AF788680

Advanced Lectures on

The STARBURST-AGN CONNECTION

Advanced Lectures on
The STARBURST-AGN CONNECTION

Tonantzintla, Puebla, Mexico June 26-30, 2000

Editors

Itziar Aretxaga

Instituto Nacional de Astrofísica, Óptica y Electrónica, Mexico

Daniel Kunth

Institut d' Astrophysique de Paris, France

Raúl Mújica

Instituto Nacional de Astrofísica, Óptica y Electrónica, Mexico

World Scientific

Singapore • New Jersey • London • Hong Kong

Published by

World Scientific Publishing Co. Pte. Ltd.

P O Box 128, Farrer Road, Singapore 912805

USA office: Suite 1B, 1060 Main Street, River Edge, NJ 07661

UK office: 57 Shelton Street, Covent Garden, London WC2H 9HE

British Library Cataloguing-in-Publication Data
A catalogue record for this book is available from the British Library.

Cover illustration: 'Señales en el jardín' (fragment). José Villalobos. Oil and encaustic on canvas. 1999.

THE STARBURST–AGN CONNECTION

Copyright © 2001 by World Scientific Publishing Co. Pte. Ltd.

All rights reserved. This book, or parts thereof, may not be reproduced in any form or by any means, electronic or mechanical, including photocopying, recording or any information storage and retrieval system now known or to be invented, without written permission from the Publisher.

For photocopying of material in this volume, please pay a copying fee through the Copyright Clearance Center, Inc., 222 Rosewood Drive, Danvers, MA 01923, USA. In this case permission to photocopy is not required from the publisher.

ISBN 981-02-4616-1

Printed in Singapore.

To Solal, born Jan 25 2001, to whom some of us are somehow indebted for a crazy but enjoyable conference.

Advanced Lectures on
"The Starburst-AGN Connection"

June 26-30, 2000. Tonantzintla, Pue., Mexico

Scientific Organizing Committee

D. Kunth (IAP, France) - chair
I. Aretxaga (INAOE, Mexico)
H. Dottori (IFUFRGS, Brazil)
J. Franco (IAUNAM, Mexico)
J. Melnick (ESO, Chile)
A. Serrano (INAOE, Mexico)
G. Tenorio-Tagle (INAOE, Mexico)
R.J. Terlevich (IoA, UK)

Local Organizing Committee

R. Mújica (INAOE) - chair
E. Benítez (IAUNAM)
V. Chavushyan (INAOE)
J. Guichard (INAOE)
F. Legrand (INAOE)
S. Linder (INAOE)
D. Mayya (INAOE)
S. del Río (INAOE)
J.R. Valdés (INAOE)

Contents

Preface

The Advanced Lectures on the Starburst–AGN Connection were organized at the *Instituto Nacional de Astrofísica Óptica y Electrónica* (INAOE), in Tonantzintla, Puebla (Mexico) on June 26-30, 2000, sponsored by the Guillermo Haro Program (`http://www.inaoep.mx/~progharo`). Seven selected experts on accretion and stellar activity delivered a set of 35 hour lectures to an audience of about 100 people, comprising students, post-docs and young faculty, experts in the fields of AGN and/or starbursts in their own right. The idea behind having a selected set of speakers to cover AGN and starburst phenomenology and physics was to present in depth state-of-the-art reviews that could enlighten the discussion about the way in which stars and accretion onto supermassive black holes convive and influence each other in AGN.

This volume is the collection of lecture notes that the teachers of this advanced school delivered. We wish to express our gratitude to our lecturers, Brian Boyle, Suzy Collin, Françoise Combes, Robert Goodrich, Hagai Netzer, Bradley Peterson and Roberto Terlevich, for providing the grounds of a very vivid debate. A separate electronic edition compiles the posters presented by the participants: in 'The Staburst–AGN Connection', Eds. R. Mújica, I. Aretxaga & D. Kunth, INAOE Electronic Edition `http://www.inaoep.mx/~agn00/posters.html`.

We are indebted to several of our colleagues at INAOE for their collaboration in this volume: Gaby López, Choco and Germán de Santos, from the graphics Dept., for their help in modifying and taking figures and photographs; Rosario Sánchez, our secretary, for her efficient administrative work; David Hughes and Bill Wall, from the Astrophysics Dept., for many suggestions on how to improve the scientific English in the book; and Esperanza Carrasco, also from the Astrophysics Dept., for selecting and arranging for us to have the cover illustration, a reproduction of *'Señales en el jardín"* by the Mexican painter José Villalobos.

Itziar Aretxaga, Daniel Kunth, & Raúl Mújica
January, 2001.

List of Participants

Alexander David (Penn State Univ., USA)
Aretxaga Itziar (INAOE, Mexico)
Argote Mauricio (IA-UNAM, Mexico)
Baliyan Kiran Singh (PRL, Ahmedabad, India)
Benítez Erika (IA-UNAM, Mexico)
Binette Luc (IA-UNAM, Mexico)
Boisson Catherine (Obs. de Meudon, France)
Boyle Brian (AAO, Australia)
Brandt William (Penn State Univ., USA)
Caccianiga Alessandro (OAL, Portugal)
Carramiñana Alberto (INAOE, Mexico)
Carrasco Esperanza (INAOE, Mexico)
Chaterjee Tapan Kumar (UAP, Mexico)
Chavushyan Vahram (INAOE, Mexico)
Chelouche Doron (Tel-Aviv Univ., Israel)
Cid Fernandes Roberto (Johns Hopkins Univ., USA)
Clénet Yann (Obs. de Paris, France)
Collin Suzy (Obs. de Meudon, France)
Collier Stefan (Ohio State Univ., USA)
Combes Françoise (Obs. Meudon, France)
Constantin Anca (Ohio Univ., USA)
Contini Thierry (ESO-Garching, Germany)
Corbett Elizabeth-Ann (RSAA, ANU, Australia)
del Río Soledad (INAOE, Mexico)
Dietrich Matthias (Univ. of Florida, USA)
Filho Mercedes (Kapteyn Inst., The Netherlands)
Franco Alfredo (IA-UNAM, Mexico)
Franco José (IA-UNAM, Mexico)
Gallagher Sarah (Penn State Univ., USA)
Gallardo Agustín (INAOE, Mexico)
Ganguly Rajib (Penn State Univ., USA)
Garcia Aurea (Univ. Sao Paulo, Brazil)
García-Barreto J. Antonio (IA-UNAM, Mexico)
Goodrich Robert (Keck Obs., USA)
González Dulce Isabel (IA-UNAM,Mexico)
Guichard José (INAOE, Mexico)
Gu Qiu Sheng (IA-UNAM, Mexico)

Hamann Frederick (Univ. of Florida, USA)
Hatziminaoglou Evanthia (Obs. Midi-Pyrenee, France)
Hooper Eric (CfA Harvard, USA)
Hornschemeier Ann (Penn State Univ., USA)
Jahnke Knud (Hamburger Sternwarte, Germany)
Janiuk Agnieszka (Copernicus Astro. Center, Poland)
Jimenéz Bailón Elena (LAEFF-INTA, Spain)
Joly Monique (Obs. de Meudon, France)
Kaspi Shai (Penn State Univ., USA)
Kawaguchi Toshihiro (Kyoto Univ., Japan)
Kloeckner Hans-Rainer (Kapteyn Inst., The Netherlands)
Komossa Stefanie (MPE-Garching, Germany)
Krongold Yair (IA-UNAM, Mexico)
Kunth Daniel (IAP, France)
Legrand François (INAOE, Mexico)
Levenson Nancy (Johns Hopkins Univ., USA)
Lira Paulina (Leicester Univ., UK)
Loaring Nicola (Univ. of Oxford, UK)
Maiolino Roberto (Oss. di Arcetri, Italy)
Marcha Maria (Obs. de Lisboa, Portugal)
Marinoni Christian (Univ. of Waterloo, Canada)
Matute Israel (Univ. Roma Tre, Italy)
Mayya Divakara (INAOE, Mexico)
Moultaka Jihane (Obs. de Meudon, France)
Mújica Raúl (INAOE, Mexico)
Nelson Charles (UNLV, USA)
Netzer Hagai (Wise Obs., Israel)
Ochoa Manuel (IA-UNAM, Mexico)
Origlia Livia (Oss. di Bologna, Italy)
Örndahl Eva (Uppsala Obs., Sweden)
Ortiz Martha (INAOE, Mexico)
Ohad Shemmer (Tel-Aviv Univ., Israel)
Oshuga Ken (Univ. of Tsukuba, Japan)
Oyabu Shinki (Univ. of Tokyo, Japan)
Peterson Brad (Ohio State Univ., USA)
Piccinelli Gabriela (IA-UNAM, Mexico)
Pineda Leopoldo (IA-UNAM, Mexico)
Ramírez Antonio (IA-UNAM, Mexico)
Ravindranath Swara (Carnegie Obs., USA)
Rodríguez Ari (INAOE, Mexico)

Rodríguez Mónica (INAOE,Mexico)
Romano Emilio (INAOE, Mexico)
Romano Roberto (INAOE, Mexico)
Rosa Daniel (INAOE, Mexico)
Ruiz Milagros (Univ. Of Hertfordshire, UK)
Salamanca Isabel (Leiden Obs., The Netherlands)
Shields Joe (Ohio University, USA)
Schmitt Henrique (STScI, USA)
Sohn Erika (IA-UNAM, Mexico)
Steffen Wolfgang (Univ. Guadalajara, Mexico)
Steinitz Raphael (Ben Gurion Univ., Israel)
Takamiya Tsutomu (Univ. Of Tokyo, Japan)
Tenorio-Tagle Guillermo (INAOE, Mexico)
Terashima Yuichi (NASA-GSFC, USA)
Terlevich Elena (INAOE, Mexico)
Terlevich Roberto (IoA, UK)
Tesch Frank (Hamburger Sternwarte, Germany)
Toledano Oswaldo (IA-UNAM, Mexico)
Trevese Dario (Univ. Roma "La Sapienza", Italy)
Treyer Marie (MRS-Marseille, France)
Uscanga Lucero (IA-UNAM, Mexico)
Vagnetti Fausto (Univ. Roma Tor Vergata, Italy)
Valdés José Ramón (INAOE, Mexico)
van Bemmel Ilse (ESO, Germany)
Vega Olga (INAOE, Mexico)
Vestergaard Marianne (The Ohio State Univ., USA)
Villa-Vargas Jorge (IA-UNAM, Mexico)
Yuan Juntao (U. of Texas at Austin, USA)
Zahn Jean Paul (Obs. de Meudon, France)
Zezas Andreas (CfA Harvard, USA)

A MEXICAN ELECTION - SPECIAL REPORT

J. RYAN O'BIBLE
Political Correspondent, INAOE

From the political posters (see Fig 1) in the streets of Tonantzintla, no attendee at the summer school can have failed to notice that there is an election taking place. The opening of this advanced school seems a good time to provide the background to this election.

There are two main political parties in this election; the SMBH and cSNR. The SMBH are the traditional party of power and have held sway for over 30 years. Despite being led by its charismatic South American-born leader, the cSNR party have never achieved more than a minority vote although the level of support for at least some of this party's policies (see below) has noticably grown in recent years.

Manifestos	
SMBH	cSNR
Accretion Disks	High Metallicities
Photoionisation	High densities
Radio Jets	Shocks
X-rays	Stellar absorption lines
Rapid Variability	Blue/UV continuum

The SMBH party's slogan *"Black holes are out-of-sight"* clearly betrays the party's origins in the 1960s. However, it has perhaps gained new meaning in recent years with the realisation that a large fraction of this party's policies may indeed be hidden. The cSNR party slogan *"A supernova a day keeps the black holes away"* has largely fallen into disfavour with its leadership, but continues to strike a chord with many of its supporters.

What then is the current state of the parties? Based on the support expressed for the parties as measured by the opinion po(ll)sters outside this lecture theatre we have the following outcome:

SMBH	40%
cSNR	27%
Don't know	33%

If this result were reproduced on June 31, it would translate to a clear win for the SMBH party but with a reduced majority. Indeed, the large number of

Figure 1. Election posters presented during the closing session of the Advanced Lectures.

floating voters might suggest to the SMBH that they would be best advised to accept some of the less-contentious cSNR policies. Political expediency may yet force a form of power-sharing coalition between the parties [a]

[a] *Editorial Note*: The outcome of the real Mexican election held on July 2 resulted in a win for the PAN, previously the minority party in Mexico, ousting the PRI party which had held power for over 70 years in Mexico. SMBH supporters take heed!

VARIABILITY OF ACTIVE GALACTIC NUCLEI

BRADLEY M. PETERSON
Department of Astronomy, The Ohio State University, 140 West 18th Avenue, Columbus, OH 43210, USA
E-mail: peterson@astronomy.ohio-state.edu

Continuum and emission-line variability of active galactic nuclei provides a powerful probe of microarcsecond scale structures in the central regions of these sources. In this contribution, we review basic concepts and methodologies used in analyzing AGN variability. We develop from first principles the basics of reverberation mapping, and pay special attention to emission-line transfer functions. We discuss application of cross-correlation analysis to AGN light curves. Finally, we provide a short review of recent important results in the field.

1 Introduction

The study of multiwavelength variability of active galactic nuclei (AGNs) is now a major branch of the field, enabled largely by the availability of suitable facilities for long-term studies of faint sources at many wavelengths; some of the scientific arguments for development of these facilities have been based on exactly these programs, which are now leading to improved understanding of the AGN phenomenon. It is only within the last few years that the evidence for supermassive black holes in both active and non-active galaxies has gone from circumstantial to compelling, and the potentially most powerful technique for measuring black-hole masses in AGNs is through study of broad emission-line variability. The existence of accretion disks in AGNs is still far from proven, but the evidence for them is improving, again as a result of variability studies. Multiwavelength monitoring observations are beginning to show the relationships between variability in different bands, and the hope is that once the phenomenology is better known, our understanding of the physics will follow.

In this contribution, we will cover the basic characteristics of AGN variability and provide what we hope is some relevant historical background. Because of the current importance of emission-line variability studies, we will develop the theory of reverberation mapping from first principles. One of the most powerful and widely used tools in the analysis of emission-line and continuum variability data is the technique of cross-correlation, and we will therefore describe in some detail the application of this method to AGN data. Throughout this contribution, we will concentrate almost exclusively on non-blazar AGNs, those for which we believe most of the observed UV/optical emission originates in an accretion disk rather than in a relativistic jet: many of the techniques described here are also applicable to blazars, however. For a fairly recent discussion of blazar variability results, we refer the reader to the fine review by Ulrich et al.[86]. We emphasize that we intend for this contribution to be primarily instructional; this should not be misconstrued as a comprehensive review of the state of the field. Our intent is to provide both students and researchers who already have some familiarity with AGNs with enough background to read critically the current literature on AGN variability and understand the strengths and weaknesses of the method.

2 Background and Basic Phenomenology

AGNs show flux variations over the entire electromagnetic spectrum. Indeed, variability was one of the first recognized properties of quasars[51,82]. Early investigations established that significant variations ($\gtrsim 0.1\,\mathrm{mag}$) in the optical brightness of quasars could occur on time scales as short as days.

Detection of rapid variability in quasars was a remarkable discovery at the time because it implies that the size of the continuum-emitting region must be of order light days (1 lt-day $= 2.6 \times 10^{15}\,\mathrm{cm}$), based on source coherence arguments: for a source to vary coherently, the entire emitting region must be causally connected, which implies a maximum size for the source based on light-travel time. Suppose, for example, that the brightness of the source doubles in a week; we can immediately conclude that the emitting region must be no larger than one light week in radius on the basis of causality. We might suppose that in fact there are multiple emitting regions varying at random, but the number of such regions must be limited or the stochastic variations would be averaged out. Individually, then, the purported independent regions face a similar size limit from causality, and the conclusion that the emitting regions are very small cannot be avoided. Historically, this is the quasar problem: how can so much energy, the equivalent of as much as trillions of stars, be produced in a region that is about the size of the Solar System?

The detection of quasar variability was a critical part of the argument that AGNs are powered by supermassive black holes. The original arguments for supermassive black holes in AGNs were based on mass constraints from the Eddington limit and size constraints from variability[63]. The Eddington limit is the requirement that gravitational forces on an ionized gas exceed outward radiation pressure, which translates to a requirement that

$$M \geq M_{\mathrm{Edd}} \equiv 8 \times 10^{5} \left(\frac{L}{10^{44}\ \mathrm{ergs\ s^{-1}}} \right) \mathrm{M}_{\odot} \quad . \qquad (1)$$

Rapid variations, in some cases on time scales as short as a day, require an emitting region less than a light day in radius, which for a 10^{44} ergs s^{-1} AGN corresponds to $\sim 10^4 R_{\rm grav}$, where $R_{\rm grav} = GM/c^2$ is the gravitational radius. Later arguments for supermassive black holes in AGNs focussed on how to power AGNs by gravitational accretion[81]. The deep gravitational potential leads to an accretion disk that radiates most strongly across the UV/optical spectrum, and for AGN masses above the Eddington limit, the thermal emission should peak in the near UV. Indeed, then, the broad UV/optical feature known as the "big blue bump" can plausibly be identified with accretion-disk emission. Furthermore, the intense magnetic fields expected in disks could provide a mechanism for jet collimation.

2.1 Basic Characteristics of Variability

AGNs have been found to be variable at all wavelengths at which they have been observed. The variations appear to be aperiodic and have variable amplitude. While variability in high-luminosity AGNs (quasars) was reported soon after their discovery, variability in lower-luminosity AGNs (Seyfert galaxies) was not reported[25] until 1967, and was less dramatic. The reason for this is probably quite simple: most of the quasars that were monitored are now known to be the jet-dominated sources known as "blazars". i.e., BL Lac objects and optically violent variables (OVVs). The optical identifications of quasars were based on coincidence with radio-source positions, which naturally led to biases towards radio-loud quasars, blazars in particular. Indeed, the original arguments about size and variability time scales in retrospect apply to the jets, not necessarily what we now identify as thermally emitting accretion disks. Nevertheless, the original conclusions about AGN sizes proved to be generally correct for both blazars and non-blazars.

UV/Optical Variability. Fig. 1 shows a light curve for a typical Seyfert 1 galaxy, NGC 5548, which will serve as a continuing example through this chapter as it is one of the best-studied objects of this class. While no periodic behavior has been identified, there are some basic parameterizations that allow us to characterize the variability. An example is shown in Fig. 2. We compare each flux measurement at an arbitrary time t_i with the flux at every later time t_j. The quantity we show is the flux ratio as a function of the time interval between observations $\Delta t = t_j - t_i$, i.e., $\Delta \log F = \log F(t_j)/F(t_i)$, where in each case the contaminating flux due to starlight in the host galaxy (as shown in Fig. 1) has been first removed. Fig. 2 shows that NGC 5548 shows little variability on time scales shorter than a few days, but on time scales of several

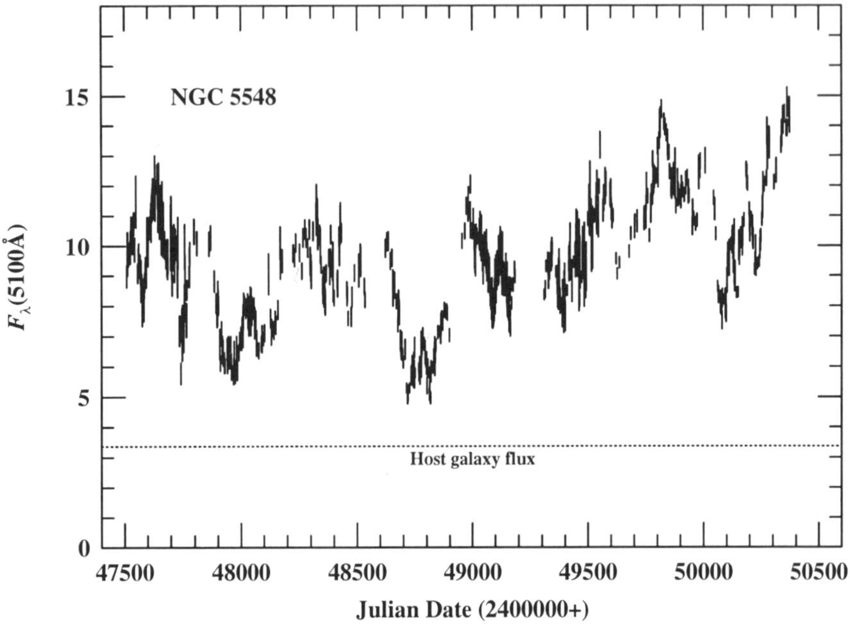

Figure 1. Optical (5100 Å) light curve of NGC 5548 from late 1988 to late 1996. The horizontal line indicates the constant contribution from starlight through the standard aperture used, which projects to $5'' \times 7''.5$. From Peterson et al.[71] ©1999 AAS.

weeks or months, very large variations can be observed. We note in passing that the quantity shown here is closely related to the "structure function". The structure function is simply the mean absolute value of $2.5\Delta \log F(\Delta t)$, i.e., the mean difference in magnitudes between observations separated by times Δt.

A common parameter to characterize variability is the mean fractional variation,

$$F_{\rm var} = \frac{\sqrt{\sigma^2 - \delta^2}}{\langle f \rangle} \quad , \tag{2}$$

where the quantities are (a) the mean flux for all N observations,

$$\langle f \rangle = \sum_{i=1}^{N} f_i \quad , \tag{3}$$

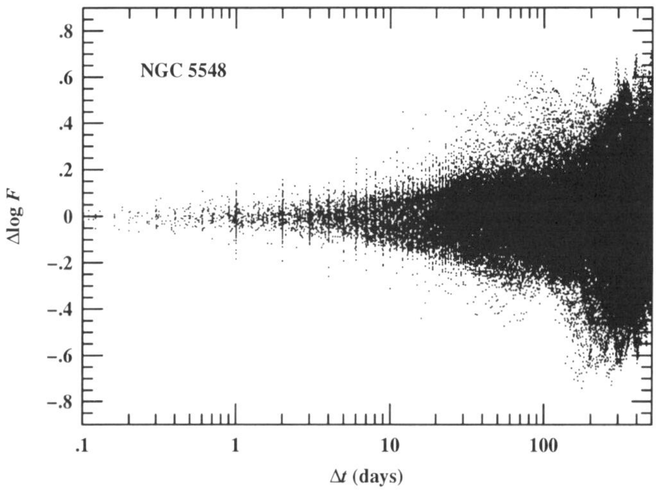

Figure 2. Difference in optical flux as a function of interval between observations Δt for the light curve shown in Fig. 1. The starlight contribution as shown in Fig. 1 has been subtracted off.

(b) the variance of the flux (as observed),

$$\sigma^2 = \frac{1}{N}\sum_{i=1}^{N}(f_i - \langle f \rangle)^2 \quad , \tag{4}$$

and (c) the mean square uncertainty of the fluxes

$$\delta^2 = \frac{1}{N}\sum_{i=1}^{N}\delta_i^2 \quad . \tag{5}$$

Even if the continuum is constant, there will still be apparent flux variations simply due to measurement uncertainties (noise); the virtue of $F_{\rm var}$ is that it adjusts the fractional variation downwards to account for the effect of random errors. The parameter $F_{\rm var}$ is thus sometimes referred to as the "excess variance." In Fig. 3, we show $F_{\rm var}$ as a function of time interval UV and optical variations in NGC 5548. This shows that UV continuum variations are typically around 10–20% on time scales of about a month. The fractional variations in the optical are less pronounced, at least in part because of contamination of the fluxes by a large constant contribution from starlight in the host galaxy.

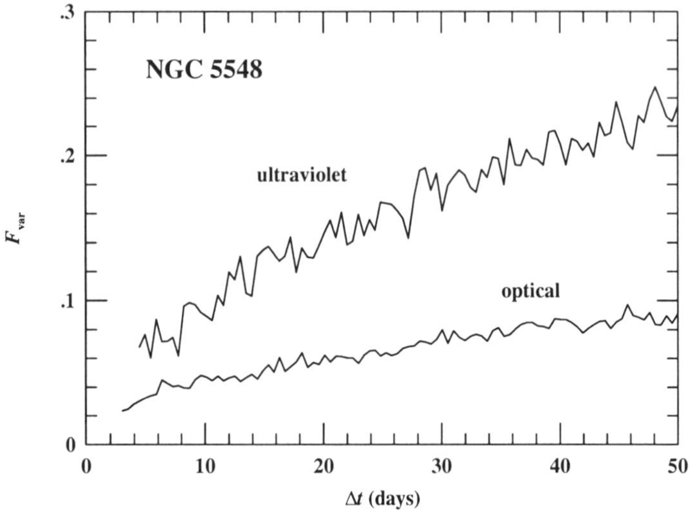

Figure 3. Variability parameter $F_{\rm var}$ from ultraviolet and optical continuum measurements of NGC 5548, as a function of interval between observations; the optical data used are those shown in Fig. 1. Courtesy of S. Collier.

X-Ray Variability. Rapid X-ray variability is a hallmark of AGNs (for a review, see Mushotzky et al.[54]). It is of historical importance since it effectively eliminates alternatives to supermassive black holes, e.g., massive stars or starbursts, as competing explanations for the high luminosities of AGNs. X-rays are expected to arise near the event horizon of the black hole, so the shortest time-scale X-ray variability is expected on a few times the crossing time

$$t_{\rm crossing} = \frac{R_{\rm grav}}{c} = 10 \left(\frac{M}{10^6\ {\rm M}_\odot} \right)\ {\rm s}. \tag{6}$$

Prior to the 1990s and the advent of RXTE, the best X-ray monitoring data was from EXOSAT, which was in a high-Earth orbit that allowed up to 80 hours of uninterrupted observations. The EXOSAT "long-looks" at variable AGNs established that variability is most rapid in low-luminosity systems.

A useful way to characterize variability is in terms of the "power density spectrum" (PDS), which is the product of the Fourier transform of the light curve and its complex conjugate. The PDS for AGNs is often parameterized as a power law,

$$P(f) = f^{-\alpha}\ . \tag{7}$$

The EXOSAT data[52] showed that AGN X-ray variations can be characterized by PDS indices in the range $1 \lesssim \alpha \lesssim 2$ over time scales of hours to months. The total power in the variations is given by integrating the PDS over all frequencies. Thus, the PDS must turn over at low frequencies (i.e., α as defined above must become less than unity) to prevent divergence in the total power. Such breaks in the PDS are observed in stellar-mass X-ray sources, and the turnover frequency correlates inversely with mass, though the fundamental reason for this is not understood. The basic idea is that the mass of the black hole can be inferred, since $M_{\rm BH} \propto R_{\rm grav} \propto t_{\rm crossing}$. If we scale AGNs relative to stellar-mass systems (which have turnover frequencies $f \approx 0.1\,{\rm Hz}$), we expect that the turnover frequencies for AGNs will occur at about 10^{-8} Hz. In only one case, NGC 3516, has there been a plausible detection of the turnover frequency. Edelson & Nandra[21] find that for this Seyfert 1 galaxy the turnover frequency is $f \approx 4 \times 10^{-7}$ Hz, which corresponds to a time scale of about one month. The mass inferred, again scaling relative to stellar-mass systems, is in the range 10^6–$10^7\,{\rm M}_\odot$.

Periodicities in X-ray light curves, which might reflect orbital or precession periods, have been searched for, but never found. As a historical footnote, however, it is worth mentioning that one such detection was claimed[53], namely a 12,000 s period in NGC 6814. However, ROSAT observations revealed that the variable source is in fact a foreground Galactic binary in the same field as the AGN[47].

2.2 Origin of the Variations

At a fundamental level, the physical origin of variations is not known, although accretion-disk instabilities are probably involved. For example, Kawaguchi et al.[37] show that continuum variations with a PDS of the form of Eq. (7) can be explained by magnetohydrodynamic instabilities, specifically disconnection events, within the disk. Variable accretion rates have also been considered. In some specific atypical cases, variations have been attributed to variable obscuration of the nuclear source and to microlensing due to stars in the host galaxy.

There are a number of important physical time scales that might be associated with variability. We include them here in the convenient form given by Edelson & Nandra[21]. The first of these is the crossing time already mentioned in Eq. (6), which we rewrite here as

$$t_{\rm crossing} = 0.011 M_7 \left(\frac{r}{10 R_{\rm grav}} \right) \ {\rm days}, \tag{8}$$

which is the time it takes a radiative signal to cross the X-ray emitting region (assumed to be at $r \lesssim 10R_{\rm grav}$). Here M_7 is the black-hole mass in units of $10^7\,{\rm M}_\odot$. Variations might also be expected on the time scale of the orbital period,

$$t_{\rm orbital} = 0.33 M_7 \left(\frac{r}{10R_{\rm grav}}\right)^{3/2} \ \text{days}. \tag{9}$$

Thermal instabilities might also cause variations on the time scale for their development,

$$t_{\rm thermal} = 5.3 \left(\frac{\alpha}{0.01}\right)^{-1} M_7 \left(\frac{r}{10R_{\rm grav}}\right)^{3/2} \ \text{days}, \tag{10}$$

where α is the viscosity parameter. Mechanical instabilities may propagate as acoustic waves, which will travel at the sound speed and thus cross the disk on a time scale

$$t_{\rm sound} = 33 \left(\frac{r}{100H}\right) M_7 \left(\frac{r}{10R_{\rm grav}}\right)^{3/2} \ \text{days}, \tag{11}$$

where H is the disk thickness. And finally, the time scale over which the effects of variations in the accretion rate will propagate through the disk is given by the drift speed,

$$t_{\rm drift} = 53000 \left(\frac{r}{100H}\right)^{2} \left(\frac{\alpha}{0.01}\right)^{-1} M_7 \left(\frac{r}{10R_{\rm grav}}\right)^{3/2} \ \text{days}. \tag{12}$$

2.3 Blazar Variability

Variability properties of blazars are distinct from those of other non-beamed AGNs and should be mentioned separately. Blazars are characterized by extreme variability at all wavelengths. Unlike radio-quiet quasars or Seyfert galaxies, significant infrared and radio continuum variability is observed in blazars. Furthermore, the polarization of the continuum radiation is also significant (i.e., greater than a few percent), and the degree of polarization and amplitude of variability correlate with luminosity, which is the opposite case for non-blazars. Blazars are also the only sources detected at TeV energies, and the TeV fluxes can vary by as much as a factor of 10 in one day. All of these properties indicate that the continuum is dominated by emission from relativistic jets, as these characteristics suggest a non-thermal (synchrotron or inverse Compton) origin.

2.4 Emission-Line Variability

The broad emission lines in AGN spectra can vary both in flux and in profile. Over time scales of months and years, the changes can be very dramatic, but on shorter time scales they are more subtle. The first detection of emission-line variations was by Andrillat & Souffrin[2], based on photographic spectra of the Seyfert 1 galaxy NGC 3516. There were a few subsequent reports[83,76], but these cases seemed to be widely regarded as "curiosities" that did not generate much follow-up work. The basic problem was that only very large changes could be detected photographically or with the intensified television-type scanners that were commonly used in AGN spectroscopy from the mid-1970s to mid-1980s. In the few cases where clear variations were detected, the changes were often dramatic; there were sometimes claims of Seyferts changing "type" as broad components of emission lines appeared or disappeared.

Pronounced variability of broad emission-line profiles was detected in the early 1980s by a number of investigators. Profile variations were originally thought to be due to excitation inhomogeneities; excitation pulses propagating through a broad-line region (BLR) with an ordered velocity field would produce features that could propagate across the profile with time. This concept led to the development of reverberation mapping, which is described in detail in Sec. 3. Peterson[61] reviews early work on emission-line variability.

A useful way to isolate the variable part of an emission line is shown in Fig. 4. The upper panel shows a mean spectrum formed from 34 individual HST spectra of NGC 5548. The lower panel shows the "root-mean square" (rms) spectrum which is formed from the same data simply by computing the rms flux at each wavelength. Constant features, such as narrow emission lines and host-galaxy flux, do not appear in the rms spectrum.

2.5 The First Monitoring Programs

The early 1980s saw the first attempts to monitor the UV/optical continuum and emission-line variations in Seyfert 1 galaxies. There were two reasons this happened when it did. First, there was a realization that variability afforded a powerful probe of the structure and kinematics of AGNs on projected scales of microarcseconds. Variability was recognized as an important new tool with which to study enigmatic quasars. Second, the right technology for such investigations became widely available: it was possible to attempt such programs on account of (a) IUE, which for the first time allowed precision UV spectroscopy of low-redshift extragalactic objects, and (b) the proliferation of linear electronic detectors (first Reticons and Image Dissector Scanners, and later CCDs) on moderate-size (1–2m) ground-based telescopes.

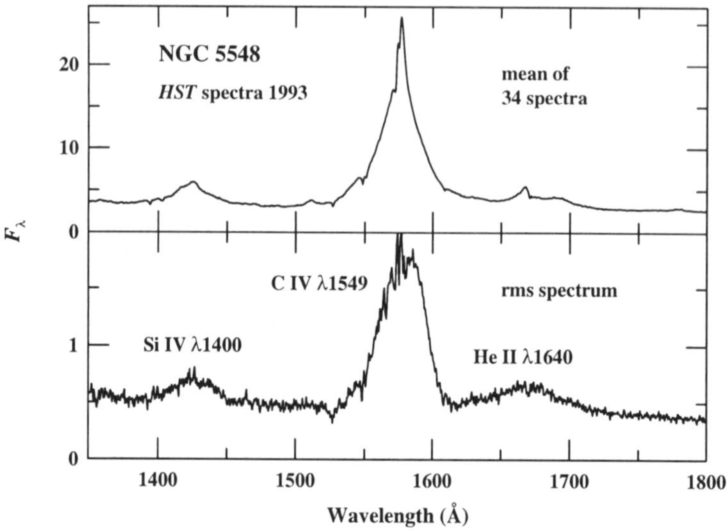

Figure 4. The top panel shows the mean spectrum computed from 34 HST spectra of the variable Seyfert 1 galaxy NGC 5548[40]. The lower panel shows the rms spectrum based on variations around this mean. The rms spectrum thus isolates the variable components of the spectrum. Fluxes are in units of 10^{-15} ergs s^{-1} cm^{-2} $Å^{-1}$.

One of the first significant monitoring programs was a multiple-year IUE-based program on NGC 4151, which was carried out by a European consortium led by M.V. Penston and M.-H. Ulrich[84]. Ultraviolet spectra were obtained with a typical sampling interval of 2–3 months. The program showed that variations in the UV and optical continua were closely coupled. It also revealed that the emission-line flux variations are correlated with continuum variations, but that different lines respond in different ways, both in amplitude and in time scale. These data also showed a complicated relationship between UV and X-ray variations and led to the discovery of variable absorption lines in the ultraviolet.

The galaxy NGC 4151 was also monitored spectroscopically in the optical at Lick Observatory by Antonucci & Cohen[4]. They found that the Balmer lines seemed to respond to continuum variations on a time scale less than around one month (their typical sampling interval). They also found that relative to Hβ and Hα, the higher-order Balmer lines and He II λ4686 varied with higher amplitudes.

Arakelian (Akn) 120 was the first higher-luminosity Seyfert that was monitored fairly extensively in the optical[64,65] as a result of dramatic Balmer-line profile changes that had been detected earlier[26,38,80]. Peterson et al.[65] found that the time scale for the response of Hβ to continuum variations suggested a BLR size of less than 1 light month across. This was a surprising result as it suggested that there was a serious problem with existing estimates of sizes of the BLR that were based on photoionization equilibrium modeling, as these indicated the BLR should be about an order-of-magnitude larger than this. The upper limit on the BLR size was similar to that obtained by Antonucci & Cohen[4] for NGC 4151, but because Akn 120 is a higher-luminosity source, the monthly sampled data provided a more critical challenge to BLR models.

Not surprisingly, the results from these earlier monitoring programs were controversial. Several observational problems could be identified:

1. *Undersampling of the variations.* The variations tended to be undersampled because the original programs for monitoring Seyfert galaxies were designed for BLRs that were thought to be many light months in size. For example, in the case of Akn 120, Peterson et al.[65] were looking for profile structures that were expected to cross the line profile on a time scale of a year or so, and monthly observations should have been sufficient to carry out this program. There was certainly at the time no reason to believe that higher sampling rates should be required; indeed, proposals to observe AGNs as often as once per month were sometimes deemed to be "oversampled" by telescope allocation committees! There is no obvious algorithm to determine whether or not the variations that have been observed are undersampled, but it is quite obvious that if the results depend on individual data points, the light curve is almost certainly undersampled and any conclusions drawn must be eyed with suspicion. A very simple operational criterion for adequate sampling is that if the results do not change much when individual points are removed from the light curve, the light curve is probably not seriously undersampled. A nice simple test is to divide a light curve into two parts, one comprised of the even-numbered points (i.e., second, fourth, etc., in the time-ordered series) and one comprised of the odd-numbered points. If the two light curves are still very similar, i.e., the important features appear in both light curves, then the original light curve is probably adequately sampled.

2. *Low S/N of the light curves.* If the detected variations are not large compared to the signal-to-noise ratio (S/N) of typical data, then spurious results can be obtained. Stated another way, $F_{\rm var}$ (Eq. (2)) must be significantly greater than zero. This was a serious issue in the case of some of the earlier data obtained with image dissector scanners and

Reticon arrays, for which uncertainties in AGN line and continuum fluxes were typically around 8–10%. To a large degree, this problem has been obviated by use of CCDs, for which typical errors in the 1–3% range are routinely achieved.

3. *Systematic errors.* There are two sometimes-related types of insidious errors that can adversely affect time series analysis: (a) correlated continuum/line errors, and (b) aperture effects.

   Correlated errors are due to systematic flux-calibration errors. Basically, if the flux calibration of a spectrum is incorrect, both the continuum and emission-line fluxes measured from it will be in error in the same sense; if the calibration is too high, both the emission-line and continuum fluxes will be too high. This introduces an artificial correlation between the line and continuum at zero lag, and can thus bias the measurement of the true lag between them to artificially small values.

   Aperture effects occur when the amount of flux entering a spectrograph is not fixed, on account of pointing or guiding errors, or variations in seeing in the case of ground-based observations. In point-like sources like stars, this affects only the overall photometric accuracy. In nearby AGNs, however, both the narrow-line region (NLR) and host-galaxy are spatially resolved, and the aperture geometry, centering and guiding, and seeing variations can lead to apparent spectral variations. Depending on their nature, aperture effects can cause either correlated, uncorrelated, or even anticorrelated errors in the line and continuum fluxes.

2.6 Spectrophotometric Flux Calibration

In this section, we will outline some of the important considerations for flux calibration of spectroscopic monitoring data, directed primarily towards ground-based optical observers.

In AGNs, aperture effects arise because the source is comprised of multiple components, some of which have angular structure on scales similar to the width of the point-spread function (PSF). The basic requirement for accurately flux-calibrating AGN spectra is that stable fractions of light from the AGN continuum source and BLR (both point-like at even the $0''.01$ level) and the NLR and the host galaxy (both extended even at arcsecond levels) must enter the aperture. We note that even in the case of space-based observations, there is a trade-off between aperture size and pointing uncertainty: the goal is to minimize the amount of host-galaxy starlight entering the aperture (arguing for a smaller aperture), while ensuring that the amount of admitted starlight is constant (arguing for an aperture large relative to the pointing accuracy of the telescope). In the UV, however, this is a less-significant problem because the host galaxy is so much fainter than the AGN itself.

Mitigation of aperture effects is one of the most important considerations for a ground-based monitoring program. This is important to keep in mind, as most AGN observers are used to background-limited observations of point sources, and this calls for adjusting the slit width for variations in seeing in order to optimize the S/N of the data. However, *this is exactly the wrong thing to do if you are monitoring Seyfert galaxy variations*, since calibration accuracy is almost always determined by systematics rather than photon statistics. Most observers *do* know that you need to open up the aperture for absolute spectrophotometry, however, and this is precisely what needs to be done in monitoring programs, as we will explain below.

The standard method of absolute flux calibration is to determine from observations of spectrophotometric standard stars how counts per second per pixel on the detector translates to flux per unit wavelength. However, standard absolute spectrophotometry is far too inaccurate for ground-based AGN monitoring; the typical accuracy that can be achieved on photometric nights is about 5%. Furthermore, at most observing sites, only a relatively small fraction of nights are of sufficient quality and stability for absolute spectrophotometry to be useful. Even at a good site like Kitt Peak, fewer than 1/3 of the nights are photometric. However, more than 2/3 of all nights (including bright time) are good enough to get quality spectra of bright AGNs with a 1–2-m telescope with a typical photometric accuracy about 15%. Since we need a high rate of sampling and do not wish to discard more than half of the potentially usable nights, secondary higher-accuracy calibration methods must be used. In ground-based AGN monitoring, there are two commonly used high-accuracy flux-calibration methods, which we outline below:

1. *Relative spectrophotometry.* This method involves simultaneous observation of a nearby field star (that must be shown to be non-variable!) in long-slit mode. Absolute calibration can be achieved by calibrating spectra of the comparison star in the usual fashion described above on nights of photometric quality. This method has been used to attain internal (relative) accuracies of $\sim$ 1–2%[16,36]. The limitations of this method are (a) an extremely accurate slit response function (i.e., relative sensitivity as a function of position in the slit) is required, (b) a large slit and high pointing and guiding accuracy is required to avoid aperture/seeing effects, and (c) this method is difficult to use shortward of $\sim$ 4000 Å because the requirement of fixed position angle (to observe the AGN and comparison

star simultaneously) leads to atmospheric dispersion problems.

2. *Internal calibration.* This is based on using a constant-flux component in the spectrum (e.g., narrow emission lines) to refine absolute calibration. In this way, internal accuracies of $\sim 2\%$ have been routinely achieved[69, 71]. The [O III] $\lambda\lambda$4959, 5007 doublet is excellent for this method because these lines are strong, relatively unblended, and close in wavelength to Hβ and He II λ4686. Unlike the broad lines, the narrow lines are almost always expected to be constant in flux. The light-crossing time for the NLR is large (typically 100–1000 years) as is the recombination time (about 100 years; see Eq. (17) in Sec. 3), so any short-term variability is smeared out. This method also has limitations: (a) most narrow lines are blended and/or weak, (b) the method works best at reasonably high spectral resolution, when the narrow-lines are at least partially resolved, and (c) in nearby AGNs, the NLR is often extended on scales of arcseconds, which introduces the possibility of aperture effects.

The last of these problems deserves serious consideration. If the NLR is extended, measuring a consistent flux is difficult due to aperture effects, as described above; either centering and guiding errors or differences in the seeing between observations can alter the calibration of the spectrum. However, it is possible to understand the nature of these effects so that in principle one can correct for them, or at minimum one can pursue mitigation strategies to reduce their influence. To do this, one must know (a) the surface brightness distributions of the extended components (the host galaxy and the NLR) and (b) the PSF, including the seeing component, for the instrument at the time of the observations.

With a good model for the surface brightness distributions, we can still use the narrow-lines for flux calibration, and then apply a seeing-dependent flux correction[89,68]. Even if the seeing is not measured, such models can be used to give us an idea of how badly our observations might be adversely affected by seeing variations. As an example, the corrections to the point-source fluxes (broad lines and AGN continuum) and host galaxy flux that need to be applied in the case of NGC 4151 are shown in Fig. 5 for different aperture geometries. We can readily conclude from these diagrams that seeing-dependence of the correction factors is smaller for larger apertures, as can be expected intuitively. A useful generalization is that the corrections for seeing effects are generally not necessary if the aperture in both dimensions is larger than the worst seeing, as measured by the full-width at half-maximum (FWHM) of a stellar image.

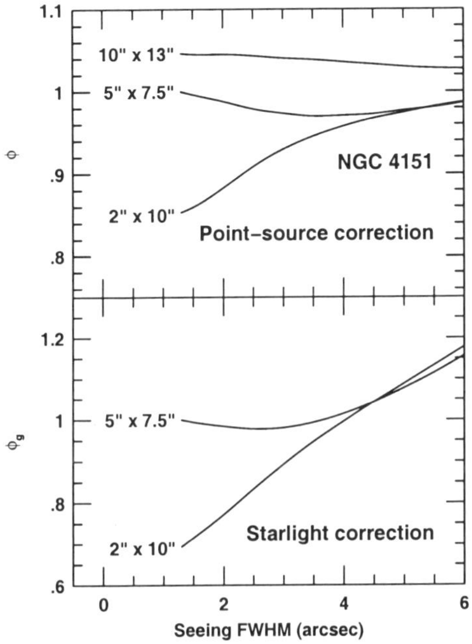

Figure 5. Aperture corrections (as defined by Peterson et al.[68]) for NGC 4151. These are multiplicative adjustments that must be made to (a) the broad emission-line and AGN continuum fluxes (top panel) and (b) the host-galaxy starlight contribution (bottom panel) as a function of seeing when the spectra are calibrated by assuming a constant flux in [O III] $\lambda\lambda$4959, 5007. The reference point is a $5'' \times 7\farcs5$ aperture and seeing FWHM = $1\farcs3$. For sufficiently large apertures, the correction is negligible under even quite poor observing conditions; for small apertures, seeing effects can produce apparent variability. The starlight correction factor for the $10'' \times 13''$ aperture has a value of ~ 1.6 and varies only weakly with seeing.

2.7 Requirements for a Successful Monitoring Program

In order to avoid misleading systematic errors of the sort discussed thus far, care must be taken in designing an observational monitoring program. The fairly obvious basic requirements that we should keep in mind are:

1. *The monitoring program must have sufficient sampling.* Specifically, the observations should be closely spaced in time relative to physical time scales of interest. In terms of total duration, the program should be at least three times as long as the longest time scale to be investigated.

2. *The S/N of the light curve must be high enough to detect variations on the shortest interesting physical time scale.* Inspection of Fig. 2, for example, demonstrates that if you can measure continuum fluxes to only 10% accuracy, 3σ detections of continuum variability in Seyfert 1 galaxies will be extremely rare on time scales as long as months.

3. *Systematic effects must be obviated.* The data must be as homogeneous as possible. The easiest way to do this is to acquire the data with a single instrument in a stable configuration that is suitable for all the observing conditions expected over the duration of the program.

3 Theory of Reverberation Mapping

After emission-line variability was detected, it became clear to a number of investigators[5,23,7,10,3] that the kinematics and the geometry of the BLR can be tightly constrained by characterizing the emission-line response to continuum variations. The time delay between continuum and emission-line variations are ascribed to light travel-time effects within the BLR; the emission lines "echo" or "reverberate" to the continuum changes. The Blandford & McKee paper[7], regarded as the seminal paper in the field, first introduced the term "reverberation mapping" to describe this process. Reviews of progress in reverberation mapping are provided by Peterson[62] and Netzer & Peterson[58].

Because the emission lines vary with the UV/optical continuum (with some small time delay), we can immediately draw several important conclusions:

1. The line-emitting clouds are close to the continuum source, i.e., the light-travel time across the BLR is small.

2. The line-emitting clouds are optically thick. If they were optically thin, the line emission from them would change little as the continuum varied.

3. The observable UV/optical continuum variations are closely related to variations of the ionizing continuum ($\lambda < 912\,\text{Å}$, $h\nu \geq 13.6\,\text{eV}$).

This information provides the crucial underpinning for the assumptions detailed below. But before we proceed any further, we need to remind ourselves of some of the basic characteristics of the line-emitting gas, much of which we derive by comparing the observed line spectra with the predictions of photoionization equilibrium computer codes (see the contribution by Netzer for a more complete discussion). In general, photoionization equilibrium models of the line-emitting clouds are parameterized by the shape of the ionizing continuum, elemental abundances, and an "ionization parameter"

$$U = \frac{Q(\mathrm{H})}{4\pi r^2 c n_e} \quad , \tag{13}$$

where

$$Q(\mathrm{H}) = \int_{\nu_1}^{\infty} \frac{L_\nu}{h\nu}\, d\nu \tag{14}$$

is the number of hydrogen-ionizing ($h\nu_1 = 13.6\,\mathrm{eV}$) photons emitted per second by the central source.

Essentially, U characterizes the ionization balance within the cloud, as $Q(\mathrm{H})/r^2$ is proportional to the number of photoionizations occurring per second at the incident face of the cloud, and n_e is proportional to the recombination rate. For a typical Seyfert 1 galaxy (NGC 5548)

$$Q(\mathrm{H}) \approx 10^{54}\, h_0^{-2} \text{ photons s}^{-1} \quad , \tag{15}$$

where h_0 is the Hubble constant in units of $100\,\mathrm{km\ s^{-1}\ Mpc^{-1}}$. By taking the electron density to be $n_e \approx 10^{11}\ \mathrm{cm^{-3}}$and the distance to the central source to be 10 light days, we find $U \approx 0.1$. The line-emitting clouds are ionized to the "Strömgren depth",

$$R_S = \frac{Uc}{\alpha_B n_e} \approx 10^{11}\ \mathrm{cm} \quad , \tag{16}$$

at which point virtually all of the incident ionizing photons have been absorbed. In BLR clouds, the region beyond the Strömgren zone is partially ionized, $n(\mathrm{H}^+) \approx 0.1 n(\mathrm{H}^0)$, and heated primarily by far infrared radiation[24] and partly by high-energy X-rays[45]. In Eq. (16), α_B is the Menzel–Baker case B recombination coefficient[59].

3.1 Reverberation Mapping Assumptions

There are a number of simplifying assumptions that we can now make:

1. The continuum originates in a single central source. The size of an accretion disk around a supermassive (say, $10^{7-8}\,\mathrm{M_\odot}$) black hole is of the order of 10^{13-14} cm. A typical size for the BLR in the same system would be of order a few light days, i.e., $\sim 10^{16}$ cm. Note that it is specifically not required that the continuum source emits radiation isotropically, though this is a useful starting point. The "point-source" assumption greatly simplifies the reverberation process. However, we should mention that the point-source assumption is probably not applicable to X-ray reverberation, as the Fe Kα emission and the hard X-rays that drive this line probably arise in regions that are of similar size, and perhaps co-spatial[79].

2. Light-travel time is the most important time scale. We assume specifically that emission-line clouds respond instantaneously to changes in the continuum flux. The time scale to re-establish photoionization equilibrium is the recombination time,

$$\tau_{\rm rec} = (n_{\rm e}\alpha_{\rm B})^{-1} \approx 40 \left(\frac{n_{\rm e}}{10^{11}\ {\rm cm}^{-3}}\right)^{-1} {\rm s}. \qquad (17)$$

The time it takes a Lyman α photon to diffuse outward from the Strömgren depth is about 20 times the direct light-travel time[35],

$$\tau_{\rm diff} = 20\frac{R_{\rm S}}{c} \approx 20\frac{U}{n_{\rm e}\alpha_{\rm B}} \approx 60 \left(\frac{n_{\rm e}}{10^{11}\ {\rm cm}^{-3}}\right)^{-1} {\rm s}. \qquad (18)$$

We also need to carry out our reverberation-mapping experiment on a time scale short enough that the structure of the BLR can be assumed to be stable. The dynamical, or cloud-crossing, time scale for the BLR is typically

$$\tau_{\rm dyn} = \frac{r}{V_{\rm FWHM}} \approx 3 - 5 \text{ years}, \qquad (19)$$

where $V_{\rm FWHM}$ is the Doppler width of the broad line for which the response time $\tau = r/c$ has been measured. Any reverberation-mapping experiment has to be short relative to the crossing time or the structural information might be washed out by cloud motions.

3. There is a simple, though not necessarily linear, relationship between the observed continuum and the ionizing continuum.

3.2 The Transfer Equation

Under these assumptions, the relationship between the continuum and emission lines can be written in terms of the "transfer equation",

$$L(V_z, t) = \int_{-\infty}^{\infty} \Psi(V_z, \tau)\, C(t - \tau)\, d\tau \ . \qquad (20)$$

Here $C(t)$ is the continuum light curve, $L(V_z, t)$ is the emission-line flux at line-of-sight velocity V_z and time t, and $\Psi(V_z, \tau)$ is the "transfer function" at V_z and time delay τ. Inspection of the transfer equation shows that the transfer function is simply the time-smeared emission-line response to a δ-function outburst in the continuum. Note that causality requires that the lower limit on the integral is $\tau = 0$.

Solution of the transfer equation to obtain the transfer function is a classical inversion problem in theoretical physics: the transfer function is essentially the Green's function for the system. Unfortunately, it is difficult to find a stable solution to such equations, especially when the data are noisy and sparse, as they usually are in astronomical applications.

In practice, most analyses have concentrated on solving the velocity-independent (or 1-d) transfer equation,

$$L(t) = \int_{-\infty}^{\infty} \Psi(\tau)\, C(t-\tau)\, d\tau \quad , \tag{21}$$

where both $\Psi(\tau)$ and $L(t)$ represent integrals over the emission-line width.

The transfer equation is a linear equation. In reality, however, the relationship between the observed continuum and the emission-line response is likely to be nonlinear. We therefore approximate $C(t) = \bar{C} + \Delta C$ and $L(t) = \bar{L} + \Delta L$, where $\bar{C}$ and $\bar{L}$ represent constants, usually the mean value of the continuum and line flux, respectively. We can then treat deviations from the mean as linear perturbations and the equation we actually solve is

$$\Delta L(t) = \int_{-\infty}^{\infty} \Psi(\tau)\, \Delta C(t-\tau)\, d\tau \quad . \tag{22}$$

Most existing data sets are inadequate for transfer-function solution, and simpler analyses are used. The most commonly used tool in analysis of AGN variability is cross-correlation, which will be discussed in detail in Sec. 4.

3.3 Isodelay Surfaces

Suppose for the moment that the BLR consists of clouds in a thin spherical shell of radius r. Further suppose that the continuum light curve is a simple δ-function outburst. Continuum photons stream radially outward and after travel time r/c, about 10% of these photons (using a typical "covering factor") are intercepted by BLR clouds and are reprocessed into emission-line photons. An observer at the central source will see the emission-line response from the entire BLR at a single instant with a time delay of $2r/c$ following the continuum outburst. At any other location, however, the summed light-travel time from central source to line-emitting cloud to observer will be different for each part of the BLR. In the case of a δ-function outburst, at any given instant, the parts of the BLR that the observer will see responding are all those for which this total path length is identical; at any given time delay, the part of the BLR that the observer sees responding is the intersection of the BLR distribution and an "isodelay surface." Astronomers, on account of their familiarity with conic sections, can readily recognize that the shape of the isodelay surface is an ellipsoid with the continuum source at one focus and

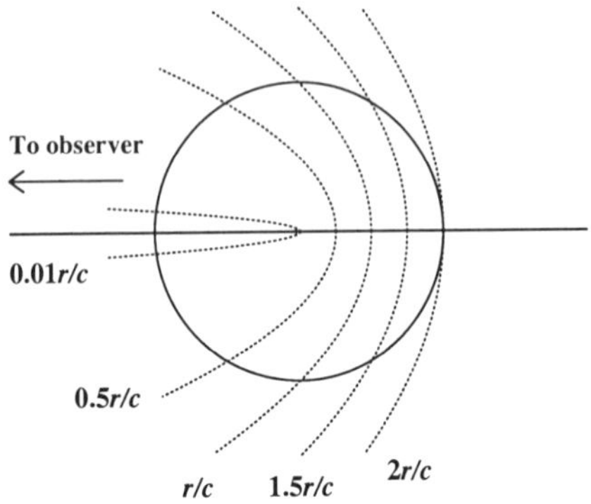

Figure 6. The circle represents a cross-section of a shell containing emission-line clouds. The continuum source is a point at the center of the shell. Following a continuum outburst, at any given time the observer far to the left sees the response of clouds along a surface of constant time delay, or isodelay surface. Here we show five isodelay surfaces, each one labeled with the time delay (in units of the shell radius r) we would observe relative to the continuum source. Points along the line of sight to the observer are seen to respond with zero time delay. The farthest point on the shell responds with a time delay $2r/c$.

the observer at the other; the light-travel time from central source to BLR cloud to observer is constant for all points on the ellipsoid. Since the observer is virtually infinitely distant from the source, the isodelay surface becomes a paraboloid, as shown schematically in Fig. 6. The figure shows the BLR as a ring intersected by several isodelay surfaces, labeled in terms of their time delay in units of r/c. Relative to the continuum, points along the line of sight to the observer are not time delayed (i.e., $\tau = 0$). Points on the far side of the BLR are delayed by as much as $2r/c$, the time it takes continuum photons to reach the BLR plus the time it takes line photons emitted towards the observer to return to the central source on their way to the observer.

Essentially, the transfer function measures the amount of line emission emitted at a given Doppler shift in the direction of the observer as a function of time delay τ. The value of the transfer function at time delay τ is computed by summing the emission in the direction of the observer at the intersection of the BLR and the appropriate isodelay surface. For a thin spherical shell, the intersection of the BLR and an isodelay surface is a ring of radius $r \sin\theta$, where the polar angle θ is measured from the observer's line of sight to the central source, as shown in Fig. 7. The time delay for a particular isodelay

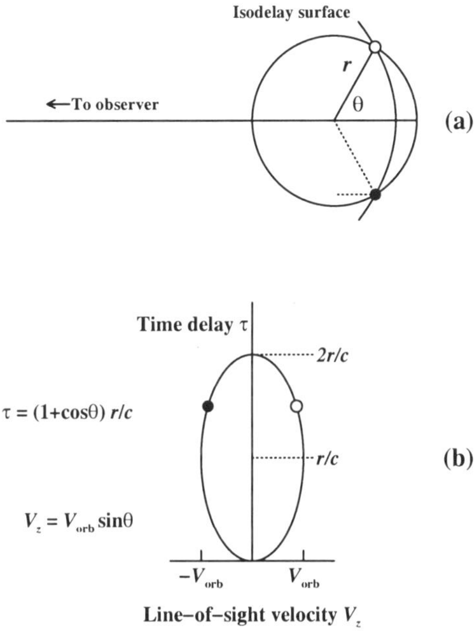

Figure 7. The upper diagram shows a ring (or cross-section of a thin shell) that contains line-emitting clouds, as in Fig. 6. An isodelay surface for an arbitrary time is given; the intersection of this surface and the ring shows the clouds that are observed to be responding at this particular time. The dotted line shows the additional light-travel time, relative to light from the continuum source, that signals reprocessed by the cloud into emission-line photons will incur (Eq. (23)). In the lower diagram, we project the ring of clouds onto the line-of-sight velocity/time-delay (V_z, τ) plane, assuming that the emission-line clouds in the upper diagram are orbiting in a clockwise direction (so that the cloud represented by a filled circle is blueshifted and the cloud represented by an open circle is redshifted).

surface is the equation for an ellipse in polar coordinates,

$$\tau = (1 + \cos\theta) r/c \quad , \tag{23}$$

as is obvious from inspection of Fig. 7. The surface area of the ring of radius $r \sin\theta$ and angular width $r\,d\theta$ is $2\pi(r\sin\theta)\,r\,d\theta$, and assuming that the line response per unit area on the spherical BLR has a constant value ε_0, the response of the ring can be written as

$$\Psi(\theta) = 2\pi\varepsilon_0 r^2 \sin\theta\, d\theta \quad , \tag{24}$$

where $0 \leq \theta \leq 2\pi$. From Eq. (23), we can write

$$d\tau = -(r/c)\sin\theta\, d\theta \quad , \tag{25}$$

so putting the response in terms of τ rather than θ, we obtain

$$\Psi(\tau)\, d\tau = \Psi(\theta) \left| \frac{d\theta}{d\tau} \right| d\tau = 2\pi\varepsilon r c\, d\tau \tag{26}$$

for values from $\tau = 0$ ($\theta = 2\pi$) to $\tau = 2r/c$ ($\theta = 0$). The transfer function for a thin spherical shell is thus constant over the range $0 \leq \tau \leq 2r/c$.

3.4 Transfer Functions for a Variety of Simple Models

The simple analytic calculation in the last section was intended to be illustrative and serve as a reference point. We will now expand on this with more general geometries, and also incorporate information from Doppler motion along the line of sight. We will derive some transfer functions for other simple models, focusing on three: (a) systems of clouds in circular Keplerian orbits, illuminated by an isotropic continuum, (b) biconical outflows, and (c) disks of random inclination. All these are physically plausible, and can produce "double-peaked" emission-line profiles, which are sometimes seen in AGNs, though not all of these models necessarily do this. We will start with the simplest models and progress to more complicated models. Much of this discussion is drawn from discussions of transfer functions in the literature[93,60,24,30,31].

Suppose line-emitting clouds are on a circular orbit at inclination $i = 90°$; imagine that the circle in Fig. 7a represents this orbit seen face on. The line response from the clouds at the intersection of an arbitrary isodelay surface and the circular orbit will be at time delay $\tau = (1 + \cos\theta)r/c$ and line-of-sight velocities $V_z = \pm V_{\rm orb}\sin\theta$, where $V_{\rm orb} = (GM/r)^{1/2}$, the circular orbital speed. It is easy to see that the circular orbit projects to an ellipse in the line-of-sight velocity/time-delay (V_z, τ) plane with semiaxes $V_{\rm orb}$ and r/c, as shown in Fig. 7b. This simple example is important because it is straightforward to generalize it to both disks (rings of different radii) and shells (rings at different inclinations).

First we consider the generalization to a shell. We can construct a shell from a distribution of circular orbits, with inclinations ranging from $i = 0°$ to $i = 180°$. As we decrease the inclination of the circular orbit in Fig. 7a from $i = 90°$, we see that the range of time delays will decrease from $[0, 2r/c]$ to $[(1 - \sin i)r/c, (1 + \sin i)r/c]$, and similarly the line-of-sight velocity range will decrease from $[-V_{\rm orb}, +V_{\rm orb}]$ to $[-V_{\rm orb}\sin i, +V_{\rm orb}\sin i]$, as we show schematically in Fig. 8. At the limiting case $i = 0°$, the time delays all contract

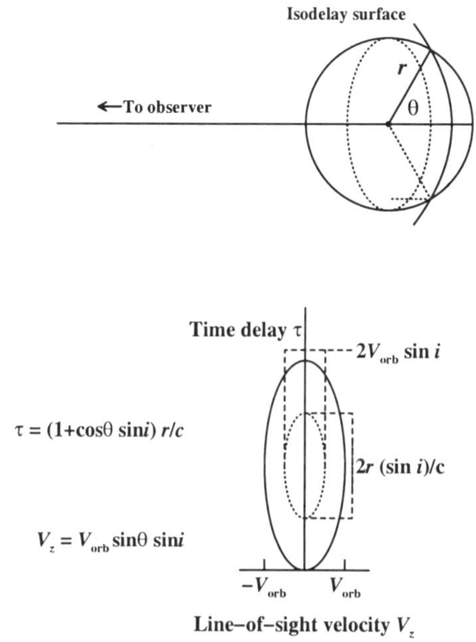

Figure 8. Here we take the circular ring shown in Fig. 7 and show how its projection on V_z, τ changes as the inclination is decreased from 90°; in the time-delay direction, the axis contracts to $2r \sin i/c$, and in the velocity direction it decreases to $2V_{\rm orb} \sin i$.

to r/c, since the light travel-time paths for all points on a face-on ring are the same, and the velocities all contract to $V_z = 0$, because the orbital velocities are now perpendicular to the line of sight. Thus, the transfer function looks like a series of ellipses as in Fig. 8b that with decreasing inclination contract down to a single point at $V_z = 0$ and $\tau = r/c$ when $i = 0^\circ$. We can construct such a transfer function by using a Monte Carlo method that places BLR clouds randomly across the surface of the shell, and the result we get is shown in Fig. 9, which corresponds to a thin shell of radius 10 light days and a central mass of $10^8\ M_\odot$. We have also integrated this transfer function over τ and V_z to obtain the emission-line profile and the one-dimensional transfer function, respectively. In the particular case of a thin spherical shell, we see that both of these are simple rectangular functions, as we showed analytically for the one-dimensional transfer function (Eq. (26)).

The first complication that we might consider is anisotropic line emission

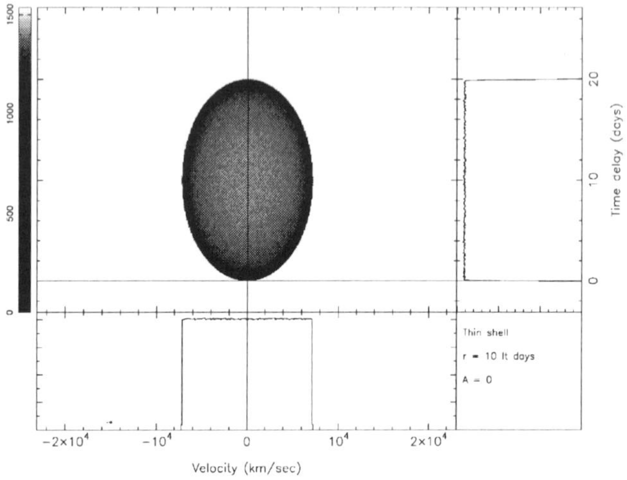

Figure 9. Transfer function for a thin spherical shell. The upper left panel shows in grey scale the two-dimensional transfer function, i.e., the observed emission-line response as a function of line-of-sight velocity V_z and time delay τ. The upper right panel shows the one-dimensional transfer function, i.e., the two-dimensional transfer function integrated over V_z, which is the response of the total emission line as a function of time (cf. Eq. (26)). The lower left panel shows the emission-line response integrated over time delay; this is the profile of the variable part of the line. The specific model here is comprised of emission-line clouds in circular orbits of radius $r = 10$ light days and random inclinations ($0^o \leq i \leq 180^o$), orbiting around a central black hole of mass $10^8\,M_\odot$. In this model, the line emission from each cloud is isotropic ($A = 0$).

by the BLR clouds. Physically, this will occur if the BLR clouds themselves are optically thick in the lines as well as the continuum. In this case, most of the line radiation emitted by the clouds will be from the side of the cloud facing the continuum source, i.e., the line emission is directed back towards the continuum source. A simple parameterization of asymmetric line emission is to describe the apparent emissivity of a cloud as

$$\varepsilon(\theta) = \varepsilon_0\,(1 + A\cos\theta) \quad , \tag{27}$$

where ε_0 is constant and the parameter $A = 0$ for isotropic emission and $A = 1$ for completely anisotropic emission; the latter case is appropriate for spherical clouds with inward-facing surfaces that are uniformly bright. In principle, A can be estimated by photoionization modeling, though in practice the values are highly uncertain on account of limitations in the accuracy of the radiative transfer codes (see the contribution by Netzer). It is certainly expected that

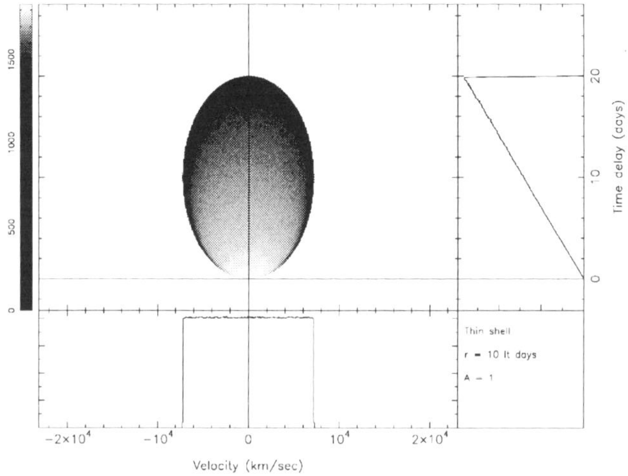

Figure 10. Transfer function for a thin spherical shell, as in Fig. 9, except with completely anisotropic ($A = 1$) line emission, i.e., all of the emission line flux from each cloud is directed back towards the continuum source.

$A \approx 1$ for Lyα, at least, and models suggest approximate values $A \approx 0.7$ for C IV and $A \approx 0.8$ for Hβ[24]. The main effect of anisotropic emission is to increase the measured lag for a given geometry because the apparent response of the near side of the BLR is suppressed. For a thin shell of radius r, the mean time delay is $\tau = (1 + A/3)r/c$. Fig. 10 shows the transfer function for the same thin spherical shell model of Fig. 9, but in this case with highly anisotropic ($A = 1$) line response.

In addition to anisotropic line response, we can also consider anisotropic illumination of the BLR clouds by the continuum source. As an example, consider the case where BLR clouds are illuminated by biconical beams of semi-opening angle ω; the limiting case as ω approaches zero would be a narrow jet-like pencil beam, and the case $\omega = 90^{\circ}$ corresponds to an isotropic continuum. We start by examining the response of an edge-on ($i = 0^{\circ}$) circular ring, as we show in Fig. 11, which is exactly like Fig. 7, but with only part of each orbit illuminated by the continuum source. We earlier generalized the result for a shell, going from Fig. 7 to Fig. 8 as we decreased the inclination of the ring; doing this again, we see in Fig. 12 how the two-dimensional transfer function is altered by anisotropic illumination. We note specifically the absence of any response near $\tau = 0$ and $\tau = 2r/c$ since the bicones shown

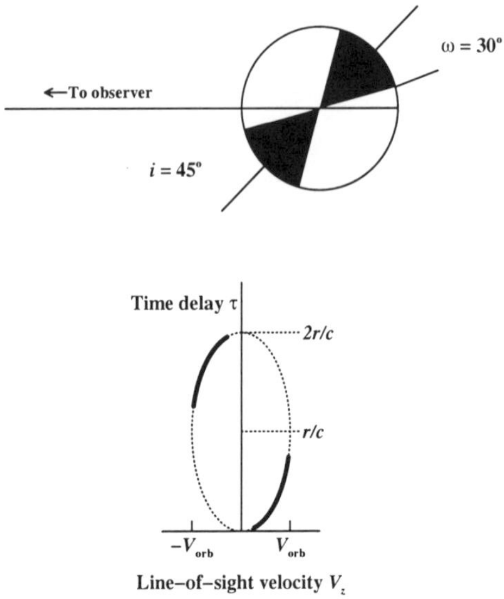

Figure 11. A ring-like line-emitting region and its projection into the V_z, τ plane is shown, as in Figs. 7 and 8, though in this case with a biconical continuum geometry. In the case shown here, the beam semi-opening angle is $\omega = 30^{\rm o}$, and the beam is inclined to the observer by $i = 45^{\rm o}$. Again assuming that clouds on the ring are orbiting in the clockwise direction, we highlight in the lower diagram the regions in the (V_z, τ) plane that will produce an emission-line response; there is no emission-line response from clouds not illuminated by the continuum beam.

do not illuminate BLR clouds directly along the observer's axis. Similarly there is also no response around $\tau = r/c$ due to the absence of response of clouds at $\theta \approx 90^{\rm o}$.

So far we have restricted our attention to "thin" geometries, i.e., single orbits and shells. Generalization to "thick" geometries, e.g., disks and shells with different inner and outer radii, is trivial: the response of a disk can be computed by integrating over a series of circular orbits, and the response of a thick shell is obtained by integrating over a series of shells. In Fig. 13, we illustrate this concept by showing the response from two rings. Thus far, the free parameters we have dealt with are the radius r, the line anisotropy factor A, and in the case of non-spherically symmetric systems, the inclination i and, in the case of biconical illumination, the beam half-angle ω. Thick geometries now introduce inner and outer radii, $r_{\rm in}$ and $r_{\rm out}$, respectively, and a distance-

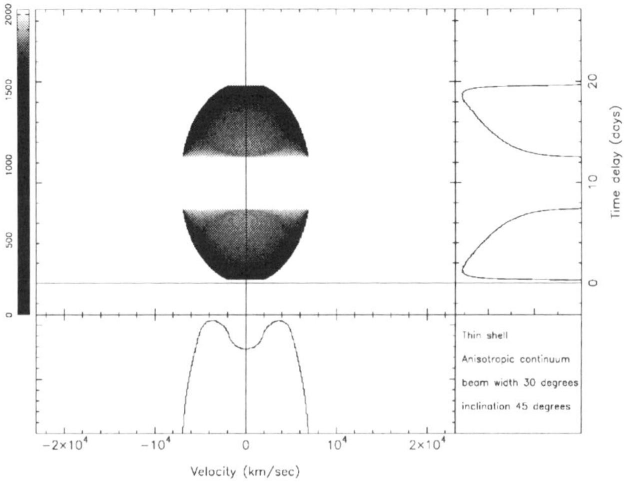

Figure 12. Transfer function for a thin spherical shell, as in Fig. 9, except with an anisotropic continuum source. In the example shown, the beam opening angle is $\omega = 30^o$ and the inclination is $i = 45^o$, as in Fig. 11.

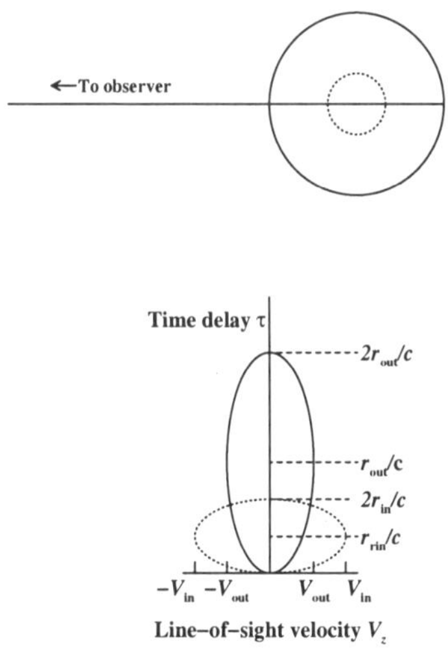

Figure 13. Two ring-like regions (as in Fig. 7), and their projected response on the (V_z, τ) plane.

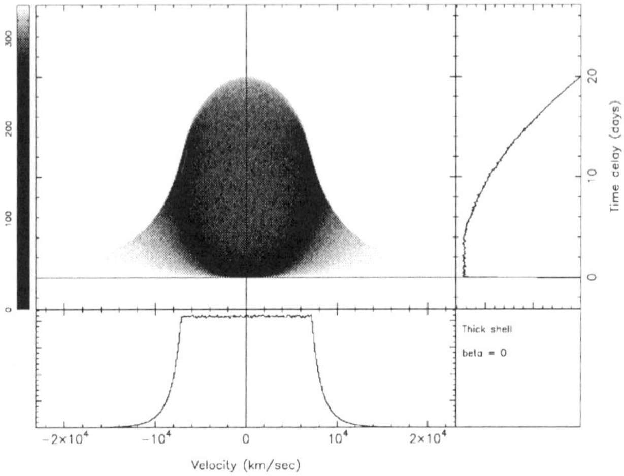

Figure 14. Transfer function for a thick spherical shell with $r_{\rm in} = 2$ lt-days, $r_{\rm out} = 10$ lt-days, and radial responsivity index $\beta = 0$.

dependent responsivity per unit volume (or per unit area, for a disk), which we can parameterize as $\varepsilon(r) \propto \varepsilon_0 r^{\beta}$. The index β allows us to condense model-dependence into a single parameter that will account for effects due to geometrical r^{-2} dilution of the continuum, a distance-dependent covering factor, etc. In Figs. 14 and 15 we show transfer functions for thick spherical shell systems with $A = 0$ and identical values of $r_{\rm in}$ and $r_{\rm out}$, but differing radial-response indices β; the effect of increasing β is to enhance the relative response of material at larger radii; the limiting cases where $\beta \to -\infty$ and $\beta \to +\infty$ correspond to the response functions of thin shells of radius $r_{\rm in}$ and $r_{\rm out}$, respectively. We will show additional examples below.

We consider now a thick shell system that is illuminated by an anisotropic continuum source. Again, we assume that the line-emitting clouds are in circular Keplerian orbits of random inclination. We show examples that are identical except for the continuum beam width and inclination in Figs. 16 and 17. An important thing to notice is that in one case (Fig. 16) the observer is outside of the continuum beam (i.e., $i > \omega$) and in the other case (Fig. 17) the observer is inside of the continuum beam (i.e., $i < \omega$); when the observer is *inside* the beam, the line profile is *single-peaked*, and when the observer is *outside* the beam, the line profile is *double-peaked.*

Another simple kinematic model for the BLR consists of clouds in spher-

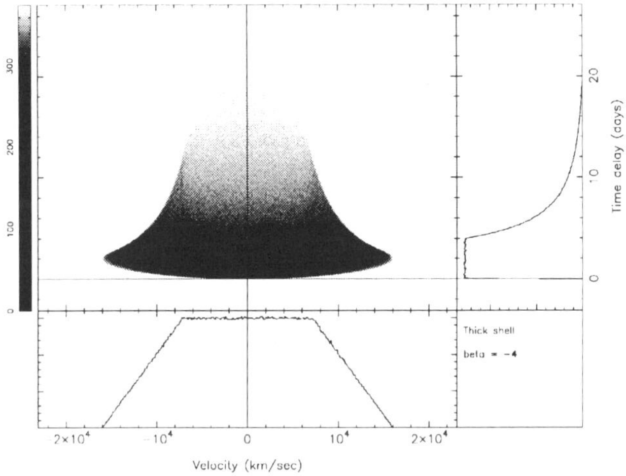

Figure 15. Transfer function for a thick spherical shell exactly as described in Fig. 14, except with radial responsivity index $\beta = -4$.

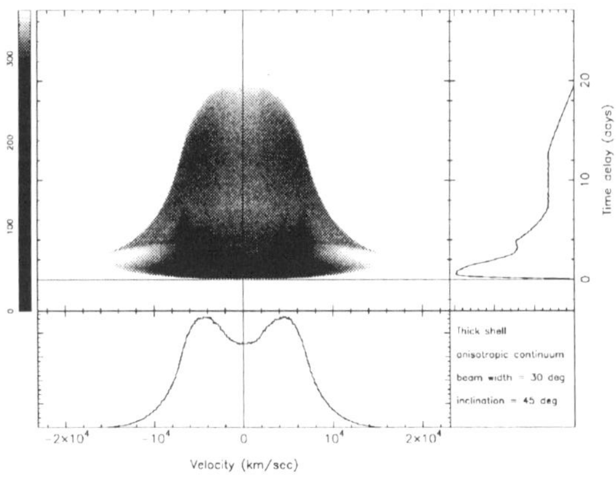

Figure 16. Transfer function for a thick spherical shell with $r_{in} = 2$ lt-days, $r_{out} = 10$ lt-days, radial responsivity index $\beta = -2$, and isotropic line response $A = 0$. In this model, the shell is illuminated by a biconical continuum with semi-opening angle $\omega = 30^o$ and inclination $i = 45^o$, i.e., the observer is outside the continuum beam. Note the double-peaked line profile, and contrast this with the model shown in Fig. 17.

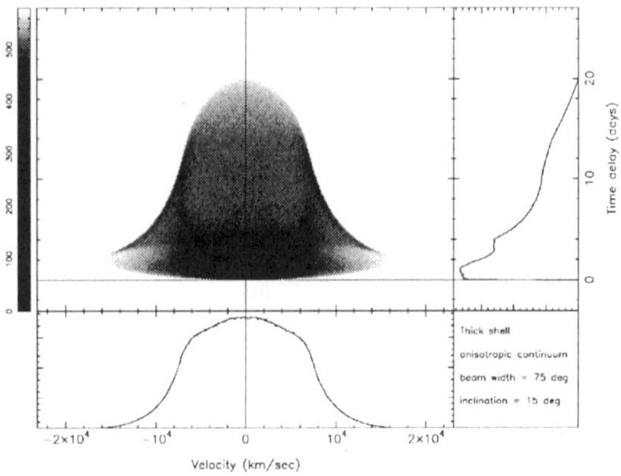

Figure 17. Transfer function for a thick spherical shell exactly as described in Fig. 16, except that in this model the shell is illuminated by a biconical continuum with semi-opening angle $\omega = 75^{\circ}$ and inclination $i = 15^{\circ}$, i.e., the observer is well inside the continuum beam. Note the single-peaked line profile, and contrast this with the model shown in Fig. 16.

ical outflow. The transfer functions for such cases are quite distinctive; an example of a two-dimensional tranfer function for a kinematic field with radial velocity $V_r \propto r$ for r less than some maximum distance $r_{\rm out}$ is shown in Fig. 18. This velocity field corresponds to either a ballistic outflow or a flow undergoing constant acceleration. At $\tau = 0$ we see response from all the material along the line of sight to the continuum source, which ranges from $V_z = 0$ for the material closest to the central source to $V_z = -V(r_{\rm out})$ for the gas farthest from the continuum source. As τ increases, we begin to see response from the far side of the BLR where the line of sight velocities V_z become positive. The range of line–of-sight velocities decreases as the isodelay surfaces get farther from the line of sight. At $\tau = r_{\rm out}/2$, the isodelay surface no longer crosses any clouds moving towards the observer, and the gas moving fastest away from the observer is that along the line of sight ($\theta = 0$) with $V_z = V(r_{\rm out})/2$. At $\tau = 2r_{\rm out}/c$, only the response from the antipodal point is seen, and the transfer function is contracted to the single point at $[2r_{\rm out}/c, V(r_{\rm out})]$.

The case of biconical outflows (which are possibly relevant, as they are certainly seen in the NLR and might well apply to at least a component of the BLR) can be dealt with by restricting the response to certain values of θ; an example is shown in Fig. 19.

We have now seen that both orbital and outflow models can produce similar emission-line profiles; if the line-emitting gas is confined to a bicone, either because of the gas distribution or the ionizing-photon distribution, one can get a single-peaked or double-peaked line profile. The two situations can be easily distinguished, however, by their two-dimensional transfer functions (or equivalently, the combination of their one-dimensional transfer functions and their line profiles). In Fig. 20, we directly compare the one-dimensional transfer functions and line profiles for two thick-shell models: (1) emission-line clouds in a biconical outflow and (2) clouds in circular Keplerian orbits of random inclination, illuminated by a biconical continuum source. In both cases, the beams (one radiation, one matter) have the same half-opening angle ($\omega = 40^{\circ}$) and two different inclinations are shown; $i = 25^{\circ}$ in the top row (i.e. the observer is in the beam, as indicated in the left-hand column illustrations of the geometry), and $i = 65^{\circ}$ in the bottom row (i.e., the observer is out of the beam). The distribution of line-emitting clouds is the same, regardless of how the clouds are moving, so in these two cases the one-dimensional transfer functions ought to be very similar, which is indeed the case, as seen in the middle column of Fig. 20. The right-hand column shows the line profiles, which are however very different. Consider the case $i = 25^{\circ}$ (top row): in the case of outflow, the line-emitting material in the beam is moving radially outward, giving relatively highly blueshifted (near side) and redshifted (far side) emission, but little emission near $V_z = 0$, since there is no line-emitting material moving transverse to the line of sight. In the case of clouds in circular orbits illuminated by an anisotropic beam, the cloud motions are perpendicular to their radial vectors, so most of the line-emitting material is at low Doppler shift as the gas motions through the beams are predominantly transverse. Now consider the higher-inclination case ($i = 65^{\circ}$; bottom row): in the case of radial outflow, the gas motions in this case are now primarily transverse so the Doppler shifts are smaller. However, in the case of circular orbits, the material in the beam is moving primarily along the line of sight, and there is a deficiency of material at $V_z \approx 0$. Note that either kinematic field can give either double-peaked or single-peaked profiles: a simple generalization is that double-peaked profiles are produced in outflow models when the observer's line-of-sight is *in the beam* (i.e., $i \lesssim \omega$) and in orbital models when the line of sight is *out of the beam* ($i \gtrsim \omega$). Neither the profiles nor the one-dimensional transfer functions individually tell us much about the BLR kinematics and velocity field, but together they can tell us a lot. Information on both τ and V_z, i.e., the two-dimensional transfer function, greatly reduces the ambiguities.

Finally, for completeness, we consider the case of an inclined disk, as this is the classic geometry for producing a double-peaked line profile. The response

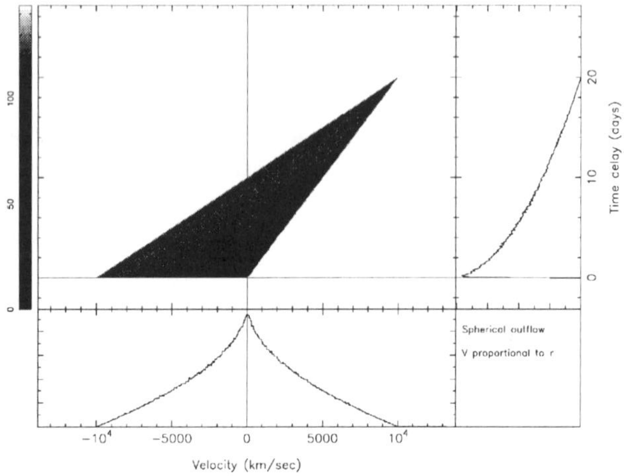

Figure 18. Transfer function for a spherical outflow, with outflow velocity $V \propto r$. For this model, $r_{\rm in} = 0.1$ lt-days, $r_{\rm out} = 10$ lt-days, $V(r_{\rm out}) = 10000\,{\rm km\ s^{-1}}$.

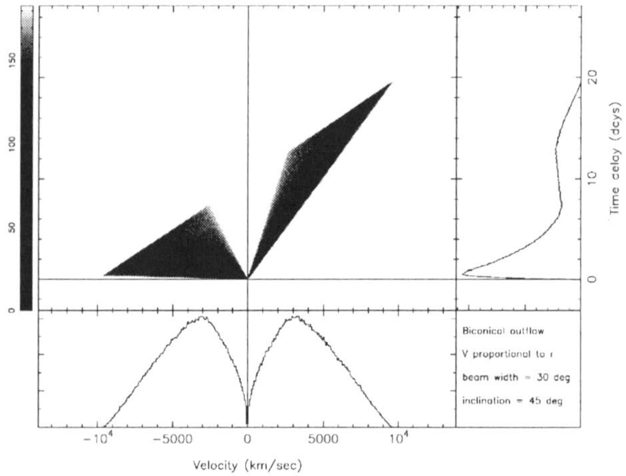

Figure 19. Transfer function for a biconical outflow, with parameters as in Fig. 18 except that the outflowing gas is confined to a bicone of half-width $\omega = 30°$ at inclination $i = 45°$.

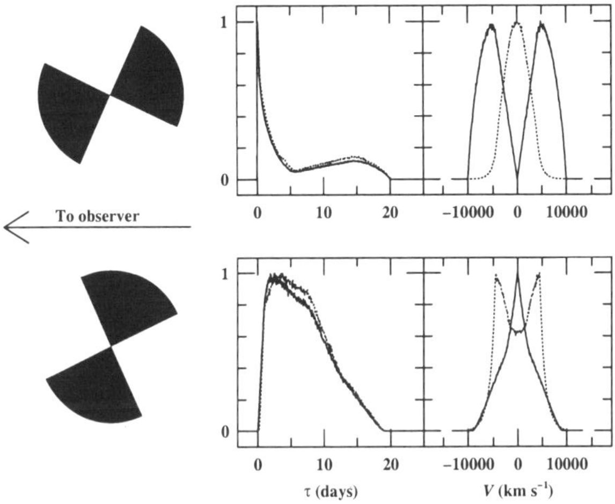

Figure 20. Response of two biconical models with different kinematics. The left hand column shows the two geometries considered, both with beam half-opening angles $\omega = 40^o$; the top row is at inclination $i = 25^o$ (i.e., the observer is in the beam) and the bottom row is $i = 65^o$ (i.e., the observer is out of the beam). One-dimensional transfer functions (middle column) and line profiles (right column) are shown. The solid line is a biconical outflow model, and the dashed line represents circular Keplerian orbits of random inclination. The one-dimensional transfer functions are identical, but the profiles are greatly different, as described in the text.

of a disk-like BLR can be computed by integrating the response of rings of different radii. In Fig. 21, we show the transfer function and line profile for a disk at intermediate inclination ($i = 45^o$). Identical line profiles can be obtained for different inclinations simply by suitably adjusting the central mass, but the transfer function allows us to reduce the ambiguity between possible disk models.

3.5 Transfer Function Recovery

With data of sufficient quality, transfer functions for different plausible models are distinguishable from one another. This is of course, why we place so much emphasis on them: if we can determine the transfer function for a particular line, we have very strong constraints on the geometry and kinematics of the responding region, and if the BLR has a spherically or azimuthally symmetric structure, it may be possible to determine the BLR kinematics and structure with fairly high accuracy and strongly constrain the emission-line reprocessing

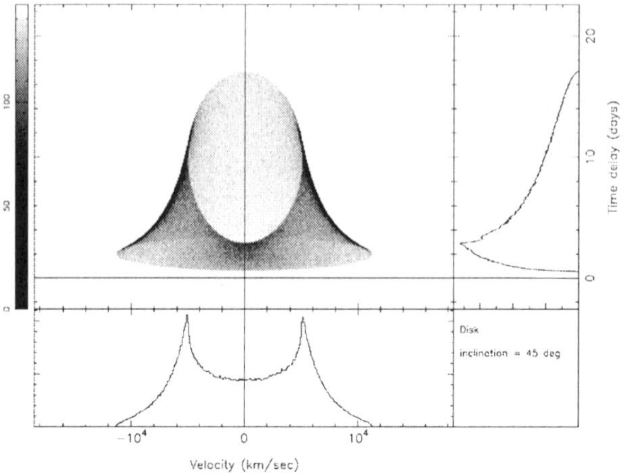

Figure 21. Transfer function for a thin disk at intermediate inclination ($i = 45^o$). Note that the line profile is double peaked, and the one-dimensional transfer function is similar to those for spherical shells.

physics. The operational goal of reverberation mapping experiments is thus to determine the transfer functions for various emission lines in AGN spectra.

How can we determine the transfer functions from observational data? Inspection of the transfer equation (Eq. (20)) immediately suggests Fourier inversion (i.e., the method of Fourier quotients), which was the formulation outlined by Blandford & McKee[7]. We define the Fourier transform of the continuum light curve $C(t)$ as

$$C^*(f) = \int_{-\infty}^{+\infty} C(t)\, e^{-i2\pi ft}\, dt \tag{28}$$

and similarly define the Fourier transform of the line light curve. By the convolution theorem[9], the transfer function becomes

$$L^* = \Psi^* \times C^* \quad , \tag{29}$$

so that $\Psi^* = L^*/C^*$, and the transfer function is obtained from

$$\Psi(\tau) = \int_{-\infty}^{+\infty} \Psi^*(f) e^{i2\pi ft}\, dt \quad . \tag{30}$$

In practice, however, Fourier methods are not used as they are inadequate when applied to data that are relatively noisy (i.e., flux uncertainties are only a factor of a few to several smaller than the variations) and which are limited in terms of both sampling rate, which is in any case usually irregular, and duration. Simpler methods, like cross-correlation (Sec. 4), can be applied with success, though with limited results. Cross-correlation, we will see, can give the first moment of the transfer function, but little else.

In principle, more powerful methods can be used to attempt to recover transfer functions. The most commonly used is the Maximum Entropy Method (MEM)[33]. MEM is a generalized version of maximum likelihood methods, such as least-squares. The difference is that in the method of least squares, a parameterized model is fitted to the data, whereas MEM finds the "simplest" (maximum entropy) solution, balancing model simplicity and realism. Examples of MEM solutions will be shown in Sec. 5. Other methodologies that have been employed for transfer-function solution include subtractively optimized local averages (SOLA)[77] and regularized linear inversion[44].

4 Cross-Correlation Analysis

4.1 Fundamentals

Cross-correlation analysis is the tool most commonly used in the analysis of multiple time series. Because its application to astronomical time series is often misunderstood and has historically been rather contentious, it merits special attention. Important steps in the development of cross-correlation analysis as applied to AGN variability studies can be found in the literature[27,28,20,49,39].

Cross-correlation analysis is basically a generalization of standard linear correlation analysis, which provides us with a good place to start. Suppose we obtain repeated spectra of one of the brighter Seyfert galaxies, and we want to determine whether or not the variations in the Hβ emission line and the optical continuum are correlated (which was an interesting question 20 years ago, even before emission-line time delays were considered). The first thing you would do is plot the Hβ flux against the continuum flux, as in the left-hand panel of Fig. 22, which shows that the two variables are indeed correlated. A measure of the strength of the correlation is given by the *correlation coefficent*,

$$r = \frac{\sum_{i=1}^{N}\left(x_i - \bar{x}\right)\left(y_i - \bar{y}\right)}{\left(\sqrt{\sum_{i=1}^{N}\left(x_i - \bar{x}\right)^2}\right)\left(\sqrt{\sum_{i=1}^{N}\left(y_i - \bar{y}\right)^2}\right)}\,, \tag{31}$$

where there are N pairs of values (x_i, y_i) and their respective means are $\bar{x}$ and $\bar{y}$. When the two variables x and y are perfectly correlated, $r = 1$. If they

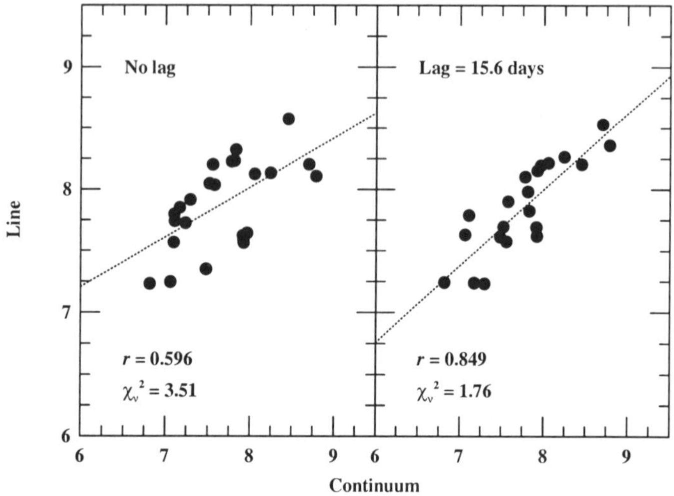

Figure 22. The left-hand panel shows simultaneous measurements of Hβ emission-line and optical continuum fluxes for the Seyfert galaxy Mrk 335[69]. These consist of 24 measurements made on an approximately weekly basis over one observing season. The right-hand panel shows the same emission-line fluxes paired with continuum values from 15.6 days earlier; i.e., the emission-line fluxes are better correlated with *earlier* rather than *current* continuum values on account of light travel-time delays. When the time lag is taken into account, the fit improves.

are perfectly anticorrelated, $r = -1$. If they are completely uncorrelated, $r = 0$. For the data shown in the left panel of Fig. 22, $r = 0.596$; for 24 pairs of points, as shown here, this means that the correlation is significant at the $\gtrsim 99.8\%$ confidence level (i.e., the chance that the two variables are in fact completely uncorrelated and the correlation we find is spurious is less than 0.02%. Confidence levels for linear correlation can be found in standard statistical tables[6])

While this is quite a good correlation, we see something more remarkable if we plot both variables as functions of time (i.e., as *light curves*), as seen in Fig. 23. We see that the patterns of variation are very similar, except that the emission-line light curve is delayed in time, or "lagged," relative to the continuum light curve. It is obvious that the correlation between the continuum and emission-line fluxes would be even better if we allowed a linear shift in time between the two light curves in order to line up their prominent maxima and minima. This is what cross-correlation does.

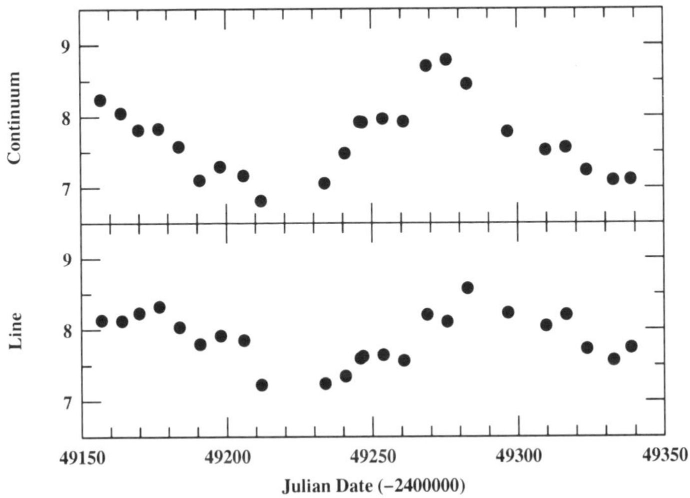

Figure 23. The Hβ emission-line and optical continuum fluxes for Mrk 335, as shown in Fig. 22, are plotted as a function of time. It is clear from the figure that the continuum and emission-line fluxes are well-correlated, and that the correlation can be improved by a linear shift in time of one time series relative to the other. The optimum linear correlation occurs by shifting the emission-line light curve backwards by 15.6 days.

The first operational problem in computing a cross-correlation is also immediately apparent: since each point in one light curve must be paired with a point in the other light curve, it is obvious that the data should be regularly spaced. The cross-correlation is then evaluated as a function of the spacing between the interval between data points Δt using the pairs $[x(t_i), y(t_i+N\Delta t)]$ for all integers N. Unfortunately, regularly sampled data are almost never found in Astronomy; ground-based programs have weather to contend with, and even satellite-based observations are almost never regularly spaced in time. The essence of the cross-correlation problem in Astronomy is dealing with time series that are not evenly sampled. Moreover, the light curves are often limited in extent and are noisy.

For well-sampled series as in Fig. 23, the sampling problem can be dealt with in a straightforward fashion. The simple, effective solution is to interpolate one series between the actual data points, and use the interpolated points in the cross-correlation. We illustrate this schematically in Fig. 24. We can then compute the cross-correlation function CCF(τ), as shown in Fig. 25, and the step size we use for τ is now somewhat arbitrary. At each value of the

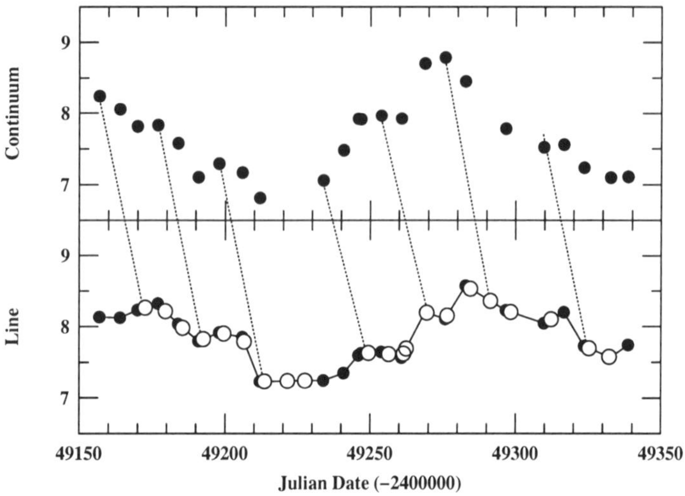

Figure 24. Continuum and emission-line light curves for Mrk 335, as in Fig. 23. This illustrates the interpolation method commonly used in cross-correlation. In this figure, the emission-line light curve is made continuous through linear interpolation between data points. Actual continuum observations are then paired with interpolated emission-line values to compute the correlation coefficient for a particular time delay. In this example, we show interpolated emission-line fluxes that are time-delayed relative to the continuum by 15.6 days, which is the value at which the cross-correlation function peaks. As a visual aid, dotted lines join a few of the data pairs. Notice how the first few points of the emission-line series and the last few points of the continuum series remain unused.

lag τ we compute r as in Eq. (31). For the example we have been using, we find that the CCF is maximized when points in the continuum light curve are matched to those in the emission-line light curve with a delay of 15.6 days. If we plot the shifted emission-line values versus the continuum values (as we have done in the right-hand panel of Fig. 22) and again perform a linear correlation analysis, we find that the fit has improved, with $r = 0.849$ and $\chi^2_\nu = 1.76$.

4.2 Relationship to the Transfer Function

It is useful at this point to examine how the CCF is related to more fundamental quantities. Defined as a function of the continuous variable τ, the

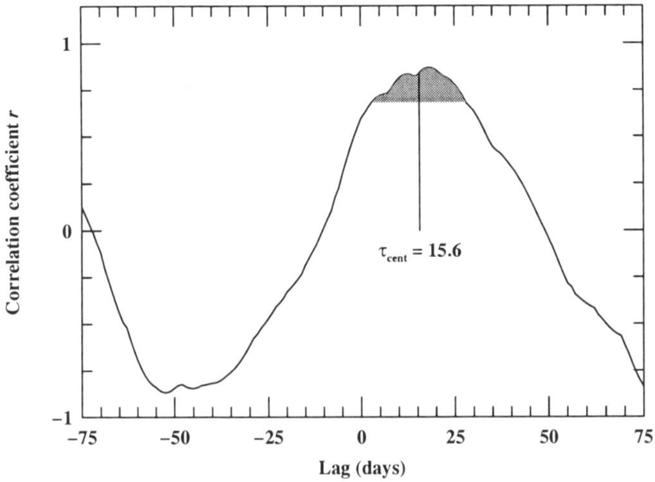

Figure 25. The interpolation cross-correlation function for the Mrk 335 data shown in the previous figures. The shaded area indicates points with values $r \geq 0.8 r_{\rm peak}$, which are the points used in computing the centroid, which is also indicated.

formal definition of the cross-correlation function is

$$\mathrm{CCF}(\tau) = \int_{-\infty}^{\infty} L(t) C(t-\tau)\; dt \quad . \tag{32}$$

If we use Eq. (21) to replace $L(t)$ in this equation, we obtain

$$\begin{aligned} \mathrm{CCF}(\tau) &= \int_{-\infty}^{\infty} C(t-\tau) \int_{-\infty}^{\infty} \Psi(\tau') C(t-\tau')\; d\tau'\; dt \\ &= \int_{-\infty}^{\infty} \Psi(\tau') \int_{-\infty}^{\infty} C(t-\tau') C(t-\tau)\; dt\; d\tau' \quad . \end{aligned} \tag{33}$$

Comparing the inner integral with Eq. (32), we see that it is the cross-correlation of the continuum light curve with itself, i.e., the "autocorrelation function",

$$\mathrm{ACF}(\tau) = \int_{-\infty}^{\infty} C(t) C(t-\tau)\; dt \quad . \tag{34}$$

Thus, we see that the cross-correlation function is the convolution of the transfer function with the continuum autocorrelation function,

$$\mathrm{CCF}(\tau) = \int_{-\infty}^{\infty} \Psi(\tau') \mathrm{ACF}(\tau-\tau')\; d\tau' \quad . \tag{35}$$

The centroid of the CCF is a usually quoted quantity,

$$\tau_{\rm cent} = \frac{\int \tau \, {\rm CCF}(\tau) \, d\tau}{\int {\rm CCF}(\tau) \; d\tau} \;, \tag{36}$$

which is related to the centroid of the transfer function

$$\tau^*_{\rm cent} = \frac{\int \tau \, \Psi(\tau) \, d\tau}{\int \Psi(\tau) \; d\tau} \;. \tag{37}$$

It is important to note that these two quantities are the same only in the limit where the light curves extend to infinity. Operationally, the centroid is computed using only points around the most significant peak, usually those points for which ${\rm CCF}(\tau) \geq 0.8 r_{\rm peak}$, where $r_{\rm peak}$ is the maximum value of the CCF. Sometimes the location of the peak $\tau_{\rm peak}$ (${\rm CCF}(\tau_{\rm peak}) = r_{\rm peak}$) is the statistical quantity used.

Cross-correlation does not necessarily give a clean and unambiguous measure of the relationship between two time series. In particular, the CCF is a convolution of the transfer function with the continuum ACF, so the CCF depends on the continuum behavior. To illustrate this, we show in Fig. 26 examples of light curves for various power-law PDSs (Eq. (7)) and the very different ACFs computed for each example.

4.3 Discrete Correlation Methods

There are some circumstances under which one might not be able to reasonably interpolate between gaps in data. This can occur (a) when there are a few large gaps in otherwise well-sampled series or (b) when there is reason to believe that the variations might be at least somewhat undersampled. In these cases, interpolation might be highly misleading, and another methodology needs to be employed. The "discrete correlation function" (DCF)[20] method is one where no assumption about light curve behavior needs to be made. The DCF method deals with irregularly sampled data by binning the data in time, as illustrated schematically in Fig. 27. This is an alternative approach to the irregular sampling requirement: instead of requiring that points contributing to ${\rm CCF}(\tau)$ are separated in time by exactly the interval τ, we time-bin the data by pairing points with time separations in the range $\tau \pm \Delta\tau/2$, where $\Delta\tau$ is the width of one time bin. Choice of the binning window is a free parameter, and two examples are shown in Fig. 28.

The principal virtues of the DCF method are (a) that only actual data points are used and (b) that it is possible to assign a statistical uncertainty to the value of the correlation coefficient in each bin. The relative weakness of the DCF is that the data are in some ways underutilized, as is evident in

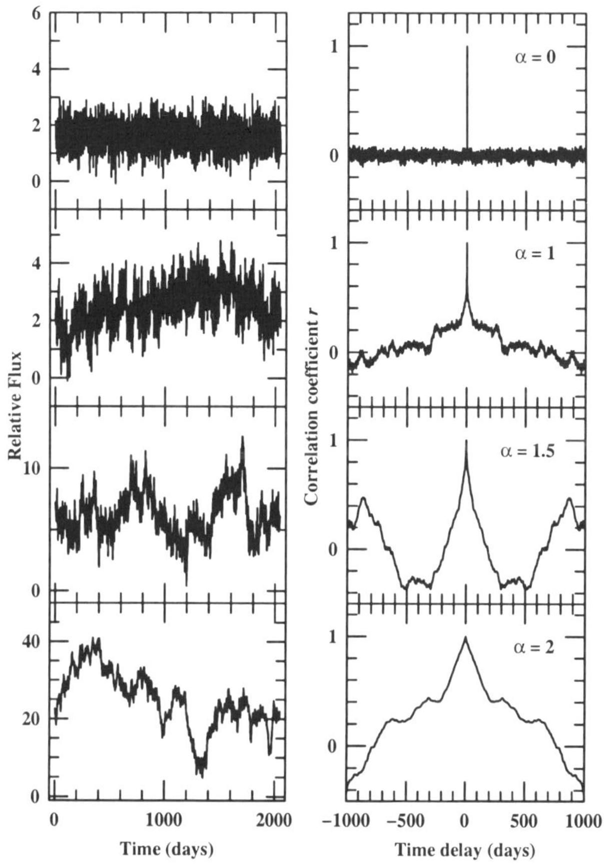

Figure 26. The left-hand column shows artificial light curves generated with different power-law indices α for the power-density spectra, $P(f) \propto f^{-\alpha}$. All light curves consist of 2048 points, normalized by setting $F_{\rm var} = 0.3$ (Eq. (2)). The right-hand column shows the autocorrelation functions for these light curves. Cases shown, top to bottom, are: (a) $\alpha = 0$. There is no correlation from one point to the next, as seen clearly in the ACF, which is essentially zero except at $\tau = 0$. This is known as "white noise." (b) $\alpha = 1$. This is the power spectrum characteristic of a random walk; each continuum point now has some dependence on history. This case is sometimes called "flicker noise" (see Press[78] for further interesting discussion of this case). (c) $\alpha = 1.5$. This appears to be rather typical of AGN variability. (d) $\alpha = 2$. This power spectrum is characteristic of a superpositioning of independent events, which is often referred to as "shot noise."

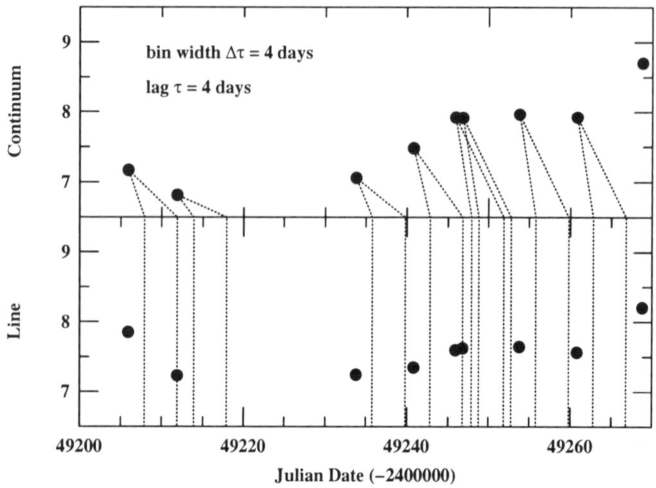

Figure 27. Part of the light curves from the previous figures, expanded to show DCF bins. The particular example shown is for a time-lag bin width $\Delta\tau = 4$ days and shows the location of the bins, pairing real continuum points with emission-line points within a 4-day window, centered at a shift of 4 days (i.e., emission-line data points shifted by 2–6 days from the continuum points). Note that in this expanded region there are only three emission-line points that fall into the bins; in the range shown here, only two continuum and three emission-line points contribute to the computation of the correlation coefficient.

Fig. 27; for a small data set, the DCF method might completely miss a real correlation, although it is less like to find a spurious correlation than is the interpolation method[94].

One difficulty of the DCF method is that the number of points per time bin can vary greatly, as can be easily inferred from inspection of Figs. 27 and 28. One solution to this is to vary the width of the time bins to ensure that there are a statistically meaningful number of points in each bin. A method for accomplishing this is the "Z-transformed DCF (ZDCF)"[1].

4.4 Computational Issues

In the foregoing discussion, we have concentrated primarily on conceptual issues without paying much attention to a number of computational issues that must be dealt with in actually computing CCFs[94,92,87].

1. *Edge effects.* Only at zero lag do all the points of the time series enter into the calculation of the correlation coefficient; at any other lag, points

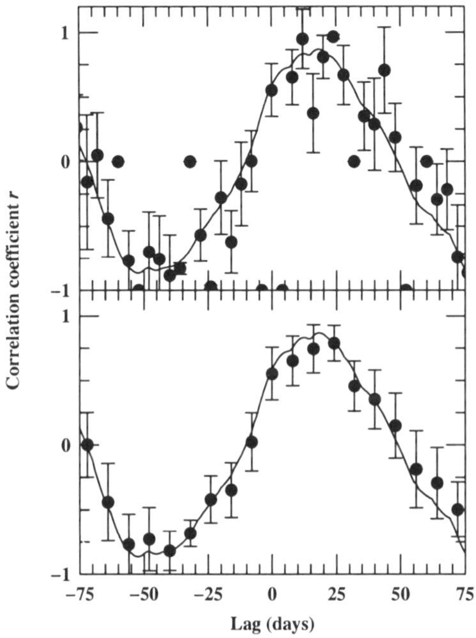

Figure 28. Cross-correlation functions for the Mrk 335 light curve shown in Fig. 23. The DCF values are shown as points with error bars, and the interpolation CCF, as in Fig. 25, is shown as a solid line. The upper panel shows the DCF with a bin width of $\Delta\tau = 4$ days, and the lower panel shows a bin width of $\Delta\tau = 8$ days; with $\Delta\tau = 4$ days, some bins have no data (those with values set to $r = -1$), but with $\Delta\tau = 8$ days, the DCF is somewhat underresolved.

at the ends of the time series drop out of the calculation since there are no points in the other series to which they can be matched (see Fig. 24). This means that at larger and larger lags, fewer and fewer points contribute to the calculation of r. This has two significant consequences:

(a) *Normalization.* Correct normalization of the CCF requires use of the correct value of the mean and standard deviation for each series, as can be seen from the definition of the correlation coefficient r (Eq. (31)). Since AGN time series are limited in duration, as points near the edges drop out of the calculation, the mean and standard deviation of the series change; in statistical language, this means that the series are "non-stationary." Thus, the mean and standard deviation of the series needs to be recalculated for every lag, using only the data points that actually contribute to the calculation. In the case where interpolated values from one series are used, the mean and standard deviation should be those of the interpolated points, not the original points.

(b) *Significance.* Once the peak value of the CCF $r_{\rm peak}$ has been found, we want to know whether or not the correlation found is "statistically significant", i.e., is it likely to be real or spurious? The statistical significance of the correlation depends on the number of points that contribute to the calculation of $r_{\rm peak}$, not to the total number of points in the series. Suppose, for example, that we have two time series consisting of $N = 50$ points, but that the maximum value of the CCF (say, $r_{\rm peak} = 0.45$) occurs at a such a large lag that only $N = 30$ of the points are actually contributing at this lag. For a linear correlation coefficient of $r = 0.45$ and $N = 30$, the correlation is significant at about the 98% confidence level. However, if we erroneously use $N = 50$, we would conclude that the level of significance is about 99.9%, clearly a major difference. If we fail to account for the correct number of points contributing to the calculation of $r_{\rm peak}$ and simply use the number of points in the series, we will *overestimate* the significance of the correlations we detect.

2. *Interpolation scheme.* There are a couple of issues that arise in this regard:

   (a) *Which series?* In the examples shown in Fig. 24, we have interpolated in the emission-line light curve. Is there any particular reason to choose one series over the other when doing the interpolation? In general there does not seem to be, unless, for example, the emission-line light curve is markedly smoother than the continuum light curve (on account of the time-smearing effect) or one light curve is much better sampled than the other. Usually, one computes the CCF by computing the CCF twice, interpolating once in each series, and then averaging the results[28].

   (b) *Interpolating function.* Here we have considered only point-to-point linear interpolation, which results in first-derivative discontinuities in the light curves and the CCFs. Is there any advantage to using higher-order interpolation functions in the time series? Generally, no, higher-order functions don't improve the CCFs in any sense, and can be grossly misleading, as higher-order fits based on only a few data points become hard to control.

3. *Resolution of the CCF.* Can the accuracy of a cross-correlation result be better than the typical sampling interval? Yes, it can, as long as the functions involved are reasonably smooth. The analogy that is usually drawn is that one can measure image centroids to far higher accuracy than the image size, which is true because both stellar images and instrumental point-spread functions are generally smooth and symmetric. Statistical tests as described below suggest that uncertainties of about half the sampling interval are routinely obtained.

4.5 Uncertainties in Cross-Correlation Lags

Although cross-correlation techniques have been applied to AGN time series for about 15 years, there is still no obvious or even universally agreed-upon way to assess the uncertainties in the lag measurements obtained. At present, the most effective technique seems to be a model-independent Monte-Carlo method known as FR/RSS (for "flux redistribution/random subset selection")[70].

FR/RSS is based on a computationally intensive statistical method known as a "bootstrap". The bootstrap works as follows: suppose that you have a set of N data pairs (x_i, y_i) and that linear regression yields a correlation coefficient r. How accurate is r? In particular, how sensitive is it to the influence of individual points? One can assess this by a Monte Carlo process where one selects at random N points from the original sample, without regard to whether or not any point has been selected previously. For the new sample of N points (some of which are redundant selections from the original sample, while some points in the original sample are missing), the linear correlation coefficient is recalculated. When this is done many times, a distribution in r is constructed, and from this, one can assign a meaningful statistical uncertainty to the original experimental value of r.

This process can also be assigned to time series, except that the time tags of the points have to be preserved. In effect, then, this means that redundant selections are overlooked; the probability that in N selections of N points a point will be selected zero times is $1/e$, so the new time series, selected at random, has typically fewer points by a factor of $1/e$ (hence the name "random subset selection"). Welsh[92] suggests that this should be modified in the sense that the weighting of each selected point should be proportional to

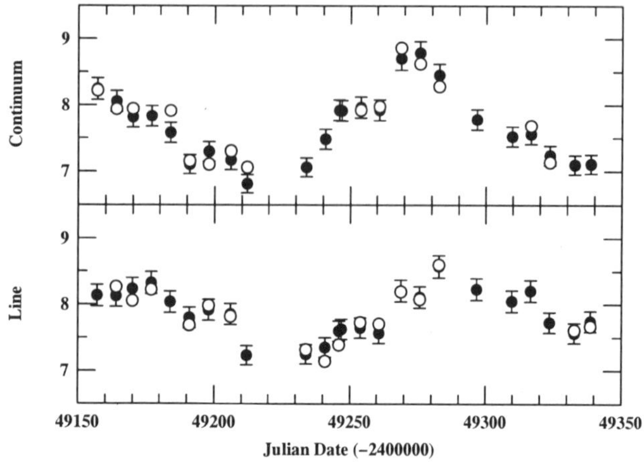

Figure 29. The filled circles show the Mrk 335 light curve from Fig. 23, but now with error bars shown. The open circles show a single FR/RSS Monte Carlo realization; the points are selected from the total subset at random, and the fluxes are adjusted as Gaussian deviates. Note that some of the original data points are not seen because the selected flux-adjusted points cover them. The realization shown here gives $\tau_{\rm cent} = 17.9$ days, compared to $\tau_{\rm cent} = 15.6$ days for the whole data set.

$\sqrt{n_i}$, where n_i is the number of times the data point (x_i, y_i) is selected in a single realization. This is closer in philosophy to the original bootstrap, but it has not been rigorously tested yet.

The other part of the process, "flux redistribution," consists of changing the actual observed fluxes in a way that is consistent with the measured uncertainties. Each flux is modified by a random Gaussian deviate based on the quoted error for that datum (i.e., after a large number of similar modifications, the distribution of flux values would be a Gaussian with mean equal to the data value and standard deviation equal to the quoted error).

A single sample FR/RSS realization is shown schematically in Fig. 29. For each such realization, a cross-correlation is performed and the centroid is measured. A large number of similar realizations will produce a "cross-correlation peak distribution" (CCPD)[49], as shown in Fig. 30. The CCPD can be integrated to assign formal uncertainties (usually $\pm 1\sigma$) to the value of $\tau_{\rm cent}$ measured from the entire data set.

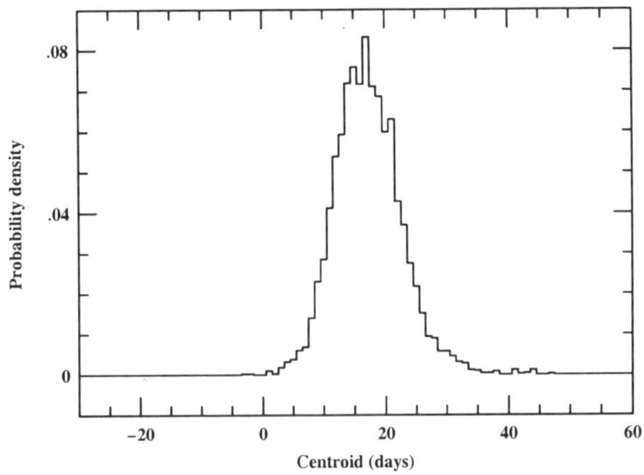

Figure 30. Multiple Monte-Carlo realizations such as those in Fig. 29 are used to build up a cross-correlation peak distribution (CCPD). Relative to the measured CCF centroid for the whole data set ($\tau = 15.6$ days), the $\pm 1\sigma$ width of this distribution is $+7.2$, -3.1 days.

5 Observational Results

With the background provided in the previous sections, we can discuss some of the more important observational results. We will begin with emission-line variability, since the results obtained thus far have been relatively less ambiguous than the results on interband continuum variations.

5.1 Size of the Broad-Line Region

By the late 1980s, the potential power of reverberation mapping was widely understood, but the high-intensity monitoring programs required to extract this information had not been carried out, mainly on account of sociological barriers: the sampling requirements, in terms of time criticality and number of observations, were far in excess of anything that had been done in the past on oversubscribed shared facilities. In late 1987, a large informal consortium known as the *International AGN Watch*[a] was formed in an attempt to obtain sufficient observing time with IUE and various ground-based observatories to measure emission-line lags for the Seyfert 1 galaxy NGC 5548, even at

[a] More information about the International AGN Watch and all papers and data obtained are available on the AGN Watch website at http://www.astronomy.ohio-state.edu/~agnwatch/.

that time one of the best-studied variable AGNs. This turned out to be an enormously successful project[14,66,67] that provided impetus for additional similar projects by the AGN Watch and other groups (see Peterson[62] for a review of the field in the wake of the first monitoring project on NGC 5548). Some of the light curves and CCFs from the original NGC 5548 project are shown in Fig. 31.

Since this initial successful campaign, a handful of AGNs have been well-monitored in the UV and optical (NGC 3783, NGC 4151, NGC 7469, 3C 390.3, Fairall 9, in addition to NGC 5548) and nearly three dozen Seyfert galaxies and low-luminosity quasars have been closely monitored in the optical[88,36]. The most important general conclusions reached from these monitoring programs are:

1. The UV/optical continua vary together, to within a couple of days.
2. The highest ionization emission lines respond most rapidly to continuum changes, implying that there is ionization stratification of the BLR.
3. The BLR gas is not in pure radial motion, since no unambiguous differences in the time scale of the response of the blue and red wings of the emission lines have been detected.

Cross-correlation results show that the size of the Hβ-emitting region scales with average continuum luminosity as[36]

$$r(\mathrm{H}\beta) \approx 0.22 \left(\frac{L_{\mathrm{bol}} h_0^2}{10^{46}\,\mathrm{ergs\ s^{-1}}} \right)^{0.7} \quad \mathrm{pc}, \tag{38}$$

and that, in general, sizes derived from other lines usually scale as $r(\mathrm{He\,II}, \mathrm{N\,V}) \approx 0.2 r(\mathrm{H}\beta)$ and $r(\mathrm{Ly}\alpha, \mathrm{C\,IV}) \approx 0.5 r(\mathrm{H}\beta)$. A slightly different version of the BLR radius-luminosity relationship is shown in Fig. 32, in which $r \propto L^{0.6\pm0.1}$. This is close to the slope of the relationship predicted by the most naive sort of scaling: since to lowest order AGN spectra all look alike, the ionization parameters (Eq. (13)) and particle densities must be the same in all of them. Thus, inverting Eq. (13), we predict that

$$r = \left(\frac{Q(\mathrm{H})}{4\pi c n_{\mathrm{e}}} \right)^{1/2} \propto L^{1/2} \ , \tag{39}$$

where in the last step we have also assumed that the shape of the ionizing continuum is not a function of luminosity. Despite these gross assumptions, the agreement with the data is not bad, although the large scatter alone tells us that this is not the whole story.

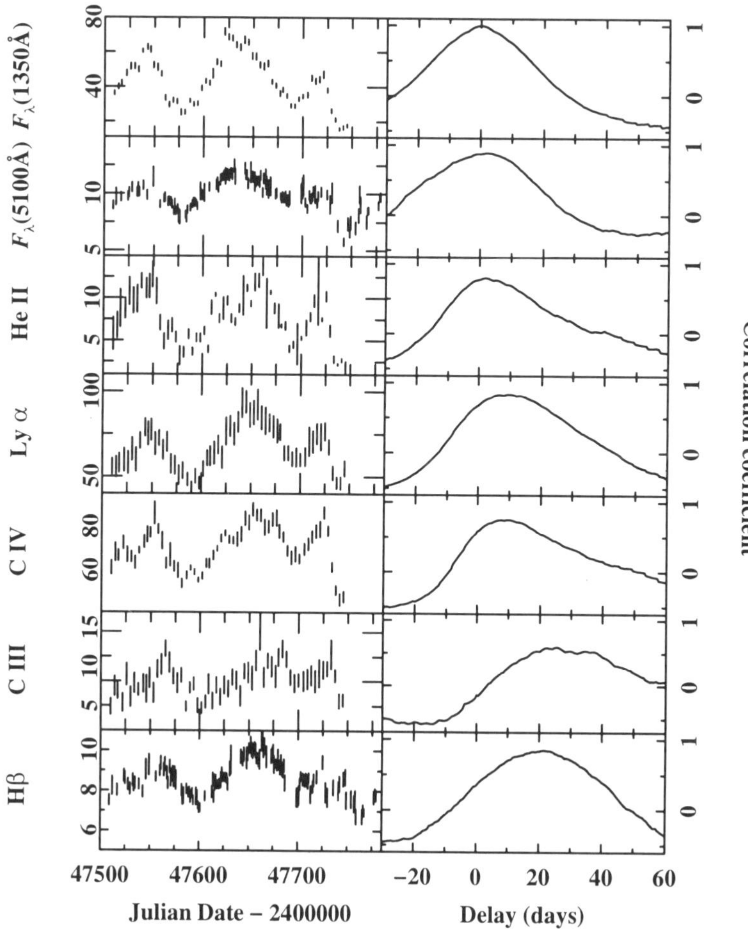

Figure 31. Light curves (left-hand column) and CCFs (right-hand column) for NGC 5548, from the first International AGN Watch monitoring project in 1988–89. From the top, the light curves are the UV continuum at 1350 Å, the optical continuum at 5100 Å, He II λ1640, Lyα λ1215, C IV λ1549, C III] λ1909, and Hβ λ4861. The continuum fluxes are in units of 10^{-15} ergs s^{-1} cm^{-2} Å^{-1}, and the line fluxes are in units of 10^{-13} ergs s^{-1} cm^{-2}. The CCFs are computed relative to the UV continuum at the top; the top panel is the UV continuum ACF. Data from Clavel et al.[14] and Peterson et al.[67].

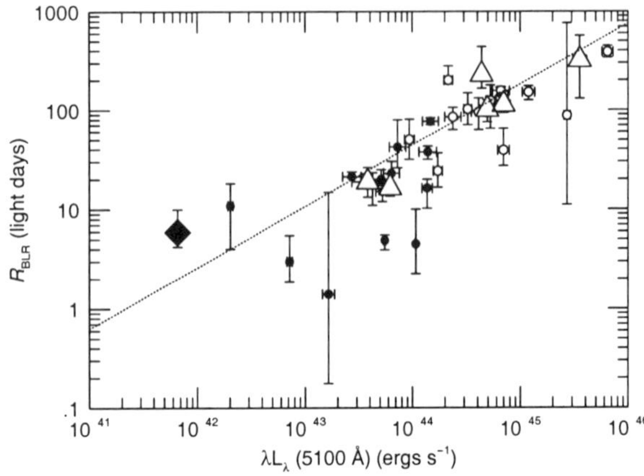

Figure 32. The BLR radius, as measured by the Hβ lag, as a function of average optical continuum luminosity. The open symbols are Palomar–Green quasars from Kaspi et al.[36], the filled circles are Seyfert 1 galaxies from Wandel et al.[88]. The large symbols are narrow-line ($V_{\rm FWHM} < 2000\,{\rm km\ s^{-1}}$) AGNs, and the filled diamond is NGC 4051[75]. The dotted line is the best fit to all the data except NGC 4051.

5.2 *AGN Black-Hole Masses*

Reverberation mapping is one method of measuring AGN central masses via the virial relationship

$$M = \frac{fr\sigma^2}{G} \quad , \tag{40}$$

where f is a dimensionless factor of order unity that depends on the geometry and kinematics of the BLR, σ is the emission-line velocity dispersion, and r is the size of the emitting region. Measurement of the emission-line time lags provides the ingredient that has been missing since the first attempts to understand the basic physics of AGNs[95]. Relative to other dynamical estimators, advantage of using the BLR to provide an estimate of the mass of the central source is that it is located very close to the central source (within $\sim 10^3 R_{\rm grav}$), leaving little doubt that the central mass is in fact a black hole. On the other hand, the kinematics of the BLR are not yet understood (see below), and non-gravitational forces might have a strong effect on gas motions. For the virial method to be applicable, the BLR kinematics must be

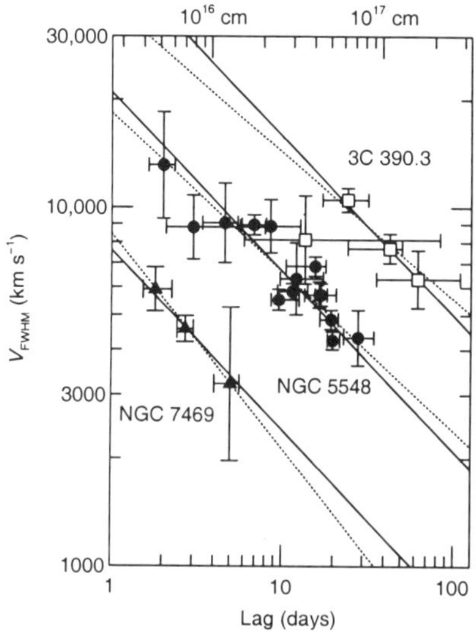

Figure 33. Line width in the rms spectrum plotted as a function of the distance from the central source (upper horizontal axis) as measured by the emission-line lag (lower horizontal axis) for various broad emission lines in NGC 7469, NGC 5548, and 3C 390.3. The dashed lines are best fits of each set of data to the relationship $\log V_{\mathrm{FWHM}} = a + b \log c\tau$. The solid line shows the best fit to each set of data for fixed $b = -1/2$, yielding virial masses of $8.4 \times 10^6\, M_\odot$, $5.9 \times 10^7\, M_\odot$, and $3.2 \times 10^8\, M_\odot$ for the three respective galaxies. From Peterson & Wandel[72] ©2000 AAS..

dominated by gravity. Even without understanding the detailed geometry and kinematics of the BLR, we can test the virial hypothesis by comparing lags and line widths measured in a single AGN: all lines must give the same virial mass, even though not all the line-emitting material needs to have common kinematics. The three AGNs for which this can be easily tested are shown in Fig. 33.

Even if this were not true for all lines, it may be true for some lines, and a given line must always yield the same mass. Only in the case of NGC 5548 is there sufficient information on the long-term behavior of a single line (Hβ) for this test to be applied, and the data seem to be consistent with the virial relationship[72]. We expect, then, that as the continuum brightens, the emission-line lag increases (see Fig. 34), and the emission-line becomes narrower. This does seem to be the case.

Virial masses based on Hβ line reverberation as a function of optical lumi-

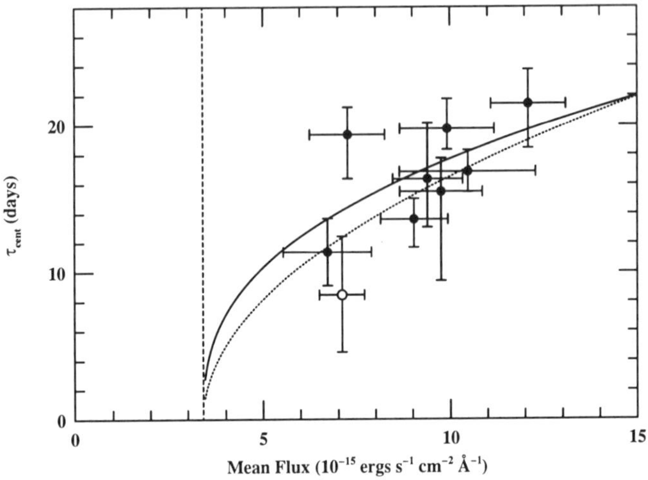

Figure 34. The Hβ emission-line lag as a function of optical continuum flux for the Seyfert 1 galaxy NGC 5548. The solid line shows the best power-law fit to the data, $\tau \propto F^{0.4\pm0.2}$, and the dotted line shows the best fit to the naive theoretical prediction $\tau \propto F^{1/2}$ (Eq. (39)). The dashed vertical line shows the estimated starlight contribution, as in Fig. 1.

nosity are shown in Fig. 35. There is considerable scatter in the relationship, but it is nevertheless clear that higher-mass black holes are found in higher-luminosity AGNs. Some of the scatter in this relationship may be attributable to differences in accretion rate or radiative efficiency: the lower end of the envelope, for example, seems to be dominated by narrow-line Seyfert 1 galaxies, which are thought to have relatively high accretion rates (and thus luminosities) for their mass. Additional factors, such as inclination of the system, may also contribute to the scatter. But we are, finally, beginning to see the first indications of a mass-luminosity relationship for AGNs.

5.3 Emission-Line Transfer Functions

While existing AGN monitoring data have been of sufficient quality and quantity to obtain cross-correlation lags, recovery of transfer functions has, not surprisingly, proven to be far more difficult. Existing transfer function solutions

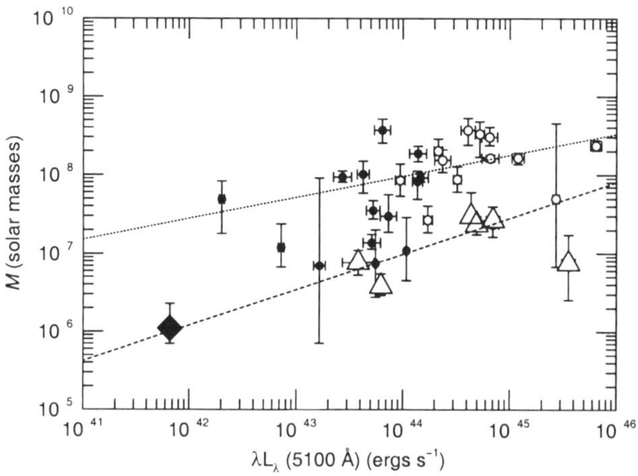

Figure 35. The central mass of AGNs, as measured from Hβ line reverberation, as a function of optical continuum luminosity. Symbols are as in Fig. 32. From Peterson et al.[75] ©2000 AAS.

tend to be very noisy and ambiguous.

In Fig. 36, we show a sample transfer function for the Hβ emission line in NGC 5548; this transfer function is based on data extending over more than a crossing time, so the reader is cautioned against concluding too much from this example, since it is based on data that span a long interval compared to the BLR dynamical time scale (Eq. (19)). The structures seen in this transfer function do not correspond to those seen in any of the simple models that we have described earlier. Note in the one-dimensional transfer function the low amplitude of the response at zero lag, first noticed even with just the first year of AGN Watch data[34]; this indicates that there is little response due to material along our line-of-sight to the continuum source, suggestive of either a low-inclination disk (i.e., there is little gas along the line of sight) or anisotropic line response[24]. Whether or not other lines have small response at small lag is less certain[43,62]; this effect may be seen clearly only in Hβ on account of the large lag.

Fig. 37 shows an attempt to recover the C IV transfer function from 39 daily observations of NGC 5548 made with HST in 1993. These data have been used in several independent analyses[90,19,11,8] with no consensus on the

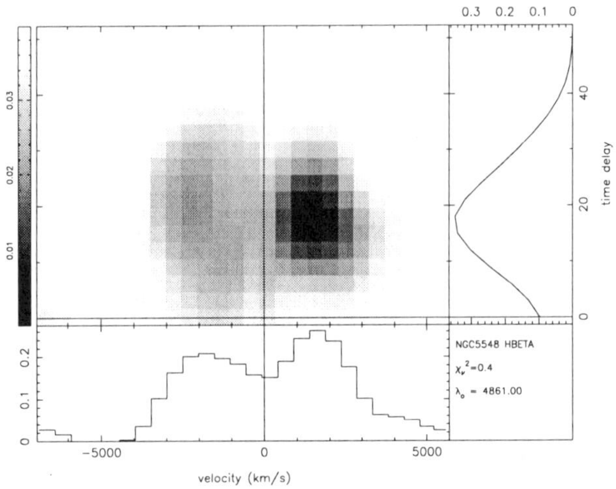

Figure 36. A transfer function for Hβ in NGC 5548, based on AGN Watch data from 1988–1996. Courtesy of S. Collier.

interpretation, and indeed with quite contrary conclusions about the kinematics: Wanders et al.[90] favor no radial motion, Done & Krolik[19] find a hint of radial infall (also previously suggested by Crenshaw & Blackwell[18] on the basis of the first year of IUE data), and Chiang & Murray[11] and Bottorff et al.[8] fit the data with different radial-outflow models.

A crude measure of the velocity field might be obtained by cross-correlating parts of the emission line: for example, comparison of the response times for the red and blue wings could be used to detect radial infall/outflow. Similarly, one could compare the response times for the line wings with the line core in an attempt to detect virial motion, i.e., $V \propto r^{-1/2}$. There have been a number of reports that the emission lines wings respond faster than core[13], as expected for virial motion, but detection is always weak.

A two-dimensional transfer function for the C IV–He II spectral region in NGC 4151 is shown Fig. 38. Ulrich & Horne[85] argue that there is a hint that the red wing responds slightly more rapidly than the blue wing, which is the expected signature for radial infall.

5.4 Emission-Line Profile Variability

If the BLR velocity field is ordered (i.e., $V = f(r)$), then the emission-line profiles should vary in response to continuum changes: the propagation of

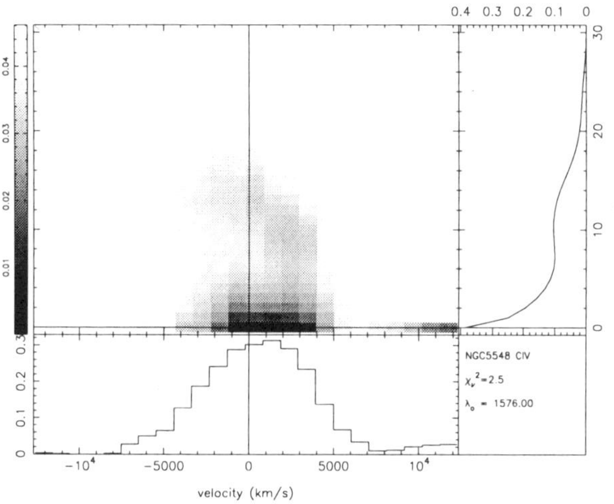

Figure 37. A transfer function for C IV λ1549 in NGC 5548, HST observations[40] from 1993. Courtesy of S. Collier.

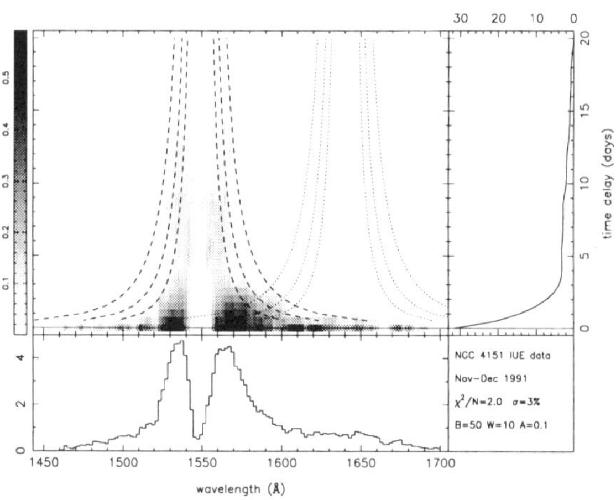

Figure 38. A transfer function for the C IV–He II region of NGC 4151 based on IUE spectra. The strong double-peaked appearance of the C IV line is due to a deep slightly blueshifted absorption feature in this line. From Ulrich & Horne[85] ©1996 Blackwell Sci.

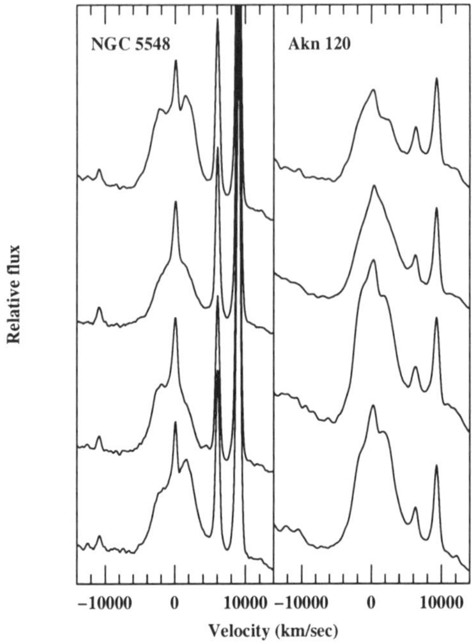

Figure 39. Emission-line profiles for Hβ in NGC 5548 (left-hand column; International AGN Watch data) and Akn 120 (right-hand column; data from Peterson et al.[69]). These are representative profiles obtained over several years of monitoring. The Hβ narrow-line component is particularly noticeable in the spectra of NGC 5548.

excitation inhomogeneities through the BLR should result in irregularities in line profiles that evolve on time scales comparable to the emission-line lags. Peterson et al.[73] have examined profile variations in several Seyfert galaxies that were monitored fairly intensively over several years. They conclude that profile variations are not obviously correlated with either the continuum or the total emission-line flux; profile variations are not reverberation effects. They also find that most Hβ profiles can be modelled fairly well with three Gaussian components that are fixed in location and width and vary only in relative flux, which thus allows multiple-peaked profiles and variable asymmetry. They also conclude that profile irregularities can occur quickly, appearing on time scales as short as days or weeks, but they can then persist for a very long time, months or even years. Sample Hβ profiles for two well-studied Seyferts are shown in Fig. 39.

5.5 Multiwavelength Continuum Variability

As noted earlier, one of the important conclusions reached from the original AGN Watch monitoring program was that the UV and optical continua vary with no apparent time delay between them, at least to the accuracy of the experiment ($\lesssim 1$ day in the best cases). An important consequence of this is that AGN continuum variability is not due either to mechanical accretion-disk instabilities or to variations in the accretion rate, since such effects would be expected to propagate through the disk (and thus across the spectrum) on much-longer sound-crossing (Eq. (11)) or drift time scales (Eq. (12)), respectively. The suggestion was made[17,15] that the UV/optical variations might in fact be driven by X-ray variations, possibly in a manner consistent with the X-ray reprocessing models that were being developed to account for the 10 keV reflection hump and the equivalent width of the 6.4 keV Fe Kα line[32,46,29]; both of these features suggest that something like half of the emitted hard X-rays in AGN interact with "cold" (not highly ionized) matter that covers much of the sky as seen from the source (and which might in fact be the accretion disk itself). If it is supposed that hard X-rays are produced above the disk plane near the axis of the accretion disk, then hard X-ray radiation striking the disk should be reprocessed into lower-energy continuum photons. Moreover, the UV/optical variations should follow those in the X-rays, with the shortest time delays for the higher-energy photons that are produced predominantly in the central regions of the accretion disk. The temperature structure of a classical thin accretion disk is given by

$$T^4 = \frac{3GM\dot{M}}{8\pi\sigma r^3}\left[1 - \left(\frac{r}{R_{\rm min}}\right)^{-1/2}\right] \,, \tag{41}$$

and assuming that AGNs are accreting close to the Eddington rate with about 10% efficiency, the radial temperature structure becomes

$$T(r) \approx 6.3 \times 10^5 \left(\frac{\dot{M}}{\dot{M}_{\rm Edd}}\right)^{1/4} \left(\frac{M}{10^8\,{\rm M}_\odot}\right)^{-1/4} \left(\frac{r}{R_{\rm S}}\right)^{-3/4} {\rm K} \,, \tag{42}$$

from which we expect a peak in the spectral energy distribution in the UV/soft X-ray region, which as noted earlier might in fact be the origin of the big blue bump[81,48].

If we suppose that a thin accretion disk is irradiated by an X-ray source on the disk axis, we should see the inner, hotter part of the accretion disk respond before the outer, cooler parts. From Wien's Law, the radiation at wavelength λ arises primarily at a particular temperature corresponding to a

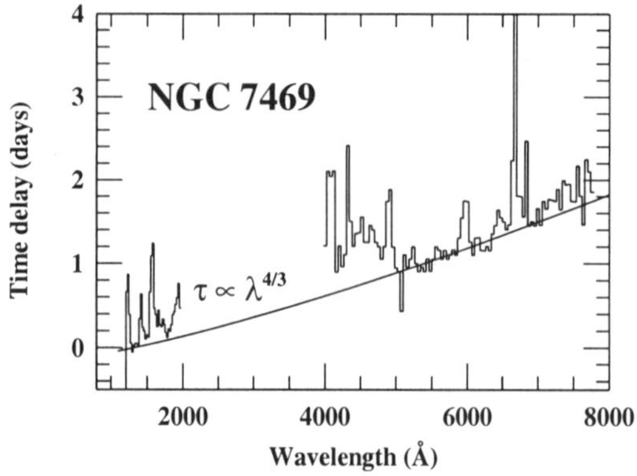

Figure 40. Lag as a function of wavelength for NGC 7469. The average flux in each wavelength band is cross-correlated with the shortest-wavelength UV (1315 Å) continuum, yielding a "lag spectrum". The expected dependence for an irradiated thin accretion disk, $\tau \propto \lambda^{4/3}$, is also shown. The broad emission lines are prominent because they have longer response times than the adjacent continuum.

particular location in the disk, i.e., $\lambda \propto T^{-1} \propto r^{3/4}$, so the difference in the response times at different wavelengths should be naively

$$\Delta\tau \propto \lambda^{4/3} \quad . \tag{43}$$

It is straightforward to show that this relationship holds for an irradiated disk that is heated locally by hard X-rays.

The predicted wavelength-dependent continuum time delays have been detected reliably in the case of NGC 7469, which was monitored intensively in 1996[91,16,42]. Fig. 40 shows the result of measuring the UV and optical fluxes across the entire monitored UV and optical spectrum, and cross-correlating each of the resulting light curves with the shortest-wavelength UV continuum. Relative to the short-wavelength UV, continuum lags in other line-free continuum bands are detected at no less than 97% confidence throughout the UV/optical region.

While the evidence shows that there are in fact wavelength-dependent continuum lags in NGC 7469, this is not an unambiguous detection of a classical thin accretion-disk structure. There are several complications:

1. Similar wavelength-dependent lags have not been reliably detected in other sources. This effect may be marginally present in the case of NGC 4151[70], but it is not found in NGC 3516[22]. However, the sampling rates

obtained for NGC 7469 are far superior to those of most other experiments, and the upper limits on UV/optical continuum lags for other AGNs seem to be consistent with predictions based on (highly uncertain) scaling of the other sources to the NGC 7469 result, except possibly in the case of NGC 3516.

2. Korista & Goad[41] have suggested wavelength-dependent lags are consistent with the expectations for diffuse emission from the emission-line clouds (Balmer and Paschen continuum). Observations over a wider wavelength range are needed to distinguish between an accretion-disk and a diffuse-continuum origin (specifically, the NGC 7469 observations did *not* cover the crucial wavelength range around the Balmer edge ~ 3650 Å).

3. The hard X-ray flux variations in NGC 7469 are not correlated with those in the UV/optical[57]. The hard X-ray spectrum was monitored with RXTE at the same time as the UV/optical monitoring campaign. Comparison of the light curves shows that the minima occurred at about the same time, but the hard X-ray maxima occurred approximately 4 days *later* than the corresponding maxima in the UV continuum light curve. This is a *prima facie* argument against the reprocessing scenario. However, a more complete spectral analysis[56] enabled by an improved background model for RXTE finds that the UV flux variations are correlated with variations in the X-ray spectral index, in the sense that the X-ray spectrum is *softer* when the UV is brighter. This suggests that the soft X-rays are strongly correlated with the UV; an increase in the number of UV seed photons cools the Comptonizing corona above the accretion disk, softening the spectrum.

By way of contrast, a recent study of high-energy variability in NGC 5548 shows clear relationships between the hard X-rays and the lower-energy photons[12]: the extreme ultraviolet (EUV) variations lead those in the soft X-ray region by about 3.5 hrs and those in the hard X-ray by about 10 hrs. The temporal order and time scales are consistent with production of both the soft and hard X-rays by Compton upscattering of lower-energy photons.

If the X-rays are indeed Comptonized lower-energy photons, are the seed photons in the EUV (~ 100 Å) or UV (~ 1000 Å)? The only good comparison available is based on simultaneous UV and EUV monitoring of NGC 5548 in 1993[50]. Cross-correlation of the UV and EUV light curves shows that the UV leads, with $\tau_{\rm cent} = 0.1^{+0.7}_{-0.2}$ days, which is consistent with zero lag and leaves the question unanswered.

A final recent observation that should be mentioned is based on a three-

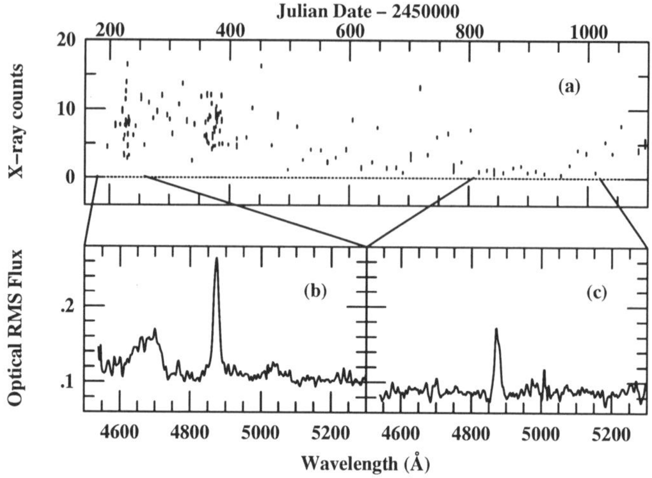

Figure 41. Comparison of hard X-ray and optical spectral variations in NGC 4051 in different X-ray states. Panel (a) shows the 2–10 keV flux measured with RXTE as a function time for the three-year period 1996–1998. Panel (b) shows the rms optical spectrum during an X-ray active period in 1996, Panel (c) shows the rms optical spectrum during an X-ray quiescent period in 1998. Note the absence of strong He II λ4686 emission in panel (c). From Peterson et al.[75] ©2000 AAS.

year combined X-ray (RXTE) and optical monitoring program on the narrow-line Seyfert 1 galaxy NGC 4051[75]. During the first two years of this program, NGC 4051 behaved in a "normal" fashion for narrow-line Seyfert 1s, with rapid, violent X-ray variability and much lower-amplitude optical variations. During the third year, NGC 4051 went into a very low X-ray state. In Fig. 41, the top panel shows the X-ray light curve. The bottom two panels show rms optical spectra, which isolate the variable parts of the spectrum, obtained during the two periods indicated, one during a high X-ray state and one during the low X-ray state. They are remarkably different. The high-state rms spectrum (lower left) shows that the optical continuum and the Hβ and He II λ4686 emission lines were all varying strongly. However, the low-state rms spectrum (lower right) shows that the optical continuum and the Hβ emission line are still present and variable, but the He II line has vanished from the rms spectrum, meaning that it is absent or constant. Whether or not it has completely vanished is difficult to determine on account of the blending of He II with strong Fe II emission in the mean spectrum, but an attempt to remove the Fe II emission by subtraction of a template indicates that most of the He II emission must have vanished during 1998. This indicates that not only has the X-ray flux dramatically decreased, but the EUV flux (which drives the He II variations) must have also decreased by a significant amount. This may suggest that the entire inner accretion disk has undergone a transition to a low-radiation state, such as an "advection-dominated accretion flow" [57].

5.6 Conclusion

The few examples cited in this section give an indication of the power of coordinated multiwavelength observations for understanding the physical mechanisms at work in AGNs on angular scales far too small to be resolvable with any current or near-term technology. These are difficult programs to implement, but the scientific return on them is potential very large, especially as we can build on the significant progress that has already been made.

Acknowledgments

I am grateful for support of AGN variability studies at The Ohio State University by the US National Science Foundation through grant NSF-9420080 and by NASA through LTSA grant NAG5-8397. I would like to thank the other lecturers, especially S. Collin and H. Netzer, and my colleagues S.J. Collier, S. Mathur, R.W. Pogge, J.C. Shields, and M. Vestergaard for suggestions on the lectures and the manuscript. I am very grateful for the hospitality extended to me at INAOE, particularly by I. Aretxaga, J. Franco, D. Kunth, and R. Mújica.

References

1. Alexander T., in Astronomical Time Series, ed. D. Maoz, A. Sternberg, & E.M. Leibowitz (Dordrecht: Kluwer), p. 163, 1997.
2. Andrillat Y., & Souffrin S., Ap. Lett., 1, 111, 1968.
3. Antokhin I.I., & Bochkarev N.G., Soviet Astr., 27, 261, 1983.
4. Antonucci R.R.J., & Cohen R.D., ApJ, 271, 564, 1983.
5. Bahcall J.N., Kozlovsky B.-Z., & Salpeter E.E., ApJ, 171, 467, 1972.

6. Bevington P.R., Data Reduction and Error Analysis for the Physical Sciences, (New York: McGraw–Hill), 1969.
7. Blandford R.D., & McKee C.F., ApJ, 255, 419, 1982.
8. Bottorff M., Korista K.T., Shlosman I., & Blandford R.D., ApJ, 479, 200, 1997
9. Bracewell R., The Fourier Transform and Its Applications, (New York: McGraw–Hill), p. 108, 1965
10. Capriotti E.R., Foltz C.B., & Peterson B.M., ApJ, 261, 35, 1982.
11. Chiang J., & Murray N., ApJ, 466, 704, 1996.
12. Chiang J., Reynolds C.S., Blaes O.M., Nowak M.A., Murray N., Madejski G., Marshall H.L., & Magdziarz P., ApJ, 528, 292, 2000.
13. Clavel J., in Variability of Active Galactic Nuclei, ed. H.R. Miller & P.J. Wiita (Cambridge: Cambridge Univ. Press), p. 301, 1991.
14. Clavel J., et al., ApJ, 366, 64, 1991.
15. Clavel J., Nandra K., Makino F., Pounds K.A., Reichert G.A., Urry C.M., Wamsteker W., Peracaula-Bosch M., Stewart G.C., & Otani C., ApJ, 393, 113, 1992.
16. Collier S., et al., ApJ, 500, 162, 1998.
17. Collin-Souffrin S., A&A, 249, 344, 1991.
18. Crenshaw D.M., & Blackwell J.H. Jr., ApJL, 358, L37, 1990.
19. Done C., & Krolik J.H., ApJ, 463, 144, 1996.
20. Edelson R.A., & Krolik J.H., ApJ, 333, 646, 1988.
21. Edelson R., & Nandra K., ApJ, 514, 682, 1999.
22. Edelson R., et al., ApJ, 534, 180, 2000.
23. Fabrika S.N., Soviet Astr. Lett., 6, 293, 1980.
24. Ferland G.J., Peterson B.M., Horne K., Welsh W.F., & Nahar S.N., ApJ 387, 95, 1992.
25. Fitch W., Pacholczyk A.G., & Weymann R.J., ApJL, 150, L67, 1967.
26. Foltz C.B., Peterson B.M., Capriotti E.R., Byard P.L., Bertram R., & Lawrie D.G., ApJ, 250, 508, 1981.
27. Gaskell C.M., & Sparke L.S., ApJ, 305, 175, 1986.
28. Gaskell C.M., & Peterson B.M., ApJS, 65, 1, 1987.
29. George I.M., & Fabian A., MNRAS, 249, 352, 1991.
30. Goad M.R., O'Brien P.T., & Gondhalekar P.M., in Reverberation Mapping of the Broad-Line Region in Active Galactic Nuclei, ed. P.M. Gondhalekar, K. Horne, and B.M. Peterson (Astronomical Society of the Pacific, San Francisco), p. 315, 1994.
31. Goad M., & Wanders I., ApJ, 469, 113, 1996.
32. Guilbert P.W., & Rees M.J., MNRAS, 233, 475, 1988.
33. Horne K., in Reverberation Mapping of the Broad-Line Region in Active

Galactic Nuclei, ed. P.M. Gondhalekar, K. Horne, and B.M. Peterson (Astronomical Society of the Pacific, San Francisco), p. 23, 1994.
34. Horne K., Welsh W.F., & Peterson B.M., ApJL, 367, L5, 1991.
35. Hummer D.G., & Kunasz P.B., ApJ, 236, 609, 1980.
36. Kaspi S., Smith P.S., Netzer H., Maoz D., Jannuzi B.T., & Giveon U., ApJ, 533, 631, 2000.
37. Kawaguichi T., Mineshige S., Machida M., Matsumoto R., & Shibata K., PASJ, 52, L1, 2000.
38. Kollatschny W., Fricke K.J., Schleicher H., & Yorke H.W., A&A, 102, L23, 1981.
39. Koratkar A.P., & Gaskell C.M., ApJS, 75, 719, 1991.
40. Korista K.T., et al., ApJS, 97, 285, 1995.
41. Korista K.T., & Goad M.R., in Probing the Physics of Active Galactic Nuclei by Multiwavelength Monitoring, ed. B.M. Peterson, R.S. Polidan, & R.W. Pogge (San Francisco: Astronomical Society of the Pacific), in press.
42. Kriss G.A., Peterson B.M., Crenshaw D.M., & Zheng W., ApJ, 535, 58, 2000.
43. Krolik J.H., Horne K., Kallman T.R., Malkan M.A., Edelson R.A., & Kriss G.A., ApJ, 371, 541, 1991.
44. Krolik J.H., & Done C., ApJ, 440, 166, 1995.
45. Kwan J., & Krolik J.H., ApJ, 250, 478, 1981.
46. Lightman A.P., & White T.R., ApJ, 335, 57, 1988.
47. Madejski G.M., Done C., Turner T.J., Mushotzky R.F., Serlemitsos P., Fiore F., Sikora M., & Begelman M.C., Nat, 365, 626, 1993.
48. Malkan M.A., & Sargent W.L.W., ApJ, 254, 122, 1982.
49. Maoz D., & Netzer H., MNRAS, 236, 21, 1989.
50. Marshall H.L., et al., ApJ, 479, 222, 1997.
51. Matthews T.A., & Sandage A.R., ApJ, 138, 30, 1963.
52. McHardy I., Mem. Soc. Astron. Ital., 59, 239, 1988.
53. Mittaz J.P.D., & Branduardi-Raymont G., MNRAS, 238, 1029, 1989.
54. Mushotzky R.F., Done C., & Pounds K.A., ARAA, 31, 717, 1993.
55. Nandra K., Clavel J., Edelson R.A., George I.M., Malkan M.A., Mushotzky R.F., Peterson B.M., & Turner T.J., ApJ, 505, 594, 1998.
56. Nandra K., Le T., George I.M., Edelson R.A., Mushotzky R.F., Peterson B.M., & Turner T.J., ApJ, in press.
57. Narayan R., Mahadevan R., Grindlay J.E., Popham R.G., & Gammie C., ApJ, 492, 554, 1998.
58. Netzer H., & Peterson B.M., in Astronomical Time Series, ed. D. Maoz, A. Sternberg, & E.M. Leibowitz (Dordrecht: Kluwer), p. 85, 1997.

59. Osterbrock D.E., Astrophysics of Gaseous Nebulae and Active Galactic Nuclei (Mill Valley: University Science Books), 1989.
60. Pérez E., Robinson A., & de la Fuente L., MNRAS, 256, 103, 1992.
61. Peterson B.M., PASP, 100, 18, 1988.
62. Peterson B.M., PASP, 105, 247, 1993.
63. Peterson B.M., An Introduction to Active Galactic Nuclei, (Cambridge: Cambridge University Press), 1997.
64. Peterson B.M., Foltz C.B., Miller H.R., Wagner R.M., Crenshaw D.M., Meyers K.A., & Byard P.L., AJ, 88, 926, 1983
65. Peterson B.M., Meyers K.A., Capriotti E.R., Foltz C.B., Wilkes B.J., & Miller H.R., ApJ, 292, 164, 1985.
66. Peterson B.M., et al., ApJ, 368, 119, 1991.
67. Peterson B.M., et al., ApJ, 392, 470, 1992.
68. Peterson B.M., Pogge R.W., Wanders I., Smith S.M., & Romanishin W., PASP, 107, 579, 1995.
69. Peterson B.M., Wanders I., Bertram R., Hunley J.F., Pogge R.W., & Wagner R.M., ApJ, 501, 82, 1998.
70. Peterson B.M., Wanders I., Horne K., Collier S., Alexander T., & Kaspi S., PASP, 110, 660, 1998.
71. Peterson B.M., et al., ApJ, 510, 659, 1999.
72. Peterson B.M., & Wandel A., ApJL, 521, L95, 1999.
73. Peterson B.M., Pogge R.W., & Wanders I., in Structure and Kinematics of Quasar Broad-Line Regions, ed. C.M. Gaskell, W.N. Brandt, M. Dietrich, D. Dultzin-Hacyan, & M. Eracleous (San Franciso: Astronomical Society of the Pacific), p. 41, 1999.
74. Peterson B.M., & Wandel A., ApJL, 540, L13, 2000.
75. Peterson B.M., et al., ApJ, 542, 161, 2000.
76. Phillips M.M., ApJS, 38, 187, 1978.
77. Pijpers F.P., & Wanders I., MNRAS, 271, 183, 1994.
78. Press W., Comments on Ap., 7, 103, 1978.
79. Reynolds C.S., in Probing the Physics of Active Galactic Nuclei by Multiwavelength Monitoring, ed. B.M. Peterson, R.S. Polidan, & R.W. Pogge, (San Francisco: Astronomical Society of the Pacific), in press
80. Schulz H., & Rafanelli R., A&A, 103, 216, 1981.
81. Shields G.A., Nat, 272, 706, 1978.
82. Smith H.J., & Hoffleit D., Nat, 198, 650, 1963.
83. Tohline J.E., & Osterbrock D.E., ApJL, 210, L117, 1976.
84. Ulrich M.-H., Boksenberg A., Bromage G., Clavel J., Elvius A., Penston M.V., Perola G.C, & Snijders M.A.J., ApJ, 382, 483, 1991.
85. Ulrich M.-H., & Horne K., MNRAS, 283, 748, 1996.

86. Ulrich M.-H., Maraschi L., & Urry M., ARAA, 35, 445, 1997.
87. Vio R., & Wamsteker W.W., PASP, in press
88. Wandel A., Peterson B.M., & Malkan M.A., ApJ, 526, 579, 1999.
89. Wanders I., Peterson B.M., Pogge R.W., DeRobertis M.M., & van Groningen E., A&A, 266, 72, 1992.
90. Wanders I., Goad M.R., Korista K.T., Peterson B.M., Horne K., Ferland G.J., Koratkar A.P., Pogge R.W., & Shields J.C., ApJ, 453, L87, 1995.
91. Wanders I., et al., ApJS, 113, 69, 1997.
92. Welsh W.F., PASP, 111, 1347, 2000.
93. Welsh W.F., & Horne K., ApJ, 379, 586, 1991.
94. White R.J., & Peterson B.M., PASP, 106, 879, 1994.
95. Woltjer L., ApJ, 130, 38, 1959.

Brad Peterson (*left*) chatting at the INAOE campus with several participants of the Advanced Lectures (left to right): Marianne Vestergaard, Stefan Collier, Eva Örndahl and Knud Jahnke.

Top: Bob Goodrich addressing some points raised by the audience.
Bottom: Daniel Kunth chairing one of Bob's sessions.

UNIFICATION OF AGN

ROBERT W. GOODRICH
The W. M. Keck Observatory, 65-1120 Mamalahoa Highway, Kamuela, HI 96743 USA
E-mail: rgoodrich@keck.hawaii.edu

In recent years researchers have begun to realize that the classification system used for active galaxies requires some changes. "Unification" refers to the merging of two or more AGN subclasses into a single subclass, usually by understanding that inclination effects or some other relatively uninteresting parameter controls how the object is classified. Once these superficial effects are understood we can get past them to the real physics of the AGN. This chapter describes the current state of unification of several common subclasses of AGN, with a view to showing what researchers can look for and avoid in classifying and understanding these objects.

1 Classification Systems

Our research is inevitably colored by the labels we use to discuss the objects we see in the Universe. We can discuss "NGC 4348" with a colleague, for example, but to put a specific object into a larger context we are inexorably drawn into creating a classification system, or taxonomy, of the objects we observe. It is useful to understand how such taxonomies evolve, and what their limits are, in order to better understand how we can use our observations to extract truly interesting physics from extraneous parameters such as viewing angle, or limitations on data quality.

We start with the discovery of a class of objects, such as active galaxies. This really was the discovery of a small number of related classes, whose intimate relationship was unclear at the time: quasars, and their lower-luminosity kin Seyfert galaxies, for instance. In the early stages, a classification system relies on easily observed keys. For AGN the original paper describing Seyfert galaxies[39] described two types differentiated primarily by line width. Seyfert 2s show strong emission from both forbidden and permitted lines (e.g. [O III] $\lambda\lambda$4959,5007 and Hβ), with line widths of hundreds of km s^{-1}. Seyfert 1s, on the other hand show an additional, "broad-line" component, seen only in the permitted emission lines such as the Balmer series, and with line widths of thousands of km s^{-1}.

Once a class is defined, further study often suggests the need for further subdivisions. Among Seyferts, the subtypes 1.2, 1.5, 1.8, and 1.9 have been used to indicate the relative strengths of narrow-line, broad-line, and continuum components in the optical spectra. Often the classification was heavily

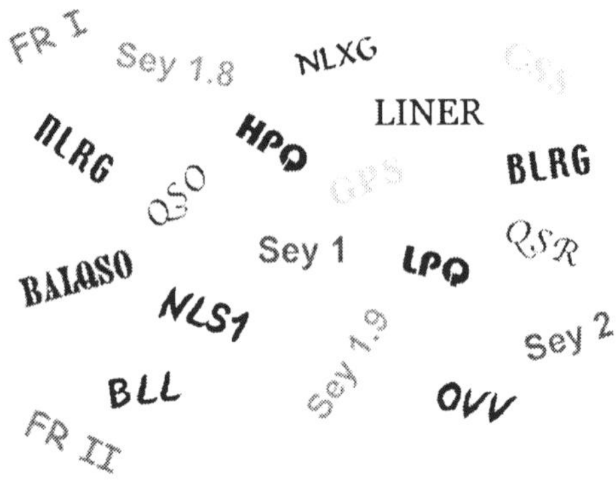

Figure 1. It is a challenge for even experts to sort out the "alphabet soup" of AGN acronyms. Or is this a NASA organizational chart?

dependent on the quality of the spectra used. Similarly, some objects were found that showed the emission line ratios of Seyfert 1s, yet with the relatively smaller line widths characteristic of Seyfert 2s. These were dubbed "narrow-line" Seyfert 1s, or NLS1s.

Observations at different wavebands also suggest the need for different subclasses. Hence from radio surveys we developed subclasses like GPS (GigaHertz-peaked spectrum), CSS (compact steep spectrum), FR I and FR II (Fanaroff-Riley I and II) sources. GPS and CSS are primarily spectral classes while FR I and FR II are based more on radio morphology. From the X-rays we defined classes such as NLXG (narrow-line X-ray galaxies). From the mid- and far-infrared data of IRAS we have ULIRGs ("Ultraluminous Infrared Radio Galaxies").

Soon the list of different AGN subclasses became an "alphabet soup" of acronyms, providing a rather confusing road map to the study of AGN physics (Fig. 1). As more and more is learned, however, it becomes clear that some objects with different classifications are in reality the same physical object, altered by some relatively uninteresting parameter such as the orientation to our line of sight. Lines between subclasses often blur or dissolve, and we achieve what we call "unification". In some cases (e.g. Seyfert 1.2s, and the so-called "N" galaxies) classes fall out of favor and are merely discarded. In other cases (Seyfert 1 and 2) so far we have retained the original classes but must keep in mind that they may be the same objects seen only from different angles.

Ideally, one would like to end up with a classification system based on readily-measurable observational parameters which have a direct relationship to physically meaningful conditions, such as the central black hole mass, its spin, the accretion rate, the star formation rate, metallicity, etc. The analogy with stellar spectral classification is useful; once it was realized that the spectral classification system for stars would make more sense if it were based on effective temperature, stellar astrophysicists threw out some subclasses and rearranged the remainder to form the usual OBAFGKM... sequence we use today. AGN taxonomy is very much a multidimensional system, so it likely cannot be reduced quite so simply by rearranging the existing subclasses. In any event the desire for a simple relationship between observables and physics is rarely met.

In the following we will visit the status of some "unifications" including attempts to unify Seyfert 1s and 2s, unify various types of radio-loud AGN, and unify broad-absorption line QSOs. The morphological nature of the topic leads to a phenomenological approach, so theory will remain relatively untouched. Observational tricks, techniques, pitfalls, and biases will be uncovered along the way.

2 Polarimetry Primer

Speaking of observational tricks, we start with a short primer on polarization and what it can tell us about astronomical objects in general. The reason for discussing polarimetry is that it has become a cornerstone in our attempts to understand AGN. Yet few astronomers understand the details of polarimetry observations and their interpretation.

Polarization originates in some form of asymmetry. Often this is caused by scattering. The Sun reflected off a lake is polarized because the scattering geometry defines a certain direction: the plane defined by the Sun/lake/observer angle.

Similarly magnetic fields define a direction in space and can lead to polarization. As electrons spiral around the magnetic field lines they typically produce circular polarization with some linear polarization mixed in. In stars with large magnetic fields, polarization effects are readily observable in the atmospheric lines, and in extreme cases cyclotron radiation features are seen in the continuum spectrum. For relativistic particles, such as are present in radio galaxy jets, synchrotron emission is produced, with corresponding high linear polarizations.

Another source of polarization is dichroic absorption. Some asymmetry in a crystal or some other organized structure can favor absorption of one sense of polarization over another. Transmission of light through such a medium leaves a net polarization. An example is the interstellar medium, where elongated dust grains are aligned in a magnetic field[a]. A single aligned grain can absorb light polarized along its long axis better than it can absorb light polarized perpendicular to that axis, hence produces polarization by dichroic absorption. Since all of the dust grains are more or less aligned with each other in the magnetic field, together they reinforce each other and produce a net polarization. Were the grains not aligned by the magnetic field, they would all polarize in random directions and the net result would be a cancellation of the polarization from a single grain.

In Astronomy, as in most of Nature, we predominantly deal with linear polarization. One of the exceptions to this is the study of magnetic stars, which often produce circular polarization. The two basic parameters we get from linear polarization measurements are the strength of polarization, represented by P and varying from 0 to 100%, and the angle of polarization, represented by θ. P provides information on both the strength of the asymmetry and the viewing geometry. The two are usually entangled and one must take care not to overinterpret either one or the other. θ gives information on the orientation of the asymmetry, which in the case of scattering means the location of the scatterers relative to the original light source and the observer. Other representations of the polarization are possible, and in fact expert polarimetrists use the linear Stokes parameters, Q and U, to do all of their calculations, converting to (P, θ) only at the end to aid nonexperts in interpretation.

Wavelength information can be used to separate kinematic or spatial components, hence spectropolarimetry is a particularly powerful tool. Moreso than imagers, spectroscopic elements in instruments tend to add their own polarization signatures. For example, reflection gratings are in some sense mirrors, and polarize light accordingly. At some wavelengths some gratings can produce up to 50% polarization or more! Hence in any polarimetry, but especially in spectropolarimetry, great attention must be given to accurate and precise calibrations and (very important, but often overlooked!) reduction and analysis techniques. For the most part polarimetry is done as a "null experiment" in which small differential measurements are made. Null experiments tend to have greater accuracy than other experiment types.

I often compare polarimetry to building an instrument. You find a large telescope. You create efficient optics and detectors to gather as many photons

---

[a] In this case, while the magnetic field provides a source of asymmetry, it does not directly influence the polarization. Spherical or nonmagnetic grains in a magnetic field would not produce significant polarization

as possible. Then you throw out 99% of the photons (for 1% polarization)! An instrument builder who did this purposefully in his/her designs would soon be looking for another job! But the trick with polarimetry is to throw out the *right* 99% of the photons. After all, what we see is light combined from any number of sources: stars in the host galaxy, light from the accretion disk, emission from the broad-line and narrow-line regions, and possibly scattered light from these sources or other, hidden sources. We can use polarimetry to isolate only one or two of these sources, and study them without the other "contaminating" sources of light.

Another advantage is found in the search for geometrical information about objects. In Astronomy we are almost always limited in the spatial scales we see. This is one of the great driving forces for putting telescopes in space above the smearing effects of the Earth's atmosphere, but even in space we are limited by the physics of optics and diffraction. Hence we search for alternate, less direct means of getting information on the geometry of distant or tiny objects. The angular information in θ provides at least a first order look at the geometry of unresolved regions, and P can add to this if interpreted in a broader framework.

Having now sold you on the power of polarimetry, and spectropolarimetry in particular, I mention some important surveys that have been done in the field of active galaxy research. Seyfert 2s[32] and Seyfert 1s[21] have been studied, along with radio galaxies.[8] I mention high-z radio galaxies[46] separately as they are being used for slightly different purposes than their more local brethren. Broad-absorption line QSOs[35] have recently begun to reveal some of their secrets to the technique; we will discuss them towards the end of this chapter. Intermediate Seyferts (Seyfert 1.8s and 1.9s) have been studied[14] as have the so-called "narrow-line" Seyfert 1s.[13] Recently low-luminosity AGN have been studied.[3] So at least in the field of active galaxies researchers are beginning to aggressively tap the potential of polarimetry, and finding it very... well, illuminating.

3 Seyfert 1 and 2 Galaxies

One of the most important driving forces in unification in recent years was the discovery by Antonucci & Miller[1] that the canonical Seyfert 2 galaxy NGC 1068 looks like a Seyfert 1 when viewed in polarized light. Goodrich & Miller[20] followed this by showing that several Seyfert 2s look like Seyfert 1s in polarized light, showing Balmer lines thousands of km s^{-1} wide, and permitted Fe II lines, both characteristic of Type 1 AGN, not Type 2s.

Shown in Fig. 2 is Mrk 348, an example of a Seyfert 2 showing a hidden

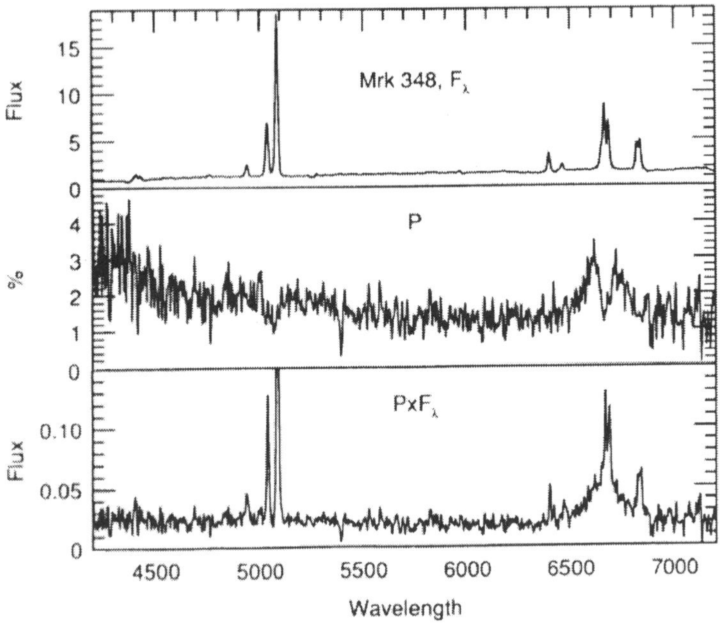

Figure 2. Spectropolarimetry of Mrk 348. Normal spectroscopy (top panel) shows the narrow lines typical of a Seyfert 2. But the polarized flux (bottom panel) shows broad Hα, typical of a Seyfert 1. From Miller & Goodrich[32] ©1990 AAS.

Seyfert 1 spectrum. Note the total flux spectrum at the top of the figure. It shows the spectrum that a normal spectrograph would produce. Relatively narrow forbidden emission lines like [O III] λ4959, 5007, [O I] $\lambda\lambda$6300, 6363, [N II] $\lambda\lambda$6548, 6583 and [S II] $\lambda\lambda$6716, 6731 dominate the spectrum. The central panel shows the fractional polarization as a function of wavelength. Note the drop in P at the strong [O III] lines, and the large rise on either side of the Hα+[N II] blend.

Looking only at fractional polarization, these features are not easy to interpret, but by multiplying the total flux and fractional polarization we get the bottom panel, the polarized flux. This in effect shows the flux spectrum of the light that is scattered, free from most of the direct light such as starlight. (In practice some of the other light sources have *some* small polarization and end up contaminating even the polarized flux.) In particular note the very broad base beneath the Hα+[N II] blend around 6560 Å. This is typical of the broad component of Hα seen in Seyfert 1 galaxies. In fact the entire polarized flux spectrum of Mrk 348 looks exactly like a Seyfert 1!

NGC 1068 was once one of the prototypical Seyfert 2 galaxies, and has become the prototypical "hidden Seyfert 1." It is easily the best studied Seyfert 2, being bright and relatively nearby. The luminosity of the hidden nucleus has been measured and shown to be similar to other Seyfert 1 galaxies, approaching QSO luminosities. Scattering regions are seen in the body of the host galaxy, far enough to be spatially resolved even from the ground. This has allowed us to show that scattering by dust occurs at these large radii, while in the nucleus electron scattering dominates. Ultraviolet spectropolarimetry supports this interpretation, and other HST data show that the nuclear scattering region is extended and effectively over-resolved by HST.

3.1 Ionization Cones and Photon Deficits

In a number of Seyfert 2s images taken in the light of highly ionized emission lines show a more or less conical morphology. These "ionization cones" indicate that the ionizing sources do not radiate isotropically. An example[37] is shown in Fig. 3.

Another indication of a hidden power source can be found when you try to estimate the number of ionizing photons reaching these emission line regions. Generally this is done by measuring a line flux, e.g. from Hβ, and calculating the number of photons, $N_{\mathrm{H}\beta}$, needed to create that flux,

$$N_{\mathrm{H}\beta} = 2.1 \times 10^{52}\ (L_{\mathrm{H}\beta}/10^{40}\ \mathrm{erg\ s^{-1}})\ \mathrm{photons\ s^{-1}} \quad . \tag{1}$$

This number is compared to the number of ionizing photons, N_{i}, in the observed spectrum,

$$N_{\mathrm{i}} = \int_{\nu_1}^{\nu_2} L_\nu / h\nu \, d\nu \quad . \tag{2}$$

In reality some assumptions must be made about the shape of the continuum between ν_1 and ν_2 as it is not observed directly. With any reasonable assumptions, however, Seyfert 2s typically show a deficit in observed photons.[49] Apparently the narrow line region (NLR) receives more ionizing flux than we see from the Earth. Again, this indicates anisotropy in the nuclear flux. This effect is seen in radio galaxies as well.

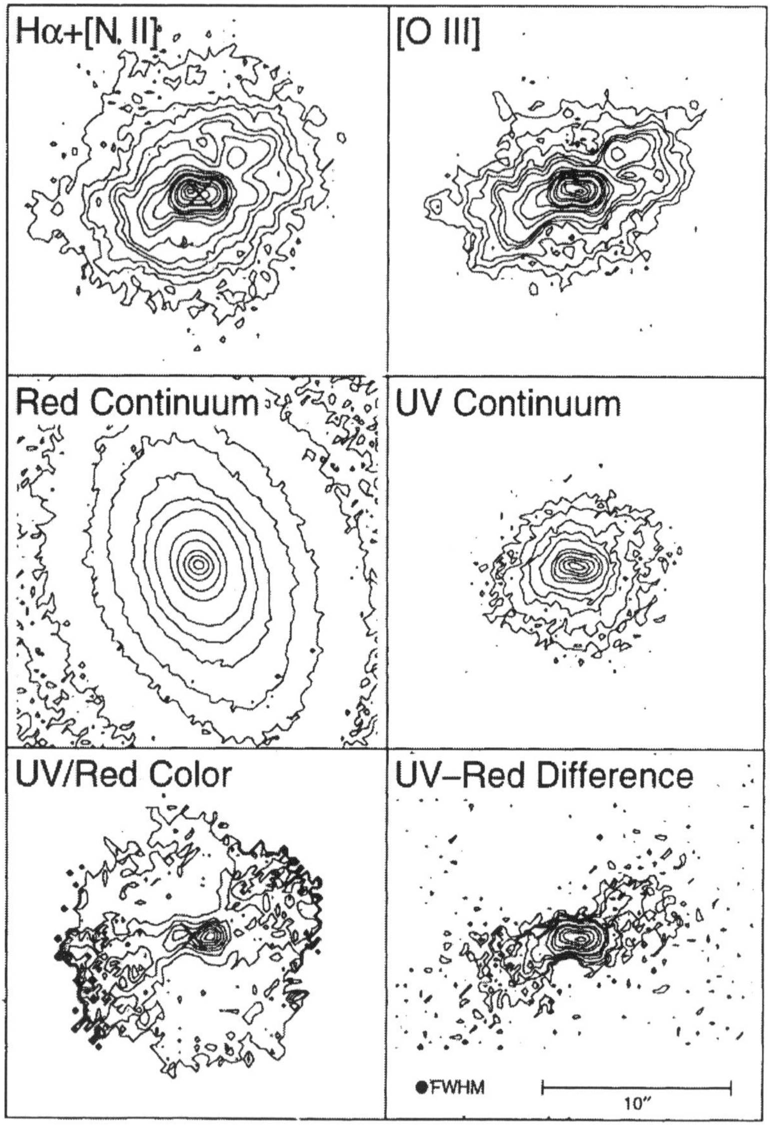

Figure 3. Ionization cones in Mrk 3. The red continuum shows the normal starlight in the galaxy. The top two panels show emission lines, with a very different morphology from than the stars, and forming two opposing fans. The UV light looks similar to the emission lines, and is likely mostly scattered nuclear continuum. Mrk 3 is one of the high-P Seyfert 2s which have been shown to have polarized broad Balmer lines. From Pogge & de Robertis[37] ©1993 AAS.

3.2 *Infrared Spectroscopy*

The picture that has emerged is one of an optically thick, geometrically thick torus surrounding the central source and causing the anisotropy seen at larger radii. Polarimetry sometimes lets us use surrounding material as a cosmic inspection mirror to see into the hidden regions. To see *through* the torus one can go to longer wavelengths, where the optical depth of dust drops, or into the X-rays, where again the optical depth drops but we must worry about gas absorption.

Near-IR spectroscopic studies of Seyfert 2s have searched for broad hydrogen emission. One of the goals is simply to find more evidence for hidden broad line regions (BLRs). Eventually we can get statistical information on the optical depth cross-section of the "mean torus." This is important for determining the physics of formation and maintenance of the thick torus. It is still unclear how the torus maintains its geometric height when dissipation should make it collapse into a thin disk in a relatively short time.

Typical near-IR studies look for broad emission in the Paschen or Brackett series[22]. Longer wavelengths probe higher optical depths; hence Paβ can be seen if the optical extinction, A_V, is less than roughly 11 magnitudes, Brγ if $A_V \leq 26$, and Brα, $A_V \leq 68$. Unfortunately, at higher wavelengths the emission lines can sometimes be overwhelmed by strong thermal emission from dust.

Shown in Fig. 4 is a spectrum of MCG −05-23-16, classified as a Seyfert 2 in the optical. A broad component of Paβ is clearly visible, dominating the narrow component at this wavelength.

3.3 *X-rays*

Comparison of the X-rays and near-IR optical depths is also instructive. The near-IR measures the dust optical depth, while absorption in the X-rays measures the gas optical depth.

Soft X-ray spectra have been used[5] to estimate gas column depths (Fig. 5). The columns are spread around $\log N_{\rm H} = 23$, implying Thomson optical depths close to unity.

In some cases, like NGC 1068, no absorption is seen, although the X-ray flux is much lower than expected from the optical continuum and typical Seyfert 1 spectral shapes. These are likely to be objects which are only showing us the *scattered* light, which is not heavily absorbed. The direct view of the nucleus in NGC 1068 is so heavily absorbed that we don't even see the soft X-rays. This allows us to estimate a lower limit to the column depth, as shown in Fig. 5 by the shaded regions.

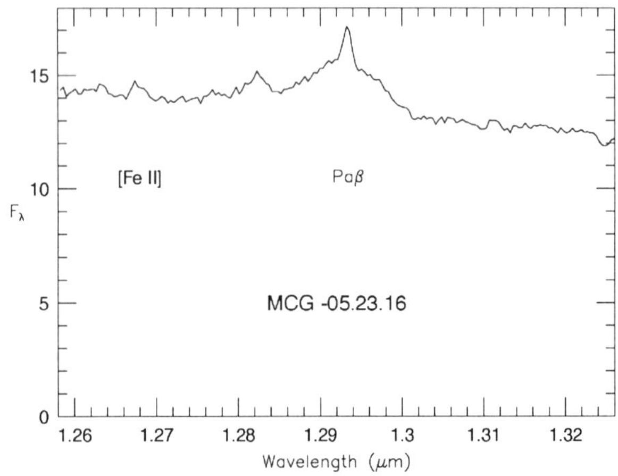

Figure 4. MCG −05-23-16 shows broad Paβ emission in near-IR spectra. From Goodrich et al.[22] ©1994 AAS

3.4 Another, Unsuspected Continuum!

In the total flux spectra of Seyfert 2s it is generally difficult to separate the broad, scattered emission-line components from the narrow-line components, but not impossible. Knowing the broad-line profile from the polarized flux helps immmensely. Once the broad-line wings are identified, it allows us to calculate the line polarization. This is typically reasonably high, 10–35% or more, as expected from simple models of optically-thin scattering. However, when the continuum polarization is measured (after removing the unpolarized starlight contribution, which is identified by its stellar absorption features), it is often found to be significantly lower. One would predict that the continuum and emission line regions would be seen more or less the same way by the scatterers, and hence would have the same polarization. In fact, if anything, the emission line region would be seen as more extended by the scatterers, and hence could be polarized *less* than the continuum.

Since the higher line polarizations are actually predicted by models, it seems likely that the problem lies in an interpretation of the continuum. Specifically, it is as if there is an extra, unpolarized continuum source, diluting

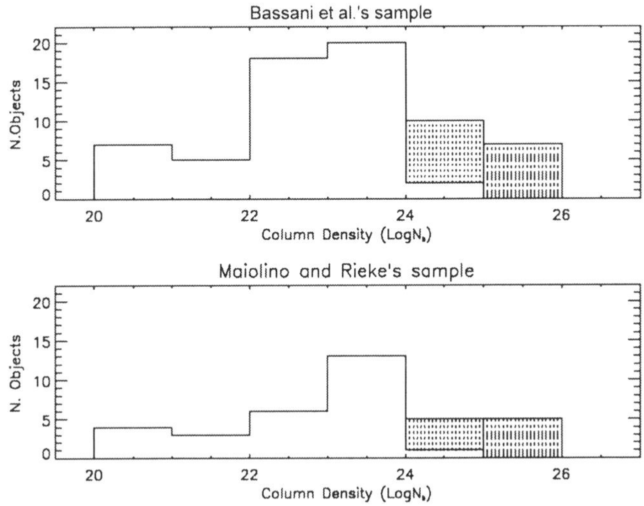

Figure 5. X-ray observations of Seyfert 2s show large absorbing columns of gas in our line of sight. The shaded regions on the right-hand side of the histograms represent lower limits. For scale in the optical, $\log_{10} N_{\rm H} = 21.3$ corresponds roughly to a visual extinction of $A_V = 1$. From Bassani et al.[5] ©1999 AAS.

the total polarization. Tran[44] coined the term "FC2" for "featureless continuum 2" for this second source. He suggested that it was free-free emission from the scattering electrons, although one then predicts that the scatterers produce a large emission line flux as well. In some cases the emission line flux predicted from the scattering region is comparable to the observed narrow-line flux, leaving little flux that could be produced from the "normal" NLR.

Another possibility, and one that almost surely applies in some cases, is that FC2 is not truly featureless, but is starlight coming from hot stars. Absorption features from hot stars can effectively hide in the spectra of AGN, swamped by cooler starlight, the emission line regions, and the featureless continuum from the AGN itself.

3.5 HST Imaging

Current ground-based and space-based spatial resolutions are not sufficient to resolve the very inner regions of AGN, such as the BLR and the accretion disk, but they are sufficient to yield interesting information on somewhat

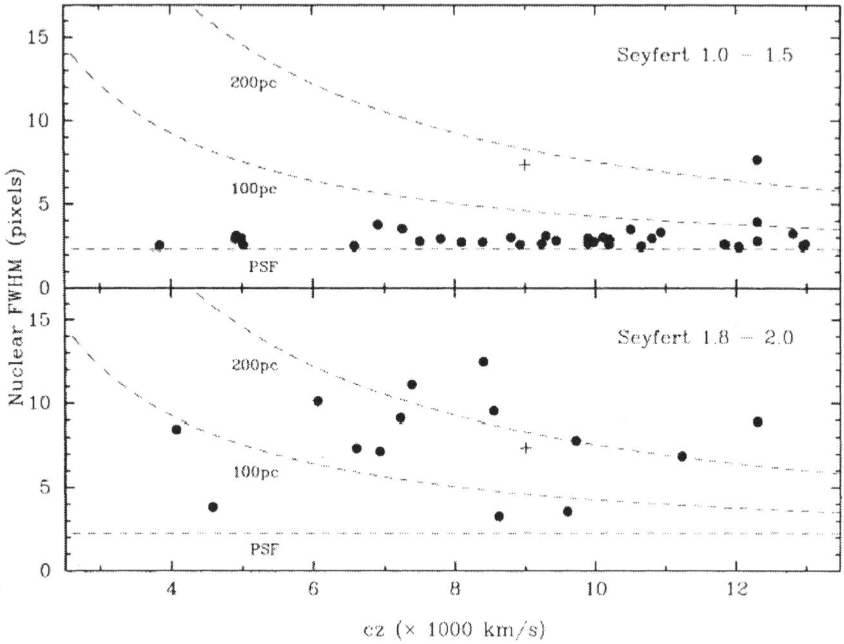

Figure 6. HST imaging of nearby AGN indicate that Seyfert 2s (bottom) do not show central point sources, while Seyfert 1s (top) do. From Nelson et al.[34] ©1996 AAS.

larger regions, including the scattering region. An HST survey by Nelson et al.[34] (Fig. 6) looked at a number of Seyfert 1s and 2s, fitting the photometric profiles with a stellar bulge component plus a more highly concentrated source. Their conclusions were that most Seyfert 1s show a point source in their centers, while in general the Seyfert 2s show more diffuse central sources. This is expected in the unified model, since we see the small continuum region directly in Seyfert 1s (hence it appears point-like), while in Seyfert 2s we see only an extended scattering region.

3.6 H_2O Megamasers

One of the most interesting recent discoveries has been that the strong water masers seen in other galaxies are in some cases related to disk-like structures around the nuclei. The best-studied case[6] is NGC 4258, shown in Fig. 7. The

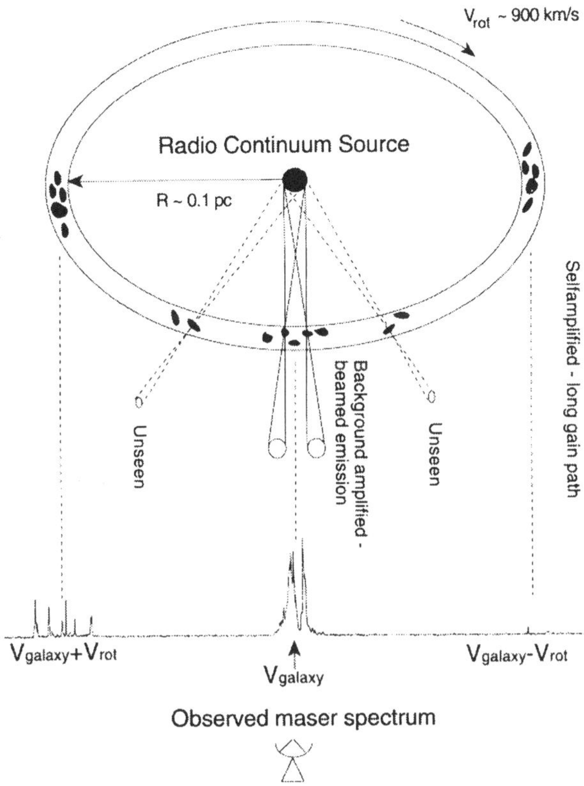

Figure 7. Data (radio spectrum at bottom) and a model of the H_2O megamaser in NGC 4258, after Greenhill et al.[27] ©1995 AAS.

maser observations at the bottom show double-peaked emission at relatively low velocities surrounding the galaxy's systemic velocity, and a number of lines at very high positive and negative relative velocities.

The schematic at the top of the figure presents the interpretation. A rotating ring of material surrounds the nucleus, which is a radio source, and has the right physical parameters to become a water maser. Along the line of sight the radio continuum from the nucleus is amplified. Along the disk edges the path through the material is long enough for self-amplification.

The great thing about the masers is that they provide very precise values of several fundamental parameters, such as velocities and positions. From this comes an accurate dynamical mass, evidence for a warped, but geometrically thin disk, and accelerations. The latter can give a direct measure of the distance to the source, giving an independent estimate of the Hubble constant.

Apart from NGC 4258, there are 20+ other megamasers, including NGC 1068. These are mostly in LINERs or Seyfert 2s, and when resolved the masing structures are perpendicular to the inner radio jets. Both of these characteristics are expected in the unification scheme. Only a few AGN show motions of the maser spots. The others may need observations over a longer time scale. In general the masers show evidence for geometrically thin, mildly warped disks with clumping of the gas. There is no sign of the geometrically thick disk postulated in unification, but remember that we cannot replace the thick disk with a thin, mildly warped disk in every case, because then such details as ionization cones, and the observed ratio of Seyfert 2s to Seyfert 1s would not be explained.

3.7 A Section on Selection

It is worthwhile here driving home an important point about selection effects. Let us approach this with a well-known example. Take the radio luminosity of Seyferts. The radio is more or less isotropic, meaning that it should not be dependent on the viewing angle. Hence one would expect that the radio fluxes from Seyfert 1s and 2s would be similar. But you have to be careful in making this comparison. If you choose a flux-limited sample in the optical, you will be limiting yourself to the Seyfert 2s with only the most luminous nuclei, since their optical light is blocked by the torus. So you will end up comparing luminous Seyfert 2s to more normal Seyfert 1s. On average the objects which are most luminous in one wave band are also more luminous in other bands, so these Seyfert 2s would then be expected to be *brighter* than the Seyfert 1s from the same sample in the radio.

Similar selection effects crop up all over AGN research. When possible we try to avoid them (like using a volume-limited sample in the above radio comparison), and when we can't do this we try to correct for them, and finally we sometimes must be satisfied with simply acknowledging them.

3.8 Summary

The picture that has emerged from this has been a central black hole and continuum source, surrounded by a broad-line region, then by a geometrically thick and optically thick torus which blocks the view of the inner regions from some lines of sight. In at least some cases thinner, warped maser disks are present. If we view the AGN along one of the blocked lines of sight we classify it as a Seyfert 2, while the same AGN viewed along an unobscured line of sight would be classified as a Seyfert 1.

Further out is a scattering region (at least sometimes), the NLR, and the rest of the galaxy. The source of FC2 must be relatively unobscured, and so must be at radii similar to the scatterers or the NLR. In some cases the inner parts of the NLR can be obscured, and possibly scattered. The evidence for this is stronger in radio galaxies, as we shall discuss below.

The evidence for unification in Seyfert galaxies seems overwhelming. All or nearly all Seyferts contain BLRs. The devil is in the details, and Nature always shows us that the details can be complicated, including allowance for ranges in disk height, optical depth, and distance from the center. There are also variable mixes of starburst/young star and AGN components. In almost all cases reported discrepancies between observations and the unification theory are explained as selection effects.

Having successfully unified (according to some!) Seyfert 1 and 2 galaxies, we look towards other AGN classes. Let us start with the radio-loud equivalents of Seyferts and their brethren.

4 Radio Galaxies

There are a large number of different types of radio galaxies, classified from the radio (at different frequencies) and from the optical. Typically their radio structure contains one or more of the following components (Fig. 8): a compact core with a steep radio spectrum, radio jets extending from the core in one, sometimes both directions, and extended, large scale radio lobes. The radio jets seen in many radio AGN are manifestations of a highly collimated relativistic particle outflow that originates in the depths of the active nucleus.

There truly seems to be a dichotomy[33] between radio-loud and radio-quiet AGN (Fig. 9). This shows up in total radio flux, as in the figure, but often it is shown in plots of $\log R$, the logarithm of the ratio of radio flux to optical flux. Astronomers often prefer to "normalize" fluxes in this manner to (a) reduce the dynamic range of their data to a more manageable level, and (b) minimize the selection effects that plague flux-limited samples. However, even looking at total radio luminosities we see a break in the distribution of values. Throughout this section we use the term "quasar" to imply a luminous radio-loud AGN, and "QSO" ("quasistellar object") to imply a luminous radio-quiet AGN. Typically $B < 23$ is used as a luminosity criterion. Be aware, however, that in the literature these terms are less rigorously used.

Later, in the section on broad-absorption line QSOs we will return to the use of $\log R$ to distinguish "radio-loud" and "radio-quiet" AGN. The problem

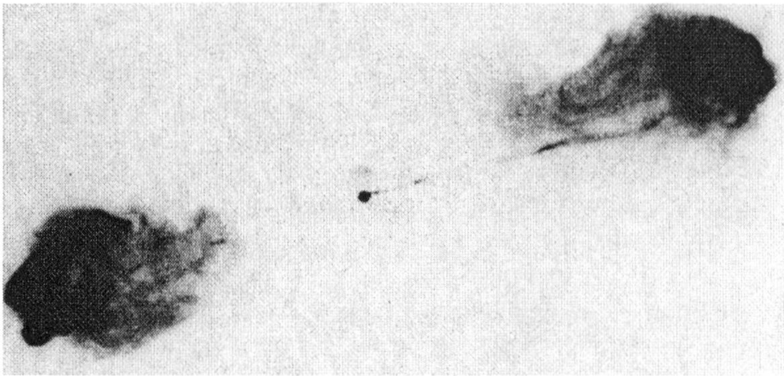

Figure 8. A radio map of Cyg A shows a classical FR II morphology, with dominant extended lobes. This very high dynamic range VLA composite also shows the weak inner jet, extending from the nucleus at center to the right lobe. The nucleus and jet would dominate the radio image were this an FR I source. From Perley et al.[36] ©1984 AAS.

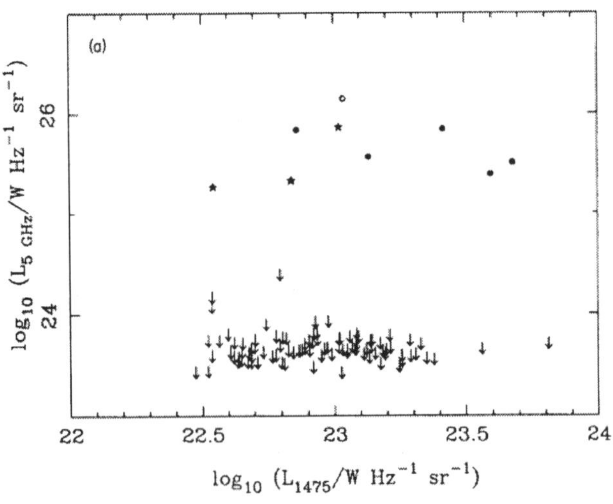

Figure 9. When radio luminosity is plotted vs. UV or optical luminosity, AGN split distinctly into two classes: radio-loud and radio-quiet objects. Upper limits to the 5 GHz radio luminosity are shown here as downward-pointing arrows. From Miller et al.[33] ©1990 Blackwell Sci.

is that if, for some reason, the optical flux is affected by something (typically extinction), then the value of $\log R$ may be compromised, giving a misleading result.

4.1 Blazars and OVVs

One of the earliest classes of radio-loud AGN was the "BL Lac" class. Their radio spectra tend to be steep, and highly variable. They also are associated with another radio class, FR I, or "Fanaroff-Riley 1" objects. This is a classification based on radio morphology. Radio images of FR Is are dominated by a bright central core. Images of FR IIs, on the other hand, are dominated by large-scale "lobes", usually one on either side of the AGN. BL Lacs and their cousins, OVVs (for "optically violently variable") also show strong, variable optical continuum, with relatively weak emission lines (Fig. 10). Further work showed that, in general, BL Lacs tend to be lower luminosity objects, but having emission lines of smaller equivalent width (< 5 Å). OVVs are true quasars which often, especially in their faintest states, show broad emission lines. BL Lacs and OVVs have been lumped into the class called "blazars."

Some other interesting characteristics of blazars are worth mentioning. Their spectra tend to be highly polarized, and highly variable not only in total flux but also polarization fraction and angle. High spatial resolution maps show bright features moving out from the cores, and when their redshift is taken into account, together with the angular rate of separation of the features from the nucleus, a naive calculation indicates that the features must be moving at greater than the speed of light. This so-called "superluminal" motion is an illusion explained by relativistic effects when highly relativistic particles outflow towards the observer. Basically, the particles are moving nearly fast enough to keep up with the light they emit!

Most of the characteristics of blazars—the strong continuum, dominance of the core in the radio, high polarization, high variability, and superluminal motion—are in fact the result of this highly relativistic outflow. When a stream of relativistic particles emits light (synchrotron radiation), the light is not isotropic, but highly beamed towards the direction of motion. This means that an observer along this line of sight will see more radiation than one looking from the side. This explains the relative dominance of the continuum at most or all wavelengths, including the core dominance in the radio. It also explains the high polarization; synchotron radiation is highly linearly polarized. The variability can be explained by slightly changing viewing angles. If the relativistic stream meanders about by small amounts, its relativistic beaming, or "Doppler-boosting," toward any given direction can change dramatically.

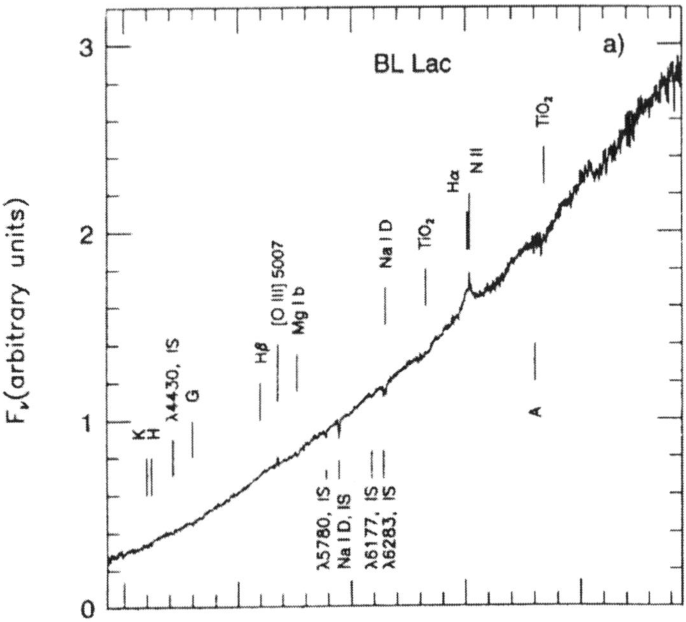

Figure 10. The prototypical blazar, BL Lac, shows a strong continuum that overwhelms many of the emission and absorption features in its spectrum. This particular spectrum is actually rather historical. It was taken at a time when the blazar continuum was weakened so much that BL Lac would not have been classified as a BL Lac object! (Talk about your classification dilemma!) Normally, however, the continuum is even stronger than shown here. From Vermeulen et al.[45] ©1995 AAS.

Viewed from the side a slight change in angle produces almost no change in flux, but viewed along the direction of motion a small change can increase or decrease the observed flux by a large factor.

4.2 Blazars, Quasars, BLRGs, and NLRGs

We have unified blazars with quasars, but what about the radio-loud equivalents to Seyfert 1 and 2 galaxies? QSOs are the more luminous versions of Seyfert 1s. Is there a parallel series among radio-loud objects? And is there a similar dichotomy between narrow-lined and broad-lined objects? Indeed there is, with Seyfert 1s being shadowed by the aptly named "broad-line radio galaxies" (BLRGs), and Seyfert 2s by (what else?) "narrow-line radio galax-

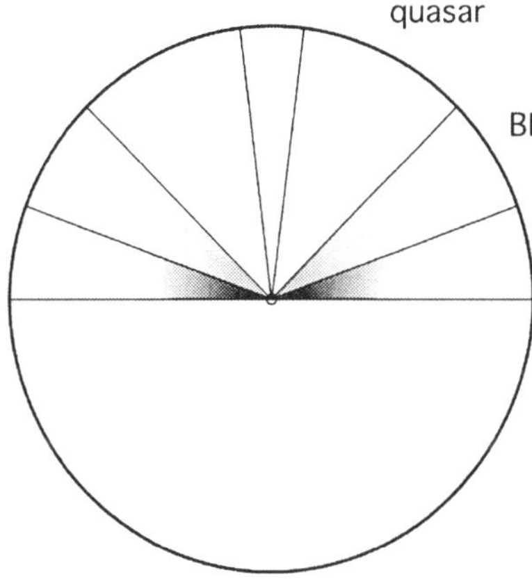

Figure 11. Cartoon showing how blazars, quasars, BLRGs, and NLRGs can all be unified by a change in aspect. The vertical is the symmetry axis of the system, and the path that the relativistic particle flow takes out of the nucleus. Observers viewing close to this axis see a blazar: a BL Lac or OVV. Observed from slightly higher angles the same object would be classified as a quasar. At even higher angles we start to see some degree of absorption, and the object would be classified as a BLRG. (Some BLRGs are undoubtedly lower-luminosity objects seen from the "quasar" angle.) Seen from near the equatorial regions the object is extincted so much that the broad-line and continuum regions are not seen directly, and we classify the object as an NLRG.

ies" (NLRGs). However, there appear to be few, if any, radio-quiet equivalents to "blazars," implying a close connection between the relativistic particle flow and radio emission.

In fact all four radio classes—blazars, quasars, BLRGs, and NLRGs—can be related in an extension to the scheme often attributed to Barthel[4]. In this picture (Fig. 11), blazars are seen down the "throat" of the radio outflow. Doppler beamed and boosted light dominates at almost every wavelength. For a relativistic Lorentz factor of γ the angle at which these effects are strong is $1/\gamma$, or typically a few degrees.

Quasars are seen at slightly higher angles, where the relativistic effects are not so strong. BLRGs are seen at even higher angles, where they are commonly seen through a relatively modest amount of dust. Finally, NLRGs are seen through significant columns of dust and gas, similar to Seyfert 2s. Let's invetsigate what hard evidence we have for the unification of these latter classes.

4.3 Do NLRGs Hide Their BLRs?

The first good clues to the BLRG/NLRG unification came from a near-IR spectroscopic study of a complete sample of radio galaxies by Hill, Goodrich, & Depoy[29]. Of eleven objects in this sample, eight were NLRGs, the other three either BLRG or, in one case, a quasar (3C 273). Of the eight NLRGs, three showed broad Paα emission. This indicated that at least 6/11 of radio objects had BLRs, probably hidden by substantial dust columns. Note that the state-of-the-art in infrared spectroscopy has progressed greatly since this early study was conducted.

In fact evidence has been put forth that NLRGs also hide parts of their narrow-line regions. Jackson & Browne[30] showed that radio galaxies seem underluminous in [O III] compared to quasars of similar low-frequency radio luminosities. However, Hes, Barthel, & Fosbury[28] showed that this is *not* the case with [O II], which comes from larger radii than [O III]. In [O II] radio galaxies and quasars are of similar brightnesses. Originally it was thought that both low-frequency (lobe-dominated) radio and [O III] emission are more or less isotropic. The interpretation now is that even some of the high-density, high-ionization inner parts of the NLR are blocked from direct view.

4.4 Do Radio Galaxies Have Mirrors?

Cygnus A is a classical double-lobed radio galaxy (Fig. 8). It is an FR II from a radio morphology classification, and an NLRG from an optical spectral classification. Early infrared studies[10] showed that the nucleus is more compact at longer wavelengths, hinting at a hidden nucleus. Indeed, polarization imaging[42] showed a vaguely centro-symmetric polarization pattern, as expected from a hidden central source shining off dust or electrons further out in the host galaxy.

However, attempts to find broad infrared emission lines turned up nothing. The imaging polarimetry showed polarization, yes, but it was quite low [19] ($\sim$ 1%), and Cyg A lies behind the Cygnus spiral arm. Much of the observed polarization could be from Galactic dust. Early attempts at spectropolarimetry[31] also showed no signs of broad lines.

At first, the unification of NLRG and BLRG seemed to be filled with quandaries and questions.

Finally, the spectropolarimeter at Keck cracked the problem of polarized broad lines in radio galaxies[8]. The problem in the past was two-fold. First, since radio galaxies are rarer than radio-quiet AGN, the nearest ones tend to lie further away than the nearest Seyferts. They are thus fainter, and in a given aperture the host galaxy contributes relatively more starlight. Second, the

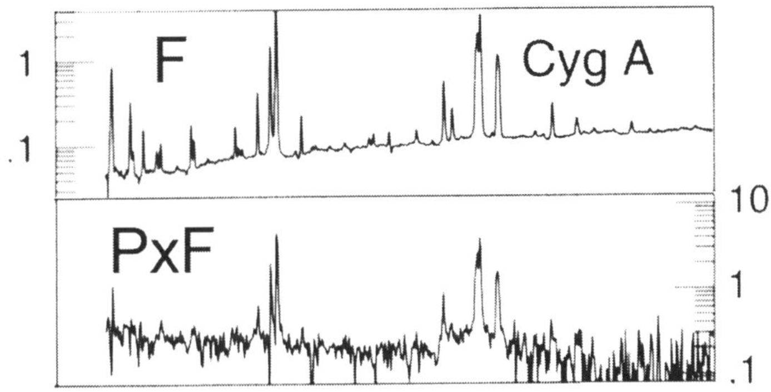

Figure 12. Keck spectropolarimetry of Cyg A, showing broad polarized emission lines in the bottom panel (polarized flux). From Cohen et al.[8] ©1999 AAS.

best-studied object, Cyg A, happened to have extraordinarily broad polarized lines, plus an unpolarized continuum that probably consists mostly of hot stars. The broader lines were harder to detect than narrower lines of the same intensity would have been.

Keck spectropolarimetry of Cygnus A is shown in Fig. 12. At top is the total flux spectrum, dominated by the NLR. At bottom is the polarized flux, in which you can see a very broad Hα emission line.

The Keck survey of radio galaxies used a very inhomogeneous sample, so while it found seven of eight NLRGs to be polarized, and six of these to show polarized broad lines, this is not directly useful in a statistical sense. They did have a number of objects in common with the sample of Hill et al.[29], and adding this new information to the latter's complete sample, six of eight NLRGs have hidden BLRs, seen either in the near-IR or via polarimetry.

Some of the Keck data, which include BLRGs, are shown in Fig. 13. A well-studied object is 3C 234, which, although originally classified as a BLRG, some argue is actually an NLRG with extremely bright scattered light. The argument is that the broad lines are highly polarized, and consist completely of scattered light. Hence, the *direct* light from 3C 234 would look like an NLRG, with the scattered broad lines superposed. 2217+259 appears to be a similar object.

The Keck survey also includes imaging polarimetry. A sample of NLRGs is shown in Fig. 14. The top row shows the total flux, the bottom panels

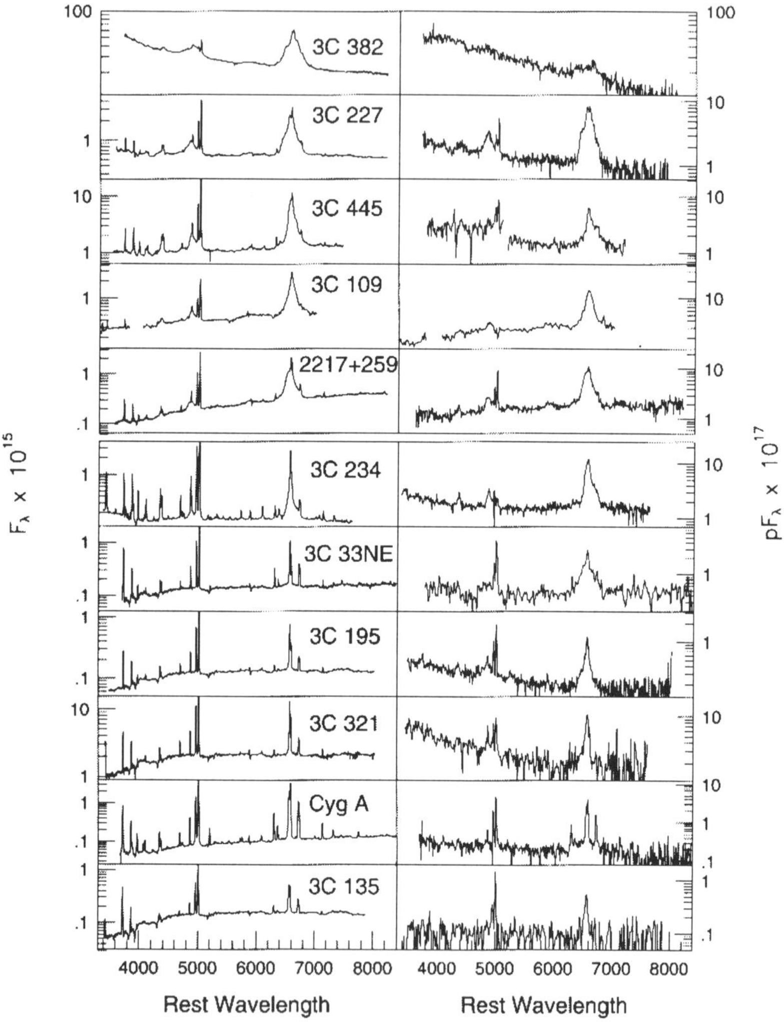

Figure 13. Spectra from the Keck polarimetric survey of low-z radio galaxies. Total flux is shown on the left, and polarized flux on the right in semilog plots. The upper three objects show broad line emission that is seen directly. The next three show highly polarized broad emission that either is due to dust obscuration or is 100% scattered light. The bottom five are NLRGs, four of which show broad lines in polarized flux. 3C 135 has very low polarization. It either has no BLR or has insufficient scattering particles in our view to overcome the starlight from the host galaxy. From Cohen et al.[8] ©1999 AAS.

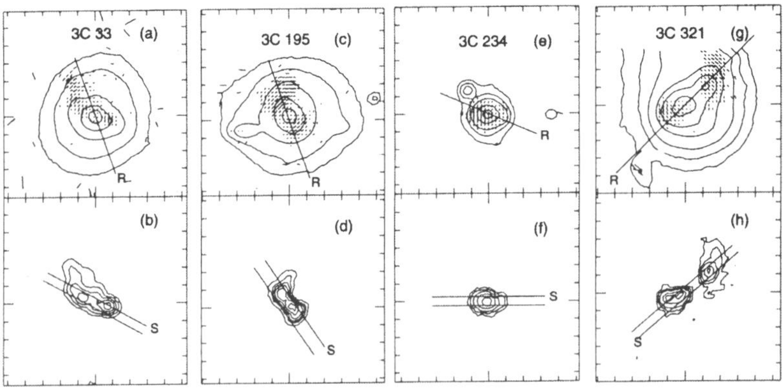

Figure 14. Imaging polarimetry of three NLRGs from the Keck survey. At top is the B-band total flux image, with the radio axis indicated by the line marked "R". Fractional polarization vectors are superposed. The bottom panels show the polarized flux. The parallel lines in these panels show the slit positions used for spectropolarimetry. From Cohen et al.[8] ©1999 AAS.

the polarized flux. Note that in general the polarized flux lines up more or less along the radio axis, often forming two opposing lobes. Since the radio axis is thought to be the symmetry axis of the AGN, this implies light leaking out of the poles of the system. The polarization is perpendicular to both the extensions and the radio axes. (Recall that polarization is perpendicular to the scattering of plane, and since we only see light scattered along the extensions, we *expect* the polarization from these regions to be perpendicular.)

These galactic-sized reflection regions are reminiscent of stellar reflection nebulae in our own Galaxy, when the central star is hidden from direct view. Note, too, that the sizes of these scattering regions are larger than the equivalent regions in Seyfert galaxies. This may simply have something to do with the intrinsic luminosities of the more distant radio galaxies, or it may indicate a difference in environment. Investigation of matched samples of radio-quiet and radio-loud AGN of similar intrinsic optical luminosity may shed more light on this.

4.5 Other Unification Keys

Radio AGN have a distinct advantage over radio-quiet AGN. The radio axes indicate the symmetry axis of the nucleus, and the strength of the Doppler-beamed core and jets relative to the lobes can give a rough idea of the angle of that axis with repect to the plane of the sky.

With this in mind, we can predict that the projected linear sizes of radio lobes in quasars should be smaller, due to foreshortening, than the sizes of radio lobes in BLRGs and NLRGs. This is indeed found[25]. A similar result among Seyferts was shown by Schmitt at INAOE[38].

Several authors have shown that in many radio galaxies, as in many Seyfert 2s, there is an ionizing photon deficit. Discussed in Sec. 3.1 for Seyferts, basically there appear to be more photons lighting up the NLR and other extended emission regions than can be accounted for in the observed spectral energy distribution.

4.6 Radio Galaxies vs. Seyferts

It is interesting to briefly note some of the differences between radio galaxies and Seyfert galaxies. The scattering appears to be on larger scales in the radio galaxies. For some time it has been claimed that radio-loud AGN tend to reside in elliptical galaxies, while radio-quiet AGN live in spirals. At some level this may be true, but the ellipticals hosting radio-loud objects are often rather dusty, and look disturbed. Perhaps a recent merger has disrupted the gas and dust normally quiescent in these ellipticals, providing a fresh gas supply to the central AGN. In any case, these are clearly *not* normal ellipticals! To further this point, Aretxaga et al.[2] in this conference showed that a number of radio galaxies show the optical signs of young stellar populations. Normally ellipticals are thought of as quiet galaxies with aged stellar populations. Brian Boyle's chapter also discusses this point.

4.7 Luminosity Effects

The character of the obscuring matter in AGN may also show a dependence on luminosity. The question is often asked, "Where are the narrow-lined QSOs, or QSO 2s?" Among higher-luminosity objects it is rare to find one with only narrow lines in its spectrum. An often-quoted answer to the question is that ULIRGs are the narrow-line equivalents to broad-line QSOs. However, it has not been shown that QSO 1s would often be classified as ULIRGs if their optical light were blocked from our view. It may instead be the case that ULIRGs are a qualitatively different class (perhaps unified through evolution) that has a large covering factor. This implies that most ULIRGs would be seen as narrow-line objects, with an occasional broad-line case, but *not* that all QSOs would appear as ULIRGs from the right viewing angle.

Hill et al.[29] proposed an alternative interpretation, called the "receding torus." In this model (Fig. 15), the obscuring dust has a more or less con-

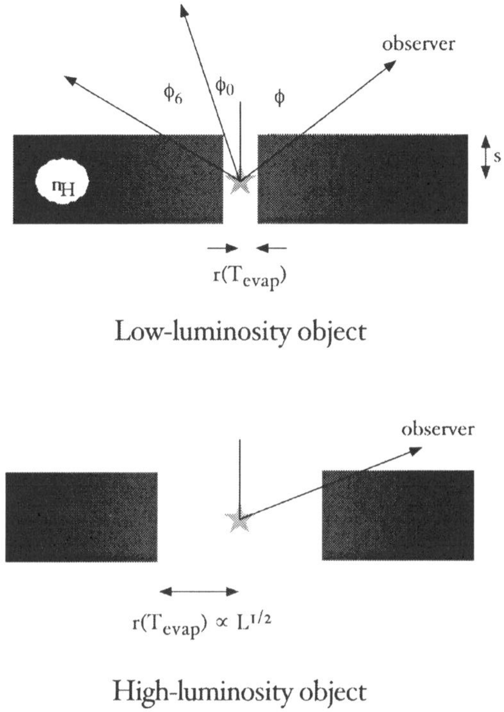

Figure 15. The "receding torus" model. More luminous sources will evaporate dust grains to a larger radius. If scale heights of the dust disks are similar at different luminosities, the probability of viewing the high luminosity object through the dust disk is smaller than the probability for the object of lower luminosity. From Hill et al.[29] ©1995 AAS.

stant scale height above the symmetry plane. Dust evaporates at a given temperature, 1100–1700 K depending on the grain composition. For a given luminosity this defines a "sublimation radius," inside of which the grain temperatures are above the sublimation temperature. Higher luminosity objects will have a larger sublimation radius. This larger radius leads to a smaller angular covering factor of the dust layer, hence fewer lines of sight which may obscure the BLR.

A third interpretation comes from a simple scaling argument, discussed by Hagai Netzer at this school. In low-luminosity AGN the narrow lines are produced in regions hundreds of parsecs from the nucleus. It seems reasonable to assume that this scales with luminosity, so that objects with a factor of

100 times the luminosity might have radii to their NLR which are actually *outside* the host galaxy! The lack of gas and dust at these distances means that only extremely weak NLRs could form.

4.8 Summary

In this section we have unified at least four (or more!) classes of radio galaxies. The blazar/quasar unification is relatively well understood. The quasar/BLRG/NLRG unification is similar to that of Seyferts, and is studied using many of the same techniques. Some of the details are different, which may give us a handle on the source of the radio loud/radio quiet dichotomy.

There are still many other radio classes we have not discussed, for want of space. Some, like compact steep-spectrum sources (CSS), may be unified through age or environment. Let us press forward, however, leaving exploration of those spoonfuls of the alphabet soup for the intrepid reader.

5 Seyfert 1.8 and 1.9 Galaxies

Let us return to Seyfert galaxies for a bit. Often looking at extreme examples of a class can tell us something useful about one or more physical parameters. The extreme character of the example may be due to an extreme in a physical parameter.

With this in mind, let's consider Seyfert 1.8 and 1.9 galaxies. These objects have broad-line regions, but they are weak relative to the NLR. The Seyfert 1.8s show detectable broad Hβ and Hα, while in Seyfert 1.9s no broad Hβ is visible (Fig. 16). In one sense these objects are intermediate between Seyfert 1s and 2s, but within the Seyfert 1 class they are at the extreme in terms of strength of the broad lines. Is this because there is less gas in the BLR? Or because the BLR is only partially hidden?

One clue comes from the Balmer decrement, which is generally high (Hα/Hβ = 10–15). This ratio can be attained in some corners of the physical parameter space of the BLR gas (density, ionization parameter, column depth, etc.). In particular high gas density and high optical depth can produce large decrements.

An alternate theory is that the BLR is reddened. This has the additional advantage that the emission lines are weakened, since extinction always accompanies reddening. Since the narrow emission lines do not show these anomalous line ratios, they must not be affected by the dust. The dust must therefore be between the BLR and the NLR.

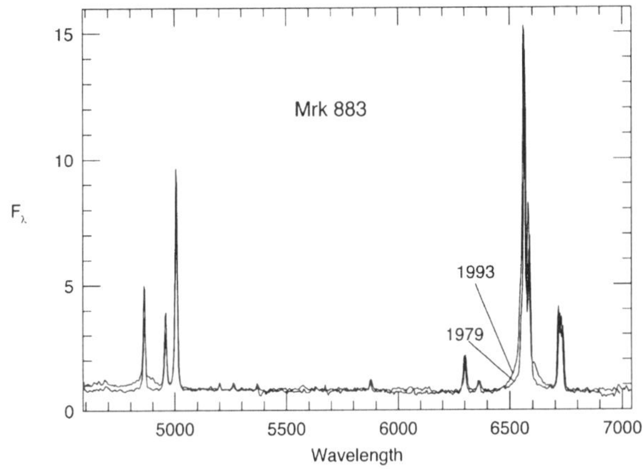

Figure 16. Mrk 883 is a good example of the Seyfert 1.8/1.9 spectral classification. While at first glance resembling a Seyfert 2, with dominant narrow emission lines, closer inspection reveals weak broad bases to the Balmer lines. If broad Hα is visible but broad Hβ is not, as in 1979, a classification of 1.9 is assigned. If broad Hβ is also visible, as in 1983, it is classed a Seyfert 1.8. From Goodrich[16] ©1995 AAS.

5.1 Variability

Most Seyferts with broad lines vary, and the Seyfert 1.8s and 1.9s are no exception, as Fig. 16 shows. In fact, some of the most extreme variability has been seen in objects which in their low state (weak lines) have the character of Seyfert 1.8s or 1.9s.

One example is shown in Fig. 17. Tohline & Osterbrock[43] documented the change of NGC 7603 from a Seyfert 1 to a Seyfert 1.8 and then back toward a Seyfert 1. Goodrich[14,16] followed the spectrum to later times. Surprisingly, this latter work, which also analyzed the variability of several other Seyfert 1.8/1.9s found that the emission line fluxes were consistent with changes in the dust column depth towards the BLR. Qualitatively, as the line fluxes increase, the Hα/Hβ ratio decreases.

This deserves some more discussion. The chapter on variability by Brad Peterson deals almost exclusively with "line reverberation," in which the emission lines change in response to a change in the continuum flux and (possibly) shape. Different emission lines will change by different amounts. Specifically,

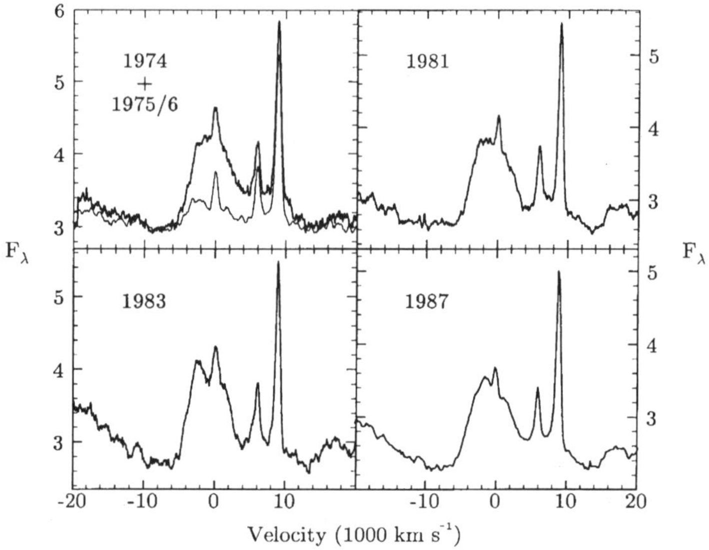

Figure 17. Hβ variability in NGC 7603. Comparing line flux measurements with continuum and Hα line measurements provides strong evidence that the bulk of the variability is due to changes in the dust optical depth towards the nuclear region. From Goodrich[14] ©1989 AAS.

if the state of the gas in Seyfert 1.8/1.9s produce extremes in the Hα/Hβ ratio, then any departures from that state (caused, e.g., by a change in ionizing radiation) will move that ratio away from the extremum. In other words, we would expect that if the line changes are responses to continuum changes, the Balmer ratio has nowhere to go but down.

However, reddening changes predict a specific path in the F(Hα) vs. F(Hβ) diagram. Reverberation could follow any number of paths, which are generally not the same path as for reddening changes. In addition, they would not be expected to follow the reddening path for very long. *For large flux changes, following the predicted reddening path can then be taken as circumstantial evidence for a change in reddening.* The larger number of emission lines that show consistency with a reddening change, the less likely that it is a coincidence. For example, in NGC 7603 the He I λ5876 line shows changes consistent with the other spectral variability when interpreted as a reddening change.

How does the continuum change? The continuum is produced at smaller radii than the BLR. Any dust that covers the BLR might also be expected to cover the continuum, although geometries can be constructed in which this is not true. Again, specific changes in the spectral shape of the continuum are expected to accompany absolute flux changes. These may be closely related to the emission line changes, although deviations from the very simplest model may change this.

What about emission line reverberation? The continuum flux changes presumably are not closely related to the spectral shape. Qualitatively one might expect that an increase in continuum flux would increase the emission line flux, but in fact this is not necessarily true. The continuum variability can change the ionization state of the gas, and actually make some lines decrease in flux. Again, a coincidental agreement between the continuum changes predicted by reddening and those seen can be taken as more circumstantial evidence for the reddening hypothesis.

So, what do we see? In some cases we do see changes consistent with changes in reddening to the BLR. The case of NGC 2622 is shown in Figs. 18 and 19. Fig. 18 shows the spectra, with starlight subtracted, in the low state (1981) and a high state (1986). The 1993 spectrum shows the continuum falling back towards the low state.

Fig. 19 shows the reddening estimated by the change in absolute flux values from two emission lines (Hα and Hβ) and the continuum near those lines. The line fluxes measured in 1981 were used as a baseline [$\Delta E(B-V) = 0$]. Each subsequent epoch provides four new, independent measurements (two line and two continuum points). Each of these points is divided by its baseline value, and four separate reddening estimates are produced. This differs from using the Hα/Hβ flux ratio to estimate a reddening at each epoch. Here we are using *absolute* fluxes to independently measure reddening for the lines (and continuum) independently.

As the figure shows, the data are reasonably consistent, with some caveats. The first is the usual uncertainty involved in absolute flux calibration[b]. The chapter on variability covers this in greater detail. A second uncertainty is in the starlight subtraction, which is never straightforward. Note, in fact, that the continuum reddening estimates are consistently slightly below the emission line estimates. This is consistent with a constant flux from a component of hot stars that was not adequately removed by the starlight template used. A third uncertainty lies in the initial line flux measurements. Any error in those initial line fluxes contributes in a systematic way in each of the reddening estimates.

---

[b]In reality, the "absolute fluxes" used here were actually calculated with respect to the [O III] line fluxes. While somewhat more accurate, this can still be problematic.

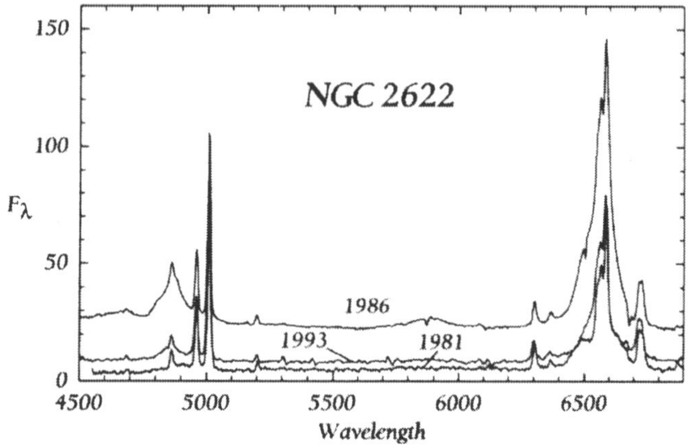

Figure 18. Spectral changes in NGC 2622. In 1981 the spectrum was classified as Seyfert 1.8, but by 1986 it was a full-fledged Seyfert 1. The continuum and broad lines dropped in brightness by 1993. From Goodrich[16] ©1995 AAS.

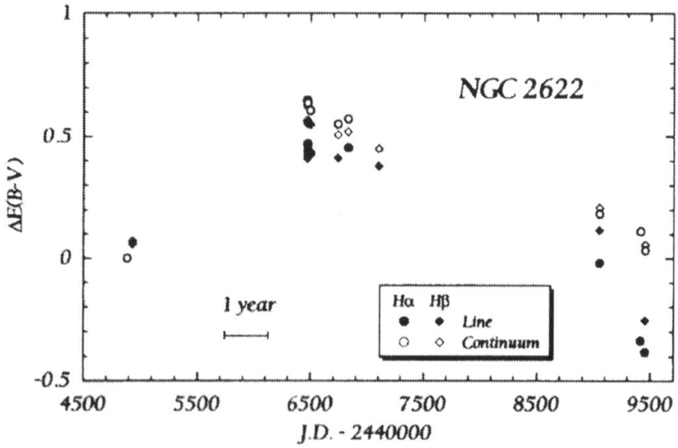

Figure 19. Reddenings estimated from absolute flux changes of different lines and continuum wavelengths in NGC 2622. From Goodrich[16] ©1995 AAS.

Finally, we have assumed only a very simple model: a foreground dust screen that completely covers the BLR and the continuum, and no "extra" lines of sight to those regions. In fact the theory appears to fall apart in 1993, when the continuum and the emission line estimates start to diverge. However, spectropolarimetry taken at that time indicate *at least* two separate polarized lines of sight.[16] One of these is clearly scattered, and the other is probably due to dust transmission polarization. This latter could be the view that shows up strongly as the reddening to the BLR decreases.

Not all Seyfert 1.8/1.9s have been seen to vary. Hence this variability test is not useful in all cases. Other consequences of the two competing theories must be sought. One idea is to move to a different wavelength. Goodrich[15] looked at Paβ spectroscopy. In the reddening theory Paβ should be quite strong, while in the other theory it should be a factor of five to ten times weaker, a clear cut signature.

What the near-IR spectroscopy showed was that Seyfert 1.8/1.9s are a mixed bag. Those objects that had varied substantially showed strong Paβ, in agreement with the reddening theory. So did a number of other objects. The rest showed an upper limit to Paβ which effectively rules out the simplest reddening theories. These objects are probably not explicable by dust.

At Lyα the case is even more clear cut. The predictions of the two theories at Lyα differ by factors of ten to 100. HST spectra[24] confirm the results of the Paβ survey on a case-by-case basis; some intermediate Seyferts show their character because of reddening (no Lyα detected) and some because of high gas optical depth and density (detected Lyα).

5.2 Time Scales

It has been thought for some time that dust cannot survive within the BLR because of the extreme conditions there. Dust sublimates at temperatures higher than 1700 K, so a dust grain close enough to the central source to have an equilibrium temperature of this value or higher should be destroyed quickly.

The highly variable intermediate Seyferts give us an estimate on where this dust lies. If the time scale for variation is given by T, and the BLR size is given by D, then to cover the entire BLR,

$$\frac{v}{c} = \frac{T}{D} \quad . \tag{3}$$

Some numbers: $T \sim 1 - 4$ yr, $D \sim 40$ light days, hence $v \sim 15,000$ km s^{-1}. This is a BLR velocity! This seems to belie the assumption that dust cannot exist within the BLR region, but remember that this dust has to cover most of the BLR emission region. So really we are talking about a region just outside the BLR, but still well inside the NLR.

More intensive, monthly monitoring of intermediate Seyferts may provide a tighter constraint on T. Until now spectra have generally been taken once a year at most.

Can the dust survive at radii less than the sublimation radius? Shielding from the harsh radiation field can help. If the dust is buried deep within a cloud, the surrounding gas may keep the radiation from penetrating far enough to destroy the grains on the estimated time scale.

Nonequilibrium processes may also be implicated. Clouds infalling from beyond the sublimation distance can contain dust, and these dusty clouds can accelerate to high velocity before the dust is fully destroyed. It would be interesting to search for signs of nonequilibrium processes, such as a shift in the mean dust grain size as the smaller grains are destroyed more rapidly.

5.3 Summary

Some Seyfert 1.8/1.9s exhibit distinctively weak broad lines and high Balmer decrements due to extreme physical parameters. These would be unified with normal Seyfert 1s, but in one corner of parameter space.

Other Seyfert 1.8/1.9s show signs of high reddening. Their spectral variability can be extreme. The dust is typically moving at BLR-like velocities, although in at least some cases it must be outside, or at least at the very outer edges, of the BLR.

The rarity of this latter type of object might be telling us something about the geometry of the region in which these dust clouds are formed and/or destroyed. One hypothesis is that these dusty clouds are being ripped off the edge of the torus, and are hence seen only over a small range in viewing angles. Alternatively, there may be some other physical parameter which causes a small fraction of AGN to display this type of variability, even if those rare objects look similar from most lines of sight.

We turn our attention now from dusty, low-luminosity objects to some more spectacular examples of absorbed AGN.

6 Broad-Absorption Line QSOs

If you like strange optical spectra, you'll love the class of radio-quiet AGN called "Broad-Absorption Line QSOs", or BAL QSOs. Their rest-frame ultraviolet spectra show deep, blue-shifted absorption features associated with

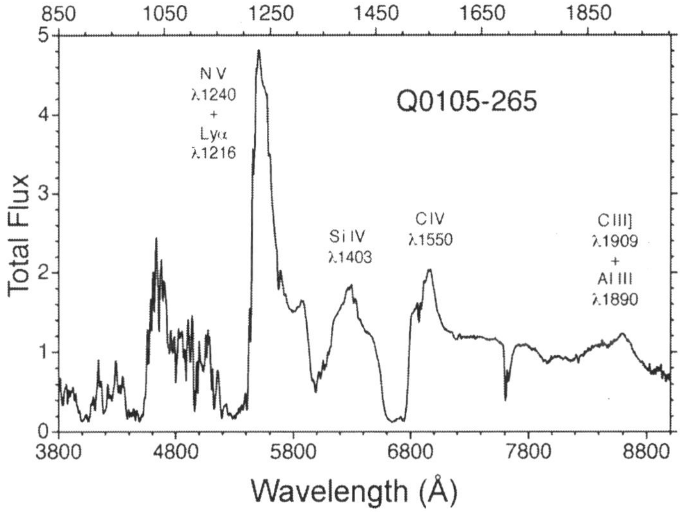

Figure 20. The strong, deep, blueshifted absorption features typical of BAL QSOs shows in this UV (rest frame) spectrum of Q0105−265. The absorption occurs predominantly in the resonance lines. Observed wavelengths are at the bottom, rest-frame wavelengths at the top in this figure. From Ogle et al.[35] ©1999 AAS.

the resonance lines such as C IV, Si IV, and N V. The spectrum of the BAL QSO Q0105−265 is shown in Fig. 20. Note that because the absorption is seen in the ultraviolet, BAL QSOs are predominantly found at high redshift ($z > 1.5$).

It is thought that BAL QSOs can be unified with non-BAL QSOs by some viewing angle. The mean emission spectrum of BAL QSOs is nearly identical to the mean spectrum of non-BAL QSOs.[48] There are two exceptions: BAL QSOs tend to have stronger N V $\lambda1240$ emission, and at least the low-ionization BAL QSOs (those with absorption at Mg II $\lambda2800$) are redder.[41] The IR properties of BAL and non-BAL QSOs are similar. One big difference is in the X-ray spectra, with BAL QSOs being much weaker relative to their optical/UV flux.

One final difference lies in the polarization properties of BAL and non-BAL QSOs. BAL QSOs tend to be more highly polarized. Spectropolarimetry of a number of them has shown that not only are the continua polarized significantly, but the polarization within the absorption troughs is very often

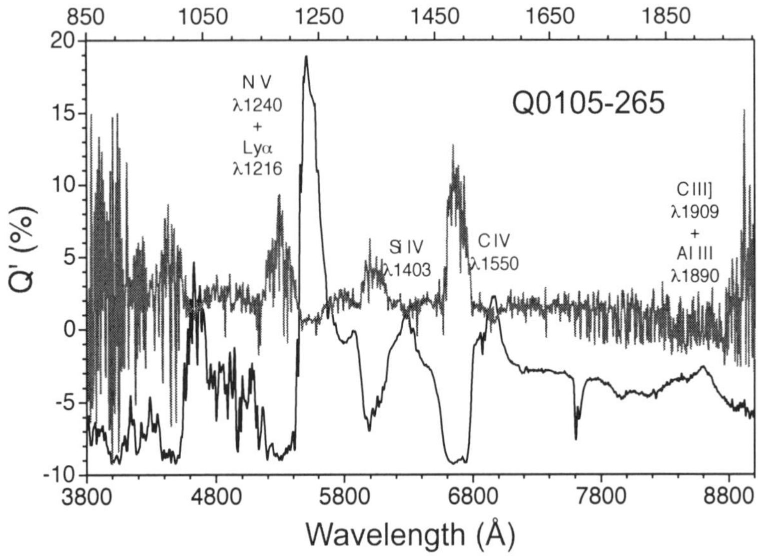

Figure 21. Polarization data (the lighter, noisier trace) added to Fig. 20 reveals high polarization in the absorption troughs, lesser but still significant continuum polarization. Note, too, the rather more subtle drops in polarization across the emission lines. From Ogle et al.[35] ©1999 AAS.

quite high. This is shown in Fig. 21, combining the total flux spectrum from Fig. 20 with polarization data. Polarization as high as 10% is seen in the C IV absorption trough.

Recall that multiplying the fractional polarization by the total flux gives a representation of the spectrum of the scattered light. Shown in Fig. 22 is the polarized flux of Q0105−265; compare to Fig. 20, the total flux. Note that the emission lines are weak or nonexistent in polarized flux; they are essentially unpolarized in most BAL QSOs. Note, too, that the absorption features are weaker (their polarizations are higher), but still present.

The polarization originates in light scattered around most of the absorption region. It apparently does not scatter completely around the BAL region (BALR) since there is still some absorption in the scattered flux. Note that in total flux one sometimes sees flat-bottomed troughs (e.g. C IV in Q0105−265), a sign of saturated absorption. However in these BAL QSOs the absorption does not go to zero flux, as one would expect. The remaining flux is the scattered light. This points out a problem with the technique of

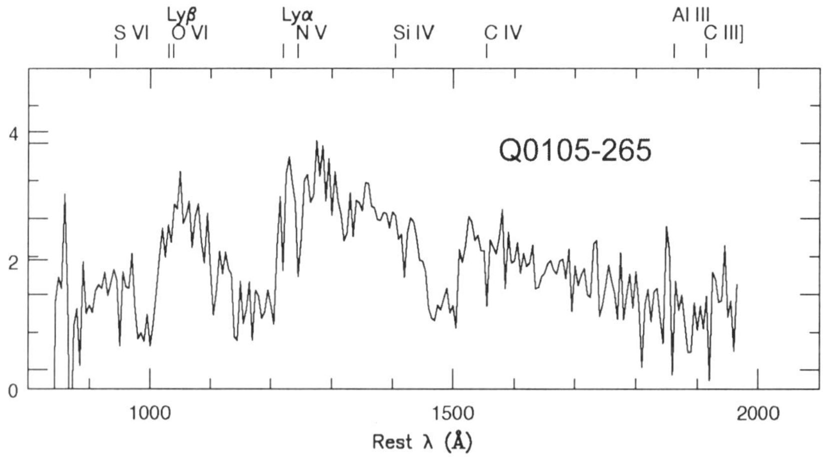

Figure 22. The polarized flux of Q0105−265 reveals the spectrum of scattered light. While the absorption troughs are still present, they are shallower. The emission lines have disappeared from the spectrum, indicating that they have essentially zero polarization. From Ogle et al.[35] ©1999 AAS.

estimating column densities as a function of velocity. What is measured in total flux is the sum of two components: one a direct, heavily absorbed view of the nucleus, and one a more lightly obscured scattered view. Unless this is taken into account, column densities will be estimated incorrectly. This affects everything from estimates of the ionization state of the absorbing gas to abundance estimates.

Unification of BAL and normal QSOs would state that most or all QSOs have this absorbing region, but most objects do not show absorption because the BAL gas does not lie along the line of sight to the continuum and emission regions.

A seemingly innocuous observation can sometimes turn out to have far-reaching consequences. Such is the case for the statement that BAL QSOs are more highly polarized than normal QSOs[17]. If all QSOs have this scattering region, then why do no *normal* QSOs show high polarization? Some non-BAL QSOs should be seen with a line of sight just missing the BALR. The small geometry change from BAL to non-BAL would not change the polarization of the scattered light significantly.

The only thing that seems to make sense is that the scattered light seen

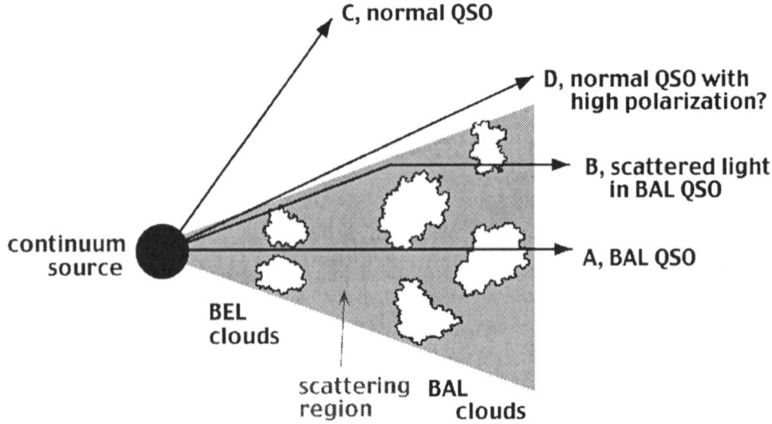

Figure 23. A schematic of the BAL QSO geometry. A direct view of the continuum seen through the BAL clouds (A in the diagram) is complemented by a polarized view from light scattering around much of the absorption (B). If we were to observe such an object from a slightly different angle (D), we would see the continuum directly, plus the scattered light. Geometry will not change the polarization of the scattered component greatly, so we expect to see some non-BAL QSOs with high polarization. We don't, which presumably means that the continuum part of the flux we see must be much stronger in non-BAL QSOs than it is in highly polarized BAL QSOs. From Goodrich & Miller[23] ©1995 AAS.

in BAL QSOs *must* exist in non-BAL QSOs, and often at about the same polarization level. This means that the unpolarized light must be less in high-P BAL QSOs. The unpolarized light is thought to be the direct view of the nuclear continuum, hence either the continuum itself must be intrinsically anisotropic, or *the continuum must be attenuated*, at least in highly polarized BAL QSOs.

That the continuum might be attenuated when there is known line absorption may not seem surprising, but continuum optical depths are typically 10^5 times less than line optical depths, so this is a very strong statement indeed! The direct line of sight must be close to Compton thick, implying a column density of $\sim 10^{23}$ cm^{-2} or greater.

Fig. 23 shows a schematic of the situation, with four lines of sight labeled A, B, C, and D. View A represents the direct view of the central source in a BAL QSO. B represents the scattered light (presumably seen by all lines of sight, not just A). C represents a sight line that would show a normal QSO.

Figure 24. A sunset analog of a BAL QSO, in which the direct view of the light source (the Sun) is heavily obscured. Scattered, polarized light dominates the scene.

D would still show a normal QSO, but is so close to being an absorbed line of sight that all other things being equal we would expect some of these objects to show high polarization. The fact that we do not[c] indicates either that in normal QSOs we do not see the scattered light, or that the direct view is much more intense.

A good analogy can be seen in the following photos of a Hawaiian sunset (Figs. 24 and 25). In the first photo the Sun is behind some clouds. There is little direct sunlight reaching the camera; most of the light is scattered, and hence polarized. The net polarization of this scene is relatively high. This scene is analogous to a high-polarization BAL QSO, with the direct view to the light source blocked.

In the second photo the Sun has come out from behind the cloud. The direct, unpolarized sunlight dominates this scene, even moreso than may be

[c]There is actually one case of a high-P non-BAL QSO known, OI 287. This is more likely analogous to 3C 234, where the emission lines are highly polarized as well, indicating that we really see none of the continuum or BLR directly, but only by scattered light.[18]

Figure 25. A sunset analog of a non-BAL QSO, in which we see the light source directly. Even a small amount of direct light will drop the total polarization of this scene to a low level.

apparent from the photographic reproduction. (Don't try looking at the Sun when it is not behind a cloud, or you will find out just how *much* more direct light there is!). This view is analogous to a non-BAL QSO. The slight change in scattering geometry between the two photos is not large enough to change the polarization or strength of the scattered component. It is only the appearance of a direct view that can drop the total polarization of the scene by a large amount.

Of course, the "weather" in QSOs may not resemble the weather here on Earth. However, other lines of evidence are consistent with this interpretation of an attenuated direct view of the continuum. Francis, Hooper, & Impey[11] showed that while the BAL QSOs in their PG sample qualify as "radio-quiet," they are at the radio-loud end of the radio-quiet distribution (Fig. 26). This is readily explained as attenuation of the optical/UV continuum. Their criterion was based on $\log R$, the ratio of radio to optical fluxes. If the optical flux is attenuated, as we are postulating, the $\log R$ will be calculated higher than it

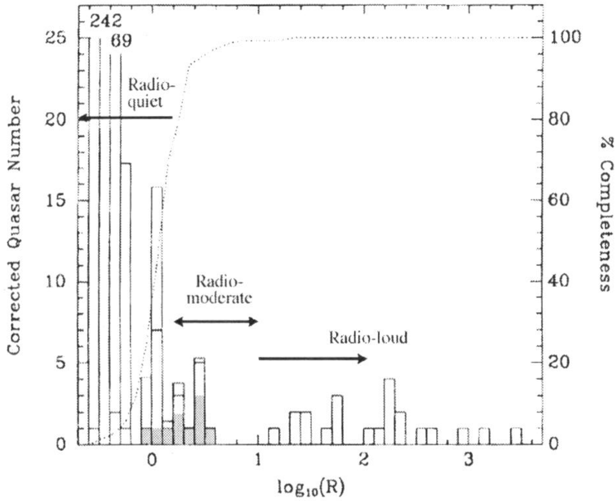

Figure 26. A survey of Palomar-Green (PG) QSOs showed that while the small number of BAL QSOs in the sample (shaded regions) were radio-quiet, they tended to lie at the radio-loud end of the radio-quiet distribution. This is consistent with the BAL QSOs having relatively fainter visual magnitudes due to attenuation of the direct view of the continuum source. From Francis et al.[11] ©1993 AAS.

is intrinsically.

6.1 *Missing BAL QSOs*

Almost every QSO known was discovered in some type of survey which had a limiting flux associated with it. The steep luminosity function of QSOs implies that a slight change in that detection limit can have a profound change in the number of QSOs found.

If some BAL QSOs are attenuated, many of those that would normally be detected in a given survey will now fall below the detection limit. We would see only the intrinsically brightest objects. Let's try to put this on a quantitative basis.

The number, N, of observed QSOs within a given redshift and luminosity range is

$$N(L_1, L_2) = \int_{z_1}^{z_2} \int_{L_1}^{L_2} \Phi(z, L)\, dL\, dz \quad . \tag{4}$$

If the BAL QSOs are "attenuated"[d] by a factor of A, then they really come from A times higher in luminosity. To recreate the original, intrinsic luminosity function, N_t,

$$N_t(L_1, L_2) = (1 - f_{obs})N(L_1, L_2) + f_{true}N(AL_1, AL_2) \quad , \tag{5}$$

hence,

$$\frac{f_{true}}{f_{obs}} = \frac{N(L_1, L_2)}{N(AL_1, AL_2)} \quad , \tag{6}$$

where f_{obs} is the observed fraction of attenuated objects and f_{true} is the true intrinsic fraction.

Using a typical QSO luminosity function, for $A = 2, 3, 4$ we calculate $f_{true}/f_{obs} = 4.3, 11.2, 23.4$. These numbers mean that if the attenuation is high enough, BAL QSOs could actually *dominate* the QSO population, rather than being a minor constituency!

A picture is worth a thousand words. Look at the luminosity function[47] shown in Fig. 27. First we clone the curve and drop it, in the figure by a factor of ten. This represents the subset of objects which are attenuated. In this case we picked the whole subset of all QSOs which show BALs. Next, we shift them over by a factor of A, for which we have used $A = 4$. With these particular parameters we would predict that there are actually more BAL QSOs than known non-BAL QSOs.

What is A, however? Two estimates have been made. One comes from matching the polarization distributions of non-BAL and BAL QSOs. This requires $A \sim 4$. Another estimate comes from the distribution of radio-to-optical flux ratios: $A \sim 3$. Both of these estimates are rather crude.

If we assume that all BAL QSOs are attenuated by factors of 3–4, we start running into trouble, not the least of which is that we predict more BAL QSOs than non-BAL QSOs. In the simplest scheme this also implies a large covering factor for the BALR. Assuming a relationship between the geometry of the scattering region and the BALR (this is not a given!) a scattering region

[d]Here we use "attenuation" to mean the ratio of flux seen by an observer looking at the object from non-BAL QSO lines of sight divided by that seen from BAL QSO lines of sight. This covers cases in which the continuum is inherently anisotropic as well.

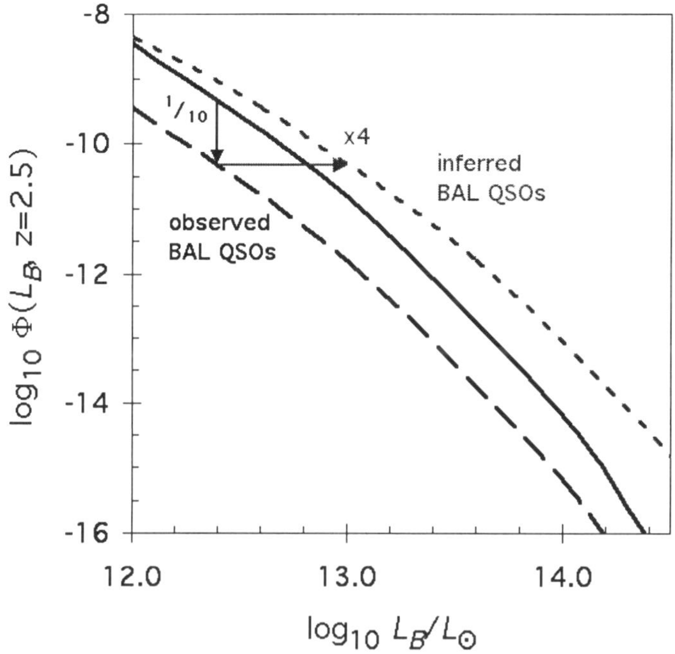

Figure 27. The observed QSO luminosity function from flux-limited surveys. BAL QSOs are observed to be roughly 10% of the observed population. This is the line shifted downward. If these BAL QSOs are actually attenuated by a factor of, say, 4, their true numbers are represented by the line shifted to the right. From Goodrich[17] ©1997 AAS.

with a large covering factor can only produce a relatively low polarization. A covering factor of 100% implies spherical symmetry, which cannot produce polarization.

Instead of assuming that *all* BAL QSOs are heavily attenuated, let's take a conservative approach and estimate the minimum number. The observed fraction is about 10%: $N(\text{BAL}) = 0.10\ N(\text{QSO})$. The minimum subset that must be attenuated is the fraction which are highly polarized, about 25%. Hence $\sim 0.02\ N(\text{QSO})$ objects are attenuated. Correcting these for $A = 3$, then intrinsically $\sim 0.22\ N(\text{QSO})$ objects are attenuated BAL QSOs. Adding back in the $0.08\ N(\text{QSO})$ which are nonattenuated BAL QSOs, we estimate that there are at least $0.30\ N(\text{QSO})$ objects which are intrinsically BAL QSOs, or more precisely, seen along lines of sight that intercept the BALR. Note, too, that roughly 3/4 of the BAL QSOs are heavily attenuated. Most of these objects are, however, missed in the surveys because the attenuation drops their fluxes below the detection limit.

6.2 X-rays

Recall that when the spectra of BAL and non-BAL QSOs were compared it was mentioned that there were two differences: the strength of N V λ1240 and the strength of the X-rays. The latter characteristics plays into the discussion on attenuation, since any material along the line of sight is expected to show some sort of signature in the X-rays.

Indeed, in the soft X-rays, where absorption from gas is expected to be strongest, BAL QSOs seem to be weak. The faintness of most BAL QSOs (due to their greater redshifts, which in turn are due to the way they are identified, through rest-frame UV features) has made it difficult in the past to study individual objects in any great detail. Still, early results[40] on PHL 5200 indicated that there was significant absorption in the soft X-rays. Other observations of individual objects generally show high absorption. One exception of absorption from a lower column was presented by Gallager & Brandt at this school[12].

Statistical approaches are also used. Summing up X-ray images that include various BAL QSOs allows for a very deep image of what might be called an "average" BAL QSO. Green & Mathur[26] found that such an average BAL QSO must be very faint in the soft X-rays, consistent with absorption from columns at or above 10^{24} cm^{-2}.

Even if the X-rays are detected at a weak level, they may not show the characteristic shape of absorbed gas, since it may be the *scattered* X-rays we are detecting. This is seen in some Seyfert 2s, where the signal-to-noise can be high in spite of the weakness of the X-rays relative to the optical. Brinkmann et al.[7] report that PG 1411+442 seems to show scattered X-ray flux.

More progress on the X-ray spectra of BAL QSOs requires higher sensitivity and wider spectral coverage, which fortunately we are getting with the new generation of X-ray instruments.

6.3 Polar or Equatorial Scattering?

It is of more than passing interest to determine the geometry of the scattering material in BAL QSOs. Is it in the equatorial plane of the system, or along the polar regions. Material along the poles has some attraction as it may be a symptom of the radio-quietness of these objects. On the other hand, material naturally tends to accumulate in the equatorial regions of systems that form through a dissipative process.

Recall that with polarimetry the symmetry direction (projected onto the plane of the sky) is revealed by the polarization's position angle. In almost any BAL QSO a large telescope like Keck can measure the polarization p.a. in the absorption troughs. As mentioned, the cleanest way to determine the symmetry axis of an AGN's inner regions is by determining the radio axis. Now, radio-quiet objects like BAL QSOs have two problems. First, and obviously, their radio emission is weak. Second, even when the radio emission is detected, they are found to be *compact*. So it turns out to be very difficult to determine the radio axis of a radio-quiet AGN. Even the recent objects which are claimed to be "radio-loud" BAL QSOs show a compact morphology. The jury is still out on whether the polarization is perpendicular to the radio axis (implying equatorial scattering) or parallel to it (polar scattering).

6.4 Low-ionization BAL QSOs

While most BAL QSOs show absorption only from high ionization resonance lines like C IV, N V, O VI, and Si IV, a number of them also show absorption in Mg II $\lambda 2800$. These are the so-called "low-ionization BAL QSOs." Sprayberry & Foltz [41] found that these objects also appeared to be more highly reddened than normal QSOs. They also appear to have a much higher fraction of high-polarization objects. Some are considered radio-loud, but again with the caveat that a definition based on $\log R$ is problematic.

In any case it seems clear that these are examples in which we see the central continuum and emission sources through attenuation, but in this case attenuation which includes some dust. Some nearby examples of low-ionization BAL QSOs, such as Mrk 231, show extraordinarily weak NLR emission, which can be interpreted as a high covering factor of absorbing gas and dust. With a high covering factor little ionizing radiation escapes to light up the gas that would normally form the NLR. A study of near-IR spectra of low-ionization BAL QSOs, comparing them with non-BAL QSOs, would be enlightening, and within reach of the new generation of large telescopes and their infrared spectrographs. In particular, a low [O III]/Hβ ratio might be expected. Note that high luminosity QSOs in general show weaker NLR emission, probably because the NLR radius is pushed out beyond the confines of the galaxy, as discussed elsewhere in these pages.

6.5 *The Link to Lower Luminosity*

What about low-luminosity AGN? Crenshaw et al.[9] found that 50% of Seyfert 1 galaxies show UV absorption, but at lower velocity. This would seem to be a natural extension of the BAL phenomenon to low luminosity. But one is left with the velocity differences, which are too large to be explained as a consequence of the deeper potential wells of the higher-luminosity objects. If that were the explanation the ratio of maximum absorption velocity would scale with the emission line widths, which it does not. Presumably the winds that form the absorbing region do not get to some high-velocity phase in the low-L AGN.

As a sidebar, I might turn the question around, and ask whether there is evidence in QSOs for the warm absorbers seen in Seyferts. As Hagai Netzer mentions in his chapter, the evidence in the X-rays is ambiguous. In the UV QSOs show less associated absorption than do Seyferts. But the connection between the X-ray and UV absorbers is still tenuous. The latest generation of X-ray satellites will study this connection in much more detail, and resolve some of our questions.

6.6 *BAL Winds*

I have casually tossed out the term "wind" to describe the BAL region. We don't really know whether the BALR has discrete clouds or a more continuous wind, but it seems likely that "wind" is a better description. How is the outflow formed? Hagai Netzer mentions several physical processes that are important in AGN. Most of the work and discussion has concentrated on photoionization, the transfer of energy between photons and ions. Some mention has been made about shocks, the transfer of kinetic energy between particles. Indeed if you scour the literature it is easy to find many papers discussing shocks. But what has been hardly mentioned at all, and what is probably the least understood, is magneto-hydrodynamics, the transfer of energy between particles and magnetic fields. Unfortunately for us, winds are often very much a magneto-hydrodynamic problem.

I won't go into details about the models here, because that would add another 10 or 20 pages, and because many of the details are poorly understood or even completely misunderstood. Clever theorists teamed with ever more powerful computers and recent ideas on AGN winds are making progress, however.

7 State of the Union—A Summary

In my concluding remarks I would like to give a quick review of some of the unification questions discussed here. Seyfert 2 is likely a "false" class. Probably all or nearly all low-luminosity AGN harbor a Seyfert 1 nucleus. IR, radio, and statistical studies are working to reveal the character (and its range) of the obscuring torus. Some Seyfert 1.8/1.9 galaxies are Seyfert 1s with heavily reddened views of the BLR and continuum. Other subclasses, such as narrow-line Seyfert 1s (NLS1s) are probably extreme in some physical parameter, such as the black hole mass (for NLS1s).

Future progress on Seyferts will come from extensive near-IR surveys with 8-m and 10-m class telescopes, mid-IR and millimeter wavelength measurements of dust emission, more work and monitoring of H_2O masers, and the extension of much existing work on Seyferts to higher luminosities. One relatively untapped field is that of adaptive optics and even (on the horizon) optical and near-IR interferometry.

Radio galaxies have many more subclasses than Seyferts, but in general they seem to follow the same Seyfert unification scheme. There is mounting evidence that NLRGs, like Seyfert 2s, harbor BLRGs, which are analogous to Seyfert 1s. So "NLRG" is a false class, too. BLRGs themselves are likely either slightly reddened radio-loud quasars or in some cases low-luminosity analogs of quasars. And blazars are clearly pole-on quasars.

Similar to Seyferts, more extensive near-IR surveys with large telescopes is an obvious area of future work. As Brian Boyle covers in his chapter, the evolution of radio-loud AGN differs from the evolution of radio-quiet AGN. So there is a lot of room for statistics and analysis of change at high redshift. This also has ties to early star formation history, as well as galaxy formation.

At the high-luminosity end of radio-quiet AGN, there appears to be no analogs to the blazar classes of radio-loud quasars. The high-luminosity equivalents of Seyfert 2s also seem to be missing, although some argument can be made for ULIRGs as "QSO 2s". Some multicolor surveys are starting to identify sets of objects which may turn out to be high-luminosity QSO 2s as well. But keep in mind the argument that as luminosity increases so does the radius of all of the associated emission line regions, and at some point the NLR may be located almost entirely outside the galaxy. Does this mean that at some high redshift NLRs will start to become important again, as the circumgalactic environment becomes more gas-rich?

It is also worth speculating on some of the outstanding questions. We still do not have a good handle on what causes the radio-loud/radio-quiet dichotomy in AGN. One key is what causes the relativistic particle outflow, and a related key may be what causes the lower velocity BAL outflow in radio-quiet objects. Does generation of the BAL flow inhibit generation of the relativistic flow or vice versa? The different evolution of radio-loud and radio-quiet objects also seems to be a key. Radio-quiet objects seem to track what we consider normal galaxies; what do radio-loud AGN track?

Throughout the summer school and this book we have also confronted the relationship between starbursts and supermassive black holes. At times they are difficult to tell apart, but there seems little doubt that they can be related. How closely? Are they two manifestations of the same trigger effect? Or is it just that they are more likely in certain types of environment, hence statistically tied together?

As our understanding of active galaxies progresses, it is clear that unification is inevitable. However, other effects are also at work and must be kept in mind: evolution, environment, fueling. Larger telescopes and advances in instrumentation are providing better surveys, and more in depth analysis of individual objects. In upcoming years one of the greatest boons will be higher quality X-ray spectra. It is indeed an exciting time to be in the business of studying active galaxies and starbursts.

Acknowledgments

I would like to thank the organizing committee of the Summer School not only for inviting me to participate, but also for inviting so many bright and talented young astrophysicists. I look forward to their future research.

References

1. Antonucci, R. R. J., & Miller, J. S., ApJ, 297, 621, 1985.
2. Aretxaga I., Terlevich E., Terlevich R.J., Cotter G., Díaz A.I., MNRAS, submitted.
3. Barth, A. J., Filippenko, A. V., & Moran, E. C., ApJ, 525, 673, 1999.
4. Barthel, P. D., ApJ, 336, 606, 1989.
5. Bassani, L., Dadina, M., Maiolino, R., Salvati, M., Risaliti, G., della Ceca, R., Matt, G., & Zamorani, G., ApJS, 121, 473, 1999.
6. Bragg, A. E., Greenhill, L. J., Moran, J. M., & Henkel, C., ApJ, 535, 73, 2000.
7. Brinkmann, W., Wang, T., Matsuoka, M., & Yuan, W. A&A, 345, 43, 1999.
8. Cohen, M. H., Ogle, P. M.., Tran, H. D., Goodrich, R. W., & Miller, J. S., AJ, 118, 1963, 1999.

9. Crenshaw, D. M., Kraemer, S. B., Boggess, A., Maran, S. P., Mushotzky, R. F., & Wu, C.-C., ApJ, 516, 750, 1999.
10. Djorgovski, S., Weir, N., Matthews, K., & Graham, J. R., ApJL, 372, 67, 1991.
11. Francis, P. J., Hooper, E. J., & Impey, C. D., ApJ, 106, 417, 1993.
12. Gallagher S.C., Brandt W. N. , Laor A., Elvis M., Mathur S., Wills B.J., Iyomoto N., ApJ, in press (astro-ph/0007384), 2001.
13. Goodrich, R. W., ApJ, 342, 224, 1989.
14. Goodrich, R. W. ApJ, 340, 190, 1989.
15. Goodrich, R. W. ApJ, 355, 88, 1990.
16. Goodrich, R. W. ApJ, 440, 141, 1995.
17. Goodrich, R. W., ApJ, 474, 606, 1997.
18. Goodrich, R. W., & Miller, J. S., ApJ, 331, 332, 1988.
19. Goodrich, R. W., & Miller, J. S., ApJL, 346, 21, 1989.
20. Goodrich, R. W., & Miller, J. S., ApJ, 355, 456, 1990.
21. Goodrich, R. W., & Miller, J. S., ApJ, 434, 82, 1994.
22. Goodrich, R. W., Veilleux, S., & Hill, G. J., ApJ, 422, 521, 1994.
23. Goodrich, R. W., & Miller, J. S., ApJ, 448, L73, 1995.
24. Goodrich, R. W., Eracleous, M., & Koratkar, A. , , in prep, 2000.
25. Gopal-Krishna, Kulkarni, V. K., & Wiita, P. J., ApJ, 463, L1, 1996.
26. Green, P. J., & Mathur, S. A&A, 462, 637, 1996.
27. Greenhill, L. J., Jiang, D. R., Moran, J. M., Reid, M. J., Lo, K. Y., & Clausen, M. J., ApJ, 440, 619, 1995.
28. Hes, R., Barthel, P. D., & Fosbury, R. A. E., Nat, 362, 326, 1993.
29. Hill, G. J., Goodrich, R. W., & DePoy, D. L. ApJ, 462, 163, 1995.
30. Jackson, N., & Browne, I. W. A., Nat, 343, 43, 1990.
31. Jackson, N., & Tadhunter, C. N., A&A, 272, 105, 1993.
32. Miller, J. S., & Goodrich, R. W., ApJ, 355, 456, 1990.
33. Miller, L., Peacock, J. A., & Mead, A. R. G., MNRAS, 244, 207, 1990.
34. Nelson, C. H., MacKenty, J. W., Simkin, S. M., & Griffiths, R. E., ApJ, 466, 713, 1996.
35. Ogle, P. M., Cohen, M. H., Miller, J. S., Tran, H. D., Goodrich, R. W., & Martel, A. R., ApJS, 125, 1, 1999.
36. Perley R.A., Dreher J.W., Cowan J.J., ApJL, 285, L35, 1984.
37. Pogge, R. W., & de Robertis, M. M., ApJ, 404, 563, 1993.
38. Schmitt H. et al., in 'The Staburst-AGN Connection', Eds. R. Mújica, I. Aretxaga & D. Kunth, INAOE Electronic Edition (`http://www.inaoep.mx/~agn00/posters.html`), p. 81, 2001.
39. Seyfert, C. K., ApJ, 97, 28, 1943.
40. Singh, K. P., Westergaard, N. J., & Schnopper, H. W., A&A, 172, L11,

1987.
41. Sprayberry, D., & Foltz, C. B., ApJ, 390, 39, 1992.
42. Tadhunter, C. N., Scarrott, S. M., & Rolph, C. D., MNRAS, 236, 163, 1990.
43. Tohline, J. E., & Osterbrock, D. E., ApJ, 210, L117, 1976.
44. Tran, H. D. ApJ, 440, 597, 1995.
45. Vermeulen R.C., Ogle P.M., Tran H.D., Browne I.W., Cohen M.H., Readhead A.C.S., Taylor G.B., Goodrich R.W., ApJ, 452, L5, 1995.
46. Vernet, J., Fosbury, R. A. E., Villar-Martín, M., Cohen, M. H., Cimatti, A., di Serego-Alighieri, S., & Goodrich, R. W., A&A, , submitted, 2000.
47. Warren, S. J., Hewett, P. C., & Osmer, P. S., ApJ, 421, 412, 1994.
48. Weymann, R. J., Morris, S. L., Foltz, C. B., & Hewett, P. C., ApJ, 373, 23, 1991.
49. Wilson, A. S., Ward, M. J., & Haniff, C. A., ApJ, 334, 121, 1988.

Hagai Netzer (*left*) discussing with Toshihiro Kawaguchi in the poster area.

PHYSICAL PROCESSES IN STARBURST AND ACTIVE GALAXIES

HAGAI NETZER
School of Physics and Astronomy, Tel Aviv University, Tel Aviv 69978, Israel
E-mail: netzer@wise.tau.ac.il

The physics of the ionized gas in active galactic nuclei (AGN) is governed by the source luminosity, the spectral energy distribution (SED) and the gas density and distribution. Photoionization is the dominant heating and excitation mechanism throughout the central ~ 1 kpc except for nuclear star-forming regions that are dominated by local heating processes. The physics of "pure" star-forming regions is very different. The energy input is mostly mechanical and the micro-physics of the ionized gas is governed by collisional processes. This review gives a short account of the steps required to construct photoionization and collisional models and apply them to three observed "regions" in AGN: the broad line region (BLR), the narrow line region (NLR) and the highly ionized gas (HIG). It also discusses the physical conditions in a typical starburst region. Specifying the gas distribution and properties in AGN reproduces the observed properties of all three regions, although extra ingredients, such as dust, are sometimes required. The emergent model is that of a "messy" AGN, with gas of different properties at almost all regions. The last part of this review deals with the AGN-starburst connection. It shows that both are intimately related in luminous AGN as well as in several ultra luminous IR galaxies (ULIRGs). New spectral tools are required to understand the physics of such objects. The most useful diagnostics for distinguishing accretion dominated from starburst dominated sources are likely to be found in the X-ray part of the spectrum.

1 The AGN Principle

Active galactic nuclei (AGN) are thought to be powered by accretion onto a massive black hole. The exact fraction of AGN among galaxies is not accurately known but recent studies suggest that it must be very large, especially if low luminosity AGN, such as LINERs, are taken into account. It is therefore tempting to present the *AGN Principle* (or, perhaps, the weak AGN principle) which states that *all galaxies contain active or dormant massive black holes.* Such a strong statement requires a proof. The present review is an attempt to look for part of this proof by studying the AGN-starburst connection.

The black hole mass is only one of the fundamental parameters determining the AGN properties. Mass accretion rate is a second important parameter. It is probably the main reason for the different appearance of LINERs, narrow line Seyfert galaxies and luminous quasars. Besides these two, there are various others that determine the observed AGN properties: the amount and geometry of obscuring gas, the dust content and the size of the various emission regions are probably the most important of those. Any complete analysis must take all those into account. This review gives only a few of the more important manifestations.

No AGN review is complete without some terminology. I will only use the most common abbreviations such as the narrow line region (NLR), the broad line region (BLR), and the highly ionized gas (HIG). Only the simplest classification will be used throughout the chapters, avoiding, as much as possible, names such as Seyfert 1s, Seyfert 2s, LINERs and NLRGs. Instead I shall refer to type-I AGN (those objects showing broad, strong optical-UV emission lines in their spectrum), and type-II AGN (those showing prominent narrow emission lines and very faint, if any, broad lines).

The spectral energy distribution (SED) of AGN is complex, with significant differences between objects. Much of the complication can be avoided by describing it, over a restricted energy band, as a power-law in energy,

$$L_\nu = a\nu^{-\gamma} , \tag{1}$$

where a is a constant. This approximation is used for various energy bands by adding the relevant subscript to the power-law index γ, e.g. γ_X.

2 The Physics of Ionized Gas

The physical properties of gas exposed to a strong radiation field are well described by photoionization theory. The time-dependent level of ionization, the time-dependent thermal structure, and the spectrum of the gas, are all well understood if the density is lower than about 10^{11} cm^{-3} and the emission lines are optically thin. Radiative transfer and high density processes are more difficult to treat and present-day calculations fall short of including all those in realistic AGN models. There are several books and comprehensive reviews that describe these processes, in particular Mihalas[50] , Osterbrock[58], Netzer[53] and Emerson[18]. Star-forming regions are thought to be powered by fast stellar winds and supernovae expanding shells. The spectrum is dominated by collisional processes and the time-independent equations, in the low density low optical depth limits, are easy to solve. Optically thick, high density collisional plasma is more difficult to treat and the time-dependent spectrum is still not well understood.

The following is a brief account of some of the principles, and the more important details necessary to analyze and compute the spectrum of starburst and active galaxies. The starburst part is less complete because it is covered in other chapters. The reader is referred to the above mentioned references, and other reviews, for a more comprehensive discussion and additional details.

2.1 *Photoionization*

The fundamental time-dependent equation that describes the conditions in a gas situated at a distance r from a strong radiation source of monochromatic luminosity L_ν, is

$$\frac{dN_X}{dt} = -N_X[I_X + R_{X-1}] + N_{X-1}I_{X-1} + N_{X+1}R_X \ , \tag{2}$$

where N_X is the number density of ion X,

$$I_X = \int_{\nu_X}^{\infty} \frac{(L_\nu/h\nu)\sigma_\nu e^{-\tau_\nu}}{4\pi r^2} d\nu \tag{3}$$

is the ionization rate per particle, ν_X is the threshold ionization frequency of ion X, and

$$R_X = \alpha_X(T)N_e \tag{4}$$

is the recombination rate per particle. Auger ionization, collisional ionization, three-body recombination, and charge exchange reactions, were neglected for the time being. The additional requirement is

$$\Sigma N_X = N_{\rm element} \ , \tag{5}$$

where $N_{\rm element}$ is the total abundance.

The steady state solution is

$$\frac{N_{X+1}}{N_X} = \frac{I_X}{R_X} \ . \tag{6}$$

This leads to the definition of two important location dependent time scales: the recombination time, $1/R_X$, and the ionization time, $1/I_X$. The first gives the typical time required for ion $X+1$ to recombine to ion X, when the radiation source is suddenly turned off. The second is the time required for ion X to get ionized to ion $X+1$ after a large increase in the ionizing flux. The two can be quite different, depending on the radiation field and the gas density. Moreover, the coupling between the various stages of ionization suggests that, under various conditions, the fractional abundance of ion X may change very slightly, or remains almost constant, despite large changes in the radiation field (for more on these issues see Krolik & Kriss[43]). More interesting, perhaps, are the global recombination and ionization time scales, i.e. the time it takes for the entire atmosphere to get completely neutral or

Table 1. Various ionization parameters

Ionization parameter	E_1	E_2
U(hydrogen)	13.6 eV	∞
U(helium)	54.4 eV	∞
U(X − ray)	0.1 keV	10 keV
U(oxygen)	0.54 keV	10 keV

the time it takes for the ionization front to reach its equilibrium location. Both are close to $1/R_X$ where X is the most abundant ion of the element in question.

The level of ionization of photoionized gas can be described by the location-dependent value of the ionization parameter

$$U = \int_{E_1}^{E_2} \frac{(L_E/E)dE}{4\pi r^2 c N_{\rm H}} , \tag{7}$$

where c is the speed of light introduced to make U dimensionless. U is relevant only if E_1 is close to the minimum energy of photons that can significantly affect the level of ionization. Thus, for gas that is ionized by a soft-UV radiation, E_1 must be related to the UV ionizing field. This is normally taken at the Lyman limit and the appropriate name is U(hydrogen). On the other hand, the most suitable choice for gas whose level of ionization is dominated by a strong soft X-ray source is $E_1 \sim 1$ keV. The UV radiation of such a source may be very intense but the gas is transparent to such photons and U(hydrogen) has no real influence on the gas level of ionization. A good choice in this case is $E_1 = 0.54$ keV, corresponding to the K-shell threshold ionization of oxygen, the most abundant element controlling the conditions in highly ionized solar-metallicity gas. A suitable name is U(oxygen). Table 1 gives a list of four ionization parameters that are useful in various situations and Fig. 1 shows a typical SED for AGN with the various energies (E_1) used to define these ionization parameters. As a rule, using the "right" ionization parameter, we normally find that $U \sim 0.1$ corresponds to gas that produces strong emission lines in the energy range under question. For example, $U(\text{oxygen}) = 0.1$ results in strong 0.5-3 keV emission lines. A much smaller U results in neutral gas and a much larger U in gas that is so highly ionized that line emission, over the relevant energy range, is not significant.

The thermal structure of the gas is governed by a similar set of equations that involves energy terms like $(E - E_{\rm threshold})$. These are not given here and can be found in the various references mentioned earlier. The ionization and

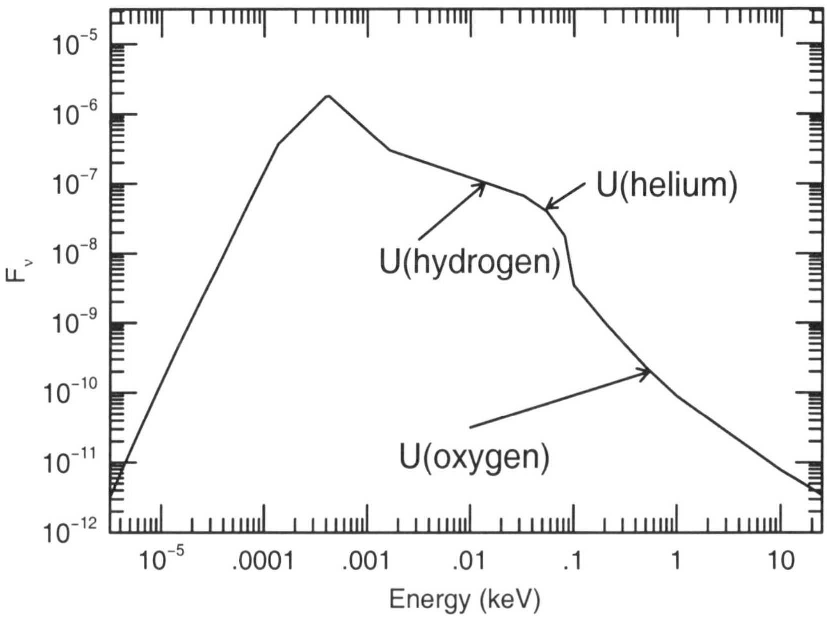

Figure 1. Theoretical spectral energy distribution (SED) for AGN and the various values of E_1 used to define the different ionization parameters.

thermal equations are, of course, coupled through the temperature dependence of the various atomic processes, such as the recombination and collisional ionization and excitation rates.

2.2 Collisional Processes and Shock-Wave Heating

In AGN, most of the emitting gas is photoionized by the central radiation field and collisions with fast particles do not play an important role in determining its level of ionization. This is very different from the conditions in X-ray bright starburst regions where the temperature can exceed several million degrees and most of the ionization is due to collisions with the fast electrons (see however comments in Sec. 6.4. regarding the photoionization of star-forming regions in AGN). In this case, collisional ionization and radiative recombination are the dominant processes.

There are other situations where collisional processes play important roles in starburst and active galaxies. Small-scale jets in type-II AGN (e.g. the nuclear jet in NGC 1068) can transfer a large amount of mechanical energy into a very small region, producing strong shocks and bright emission from collisionally excited plasma[16]. Collisional ionization and three-body recombination, are affecting the level of ionization of the high density gas in the central accretion disk. Finally, the innermost part of the line emission region in AGN can attain such high densities that collisional ionization and three-body recombination cannot be neglected.

A comment about the general energetics of shock waves in AGN narrow line regions is in order. Fast moving shocks are capable of producing bright emission line condensation in the NLR[79]. Jets can be the most efficient in producing such shocks and, indeed, some models of the NLR[10] argue that the narrow line emission in Seyfert galaxies is powered entirely by radio-emitting jets. Such "photoionizing shocks" are powerful sources of ionizing radiation and the resulting spectrum is significantly different from the pure, single temperature collisional plasma spectrum.

As argued by Laor[44], the mechanical energy produced by the shock is roughly $(v^2/2c^2)m_{\rm shock}$ and the one produced by the black hole accretion is about $0.1m_{\rm accretion}c^2$. Observed NLR velocities are of the order of 500 km s^{-1} thus the ratio of the two is approximately $10^{-5}(m_{\rm accretion}/m_{\rm shock})$. For shock excitation to be energetically significant, the amount of shocked material must exceed, by many orders of magnitude, the amount of accreted material. However, much of the NLR gas seems to be gravitationally bound (see Sec. 4.3) and the loss of kinetic energy must also be associated with a loss of angular momentum. This must result in an inflow of the shocked gas into the center of the AGN and, eventually its accretion. This would lead to accretion luminosity which is orders of magnitude larger than the shock produced luminosity.

The argument can be applied, in a more quantitative way, to specific emission lines, e.g. the [O III] λ5007 line. A certain fraction of the shock energy is radiated as photons of this line and a certain fraction of the central source radiation is absorbed and re-radiated as 5007Å line photons. Comparing the two, leads to a similar conclusion: shock heating cannot contribute, significantly, to the emission lines in large accretion rate sources, unless most of the shocked gas is flowing out of the system. The accretion rate in faint AGN, in particular in LINERs, may be much smaller while the gas velocity is basically the same. Shock heating can play a more important role in such sources.

2.3 Additional Atomic Processes

The list of atomic processes mentioned so far is by no means complete. The following is a short reminder of other important processes.

Low temperature dielectronic recombination is important in gas whose level of ionization is determined by the UV radiation field. High temperature dielectronic recombination is important in collisional plasma. X-ray ionization from inner shells (Auger ionization) is important for ions with three or more electrons in strong X-ray radiation fields. The process couples non-adjacent stages of ionization and results in X-ray fluorescence lines that are commonly observed in the spectrum of AGN. Heating and ionization by secondary electrons must be considered when the gas is almost completely neutral and the radiation field contains hard photons. It increases the level of ionization (and the intensity of some lines, like Lyα) and reduces the cooling of the gas. Finally, charge exchange reactions, mostly with neutral hydrogen, are important in low ionization gas.

2.4 The Spectrum of Ionized Gas

The main "product" of photoionization and collisional plasma calculations is the predicted spectrum. This requires the solution of the time-dependent level population equations at every point in the cloud. For a two-level atom whose level-2 population is determined by recombination with rate $\alpha_{\rm eff}$, collisional excitation and de-excitation with rates q_{12} and q_{21}, and line (but no continuum) photon emission and absorption, the equations to solve are

$$\frac{dn_2}{dt} = n_1 q_{12} - n_2(A_{21}\beta_{21} + q_{21}) + N_{X+1} N_e \alpha_{\rm eff} \tag{8}$$

and

$$n_1 + n_2 = N_X \ , \tag{9}$$

where N_X itself is time-dependent (Eq. 2).

In Eq. 8, β_{21} is the "local escape probability", i.e. the probability of line photons to leave that location without being absorbed by level 1. This escape probability treatment replaces the more rigorous, full radiative transfer solution. It is introduced to enable a full local solution while general radiative transfer couples different locations and is, therefore, more complicated. A reasonable approximation is $\beta = 1/(1 + b\tau_{12})$ where b is a constant of order unity.

Having solved for n_1 and n_2, we can now express the line emissivity as

$$\epsilon_{21} = n_2 \beta_{21} A_{21} h\nu_{21} \ , \tag{10}$$

where, in general, ϵ_{21} is time dependent.

In AGN and starburst galaxies, emission lines are normally classified as either *recombination lines* or *collisionally excited lines.* Most of the observed hydrogen and helium lines belong to the first category, although collisional contributions to the Lyman and Balmer hydrogen lines, in high density AGN clouds, are known to be important. Most metal lines are collisionally dominated in photoionized and collisional plasmas. The exception is the soft X-ray (0.5–3 keV) lines in AGN that originate in photoionized gas (see below). The temperature in such a gas is far below the plasma equilibrium temperature and strong metal lines, like O VIII 653 eV, are mostly due to recombination (see however Nicastro et al.[57], regarding time dependent collisional processes in such a gas).

The equivalent width (EW) of recombination lines is used to deduce the *covering fraction*, C_f, measured in units of $\Omega/4\pi$, i.e. the part of the source covered by optically thick gas. This is easy to demonstrate by considering Lyα, one of the strongest emission lines in the spectrum. Under most circumstances, the number of Lyα photon is comparable (and in the simplest, recombination Case B conditions, identical) to the number of hydrogen ionizing photons. For a power-law continuum (Eq. 1) extending from the Lyman limit to some high cut-off frequency $\nu_{\text{cut-off}}$, this results in

$$\mathrm{EW}(L_\alpha) = \frac{1215}{\gamma}(\frac{3}{4})^\gamma[1 - \nu_{\text{cut-off}}^{-\gamma}]C_f \text{ Å} . \quad (11)$$

A comparison with the observed equivalent widths of broad and narrow recombination lines suggests that, for the low temperature AGN gas, $C_f \sim 0.1$. For the HIG, the number is much larger and C_f is closer to unity.

Two additional processes leading to line emission need to be considered. The first is *continuum fluorescence*, sometimes referred to as photo-excitation or line scattering. This process is the result of populating level n_2 by absorbing the incident continuum. The process, that is not included in the simple two-level scheme of Eq. 8 (see Netzer[53] for the required modification) results in line emission whose intensity depends on the overall geometry. For example, in a static spherical configuration, or in a static spherical shell, where the continuum source is in the center, the emission due to this process is completely balanced by line of sight absorption, provided there is no additional opacity sources (e.g. a large disk or a large gas cloud around the center). The net emission in this case is zero. In a moving atmosphere, the emitted photons are Doppler shifted relative to the gas rest frame, and absorption and emission appear at different energies. This results in both emission and absorption whose equivalent widths are the same. Well known examples are

P-Cygni profiles in expanding stellar atmosphere and the X-ray spectrum of several type-I AGN.

Simple arguments can be used to estimate the relative intensity of continuum fluorescence lines and line emission due to radiative recombination. Both are related because, for a given ion and a strong resonance line, the bound-bound absorption cross section integrated over the line profile, and the bound-free absorption cross section integrated over the ground level continuum, are comparable. Since most recombinations lead to line emission, we find that for optically thin lines and continuum (this is the situation where continuum fluorescence is the most efficient), the relative number of line photons produced by the two processes is

$$\frac{\mathrm{N(continuum\ fluorescence)}}{\mathrm{N(recombination)}} = \frac{(\nu_l/\nu_o)^{-(\gamma+1)}(\pi e^2/mc)f_{12}}{\int_{\nu_0}^{\infty}(\nu/\nu_0)^{-(\gamma+1)}\sigma_\nu d\nu} \simeq$$
$$\simeq \frac{3+\gamma}{2}[\nu_{12}/\nu_0]^{-(\gamma+1)} \quad (12)$$

For H-like and He-like ions, and a typical $\gamma \simeq 1$ continuum, the ratio is about 3. An important deviation is a situation where the line in question becomes optically thick and continuum fluorescence drops exponentially. Continuum fluorescence excitation of emission lines is thus strongly column-density and line-width dependent.

The second process is line fluorescence, i.e. line emission following the ejection of an Auger electron. The fluorescence efficiency depends on Z^3 thus the most abundant elements, like oxygen, do not produce strong fluorescence lines, compared with iron, despite the order of magnitude difference in abundance. This difference can partly be compensated by the larger incident flux around 0.54 keV (the K-shell threshold for neutral oxygen), compared with the flux near 7.1 keV (the K-shell threshold of neutral iron) in typical AGN.

Simple photon counting arguments, like those introduced for calculating the covering fraction and EW(Lyα) (Eq. 11), can be used to estimate the EW of fluorescence lines relative to the incident continuum. Thus, the equivalent width of the 6.4 keV Kα line of iron can be calculated for a given SED and iron abundance. There are two practical differences. First, the column density required for complete absorption of the incident radiation is very large and, in some cases, a considerable part of the radiation is not absorbed by the gas. Second, the absorption at the relevant energy band (e.g. beyond the 7.1 keV threshold of neutral iron) is never due to only one element and Compton scattering is non-negligible too. Thus, not all of the absorbed photons, beyond a certain threshold, result in the emission of the fluorescence line in question.

The situation is much simpler when estimating the EW relative to the *scattered* continuum. The reason is that, both the number of line photons, as well as the number of scattered continuum photons (against which the lines are measured) depends on the column density of the gas, provided it is Compton thin. For example, the EW of the strong 6.4 keV Kα line observed in AGN, assuming a power-law continuum with energy slope γ_X, is

$$\mathrm{EW}(6.4\ \mathrm{keV}) \simeq 3.2[\frac{1.11^{-\gamma_X}}{3+\gamma_X}][\frac{\mathrm{F_Y}}{0.3}][\frac{n_{\mathrm{Fe}}/n_{\mathrm{H}}}{4\times 10^{-5}}]\ \mathrm{keV}\ , \tag{13}$$

where $\mathrm{F_Y}$ is the fluorescence yield (of the order of 0.3 for low ionization iron). There are small uncertainties on this estimate, to do with the metal abundances. There is also a viewing angle dependence[49] of EW(6.4 keV) that must be taken into account. However, the approximation is good enough to deduce the iron abundance (n_{Fe} in the above expression) in type-II AGN whose 6–7 keV continuum is thought to be dominated by reflection. The expression fails in situations where the Compton depth is large, and angle-dependent multiple scattering of the continuum photons must be taken into account[49,56] (Eq. 2).

2.5 A Cloud or not a Cloud?

Real computations require the definition of the basic line emitting entity which is normally referred to as a *cloud*. The assumption of clouds, or more generally condensations, is reasonable in view of spatially resolved observations of galactic HII regions. It is also required by line intensity considerations, mainly the observation that the typical line widths, in the BLR, are similar for low and high ionization lines. This is however problematic since such entities are either gravitationally bound or must be confined by external pressure. Self gravitating clouds must contain large amounts of gas. Most likely candidates are stars in the central cluster. These will be dealt with later under the name *bloated stars* (BS). Much lower mass clouds are normally assumed in the BLR. They can be as small as 10^{12} cm in diameter and contain as little as 10^{25} grams.

Confined clouds have been addressed in many papers but there is no real understanding of the confining medium or agent. The most likely possibilities are either thermal confinement, by a low density hot inter-cloud medium (HIM), or confinement by magnetic fields. The first has been discussed in numerous publications[42]. The idea is of a two-phase medium, where the hot and cold components are in pressure equilibrium. The HIM temperature is determined by Compton heating and cooling by the central radiation source. As pointed out by Mathews & Ferland[46] and many papers since, this results in a low temperature ($T \sim 10^7$ K) HIM with a large Compton depth, especially in high luminosity AGN. The resulting optical depth is too large to be compatible with the observed variability of such sources. Relativistic HIM has also been suggested. In this case, the confining gas is of much higher temperature and the Compton optical depth significantly smaller.

An alternative and likely solution is magnetic pressure confinement[64]. In this case, $B^2/8\pi \geq NkT$ and the required magnetic field is of the order of one gauss for the BLR and much smaller for the NLR. Such magnetic fields are likely to occur near the central accretion disk and, perhaps, throughout the emission line region.

2.6 A System of Clouds

Adopted the cloud concept, we proceed in two stages. First to calculate the ionization and thermal structure for individual clouds. This results in the location-dependent cloud emission in the various emission lines, $j_{\rm line}(r)$ (erg s^{-1}). We then integrate $j_{\rm line}(r)$ over volume taking into account the spatial distribution in properties (e.g. cloud size and shape) as well as absorption and obscuration by other clouds and by inter-cloud dust. In a spherical system, extending from $r_{\rm in}$ to $r_{\rm out}$, the total line emission is

$$E_{\rm line} = 4\pi \int_{r_{\rm in}}^{r_{\rm out}} n_{\rm c}(r) j_{\rm line}(r) \exp[-\tau({\rm line}, r)] r^2 dr \ , \qquad (14)$$

where $n_{\rm c}(r)$ is the radial dependent cloud number density and $\exp(-\tau({\rm line}, r))$ the absorption and extinction continuum optical depth, at the line energy, between location r and the observer. The computed $E_{\rm line}$ can be directly compared with the observations.

The above process must also include the computation of the bound-free and free-free emissions . In some situations (e.g. the Balmer continuum of the BLR clouds and the O VII and O VIII bound-free X-ray continua in the HIG) these provide additional diagnostics.

2.7 Diagnostic Diagrams

The observed line intensities can be used, in combination with the results of the calculations, to deduce the important physical parameters such as the ionization parameter, the gas temperature (in a collisional plasma) and composition. This is done by constructing line ratio or line diagnostic diagrams. The first give the line intensity as a function of ionization parameter (in photoionized gas) or temperature (in collisional plasma). The second convey the same information by plotting one line ratio against another. Fig. 2 shows an example that can be applied to X-ray lines in photoionized AGN gas.

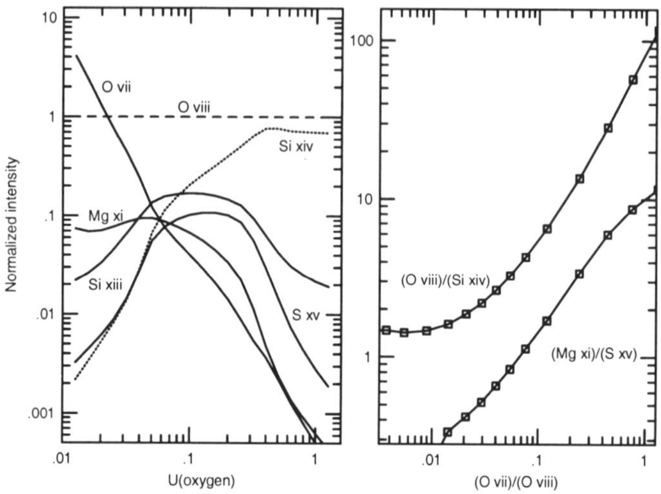

Figure 2. Left: An X-ray line ratio diagram, as a function of U(oxygen), for a low density ($N_{\rm H} = 10^5$ cm^{-3}) solar metallicity, low opacity photoionized gas. Right: Two diagnostic (line ratio) curves for the same gas. The combination of several such diagrams can be used to solve for both the ionization parameter and the gas composition.

2.8 Limitations and Uncertainties

While present photoionization computations are very advanced, there are several known limitations:

Reliable and unreliable line calculations: Most strong broad AGN lines are very optically thick. Such lines are normally treated with the escape probability formalism even in the most advanced photoionization codes. As explained, this treatment is local in nature and line transfer is not treated in detail. The local escape probability method is probably adequate for treating the high ionization lines like C IVλ1549 and N Vλ1240, the recombination lines like He IIλ1640, intercombination lines like C III]λ1909, and perhaps also for Lyα. It is definitely inadequate for the hydrogen Balmer lines and, perhaps, also for calculating the intensity of lines of low ionization ions like Fe^+ and Mg^+. It is also limited in its treatment of the angular emission pattern of large optical-depth lines. Full radiative transfer methods, combined with detailed ionization and thermal treatment, must be used to verify the accuracy of the present calculations. Such codes are not yet available.

Time-dependent spectra: While time dependent line emission in AGN must be important, because of the large continuum variations, a full time-dependent treatment of the spectrum is still lacking (see however a time-dependent treatment of the NLR gas[11,20]).

Cloud distribution: Cloud confinement is still a major problem and the cloud distribution is still unknown. Regarding the latter, two different possibilities have been considered. The first[53] is a simple spherical cloud system, where there is one type of cloud at every distance. An example is given in the following section. The second is the locally optimally emitting clouds (LOC) model[9,39], that assumes numerous clouds, with a range of density, column density and covering fraction, at any given location. The main idea of the model is that the local ionizing flux, and the gas density, produce a large range of ionization parameter at any given location. Lines from high and low ionization species are produced, with different efficiencies, at all radii, and the local spectrum reflects the range of physical properties. The $1/r^2$ central-source flux variation is the main reason for the different contributions to the various emission lines at different locations. In the LOC model, the general tendency is for lines of higher ionization and higher critical density to be emitted closer to the center.

2.9 The Motion of Ionized Gas

The equation of motion for ionized gas includes three major terms, gravity, $g(r)$, radiation pressure acceleration, $a_{\rm rad}(r)$, and drag-force. Gravity dominates in almost all cases of low ionization, large column density gas. However, for highly ionized gas in strong radiation fields, acceleration by radiation pressure force can be very efficient and dominates over gravity and drag forces.

The radiation pressure force is obtained by summing over the contributions from various ions. The contribution due to ion X is

$$a_{\rm rad}(r, X) = \frac{N_X}{\rho(r)c} \int_{\nu_X}^{\infty} \frac{L_\nu \chi_\nu e^{-\tau_\nu}}{4\pi r^2}, \tag{15}$$

where χ_ν is the total absorption cross section. Assuming an average ionizing energy $\overline{h\nu}$, an absorption cross section which is dominated by bound-free processes, and using the definition of I_X from Eq. 3, we find $a_{\rm rad}(X) \propto \overline{h\nu} N_X I_X / N_e$. For a steady state gas, photoionization is exactly balanced by radiative recombination and $N_X I_X = R_X N_{X+1}$. Thus $a_{\rm rad}$ is proportional to N_{X+1} and for the specific case of hydrogen, $a_{\rm rad}(\text{total}) \propto N_e$.

It is customary to introduce the "force multiplier", $M(r)$, which is the ratio of the total radiation pressure to the Compton radiation pressure

$$M(r) = \frac{a_{\rm rad}(\text{total}, r)}{a_{\rm rad}(\text{Compton}, r)} , \tag{16}$$

where

$$a_{\rm rad}(\text{Compton}, r) = \frac{N_e(r)\sigma_{\rm T} L_{\rm total}}{4\pi r^2 \rho(r) c} , \tag{17}$$

and $\sigma_{\rm T}$ is the Thomson cross section. Obviously, $M(r) \geq 1$ by definition.

The main contributions to $M(r)$ are from bound-bound and bound-free transitions. The former dominates in all cases where the line optical depths are small. In such cases it contributes between 50 and 90% of the total radiation pressure. In other situations, some or all of the resonance lines are optically thick and the bound-bound contribution is reduced with the increased line opacity. Fig. 3 shows an example of the various contributions to the force multiplier in a highly ionized AGN gas[13].

3 The Broad Line Region

Having established the general principle, we can now examine different "regions", i.e. various combinations of cloud location, gas density and column density. A complete description of the specified region includes also its boundaries, the gas dynamics and additional components (e.g. dust) that can be found within these boundaries. Such regions may or may not exist in real AGN and observations must provide the ultimate test.

3.1 The BLR Spectrum

First we consider a situation where the gas density is of order 10^{10} cm^{-3}, the column density is 10^{22-23} cm^{-2}, and the location a few light-days from the center. For a low luminosity AGN, this would imply $U(\text{hydrogen}) \sim 10^{-2} - 10^{-1}$ which means that the dominant carbon ion is C IV, the dominant oxygen ions are O VI, O V and O VI, etc. Such clouds must emit strong optical and UV lines of moderately ionized carbon, nitrogen and oxygen, as well as strong Lyman and Balmer hydrogen lines. The large column density also suggests that, if the internal gas motion is mostly thermal, all metal lines, as

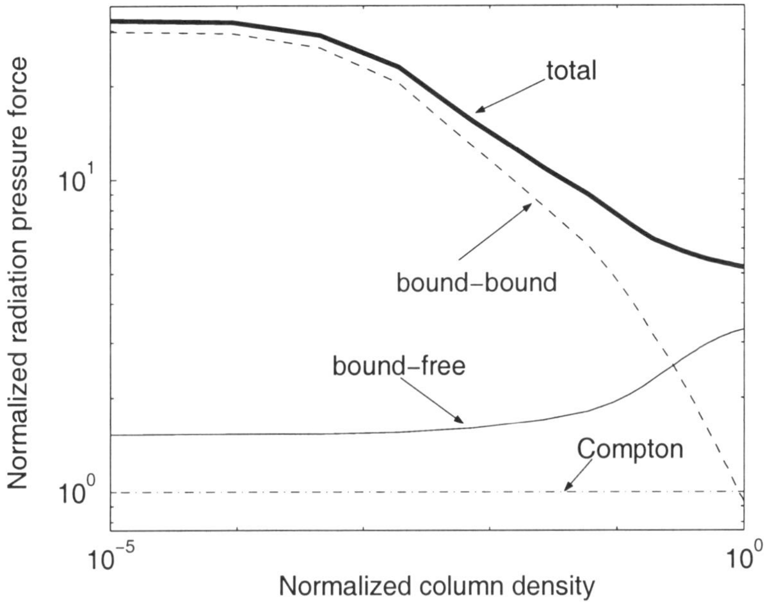

Figure 3. Contributions to the radiation pressure force at various depths inside a highly ionized AGN cloud of column density 10^{22} cm^{-2} and U(oxygen) = 0.01 at the illuminated face. From Chelouche & Netzer[13] ©2001 Blackwell Sci.

well as the Lyman and Balmer lines, have large optical depths. Combined with the large Keplerian velocities corresponding to this location, the predicted spectrum is of broad strong emission lines. This is, indeed, observed in AGN and is known as the BLR. Fig. 4 shows a composite BLR spectrum.

The general BLR properties have been reviewed, in great details, in several books and articles[53,29,61,41,67] and the following is only a short account of some of the most important issues.

3.2 Line Intensity and Line Profiles

Present day photoionization models are quite successful in explaining many of the observed properties of the BLR gas. In particular, there is a good agreement between the mean value of U(hydrogen) required to explain the observed line intensities and the emissivity-weighted-radius of the BLR which,

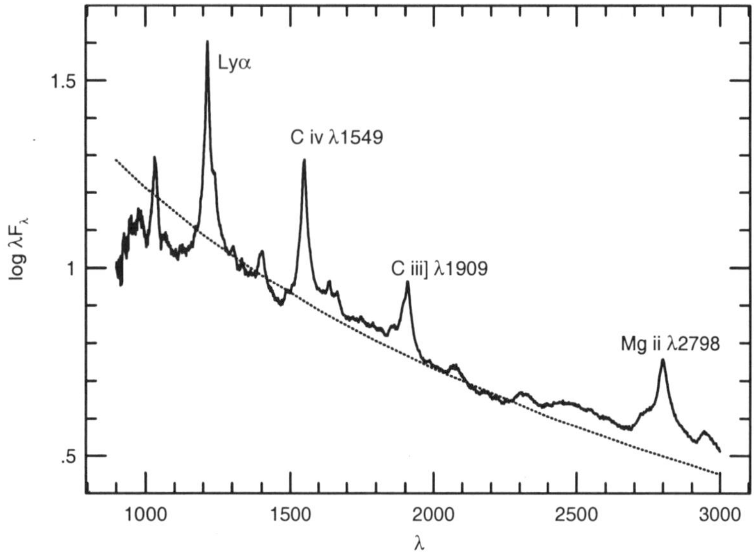

Figure 4. A composite AGN spectrum obtained by adding spectra of several quasars. The dotted line is a power-law continuum with $\gamma_{\rm UV} = 0.6$. Note the turnover of the continuum, at short wavelengths, relative to the power-law. Courtesy of A. Laor.

for the Balmer lines is given by

$$R_{\rm BLR}({\rm Balmer}) \simeq 0.15 L_{46}^{0.7}\ {\rm pc}, \tag{18}$$

where $L_{46} = L_{\rm ionization}/10^{46}$ erg s^{-1}. This expression, which is obtained from the reverberation mapping of 34 AGN[35] with the additional assumption that $L_{\rm Bolometric} = 9\lambda L_\lambda(5100\text{Å})$, suggests that the physical conditions in BLR clouds do not scale linearly with $L^{1/2}$. In particular, it suggests that, in the BLR,

$$U({\rm hydrogen}) \propto L_{\rm ionization}^{-0.4}\ , \tag{19}$$

which means that, on the average, the BLR gas in low luminosity AGN is more ionized compared with the gas in higher luminosity sources. This interesting conclusion must be tested for high ionization lines, like C IV $\lambda1549$, since the result only applies to the hydrogen Balmer lines[35] .

Another success of the models is the realization of a thick geometry, i.e. the understanding that $r_{\rm out} \gg r_{\rm in}$. A detailed mapping of the BLR requires specific assumptions about the gas distribution, as illustrated below. A notable exception is the relative intensity of the Balmer and Lyman hydrogen lines which cannot be explained by present day models. This problem, first raised by Baldwin[8] and discussed extensively in the literature, is related to the large observed intensity of the Balmer lines, far in excess of their predicted intensity.

3.3 Energy Budget and SED

Among the few open issues regarding the BLR spectrum are the *energy budget* and the SED problems. Both are related to the fundamental assumption that the observed Lyα flux represents, to a high degree of accuracy, the total number of ionizing photons (Eq. 11). Similar arguments suggest that the observed flux of the HeII lines is a good measure of the $E > 4$ Rydberg continuum since most of the radiation at those energies is absorbed by He^+. Thus, several observed lines give indications for the number of unobserved Lyman continuum photons as well as on the shape of the unobserved UV continuum. Given this assumption, the covering factor can be deduced from the measured equivalent widths of the hydrogen and helium lines.

However, years of observations and great improvement in AGN modeling cannot reconcile the discrepancy between the observed SED and the equivalent widths of the recombination lines. The observed SED seems to be under luminous to explain the observed line intensities and too soft to explain the observed line ratios. It seems that, either the far UV continuum is very different from the one obtained from the extrapolation of the optical-UV continuum (which is a problematic assumption by itself since, in some high redshift sources, much of the Lyman continuum is directly observed) or there is a fundamental flaw in the models.[52,38,81].

3.4 Gas Composition

An interesting development of the last decade[29] is the realization that AGN metallicity, that was suspected to be high, as expected in galactic nuclei, is indeed unusual at least in high luminosity sources. The unusually high abundances are deduced from comparing strong UV lines, such as N V $\lambda1240$, with recombination lines such as He II $\lambda1640$. The enhancement is more noticeable for secondary elements like nitrogen. The recent study of lower luminosity

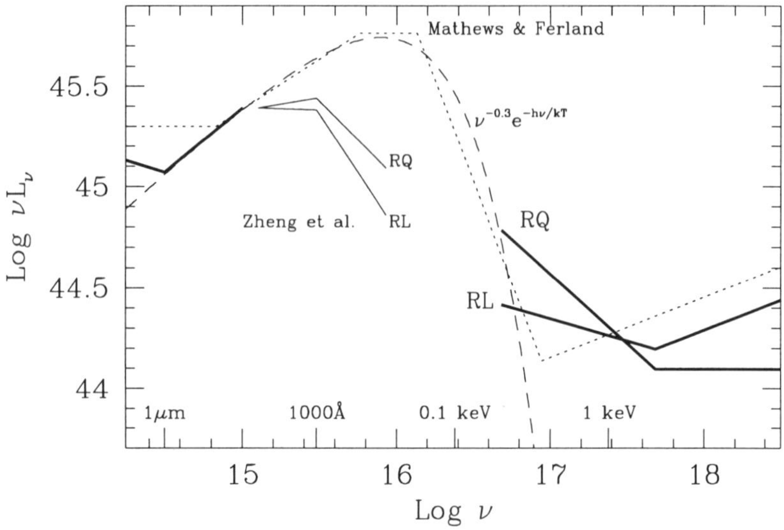

Figure 5. Several theoretical SEDs compared with real observations at optical-UV and X-ray energies. The extrapolated X-ray flux meets the observed UV continuum and does not require the "big blue bump" marked by the dashed lines. Courtesy of A. Laor.

PG quasars[77] suggests a similar behavior in those sources and, perhaps, a luminosity dependence of the composition of the line emitting material. All this hints to the suggestion that luminous AGN are related to intense star-forming activity. Obviously, there are interesting implications to galaxy and black hole evolution.

3.5 Mapping the BLR

The great advance in applying reverberation mapping to AGN observations (see Brad Peterson's review in this book), enables, for the very first time, a meaningful attempt to map the ionization and mass distribution in the BLR in several low luminosity AGN. The following is a brief account of the latest results based, primarily, on Kaspi & Netzer[33] and Korista & Goad[39].

A recent work by Kaspi & Netzer[33] presents an attempt to model, in detail, the time-dependent line and continuum variations across the BLR, in the well studied case of NGC 5548. The observations of this galaxy are particularly useful since they include well sampled light curves of the optical and UV continuum, as well as about eight isolated emission lines. This is essential for checking the consistency of the emission line models since various emission lines respond differently to the ionizing continuum variations. The model assumes a simple radial distribution of spherical, constant density clouds and requires seven parameters: the inner and outer radii, the radial particle-density parameter, s, defined by

$$N_{\rm H}(r) \propto r^{-s} \ , \tag{20}$$

the cloud number density parameter, p, defined by

$$n_{\rm c}(r) \propto r^{-p} \ , \tag{21}$$

the normalization of the density and column density at a given radius, and the global normalization of the covering factor (i.e. the total line emission). A modified version includes, also, a variable-shape SED which adds one or two more parameters. Since the clouds are assumed to be spherical, and the cloud mass is conserved, it is easy to obtain also the specific radial dependences of the column density, the cloud radius, the ionization parameter and the covering factor that are all related to s and p.

The calculations[33] do not attempt to fit the Balmer line light curves because of the uncertainty in calculating the transfer of these photons. Similarly, there is no attempt to calculate the FeII and MgII line intensities. There are four UV lines that are modeled (Lyα, He II λ1640, C III]λ1909 and C IVλ1549) and the results are in good agreement with the observations provided $1 \lesssim s \lesssim 1.5$ and $p \gtrsim 1.5$. Other important results include a lower limit on the column density of the BLR clouds, a well defined range of acceptable densities. The results of the fitting to the light-curves of the four emission lines are shown in Fig. 6.

In a parallel effort, Korista & Goad[39] constructed a LOC model in attempt to fit the observed light curves of the very same source. As explained, the LOC model assumes a very different gas distribution with a large range of properties at any given location. The current version of the model requires eight parameters (some of which are fixed and others allowed to vary) that describe the density, column density and covering fraction at any chosen location. The model fits reasonably well the observed time-dependent fluxes of the four lines modelled by Kaspi & Netzer[33], and fails, like the other model, to explain the Mg IIλ2798 line intensity.

Given the above two attempts, and various other works underway to improve the models, and to fit the spectrum of other sources, we conclude

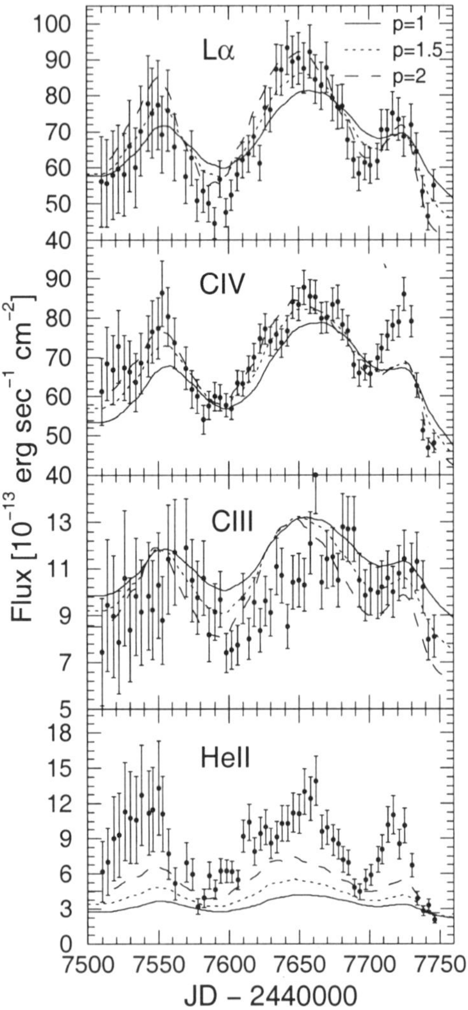

Figure 6. Observed (points with error bars) and calculated (smooth curves) light curves for NGC 5548, under various assumed conditions on the cloud number density parameter p. From Kaspi & Netzer[33] ©1999 AAS.

that AGN-BLRs are reasonably well understood although outstanding issues still remain.

3.6 Bloated Stars as BLR Clouds

The idea that stars in the (hypothetical) nuclear cluster can absorb and reprocess the AGN continuum radiation, is not new. It has been discussed in various papers[17,60,71,36] since the early 1980s and studied, in much greater details recently[2,3,1]. The most recent version of the model is based on the suggestion that the broad emission lines are emitted from the winds or the mass-loss envelopes of giant stars in the central cluster. These have been named *bloated stars* (BS) and while their origin and formation is not entirely understood, there are reasons to believe that they can evolve in the extreme environment near the central black holes. Models involving the emission of strong lines from the cometary tails of such stars, or their tidally disrupted atmospheres, have been discussed in various levels of detail.

The challenge of the BS model is to explain the various observed BLR properties. In particular, the observed broad line intensities, profiles and variability must be reconstructed by the model. Several papers[2,3,1] go a long way in this direction, although significant problems remain. A major advantage of the model is the removal of the confined cloud hypothesis. A major disadvantage is the predicted small number of BSs. The reason is that, such highly evolved stars cannot represent more than a tiny part of the nuclear stellar populations. Since the number of BSs can be estimated from the observed line intensity and covering fraction, there are strong limits on the total stellar population. Stellar collisions are expected too (this presents a problem for all cloud models) and there are other difficulties. A fundamental difference between the BS model and the cloud model is the total mass. A single BS can be more massive than 10^8 "standard" BLR clouds. The BS model requires, therefore, much more mass in the BLR.

3.7 The Number of BLR Clouds

High spectral resolution observations suggest that, in general, the observed emission line profiles are smooth over thousands of km s^{-1}. For example, the very detailed observations of low luminosity sources like NGC 4151, clearly demonstrate that, if the line profiles are produced by small individual clouds, the deviation from a smooth flow or motion is extremely small[7]. The observed deviations from a smooth profile, with assumptions on the velocity field, can be used to deduce the number of clouds. Arav and collaborators[7] have used such arguments to argue that, if individual clouds emit pure thermal line profiles, the number of clouds is huge. For the specific case of NGC 4151, their Monte-Carlo simulations suggest at least 10^7 clouds.

While the smooth line profiles are problematic for all cloud models, this is specifically noticeable in the bloated star model where the expected number of stars is only about 10^{4-5}. There are several suggestions of how to reconcile the observed profiles with a much smaller number of clouds. One possibility is that internal motions (the so-called microturbulences) are super-sonic, reaching[12], perhaps, several hundred km s^{-1}. For BSs, the suggestion is that cometary tails, resulting from tidal forces in the vicinity of the central black hole, are associated with large internal differential velocities. This helps to reduce the number of clouds. It also introduces other difficulties, such as the dramatically reduced line optical depth in the presence of large turbulent motions.

4 The Narrow Line Region

Next we consider lower density gas, $N_{\rm H} \sim 10^{3-5}$ cm^{-3}, at larger distances, up to about 1 kpc, with typical dimensions of several hundred parsecs. This would give an ionization parameter which is similar, or perhaps slightly smaller, than the one deduced for the BLR. Scaling the Keplerian velocities with the increased distance, we anticipate narrower emission lines of FWHM= few$\times$100 km s^{-1}. The appropriate name would be the *narrow line region* (NLR).

4.1 The NLR Spectrum

The main expected spectral difference between the BLR and the NLR (except, of course, for the line width) is the presence, in the NLR, of strong forbidden lines, as a result of the lower density. The relative intensity of the NLR and BLR lines, assuming ionization bounded clouds in both regions, reflects the relative covering factor of the two regions. A rough estimate can be obtained by comparing the relative intensity of the broad and narrow Hβ line components. In many type-I AGN, the narrow component of the Hβ line is an order of magnitude weaker than the broad Hβ component. We expect this to be also the ratio of the covering factors in the two regions. As explained below, this is not necessarily the case if other ingredients, such as dust, are included.

The NLR is, perhaps, the best observed region in active galaxies. In nearby sources it is spatially resolved, thus a location-dependent velocity map can be obtained. The brightness of the NLR in nearby sources allows high quality optical, UV and IR observations and, starting from 1999, also real X-ray spectroscopy. Figs. 7- 9 show the narrow line spectrum of various AGN

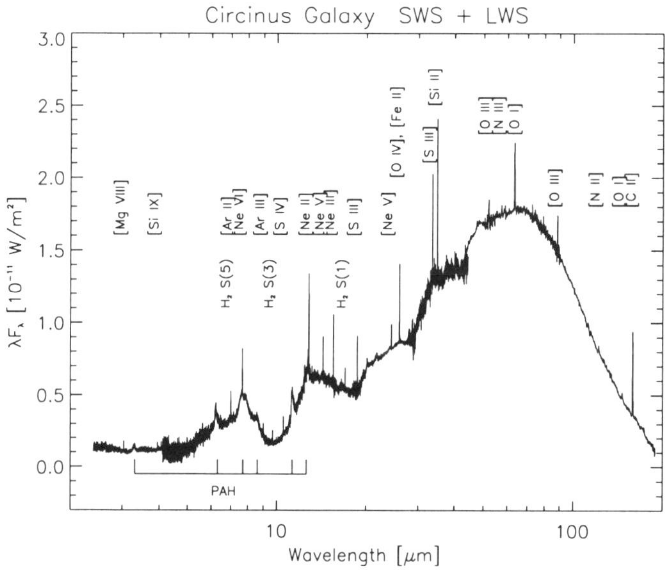

Figure 7. The ISO SWS+LWS spectrum of the Circinus galaxy. Courtesy of E. Strum.

in various wavelength bands.

Some of the most important characteristics of the spectra are worth commenting on.

- Many of the observed optical, UV and IR lines seem to originate from the same NLR clouds. Given this, we can infer the physical conditions in the gas with great accuracy because (a) the lines are easy to interpret (low density, no complicated radiative transfer); (b) we can observed many lines and hence deduce the important parameters of density, temperature and abundance using standard line ratio techniques, and (c) the required reddening corrections can be obtained by comparing reddening sensitive line pairs such as the Balmer and Paschen hydrogen lines (see however a comment below about internal and external dust). The main problem is that the spatial resolution of present day mid-IR instruments (e.g. ISO) is

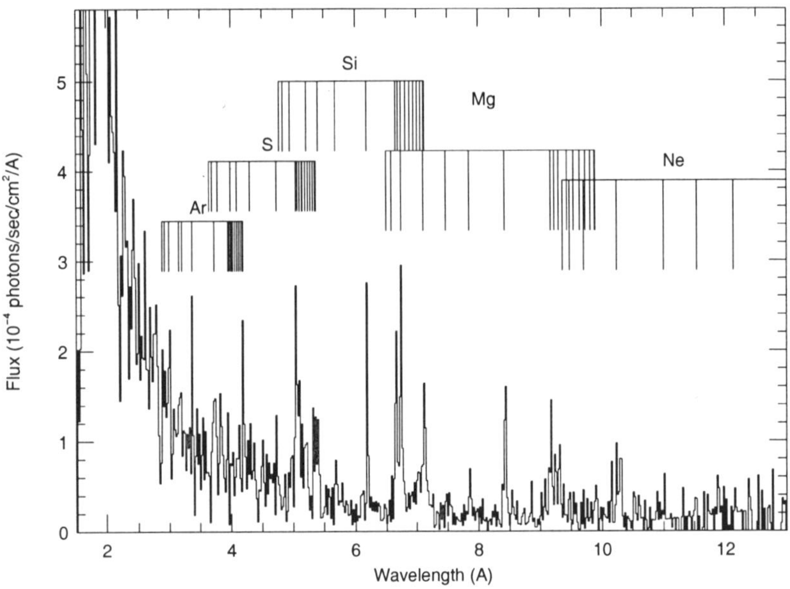

Figure 8. The *Chandra* X-ray spectrum of the Circinus galaxy. From Sambruna et al.[69] ©2001 AAS.

inferior to that of the large ground-based telescopes and the HST. Thus, only global reddening corrections can be obtained by using the mid-IR lines.

- Some mid-IR emission lines (e.g. [NeVI]7.66μm) as well as optical lines due to Fe X, Fe XI and Fe XIV, clearly originate in clouds of very high level of ionization. These *coronal lines* are probably ionized by the same central continuum, although there are alternative explanations involving collisional ionization in a much hotter environment[79]. In several well studied cases, there are clear signs of different dynamics of the coronal line region that suggest that this part of the NLR is closer to the central source and is characterized by higher than average density and velocity.

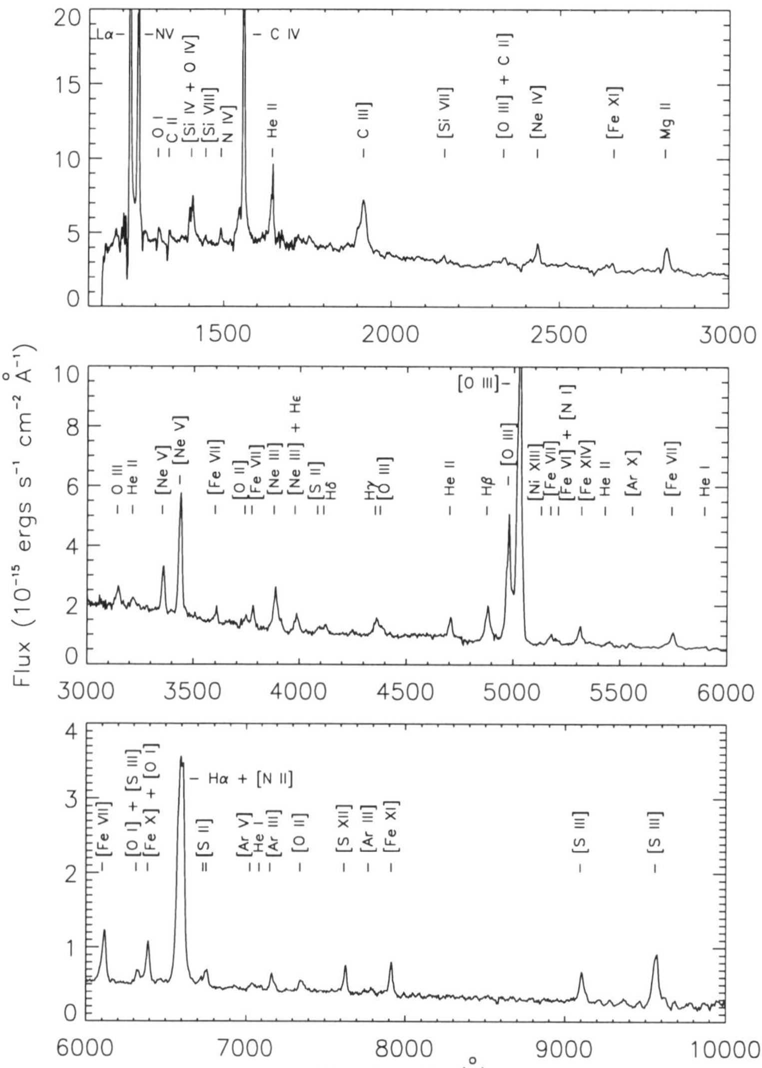

Figure 9. An HST spectrum of a hot-spot in NGC 1068. From Kraemer & Crenshaw[40]
©2000 AAS.

Thus, in at least several sources, the NLR is stratified with both density and ionization decreasing outwards. In the simple radial-dependent cloud model, this means $s < 2$ (Eq. 20 but note, again, the alternative LOC model[22]).

- The ISO observations of starburst and active galaxies[24,14] clearly show the presence of strong PAH (polycyclic aromatic hydrocarbon) features, mostly at around 7.7 and 8.6 μm. While the physics and chemistry of these small molecules is not very well understood, it is believed that they are sensitive to high energy (far UV and soft X-ray) radiation that can easily destroy them. The EW of the PAH features is argueably a good measure of the strength of the unobserved continuum radiation. Indeed EW(PAH) seems to be larger in starburst galaxies compared with AGN (see however an alternative[14] analysis). Such spectral features are useful in determining the ionizing flux in the NLR, even in highly obscured sources. Present observations are consistent with the idea that the line emitting part of the NLR is illuminated and excited by the central source in the direction where there is a clear view of the center. NLR-type clouds that are shielded from the central source (if such clouds exist) do not emit detectable narrow lines.
- The narrow X-ray lines in type-II AGN (only two published results, correct to October 2000) come from a region which seems to overlap with the NLR but with very different properties. Evidently, some of the nuclear gas is extremely highly ionized and while occupying the same volume, it is not necessarily of the same origin. This highly ionized gas (HIG) is further discussed in Sec. 5.

4.2 Dusty NLRs

Like all known HII regions, dust is likely to be present in AGN emission line regions. Such dust will scatter and absorbe some of the ionizing and non-ionizing radiation, and will affect the observed spectrum. There is clear evidence for dust in the NLR, obtained by comparing narrow emission lines of known relative intensity and very different wavelengths. However, there is no indication for dust mixed in with the BLR gas. It is likely that the harsh environment of the BLR, mainly the strong radiation field, is the reason for this deficiency. As explained below, this is also supported by other observations.

A comparison of the broad and narrow Balmer lines suggests, based on simple photon-counting arguments, that if the gas in both regions is optically thick in the Lyman continuum, the covering factor of the BLR is an order of magnitude larger than the covering factor in the NLR. Moreover, counting the number of the narrow Hβ line photons, shows them to be very small considering the expectation, only about 1% of the expected number given the NLR geometry. This is entirely inconsistent with estimates of the (hypothetical) cone opening angle (see R. Goodrich contribution) and direct mapping of the NLR, by HST, in nearby sources. A possible explanation is that the NLR gas entirely fills the open cone but is mostly optically thin and does not absorbe a large fraction of the ionizing continuum. This is in conflict with several observed line ratios as well as the emission measured in some of the lines.

An alternative explanation for the "missing" NLR gas is the presence of dust in most of the volume outside the BLR[55]. If AGN dust is of similar composition as the interstellar dust in our own galaxy, we can calculate the dust sublimation radius, i.e. the minimum radius where dust of a certain composition can survive the central radiation field and does not evaporate. For graphite dust particles, this radius is given roughly by

$$R_{\rm sub} = 0.2 L_{46}^{1/2}\,{\rm pc}\,. \tag{22}$$

Evidently, $R_{\rm sub} \simeq R_{\rm BLR}$ which suggests that there is a natural boundary to the BLR. According to this explanation, the entire volume, from a small fraction of a parsec, very close to the central black hole, to hundreds of parsecs from the center, is filled with gas whose average density is a decreasing function of distance. The inner $R_{\rm sub}$ parsecs are void of dust since any dust particle that could have formed there, evaporates on a short time scale. In this region, almost all of the ionization radiation is absorbed by gas.

The presence of dust in the NLR, just outside $R_{\rm sub}$, can substantially change the level of ionization. Such dust, referred to as *internal dust*, competes, effectively, with the ionization of the gas since the dust particles have large absorption cross sections at all energies between about 0.02μm and 10μm. The fraction of Lyman continuum photons absorbed by the dust, relative to those absorbed by the gas, is proportional to U(hydrogen) since

$$\frac{N_{\rm dust}}{N_{\rm H^0}} \propto \frac{N_{\rm H^+} + N_{\rm H^0}}{N_{\rm H^0}} \propto U({\rm hydrogen})\,. \tag{23}$$

Therefore, dust is more efficient in attenuating the ionizing radiation in highly ionized gas, where it absorbs a larger fraction of the photons capable of ionizing the gas. The different ionization structures of dusty and dust-free environments are illustrated in Fig. 10 that shows the shrinking of the HII part of the dusty gas. This involves weaker emission lines.

For AGN, the suggestion is that the gas just outside the BLR is dusty and characterized by a large ionization parameter which results in inefficient

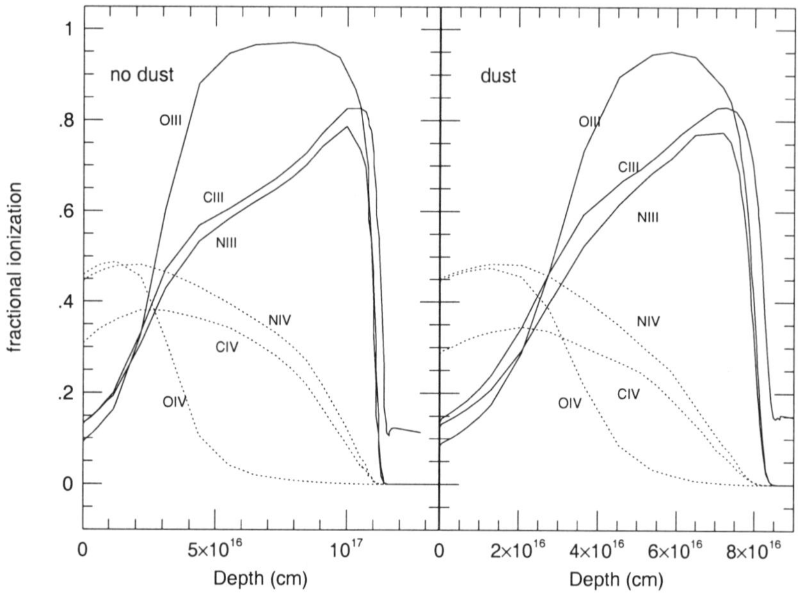

Figure 10. The effect of internal dust on the level of ionization, and the depth of the ionization front, inside a typical NLR cloud. In this case, U(hydrogen) $= 0.01$ and the gas-to-dust ratio is similar to the mean interstellar value. Note the shrinking of the HII region size by about 30% but the very similar ionization structure.

line emission. Moreover, such dust has great influence on the emissivity of Lyα and other resonance lines that suffer additional attenuation because the effective dust opacity in those wavelengths is greatly enhanced (by a factor of ~ 5) due to the large number of scattering of the line photons. The result is that highly ionized dusty regions emit weaker emission lines.

As stated earlier, this review is concerned with the consequence of having gas of certain properties at certain distances from a typical AGN continuum source. In Sec. 3 we found that such a combination produces the standard BLR and in this section we illustrated how a different combination produces the standard NLR. It is therefore interesting to note that NLR gas, mixed with dust, can explain the *absence* of *intermediate line region* in AGN (a statement challenged by some AGN observers). The combination of gas density at the location where the gas velocity is intermediate between the BLR and the NLR velocities, does exist. However, if the above explanation is correct, the emission lines produced in this region are extremely weak because of the dust. This leads to the apparent absence of such regions in AGN.

4.3 The Motion of the NLR Gas

The dynamics of the NLR gas has been studied extensively, but the results are still inconclusive[75]. Optically thick NLR clouds are likely to be dominated by the gravitational field, because of the large amount of neutral gas. Indeed, typical observed widths of narrow emission lines are similar to what is expected given the bulge mass of the parent galaxy. Radiation pressure force must be important for the dynamics of those NLR clouds with little or no neutral gas. The velocity of such a gas is likely to be dominated by a large outward component. Unfortunately, it is not simple to tell them apart.

Observations of narrow emission lines clearly indicate that some narrow line profiles, especially in lines of highly ionized species, exhibit strong blue asymmetry. This is particularly noticeable in almost all coronal lines in many AGN. The asymmetry can be due to either an expanding atmosphere whose receeding part is partly obscured, or from infalling clouds that are very faint when observed from their neutral back side (far from the central source). The most likely opacity source in such clouds is internal dust. Recent ISO observations seem to give the first clue to the origin of the blue asymmetric narrow lines. The comparison of optical and IR lines of the same ion, enables to test the effect of dust since the IR lines are hardly affected by reddening[73]. Thus if dust is present, and causing the blue asymmetry in the optical line profiles, this will not be seen in the IR line profiles. The conclusion, after studying several IR-optical line pairs in two AGN, is that the same blue asymmetry is observed in the IR lines, suggesting that the blue asymmetry is not due to internal dust. It seems that the NLR gas, at least the highly ionized clouds in those sources, is flowing outward at high velocities. A plausible (but not the only) explanation for the blue asymmetry is central obscuration, perhaps by the central accretion disk.

4.4 Recovering the SED

The combination of a large number of UV, optical and IR narrow emission lines is also useful in recovering the unobserved (far UV) part of the SED. This has been recently demonstrated[4,5] for NGC 4151 and NGC 1068.

The idea behind this work[4,5] is the fact that different ions require different ionization energies and the observed spectrum is due to some weighted mixture of all those ions. The accurate knowledge of this mixture, based on the comparison of model calculations with observations, can be used to deduce the relative ionizing flux at each energy. This is illustrated in Fig. 11 that shows the ionization potential of several ions that are known to produce strong emission lines against the unobserved part of the SED. Obviously, there are other unknowns to do with the gas distribution, composition and amount of dust. However, the superb two dimensional spectroscopy provided by HST, combined with accurate IR line measurements, produces such a large number of observed emission lines, that the results are quite secured.

An interesting result of the work of Alexander and collaborators[4,5] is that, in the two cases studied, there is clear indication of an absorption feature in the SED, just above the Lyman limit. Such an absorption can be caused, for example, by a large covering fraction of optically thin gas that is situated either in the BLR or just outside it. This absorbing gas may be related to optically thin BLR gas, or the X-ray and UV absorption systems discussed in the following section. It gives another indication that the earlier picture of two isolated regions in AGN, the BLR and the NLR, is just a very poor approximation of a very complex emission region. Real AGN seem to be full of clouds and filaments, close to the center and very far from it, that sometimes obscure the ionizing radiation affecting the conditions in some other clouds. A proper name is the *messy emission line region.*

5 The Highly Ionized Gas

The final example that is treated here in detail is that of a high ionization gas in the BLR and/or the NLR. We do not specify the gas density (mostly because it is unknown) but consider the general case where it is some two or three orders of magnitude smaller than the density of the material in that region. In the BLR, this would mean $N_{\rm H} \sim 10^{7-8}$ cm^{-3} and in the NLR $N_{\rm H} \sim 10^{1-3}$ cm^{-3}. We also assume a column density which does not exceed the column density of the already discussed components. The resulting ionization parameter is two-to-three orders of magnitude larger than the corresponding BLR or NLR ionization parameter. The gas is exposed to the AGN radiation field and we expect strong absorption and emission features in the soft X-ray part of the spectrum. The relevant ionization parameter to use in this case is U(oxygen) and the component is named the *highly ionized gas* (HIG).

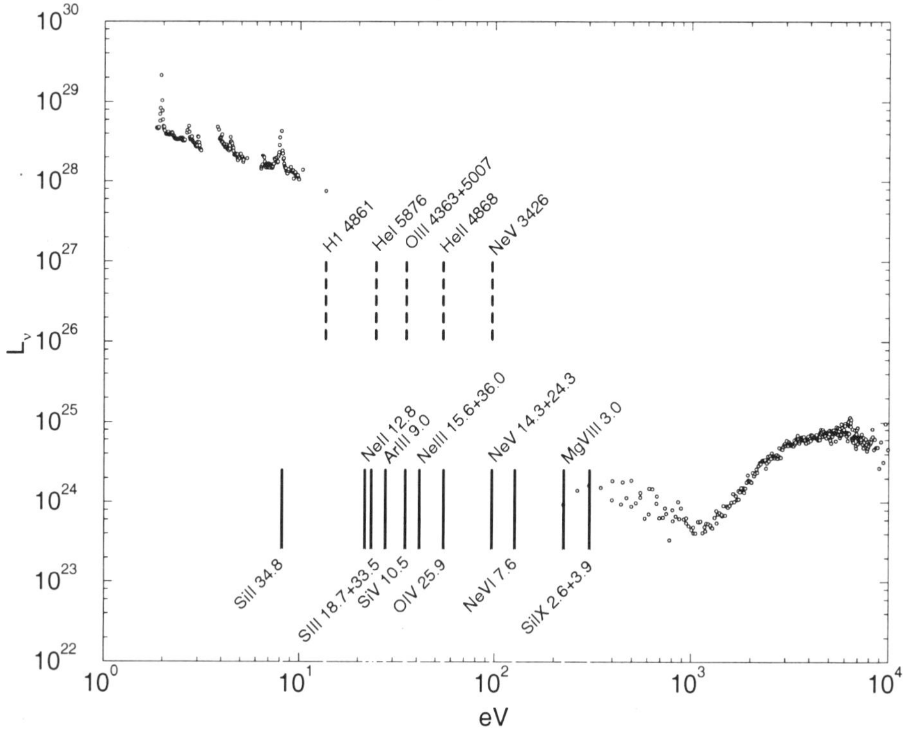

Figure 11. The unobserved part of the SED in AGN can be recovered by noting that various observed narrow lines are due to ions that require different ionization energies. The diagram shows some of the observed line ionization energies superimposed on the NGC 4151 continuum. From Alexander et al.[4] ©1999 AAS.

5.1 *The Spectrum of the HIG*

X-ray observations since the early 1990s show that many type-I AGN contain significant amounts of circumnuclear material which re-processes and modifies the intrinsic X-ray radiation. Partially-ionized X-ray absorbers were first suggested to exist in AGN in sources such as MR 2251-178 [27,59] and NGC 4151 [23,80]. Supporting evidence for such HIG (also referred to as "warm absorber") was also provided by *Ginga*, *ROSAT* and *BBXRT*. Study of the HIG spectrum has improved by the launch of *ASCA* that confirmed the already observed deficit of photons close to ~0.8 keV, soon to be interpreted as due to bound-free absorption by O VII and O VIII. Extensive studies of the feature, including its time-dependent behavior, can be found in several papers[21,66,65,25]. Such systems are very common in low luminosity AGN but seem to be less common in higher luminosity sources (a result that needs confirmation since observations of high luminosity sources are severely limited by the poor signal-to-noise ratio of the *ASCA* observations).

The X-ray spectrum of type-II AGN is completely different[74,56,26]. *ASCA* observations clearly show large EW lines (few × 100 eV) of H-like and He-like ions of neon, magnesium, silicon and sulfur, and weak X-ray continuum. The observed 0.5–10 keV luminosity is consistent with the standard model that involves obscuration by the central torus. The amount of obscuration can be directly measured in several type-II sources since the line-of-sight column is such that it is opaque to the very soft (~ 1 keV) continuum yet transparent to the medium X-ray (~ 5 keV) radiation. The material producing the large EW X-ray lines is not obscured, hence the large line-to-continuum intensity ratio. The HIG in type-II AGN is likely to have large dimensions and low density. It is not yet clear whether it is the same component observed in type-I AGN.

There are interesting suggestions about HIG which is far enough from the center to allow some dust particles to co-exist with the hot gas[37]. So far, there are only poor resolution *ASCA* data to support this idea while high resolution spectra, at around 0.5 keV (to detect the signature of neutral oxygen, the most abundant ion in the hypothetical dust) are required.

The study of HIG in active galaxies is being revolutionized by the new, high resolution spectra of the *Chandra* and *XMM* grating instruments. The handful of spectra already published shows rich emission and absorption features that confirm, in some cases, the earlier findings, but contradict others. An example is shown in Fig. 12 where the spectrum of NGC 3783, a type-I AGN with strong absorption features, is shown. The spectrum is clearly under-exposed but is good enough to illustrate the complexity of the features. Better spectra are expected soon.

Equally interesting are new *Chandra* observations of type-II sources[68,69] (see Fig 8) that enables the very first measurement of the HIG extent. In those two sources (Mkn 3 and the Circinus galaxy) the HIG extends over tens to hundreds of parsecs.

X-ray emission and absorption lines are relatively easy to interpret. In general, we expect three generic type spectra:

Absorption dominated spectra are seen when the absorbing material is covering the entire source but the emission features are very weak since

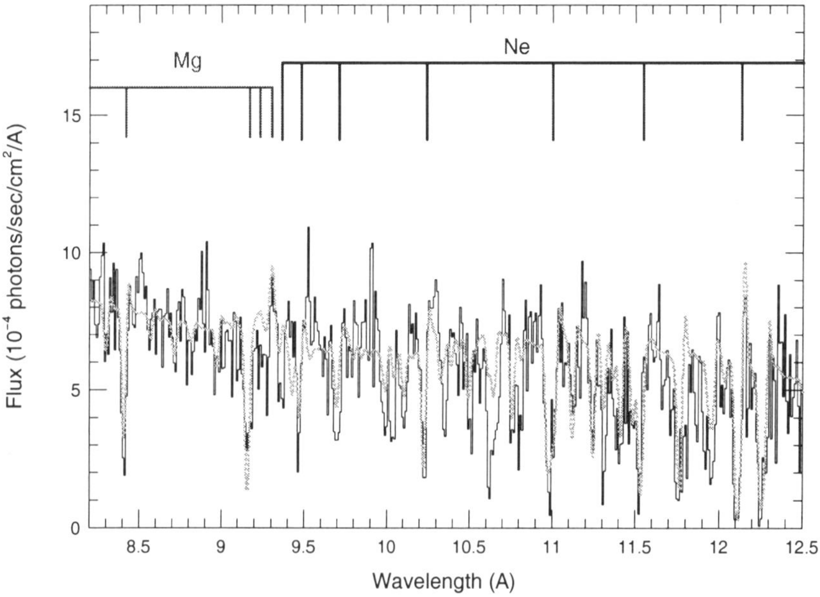

Figure 12. The spectrum of NGC 3783 as observed by *Chandra* (dark line) and a photoionization model fit to the data (light dotted line). All the absorption lines that are not marked are iron L-shell lines. Note the P-Cygni profiles of some lines. From Kaspi et al.[35]

there is little or no material on other sides (i.e. $C_f \ll 1$). An example is shown in Fig. 13.

Emission dominated spectra are observed when the emitting gas is directly observed but the ionizing continuum is obscured. The spectrum is composed of X-ray lines and reflected (or scattered) X-ray continuum. As explained, the lines are mostly due to recombination in photoionized gas and due to collisional excitation in thermal plasma. The line EWs are normally very large because of the weak scattered continuum. An example is shown in Fig. 13.

Emission and absorption spectra are seen in those cases where there is

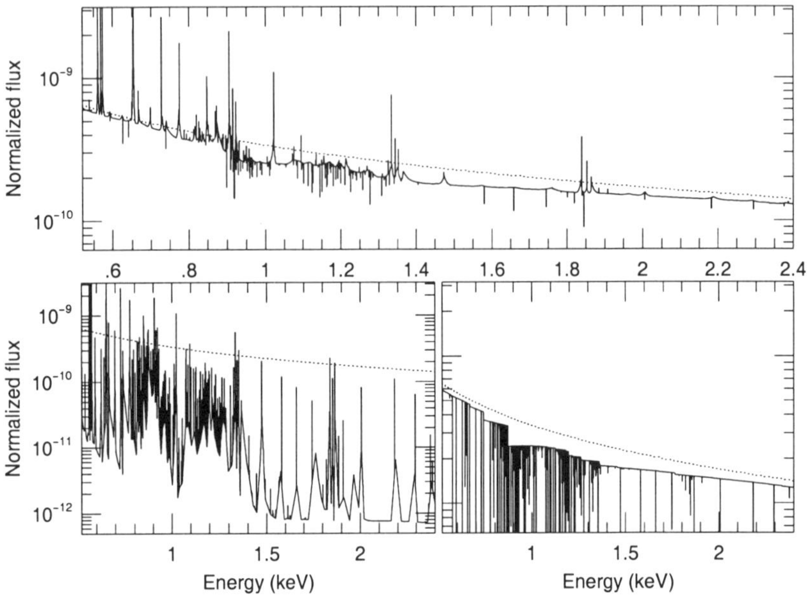

Figure 13. Pure absorption (bottom right), emission and reflection (bottom left) and combined (top) theoretical X-ray spectra of type-I AGN. The HIG is characterized by U(oxygen) = 0.03 and hydrogen column density of 10^{22} cm^{-2}. The dashed line is the incident continuum. Note the different energy scale on top and the different flux scale on the bottom left.

a direct view of the continuum, large covering fraction, large column-density HIG, and some gas on the line of sight to the center. The spectral appearance depends on various parameters. The ratio of the line-of-sight and total covering fractions determines the relative strength of the emission and absorption features. Relative motion, such as expansion, can result in P-Cygni type profiles and any azimuthal dependence in the gas properties can result in very different emission and absorption line intensities. Fig. 13 shows a combination of the absorption and emission spectra in a simple case where the HIG properties are the same on all sides of the source.

5.2 *Where is the HIG?*

The HIG is very different in its properties from the NLR and BLR gas. Its level of ionization is much higher and so is its covering fraction. It is therefore reasonable to assume that it occupies a much larger volume of space and the picture of small isolated clouds may not be valid in this case. In particular, the mass of this component may exceed, by much, the BLR and the NLR mass.

Unfortunately, so far there is no direct measurement of the HIG location and we can only parameterize its density mass and location by using the known U(oxygen) (~ 0.02) and column density. Scaling the HIG to the known BLR density, location and ionization parameter, and using the observed HIG column density, suggests that the HIG density is

$$N_{\rm HIG} \simeq 6 \times 10^6 \, L_{44}({\rm oxygen}) R_{\rm pc}^{-2} \ {\rm cm}^{-3} , \tag{24}$$

the physical thickness of a typical HIG cloud is

$$\Delta R_{\rm HIG} \simeq 5 \times 10^{-3} \, L_{44}^{-1}({\rm oxygen}) N_{22} R_{\rm pc}^2 \ {\rm pc} , \tag{25}$$

and the mass of such clouds is

$$M_{\rm HIG} \simeq 10^3 \, N_{22} C_f R_{\rm pc}^2 \ M_{\odot} , \tag{26}$$

where N_{22} is the hydrogen column density (assuming solar composition) in units of 10^{22} cm^{-2} and L_{44}(oxygen) the 0.54–10 X-ray luminosity in units of 10^{44} erg s^{-1} (for low luminosity AGN, like NGC 5548 and NGC 3783, $L_{44}({\rm oxygen}) \simeq 0.05$ and for the high luminosity quasars $L_{44}({\rm oxygen}) \simeq 1$). Given these properties, there are several likely locations for the HIG:

1. Just outside the central source, perhaps due to material blown away from the central accretion disk. In this case, the HIG density must be large, 10^{11} cm^{-3} or even larger.

2. Just outside the BLR, due to gas evaporation off condensations in the BLR. The condensations may be the high density cores of BLR clouds or bloated stars in the central star cluster. The blown-away gas, or wind, becomes transparent to the X-ray radiation and is pushed outwards by radiation pressure. Close to the outer BLR boundary, the density is about 10^8 cm^{-3}.

3. Well outside the BLR, occupying the same volume as the NLR. The density of this gas is very low and its mass extremely large (Eqs. 24 and 26). Its origin is not clear, but so is the origin of the NLR gas. Such gas can directly be observed by *Chandra* (as shown, already, in two cases).

Obviously, there is nothing to against the possibility of more than one origin or location.

Regarding the largest region, it has been suggested[62] that the HIG is, in fact, the coronal line region gas. The few available calculations suggest that the level of ionization is similar (yet not identical) and the IR and X-ray observations may represent the same phenomenon. This interesting suggestion is currently under investigation.

5.3 The Motion of the HIG

The high level of ionization makes the HIG subject to strong radiation pressure force. In this respect, there is a major distinction between this component and the colder BLR and NLR gas. We therefore expect the HIG to flow out of the system with velocities that depend on the "launching" location of the flowing gas (the attained velocity is typically of the same order of magnitude as the escape velocity at the base of the flow), the run of external pressure as a function of radius, and the micro-physics of the gas. The micro-physics determines the value of the force multiplier (see Sec. 2.9) and the local acceleration. Additional factors are the ionization level, the metal abundance and the internal motion (microturbulence) that controls the contribution of bound-bound transitions to $M(r)$. Recent calculations indicate that $M(r) \sim 10$ if the internal motion is of the order of the sound speed in the gas. This means that sources with $L/L_{\rm Eddington} > 0.1$ can efficiently accelerate HIG clouds.

It is not yet clear whether the term "cloud" used for the colder components (and required to explain the optical-UV observations) can be applied to the HIG. While high ionization clouds can form in a medium controlled by external magnetic pressure, or pressure due to very high temperature material, they may not survive long enough to show the observed spectral features. The existence of such clouds is still an open question that may be resolved by new X-ray observations. Alternatively, the HIG may be an outflowing wind that requires no confinement. This would fill large volumes of space and is less likely to produce multi-component absorption systems. Fig. 14 shows new calculations[13] for the velocity of HIG clouds under various conditions. So far, there is no detailed study of the complementary case of a wind-type HIG.

5.4 Associated UV absorbers

UV spectroscopy of type-I AGN clearly shows narrow absorption lines that are usually blueshifted with respect to the BLR velocity. These are most

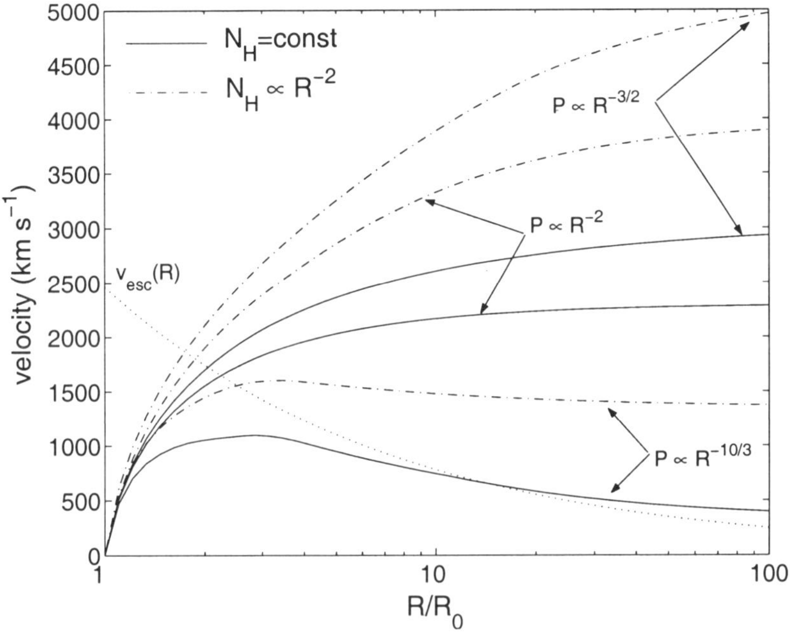

Figure 14. Calculated velocities of HIG clouds under various conditions of external pressure, P, as marked. The dotted line gives the escape velocity at the various locations. From Chelouche & Netzer[13] ©2001 Blackwell Sci.

probably caused by intrinsic outflowing gas. Another well known example is the broad absorption lines (BALs) that will not be discussed here.

Narrow intrinsic AGN absorption lines span a wide range of ionizations, from O VI and Ne VIII to C II, Mg II and Si II. Intrinsic QSO absorbers are not very common. They can have single or multiple line components, velocity shifts from near 0 km s^{-1} to many thousands km s^{-1}, and velocity widths from near the thermal Doppler speed to at least[15] several hundred km s^{-1}. Intrinsic absorption lines appear in a large fraction (~50%) of type-I AGN. The lines are typically highly variable, a few hundred km s^{-1} wide, and blueshifted by about 500–1500 km s^{-1} from the emission redshift[45,15].

Recent studies of type-I AGN led to the suggestion that the presence of UV line and X-ray continuum absorptions are correlated[47,48]. The extreme view of this model is the speculation that the two absorption phenomena occur in the same gas. Photoionization calculations have shown that, in some cases, the column densities measured from the UV and X-ray features are consistent with a one-component absorber[47,48,72,54]. In these models, the ions producing the low- to moderate-ionization UV lines, such as C IV λ1549 and N V λ1240, are just trace constituents in a gas dominated by O VII, O VIII, Ne IX and Ne X. Other cases, showing incompatible conditions[30] are known too and it is quite likely that, in many sources, both originate in very different clouds.

While it is not yet clear whether the HIG gas is related, in any way, to the coronal line region, or the gas producing the intrinsic UV absorption lines, its observation re-enforces the view presented earlier that the picture of distinct isolated regions is not valid any more. Instead, we should adopt the more realistic view of a "messy AGN" in which clouds of different types exist in many different locations. Some are of high enough density to produce a typical $T_e \sim 10^4$ K photoionized gas spectrum while others are of lower density, higher temperature and higher level of ionization. It is not yet clear what dominates the motion of the BLR and NLR gas but it is clear that the lower density HIG is moving out of the system, being pushed, most likely, by radiation pressure force.

5.5 Time Dependent HIG?

The HIG in AGN is subjected to a highly variable continuum source. Depending on the gas density and temperature, the emergent spectrum will be time dependent. We can estimate the location of the HIG where such short-time effects will become noticeable. For example, in type-I AGN, the ionization parameter is well constrained and the location and density of the gas are simply related (Eq. 24). A recombination-time vs. distance relation can be established. For example, the typical few hours time variability of many type-I AGN will result in variable X-ray lines if the HIG distance is 0.1 pc or less.

The investigation of time dependent photoionization of the HIG gas is just starting[43,57]. The observational work is also in a preliminary stage. *Chandra* and *XMM* are definitely capable of determining the HIG location using reverberation studies of X-ray emission lines, similar to the reverberation mapping of the BLR, and measuring variable absorption lines. The next couple of years is likely to see the answer to this question.

Even if the HIG is further out, and the density is smaller, much longer time scale variations (years!) are known to occur in AGN and the presently observed X-ray spectrum may reflect past very bright or very faint phases of the source. HIG models must take this into account.

Table 2. AGN components: locations and densities

Component	*location*	*density*
Accretion disk	10^{-3} pc	$\sim 10^{15}$ cm^{-3}
BLR	0.01-0.1 pc	$\sim 10^{10}$ cm^{-3}
Torus	1–10 pc	10^{3-6} cm^{-3}
Warm absorber	?	10^{2-10} cm^{-3}
NLR	100–1000 pc	10^{3-5} cm^{-3}
Starburst	$\sim$ 1000 pc	10^{0-4} cm^{-3}

5.6 Other X-ray Bright Components

The X-ray bright components in AGN are those clouds, or condensations, that are exposed to the direct radiation field of the central continuum source and where U(oxygen) is large enough for the H-like and He-like ions of most metals to be the most abundant ions. Since the location and the density of most components, except the HIG, are roughly known, we can check each of those, in turn, to see whether it can contribute to the observed X-ray spectrum. Using the previously defined L_{44}(oxygen) and assuming a typical $\gamma_X = 1$, we estimate

$$U(\text{oxygen}) \simeq 0.2 L_{44}(\text{oxygen}) R_{\text{pc}}^{-2} N_5^{-1} , \qquad (27)$$

where $N_5 = N_{\text{H}}/10^5\,\text{cm}^{-3}$. A list of components is given in Table 2. As evident from the combination of Eq. 27 and Table 2, the central molecular torus, if it exists with dimensions and density as assumed in standard AGN models[6], is a likely strong X-ray source. The level of ionization of the inner torus walls can be high enough to produce X-ray lines, and the column density large enough to reflect a large fraction of the incident continuum. The expected covering fraction of the inner torus walls exceeds 50% (see Goodrich in these proceedings) so the reflected continuum can contribute significantly to the observed X-ray flux.

The uncertainty in the inner torus density is large enough to prevent us from making detailed predictions. At the high density range, most of the gas is almost neutral and X-ray reflection is inefficient. The resulting reflected X-ray continuum is very hard and the iron Kα line is dominated by low ionization iron. The central energy of this feature is around 6.4 keV and we expect a low-contrast Compton scattering shoulder that extends the line profile to lower energies. At the low density limit, the gas is more ionized and the emission line spectrum dominated by H-like and He-like species. Obviously, the above description is highly simplified since the X-ray flux itself can affect the inner-wall density by producing winds and by causing low density material to evaporate off the inner surface.

6 The AGN-starburst Connection

6.1 Star-Forming Regions in AGN

Several type-II AGN show nuclear star-forming rings or bright nuclear HII regions. The X-ray band is probably most suitable to investigate and identify this activity, because of the large expected X-ray luminosity, as a result of fast stellar winds and supernovae explosions. Extended soft X-ray emission, which is most probably due to starburst activity, has already been observed in the inner ~1 kpc region of many such sources (e.g.[78] NGC 1068). The phenomenon is very common – indeed there are only a few sources where this is *not* observed.

The difference between "pure" starburst activity and the one observed in active galaxies is the presence of the central, powerful X-ray source. Keeping in mind the definition of U(oxygen), we can assess the influence of this source using the known location and the expected density of the star-forming regions. This is given in Table 2 and Eq. 27. Evidently, under reasonable assumptions, the central X-ray source must influence and modify the physical properties of the starburst produced gas.

Finally, the suggestion that high accretion rates in luminous AGN may be related to high column of nitrogen enriched material, whose origin is nuclear starburst[29,77] may or may not be related to this question.

6.2 The X-ray Spectrum of Star-Forming Regions

"Pure" starburst galaxies (e.g. NGC 253) show the typical X-ray signature of supernova explosions and fast stellar winds, and the high velocity flows and dusty condensation observed in star forming regions. The X-ray signature is of a hot ($kT \sim 0.3 - 2$ keV) plasma, with its typical free–free continuum and several large ($\geq$100 eV) equivalent width emission lines[63,32]. Extreme starburst activity is associated with so-called *superwinds*[31] that are observed at both optical and X-ray energies. The X-ray emitting regions of starburst regions are spatially resolved, showing typical dimensions of order one kpc. The pre-*Chandra* and *XMM* abundance determinations, using the X-ray lines, were problematic since they implied extremely low metallicities, in clear contrast to the theoretical expectations and the optical-IR-UV measurements. These may be traced to wrong identification of the low resolution instruments.

As noted in Sec. 2, the spectrum of low density, low opacity collisional plasma is determined by the gas temperature and composition. Accurate calculations of pure thermal gas are widely available and the only significant unknowns are certain atomic data (most notably, the collision strengths and dielectronic recombination rates of iron M-shell and L-shell lines). Since the level of ionization is uniquely defined by the gas temperature, the latter can be found by comparing the intensity of lines from different ions, for a given metal abundance. One can thus construct line ratio and diagnostic diagrams, similar to those shown in Fig. 2, where the independent variable is the temperature, rather than the ionization parameter. An example is shown in Fig. 15. Observing a number of X-ray lines of different species, and using such diagrams, allows to solve for both the gas temperature and composition. Obviously, this single temperature view of the starburst gas is over simplified and can only be used as a first approximation to the more complex real situation.

6.3 Starburst Galaxies and ULIRGs

The term *starburst galaxies* used here refers to one of two cases. The first is the "pure" starburst galaxy described in Sec. 6.2. Such object shows no clear central radiation source and the gas micro-physics is controlled by collisional processes. The second category includes the Ultra Luminous IR Galaxies (ULIRGs) that are characterized by very high IR luminosity ($> 10^{45}$ erg s^{-1}) and extreme extinction ($A_V \sim 50 - 100$ mag). The main power house of ULIRGs is completely obscured from view and the observed luminosity is due to thermal emission by the hot dust. Most ULIRGs show clear star forming activity but some also show signs of accretion onto a massive black hole. ULIRGs are thought to be the result of mergers[70], an idea that is clearly supported by recent HST near-IR imaging.

Galactic merger can lead to several different results. First, it can trigger violent star-forming activity including heating of the ISM to X-ray energies and large scale outflows with typical velocities[31] of ~ 1000 km s^{-1}. Second, it can result in the formation of giant black holes (or, alternatively, in fast growing of small existing black holes) during the merging event. This may eventually lead to the formation of high luminosity AGN. The two can proceed along different tracks but can also operate, side by side, in the very same merger.

The observational manifestation of a galactic merger is different for different cases and depends, also, on the stage of the merger. In particular, there is an interesting on-going debate about the main power source of several dozens of observed ULIRGs which can be either the violent star formation (producing

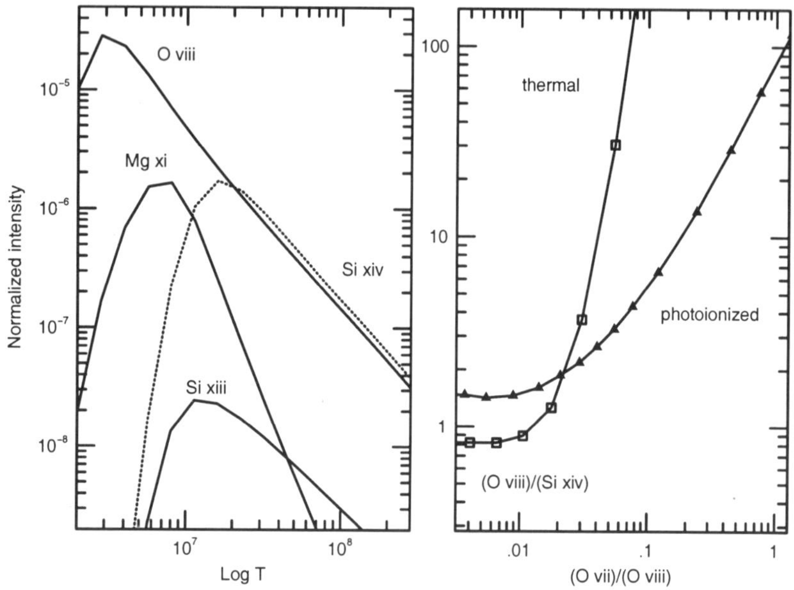

Figure 15. Left: line ratio diagram, for several intense X-ray lines, in a low density ($N_{\rm H} = 10^3$ cm^{-3}) solar metallicity, optically thin collisional plasma. Right: A diagnostic diagram showing, this time, the thermal gas line ratio (left curve with squares; the temperature is increasing from top to bottom) along side the photoionized gas curve (marked with triangles) from Fig. 2

star clusters that are more luminous than many quasars!) or the newly formed central AGN. While the spectroscopic signatures of the two are predicted to be very different, the analysis is extremely complicated since the optical-UV part of the spectrum is completely obscured by dust, and the standard line emission diagnostics cannot be used. Only two spectral bands are available for a detailed study, the hard X-ray band and the mid-to-far IR band.

Genzel et al.[24] suggested several diagnostics to identify the main source of energy. They are based on the level of ionization of the IR lines and the intensity of the PAH features. For example, only AGN type continuum can ionize oxygen beyond O^{+2} thus a strong O IV line is a clear AGN signature. As for the PAHs, such molecular features are easily destroyed by the intense, high energy radiation field of an AGN. Strong PAH is thus evidence for the absence of a powerful AGN.

There is a simple way to test for the existence of a central AGN in ULIRGs using hard X-ray observations. At this energy ($\sim 3-10$ keV), the large obscuring dusty material becomes transparent, in some sources, and the central source can be directly observed. Indeed, X-ray observations clearly show a powerful central AGN-type source in several ULIRGs, on top of soft (0.5-2 keV) nuclear sources that can be due to the starburst activity. In some cases, the power-law X-ray source is associated with IR lines from highly ionized gas. This gas (which is, in fact, the NLR) must be exposed to a strong central UV source in order to emit such lines. However, in several sources, like[56,76] NGC 6240, there is no indication of highly ionized species in the IR and the only AGN signature is in the X-ray band (in the case of NGC 6240, the X-ray source is obscured by a huge column of column of $\sim 10^{24}$ cm^{-2}). Such cases defy standard explanation. One possibility is that the obscuring material covers the entire source and there is no cone through which the NLR gas, if exists, can be ionized. The observed low excitation IR lines are emitted, in this case, by star-forming regions outside the central obscurer.

6.4 A Combined Model

The properties of the X-ray emitting gas in several ULIRGs must be governed by both large scale dynamical processes (mergers and star-forming activity), as well as by the strong radiation field of the central AGN. Modeling them as pure photoionized regions regions that are excited by a single central source, or as pure collisionally ionized plasma in extended star-forming regions, is clearly oversimplified. Understanding the sources, and the overall AGN-starburst relationship, requires more sophisticated models, combining thermal and photoionization excitations. This is true in particular in those ULIRGs where the central AGN is powerful enough and the star-forming regions are close enough to the nucleus. In such cases, collisional excitation together with photoionization is affecting the physical conditions in the observed X-ray gas. The analysis of such spectra requires new and advanced diagnostic tools.

Fig. 16 shows two spectra. One is a standard $T = 10^7$ K stationary collisional plasma and the other the spectrum of this gas when illuminated by a power-law X-ray continuum similar to the one expected from a central AGN, assuming $U(\text{oxygen}) = 0.2$. The spectral appearance of the two is

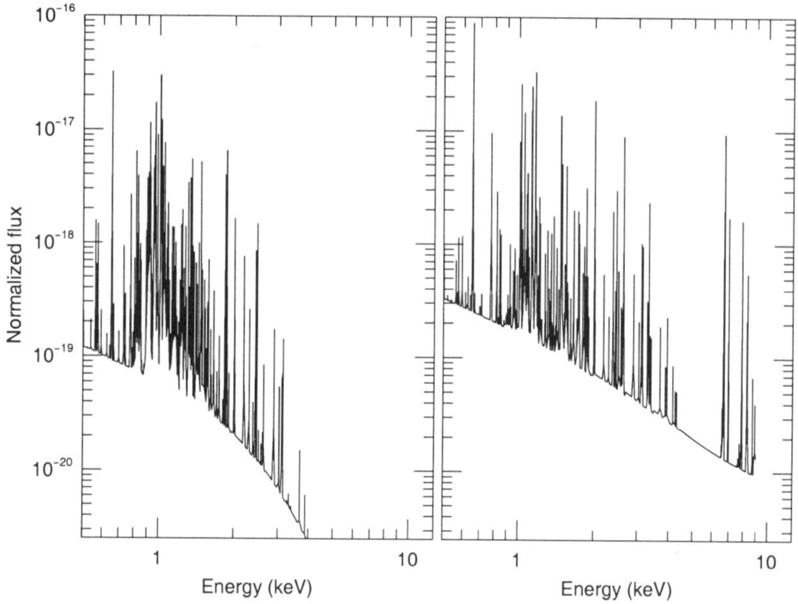

Figure 16. Left: The spectrum of a steady state $kT = 1$ keV collisional plasma. Right: The spectrum of the same $kT = 1$ keV plasma when illuminated by a central AGN source with U(oxygen)=0.2. Note the absence of strong iron lines in the pure collisional case.

very different. While the pure thermal plasma continuum is curved, and the high ionization lines of H-like and He-like iron, around 6.7-7 keV, are weak or absent, this is not the case when photoionization is taken into account. A clear signature of the photoionized gas is the flat continuum and the strong ionized iron lines. Present day X-ray spectroscopy can distinguish between the two possibilities provided adequate theoretical tools are available. The observations can supply the location of the line emitting gas, the luminosity and SED of the X-ray source and the high resolution spectra. New diagnostic diagrams, especially prepared for analyzing such cases, taking into account both excitation processes, must be developed. They will use the observed line ratios to deduce the value of U(oxygen), the gas temperature and composition, and the effect of stellar winds and central radiation pressure on the gas dynamics. X-ray observations and analysis provide a new way, perhaps the most efficient and meaningful one, to understand ULIRGs and the physics of star-forming regions near powerful, AGN-type sources.

Finally, it must be re-emphasized that the gas in starburst galaxies, in star-forming regions in AGN and in ULIRGs, may not be stationary or in steady state equilibrium. The single temperature stationary plasma models illustrated here, give only the zero order approximations for a more complex situation. New, time-dependent non-stationary models must clearly be developed to fully understand some of these objects.

7 Open questions

Throughout this review we have mentioned several outstanding questions regarding the standard AGN model, starburst galaxies and the AGN-starburst relationship. A list of this type is necessarily long and I will make no attempt to present a complete one. The following is a summary of some of the questions that were addressed here:

- What is the nature of AGN clouds? Are they bloated stars or smaller condensations confined by external pressure? What is the number of clouds and their radial distribution? What is the origin of the confining pressure?

- What is the mass budget of AGN? What is the ratio of the accreted mass to the outflowing mass?

- What is the resolution of the energy budget problem and what is the "real" SED?

- How realistic is the simple two-zone model of AGN low temperature gas (BLR and NLR)? Is there an *intermediate line region* and what is the reason for its low emissivity? What is the role of dust in all emission line regions?

- What is the global dynamics of the NLR gas? Is the outflowing model with central obscuration observed in NGC 4151 and NGC 1068, typical of other source?

- How common are central absorbers (i.e. the material between the center and the NLR)? What is their effect on the NLR spectrum?

- What is the location, the mass and the motion of the HIG in type-I AGN? Is it the same component producing the UV absorption lines and what is the relationship with the coronal line region?

- What are the properties of the HIG in type-II AGN? Is it the same component producing the ~1 keV absorption in type-I AGN? Can the apparent differences between type-I and type-II objects be fit into a single unified scheme?

- What are the physical properties of the X-ray emitting gas in starburst galaxies not hosting luminous AGN? What is the composition of this gas?

- What is the dominant excitation mechanism of the HIG in ULIRGs and starburst galaxies hosting luminous AGN: mechanical heating, photoionization by the central radiation, or both?

- Is the X-ray emitting gas in nuclear starburst region showing type-II AGN influenced mainly by the starburst activity or the central radiation field? Is the spectrum time dependent and on what time scale? Is the motion of this gas influenced by the central radiation field?

Acknowledgments

This work is supported by the Israel Science Foundation and by the Jack Adler Chair of Extragalactic astronomy.

References

1. Alexander, T., MNRAS, 285, 891, 1997.
2. Alexander, T., & Netzer, H., MNRAS, 270, 781, 1994.
3. Alexander, T., & Netzer, H., MNRAS, 284, 967, 1997.
4. Alexander, T., Strum., E., Lutz, D., Sternberg, A. Netzer, H., & Genzel, R, ApJ, 512, 204, 1999.
5. Alexander, T., Lutz, D., Strum., E., Genzel, R., Sternberg, A. & Netzer, H., ApJ, 536, 710, 2000.
6. Antonucci, R., ARAA, 31, 473, 1993.
7. Arav, N., Barlow, T., Laor, A., Sargent, W.L.W., & Blandford, R.D., 297, 990, MNRAS, 1998.
8. Baldwin, J.A., ApJ, 214, 679, 1977.
9. Baldwin,J.A., Ferland, G.J., Korista, K. & Verner, D., ApJL, 445, L119, 1995.

10. Bicknell, G.V., Dopita, M.A., Tsvetanov, Z.I., & Sutherland, R.S., ApJ, 495, 680, 1998.
11. Binette, L. & Robinson, A., A&A, 177, 11, 1987.
12. Bottorff, M. C., & Ferland, G.J, ApJ, in press.
13. Chelouche, D., & Netzer, H., MNRAS, in press.
14. Clavel, J. et al., A&A 357, 839, 2000.
15. Crenshaw, D.M., Kraemer, S.B., Bogess, A., Maran, S. P., Mushotzky, R.F., & Wu, C., ApJ 516, 750, 1999.
16. Dopita M. & Sutherland, R.S., ApJ, 455, 468, 1995.
17. Edwards, A. C. , MNRAS, 190, 757, 1980.
18. Emerson D. in "Interpreting Astronomical Spectra", John Wiley and Sons, 1997.
19. Emmering, R.T., Blandford, R.D., & Shlosman, I., ApJ, 385, 460, 1992.
20. Eracleous, M., Livio, M., & Binette, L., ApJL 445, L1, 1995.
21. Fabian, A., Kunieda, H. Inoue, S. Matsuoka, M. Mihara, T. Miyamoto, S., Otani, C. Ricker, G., Tanaka, Y., Yamauchi, M., & Yaqoob, T., PASJ, 46L, 59F, 1994.
22. Ferguson, J.W., Korista, K.T., & Ferland G.J., ApJS, 110, 287, 1997.
23. Fiore, F., Perola, G. C., & Romano, M., MNRAS 243, 522, 1990.
24. Genzel, R., et al., ApJ 498, 579, 1998.
25. George, I.M., et al., ApJS 114, 73, 1998.
26. Guanazzi M. et al., MNRAS, 310, 10, 1999.
27. Halpern, J.P., ApJ, 281, 90, 1984.
28. Hamann, F., & Ferland, G.J., ApJ, 418, 11, 1993.
29. Hamann, F., & Ferland, G.J., ARAA, 37, 487, 1999.
30. Hamann, F.W., Netzer, H., & Shields, J.C., ApJ 536, 101, 2000.
31. Heckman, T., et al., ApJ 457, 616, 1996.
32. Iwasawa, K., MNRAS 302, 961, 1999.
33. Kaspi S. & Netzer, H. ApJ, 524, 71, 1999.
34. Kaspi, S., Smith, P.S., Netzer, H., Maoz, D., Jannuzi, B.T., & Giveon, U., 1999, ApJ, 533, 631, 2000.
35. Kaspi, S., Brandt, W.N., Netzer, H., George, I.M., Chartas, G., Behar, E., Sambruna, R.M., Garmire, G., & Nousek, J.A., ApJ, submitted.
36. Kazanas, D., ApJ, 347, 74, 1989.
37. Komossa, S., & Breitschwerdt, ApJS, 272, 299, 2000.
38. Korista, K. T., Ferland, G.J., & Baldwin, J., ApJ 487, 555, 1997.
39. Korista, K. T. & Goad, M. R., ApJ, 536, 284, 2000.
40. Kraemer, S,B. & Crenshaw, D.M., ApJ, 532, 256, 2000.
41. Krolik J. "AGN" , Princeton Series in Astrophysics, 1999.
42. Krolik, J.H., McKee, C.F., & Tarter, C.B., ApJ, 249, 422, 1981.

43. Krolik J. & G. Kriss, ApJ, 447, 512, 1995.
44. Laor A., ApJL, 496, L71, 1998.
45. Maran, P.S., Crenshaw, D.M., & Mushotzky, R.F., ApJ, 465, 733, 1996.
46. Mathews, W.G., & Ferland, G., J., ApJ, 323, 456, 1987.
47. Mathur, S., ApJ, 431, L75, 1994.
48. Mathur, S., Wilkes, B., Elvis, M., Fiore, F., ApJ, 434, 493, 1994.
49. Matt, G., Brandt, W. N., & Fabian, A.C., MNRAS, 280, 823, 1996.
50. Mihalas, D., "Stellar Atmospheres" , W.H. Freemen and Company, 1978.
51. Nandra, K., Pounds, K., Nat, 359, 215, 1992.
52. Netzer, H., ApJ, 289, 451, 1985.
53. Netzer, H., in Active Galactic Nuclei, Courvoisier & Mayor eds., Berlin:Springer, p57, 1990.
54. Netzer, H. , ApJ, 473, 781, 1996.
55. Netzer, H., & Laor, A., ApJ, 404, L51, 1993.
56. Netzer, H., Turner,T.J., George,I.M., ApJ, 504, 680, 1998.
57. Nicastro, F., Fabrizio, F., Perola, G.C., & Elvis, M., ApJ, 512, 184, 1999.
58. Osterbrock D.E. "Astrophysics of Gaseous Nebulae and Active Galactic Nuclei", University Science Books, 1989.
59. Pan, H., Stewart, G.C., Pounds, K.A., MNRAS, 242, 177, 1990.
60. Penston, M.V., MNRAS, 233, 601, 1988.
61. Peterson, B.M. "An Introduction to Active Galactic Nuclei", Cambridge University Press, 1997.
62. Porquet, D, Dumont, A.M., Collin, S, & Mouchet, M., A&A 341, 59, 1999.
63. Ptak, A., Serlemitsos, P., Yaqoob, T., Mushotzky, R.., ApJS, 120, 179, 1998.
64. Rees, M.J., MNRAS, 228, 47p, 1987.
65. Reynolds, C., MNRAS 286, 513, 1997.
66. Reynolds, C.S., Fabian, A.C., Nandra, K., Inoue, H., Kunieda, H., & Iwasawa, K., MNRAS, 277, 901, 1995.
67. Robson I., "Active Galactic Nuclei", John Wiley & Sons, 1996.
68. Sako, M., Kahn, S.M., Pharels, F, & Liedahl, D.A., ApJL, 543, L115, 2000.
69. Sambruna, R. M., Netzer, H., Kaspi, S., Brandt, W. N., Chartas, G., Garmire, G. P., Nousek, G. A., Weaver, K. A., ApJL, in press.
70. Sanders, D., Mirabel, F., ARAA, 34, 749, 1997.
71. Scoville, N., & Norman, C. , ApJ, 332, 163, 1988.
72. Shields, J., & Hamann, F., ApJ 481, 752, 1997.
73. Strum, E., Alexander, T., Lutz, D., Sternberg, A. Netzer, H. & Genzel, R, ApJ 512, 197, 1999.

74. Turner, T.J., et al., ApJ, 488, 164, 1997.
75. Veilleux, S., ApJS, 75, 357, 1991.
76. Vignati, P., et al. A&A 349, l57, 1999.
77. Wills, B.J., Laor, A., Brotherton, M.S., Wills, D., Wilks, B.J., Ferland, G.J., & Shang, Z., ApJL 515, L53, 1999.
78. Wilson, A., & Elvis, M., Ap. Sp. Sci., 248, 14, 1997.
79. Wilson, A.S. & Raymond, J.C., ApJL, 513, L115, 1999.
80. Yaqoob, T., & Warwick, R. S., MNRAS, 248, 773, 1991.
81. Zheng, W., Kriss, G.A., Telfer, R.C., Grimes, G.P., & Davidsen, A.F., ApJ 475, 469, 1997.

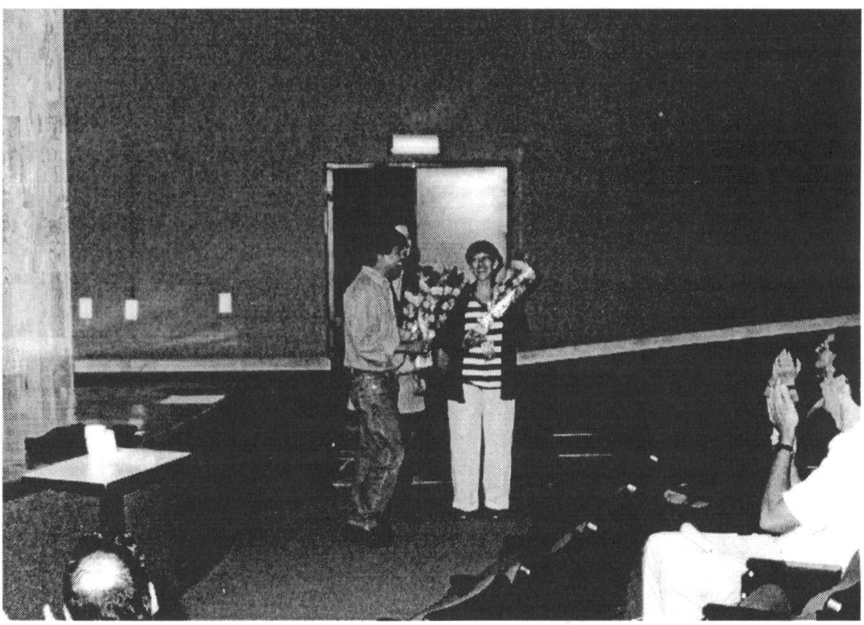

Raúl Mújica thanking the secretaries of the Advanced Lectures at the end of the meeting.

Itziar Aretxaga (*top*) and Suzy Collin (*bottom*) dancing in the party that followed the official dinner of the Advanced Lectures. It was fun!

ACCRETION AND EMISSION PROCESSES IN AGN

S. COLLIN
Observatoire de Paris-Meudon, 92195 Meudon, France
E-mail: suzy.collin@obspm.fr

The UV-X continuum, the X-ray spectral features, and the variability in these bands provide powerful tools for studying the innermost regions of AGNs from which we gain an insight into the accretion process. In this chapter the discussion focusses on luminous AGN, i.e. Seyfert galaxies and quasars. The standard accretion disk model (a stationary geometrically thin disk) is described, and vertically averaged solutions for the radial structure are given. The emission of the standard disk is discussed using different approximations, and it is compared to the observations. This leads to the conclusion that more complex models are required, such as the irradiated disk and the disk-corona models. The advantage of this last model is that it explains the overall UV-X spectral distribution. In the framework of these disk models, the profile, intensity, and variability properties of the X-ray iron line can be explained by reprocessing at the surface of the cold disk very close to the black hole (the "relativistic disk model"). An alternative possibility is discussed, where the UV-X continuum is produced by a quasi-spherical distribution of dense clouds surrounded by (or embedded in) a hot medium. In such a model the iron line profile could be due to Comptonization instead of relativistic effects.

1 A Few Introductory Words

In these lectures, I adopt the "black hole paradigm", in which AGN are fueled by accretion onto a massive black hole (BH)[100]. Besides the BH paradigm, there is another generally accepted model, i.e. that accretion takes place via a disk. A few initial comments should be made concerning this point.

One tries to account for AGN luminosity by invoking accretion with angular momentum. If the flow is rotationally supported, it will form a disk. The presence of a disk is inferred from several observational facts. The most obvious one is the existence of a privileged direction: several active galaxies display cones of ionized gas, and collimated jets are commonly observed in radio-loud objects, as well as in a few radio-quiet ones. The "Unified Scheme"[8] implies the existence of a dusty torus. Double peaked line profiles observed in a small fraction of AGN are also believed to be the signature of disk emission. Finally, large gaseous disks have been observed in some cases, for instance in NGC 4258. However, one should keep in mind that all these evidences point towards the existence of a disk *at large distances*, i.e. at $R \gtrsim 10^4 R_G$ ($R_G = GM/c^2$ is the gravitational radius, where M is the BH mass), while the bulk of the accretion luminosity is emitted by a very small region, at $\sim 10 R_G$.

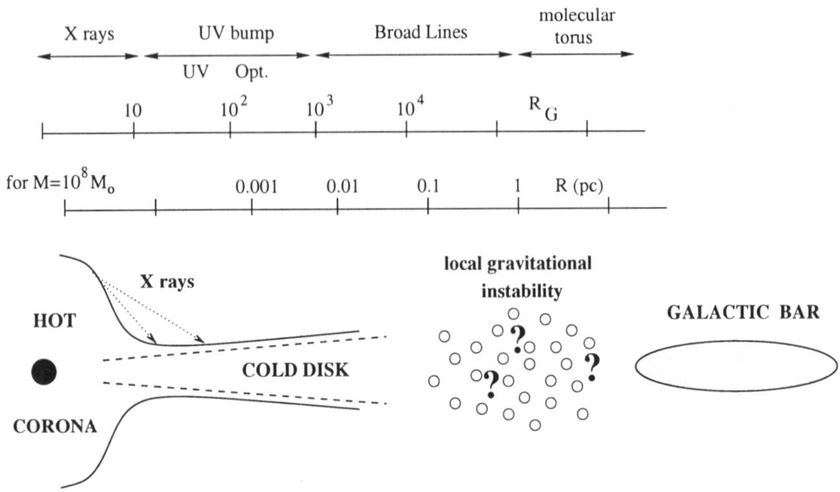

Figure 1. Schematic view of an AGN, showing the dimensions of different regions in R_G (and in pc for $M = 10^8$ $M_\odot$). The present chapter deals with the region located inside the "gravitationally unstable" part of the disk, closer than a few $10^3 R_G$ from the BH.

At such small distances the only evidence for an accretion disk is the profile of the Fe K line. Since this issue is important (as it implies the presence of a "cold disk" at only a few R_G in some cases), one should question this "proof", and examine whether there is not another possible interpretation. Moreover, there are some theoretical hints for quasi-spherical accretion. For these reasons I discuss quite extensively the "quasi-spherical" model, which does not preclude the presence of angular momentum, although this is presently not a popular model.

In order to show the physical scale of the region discussed in this chapter, Fig. 1 displays a very schematic representation of the central flow in an AGN. We shall deal with the region located inside the "gravitationally unstable" part of the disk, closer than a few $10^3 R_G$ (the distance where the broad lines are formed). There is presently no model proposed for the locally unstable region indicated in the figure, as the gravitational torques discussed in F. Combes' lectures concern the outermost parts of the nucleus, beyond 10 pc.

I have chosen to discuss extensively the most simple models, and to mention only briefly the recent more sophisticated developments. My aim is also to stress the uncertainties underlying some widely accepted and fashionable models. Finally I should mention that I focus on radio-quiet objects, where accretion processes dominate the spectrum.

The observational evidence is summarized in Sec. 2, and the basic physics of BHs are given in Sec. 3. In Sec. 4 a few results on the physical state of the medium emitting the bulk of the accretion power are deduced directly from the observations. In Sec. 5, the basis of accretion disk theory is recalled, emphasising some robust results concerning the emission spectrum, which are almost independent of the model. In Sec. 6, the standard (i.e. thin and steady) α-disks are discussed in their simplest (vertically averaged) form. Sec. 7 is devoted to the vertical structure. Sec. 8 tackles different models of "irradiated" disks, and their emission spectrum. Sec. 9 discusses the "cloud model", which is an alternative to these disks. In Sec. 10 the "thick" disk models envisioned for very low and very high accretion rates are briefly recalled. Sec. 11 reminds us that all this chapter deals with regions close to the BH, and that the accretion process is still not understood on larger scales.

2 Summary of Observations

I give here a very brief summary of some basic observational results which are crucial to constrain a model of the central engine. For more details the reader is referred to Koratkar & Blaes' excellent review[57].

2.1 The Broad-Band Spectrum

Fig. 2 displays a typical broad-band continuum, from the radio to the γ-ray range, for radio-quiet and radio-loud AGN[112]. The continuum in radio-quiet objects can be divided into three parts: an "IR-bump", a "UV-bump" (creating a gap at about 1μm), and an X-ray power-law continuum. The "1μm gap" is clearly due to the superposition of two different components which dominate the IR and the UV emission, and fade at 1μm. Since 1μm is the shortest wavelength at which hot dust can radiate, the IR-bump is attributed to hot dust radiation, at least in radio-quiet objects.

The continuum produced directly by accretion processes is restricted to the UV-X range (though the mid-IR continuum could be partly reprocessed nuclear radiation). From 1 to 10 eV the spectral index α, defined as $F_\nu \propto \nu^{-\alpha}$, lies in the range from -0.3 (for luminous quasars) to $+0.5$ (for Seyfert galaxies and low luminosity quasars). The soft X-ray continuum below 1 keV seems to be simply the extension of the UV with a spectral index close to 1.5. The continuum from 1 eV to about 100 eV is known as the *Big Blue Bump* (BBB), with a peak at 10 eV (see Fig. 5 in H. Netzer's chapter). A noticeable fact concerning the UV continuum is the almost complete *absence*

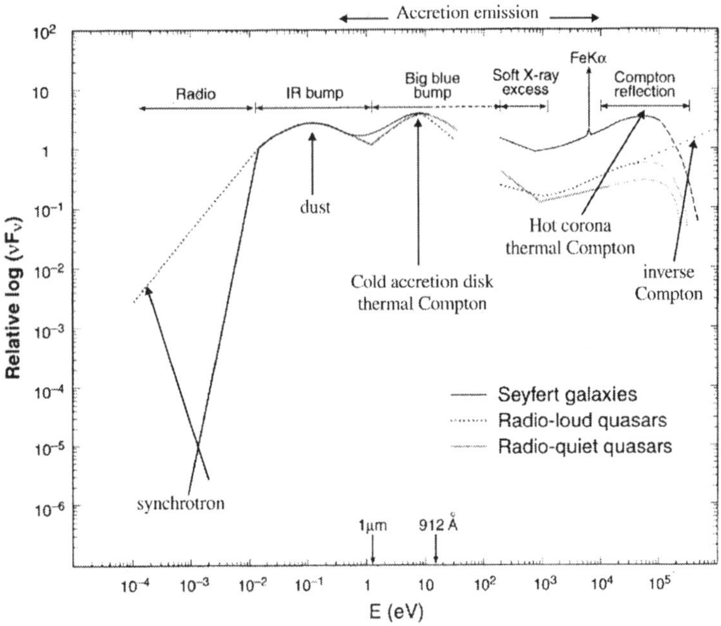

Figure 2. Typical AGN continuum of radio-loud and radio-quiet objects, and possible emission mechanisms (some of them not mentioned in the text). Figure adapted from Koratkar & Blaes[57], ©1999 Astr. Soc. of the Pacific, reproduced with permission of the editors.

of a Lyman discontinuity (a weak absorption edge is sometimes observed but it is most probably due to the Broad Line Region or to intervening intergalactic clouds with a redshift close to the emission one). I will not discuss here the "Warm Absorber", which imprints features in the soft X-ray continuum, as it is covered in H. Netzer's lecture notes.

Above 1 keV the average spectral index is about 0.7, so that in the majority of objects the continuum below 1 keV shows a "soft X-ray excess" when compared to the extrapolation of the continuum in the 1-10 keV range. Recent observations performed with Chandra show that the strength of the excess has probably been overestimated in the past. Above 1 keV the spectrum can be decomposed into an underlying power-law with a spectral index (in photon number) $\Gamma \sim 1.9$ and, superimposed on it, a broad emission line peaking at 6.4 keV, identified with the Kα line of weakly ionized iron, and a "hump" peaking at about 30 keV (for a review see Mushotzky et al.[86]). It has been shown[132] that the strength of the hump, measured by a parameter R, corre-

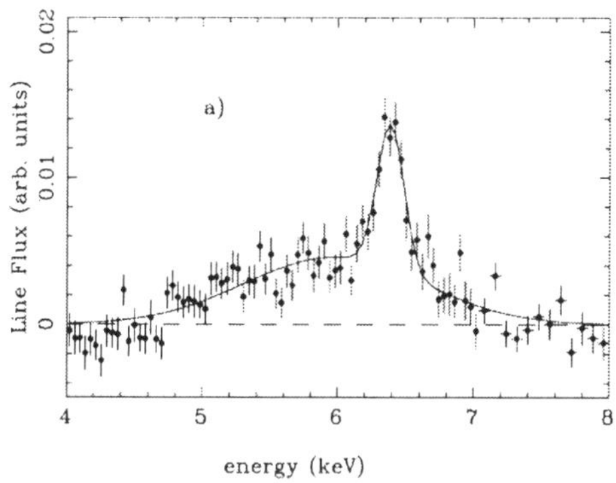

Figure 3. Mean line profile of the Fe K line for a sample of Sy 1 nuclei. The solid line is a double-Gaussian fit to the profile. From Nandra et al.[89] ©1997 AAS.

lates with Γ. The power-law continuum has a cut-off with an e-folding energy of 50 to 300 keV.

The Fe K line displays an asymmetrical profile with an extended red wing and a smaller blue wing[89] (see Fig. 3). The presence of the broad red wing in the average spectrum has been recently questioned[133]. However in the two best studied objects, MCG-6-30-15 and NGC 3516, the red wing extends down to 4 keV. The strength of the line seems also to correlate with Γ [133].

Finally, another important result is that the BBB luminosity is larger than the hard X-ray luminosity, in particular in luminous quasars.

2.2 Variability Properties

Since B. Peterson focussed mainly on the optical-UV properties, it is necessary to recall briefly some aspects of X-ray variability relevant to this chapter.

A well established result is that the characteristic time variability in the UV scales with luminosity: in Seyfert galaxies it is typically of the order of a few days, while it is several months in high luminosity quasars. The X-ray flux varies more rapidly than the UV, with typical time-scales of hours for Seyfert galaxies.

The conclusions one can draw from correlations between continuum bands are complex and sometimes controversial. For instance, NGC 7469 shows variations in the UV and X-rays with similar large amplitudes but with a time-delay of the order of 2 days, the UV leading the X-rays at maxima. The X-ray flux displays in addition short term variations not seen in the

UV[90]. Moreover, the X-ray spectral index seems to be correlated with the UV flux[91]. In NGC 3516 the delay between UV and X-bands is larger than 2 days[29]. In NGC 4051 no optical variations were observed during strong X-ray variations[25]. The UV variations drive the optical variations with a time-delay ≤ 0.2 days in NGC 4151[28], ≤ 0.15 days in NGC 3516[29], and ~ 1 day (but increasing with wavelength[129,16]) in NGC 7469. Soft X-ray variations are generally larger than hard X-ray ones, but this is not always the case[89]. In NGC 3516, they were strongly correlated with no measurable lag (≤ 0.15d)[29].

The centroid energy, the intensity, and the equivalent width of the Fe Kα line, are variable on short time-scales (down to ks). Correlations between the continuum and Fe K have been intensively looked for in a few objects with ASCA, RXTE, and now with Chandra, revealing a complex behavior (for a review see Nandra[88]). ASCA has observed rapid (1 to 10 ks) variations of Fe K in profile and intensity in MCG-6-30-15[51]. This result has been questioned[67,14] by recent observations of MCG-6-30-15 and NGC 5548 with RXTE, which show that the line flux is constant over time-scales of 50 to 500 ks, while the profile is variable and the underlying continuum displays large flux and spectral variations! The same result seems to be observed in NGC 3516. Reynolds[101] excludes a time-delay of 0.5 to 50 ks between the line and the continuum, and suggests that the line flux remains constant on time-scales between 0.5 and 500 ks in MCG-6-30-15. An analysis of a sample of Seyfert nuclei from the ASCA archive[130] finds that in most cases changes in the line do not appear to track changes in the continuum. Unfortunately, the new results found with Chandra are not yet published at the time of writing these lectures.

We summarize here the most important points that any model should be able to explain:

- Spectral features:
  - the overall shape of the continuum: the BBB (but does it include or not a soft X-ray excess?), the absence of the Lyman discontinuity, the UV/X ratio, the X-ray hump, the power-law, the break at ~ 100 keV, the Fe K line;
  - the correlation between R and Γ;
  - the very broad red wing of the iron line (at least in a few objects), and its centroid energy.
- Variability properties:
  - the rapid variations ($\leq$ ks) of the 2-10keV continuum;

- the correlation between (and only between) long-term variations of UV and X-ray fluxes;
- the very short (or absent) time-lag between optical and UV variations.

And perhaps a few other, but not so well established, variability properties: the slow variation of R; the rapid changes of the Fe K profile; and the absence of correlation of the Fe K flux with the underlying continuum.

Several other issues are not mentioned here, including the polarization properties, in particular, the abrupt increase of polarization level observed at 750 Å in quasars.

3 Black Holes, Accretion Rates, and Luminosities

It is beyond the scope of this chapter to discuss BH physics, for which excellent text books already exist[131,80]. For the purpose of this chapter it is only necessary to know that there are two types of BHs: non rotating or Schwarzschild BHs; and rotating or Kerr BHs. The innermost stable orbit of a Schwarzschild BH is equal to $6R_{\mathrm{G}}$, while it depends on the angular momentum, J, for a Kerr BH. Due to photon capture the maximum value of $cJ/GM^2 = 0.998$ and not unity, and it corresponds to a radius of the last stable orbit equal to $1.24R_{\mathrm{G}}$. Inside the innermost stable orbit, a particle falls almost radially onto the BH. The binding energy of the last stable orbit determines the efficiency of energy conversion. For Schwarzschild and Kerr BHs this is $\eta = 0.057$ and $\eta \sim 0.30$ respectively.

Another important parameter is the "gravitational redshift". This is due to the increase of proper-time with decreasing distance from the BH. Since $r = R/R_{\mathrm{G}}$ can be smaller for Kerr BHs, photons orbiting around a Kerr BH can be more redshifted than around a Schwarzschild BH.

The Eddington luminosity, L_{Edd}, describes the maximum available power due to accretion. It corresponds to the equality between the gravitational force exerted on protons and the radiative force. In a fully ionized gas, the opacity is due to Thomson scattering, and this leads to the well known expression

$$L_{\mathrm{Edd}} = \frac{4\pi\, c\, GM\mu_e}{\sigma_{\mathrm{T}}} \quad , \tag{1}$$

where σ_{T} is the Thomson cross section and μ_e is the unit mass per electron. In units appropriate for quasars, this can be written as

$$L_{\mathrm{Edd}} = 1.5 \times 10^{46} M_8 \ \ \mathrm{erg\ s^{-1}} \quad , \tag{2}$$

where M_8 is the mass expressed in 10^8 M$_\odot$. Note that this expression is valid in principle only for spherical accretion of a fully ionized medium in steady state.

The corresponding accretion rate is

$$\dot{M}_{\rm Edd} = \frac{L_{\rm Edd}}{c^2\eta} = 2.8\ M_8 \left(\frac{\eta}{0.1}\right)^{-1}\ \ {\rm M_\odot yr^{-1}}\ \ . \tag{3}$$

We will often use the "Eddington ratio" $\dot{m} = \dot{M}/\dot{M}_{\rm Edd} = L/L_{\rm Edd}$. Some people define instead $\dot{m} = c^2\dot{M}/L_{\rm Edd}$, which differs by η from our expression.

B. Peterson showed in his lectures that BHs in Seyfert nuclei have a mass range of 10^7 to 10^8 M$_\odot$, with Eddington ratios of about 0.1.

4 What Does the BBB Tell us About the Emitting Medium?

The shape of the BBB implies thermal emission from an optically thick medium radiating as a black body (BB) at temperatures from a few 10^4K (optical emission) to a few 10^5K (EUV emission). The medium can however be optically thin at frequencies larger than the peak of the local Planck curve.

If the medium radiates like a BB, the size of the emitting region, $R_{\rm BBB}$, is given by

$$L_{\rm BBB} = {\rm Min}(1,\tau)\ a\ 4\pi R_{\rm BBB}^2\ \sigma T_{\rm eff}^4\quad , \tag{4}$$

where $L_{\rm BBB}$ is the luminosity of the BBB, τ is the optical thickness, a is a factor of the order of unity that depends on the geometry of the emitting region (in the case of a disk $a = 0.5$), σ is the Stefan constant, and $T_{\rm eff}$ is the effective temperature. Since τ is larger than unity, using the definition of $R_{\rm G}$ one gets

$$r_{\rm BBB} = \frac{R_{\rm BBB}}{R_{\rm G}} \sim 30\ a^{1/2}\ T_5^{-2} M_8^{-1/2} \left(\frac{L_{\rm BBB}}{L_{\rm Edd}}\right)^{1/2}\quad , \tag{5}$$

where T_5 is the effective temperature in units of 10^5K.

One can deduce that the density of the emitting medium is high. Let us assume that the UV emitting medium "sees" the X-ray source, and that the surface layers are heated by this source. The temperature T of an X-ray heated gas, assuming a spherical geometry, is governed by a parameter Ξ defined as

$$\Xi \sim \frac{1}{n_e ckT}\ \frac{L_X}{4\pi R^2}\ , \tag{6}$$

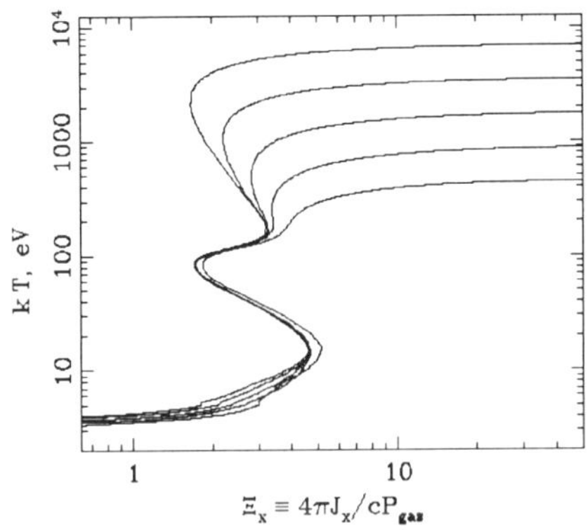

Figure 4. T vs. Ξ for an optically thin medium exposed to a mixture of power-law continuum (of index 0.8 in F_ν) and a 10 eV BB, for increasing fractions of the BB intensity from top to bottom. From Nayakshin & Kallman[96].

where L_X is the X-ray luminosity. Note that Ξ is closely related to the ionization parameter U used by H. Netzer in his lectures. The Ξ vs. T curve depends on the incident spectrum, as first shown by Krolik et al.[60]. This curve is characterized by two thermally stable solutions: a cool phase up to about 2×10^5 K, where the energy balance is due to atomic processes; and a hot phase close to the "Compton temperature" $T_{\rm Comp}$, that corresponds to the balance between Compton heating and Compton cooling. As an illustration, Fig. 4 displays the Ξ vs. T curves for an optically thin gas illuminated by a mixture of power-law continuum (index 0.8 for F_ν) and a 10 eV BB, for increasing fractions of the BB intensity. One can derive that the fraction of soft photons decreases with an increasing $T_{\rm Comp}$. For instance, the composite Laor et al[65] AGN continuum has $T_{\rm Comp} = 2.5 \times 10^6$ K, corresponding to the lower curve on the figure. We shall come back to the Ξ vs. T curve in the following sections. For the moment, we only need to know that whatever the fraction of BB emission, a temperature of 10^5K is obtained for $\Xi \sim 1 - 10$.

Since Ξ must be of the order of a few units, using Eq. (4) one gets:

$$n_e \sim 3 \times 10^{16}\ T_5^{-1}\ \ M_8^{-1} \left(\frac{L_{\rm X}}{L_{\rm Edd}}\right) \left(\frac{R}{10R_G}\right)^{-2} \ {\rm cm}^{-3} \ . \qquad (7)$$

Of course, things are more complicated if the medium is stratified and heated by non radiative processes, but this result remains basically correct.

The suggestion that this thick, dense, and relatively cold medium emitting the BBB could be identified with an accretion disk was made for the first time by Shields[115] for 3C 273, followed by subsequent papers[71,70]. There are however other alternative explanations which will be discussed at the end of the chapter. We now concentrate on the physics of "standard accretion disks" and their emission properties.

5 Thin Steady Accretion Disks: the Simplest Approximation

First we can show that *if the disk radiates like a BB at* $\sim 10^5 K$*, it is geometrically thin*, i.e. its scale height H is much smaller than its radius R. The vertical component of the gravitational acceleration at an altitude z is GMz/R^3, so $H \sim Rc_{\rm s}/V_\phi$, where $c_{\rm s}$ is the sound velocity and V_ϕ the orbital velocity. Although the temperature inside the disk can be (and actually is) larger than 10^5K, at 10-100$R_{\rm G}$ $c_{\rm s} \ll V_\phi$, which equals a fraction of the light velocity, so $H/R \ll 1$.

Since the disk is geometrically thin, the problem is reduced to two dimensions. The gas velocity has only two components, V_ϕ and V_r, the radial component. If the BH mass dominates the mass of the disk and the central stellar cluster (which is the case for $R < 10^4 R_{\rm G}$), V_ϕ is Keplerian, so the orbital frequency $\Omega_{\rm K} = (GM/R^3)^{1/2}$, and one deduces the basic equation

$$H \sim c_{\rm s}\Omega_{\rm K} \ , \tag{8}$$

which is a simplified form of the hydrostatic equilibrium equation.

At this stage we can determine the dynamical time $t_{\rm dyn}$, i.e. the time it takes for a perturbation to cross the disk vertically. According to Eq. (8),

$$t_{\rm dyn} = \frac{H}{c_{\rm s}} = \frac{1}{\Omega_{\rm K}} \ , \tag{9}$$

which written in appropriate units gives

$$t_{\rm dyn} \sim 500 M_8 r^{3/2} \ \ {\rm s.} \tag{10}$$

This is the shortest time on which one can expect a change of structure to occur at a given radius.

The surface density, Σ, and V_r are governed by the continuity equation and the equation of motion:

$$\frac{\partial \Sigma}{\partial t} + \frac{1}{R}\frac{\partial}{\partial R}(R\Sigma V_r) = 0 \ , \tag{11}$$

$$\frac{\partial}{\partial t}(R^2\Sigma\Omega_K) + \frac{1}{R}\frac{\partial}{\partial R}(R^3\Sigma\Omega_K V_r) = \frac{1}{2\pi R}\frac{\partial \mathcal{G}}{\partial \mathcal{R}} \quad , \tag{12}$$

where $\mathcal{G}$ is the torque between two adjacent rings.

To simplify the discussion, we consider a stationary disk. This is the case if the disk is fueled at a constant rate at its outer radius during the time it takes for the gas to reach the BH (the "viscous time"), and if there is no matter falling radially onto the disk at smaller radii. Instabilities can however suppress the stationarity. Under those conditions Eq. (11) becomes

$$\dot{M} = 2\pi R\Sigma V_r \quad , \tag{13}$$

where the accretion rate $\dot{M}$ is constant along the radius; and Eq. (12) becomes

$$\mathcal{G} = R^2\Omega_K\dot{M} \; + \; \text{constant} \quad , \tag{14}$$

which describes the constancy of the angular momentum flux. From Eq. (13) it is interesting to note that the thin disk assumption (i.e. $V_r \ll V_\phi$) implies a large value of the surface density,

$$\Sigma \sim 10^{22}\frac{1}{\eta}\frac{c}{V_r}\frac{\dot{m}}{r} \quad \text{g cm}^{-2}. \tag{15}$$

Since matter orbits in circles it requires a dissipation mechanism to lose energy and move towards the BH. The question is how to ensure accretion, or in other words: what is the torque $\mathcal{G}$? Microscopic viscosity is far too small, so one has to seek for a more efficient mechanism. Let us assume for the moment that there is a *local* unspecified viscosity.

The kinematic viscosity ν is related to the torque through the viscous stress tensor, giving the rate at which angular momentum is transported. For a thin disk, the only interesting coordinate of the stress tensor is

$$T_{r\phi} = \rho\nu R\frac{\partial\Omega_K}{\partial R} \quad , \tag{16}$$

where ρ is the density. The torque is thus equal to

$$\mathcal{G} = \int R d\phi \int R T_{r\phi} dz \; = \; 2\pi R^3\nu\Sigma\frac{\partial\Omega_K}{\partial R} \quad . \tag{17}$$

Note that since $\partial\Omega_K/\partial R$ is negative, the torque is negative, and angular momentum is transported outwards, as required for accretion to take place.

Finally, one can compute the rate of energy dissipated in the disk per unit surface, $D(R)$. According to our assumption of a local viscosity, the potential energy is transformed into heat, so we would get $2\pi D(R) = GM\dot{M}/R^3$. However, we also have to take into account the conservation of angular momentum and the fact that the torque cancels at R_{in}, the radius of the last stable orbit.

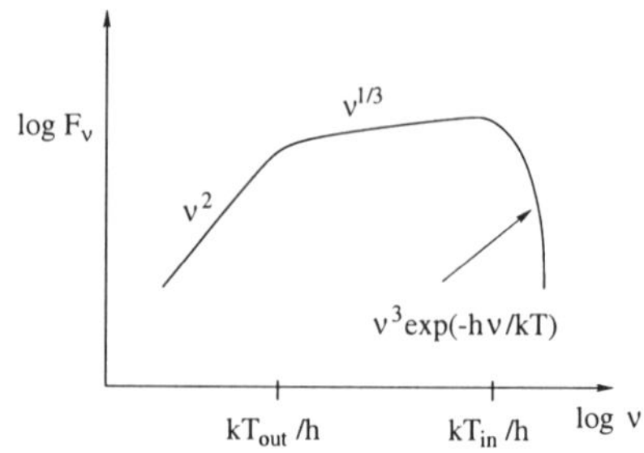

Figure 5. Disk spectrum, assuming that the disk radiates locally like a BB. Boundary corrections were not taken into account. The index 2 and the exponential cut-off correspond to the Rayleigh-Jeans and to the Wien part of the spectrum respectively.

This leads to an extra factor $(3/2)(1-\sqrt{R_{\rm in}/R})$ (for a detailed demonstration see Krolik[59]). Finally one gets

$$D(R) = \frac{3GM\dot{M}}{4\pi R^3} f(R) \ , \tag{18}$$

where $f(R) = (1 - \sqrt{R_{\rm in}/R})$. In a fully relativistic treatment the expression of $f(R)$ would be more complicated. It is important to realize that the approximate R^{-3} dependence of the dissipation leads to a luminosity roughly proportional to $1/R$. Therefore, *the bulk of the gravitational energy is released close to the BH, at a few $R_{\rm G}$, but this does not mean that the bulk of the radiation necessarily comes from this region* (cf. the "cloud model").

If the disk radiates like a BB, the temperature $T_{\rm BB}$ is equal to the effective temperature $T_{\rm eff}$. Since the energy is radiated by both faces of the disk, $T_{\rm eff} = (D/2\sigma)^{1/4}$, which translates into

$$T_{\rm eff} \sim 6 \times 10^5 \eta^{-1/4} f(R)^{1/4} \dot{m}^{1/4} M_8^{-1/4} r^{-3/4} \ \ {\rm K}, \tag{19}$$

where $\dot{m}$ is defined by Eq. (3). Integrating the Planck function over the radius and neglecting the factor $f(R)$, one obtains the spectrum shown in Fig. 5, where $T_{\rm in}$ and $T_{\rm out}$ are the temperatures at the inner and outer radii. According to the Planck law, at a frequency $\nu = kT_\nu/h$, the regions with $T < T_\nu$, i.e. $R > R(T_\nu)$, contribute negligibly to the emission. On the other hand, in the inner regions where $T > T_\nu$, the spectrum at the frequency ν is given by the Rayleigh-Jeans expression $B_\nu = 2\nu^2 c^{-2} kT_{\rm eff}$. Thus, $L_\nu \propto \int_{R_{\rm out}}^{R(T_\nu)} R^{1/4} dR$, and finally *the emission at a frequency ν is dominated by material at the radius $R(T_\nu)$.*

This spectrum should be corrected for relativistic effects close to the BH, in particular in the case of a Kerr BH, where the inner radius is small. The effect is to make the spectrum dependent on the inclination angle, as L_ν is relativistically boosted towards high frequencies for edge on disks (cf. Fig. 12).

The fact that the bulk of the emission at a given wavelength is produced at the radius R_ν has important consequences for the variations of the different spectral bands. Since $B_\nu(T)$ peaks at $\lambda = 0.51/T(\mathrm{K})$ cm, one deduces from Eq. (19), that the emission at λ_μ, expressed in microns, is produced mainly at a distance

$$r \sim 650\ \lambda_\mu^{4/3} \eta^{-1/4} f(R)^{1/4} \dot{m}^{1/4} M_8^{-1/4} \quad , \tag{20}$$

or

$$R \sim 1.2 \lambda_\mu^{4/3} \eta^{-1/4} f(R)^{1/4} \dot{m}^{1/4} M_8^{3/4} \ \ \mathrm{lt-d.} \tag{21}$$

From this we infer that the optical to near-UV emission corresponds to regions located at large distances from the BH. Comparing this result with the small time-delays between the UV and optical light curves (less than a fraction of a day), we see that it is marginally compatible with a causal link propagating at the speed of light. It is however incompatible with the dynamical time, according to Eqs. (10) and (20). But before we accept this conclusion we should question whether it is legitimate to assume that the disk is radiating locally like a BB, and therefore we have to obtain information on the optical thickness of the disk, and more generally on its physical properties. In this aim we have to specify the viscosity mechanism.

6 A Further Step: the α-Prescription, Vertically Averaged Models

6.1 Disk Structure

It is widely accepted that the mechanism for transportation of angular momentum can be identified with "turbulent viscosity" according to the "α-prescription" proposed by Shakura & Sunyaev[113] (for a detailed discussion of α-disks see Frank et al.[33]). In this prescription it is assumed that the *sizes of the turbulent eddies are smaller than the thickness of the disk and that the turbulence is subsonic*, hence

$$\nu = \alpha c_{\mathrm{s}} H \quad , \tag{22}$$

where $\alpha \lesssim 1$.

The α-prescription has proven to give satisfying results when applied to disks in cataclysmic variables, where a relatively small value (≤ 0.1) of α

is generally required. A magnetohydrodynamic instability pointed out by Chandrasekhar[13] is able to provide this viscosity[10]. Other prescriptions based on purely hydrodynamic (shear) instabilities are also under debate[105], in which the corresponding value of α can be larger and vary with $\Omega_{\rm K}$. The viscosity can be parameterized with a fixed Reynolds number[41,42,105], following an approach suggested already by Lynden-Bell & Pringle[69]. Finally the viscosity might also be a non-local mechanism, for instance in the case of an organized magnetic field anchored in the disk.

From Eqs. (13) and (14), one finds

$$V_r = \frac{3}{2}\alpha c_{\rm s}\frac{H}{R} \quad , \tag{23}$$

which shows that the radial velocity is strongly subsonic, and a fortiori strongly sub-keplerian, as expected.

The α-prescription is equivalent to the previous expression if *and only if* the disk is not self-gravitating. Eqs. (16) and (22) give

$$T_{r\phi} = \frac{3}{2}\alpha\rho c_{\rm s} H\Omega_{\rm K} \quad . \tag{24}$$

Using $c_{\rm s} = H\Omega_{\rm K}$, one gets

$$T_{r\phi} = \frac{3}{2}\alpha P \quad , \tag{25}$$

where P is the sum of the radiative and gas pressure.

If the ratio of the vertical component of the disk gravity to that of the central gravity, $\zeta = 4\pi G\rho/\Omega_{\rm K}^2$ is of the order of unity, then[48]:

$$c_{\rm s} = H\Omega_{\rm K}(1+\zeta)^{1/2} \quad \text{and} \quad T_{r\phi} = \frac{3}{2}\alpha P(1+\zeta)^{-1/2} \quad . \tag{26}$$

Actually there is a debate about the value of P one should take in the stress tensor: is it the total pressure or only the gas pressure? For the moment we adopt the use of the total pressure.

Because it is a matter of confusion, it is necessary to stress here that $T \sim T_{\rm mid-plane}$ and not $T_{\rm eff}$. If the diffusion approximation holds between the two temperatures, then $T \sim \tau^{1/4} T_{\rm eff}$, which implies that $T \gg T_{\rm eff}$ in the optically thick region.

All these expressions are *vertically averaged.* Since the disk is geometrically thin, these averaged quantities can be used to decouple the radial and vertical structures. Additional equations describe:

- The gas pressure:

$$P_{\rm gas} = \frac{\rho k T}{\mu m_{\rm H}} \quad , \tag{27}$$

where μ is the mean mass per particle.

- The radiative pressure, which in the 2-stream approximation is given by

$$P_{\rm rad} = \frac{F_{\rm rad}}{c}\left[\frac{1}{2}\tau_{\rm R} + \frac{1}{\sqrt{3}}\right] \quad , \tag{28}$$

where $\tau_{\rm R}$ is the Rosseland optical thickness of the disk.

- The flux radiated by each face of the disk[40]

$$F_{\rm rad} = \frac{4}{3}\frac{\sigma T^4}{\frac{1}{2}\tau_{\rm R} + \frac{1}{\sqrt{3}} + \frac{1}{3\tau_{\rm P}}} \quad , \tag{29}$$

where $\tau_{\rm P}$ is the Planck optical depth. Note that, in the optically thick case, $P_{\rm rad}$ tends to its LTE value and $F_{\rm rad} \approx (8/3)\sigma T^4/\tau_{\rm R}$, which is the well-known diffusion approximation. In the optically thin case, $F_{\rm rad} \approx 4\sigma T^4 \tau_{\rm P}$.

- The energy balance between the local viscous dissipation and the radiative cooling,

$$F_{\rm rad} = D(R)/2 \quad . \tag{30}$$

These equations, when solved self-consistently with the appropriate functions for $\kappa_{\rm R}(\rho, T)$, $\kappa_{\rm P}(\rho, T)$ and $\mu_{\rm P}(\rho, T)$, yield the radial disk structure in the non self-gravitating region of the disk for given values of M, $\dot{M}$, and α. One finds that the disk is divided into several regions, according to the dominating processes: radiation or gas pressure; Thomson, Kramers, or more complicated opacity; and self-gravity or no self-gravity.

To show the importance of the opacity, Fig. 6 displays the radial dependence of several physical quantities as function of the radius for an α-disk in the case of a typical AGN. In these computations self-gravity has been neglected in order to focus on the other parameters. Different approximations have been compared to a 2D computation with realistic opacities. We see that in the inner region ($R \leq 10^{16}$ cm, i.e. $r \leq 10^3 R_{\rm G}$), the opacity is better approximated by Thomson scattering, while in the outer region it is better approximated by Kramers' opacity. Note that the radiation pressure dominates in the Thomson regime ($\beta = P_{\rm gas}/P_{\rm rad}$).

Fig. 7 shows that the transition between the radiation and gas pressure dominated regimes occurs at a larger radius for a larger accretion rate in Eddington units. This radius also increases with a decreasing viscosity parameter α, which corresponds to an increasing density. For a typical AGN ($\dot{m} = 0.1$ and $M = 10^8 {\rm M}_\odot$) and $\alpha = 0.1$, $R \sim 2000 R_{\rm G}$. Note that it occurs

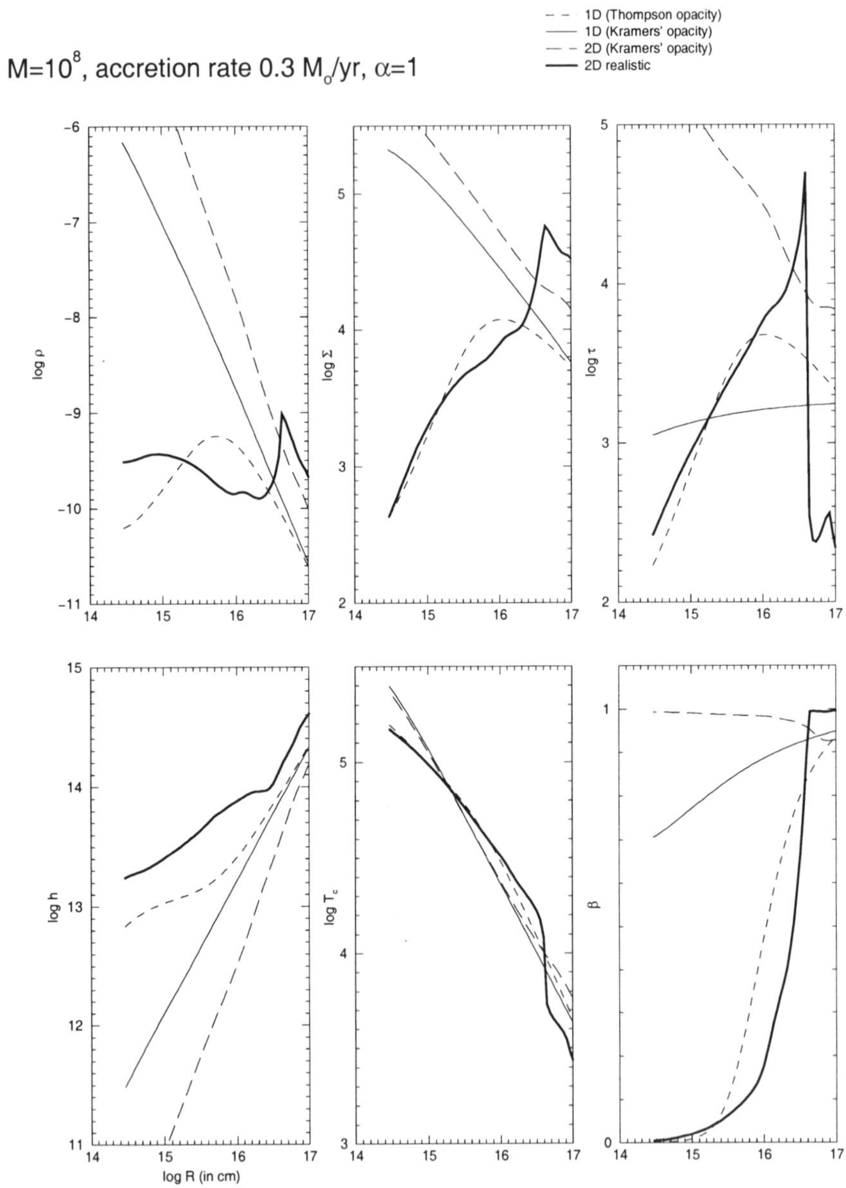

Figure 6. From left to right and top to bottom, density, surface density, Rosseland mean optical thickness, scale height, equatorial temperature and radiation to gas pressure ratio as functions of the radius for a vertically averaged α-disk. The thick solid line corresponds to a 2D model with realistic opacities. Courtesy of J.-M. Huré.

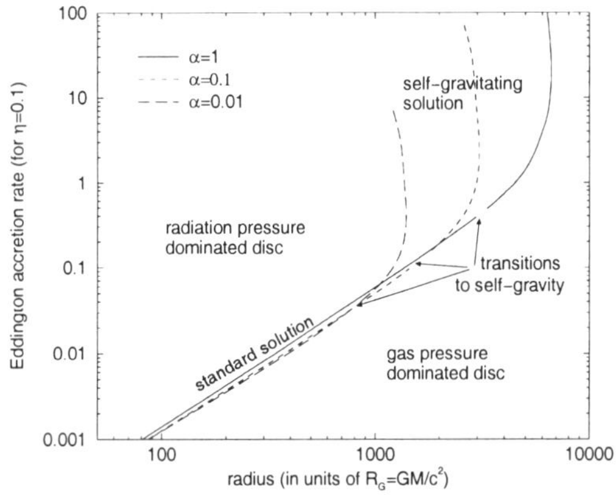

Figure 7. Transition from radiation to gas pressure regime for a vertically averaged α-disk, for $M = 10^8 \mathrm{M}_\odot$. The figure displays $\dot{m}$ versus r for different values of α. Courtesy of J.-M. Huré.

in the same region as the transition to the self-gravity regime (this is due to the rapid increase of density when the disk gravity begins to dominate).

Finally the disk becomes self-gravitating at a radius $R_{\rm sg}$ corresponding to $\zeta \sim 1$. As shown in Fig. 8, the transition occurs at $R \sim 100 - 1000 R_{\rm G}$. It is located further away for a larger value of α, owing to the corresponding smaller value of the density.

In summary *the non self-gravitating part of the disk is generally reduced to the Thomson opacity–radiation pressure dominated regime.* Neglecting the boundary function $f(R)$ the different parameters vary as:

- $\rho \propto R^{3/2}$,
- $H =$ constant ,
- Σ and $\tau \propto R^{3/2}$,
- $T \propto R^{-3/8}$,
- $\zeta \propto R^{9/2}$,
- $P_{\rm gas}/P_{\rm rad} \propto R^{21/8}$.

These relations explain why the transition from the gas pressure to the radiation pressure regime, and the transition to self-gravity, are so rapid (see the

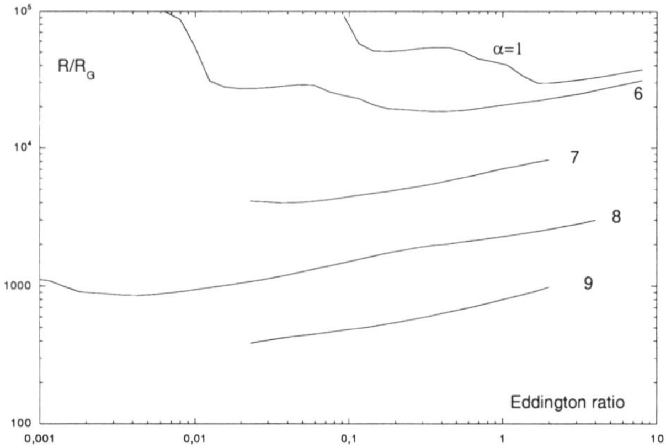

Figure 8. Transition to the self-gravity regime for a vertically averaged α-disk. The figure displays R_{sg}/R_G versus $\dot{m}$ for different values of M. All models except one are computed for $\alpha = 0.1$. The curves are labelled with $\log M$ in units of $M_\odot$.

strong variation of ζ and P_{gas}/P_{rad} as functions of the radius on Fig. 7). It is also interesting to note that the gas temperature decreases less rapidly than T_{eff} (which is proportional to $R^{-3/4}$). The scale height to radius ratio can be expressed in this region as

$$\frac{H}{R} = 6.7\frac{\dot{m}}{r} \ , \tag{31}$$

which shows that the "geometrically thin" assumption does not hold for large $\dot{m}$; and the Thomson thickness is given by

$$\tau_T \ = \ 55\alpha^{-1}\left(\frac{R}{R_G}\right)^{3/2} \ . \tag{32}$$

A few important time-scales relevant to the external regions which emit at optical wavelengths, where self-gravity is not negligible, are defined below. The "viscous time" t_{visc}, basically the time it takes for matter to move to the center, can be expressed as (using in particular Eq. (26)):

$$t_{visc} \sim \frac{R}{V_r} \ \sim \frac{1}{\alpha}\left(\frac{H}{R}\right)^{-2}\frac{1}{\Omega_K\sqrt{1+\zeta}} \ . \tag{33}$$

The "thermal time" t_{th}, i.e. the time it takes to radiate the thermal energy, can be expressed as

$$t_{\rm th} \sim \frac{PH}{D(R)} \sim \frac{\sqrt{1+\zeta}}{\alpha\Omega_{\rm K}} \quad . \tag{34}$$

The expression for the dynamical time $t_{\rm dyn}$ has already been given in Eq. (9). Correcting for self-gravity, it becomes

$$t_{\rm dyn} \sim \frac{1}{\Omega_{\rm K}\sqrt{1+\zeta}} \quad . \tag{35}$$

It is easy to see from these expressions that

$$t_{\rm visc} \gg t_{\rm th} \geq t_{\rm dyn} \quad . \tag{36}$$

For instance in the inner Thomson opacity - radiation pressure dominated region $t_{\rm visc}$ is given by

$$t_{\rm visc} = 10\ \alpha^{-1} M_8 \dot{m}^{-2} r^{5/2}\ \ {\rm s}. \tag{37}$$

6.1.1 Stability of the Disk

There are 3 main instabilities acting in the disk:

1. the thermal instability, which occurs when

$$\left(\frac{\partial \ln D}{\partial \ln T}\right)_R - \left(\frac{\partial \ln F_{\rm rad}}{\partial \ln T}\right)_R > 0 \ ; \tag{38}$$

2. the viscous instability, which occurs when

$$\left(\frac{\partial \ln \dot{M}}{\partial \ln \Sigma}\right)_R < 0 \ ; \tag{39}$$

3. the gravitational instability, which occurs when[127,34]

$$Q = \frac{\Omega_{\rm K}^2}{\pi G \rho} < 1 \ . \tag{40}$$

In the self-gravitating regime, where[46] $\rho \propto R^3$ and $\Sigma \propto R^2$, the disk shrinks, rapidly becomes strongly gravitationally unstable, and unstable fragments with a typical volume H^3 begin to collapse. Consequently, for $R > R_{\rm sg}$ the α viscosity does not hold. We stress that this gravitational instability is *local*, and does not require the disk mass to be of the order of the BH mass. The instability corresponds simply to the tidal limit $4\pi\Sigma \geq MH/R^3$, or equivalently to $M_{\rm disk} \sim MH/R$, hence $M_{\rm disk} \ll M$.

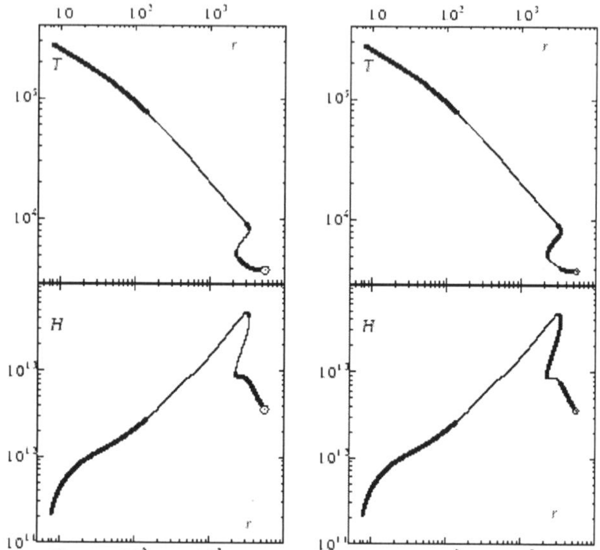

Figure 9. Viscously (left) and thermally (right) unstable regions for a vertically averaged α-disk, with $\alpha = 1$, $M = 3 \times 10^7 M_\odot$, and $\dot{m}$=0.03. The figure displays T (in K) and H (in cm) versus r. The solid line represent the unstable regions. From Huré[45].

It can be shown that the thermal and the viscous instabilities are strongly linked, so they generally occur in the same region of the disk (see Fig. 9). We see that the disk is both viscously and thermally unstable in a large fraction of the inner region. This combined instability should operate on a thermal time-scale, which is quite small in the inner regions (cf. Eqs. (10) and (34)). It should therefore induce rapid variations of EUV emission. Moreover as α-disks are also gravitationally unstable in the outer region, one deduces that *in AGN, α-disks are unstable almost everywhere!*

6.1.2 "β-Disks"

In the case where the stress tensor is assumed to be proportional to the gas pressure and not to the total pressure[15,78], the disks are known as "β-disks". Strictly speaking they should be called "$\alpha - \beta$-disks"! They are denser than α-disks, and dominated by gas pressure even in the inner regions. Consequently they are not subject to the viscous/thermal instability, but to another instability due to ionization of hydrogen, which occurs in regions where the mid-plane temperature is of the order of 10^4K. This instability induces a "limit-cycle" behavior, which is thought to explain the periodic variations of dwarf novae. In AGN it should operate on a much larger time-scale (10^5 yr), and could correspond to variations of 4 orders of magnitude in luminosity[120].

6.2 *Where a Fundamental Ingredient, Compton Scattering, Appears*

The solution of the vertically averaged disk showed that Thomson opacity dominates over free-free opacity in the inner regions. Diffusions are therefore important, and modify the equilibrium spectrum. Radiation is thermalized for a relatively small absorption optical thickness, when the effective optical thickness $\tau_{\rm eff} = \sqrt{\tau_{\rm abs}(\tau_{\rm abs} + \tau_{\rm T})} \sim \sqrt{\tau_{\rm abs}\tau_{\rm T}} > 1$, where $\tau_{\rm abs}$ and $\tau_{\rm T}$ are the absorption and diffusion optical thicknesses respectively. This occurs in the inner regions. In this case the spectral distribution is given by[111]

$$I_\nu = 2B_\nu\sqrt{\frac{\kappa_{\rm abs}}{\kappa_{\rm T}}} \ll B_\nu \ . \tag{41}$$

For instance, if absorption is dominated by free-free processes,

$$\kappa_{\rm abs} = 3.69 \times 10^8 N_e N_i Z^2 g_{\rm ff} \frac{1}{\sqrt{T}}\frac{1}{\nu^3}\left(1 - \exp\left(-\frac{h\nu}{kT}\right)\right), \tag{42}$$

and, using Eqs. (41) and (42), one deduces that the intensity is given by

$$I_\nu \propto \nu T^{1/4} \tag{43}$$

in the Rayleigh-Jeans regime, where $h\nu/kT < 1$; or

$$I_\nu \propto \nu^{3/2} \exp\left(-\frac{h\nu}{kT}\right) T^{-1/4} \tag{44}$$

in the Wien regime, where $h\nu/kT > 1$. The spectrum emitted by the innermost regions will therefore differ from a local BB.

However it is not sufficient to take into account Thomson scattering, as one can easily show that Compton scattering is also important. Since the gas temperature is high and we are interested in soft X-ray photons, the collisions between electrons and photons are not elastic, and they lose or gain energy with each scattering. The "Compton parameter" y, which is the total relative change of energy when a photon travels through the medium, is equal to the mean number of diffusions times the relative change of energy at each scattering. The mean number of diffusions, $N_{\rm s} = {\rm Max}(\tau_{\rm T}, \tau_{\rm T}^2) = \tau_{\rm T}^2$. In the case of a thermal non relativistic distribution of electrons, the relative change of energy at each scattering is given by[111]

$$\frac{\Delta h\nu}{h\nu} = -\frac{h\nu}{m_o c^2} + \frac{4kT}{m_o c^2} , \tag{45}$$

and thus one gets

$$y = \frac{4kT - h\nu}{m_o c^2} \tau_{\rm T}^2 . \tag{46}$$

In the inner regions of the disk, $\tau_{\rm T} \sim 10^4$ and $T \sim 10^6$K, so y can easily be larger than unity. If free-free dominates, then $\tau(\nu)$ and hence y increases with decreasing frequency, and there is a frequency ν_o below which $y > 1$. If $h\nu_o/kT \ll 1$, inverse Compton processes will drive soft photons to higher frequencies, and there will be an accumulation of photons at $\nu \sim 4kT/h$, at the expense of the low frequency population[22,128]. Another consequence of Compton diffusion is that the profile of a line will be asymmetrically broadened, as it is dominated by Inverse Compton diffusions for $\nu_{\rm line} < 4kT/h$, and by direct Compton diffusions for $\nu_{\rm line} > 4kT/h$ (i.e. down-scatterings leading to a broad red wing). The computation of the spectrum in this case is performed with transfer codes using the "Kompaneets diffusion equation", or with Monte Carlo codes. The effect of Compton scatterings on the spectral distribution of the disk is illustrated in Fig. 11, and at the end of the chapter we will discuss the effect on the profile of the iron line.

Another problem which has to be taken into account is the existence of an atmosphere above the disk. As a first approach[56,125], the disk was replaced by low gravity stellar atmospheres of different $T_{\rm eff}$. However, since gravity is constant in a stellar atmosphere, while it increases with height in disks, this method is not valid, and led to the erroneous result that the spectrum should always show a strong Lyman discontinuity in absorption.

7 The α-Prescription, Computation of the Vertical Structure and Vertical Transfer

The computation of an accurate emission spectrum requires a model of vertical structure with the self-consistent treatment of radiative transfer. Since the disk is still assumed to be geometrically thin, it is again possible to decouple the vertical and the horizontal structure equations. One then solves for each annulus a set of differential equations in z, instead of a set of algebraic equations. A simplification consists, at least in a first step, of decoupling the computation of the emitted spectrum from that of the vertical structure, as the disk is optically thick and the spectrum forms in the upper layers. The radiative transfer is then solved using the diffusion approximation.

7.1 Vertical Structure

The equations for the vertical structure are:

- the hydrostatic equilibrium,

$$\frac{1}{\rho}\frac{dP_{\rm tot}}{dz} = -\Omega_{\rm K}^2 z \quad ; \tag{47}$$

- the vertical variation of the energy produced by viscous dissipation,

$$\frac{dF}{dz} = \frac{9}{4}\rho\nu(z)\Omega_{\rm K}^2 \quad ; \tag{48}$$

- the variation of the mean optical thickness and of the surface density,

$$\frac{d\tau}{dz} = -\kappa\rho \;\; \text{and} \;\; \frac{d\Sigma}{dz} = \rho \quad , \tag{49}$$

  where κ is a frequency averaged opacity;

- the variation of the radiation pressure in the diffusion approximation, assuming that the energy is transported only via radiation,

$$-\frac{c_{\rm s}}{\mu\kappa}\frac{dP_{\rm rad}}{dz} = F(z) \quad . \tag{50}$$

A question is raised here: how is viscous heating deposited vertically? The original Shakura-Sunyaev prescription was introduced to describe vertically averaged quantities, and it is not clear how to extend it to a z-dependant model. According to the $\alpha - P$ prescription described above, it seems natural to assume that ν is constant vertically, which implies that the heat deposition dF/dz is proportional to ρ. If the opacity is dominated by Thomson scattering and the pressure is dominated by $P_{\rm rad}$, as is the case in the inner region of the disk, then Eqs. (47), (48) and (50) lead to the result that ρ does not depend on z. A different prescription is sometimes used for disks in dwarf novae[75].

Other ingredients can also be added to the model, such as convective transport. As an illustration, Fig. 10 displays the 2D structure of an α-disk, for $\alpha = 0.1$, taking into account convective and turbulent transport and self-gravity[47]. The parameters are those of a typical AGN ($M = 10^8 {\rm M}_\odot$, $\dot{M} = 0.1{\rm M}_\odot/{\rm yr}$, or $\dot{m} \sim 0.03$). Comparing the results of 2D and 1D computations, one finds that the results are similar for the radial structure within factors of

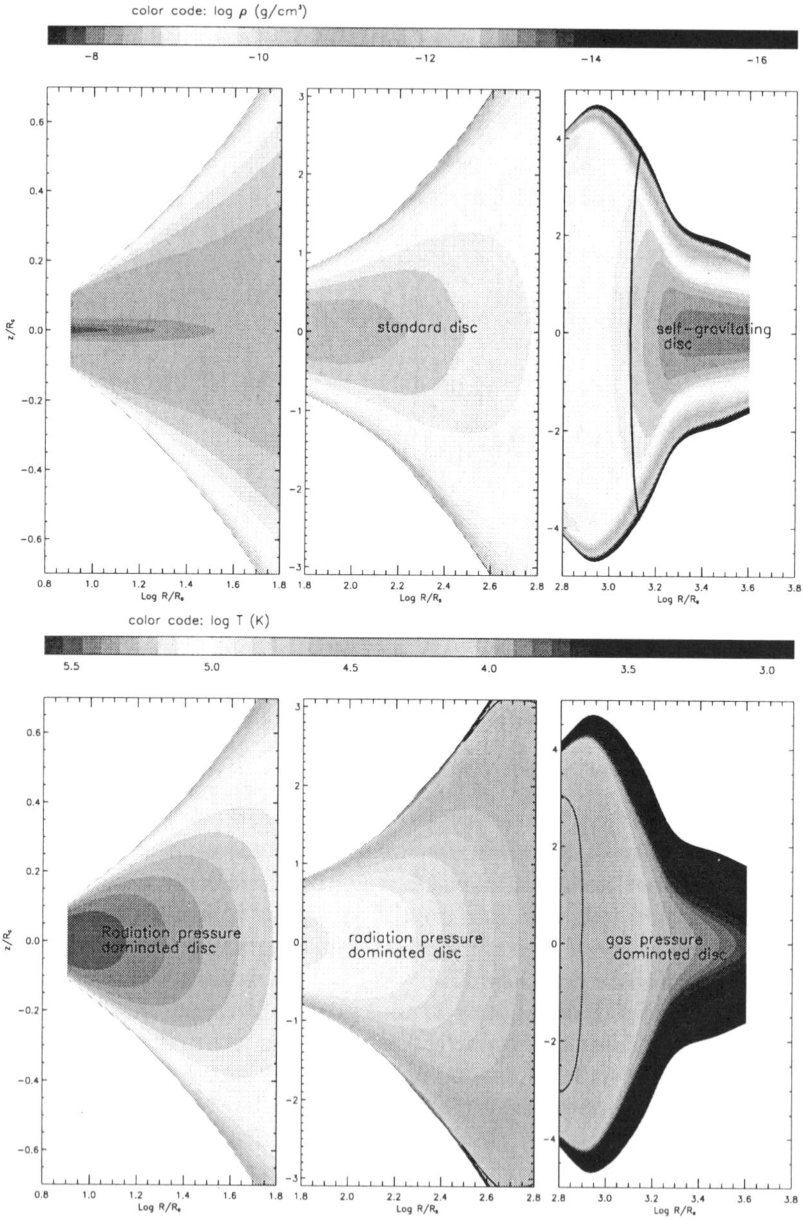

Figure 10. Temperature and density structure of an α-disk for a typical AGN ($M = 10^8 M_\odot$, $\dot{M} = 0.1 M_\odot\ \mathrm{yr}^{-1}$ or $\dot{m} \sim 0.03$, $\alpha = 0.1$). Courtesy of J-M. Huré.

a few. The temperature in the equatorial plane is accurate to within 30% and the most important difference (by a factor ≤ 4) is in the disk thickness[49]. The density and the surface density, the critical radius for self-gravity, the ratio of gas to radiation pressure ratio are correctly computed in the 1D model (cf. Fig. 6). The 2D structure avoids however some multiple solutions met in the 1D model (cf. Fig. 9).

7.2 Disk Spectrum

Vertically averaged α-disks, which have proven to be extremely useful in determining the radial structure of the disk, are not sufficient to determine the emission spectrum. This requires solving the vertical radiative transfer in a consistent way with the vertical structure of the disk.

Several authors have built consistent models of the vertical structure of AGN accretion disks using various approximations for the radiative transfer. The first computation was performed by Laor & Netzer[64], who solved the radiative transfer in the Eddington approximation, with LTE, bound-free and free-free opacities, and with different assumptions for the stress tensor as well as for vertical heat deposition laws. They showed in particular that the Lyman discontinuity can appear either in absorption or in emission according to these laws. Ross et al.[107] relaxed the LTE assumption and took into account Comptonization using the Kompaneets equation. Störzer[123] solved the radiative transfer equation in 3D, also taking into account the influence of self-gravity, but keeping the LTE approximation. Shimura & Takahara[117,118] also solved the vertical structure together with the transfer treated with the Kompaneets equation, but for a stress tensor proportional to the gas pressure and not to the total pressure. Störzer et al.[124] and Shields & Coleman[116], addressed the question of how departures from LTE influence the Lyman discontinuity. Dörrer et al.[26] solved the vertical structure self-consistently, with LTE and no bound-free opacities, and a pure-H atmosphere, but taking into account Compton scattering. Sincell & Krolik[122] solved the coupled vertical structure and radiation transfer with a two stream Eddington approximation, restricting the elements to He and H in LTE.

Fig. 11 illustrates the influence of Comptonization[107]. It shows the spectra of two (non-rotating) BHs, with different masses and accretion rates. This figure illustrates two interesting results:

1. The soft X-ray excess cannot be reproduced. Although a non-negligible fraction of the disk luminosity is radiated in the soft X-ray range as a consequence of Comptonization, it is obvious that the spectral energy distribution of AGN cannot be accommodated if one only assumes an

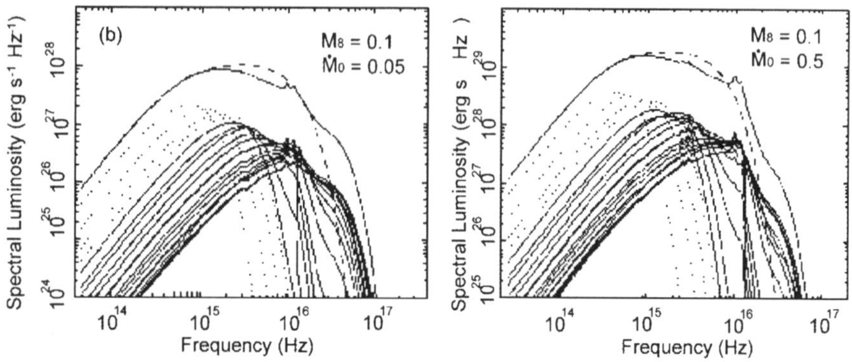

Figure 11. Spectrum of an α-disk around a Schwarzschild BH. Solid lines represent the contribution from different annuli and the total spectrum (with Componization). Dotted lines represent the local BB approximation used for the outermost annuli. Dashed lines represent the total spectrum with the local BB computation. From Ross et al.[108] ©1993 Blackwell Sci.

extra power-law component exists in the soft X-ray range. Maximally rotating BHs produce continua which are harder those of non-rotating BHs, but they do not solve the problem either.

2. The local BB gives the correct result at optical to near-UV wavelengths (this would not be true for the UV band and for higher masses[43]), and confirms *our previous conclusion that the time-delay observed between the optical and UV bands is too small compared with the smallest time-scale in the disk.*

Hubeny and collaborators[41,42,43] solved self-consistently the vertical structure with an exact radiative transfer with no diffusion and escape probability approximation, in the non LTE case, taking into account the general relativistic transfer function. However, they did not take into account Comptonization, or heavy elements, so their models probably do not correctly describe the EUV. They computed the most extensive set of models to date: a grid of non-LTE disks around maximally-rotating Kerr and Schwarzschild BHs, for a wide range of BH masses and accretion rates. Some of their results are shown in Figs. 12, 13, and 14.

Fig. 12 compares the energy distributions of disks of Schwarzschild and Kerr BHs of 10^9 $M_\odot$ viewed at 37°, for different disk luminosities.

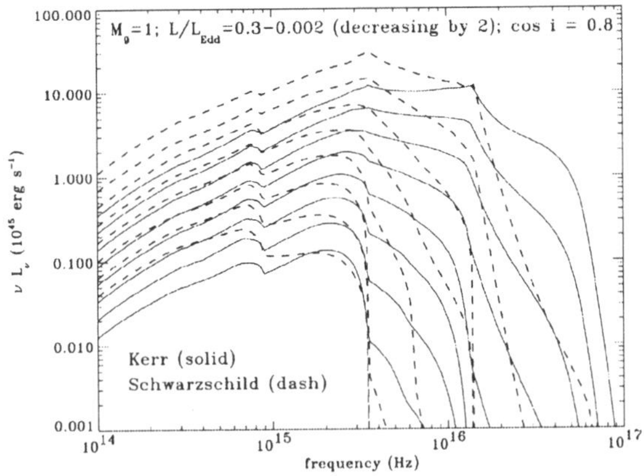

Figure 12. Spectral energy distribution of an α-disk around a maximally rotating BH (solid lines) and a Schwarzschild BH (dashed lines), with $M = 10^9$ $M_\odot$ and various values of the accretion rates. The curves correspond to $L/L_{\rm Edd}=$ 1, 1/2, and 1/4, etc., for the Kerr BH; and the corresponding values for the Schwarzschild BH are 5.613 times larger. From Hubeny et al.[43] ©2000 AAS.

Schwarzschild and Kerr BHs produce quite similar optical spectra, except that in Kerr BHs the spectrum is shifted towards higher frequencies.

Fig. 13 displays 1450 to 5050 Å spectral slopes, β (defined here as $F_\nu \propto \nu^\beta$), as a function of the Eddington ratio for Kerr BHs of different masses. The local BB models depart from real transfer models except for low accretion rates. This is due to the relatively high masses considered. Both sets of models can reproduce the mean observed colors of the LBQS sample[32]. However, one should note that *the models cannot account for the positive slopes observed in the majority of Seyfert galaxies.*

Finally, Fig. 14 shows the strength of the Lyman edge in a Kerr BH with different disk inclinations. The Lyman discontinuity is always more intense than observed, except for very low accretion rates or very large disk inclinations, not plausible in the framework of Unification. It is however possible that the addition of metal opacity (not taken into account in the computations) changes this result.

This discussion illustrates the fact that the computation of disk spectra is difficult and involves poorly known parameters. Besides, several other parameters not mentioned here complicate the problem, such as the value of α, the inner and outer radii of the disk, the different assumptions concerning heat deposition, etc. Therefore, in some cases it may not be a bad choice to use

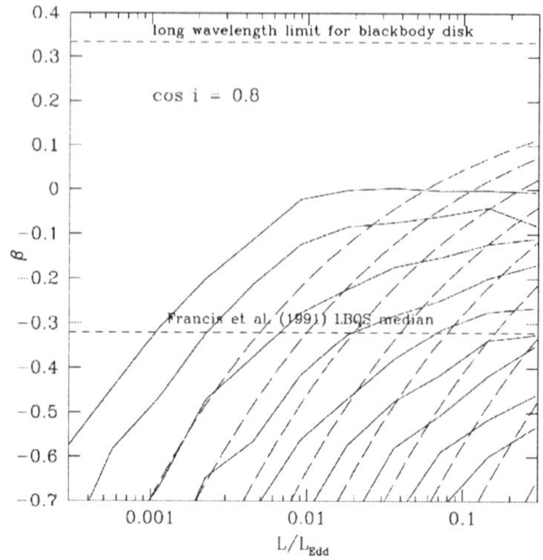

Figure 13. Optical/ultraviolet spectral slope between 1450 and 5050Å for an α-disk around a Kerr BH with 37° viewing angle, and $\alpha = 0.01$. Solid curves: computed models; dashed curves: local BB. From top to bottom, the curves are computed for 9 values of M: 1/8, 1/4, 1/2, 1, 2, 4, 8, 16, and 32 $\times 10^9 M_\odot$. From Hubeny et al.[43] ©2000 AAS.

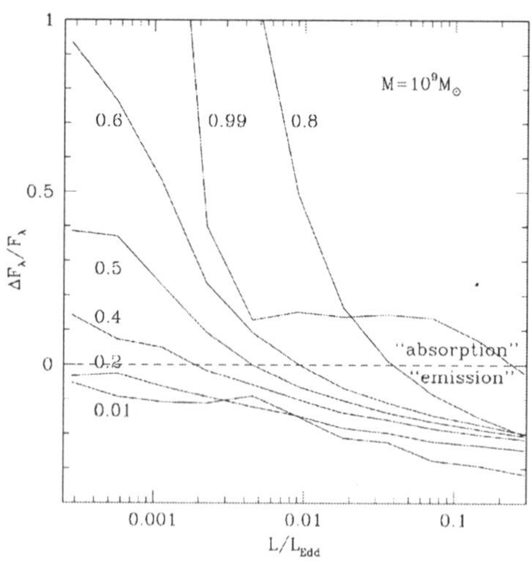

Figure 14. Lyman edge strength as a function of Eddington ratio for an α-disk around a Kerr BH with $M = 10^9$ $M_\odot$. The curves are labelled with the value of $\cos i$. From Hubeny et al.[43] ©2000 AAS.

the simplest and least model-dependent solutions. Despite these uncertainties, the following results can be considered as secure:

- the spectral energy distribution does not fit the observations in the soft X-ray range: another emission mechanism is required;
- the disk spectrum in the optical to near-UV range is too flat compared to that of many Seyfert nuclei;
- the computed Lyman discontinuity is too strong or has to be fine-tuned;
- the small or absent time-lag between the optical and UV bands is not compatible with the viscous time-scale, or even with the dynamical time;
- the disk is not emitting hard X-rays at all.

Additionally, we paid no attention to polarization properties. They are also hard to explain in the context of the standard disk model.

8 Other Disk Models

8.1 The "Lamppost" Model

The discussion on the standard disk model has shown that the regions emitting at UV and optical wavelengths are located at large distances from each other, so the small time-lags between the (well correlated) optical and UV light curves require a causal link propagating at the speed of light. This led to questioning the validity of the thin steady disk model[7,21], and to propose that the disk was actually irradiated by the X-ray continuum, and emitting as a result of both gravitational and external radiative heating[17,83,106]. The discovery of the iron line and the X-ray hump[99] implied reprocessing by a cold medium, in agreement with this model.

There then followed the era of "irradiated disks". These models included an X-ray point source located at a given height A above the center of the disk, or a spherical source of radius A around the BH. The former is now referred to as the "lamppost model" (cf. Fig. 15).

The flux falling on each face of the disk F_X can be expressed as

$$F_X = a f_X \frac{L_{bol}}{4\pi(R^2 + A^2)} \cos\theta \sim f_X \frac{L_{bol} A}{4\pi R^3} \quad , \tag{51}$$

where f_X is the fraction of the bolometric luminosity emitted by this source, θ is the angle of the light rays to the normal, and $a = 1/2$ for a point source and $a = 2/3\pi$ for a spherical source. The second approximation is valid for

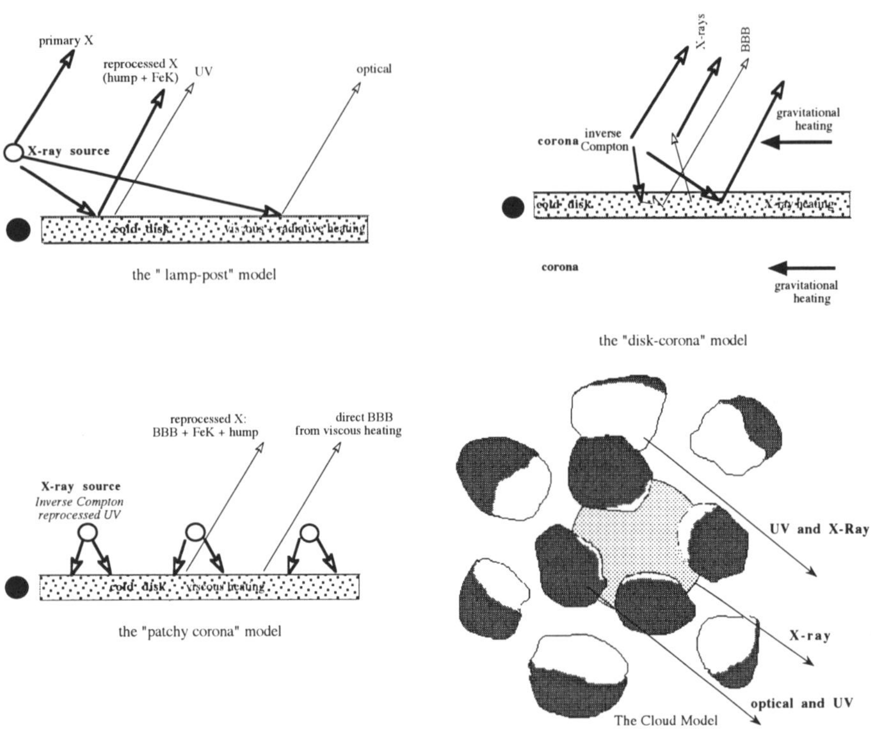

Figure 15. Sketches of several models of irradiated disks (the lamppost, the disk-corona, the patchy corona), and of the cloud model.

$R \gg A$, i.e. for the optical to near-UV emitting regions, located much further away than the X-ray source. X-ray variability studies suggest that $A \sim 10R_{\rm G}$. *Therefore, the external flux varies with the radius like the viscous flux.* This external irradiation has several effects on the disk:

- A major fraction of X-rays is absorbed by photoelectric processes in the upper layers of the atmosphere, and this creates a hot and ionized "skin" which emits the iron line.

- The remaining fraction is Compton scattered in this ionized atmosphere and gives rise to the X-ray "hump".

- The absorbed fraction is reprocessed into UV, of which about 50% is re-emitted outwards and 50% towards the interior. This latter fraction can modify the internal structure of the disk. The effect is however negligible, unless $f_X \approx 1$.

  Indeed, the condition for radiative heating to dominate the viscous heating when the disk is optically thick is[40,48]

$$F_{\rm visc}\ (1 + \frac{3}{8}\tau_{\rm R})\ \leq \frac{F_{\rm X}}{2}\ . \quad (52)$$

  Since this condition is not realized, the radial structure is not modified (unlike in the disk-corona model).

- The fraction of UV radiation re-emitted outwards is added to the emission due to the viscous flux.

The emitted spectrum has been computed using different schematizations. Except for the most recent ones[95], the models consist of a disk of constant density in the vertical direction, irradiated on the top by a power-law continuum and on the bottom by a BB (or by a Wien spectrum) to approximate the effect of viscous heating. The transfer is treated with various approximations[108,73,19,121] or with a Monte Carlo method coupled with a photoionization code[135]. An example is given in Figs. 16 and 17. Fig. 16 shows that iron is strongly ionized in the upper layers. The spectrum displayed in Fig. 17 illustrates the effects mentioned above. It is computed with the code TITAN[27], which solves the complete non-LTE transfer inside lines and continuum, with photoionization and thermal equilibrium, and is coupled to a Monte Carlo code to handle Compton diffusion. The strong UV-bump corresponds to the viscous release, to which a small ($\sim$ 10%) contribution of reflected radiation is added. It is interesting to note that the Lyman discontinuity, present here in absorption, is quite weak, though not completely absent, as in the observed spectra. In the soft X-ray range the spectrum has a slope of the order of 1.5 in flux, in agreement with the observations, and displays many lines and edges from a highly ionized medium. Such lines are observed in the Chandra spectra of NGC 3783[54], and NGC 5548[52], unfortunately strongly blended with absorption lines from the Warm Absorber, and perhaps partly produced by it. This continuum is the "reflected" spectrum, though obviously it is not due to pure reflection! In the hard X-ray range an intense iron line is produced at 6.4 keV, due to Fe XVII and less ionized species.

This model is able to solve several problems raised by the standard non-irradiated disk model, like the presence of the soft X-ray excess, the iron line

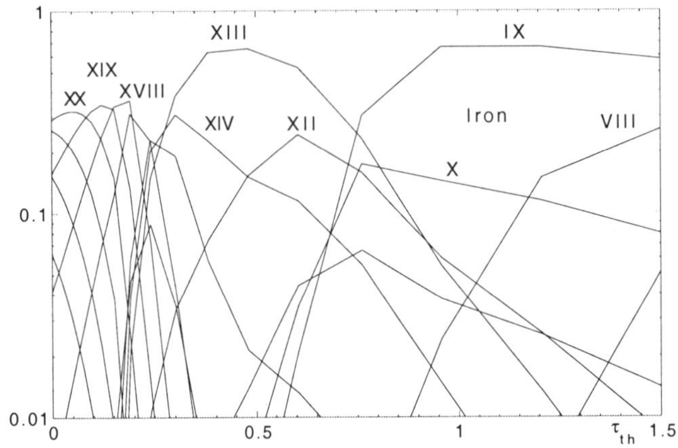

Figure 16. Ionization state of iron in the upper layers of an irradiated disk. The model is a slab of constant density 10^{14}cm^{-3}, irradiated on one side by a power-law $F_\nu \propto \nu^{-0.9}$ from 1eV to 100 keV and on the other side by a BB with $T = 10^5$K to mimic the viscous heating. The ionization parameter is $\xi = 300$, and the luminosity of the power-law is equal to the luminosity of the BB. Computations made with TITAN.

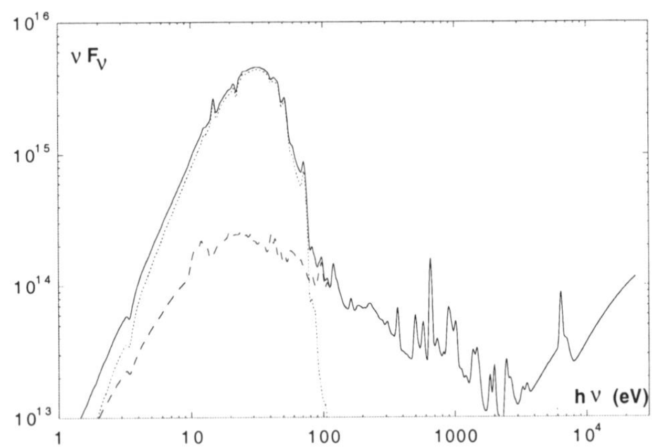

Figure 17. Emission spectrum for the same model as Fig. 16. The dotted line represents the underlying transmitted BB spectrum; the dashed line, the reflected spectrum; and the solid line, the total spectrum emitted by the disk. The spectral resolution (R) of the spectrum has been degraded to 30.

and of the "hump"; but it retains a few unsolved problems:

- Only a small fraction of the UV flux can be produced in this way as the X-ray luminosity is smaller than L(BBB). One deduces that the optical to UV emission is only slightly modified with respect to the standard disk (see also Fig. 17). Since the local BB picture holds for the optical to near-UV emission, one has $T_{\rm BB} = [(F_{\rm visc} + F_{\rm X})/\sigma]^{1/4}$, and therefore $T_{\rm BB} \propto R^{-3/4}$, except in the very central regions, where anyway the BB approximation is not valid and only EUV is emitted. Consequently to get a steep spectrum, $F_\nu \propto \nu^{-\alpha}$ with $\alpha \geq 0$ as observed in Seyfert nuclei, one should seek another geometry for the irradiation of disks.

- The disk will respond to a variation of the central source on a time-scale of $\sim 1.2\lambda_\mu^{4/3}\eta^{-1/4}f(R)^{1/4}\dot{m}^{1/4}M_8^{3/4}$ days, according to Eq. (21). Thus, rapid X-ray fluctuations will be erased by the light travel-time towards the disk, as observed. But the UV–optical light curves should always lag behind the X-ray one, while the opposite is sometimes observed, for instance in the case of NGC 7469. Also the model cannot explain the fact that the UV light curve has the same variability amplitude as the X-ray with the characteristic UV time-scale.

- The problem of the Lyman discontinuity remains, although improved with respect to non-irradiated disk.

- Finally, the model requires the presence of an ad-hoc X-ray source for which no physical emission mechanism is proposed.

8.2 What Can We Learn about the Emitting Medium from the X-ray Spectrum?: Where Compton Scattering Makes its Appearence Again

We know that the X-ray spectrum is made of a reflected part (the "cold" iron line, the hump), and of a primary spectrum. The reflection takes place on a "cold" medium, which most probably is that emitting the UV-bump. The correlated variations and the small time-lag of the UV and X-ray fluxes confirm that the two emitting media are "seeing" each other. The high energy cut-off of the primary spectrum at about 100 keV tells us that it is emitted by a hot medium with an electron temperature such that $\Theta_e = kT_e/mc_o^2 \sim 0.1$ (if the emission is mainly due to electrons, which is most likely).

Three thermal mechanisms can compete in this hot medium: free-free, cyclotron, and inverse Compton emission. The time-scale for inverse Compton cooling is smaller than the time-scale for free-free cooling when the following inequality holds

$$l = \frac{L_{\rm UV}}{R}\frac{\sigma_{\rm T}}{mc^3} \sim 10^4 \frac{L_{\rm UV}}{L_{\rm Edd}}\frac{1}{r} > 0.03\Theta_e^{-1/2} \sim 10^{-2} \quad , \tag{53}$$

where l is the "compacity" of the UV emission region.

The dimension of the UV emission region is typically $100R_{\rm G}$. One deduces that the condition is easily fulfilled, so inverse Compton cooling dominates over free-free cooling. It dominates also over cyclotron cooling unless the magnetic pressure is of the order of the density of radiation, which implies a very high value of the magnetic field. In conclusion: *the optical-UV and the X-ray emissions are probably strongly coupled through direct (in the cold medium) and inverse (in the hot medium) Compton scattering.*

In the hot Thomson thick medium there were multiple scatterings of hard photons. In the hot Thomson thin medium with soft photon input there is a small number of scatterings. This model was proposed for X-ray binaries[114].

The mean amplification per scattering is $A = 1+4\Theta_e - h\nu_0/m_ec^2$. If $\tau_{\rm T} < 1$, the probability of m scatterings is $\tau_{\rm T}^m$. After m scatters, $\nu/\nu_0 = A^m$. This creates a series of humps with decreasing intensities. One can thus approximate the Comptonized spectrum as a power-law $I(\nu) = I(\nu_0)(\nu/\nu_0)^{\ln(\tau_{\rm T})/\ln A}$. For a given energy of the soft photons, the spectral index is therefore defined by $\tau_{\rm T}$ and T_e. Moreover no photons can be scattered to an energy greater than that of the thermal electrons, and thus, one expects the power-law to be cut at $h\nu/kT_e \sim 1$.

Since the cooling of the hot gas is due to inverse Compton scattering, its temperature can be determined from an energy balance equation, provided that the heating is known, and thus one can completely determine the spectral distribution. This led to the introduction of a disk model where the release of gravitational energy occurs in a hot optically thin "corona" above the disk, through buoyancy and magnetic field reconnection[68,36,37]. If a large fraction ($f \sim 1$) of the gravitational release takes place in the corona, the disk is almost exclusively heated by the backscattered radiation coming from the corona. It reprocesses this radiation into a soft BB, and the corona upscatters this radiation while being Compton cooled by this process (cf. Fig. 15). This radiative coupling between the disk and the corona and the energy balance of the corona impose a relation between T_e and $\tau_{\rm T}$ (the smaller $\tau_{\rm T}$, the larger T_e), such that the X-ray spectral index is close to unity. Since an increase in $\tau_{\rm T}$ causes both a decrease of T_e and a steepening of the spectrum, a relation between the cut-off at high energy and the spectral index in the 2-10 keV range is predicted (cf. Fig. 18).

Since the ratio between the inverse Compton luminosity and the soft lu-

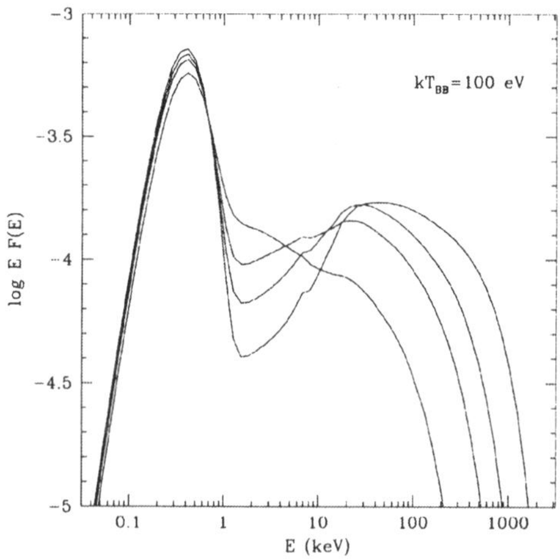

Figure 18. Comptonized spectra for different couples of (τ_T, Θ_e) [(0.63, 0.11), (0.32, 0.21), (0.2, 0.3), (0.1, 0.5)]. The soft photons correspond to a BB with kT_{BB} =100 eV. From Haardt et al.[39] ©1997 AAS

minosity is of the order of unity, the model predicts a BBB/X-ray luminosity ratio smaller than observed. This is why a variant of the disk-corona model, the "patchy corona" (cf. Fig. 15), was introduced[38]. The corona is not homogeneous, but is made of a few blobs, which could be due to the formation of magnetic loops storing energy and releasing it rapidly through reconnection like in solar flares. Below a flare, the disk is heated to a higher temperature. The rest of the disk radiates through viscous processes. An important difference with the continuous corona is that the fraction of the disk located below the "flare" is much more irradiated and consequently its upper layers are strongly modified, in particular due to the large radiative pressure.

8.3 The Two-Phase Disk

In the previous model the gravitational release takes place mainly in the corona. This implies strong structural changes of the disk, both vertically and radially, with respect to the standard model.

Disk-corona models have been studied in many papers[87,63,136,126,122,109,110]. Since the disk has no or only weak internal heat generation, it is almost isothermal and has a weak radiation pressure. It is therefore very dense and supported only by gas pressure.

Consequently *it is not subject to the viscous/thermal instability like radiation supported disks.* This is a great advantage, but on the other hand, since these disks are very dense, they are gravitationally unstable at all radii for relatively modest luminosities. Other differences with the lamppost model is that the corona pressurizes the disk, and there is conductive transport between the disk and the corona.

Generally, the corona is a two temperature plasma, where ions are heated and transfer their energy to the electrons through Coulomb collisions, and the electrons are Compton cooled. The corresponding equilibrium temperature is close to virial for the ions (10^{11}K), and of the order of 10^9K for the electrons, as already mentioned. Therefore, the gas pressure supported corona is at the limit of being geometrically thick, and sometimes it is quasi-spherical. Some people think that a collective process able to rapidly transfer energy between ions and electrons can maintain the two constituents at the same temperature.

In these models, the fraction of energy generated in the corona is a free parameter given for instance[55] by a power-law in R. By changing this fraction and the geometry of the hot region, the models can fit the observed X-ray spectra. But one should search for a physical basis for the choice.

Some models take into account mechanical transport only in the radial direction, and others add the vertical transport by conduction or evaporation between the disk and the corona. Models with the local corona strength, calculated on the basis of the disk evaporation rate, tend to predict a relatively weak corona above a disk, with possible total disk disruption in its innermost part[110,76]. These results are supported by 2D numerical time-dependent computations[44]. However, these models do not predict strong the soft X-ray excesses present in composite spectra[65], as Comptonization in the corona is too weak and the emission from the innermost single phase flow arises in the hard X-ray part of the spectrum.

8.4 The Iron Line: a Proof for Irradiated Cold Relativistic Disks?

When an X-ray photon is absorbed to eject one of the two K-shell electrons of an iron ion, this photoelectric process is followed by the de-excitation of an electron from the L-shell, and then, either by the emission of a "fluorescent" K-photon, or by a subsequent Auger ionization (escape of an electron from an upper shell without emission of radiation). The probability of these two processes is of the same order and depends slightly on the ionization state of the ion. The energy of the fluorescent photon is almost constant and equal to 6.4 keV for Fe I to XVII, then it increases to 6.9 keV for the H-like Fe XXVI. Actually this transition and the one corresponding to the He-like Fe XXV are not "fluorescent" but simply normal resonance Lα transitions. Besides, from Li-like (Fe XXIV) to H-like iron, the Auger effect cannot occur, owing to the lack of two electrons in the L-shell. The absorption threshold is 7.1 keV for photoionization of a neutral atom, and it increases to 8 keV for Fe XVII, to reach 9.3 keV for Fe XXV.

The ionization state of iron depends on the ionization parameter ξ. For constant gas density one finds[73,74]:

for $\xi \leq 100$: a "cold" line is produced at 6.4 keV by ions less ionized than Fe XVII, and the iron K-shell absorption edge is small since the contribution of the other elements to the photoelectric opacity is large.

for $100 < \xi \leq 500$: iron is in the states Fe XVII to Fe XXIII. There is a vacancy is the L-shell of the ion, so Kα photons are resonantly absorbed and trapped until they disappear by the Auger effect. Only a few line photons can escape and the Fe K line is very weak. The lower-Z elements are ionized, leading to a moderate iron absorption edge.

for $500 < \xi \leq 5000$: the line photons are still subject to resonant scattering, but since they are not destroyed by the Auger effect they can escape the disk and produce an intense "hot" iron line at 6.8 keV. The absorption edge is strong.

for $\xi > 5000$: iron and the other elements are completely ionized, so there is no line and no edge.

As we shall see in the next section, *this is no longer valid for a constant pressure medium or for a disk in hydrostatic equilibrium.*

The flux of the line and the strength of the ionization edge depend on the ionization state, but also on the iron abundance, the geometry of the reprocessing medium, and, in the case of a plane-parallel slab, on the inclination of the line of sight and the illuminating source itself. We shall not give here more details about these dependences (see Fabian et al.[31]). The observed mean intensity, centroid energy, and equivalent width, lead to the conclusion that the line is produced in a relatively "cold" medium (meaning that iron is less ionized than Fe XVII), covering about 2π steradians of the central source (therefore an infinite slab irradiated by an X-ray source is an acceptable geometry).

It is widely accepted that the line profile is a proof of the formation of the line at the surface of the accretion disk, very close to the BH. The extended red wing is then due to gravitational redshift, while the blue wing is relativisticaly boosted due to the large rotational motion. As an illustration, Fig. 19 displays

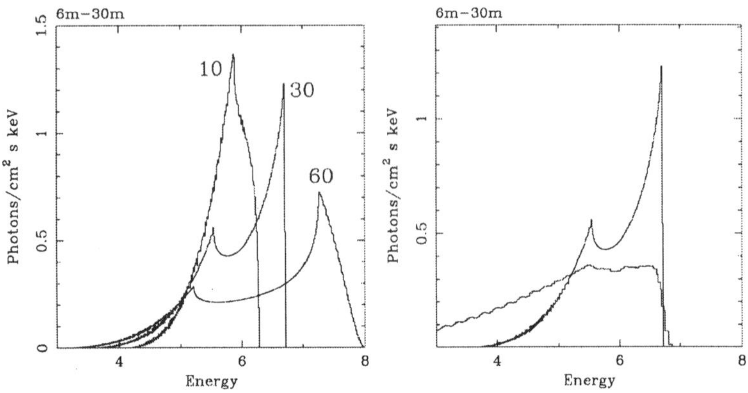

Figure 19. Profile of the iron line emitted by a disk around a Schwarzschild and a maximally rotating Kerr BH. The disk is assumed to extend down to the last stable orbit. From Fabian et al.[31], ©2000 Astr. Soc. of the Pacific, reproduced with permission of the editors.

on the left the profile of the line emitted by a disk around a Schwarzschild BH extending from 6 to $30R_{\rm G}$, as a function of the inclination of the disk. The intensity of the blue wing depends strongly on the inclination. The figure on the right compares the profile of a line formed in an accretion disk around a Schwarzschild and a Kerr BH, for an inclination of 30°. In the case of the Kerr BH, the disk is assumed to extend down to $1.25R_{\rm G}$, so the gravitational redshift is stronger and the red wing is more extended.

Fitting the profile of the line in several objects led to a number of surprising results. First the absence (or the weakness) of the blue wing corresponds to an inclination close to 30°. Second, at least in the two best observed objects, NGC 3615 and MCG-6-30-15, the disk should extend down to $2R_{\rm G}$ to account for the broad red wings. Fabian et al.[31] argued against Reynolds & Begelman[103] for a Kerr geometry, as the gas falling inside the last stable orbit on a Schwarzschild BH would be highly ionized and the spectrum would display a strong absorption edge, inconsistent with the observed data.

8.5 More Realistic Computations of the Emission Spectrum

A first attempt to compute the emission of a disk topped by a corona in a consistent way was made by Sincell & Krolik[121]. Their transfer method was already mentioned in the discussion of the standard disk. They relaxed the condition of constant density, and solved the coupled vertical structure (i.e.

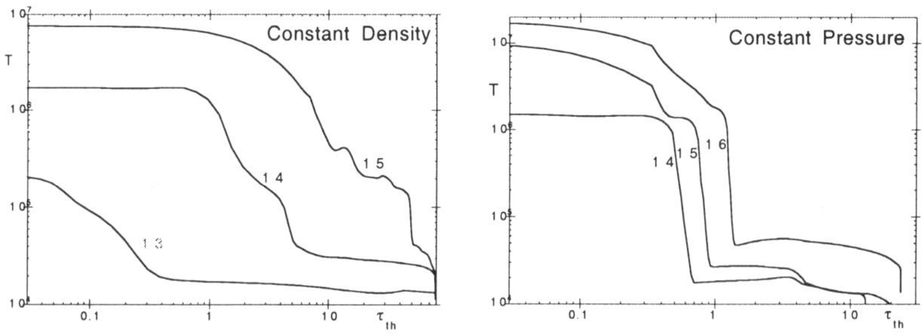

Figure 20. Comparison between the temperature structure of an irradiated layer for constant density and for constant pressure. The curves are labelled with the log of the flux incident on the layer. The constant density layer and the surface of the constant pressure layer have a density 10^{12} cm^{-3}, the incident spectrum is a power-law with a spectral index unity, from 0.1 eV to 100 keV. Computations made with TITAN.

hydrostatic equilibrium of the irradiated layer) and the radiation transfer, assuming that the energy release takes place entirely in the corona. The density in the X-ray heated skin is very high ($\sim 10^{18}$ cm^{-3}) and justifies their LTE approach. Integrating the emission over the radius, they computed the UV continuum emitted by the X-ray heated skin. Their calculations predict a strong Lyman discontinuity in emission, in strong disagreement with the observations.

Nayakshin and collaborators[95,96] also solved consistently the radiation transfer, with a complete treatment of the ionization and thermal equilibrium (not limited to H and He) and the hydrostatic equilibrium of an irradiated disk, not restricted to the case of a strongly pressurizing corona. They stressed that *hydrostatic equilibrium is extremely important for the structure of the ionized skin*, owing to the the thermal instability mechanism linked with the Ξ vs. T curve (cf. Fig. 4): when the temperature reaches the unstable part of the curve, it "jumps" to the other stable part. In contrast with the constant density case, which predicts a smooth decrease of the temperature with increasing depth inside the irradiated layer, hydrostatic equilibrium predicts a sharp transition between the hot and the cold stable phases. The same phenomenon occurs in the constant pressure case, as shown in Fig. 20. One also sees that the hot layer is thicker in the constant density case, owing to the rapid increase of density. As a consequence, for constant pressure the spectrum displays a UV-bump restricted to a one temperature BB, while for constant density the UV-bump is much more extended in frequency, and it joins smoothly the soft X-ray range (cf. Fig. 21). Fig. 22 shows that for constant density, no "cold" iron is present, unlike in the constant pressure case, where the cold layers are closer to the surface, so the soft photons emitted by these layers are able to leave the medium.

For hydrostatic equilibrium, the emissivity of the line and the spectral distribution of the continuum reprocessed in the atmosphere of an irradiated disk depends strongly on a dimensionless "gravitational parameter" A, proportional to the ratio of the gravity to the radiation pressure term[95,96,94]. Fig. 23 shows that for constant density, when the iron line is intense, it is always highly ionized. For hydrostatic equilibrium (which isi referred to as the self-consistent case), when the iron line is intense, it is a "cold" line. This might explain the predominance of "cold" iron in AGN spectra. The observed correlations are better accounted for by a strong illumination of the disk, corresponding to the "patchy" or "flaring" corona model, than by the lamppost or the continuous corona[96]. Clearly more of these self-consistent computations (unfortunately enormously time consuming) are needed to get a real physical understanding of the complex observed behavior of AGN spectra.

9 An Alternative to the "Cold Relativistic Disk"? The "Cloud Model"

The iron line and UV continuum are both due to reprocessing of the X-ray continuum, for instance in the atmosphere of an irradiated disk. Hence the regions contributing the most to the variable fraction of the UV flux should give rise to an iron line, and inversely the region emitting the iron line should produce a UV or a EUV excess of continuum emission.

We have seen that the optical-UV variable continuum is emitted at large distances from the BH (at $R \sim 100$ to $1000\ R_{\rm G}$), with the result that a fraction of the iron line is also reprocessed in this region. The iron emission will appear as a narrow line at 6.4 keV, constant on UV variation time-scales. It is not clear whether a narrow core is always observed, but we know that the broad-line stays constant on time-scales of the order of days while the underlying X-ray continuum varies. Also there seems to be a contradiction between the short time-lag of the optical and UV light curves and the relative large time-lag between the UV and the X-ray light curves. Generally speaking both the iron line and the optical-UV-X observations, in particular their variability properties, have been surprising in the context of the disk model.

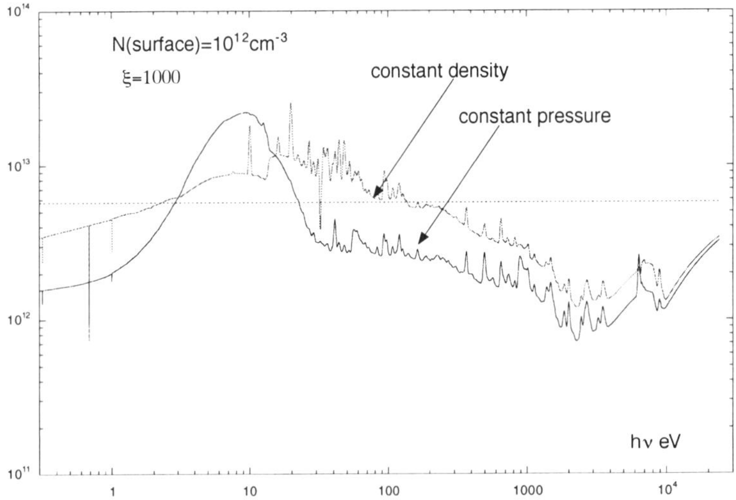

Figure 21. Comparison between the spectrum reflected by an irradiated layer for constant density and for constant pressure. Same density and incident continuum (the dotted line) as in Fig. 20.

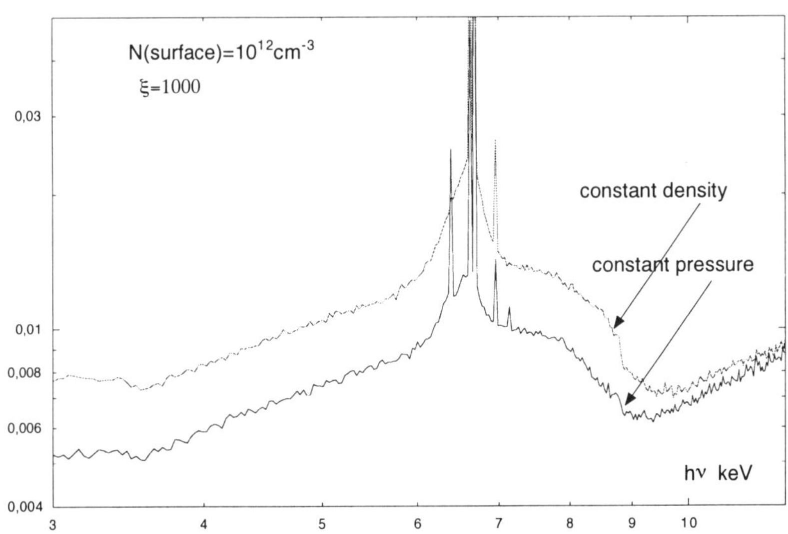

Figure 22. Zoom of Fig. 21. Computation made with TITAN and NOAR.

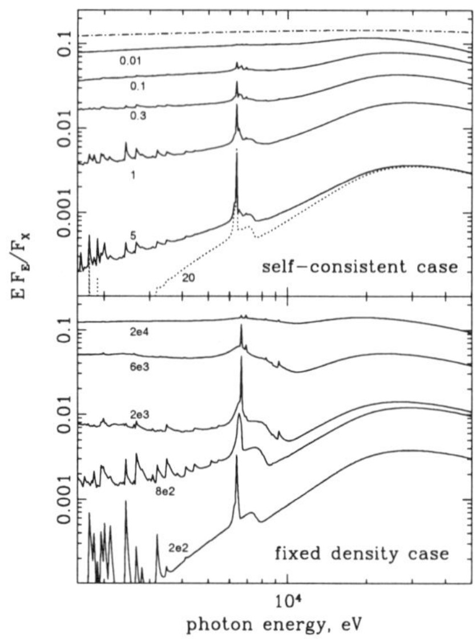

Figure 23. Comparison of the constant density case with the hydrostatic equilibrium case ("self-consistent case"). The curves are labelled with ξ in the first case and with the gravitational parameter A in the second case. One sees that the iron line is "cold" in the self-consistent case, and "hot" in the constant density case. From Nayakshin et al.[95] ©2000 AAS.

An alternative geometry has been proposed[35,12,62], where the Fe K line and the UV continuum may arise from dense clouds (optically thick for free-free absorption, but not Thomson thick) with a large covering factor embedded in a hot quasi spherical medium, rather than from a disk. A variant of this model, with Thomson thick clouds, has also been proposed[18]. Fig. 15 displays a sketch of the "cloud model". In this model Compton scattering broadens the iron line.

A physical basis for the model can be found in a new inhomogeneous equilibrium of the disk where the bulk of the mass rests in dense clumps magnetically connected in a bath of hot gas[58]. One advantage of the model is that it is thermally and viscously stable. The transport of angular momentum is insured by collisions between clumps. The requirements for dynamical and thermal equilibrium determines the parameters of the system, for example the temperature of the hot gas and the optical thickness (quite similar to that of the corona in the disk-corona model) of the clumps which are Compton thick with a covering factor close to unity.

Another possibility exists in the framework of a nuclear wind. There are several theoretical arguments for the existence of hydrodynamic winds or radiation-driven flows close to the BH. On the other hand, the fact that the UV and soft X-ray absorption lines are blueshifted is observational evidence for an outflowing medium. One can thus imagine that the quasi-spherical medium invoked in this model is simply the basis of the Warm Absorber, for instance similar to the wind envisioned by Murray & Chiang[85]. One interesting aspect of such a model is that the iron line would be easily redshifted by the Doppler effect from the position of the Fe XXV or Fe XXVI line (6.7 keV) to 6.4 keV or less, since the observer would be seeing a line emitted only from the illuminated side of the clouds.

This model is able to account for the UV-X spectrum of AGN[18,19,24]. The observed X-ray continuum is then a mixture of the primary continuum and of the continuum reflected by the illuminated side of the clouds, while the BBB is produced both by intrinsic emission of the clouds heated by the X-ray source and by reflection on the illuminated side of the clouds. Similar to the disk-corona model, the X-ray source can be the result of Compton upscatterings of the UV photons emitted by the surrounding clouds. The radius of the cold medium is $r \sim 40\dot{m} n_{14}^{-1/2} M_7^{-1/2} \xi_3^{-1/2}$, where n_{14} is expressed in units of 10^{14} cm^{-3} and ξ_3 in 10^3. Fig. 24 shows an example of the spectrum obtained for such a model.

In this closed geometry the UV photons which are not upscattered to the X-ray range by clouds in the hot medium, undergo multiple reflections which amplify the UV-bump. Consequently the clouds "see" a spectrum richer in soft photons than the observer (cf. Fig. 25). This leads to a small Compton temperature of the clouds, which is important to account for Compton downscattering of the iron line photons (cf. below).

Such a medium has the power to Compton-broaden the iron line[23]. There are actually two possibilities[84]:

- the hot medium is surrounded by a cold Compton thick medium, so the iron line and the hump are emitted by the Comptonizing medium itself (Comptonization by reflection)[3];

- the iron line is intrinsically narrow because it is emitted at $\sim 100 R_{\rm G}$ from the BH, and it is Compton broadened when crossing a surrounding ionized Compton thick medium (Comptonization by transmission)[81,82,4].

An example of the model fit of the mean AGN spectrum[89] with Comptonization by reflection is shown in Fig. 26. Two types of clouds are required:

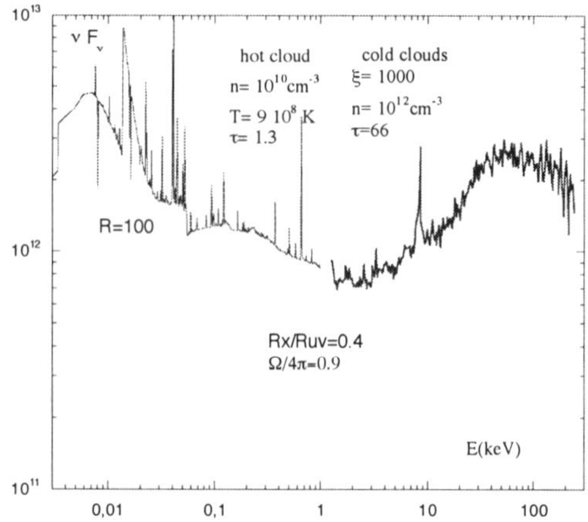

Figure 24. Spectrum emitted by a spherical shell of clouds surrounding a hot medium that upscatters the UV photons emitted by the irradiated clouds. The covering factor of the clouds is 0.9, and the ratio of the radius of the hot and cold media is 0.4. Spectrum computed with TITAN and NOAR. The noise in the right-hand side is a numerical artifact of the Monte Carlo method. The spectral resolution of the left-hand side is 100.

"cold" clouds, which emit the "cold" component of the iron line, and highly ionized clouds, which insure Comptonization of the iron line. The BBB is produced by both media[3].

An interesting aspect of Comptonization by transmission is that it can account not only for the broad red wing of the iron line, but also for the absence of a Lyman discontinuity in the UV spectrum without appealing to a fine tuning of the parameters (cf. Fig. 28).

The transmission model has been rejected[30,104] based on the size given by variability. The very high ionization state required to avoid an intense iron absorption edge implies too large a dimension, in particular in the case of MCG-6-30-15. However, assuming that the Comptonizing medium is homogeneous (the most conservative case), one deduces from Eq. (6) assuming $\xi = 10^6$ and $\tau_{\rm T} \sim 3$ a dimension of 0.03 lt-d, consistent with variability timescales. It corresponds to a distance of about $100R_{\rm G}$ if the BH in MCG-6-30-15 has a mass of a few $10^6{\rm M}_\odot$, as it is now believed (note that such a small mass implies that this object is radiating close to its Eddington luminosity, making it peculiar among Seyfert nuclei). However a serious argument against the transmission model, is that the energy cut should take place at 500 keV / τ^2, i.e. at about 50 keV, while it seems to be observed above 100 keV.

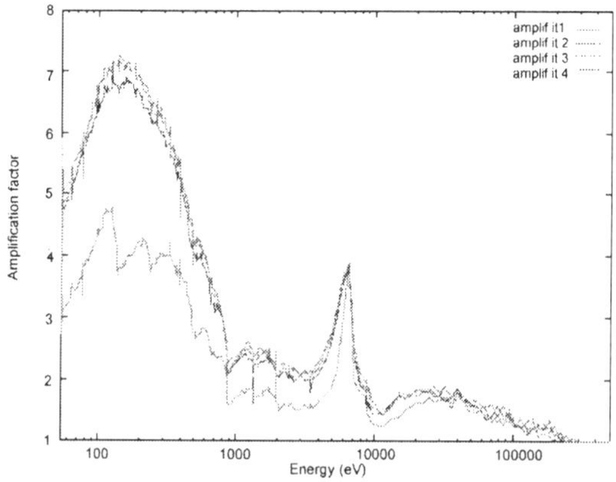

Figure 25. Amplification factor due to multiple reflections in the cloud model, assuming that the X-ray source is point-like. Starting from $\xi = 3000$, the effective primary ionization parameter illuminating the clouds reaches $\xi = 10880$. The covering factor of the clouds is 0.9, and the primary spectrum is a power-law with a spectral index equal to unity between 0.1 eV and 100 keV. From Abrassart & Dumont[3].

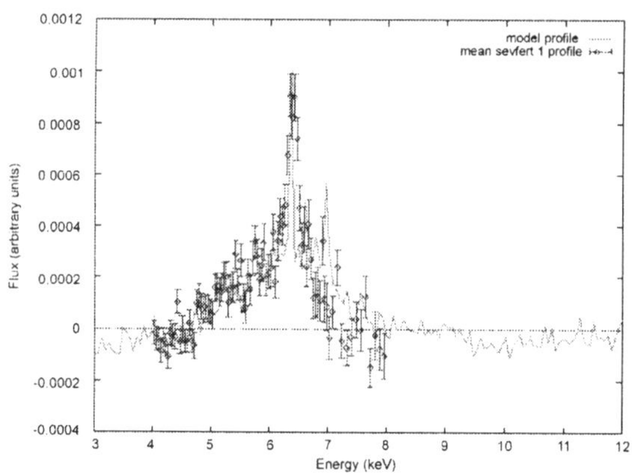

Figure 26. Fit of the mean iron line profile with the cloud model. The incident spectrum on the reprocessing shell is a power-law with unity spectral-index between 0.1 eV and 100 keV, the covering factor is 0.9, and the shell contains highly ionized ($\xi = 10000$) and nearly neutral ($\xi = 300$) clouds in equal proportion. From Abrassart & Dumont[3].

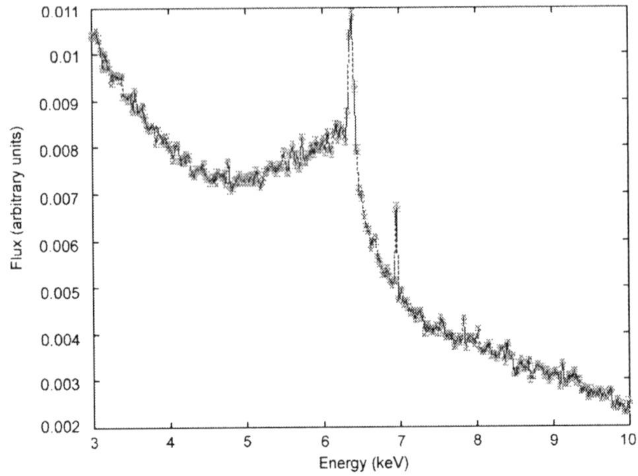

Figure 27. Profile of an intrinsic cold narrow 6.4 keV iron line, transmitted through an extended shell of Thomson depth 3, with a density varying as $1/R^2$ and $\xi = 2 \times 10^6$. The incident spectrum is a composite AGN[65]. From Abrassart & Dumont[4].

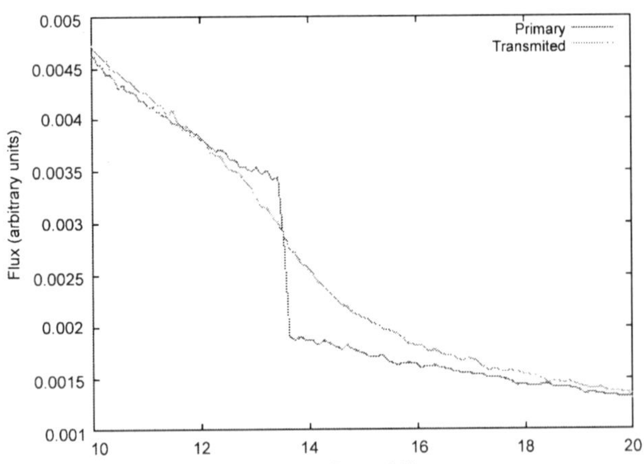

Figure 28. Same as in Fig. 27, but showing the effect on an extrinsic Lyman edge in absorption. From Abrassart & Dumont[4].

A strong argument against the reflection model is the fact that it still has not been shown able to account for a very extended red wing with no blue wing, as observed in MCG-6-30-15. However the parameter space is far from having been completely explored, as these computations are extremely time consuming.

The main difference with the disk model is that the region where the iron line is emitted has not to be located close to the BH. This is in agreement with the absence of short term variations of the line strength correlated with variations of the continuum (though the rapid variations of the profile are difficult to account for except with an ad hoc geometry). It allows also for the UV and optical emission to arise at similar distances, of the order of $100R_G$. On the other hand, in this model the clouds can randomly obscure the line of sight, thus accounting for short term variations of the primary X-ray source[5]. Another major difference with the disk model is that it also explains naturally a large L_{UV}/L_X ratio, owing to the large covering factor of the "cold" medium. Note that this model is in a sense "non local", as the cloud system can be compared to a photosphere which is reprocessing the gravitational energy that is emitted much closer to the BH.

Finally an intermediate solution has been proposed[53] where a system of quasi-spherically randomly keplerian moving clouds is located relatively close to the BH. In this scenario the iron line is broadened both by relativistic effects and by Comptonization.

10 And What Happens at the Largest and at the Smallest Accretion Rates? The Case for Thick Disks.

Up to now we have concentrated on accretion rates similar to those observed in Seyfert nuclei and modest quasars, where geometrically thin disks are a natural solution of the α-prescription. It is interesting to consider what happens at both higher and lower accretion rates, where thick disks are predicted.

In thin disks, vertical motions and radial pressure gradients are negligible, and it is therefore possible to neglect the flux advected towards the center,

$$F_{\rm adv} = \frac{\dot{M} c_{\rm s}^2}{4\pi R^2} f(R) \left(\frac{d\ln P}{d\ln R} - \frac{5}{2}\frac{d\ln T}{d\ln R} \right) \quad . \tag{54}$$

When the disk is geometrically thick, the radial velocity is comparable to the rotational velocity, and this induces strong radial pressure and temperature gradients. The advected flux must then be taken into account,

$$F_{\rm visc} = F_{\rm cool} + F_{\rm adv} \quad , \tag{55}$$

where $F_{\rm cool}$ is the heat flux radiated away. Let us consider the two extreme cases:

1. for $\dot{m} \gg$ or ~ 1: Owing to the large accretion rate, the flow is dense and optically thick, the inner regions are thus supported by the strong radiation pressure. Consequently, the disk is inflated and becomes geometrically thick. Heat advected to the BH exceeds the heat radiated away, and η decreases with $\dot{m}$ [1]. For super-Eddington accretion rates these disks have relatively small luminosities, of the order of $L_{\rm Edd}$. They radiate mainly in the soft X-ray range and it is suggested that they could account for Narrow Line Sy 1s, which probably radiate near their Eddington limit, and display large soft X-ray excess[79].

2. for $\dot{m} < \alpha^2$, i.e. typically ≤ 0.01: Owing to the small accretion rate, the flow has a low density and does not radiate efficiently. Like the coronae discussed previously, the protons remain very hot and the disk is supported by gas pressure, becoming geometrically thick and optically thin. Several cooling processes can dominate whether or not soft photons are present to induce inverse Compton cooling[50,92,2,93]. Since these disks are not efficient radiators, a large fraction of the gravitational energy will be advected towards the BH and simply "swallowed". These disks are referred to as "Advection Dominated Accretion Flows" (ADAFs). Both solutions (i.e. the standard disk and the ADAF) exist, without any understanding on how to switch between them. If the ADAF is surrounded by a cold disk the emission spectrum will depend on the location of the transition between the cold disk and the ADAF (cf. Fig. 29). In this case the hot medium constituting the ADAF can be cooled by inverse Compton scattering of the soft photons produced by the cold disk. If a nuclear radio source is present, the synchrotron photons can play the same role in cooling the hot gas.

Much recent attention has been devoted these last years to ADAFs. They are generally thought to be present in low luminosity objects, such as LINERs and the nuclei of elliptical galaxies (M 87[102]), the nucleus of NGC 4258[66], and the Galactic Center[77]. All show a non-resolved X-ray source but no optical/UV continuum, as one expects from an ADAF. The theoretical possibility of an ADAF is a matter of current debate, and observationally it is controversed (in particular for Sgr A*[6]).

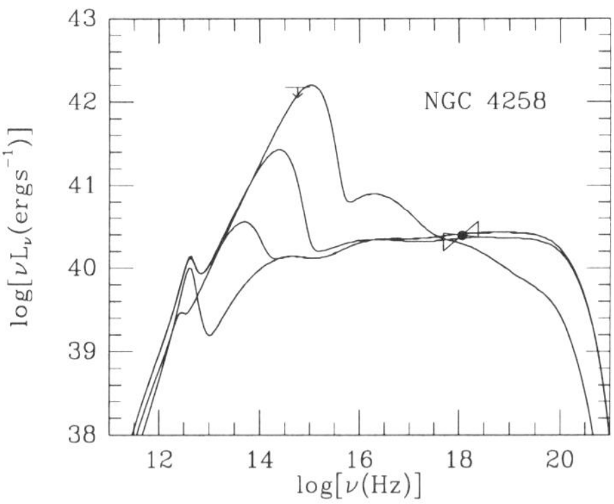

Figure 29. Spectrum of an ADAF with a cold disk. The 4 curves correspond to different positions of the transition between the ADAF and the cold disk: infinity (no cold disk, pure flat ADAF spectrum), 1000 R_G, $100R_G$, $10R_G$. The importance of the blue bump and its peak frequency increase when the transition radius decreases. Courtesy J.-P. Lasota.

11 A "No Man's Land": the Disk between 10^3 and 10^6 R_G

We have seen that the disks become gravitationally unstable beyond a few $10^3 R_G$. Moreover, the viscous time-scale given by the α-prescription for transporting the gas towards the BH at such distances is larger than the lifetime of the active nucleus. We know that further from the center, at about 10^6 R_G, the disk is globally gravitationally unstable and the supply of gas can be achieved by gravitational torques (cf. F.Combes in these lecture notes). However we know nothing about the structure of the disk in the intermediate region and how the outward transport of angular momentum is ensured.

Several solutions have been proposed. A magnetic field anchored in the disk can transport angular momentum non-locally, and possibly prevent gravitational instability. Another solution consists of a disk made from marginally unstable and randomly moving clouds with large bulk velocities, where the transfer of angular momentum is provided by cloud collisions[11]. A more elaborate model is proposed for NGC 1068[61] in which a disk of gas clumps undergoes mutual gravitational interactions. These disks would be geometrically thick, and are actually similar to the tori invoked in the Unified Scheme.

The observations of purely keplerian velocities in masers lying at about 10^5 R_G from the BH in NGC 4258[97] argue strongly for a geometrically thin warped disk. Collin & Zahn[20] have therefore adopted the view of a marginally stable disk where unstable clumps collapse until protostars are formed. (However, such a picture is only well suited for low Eddington ratios. For large Eddington ratios the number of newly formed stars is high, and the disk will become highly turbulent and inhomogeneous.) Stars embedded in an accretion disk should accrete at a high rate and rapidly acquire masses of a few tens of $M_\odot$, leading to a "starburst" of massive stars (not comparable in size with real starbursts)[9]. The mass transport can thus be driven by supernovae or by stellar outflows, which at the same time can contribute to the heavy element enrichment observed in the BLRs at high-z and in BALs. A larger scale starburst could also be induced in the shocked gas of supernova remnants. Several pieces of evidence have been discussed in these lectures for the existence of starbursts in the very central region of Seyfert galaxies.

Finally Ostriker[98] made a very interesting suggestion that has never been seriously considered. He stated that if the disk is embedded in a massive stellar system, interactions between the stars and the disk can remove angular momentum from the disk at the rate expected for an AGN.

12 Conclusion

It is not easy to conclude such a chapter because the results are really difficult to summarize. We have seen that the tremendous increase of data with the advent of X-ray missions such as ASCA, RXTE, and now Chandra and XMM, has considerably complicated the problem, and now precludes any simple interpretation. We have seen that the most detailed models, used to make observational predictions, are still based on crude approximations, such as geometrically thin, stationary and axisymmetric disks. For instance the models are generally assumed to be steady-state, despite the evidence that there exist strong variations on small time-scales. The viscosity is currently parametrized by the α-prescription, but other prescriptions are not excluded. There are heated debates about the value of α, whether it is constant or not, whether the stress tensor is proportional to the total pressure or to the gas pressure. We don't know the influence of the magnetic field. We have no idea of the geometry of the inner regions. We know that there is a quasi-spherical hot accretion flow, but at the same time we believe that the cold thin disk extends down to a few gravitational radii from the BH, and in this case we know that it should be unstable. The status of the hot flow is not clear yet. In short, one could say that the more sophisticated the models, the larger the uncertainties! To continue to make progress, we must obtain very good new observational data, and at the same time develop the basic physics and tools for the modeling. The field has still a promising future in the third millenium!

Acknowledgments

I am grateful for enlighting discussions with B. Czerny, A-M. Dumont, J-M. Huré, M. Mouchet, and J-P. Zahn and to I. Aretxaga for a careful reading of the manuscript, leading to many clarifications and improvements.

References

1. Abramowicz, M.A., Calvani, M., Nobili, L., ApJ, 242, 772, 1980.
2. Abramowicz, M.A., Chen,X., Kato, S., Lasota, J.-P., ApJ, 438, L37, 1995.
3. Abrassart, A., Dumont, A.-M., Proc. of the First XMM workshop on "Science with XMM", M. Dahlem (ed.), 1998.
4. Abrassart, A., Dumont, A.-M., Proceedings of the Bologna Conference on "X-ray Astronomy, Stellar Endpoints, AGNs and the Diffuse X-ray Background" (Astroph. Lett. and Comm.), 1999.
5. Abrassart, A., Czerny, B., A&A, 356, 475, 2000.
6. Agol, E., ApJL, 538, L121, 2000.
7. Alloin, D., Pelat, D., Phillips, M. & Whittle, M., ApJ, 288, 205, 1985.
8. Antonucci, R. R. J., Miller, J.S., ApJ, 297, 621, 1985.
9. Artymowicz, P., Lin, D.N.C., Wampler, E.J., ApJ, 409, 592, 1993.
10. Balbus, S.A., Hawley, J.F., ApJ, 376, 214, 1991.
11. Begelman, M.C., Frank, J., Shlosman, I., in "Theory of Accretion disks", eds. F. Meyer et al., Kluwer Academic Publishers, 1989.
12. Celotti, A., Fabian, A.C., Rees, M.J., MNRAS, 255, 419 , 1992.
13. Chandrasekhar, S., Proc. Nat. Acad. Sci., 46, 53, 1960.
14. Chiang J., et al., ApJ, 528, 292, 2000.
15. Clarke, C.J., Shields, G.A., ApJ, 338, 32, 1989.
16. Collier, S.J., et al., ApJ, 500, 162 , 1998.
17. Collin-Souffrin, S., A&A, 249, 344, 1991.
18. Collin-Souffrin, S., Czerny, B., Dumont, A.-M., Zycki, P., A&A, 314, 393, 1996.
19. Collin-Souffrin, S., Dumont, A.-M., "Theory of Accretion Disks - 3", Eds E. Meyer-Hofmeister & H. Spruit, Springer, 1997.
20. Collin, S., Zahn, J-P., A&A, 344, 433, 1999.
21. Courvoisier, T., Clavel, J., A&A, 288, 389, 1991.
22. Czerny, B., Elvis, M., ApJ, 321, 305, 1987.
23. Czerny, B. & Zbyszewska, M., MNRAS , 249, 634, 1991.
24. Czerny, B., Dumont, A.-M., A&A, 338, 386, 1998.
25. Done, C., et al., MNRAS, 243, 713, 1990.
26. Doerrer, T., Riffert, H., Staubert, R., Ruder, H., A&A, 311, 69 , 1996.

27. Dumont, A-M., Abrassart, A., Collin, S., A&A, 357, 823 , 2000.
28. Edelson, R., Alexander, T., Crenshaw, D.M., ApJ, 470, 364, 1996.
29. Edelson, R., et al., ApJ, 534, 180, 2000.
30. Fabian, A.C, et al. 1995, MNRAS, 277, L11, 1996.
31. Fabian A.C., Iwasawa K., Reynolds C.S., Young A.J., PASP, 112, 1145, 2000.
32. Francis, P.J., et al., ApJ, 373, 465, 1991.
33. Frank J., King A.R., Raine D.J., "Accretion power in astrophysics", 2nd Ed., Cambridge University Press, 1992.
34. Goldreich P., Lynden-Bell D., MNRAS, 130, 97, 1965.
35. Guilbert, P.W., Rees, M.J. 1988, MNRAS, 233, 475, 1965.
36. Haardt, F., Maraschi, L., ApJ, 380, L51, 1991.
37. Haardt, F., Maraschi, L., ApJ, 413, 507, 1993.
38. Haardt, F., Maraschi, L., Ghisellini, G., ApJ, 432, L95, 1994.
39. Haardt, F., Maraschi, L., Ghisellini, G., ApJ, 476, 620, 1997.
40. Hubeny, I., ApJ, 351, 632, 1990.
41. Hubeny, I., Hubeny, V., ApJL, 484, L37, 1997.
42. Hubeny, I., Hubeny, V., ApJ, 505, 558, 1998.
43. Hubeny, I., Agol, E., Blaes, O., Krolik, J.H., ApJ, 533, 710, 2000.
44. Hujeirat, A., Camenzind, M., A&A, 361, 53, 2000.
45. Huré J-M., PhD thesis, University Paris 7, 1994.
46. Hure, J., A&A, 337, 625, 1998.
47. Huré, J-M., A&A, 358, 378, 2000.
48. Huré, J-M., Collin-Souffrin S., Le Bourlot J., Pineau des Forets G., A&A 290, 19, 1994.
49. Huré, J-M., Galliano , F., A&A, in press
50. Ichimaru, S., ApJ, 214, 840, 1977.
51. Iwasawa, K., Fabian, A.C., Reynolds, C.S., MNRAS, 282, 1038, 1996.
52. Kaastra, J.S., Mewe, R., Liedahl, D.A., Komossa, S., Brinkman, A.C., A&A, 354, L83, 2000.
53. Karas, V., Czerny, B., Abrassart, A., Abramowicz, M.A., MNRAS, 318, 547, 2000.
54. Kaspi, S., Brandt, W. N., Netzer, H., Sambruna, R., Chartas, G., Garmire, G.P. & Nousek, J. A. ApJL, 535, L17, 2000.
55. Kawaguchi, T., Shimura, T., Mineshige, S., ApJ, in press (astro-ph/0008404)
56. Kolykhalov, P. I., Sunyaev, R.A., High Energy Astrophysics and Cosmology, 249, 1984.
57. Koratkar A., Blaes O., PASP, 111, 1, 1999.
58. Krolik, J.H., ApJL, 498, L13, 1998.

59. Krolik, J.H., "Active galactic nuclei : from the central black hole to the galactic environment", Princeton University Press, 1999.
60. Krolik, J. H., McKee, C.F., Tarter, C.B., ApJ, 249, 422, 1981.
61. Kumar P., ApJ 519, 589, 1999.
62. Kuncic, Z., Celotti, A., Rees, M.J., MNRAS, 284, 717, 1997.
63. Kusunose, M. & Mineshige, S., ApJ, 423, 600, 1994.
64. Laor, A., Netzer, H., MNRAS, 238, 897, 1989.
65. Laor, A., Fiore, F., Elvis, M., Wilkes, B.J., McDowell, J.C., ApJ, 477, 93, 1997.
66. Lasota, J-P., Abramowicz, M. A., Chen, X., Krolik, J., Narayan, R., Yi, I., ApJ, 462, 142, 1996.
67. Lee, J.C., Fabian, A.C., Reynolds, C.S., Brandt, W.N., Iwasawa, K., MNRAS, 318, 857, 2000.
68. Liang, E.P.T. 1977, ApJL, 211, L67, 2000.
69. Lynden-Bell, D., Pringle, J.E., MNRAS, 168, 603 , 1974.
70. Malkan M.A., ApJ, 268, 582, 1983.
71. Malkan, M.A., Sargent, W.C.W., ApJ, 254, 122, 1982.
72. Mathews, W.G., Ferland, G.J., ApJ, 323, 456, 1987.
73. Matt, G., Fabian, A.C., Ross, R.R., MNRAS, 262, 179, 1993.
74. Matt, G., Fabian, A.C., Ross, R.R., MNRAS, 278, 1111 , 1996.
75. Meyer, F., Meyer-Hofmeister, E., A&A, 287, 559, 1994.
76. Meyer, F., Liu, B.F., Meyer-Hofmeister, E., A&A, 361, 175 , 2000.
77. Mezger, P. G., Duschl, W. J., Zylka, R., A&AR, 7, 289, 1996.
78. Mineshige, S., Shields, G.A., ApJ, 351, 47, 1990.
79. Mineshige, S., Kawaguchi, T., Takeuchi, M., Hayashida, K., PASJ, 52, 499 , 2000.
80. Misner, C. W., Thorne, K.S., Wheeler, J.A., San Francisco: W.H. Freeman and Co., 1973.
81. Misra, R., Kembhavi, A.K., ApJ, 499, 205, 1998.
82. Misra, R., Sutaria, F.K., ApJ, 517, 661, 1999.
83. Molendi, S., Maraschi, L., Stella, L., MNRAS, 255, 27, 1992.
84. Mouchet, M., Abrassart, A., Porquet, D., Dumont, A-M., Collin, S, to appear in "AGN in their Cosmic Environment", Eds. B. Rocca-Volmerange & H. Sol, EDPS Conf. Series in A&A, 2000.
85. Murray, N., Chiang, J., ApJ, 454, L105, 1995.
86. Mushotzky, R.F., Done, C., Pounds, K.A., ARAA, 31, 717, 1993.
87. Nakamura, K., Osaki, Y., PASJ, 45, 775, 1993.
88. Nandra, K., Large Scale Structure in the X-ray Universe, Proceedings of the 20-22 September 1999 Workshop, Santorini, Greece, eds. Plionis, M. & Georgantopoulos, I., Atlantisciences, Paris, France, p.249, 249, 2000.

89. Nandra, K., George, I.M., Mushotzky, R.F., Turner, T.J., Yaqoob, T., ApJ, 477, 602, 1997.
90. Nandra, K., et al., ApJ, 505, 594, 1998.
91. Nandra K. et al., ApJ, in press (astro-ph/0006339), 2000.
92. Narayan,R., Yi, I., ApJ, 452, 71, 1995.
93. Narayan, R., Mahadevan, R., Grindlay, J.E., Popham, R.G., Gammie, C., ApJ, 492, 554, 1998.
94. Nayakshin, S., ApJ, 534, 718, 2000.
95. Nayakshin, S., Kazanas, D., Kallman, T.R., ApJ, 537, 833, 2000.
96. Nayakshin, S., Kallman, T.R., ApJ, submitted (astro-ph/0005597).
97. Neufeld D.A., Maloney P.R., ApJL, 447, L17, 1995.
98. Ostriker, J.P., ApJ, 273, 99, 1983.
99. Pounds, K.A., Nandra, K., Stewart, G.C., George, I.M., Fabian, A.C., Nat, 344, 132, 1990.
100. Rees M.J., ARAA, 22, 471, 1984.
101. Reynolds, C.S., ApJ, 533, 811, 2000.
102. Reynolds, C. S., di Matteo, T., Fabian, A. C., Hwang, U., Canizares, C. R., MNRAS, 283, L111, 1996.
103. Reynolds, C.S., Begelman, M.C., ApJ, 488, 109, 1997.
104. Reynolds, C.S., Wilms, J., ApJ, 533, 821, 2000.
105. Richard, D., Zahn, J., A&A, 347, 734, 1999.
106. Rokaki, E., Collin-Souffrin, S., Magnan, C., A&A, 272, 8, 1993.
107. Ross, R. R., Fabian, A.C., Mineshige, S., MNRAS, 258, 189, 1992.
108. Ross, R.R. & Fabian, A.C. MNRAS, 261, 74, 1993.
109. Różańska, A., Czerny, B., Życki, P.T., Pojmański, G., MNRAS, 305, 481, 1999.
110. Różańska, A., Czerny, B., A&A, 360, 1170, 2000.
111. Rybicki, G.B., Lightman, A.P., New York, Wiley-Interscience, 1979.
112. Sanders, D.B., Phinney, E.S., Neugebauer, G., Soifer, B.T., Matthews, K., ApJ, 347, 29, 1989.
113. Shakura, N.I., Sunyaev, R.A., A&A, 24, 337, 1973.
114. Shapiro, S.L., Lightman, A.P., Eardley, D.M., ApJ, 204, 187, 1976.
115. Shields, G.A., Nat, 272, 423, 1978.
116. Shields, G.A., Coleman, H.H., Theory of Accretion Disks - 2, Kluwer Ed., 1994.
117. Shimura, T., Takahara, F., ApJ, 419, 78, 1993.
118. Shimura, T., Takahara, F., ApJ, 440, 610, 1995.
119. Shlosman, I., Begelman, M.C., ApJ, 341, 685, 1989.
120. Siemiginowska, A., Czerny, B., Kostyunin, V., ApJ, 458, 491, 1996.
121. Sincell, M.W. & Krolik, J.H., ApJ, 476, 605, 1997.

122. Sincell, M.W. & Krolik, J.H., ApJ, 496, 737, 1998.
123. Störzer, H., A&A, 271, 25, 1993.
124. Störzer, H., Hauschildt, P.H., Allard, F., ApJL, 437, L91, 1994.
125. Sun, W., Malkan, M. A., in "Astrophysical Jets and their Engines", 125, 1987.
126. Svensson, R., Zdziarski, A.A., ApJ, 436, 599, 1994.
127. Toomre A., ApJ 139, 1217, 1964.
128. Wandel, A., Petrosian V. ApJ 319, L11, 1988.
129. Wanders, I., et al., ApJS, 113, 69, 1997.
130. Weaver, K.A., Gelbord, J., Yaqoob, T., ApJ, in press (astro-ph/0008522).
131. Weinberg, S., "Gravitation and cosmology: Principles and applications of the general theory of relativity", New York: Wiley , 1972.
132. Zdziarski A.A., Lubinski P., Smith D.A., MNRAS, 303, L11, 1999.
133. Zdziarski A.A., Lubinski P., MNRAS, submitted (astro-ph/0009017).
134. Zheng, W., Kriss, G.A., Telfer, R.C., Grimes, J.P., Davidsen, A.F., ApJ, 475, 469, 1997.
135. Zycki, P.T., Krolik, J.H., Zdziarski, A.A., Kallman, T.R., ApJ, 437, 597, 1994.
136. Zycki, P.T., Collin-Souffrin, S., Czerny, B., MNRAS, 277, 70, 1995.

Top: Opening of the Advanced Lectures.
Bottom: Francoise Combes, during one of her lectures.

FUELING THE AGN

F. COMBES
Observatoire de Paris, DEMIRM, 61 Av. de l'Observatoire, F-75 014, Paris, France
E-mail: francoise.combes@obspm.fr

Active Galactic Nuclei are fueled from material (gas or stars) that are in general far away from the gravitational influence of the central black hole, the engine thought to be responsible for their activity. The required material has a lot of angular momentum that, a priori, is quite difficult to evacuate. The various dynamical mechanisms that may play a role in this game are reviewed, including $m = 2$ perturbations (bars and spirals), $m = 1$ perturbations (spirals, warps, lopsidedness), and tidal interactions between galaxies and mergers. In the latest stages of the merger, a binary black hole could be formed, and its influence on the dynamics and fueling is discussed. Starbursts are often associated with AGN, and the nature of their particular connection, and their role in the nuclear fueling is described. Evolution of the fueling efficiency with redshift is addressed.

1 Introduction

Some galaxies are dubbed "active", which means that they radiate much more energy than can be sustained by a normal evolution rate, or average gas consumption rate, given their size and gas reservoir mass. Their activity could be due to a starburst, or to a material-accreting compact nucleus. Both are often related or simultaneous. The activity has a short duration with respect to the galaxy life-time, but can be recurrent. A key issue is to understand the mechanisms that trigger the activity, and how a large amount of material can be funneled to the central regions, to fuel this activity.

A first possibility is provided by the dense star clusters that are present near the nucleus. These nearby stars can provide gas fuel to the nucleus through their mass loss rates, or even the tidal disruption of stars can liberate their whole gas mass. We will present and discuss the various physical phenomena involved, and their respective time-scales in Sec. 2: only those stars that have low angular momentum orbits are available to fuel the activity, but these can be replenished through dynamical diffusion when depleted. A bigger problem arises when the nearby stars are not numerous enough to sustain the activity: mass has then to come from the whole galaxy in a short time-scale, and this rises the problem of angular momentum transfer, that must involve internal or externally-triggered gravitational instabilities. Note that the mass fueling can occur in two steps: a first instability drives the gas towards the center, giving rise to a nuclear starburst and the formation of dense nuclear star clusters. Then, the active nucleus can be fueled by the evolution of the dense star cluster.

The gravitational instabilities of a galaxy disk are described in the next sections. Two-fluid instabilities are considered, and the critical role of gas is emphasized (Sec. 6). Because of its dissipative character, the gas can cool down as soon as instabilities heat it, and maintain non-axisymmetric features like spiral structure. The corresponding gravity torques are the tool to transfer angular momentum. The most wide-spread instabilities are the $m = 2$ spirals and bars: their formation mechanisms, their family of orbits, their resonances, etc... are essential to better understand the dynamical gas fueling (Sec. 4). It will be shown how bars can be destroyed, or re-born, how bars within bars develop, and how material could be driven towards the nucleus more efficiently through a hierarchy of bars.

Also very frequent are the $m = 1$ instabilities, lopsidedness and off-centring. Several possible mechanisms will be described for these instabilities, that can also favor material infall to the center (Sec. 5). These include counter-rotation, warps or peculiar gaseous instabilities in a near-keplerian disk, a situation common in the neighborhood of central supermassive black holes.

Finally, external triggers through galaxy interactions and mergers should also be emphasized. Observational evidence of internal and external fueling will be discussed and compared. The role of gas and bars in minor and major mergers is investigated through numerical simulations. These external mechanisms are favored by hierarchical scenarios of galaxy formation, where massive galaxies are the result of the merging of several smaller entities. In this frame, it is expected that the central black hole mass increases in parallel to the mass of the galaxy, as observations seem to confirm.

One consequence of galaxy merging is the possibility of binary black hole formation. The evolution and history of the black hole population will be investigated through the main dynamical mechanisms (Sec. 8), and implications on the fueling will be explored. The Starburst-AGN connection (competition or collaboration) is then described, in a cosmological perspective.

2 AGN Fueling through Dense Nuclear Star Clusters

2.1 Fueling: Why is there a Problem?

To estimate more quantitatively the fueling problem, let us consider the typical luminosity and power of an active nucleus, that can be of the order or higher than 10^{46} erg/s. If we assume a mass-to-energy conversion efficiency $\epsilon \sim 10\%$ ($L = dM/dt\, c^2 \epsilon$), then the mass accretion rate should be $dM/dt \sim 1.7\ (0.1/\epsilon)\ (L/10^{46}\ \mathrm{erg/s})\ M_\odot/\mathrm{yr}$. If the duty cycle of the AGN is of the order of 10^8 yr, then a total mass up to $2 \times 10^8\ M_\odot$ should be available. It is a significant fraction of the gas content of a typical galaxy, like the Milky Way! The time-scale to drive such a large mass to the center is likely to be larger than 1 Gyr.

For the mass to infall to the center, it must lose its angular momentum. Could this be due to viscous torques? In a geometrically thin accretion disk, one can consider the gas subsonic viscosity, where the viscous stress is modeled proportional (α) to the internal pressure, with a factor $\alpha < 1$. This can only gather in 1 Gyr the gas within 4α pc typically[160,140]. This shows that viscous torques will not couple the large-scale galaxy to the nucleus, only the very nuclear regions could play a role through viscous torques.

2.2 Are Stars a Possible Fuel?

The stars themselves could provide gas to the nucleus, through their mass loss, if there is a local stellar cluster, dense and compact enough (core radius R_c of less than a pc, core mass M_{core} of the order of $10^8\ M_\odot$). However, the mass loss rate derived from normal stellar evolution gives only $10^{-11} M_\odot/\mathrm{yr}/M_\odot$, orders of magnitude below the required rate of a few $M_\odot$ /yr. The contribution will be significant, only if a massive stellar cluster ($4 \times 10^9\ M_\odot$) has just formed through a starburst[134]. A coeval cluster can liberate $10^9 M_\odot$ on 10^8 yr, since mainly massive stars evolve together in the beginning. Thus the existence of a starburst in the first place solves also the problem of the AGN fueling, as in the symbiotic model of Williams et al.[186]. The angular momentum problem is now reported to the starburst fueling, and could be solved only through large-scale dynamical processes.

There are several processes to fuel gas to the black hole, directly from the stars, and these could work for the low-luminosity end of AGNs; one can invoke:

- bloated stars, a phenomenon that makes mass loss more efficient [51,2,3];
- tidal disruption of stars[78,58];
- star collisions [167,27,39,145].

To quantify all these processes, it is important now to define the corresponding characteristic radii.

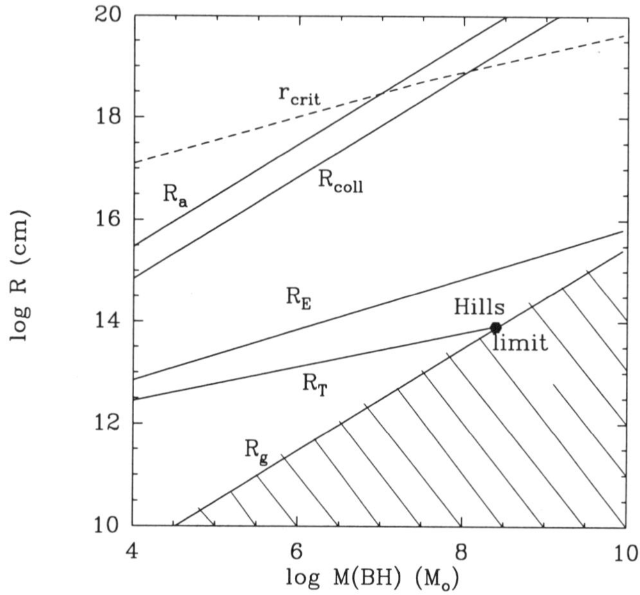

Figure 1. Characteristic radii, corresponding to the various physical phenomena, as a function of black hole mass: from top to bottom, R_a, the accretion radius, R_{coll}, the collision radius, R_E, the Eddington radius, R_T, the tidal radius, R_g, the gravitational radius (see text for definitions). Loss-cone effects are important inside the critical radius r_{crit} (see Sec. 2.4).

2.3 Characteristic Radii

First, it is obvious to define the *"accretion radius"* R_a, or the radius under which the black hole of mass $M_\bullet$ dominates the dynamics, as a function of the velocity dispersion V_∞ of the stellar component around it,

$$R_a = GM_\bullet/V_\infty^2 \sim 3 \times 10^{19}\text{cm } M_8(200/V_\infty)^2 \ , \tag{1}$$

where M_8 is the black hole mass in units of $10^8 M_\odot$. The radii[58,112] are plotted in Fig. 1.

Then the *"collision radius"* R_{coll}, is the radius under which stellar collisions are disruptive, i.e. the free-fall velocity around the black hole

$(GM_\bullet/r)^{1/2}$ is comparable to the escape speed v_* of a typical individual star $(GM_*/r_*)^{1/2}$. For solar mass stars (escape velocity of the order of 500 km/s)

$$R_{\rm coll} \sim 7 \times 10^{18} {\rm cm}\ M_8 \quad . \tag{2}$$

The *"Eddington radius"* is the radius under which a star receives more light than its Eddington luminosity,

$$R_{\rm E} \sim R_*(M/M_*)^{1/2} \sim 7 \times 10^{14} {\rm cm}\ M_8^{1/2} \quad , \tag{3}$$

for solar mass stars. The radiation pressure can then disrupt the envelope, or at least form bloated stars, more fragile with respect to mass loss.

The *"tidal radius"* has a great importance, it is the radius under which a star is disrupted by the tidal forces of the black hole (calculated like a Roche radius),

$$R_{\rm T} \sim R_*(M/M_*)^{1/3} \sim 6 \times 10^{13} {\rm cm}\ M_8^{1/3} \rho_*^{-1/3} \quad , \tag{4}$$

where ρ_* is the average density of solar mass stars. From that, we can estimate the accretion rate due to tidal disruption of stars, by integrating the mass available (in $\rho_{\rm core} R_{\rm T}^3$), divided by the dynamical time, in $\rho_{\rm core}^{-1/2}$,

$$dM/dt\,({\rm tide}) \sim 0.3\, M_\odot/{\rm yr}\ (M_{\rm core}/10^8 M_\odot)^{3/2}\ (0.01\,{\rm pc}/R_{\rm c})^{9/2} M_8 \quad . \tag{5}$$

Note that the efficiency of collisions between stars becomes larger than the tidal contribution, for large compactness of the nuclear star clusters, such as their velocity dispersion $\sigma_* > v_*$:

$$dM/dt\,({\rm coll}) \sim dM/dt\,({\rm tide})(\sigma_*/v_*)^4 \quad . \tag{6}$$

Finally, let us recall the horizon radius of the black hole (or *"gravitational radius"*) under which matter cannot escape:

$$R_{\rm g} \sim 2GM_\bullet/c^2 \sim 3 \times 10^{13} {\rm cm}\, M_8 \quad . \tag{7}$$

As the black hole horizon grows faster with $M_\bullet$ than the tidal radius, there is a limit, when $M_8 \sim 3$, above which the star disruption occurs inside the black hole, and there is no gaseous release or AGN activity (but the black hole grows even more rapidly). This is the Hills limit[78].

2.4 Black Hole Growth by Star Accretion

Let us compute the time required to reach the critical mass $M_{\rm c}$ where $R_{\rm T} = R_{\rm g}$, above which stars are swallowed by the black hole without any gas radiation ($M_{\rm c} = 3 \times 10^8 M_\odot$). When a tidal breakup of a star (of mass m, radius R) occurs, the energy required is taken from the orbital energy of the star, $E_b = 3/4Gm^2/R$, then the gas coming from the disruption will have an orbit of typical semi-major axis

$$r = (2M/3m)R/(1 - 2V^2R/3GM) \quad , \tag{8}$$

$$r = 1.5\,\text{pc}\,M_8/(1 - (< V^2 >^{1/2} /535\,\text{km/s})^2) \quad . \tag{9}$$

For our own Galaxy, with a $2 \times 10^6 M_\odot$ black hole, this means a typical radius of the gas disk of 0.03 pc.

The black hole cannot swallow the gas too fast, the maximum rate occurs when it radiates at Eddington luminosity (above which the radiation pressure prevents the material to fall in). This maximum luminosity is $L_E = 3.2 \times 10^4 (M/M_\odot)\, L_\odot$. For a mass M_c, the maximum is $10^{13} L_\odot$ (close to the peak luminosity of QSOs). Then the corresponding accretion rate, assuming an efficiency of ϵ =10–20% is $(dM/dt)_E = 1.1 \times 10^{-8} (M/M_\odot) M_\odot/\text{yr}$. This implies an exponential growth of the black hole; it takes only 1.6×10^9 yr to grow from a stellar black hole of 10 $M_\odot$ to M_c, $t_E = 9.3 \times 10^7 \ln(M_c/M)$ yr . Note that this very simple scheme would lead at $z = 2.8$ to a maximum of the number of quasars. This maximum rate, however, is not realistic, since the black hole quickly gets short of fuel, as the neighboring stars (in particular at low angular momentum) are depleted. Then it is necessary to consider a growth limited by stellar density ρ_s, $dM/dt = \rho_s \sigma V$, where σ is the accretion cross-section, and V the typical stellar velocity. The corresponding time-scale to grow from M to M_c is

$$t_D = 1.7 \times 10^{15}\text{yr} \left(\frac{\rho_s}{M_\odot \text{pc}^{-3}}\right)^{-1} \left(\frac{M}{M_\odot}\right)^{-1/3} \left(1 - \left(\frac{M}{M_c}\right)^{1/3}\right) \left\langle \left(\frac{V}{\text{km/s}}\right)^2 \right\rangle^{1/2} \tag{10}$$

Typically in galactic nuclei, $\rho_s = 10^7 M_\odot/\text{pc}^3$, $< V^2 >^{1/2} = 225$ km/s. A black hole could grow up to M_c in a Hubble time, and the luminosity at the end could be of the order of 10^{46} erg/s (see Fig. 2). More detailed considerations[58,110] introduce the loss-cone effect: the angular momentum can diffuse faster than the energy (faster than a stellar relaxation time t_R). Stars with low angular momentum, or very excentric orbits, will be swallowed first. Since the low angular momentum stars are replenished faster, the loss-cone effect increases the accretion rate by $t_l = t_R(1 - e^2)$, with e the excentricity of the orbits. This is significant inside a critical radius r_{crit}, where the loss-cone angle becomes larger than the diffusion angle $\theta_D \sim (t_{dyn}/t_R)^{1/2}$. This critical radius is also plotted in Fig. 1.

More detailed considerations also can change the above scenario, for instance when a mass spectrum for the stars is taken into account. The critical

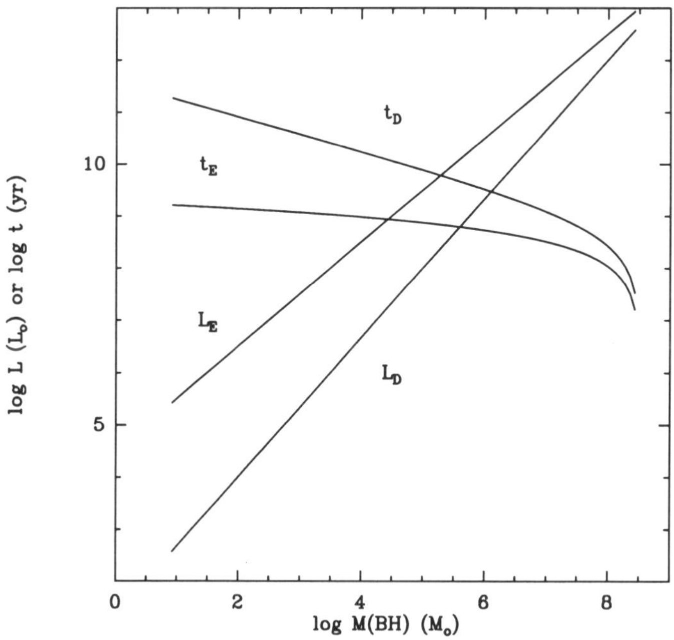

Figure 2. Growth of a supermassive black hole in two simple models: accretion at Eddington luminosity (time-scale t_E and corresponding luminosity L_E, as a function of black hole mass M_BH), and when accretion is limited by diffusion (t_D and L_D). Reprinted by permission from Nature (Hills[78]) ©1975 Macmillan Magazines Ltd.

mass can be then be higher than M_c, because of large mass stars: giants are less dense and disrupted before solar-mass stars. This leads to higher luminosities for the active nuclei. Also the presence of the supermassive black hole may form a cusp of stars in the center. Then the stellar density is much higher and it is R_coll that limits the rate of accretion. Gas is produced by the star-star collisions, and again higher masses and luminosities can be reached.

2.5 Formation of a Cusp of Stars around the Black Hole

From a numerical resolution of the time-dependent Boltzmann equation with the relevant diffusion coefficients, it can be shown that around a black hole at the center of a globular cluster, the stellar density should be of a power-law shape[7], with a slope of 7/4. The relevant two-body relaxation time t_R is dependent on the number of bodies N in the system, as $t_R/t_c = N/\log N$, where t_c is the crossing time $= r_c/V$. For a galactic center, with a volume density of stars of $10^7 M_\odot/\mathrm{pc}^3$, this relaxation time is 3×10^8 yr.

The distribution of stars around a black hole can be described, according to the distance to the center:

- For the stars not bound to the black hole, at $r > R_a$, their velocity distribution is Maxwellian, and their density profile has the isothermal power law in r^{-2}. There are also unbound stars inside R_a, but with a density in $r^{-1/2}$. This allows to compute the penetration rate of these unbound stars in the tidal or collision radius, to estimate the accretion rate. With a core stellar mass of $M_{\rm core} = 10^7 - 3 \times 10^8 M_\odot$, a density 10^7 pc^{-3}, the accretion rate by tidal disruption is $(dM/dt)_{\rm tide} = 1 M_\odot/\mathrm{yr}\ M_8^{4/3}$; and by stellar collisions, $(dM/dt)_{\rm coll} = 0.1 M_\odot/\mathrm{yr}\ M_8^3$.
- The orbits bound to the black hole $r < R_a$: due to the cusp, their density is in $r^{-7/4}$, there is an excess of stars inside $R_{\rm coll}$, that favors stellar collisions.

More refined Monte-Carlo simulations with a distribution function $f(E, L, t)$, taking into account the velocity anisotropy, the disappearance of disrupted stars, etc.. show that the stellar cluster cannot fuel the black hole indefinitely[48]. The growth rate of the black hole and its luminosity decreases as 1/time. The loss-cone theory and the simulations are in agreement: the accretion rate due to tidal disruptions is $M_{\rm core}/t_{\rm R}$, typically of $10^{-2} M_\odot/\mathrm{yr}$, with a maximum lower than 1 $M_\odot$/yr; this cannot explain the luminosity of QSOs. QSOs might be explained only when stellar collisions are included, the corresponding accretion rate is typically a hundred times higher.

Triaxial deviations from spherical symmetry of only 5% (due to a bar or binary black hole) can repopulate the loss-cone, increasing tidal disruption to QSO levels. However, $t_{\rm coll} < t_{\rm R}$, and collisions may destroy the cusp[133].

2.6 *Stellar Collisions*

There is a tight analogy between the tidal disruption of stars by the black hole and the disruption of two stars during a head-on collision. The factor $\beta = v_{\rm coll}/v_*$ ($v_* = 500$ km/s for a solar mass star) regulates the fate of collisions, and is equivalent to the factor of penetration of the star inside the tidal radius $R_{\rm T}/r_{\rm peri}$, where $\mathrm{r_{peri}}$ is the pericenter of the stellar orbit.

- if $\beta = 1$, the collision occurs near the collision radius $R_{\rm coll}$, there is coalescence of the two stars; the result may quickly explode as a supernova.

- if $\beta > 1$ a fraction of the mass is lost, and fuels the black hole.

- if $\beta >> 1$ (well inside $R_{\rm coll}$), a deformation in pancakes occurs, very similar to the tidal disruption, when $R_{\rm T}/r_{\rm peri} >> 1$. Gas and debris are bound to the black hole.

Stellar collisions help to refill the loss-cone, although they flatten the stellar cusp[145]. The collision rate is comparable to the diffusion rate, that refills the central core.

It is now generally accepted that for low density nuclei, stellar evolution and tidal disruption is the main mechanism to bring matter to the black hole; and for high density nuclei, stellar collisions dominate the gas fueling. The evolution of the stellar density through these processes is then opposite, and accentuates the differences:
- for $n < 10^7/{\rm pc}^3$ the core expands, due to heating that results from the settling of a small population of stars into orbits tightly bound to the black hole;
- for $n > 10^7/{\rm pc}^3$, the core shrinks due to the removal of kinetic energy by collisions. To give an order of magnitude[50], the nuclear density in our own Galaxy is estimated to be $10^8 M_\odot/{\rm pc}^3$.

These mechanisms produce differing power-law slopes in the resulting stellar density cusp surrounding the black hole, $-7/4$ and $-1/2$ for low- and high-density nuclei, respectively[130,145]. In simulations however[145], collisions tend to produce a flat core, instead of $r^{-1/2}$ law in Fokker-Plank studies, which imply isotropy (and are unable to treat sparse regions).

Finally, stellar collisions could explain both the high luminosities[167,27] (up to 10^{45} erg/s), and the variability of some QSOs ($\sim$ 10 yr scale) according to Courvoisier et al.[39], see Fig. 3.

But still, even with collisions, the most active quasars requiring $100 M_\odot$/yr remain to be fueled. Can the gas from stellar disruptions be stored in a disk, and suddenly poured in a burst of activity[159]? There might be tight relations with nuclear starbursts[134,137,186].

3 Large-Scale Instabilities

Since the available fuel near the black hole is not sufficient, and a large fraction of the gaseous mass of the galaxy disk must be involved, it is necessary to rely on violent large-scale instabilities, that can drive the matter inwards in a few dynamical times.

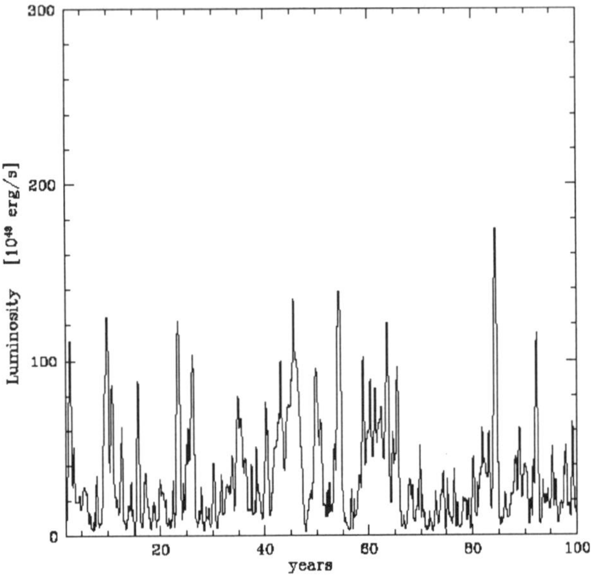

Figure 3. Light curve produced by the energy released in star collisions, with a collision rate of τ = 12 collisions per yr, provided for instance by a star cluster of mass $10^7 M_\odot$, a core radius of $R_c = 0.001$ pc, surrounding a black hole of mass $10^9 M_\odot$. From Torricelli-Campioni et al.[179] ©2000 ESO.

3.1 Gravitational Stability

The conditions of stability of a dissipation-less component like the stellar one, is relatively well known. High enough velocity dispersions are required to suppress axisymmetric instabilities and even spirals, bars or z-instabilities. The latter provide heating to the medium which becomes un-responsive. The gas is dissipative and has a completely different behavior; it is always unstable, and its velocity dispersion is fixed by regulation and feed-back.

The local stability criterion has been established by Toomre[177]: stabilization is obtained through pressure gradients (velocity dispersion c) at small scale, smaller than the Jeans length: $\lambda = ct_{ff} = c^2/G\Sigma$, where Σ is the disk surface density. At large scale, the rotation stabilizes through centrifugal forces, more precisely the scales larger than $L_{crit} = 4\pi^2 G\Sigma/\kappa^2$ (where κ is the epicyclic frequency). The Safronov-Toomre criterion is obtained in equaling

λ and L_{crit}

$$Q = \frac{\kappa c}{3.36 G\Sigma} > 1 \quad . \tag{11}$$

For one component, a radial mode (ω, k) in a linear analysis obeys the dispersion relation (where k is the wave number),

$$\omega^2 = \kappa^2 + k^2c^2 - 2\pi G\Sigma k = \kappa^2 + 4\pi^2c^2/\lambda^2 - 4\pi^2 G\Sigma/\lambda \quad , \tag{12}$$

which means that self-gravity reduces the local frequency κ. If a stability criterium is easy to derive for one component, the same is not true for a two-components fluid, since the coupling between the two makes the ensemble more unstable than each one alone. The one-component criterion can be applied separately to the stellar and gas components, where the corresponding values of Σ and c are used, leading to Q_s and Q_g. Because $c_g << c_s$ however, only a small percentage of mass in gas can destabilize the whole disk, even when $Q_s > 1$ and $Q_g > 1$. For two fluids with gravitational coupling, the dispersion relation yields a criterion of neutral stability $\omega^2 = 0$,

$$\frac{2\pi G k \Sigma_s}{\kappa^2 + k^2 c_s^2} + \frac{2\pi G k \Sigma_g}{\kappa^2 + k^2 c_g^2} = F = 1 \quad , \tag{13}$$

which gives directly an idea of the relative weight of gas and stars in the instabilities[88]. At low k (long waves), essentially the stars contribute to the instability, while at high k (short waves), the gas dominates. The neutral equilibrium requires the simultaneous solution of $d\omega^2(k)/dk = 0$, which leads to a system of 2 polynomial equations of 4th and 3rd order in k, and no analytic criterion can be derived; there are only numerical solutions in terms of a two-fluid Q_{s-g} value, which is always lower than the Q_s or Q_g values. $Q_{s-g} = 1$ defines the neutral stability, by analogy with the one-fluid treatment, at the fastest growing λ, $F = 2/(1 + Q_{s-g}^2)$. The stability depends strongly on the gas mass fraction ϵ (between 5 and 25%), and there can be sharp transitions from high to low values of λ, as for instance the mass fraction increases from $\epsilon = 0.1$ to 0.15. When Q_s is high and Q_g is low (a frequent situation, given the dissipation and cooling in the gas), the main instability is at small-scales. A gas-rich galaxy (with $\epsilon > 0.25$) is only stable at very low surface densities (this explains the Malin 1-type galaxies[85]); the center of early-type galaxies, where Q_s is high, and Q_g is low (due to low gas velocity dispersion), could be dominated by the gas wavelength, even at very low gas mass fraction: this can explain spiral arms in galaxies such as[18] NGC 2841. Also interaction of galaxies, by bringing in a high amount of gas, may change abruptly the fastest growing wavelength from λ_s to λ_g, and trigger star-formation. Even beyond the neutral stability criterium, when $Q_{s-g} > 1$, a galaxy disk can be unstable with respect to non-axisymmetric perturbations, such as bars or spirals; in this case also, the two-fluid coupling increases the instability, i.e. the disk will form a bar, even if the stars or the gas alone are stable with respect to such perturbations[88].

3.2 Feedback on the Dynamics

Many photometric and kinematic studies have computed the Q values over galactic disks, and it appears that Q does not follow big variations, but on the contrary is almost constant over the systems, as if self-regulating processes were at work. The stellar dispersion has been studied in our own galaxy and in a few external galaxies[107,20]. The velocity dispersion decreases exponentially with radius, in parallel to the stellar surface density, and Q_s is nearly constant as a function of radius, at least for large galaxies, and vary between 2 and 3 from galaxy to galaxy.

If the stellar disk can be considered as a self-gravitating infinite slab, locally isothermal (σ_z independent of z), its density obeys

$$\rho = \rho_0 \mathrm{sech}^2(z/z_0) = \frac{\rho_0}{\mathrm{ch}^2(z/z_0)} , \tag{14}$$

where z_0 is the characteristic scale-height of the stellar disk, given by

$$z_0 = \frac{\sigma_z^2}{2\pi G \Sigma_s(r)} , \tag{15}$$

where σ_z is the vertical velocity dispersion, and $\Sigma_s(r) = z_0 \rho_0$ is the surface density. The latter has a general exponential behavior[59], with a radial scale-length h. The scale-height z_0 has been observed to be independent of radius[181,182,44], and there are only small departures from isothermality[183] in z. Then, if the mass to light ratio is constant with radius for the whole stellar disk, and there is no or little dark matter within the optical disk, we expect to find a velocity dispersion varying as $e^{-r/2h}$. This is exactly what is found, within large uncertainties[20]. Since some galaxies of the sample are edge-on and others face-on, the comparison requires to know the relation between the z and r projection of the dispersion. In the solar neighborhood, $\sigma_z = 0.6\sigma_r$, and this ratio is assumed to be valid in external galaxies too.

As for the gaseous component, the vertical velocity dispersion is constant with radius in the outer parts of galaxies, where the rotation curve is flat[46,35]: $c_g \approx 6$ km/s. The behavior of $\kappa \propto 1/r$ (for a flat rotation curve) is exactly parallel to the gas surface density[19] $\Sigma_g \propto 1/r$, and therefore $Q_g \propto c_g \kappa / \Sigma_g$ is constant with radius in the outer parts. This strongly suggests a regulation mechanism, that could maintain the values of Q about constant for both stars and gas.

The mechanism could be simply gravitational instabilities coupled with gas dissipation. When the medium is cool enough, so that the Q value is too low, gravitational instabilities quickly provide heating. The stellar component cannot cool down and keeps hot, although this is somewhat moderated by the gravitational action of the gas, and the young stars born in the cool component. The gas is even more sensitive to the heating, but it can dissipate its disordered motions. The key point is that cooling encourages dynamical instabilities, and therefore produces heating, which is how the regulation works[15].

It is tempting to relate the gas stability criterium ($Q_g \sim 1$) to the threshold for star formation in galaxy disks[96]. However, things are certainly less simple, since the gas component does not form stars as soon as it is unstable and form clouds. In the outer parts of galaxies, the HI gas that extends much further than the last radius of star-formation, appears patchy, clumpy, and following some kind of spiral structure. The outer gas is unstable at any scale. This suggests that instabilities are present at the origin of cloud formation, and are the regulator[109] of the constant c_g and Q_g.

Note that there are several ways to maintain Q constant, and that the stars and gas have chosen two different ways: the stellar component keeps its scale-height constant, while its velocity dispersion is exponentially decreasing with radius; the gas keeps its velocity dispersion constant, while its scale-height increases steadily with radius (linearly, when the rotation curve is flat). This might be related to the different radial distribution. The gas does not display an exponential radial profile, may be due to continued infall or accretion.

4 Bars

Spiral waves, bars, and more generally non-axisymmetric features in the galaxy mass distribution and potential can produce torques and gas radial inflows towards the center. Bars are more long-lived and usually represent a stronger perturbation, and therefore are the privileged way to fuel the nucleus dynamically. Let us first describe the orbits and resonances in a barred potential, to better understand the existence and direction of the torques.

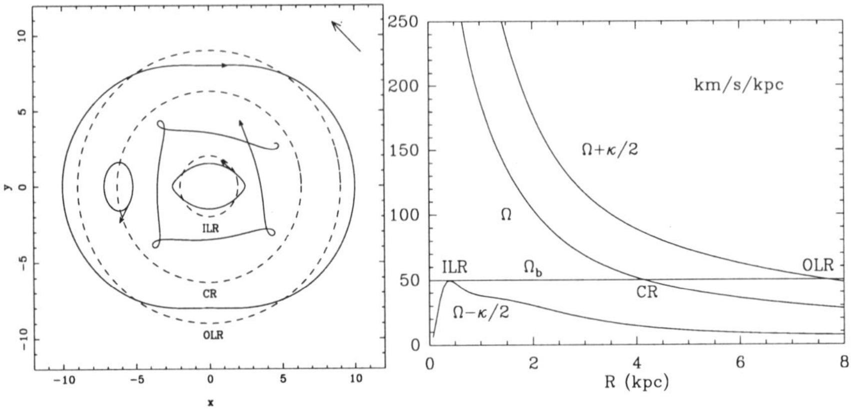

Figure 4. *Left*: various types of resonant orbits in a galaxy. At the ILR, the orbit is closed and elongated, and is direct in the rotating frame. At Corotation (CR), the orbit makes no turn, only an epicycle; at OLR, the orbit is closed, elongated and retrograde in the rotating frame. In between, the orbits are rosettes that do not close. *Right*: frequencies Ω, $\Omega - \kappa/2$ and $\Omega + \kappa/2$ in a galaxy disk. The bar pattern speed Ω_b is indicated, defining the locations of the Lindblad resonances.

4.1 Orbits and Resonances

First, let us recall the characteristics of orbits in an axisymmetric potential $\Phi(r)$ in the plane $z = 0$. A circular orbit has an angular velocity $\Omega^2 = \frac{1}{r}\frac{d\Phi}{dr}$. In linearizing the potential in the neighborhood of a circular orbit, the motion of any particle can be expressed in first order by an epicyclic oscillation, of frequency κ,

$$\kappa^2 = \frac{d^2\Phi}{dr^2} + 3\Omega^2 = r\frac{d\Omega^2}{dr} + 4\Omega^2 \quad . \tag{16}$$

The general orbit is therefore the combination of a circle and an epicycle, or a rosette, since there is no rational relation between the two periods.

The bar creates a bisymmetric gravitational potential, with a predominant Fourier component $m = 2$, which rotates in the galaxy with the pattern speed Ω_b. There is a region in the plane where the pattern speed is equal to the frequency of rotation Ω, and where particles do not make any revolution in the rotating frame. This is the resonance of corotation (cf Fig. 4).

In the rotating frame, the effective angular velocity of a particle is $\Omega' = \Omega - \Omega_b$. There exists then regions in the galaxy where $\Omega' = \kappa/m$, i. e. where the epicyclic orbits close themselves after m lobes. The corresponding stars are aligned with the perturbation and closely follow it; they interact with it always with the same sign, and resonate with it. These zones are the Lindblad resonances, sketched in Fig. 4. According to the relative values of Ω and κ in a realistic disk galaxy, and because the bar is a bisymmetric perturbation, the most important resonances are those for $m = 2$.

Periodic orbits in the bar rotating frame are orbits that close on themselves after one or more turns. Periodic orbits are the building blocks which determine the stellar distribution function, since they define families of trapped orbits around them. Trapped orbits are non-periodic, but oscillate about one periodic orbit, with a similar shape. The various families are[38]:

- The x_1 family of periodic orbits is the main family supporting the bar. Orbits are elongated parallel to the bar, within corotation. They can look like simple ellipses, and with energy increasing, they can form a cusp, and even two loops at the extremities.

- The x_2 family exists only between the two inner Lindblad resonances (ILR), when they exist. They are more round, and elongated perpendicular to the bar. Even when there exist two ILRs in the axisymmetric sense, the existence of the x_2 family is not certain. When the bar is strong enough, the x_2 orbits disappear. The bar strength necessary to eliminate the x_2 family depends on the pattern speed Ω_b: the lower this speed, the stronger the bar must be.

- Outside corotation, the 2/1 orbits (which close after one turn and two epicycles) are run in the retrograde sense in the rotating frame; they are perpendicular to the bar inside the outer Lindblad resonance (OLR), and parallel to the bar slightly outside.

4.2 Gas Flow in Barred Galaxies

In summary of the previous section, the orientation of the periodic orbits rotates by 90° at each resonance crossing, and they are successively parallel and perpendicular to the bar. The gas will first tend to follow these orbits, but the streamlines of gas cannot cross. Since periodic orbits do cross, gas clouds can encounter enhanced collisions, such that their orbits are changed. Instead of experiencing sudden 90° turns, their orbits will smoothly and gradually turn, following the schematic diagram of kinematic waves, first drawn by Kalnajs[92], and illustrated in Figs. 8 and 9.

This interpretation predicts that the arms will be more tightly wound when more resonances exist; there will be a winding over 180° with only CR and OLR, with the gas aligned with the bar until corotation. When 2 ILRs

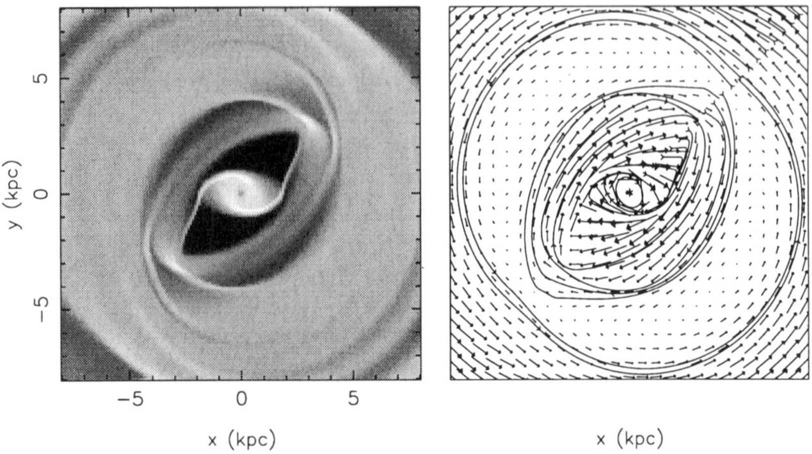

Figure 5. Response of the gas to a bar potential. The left panel shows the gas density in grey-scale and the right panel the gas flow lines and the velocity vectors in the rotating frame of the bar. The bar potential is at 45° from horizontal. From Athanassoula[4] ©1992 Blackwell Sci.

exist, the gas response can be perpendicular to the stellar bar. When there is barely one ILR, strong shocks can occur on the leading edge of the bar, corresponding to the offset dust lanes observed in barred galaxies (cf Fig. 5).

4.3 Angular Momentum Transfer

To minimize its total energy, a galaxy tends to concentrate its mass towards the center, and to transfer its angular momentum outwards[113]. It is the role of the spiral structure to transport angular momentum from the center to the outer parts, and only trailing waves can do it. This transfer, mediated by non-axisymmetric instabilities, is the motor of secular evolution of galaxies, and of the formation of bars and resonant rings.

The angular momentum transfer is due to the torques exerted by the bar on the matter forming spiral arms. There is a phase shift between the density and the potential wells, resulting in torques schematized in Fig. 6. The gas is much more responsive to these torques, since they form the spirals in a barred galaxy. The torque changes sign at each resonance, where the spiral turns by 90°. Between the ILR and corotation, the torque is negative, while between CR and OLR, the torque is positive. These torques tend to depopulate the corotation region, and to accumulate gas towards the Lindblad resonances, in

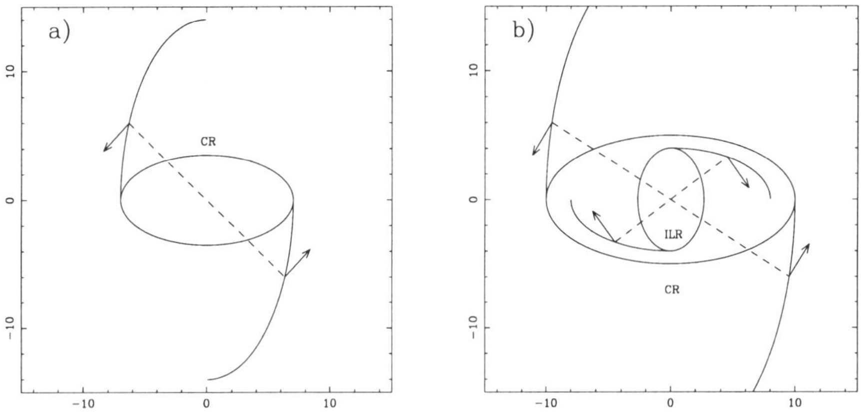

Figure 6. Schematic representation of the gravity torques exerted by the bar on the gaseous spiral: a) between CR and OLR, the gas acquires angular momentum and is driven outwards; b) between CR and ILR, the gas loses angular momentum. From Combes[31] ©1988 Kluwer.

the shape of rings. Indeed, these rings then are aligned with the symmetry axis of the bar, and no net torque is acting on them. Numerical simulations of colliding gas clouds in a barred potential show that rings form in a few dynamical times, i.e. in a few Gyrs for the outer ring at OLR[155], or in $\sim 10^8$ yr for nuclear rings at ILR[30].

This mechanism for ring formation has been confirmed by many studies, and confronted to observations, where it always encounters large success. The potential is obtained from photometry of the galaxies in NIR bands, and derived with a small number of assumptions (usually constant M/L over the disk). The simulation of the gas flow in the potential allows the derivation of the pattern speed[23] (see Fig. 7).

4.4 Fueling Nuclear Activity

The main problem to fuel the nucleus is to solve the transfer of angular momentum. Torques due to the bar are very efficient, but gas can be stalled in a nuclear ring at ILR. Other mechanisms can then be invoked: viscous torques, or dynamical friction of giant clouds (GMC) against stars. The viscosity is in general completely inefficient over the galactic disk, but the corresponding

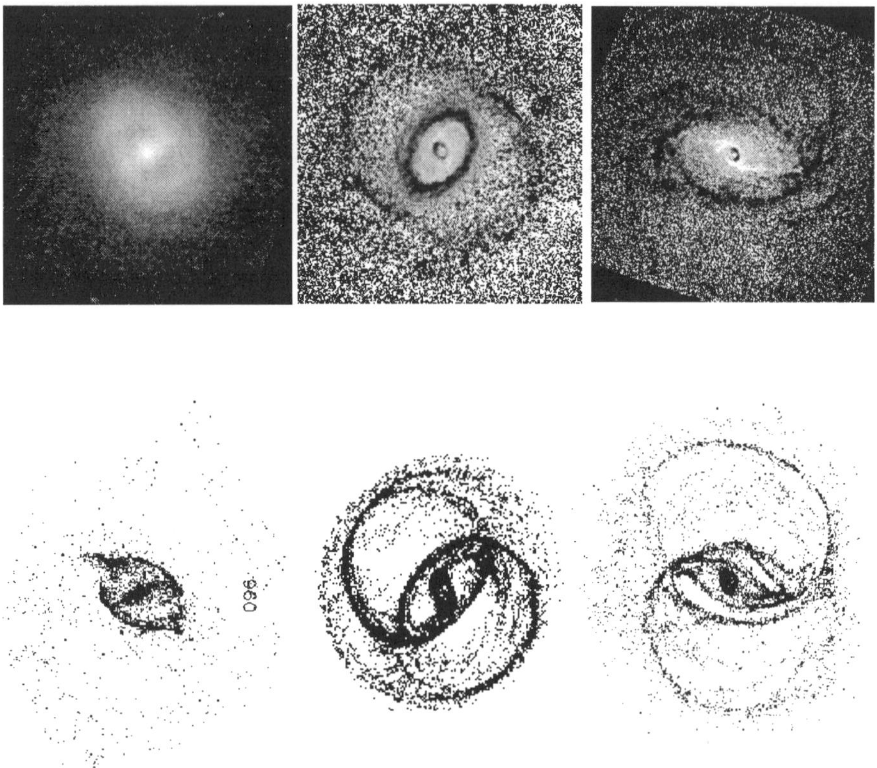

Figure 7. Comparison of observed maps (*top*) and corresponding simulations (*bottom*) for NGC 6300 (*left*), NGC 3081 (*middle*) and NGC 1433 (*right*). For NGC 3081 and NGC 1433, the top images are $B - I$ colors (blue is dark), and for NGC 6300 it is the H-band image. From Buta & Combes[24] ©2000 Astron. Soc. of the Pacific.

time-scale is decreasing with decreasing radius. Unfortunately, in the center of galaxies, the rotation is almost rigid, the shear is considerably reduced, and so are the viscous torques. The time-scale for dynamical friction becomes competitive below $r = 100$ pc from the center (about $10^7 (r/100 \text{ pc})^2$ yr for a GMC of $10^7 M_\odot$). For the intermediate scales, a new mechanism is required.

Note that if there is a supermassive black hole in the nucleus, it is easier to bring the gas to the center. Indeed, the presence of a large mass can change the behavior of the precessing rate of orbits $\Omega - \kappa/2$: instead of increasing

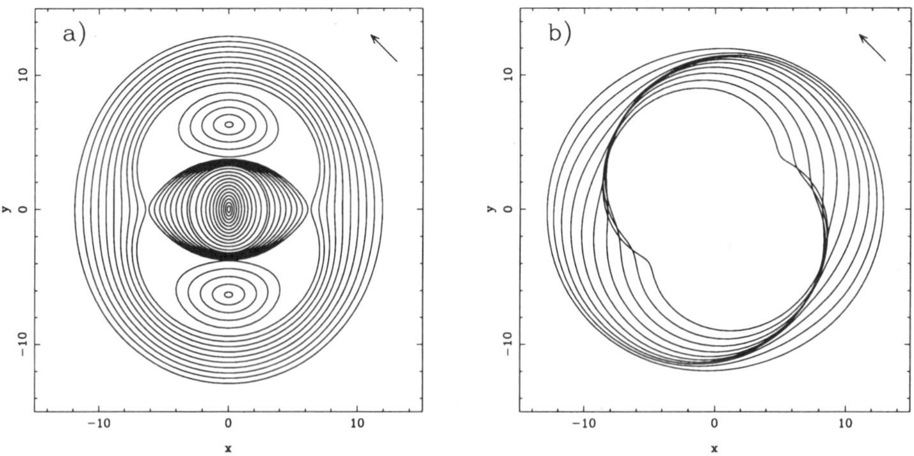

Figure 8. a) Periodic orbits in a barred galaxy ($\cos 2\theta$ potential, oriented horizontally). Their orientation rotates by $\pi/2$ at each resonance. b) The gas tends to follow these orbits, but is forced to precess more rapidly while losing energy and angular momentum, since $\Omega - \kappa/2$ is a decreasing function of radius.

with radius inside ILR (as in Fig. 4), it will decrease.

Due to cloud collisions, the gas clouds lose energy, and their galactocentric distance shrinks. Since it tends to follow the periodic orbits, the gas streams in elliptical trajectories at lower and lower radii, with their major axes leading more and more the periodic orbit, since the precession rate (estimated by $\Omega - \kappa/2$ in the axisymmetric limit, for orbits near ILR, and by $\Omega + \kappa/2$ near OLR) increases with decreasing radii in most of the disk (Fig. 8). This regular shift forces the gas into a trailing spiral structure, from which the sense of the gravity torques can be easily derived. Inside corotation, the torques are negative, and the gas is driven inwards towards the ILR. Inside ILR, and from the center, the precessing rate is increasing with radius, so that the gas pattern due to collisions will be a leading spiral, instead of a trailing one (see Fig. 9). The gravity torques are positive, which also contributes to the accumulation of gas at the ILR ring. This situation is only inverted in the case of a central mass concentration (for instance a black hole), for which the precession rate $\Omega - \kappa/2$ is monotonically increasing towards infinity with decreasing radii. Only then, the gravity torques will pull the gas towards the very center, and "fuel" the nucleus.

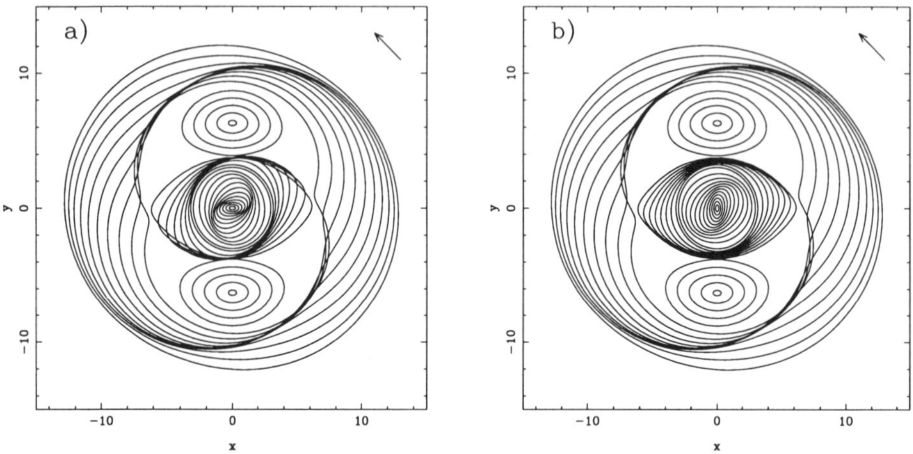

Figure 9. a) Without a central mass concentration, $\Omega - \kappa/2$ is increasing with radius in the center: the gas winds up in a leading spiral inside the ILR ring; b) with a central mass concentration, it is the reverse and the gas follows a trailing spiral structure, inside ILR.

The problem reduces to forming the black hole in the first place. We show next that the accumulation of matter towards the center can produce a decoupling of a second bar inside the primary bar. This nuclear bar, and possibly other ones nested inside like Russian dolls, can take over the action of gravity torques to drive the gas to the nucleus, as first proposed by Shlosman et al.[160].

4.5 Decoupling of a Nuclear Disk

The bar torques drive progressively more mass towards the center. This matter, gaseous at the beginning, forms stars, and gradually contributes to the formation of the bulge, since stars are elevated above the disk plane, through vertical resonances with the bar[32,143]. When the mass accumulation is large enough, then the precessing rate $\Omega - \kappa/2$ curve is increasing strongly while the radius decreases, and this implies the formation of two inner Lindblad resonances. In between the two ILRs, the periodic orbits are perpendicular to the bar (x_2 orbits), and the bar loses its main supporters. The weakening of the primary bar, and the fact that the frequencies of the matter are considerably different now between the inner and outer disk, forces the decoupling of a nuclear disk, or nuclear bar from the large-scale bar (primary bar).

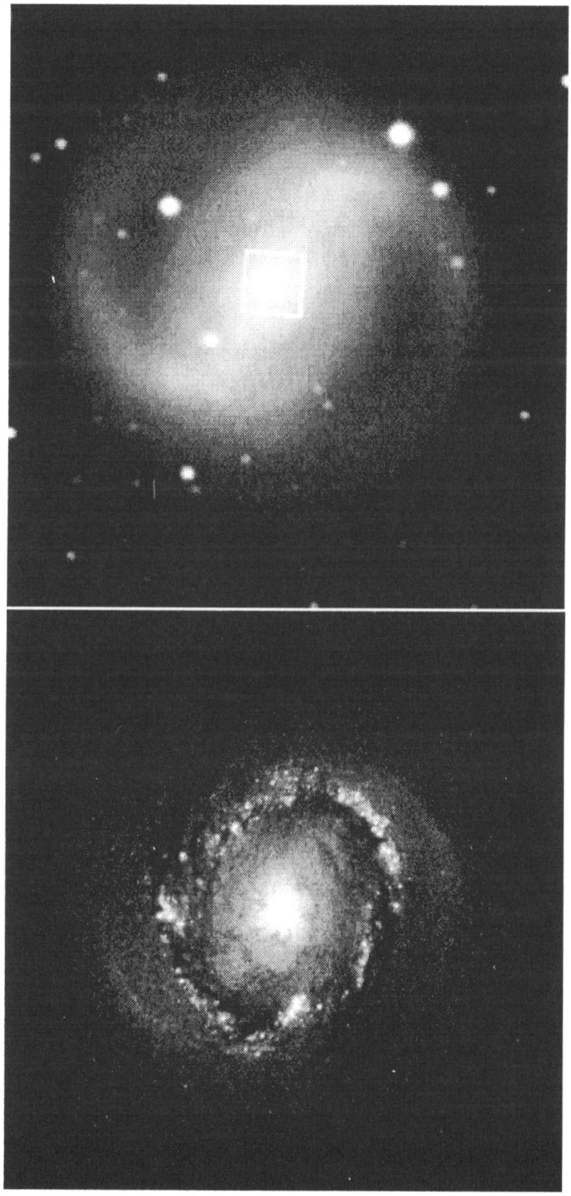

Figure 10. Photographs of the barred galaxy NGC 4314. *Top:* a general view of the bar and the spiral arms (photo from McDonald Observatory, Texas). *Bottom:* a zoom on the central parts (corresp. to the square in the first picture), with the Hubble Space Telescope. A nuclear ring can be discerned, within which a second independent spiral structure has developed. From Benedict et al.[14] ©1996 AAS.

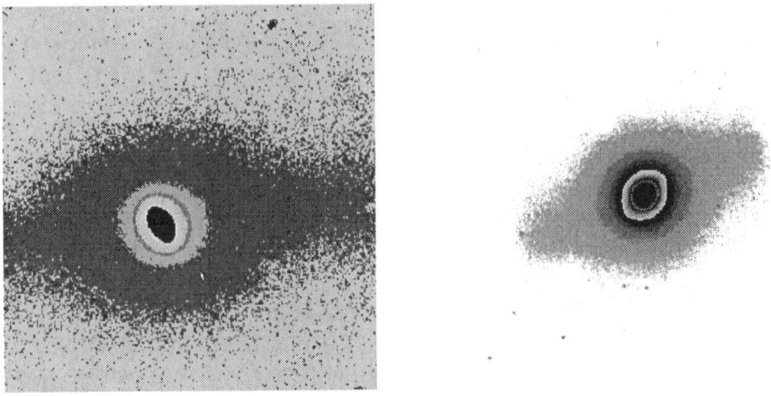

Figure 11. Examples of nuclear bars seen in NIR bands. *Left:* NGC 1433 in H. *Right:* NGC 2217 in H. The field of view of both images is 2 x 2 arcmin. From Jungwiert et al.[89] ©1997 ESO.

Nuclear disks are frequently observed, at many wavelengths: optical with the HST[10] (Fig. 10) or in CO molecules with millimeter interferometers[86]. A recent survey in the Virgo cluster[149] reveals that about 20% of the 80 spirals observed possess a decoupled nuclear disk. In nearly edge-on systems, these nuclear disks are conspicuous through large velocity gradients, like in the Milky-Way[43] or NGC 891 [63].

4.6 Bars within Bars

Secondary bars form through decoupling[60,34]. The second bars rotate with a much faster angular velocity, and are observed with a random angle from the primary bar (see Fig. 11). To avoid chaos, the two bars have a resonance in common. It is frequent that the ILR of the primary coincides with the corotation of the secondary. Multiple periodic particle orbits have been identified in such time-varying potentials[114]. It is possible that the two bars exchange energy with each other, through non-linear coupling; then $m = 4$ and $m = 0$ modes are also expected, and these have been seen in simulations[172,119]. Even then, the life-time of the ensemble is rather short, a few rotations. But the nuclear bars could help to prolong the action of the primary bar towards the nucleus (as first proposed by Shlosman et al.[160]).

4.7 Bar Destruction

The inflow of matter in the center can destroy the bar. It is sufficient that 5% of the mass of the disk has sunk inside the inner Lindblad resonance[70,139,71]. But this depends on the mass distribution, on the size of the central concentration; a point mass like a black hole is more efficient (may be 2% is sufficient). The destruction is due to the mass re-organization, that perturbs all the orbital structure: the x_1 orbits sustaining the bar for instance are shifted outwards. Near the center, the central mass axisymmetrizes the potential. Then there is a chaotic region, and outside a regular one again. When a central mass concentration exists initially, in N-body simulations, a bar still forms, but dissolves quicker. It is also possible that after a bar has dissolved, another one forms, after sufficient gas accretion to generate new gravitational instabilities: the location of the resonances will not be the same.

4.8 Nuclear Spirals in Disk Galaxies

When the mass is concentrated towards the center in a galaxy, there can be several possibilities and mechanisms to fuel the nucleus: either bars within bars, as multi-mode stellar density waves, as described previously, or an instable gas-dominated central disk[75]. Another possibility is simply a large-scale stellar bar, and a gaseous nuclear spiral by continuity. It has indeed been proposed that the observed nuclear spirals are only gaseous[55,146]. They could be acoustic waves, developing without self-gravity. Gas short waves are able to propagate inside ILR (and beyond OLR outside). The dispersion relation is (with v_s the sound speed)

$$m^2(\Omega - \Omega_\mathrm{p})^2 = \kappa^2 + k^2 v_\mathrm{s}^2 \tag{17}$$

for $m = 2$

$$k^2 v_\mathrm{s}^2/4 = (\Omega - \kappa/2 - \Omega_\mathrm{p})(\Omega + \kappa/2 - \Omega_\mathrm{p}) \tag{18}$$

if $kR >> 1$, the gas waves propagate only when $\Omega - \kappa/2 > \Omega_\mathrm{p}$, in between two ILRs, or if there is only one ILR. Multi-arms (large m) are possible. The value of Q is large, since the gas is not self-gravitating. Simulations show the gas response in a stellar bar potential, and how the morphology depends on the sound speed and on the shape of the potential[56] (cf. Fig. 12).

4.9 Bars in Early- and Late-Types

Observations reveal that bars have different properties along the Hubble sequence[52,53]. In early-type galaxies, bars have flat profiles, while in late-types, they have exponential profiles. Also, bars extend farther with respect

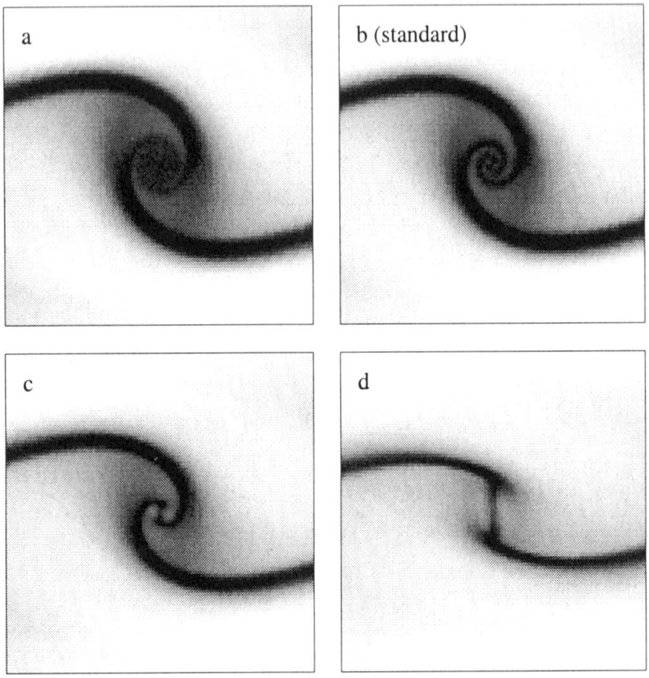

Figure 12. Gas streaming in a barred potential. The stellar bar is horizontal and extends more than twice the size of the box. The different morphologies are due only to the different values of the sound speed: (a) 7 km/s, (b) 10 km/s, (c) 14 km/s, (d) 20 km/s. From Englmaier & Shlosman[56] ©2000 AAS.

to the disk exponential scale in late-types. Numerical simulations allow to interpret these features. Due to their relatively massive bulge, and large mass concentration, early-types have[33] large precessing frequencies $\Omega - \kappa/2$ and tend to have large Ω_b. Two ILRs then exist, and bars end near their corotation. On the contrary, late-types have low $\Omega - \kappa/2$, and low Ω_b. Corotation is then far away in the disk, and lets the disk scale-length determine the bar length (see Fig. 13). The profile along the bar is therefore exponential, very similar to the disk profile. Bars in early-types continuously grow in time, since angular momentum is transferred in outer regions. But in late-types they cannot, since there are not enough stars to accept angular momentum, and they stop to grow before. Since the existence of ILRs favors radial gas inflows, fueling appears more efficient in early-type galaxies.

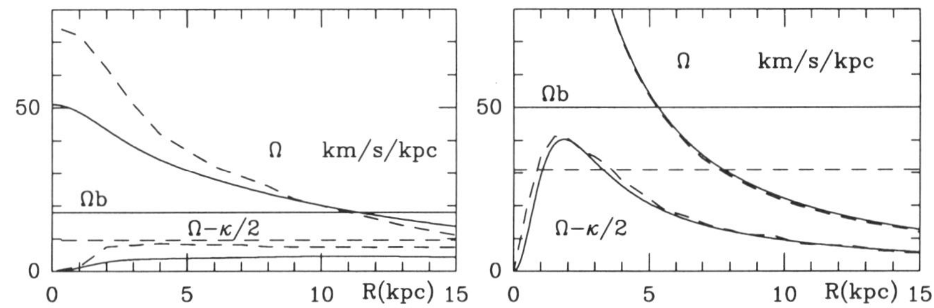

Figure 13. Rotation curves for late and early-type galaxy models, compared with the bar pattern speed. The initial values are full lines, while the final values are dashed lines. *Left:* late-type model. *Right:* early-type. From Combes & Elmegreen[33] ©1993 AAS.

4.10 Evolution along the Hubble Sequence and Growth of the Black Hole

Non-axisymmetric instabilities, like bars, force galaxies to evolve towards large mass concentrations, large bulge-to-disk ratios, and lower gas content through consumption by star formation. This means that a galaxy born as a late-type, will progressively evolve towards early-types, with a somewhat chaotic path, from barred to unbarred, and sometimes moving backwards, when the disk accretes mass. The gross lines of this evolution are sketched in Fig. 14.

To summarize this bar-driven evolution, gather the main features obtained through N-body simulations, and supported by observations, it is interesting to test a toy model, in a semi-analytical way, including:

– star formation, with a combination of a quiescent rate, proportional to the gas density, in a time scale of 3 Gyr; and a bar-driven contribution, with a threshold ($Q < 1$) and a rate equal to $(1-Q)/t_*$, where t_* is the star-formation time-scale (proportional to the dynamical time-scale for gravitational instabilities);

– radial flows: when a bar is formed, gravity torques produce gas inflow, therefore with a threshold $Q < 1$, and rate $(1 - Q)/t_{\rm vis}$, with $t_{\rm vis}$, the "gravitational viscosity" time-scale, $\sim \frac{1}{\Omega}(\frac{M_{\rm tot}}{M_{\rm d}})^2$;

– bulge formation: the inflowing gas (and stars) are assumed to form the bulge through star-formation and vertical resonances;

– death of bars: when $Q > 1$ (central concentrations, lack of gas and self-gravitating disk);

– gas infall: possibility of a continuous small infall or a periodically substan-

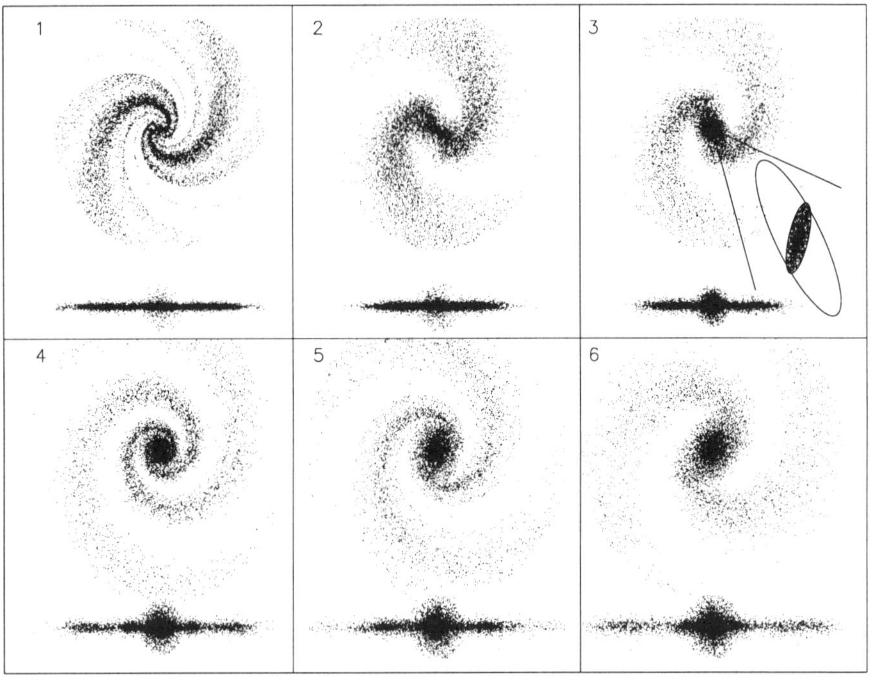

Figure 14. Cartoon of galaxy evolution along the Hubble sequence. A galaxy, with a small mass distributed mainly in a disk, without bulge (late-type), is unstable with respect to spiral and bar formation (steps 1 and 2). The bar drives the gas towards the center, and the bulge is building up (see in each frame the edge-on projection). When there is too much mass concentrated in the center, the bar is destroyed (step 4), and the gas coming from the outer parts, enrich the disk, and re-establish a larger disk to bulge ratio. Later on (steps 5 and 6), another bar will form, when the disk to bulge ratio is favorable. A secondary bar (cf step 3) may help the primary one to drive the mass towards the center. At the end, the galaxy may be classified early-type.

tial one (from companions).
– black hole formation: a fixed fraction $b_{\rm eff}$ of the radial gas flow is taken to contribute to its formation, i.e. $dM_{\rm BH}/dt = b_{\rm eff} M_{\rm g}(1-Q)/t_{\rm vis}$, with a threshold $Q < 1$.

Fig. 15 displays some results of the toy model[36]. The most striking feature is the self-regulation of the stability parameter Q towards 1. Although the galaxy initially starts almost completely gaseous, the gas mass fraction soon stabilizes to 10% of the total. Also the mass of the central concentration (or

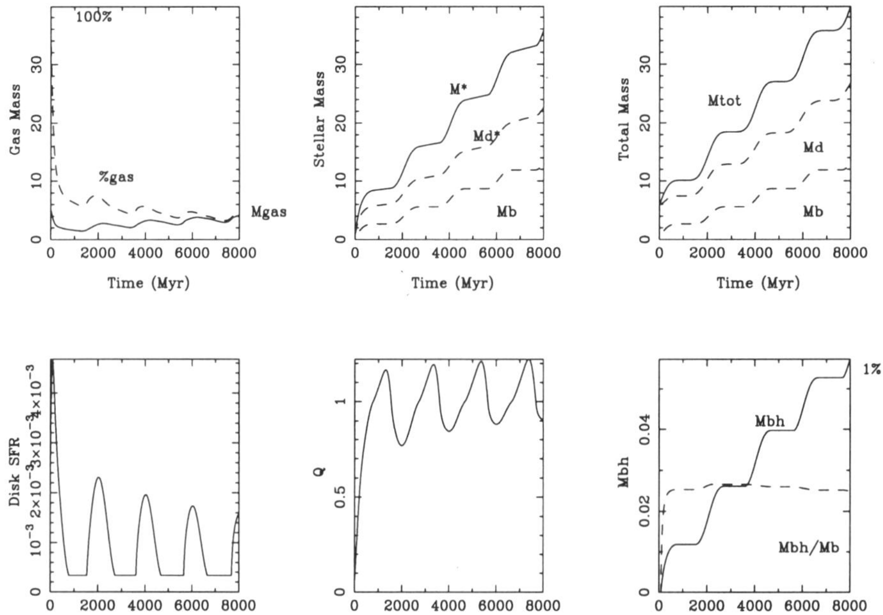

Figure 15. Model of periodic gas accretion in a galaxy with star formation, birth and death of bars, radial flows and black hole formation taken into account: *Top left:* full line, gas mass versus time; dash line, gas mass fraction. *Top middle:* full line, stellar mass; dash lines, disk stellar mass at top, and bottom bulge mass. *Top right:* full line, total mass; dash lines, total disk mass at top, and bottom bulge mass. *Bottom left:* disk star formation rate versus time. *Bottom middle:* Toomre Q parameter. *Bottom right:* Full line, mass of the central black hole; and dash line, mass ratio between the black hole and the bulge.

black hole) stabilizes to a constant fraction of the bulge mass, as observed[115].

5 $m = 1$ Perturbations

Asymmetries, lopsidedness, and one-arm spirals are also frequent in galactic disks, and may play a role in fueling the nucleus. Their physical processes are varied and different from the more familiar $m = 2$ instabilities, and these are first described briefly.

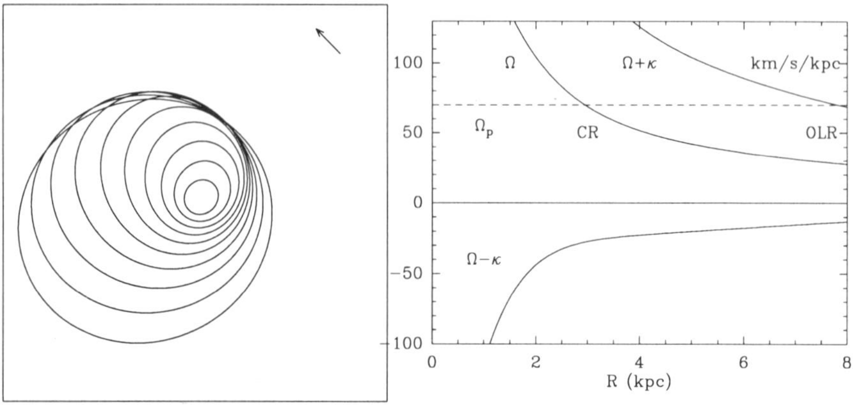

Figure 16. *Left*: pattern of lopsided ellipsoidal orbits forming a one-arm leading spiral. *Right*: frequencies Ω, $\Omega - \kappa$ and $\Omega + \kappa$ in a galaxy disk. A possible pattern speed Ω_p is indicated, allowing CR and OLR resonances.

5.1 Physical Nature of Instabilities

The density wave theory has been developed in the WKBJ ($kr >> 1$ approximation) and linear regime[108,178]. The amplification of the waves occurs at corotation, since the energy and angular momentum of the perturbation are positive outside and negative inside corotation. Waves are partially transmitted, and partially reflected at corotation, which is a zone of evanescence for the waves if $Q > 1$. The wave transmitted will carry energy and angular momentum of opposite sign of the incident wave: for conservation, the reflected wave must have increased amplitude. Waves can spontaneously develop if the corotation amplifier is coupled to a reflection at a resonance or boundary (turning point). The feedback cycle may be the *waser*[117] or the *swing*[178]. The turning points are located at the radii when $\Omega_p = (\Omega \pm \kappa/m)(1 - 1/Q^2)^{1/2}$.

For $m = 1$ perturbations, ILR and OLR cannot exist at the same time (see Fig. 16). For lopsided instabilities, developing around a central mass in a nearly-keplerian disk, another amplifier exists, which releases the need for corotation: the indirect potential, which is due to the off- centring of the central mass[1,162],

$$\Phi(r, \theta, t) = \alpha\omega^2 r \cos(\omega t - \theta) \quad . \tag{19}$$

This indirect potential creates in permanence a long-range force. The disk behaves like a resonant cavity with the off-centring constantly stimulating new long trailing waves. The central mass gains angular momentum, and also the disk outside corotation; this is not in contradiction, since in fact in a frame centered on the central mass, the angular momentum of the disk is of opposite sign (with respect to that centered on the system center of mass). While the growth rate of the mode γ must be $\sim \Omega$ for the *swing* mechanism, here $\gamma << \Omega$. This mode allows the inner disk to lose angular momentum, and to inflow on the central mass.

5.2 Lopsidedness and $m = 1$ Asymmetries

The $m = 1$ perturbation is present in most galactic disks[147], often superposed on $m = 2$ ones. These perturbations can be of different nature and origin, according to their scale (nuclear or extended disk), or whether they involve the gaseous or stellar disks.

Linear analysis, supported by numerical calculations, show that gaseous disks rotating around a central mass, are unstable for low values of the central mass[74]. The instability disappears when the central mass equals the disk mass. For a gaseous disk, which can develop acoustic waves, and subject to the indirect term (or off-centering of the center of mass) as amplifier, an $m = 1$ can form and grow[162,90].

For nuclear stellar disks, long lasting oscillations of a massive nucleus have been observed[127,171]. For extended disks, off-centered in an extended dark halo, lopsidedness could survive, if the disk remains in the region of constant density (or constant $\Omega(r)$) of the halo[106].

In the special case of a stellar nuclear disk around a massive black hole, it is possible that self-gravity is sufficient to compensate for the differential precession of the nearly keplerian orbits, and that a long-lasting $m = 1$ mode develops, or is maintained long after an external excitation[6]. This could be the explanation of the double nucleus observed[5,102] for a long time in M31, for which an eccentric disk model has been proposed[180,169]. In this $m = 1$ mode, the maximum density is obtained at the apocenter of the aligned elongated orbits. The pattern speed, equal to the orbital frequency of the barycentre of the stellar disk, is slow (3 km/s/pc, Fig. 17), with respect to the orbital frequency of the stars themselves (250 km/s/pc in the middle of the disk). The excitation of the $m = 1$ perturbation can then last more than 3000 rotation periods. This could be a frequent phenomenon around supermassive black holes.

The decoupling of the nuclear disks also implies the possibility of decoupled z-oscillations, and these nuclear regions are frequently observed to be inclined at angles that differ from that of the main disk. Jet orientations are not correlated with large-scale disk orientations in Seyfert galaxies[99], while

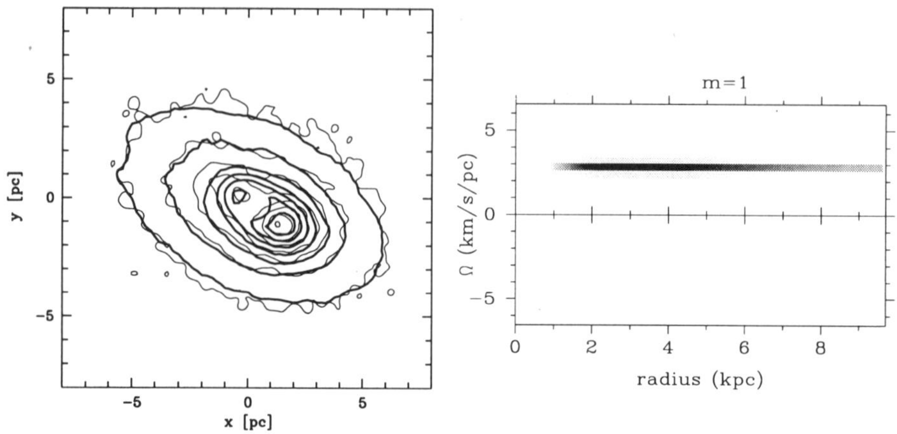

Figure 17. HST map in the visible of the nuclear disk surrounding the black hole of M31: thin lines are the observed contours, thick lines are those of the $m = 1$ model simulations (*left*). Pattern speed as a function of radius for the $m = 1$ mode obtained in the simulation (*right*). From Bacon et al.[6] ©2001 ESO.

they should be perpendicular to the accretion disk. The Galactic Center is a well-studied example, our nearest supermassive black hole. There must be fueling in action. There is a large-scale bar, a nuclear bar/spiral, an $m = 1$ off-centring, a tilted disk (warped?) etc...

5.3 Counter-Rotating Components

The phenomenon of counter-rotating components is a tracer of galaxy interactions, mass accretion or mergers. It has been first discovered in ellipticals with kinematically decoupled cores (likely to be merger remnants[9]). It has been observed also in many spirals; either two components of stars are counter-rotating, or the gas with respect to the stars, or even two components of gas, in different regions of the galaxies[62,16]. In the special case[150] of NGC 4550, two almost identical counter-rotating stellar disks are superposed along the line of sight.

These systems pose a number of questions, first about their formation scenario, but also about their stability, their life-time, etc.. Do special waves and instabilities develop in counter-rotating disks, and does this favor central gas accretion?

5.4 Stability

First, it appears that the counter-rotation (CR) can bring more stability. Even a small fraction of CR stars has a stabilizing influence with respect to bar formation ($m = 2$), since the disk has then more velocity dispersion[92]. But a one-arm instability is triggered, for a comparable quantities of CR and normal stars. This comes from the two-stream instability in flat disks, similar to that in CR plasmas[111]. They develop two $m = 1$ modes in the two components, with energies of opposite signs: the negative-E mode can grow by feeding energy in the positive-E mode.

A quasi-stationary one-arm structure forms, and lasts for about 1 to 5 periods[37]. The structure is first leading, then trailing, and disappears. The formation of massive CR disks in spirals has been studied by Thakar & Ryden[175,176].

5.5 Counter-Rotation with Gas

The presence of two streams of gas in the same plane will be very transient: strong shocks will occur, producing heating and rapid dissipation. The gas is then driven quickly to the center. But the two streams of gas could be in inclined planes, or at different radii. This is the case in polar rings, discovered in 0.5% of all galaxies[185]. After correction for selection effects (non optimal viewing, dimming, etc..) 5% of all S0 would have polar rings.

If there is only one gas stream, the problem is more similar to the two-stream instabilities of stellar disks mentioned above. However the gas is cooling, and it is not easy to stabilize against $m = 2$ components. Both $m = 1$ and $m = 2$ may be present simultaneously in these systems. This is the case of the galaxy NGC 3593, composed of two CR stellar disks[17]: in an extended stellar disk, a CR nuclear disk is embedded possessing co-rotating gas. The molecular component associated with this nuclear disk reveals both a nuclear ring and a one-arm spiral structure outside of the ring[64]. N-body simulations have shown that both structures can be explained by the superposition of $m = 1$ and $m = 2$ in the gas component, the ring being formed at the ILR of the bar (Fig. 18). In the $m = 2$ pattern, two counter-rotating bars develop (Fig. 19).

6 Interactions and Mergers

Galaxy interactions can be the most efficient way to produce strong torques, and transfer the angular momentum away. During the interaction period,

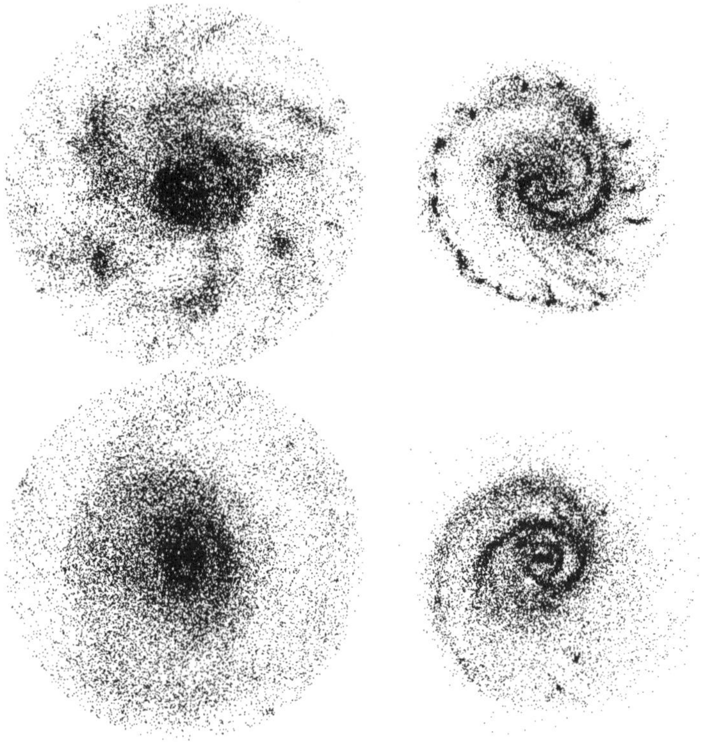

Figure 18. Particle plots of the stars (*left*) and the gas (*right*) for a model of the counter-rotating galaxy NGC 3593, at successive epochs: 200 and 400 Myr. Particles are plotted until a radius of 6.25 kpc. The majority of stars rotate in the direct sense, while the gas is retrograde. From Garcia-Burillo et al.[64] ©2001 ESO.

strong bars are triggered in the galaxy disks, and through the same mechanisms as described before, gas is driven inwards. During the merger, a complete change of geometry also brings most of the gas to the center. The main consequence is the trigger of spectacular starbursts, in particular in major mergers, i.e. merging disk galaxies of about equal masses. The star forming activity is concentrated in the nuclei, with some exceptions (as The Antennae or Arp 299, for instance). The same gas inflow can also fuel an active nucleus, either directly, or indirectly through the coeval dense nuclear clusters formed in the starburst (cf Sec. 2.2).

In these huge starbursts discovered by IRAS, CO concentrations suggest

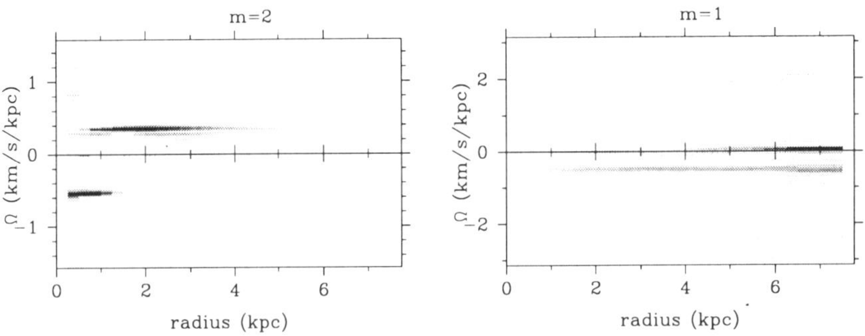

Figure 19. Pattern speed as a function of radius, in units of 100 km/s/kpc, for the $m = 2$ mode, total density (*left*) and $m = 1$ mode, gas density (*right*), for the NGC 3593 model of Fig. 18. Note the presence of two counter-rotating bars.

that the the cause of the starburst is the concentration of huge gas masses in the center; the molecular gas represents a significant fraction of the dynamical mass[157]. This gas must be brought to the center in a time- scale short enough with respect to the feedback time-scale of star-formation[103], i.e. a few 10^7yr. This requires very strong gravity torques.

6.1 Physical Processes

Numerical simulations have enlightened the dynamical processes[9]. They represent galaxies as 3-component systems: disks, bulges, and dark haloes. The latter play an important role, although they do not share most of the perturbations, they are essential to provide the dynamical friction that makes the baryonic systems decay and merge. It is the dark matter that takes most of the angular momentum away, allowing the luminous mass to fall towards the center[8].

Gas dissipation is also a key factor in merger simulations. The actual equivalent viscosity of the interstellar medium (ISM) is not well known, and it is not possible to reproduce the full multiphase medium. Two extreme models are commonly used: the ISM is represented either as a continuous fluid, submitted to pressure forces, shock waves and artificial viscosity (finite difference scheme, smooth particle hydrodynamics (SPH)); or as colliding clouds, with no pressure forces and no shocks (sticky particles). Both can reproduce the main characteristics of gas flow in galaxies, provided that true viscous torques are negligible. The star-formation rate, global laws and corresponding feed-back are also not known in most perturbed dynamical situations, and present models are rough approximations. Note that the star formation is itself fundamental for the dynamics, since it can lock the gas into the non-dissipative medium and halt for a while the mass transfer towards the center. Since stars are observed to be formed inside giant molecular clouds in our Galaxy, the latter being the result of agglomeration of smaller entities, one process could be to relate star formation to cloud-cloud collisions, in the sticky particles model[132]. A more widely used approach is to adopt a Schmidt law for the star formation rate (SFR), i.e. the rate is proportional to a power n of the gas volume density[125], n being between 1 and 2. In both cases, it was shown that interacting galaxies were the site of strong starbursts, that could be explained both by the orbit crowding in density waves triggered by the tidal interactions, and by the gas inflow and central concentrations, accumulating the gas in small and very dense regions. This depends of course on the non- linearity of the Schmidt law, and star-formation efficiency strongly depends on the power[125] n. Mihos & Hernquist[123] use hybrid-particles techniques, within SPH, to describe the effects of gas depletion and formation of a young star population.

Another peculiar feature of merging galaxies is the ability of forming giant complexes, that will soon become dense stellar clusters. This can be understood, given the larger velocity dispersion of perturbed systems. The disks are heated in tidal interactions, since the relative orbital energy is transformed into internal disordered motions. This has the consequence to increase the critical Jeans scale for gravitational instabilities, and to create giant complexes[144]. The global Jeans length is $\lambda \propto \sigma^2/\Sigma_{\rm g}$, where σ is the velocity dispersion and $\Sigma_{\rm g}$ is the gas surface density of the galactic disk. The corresponding growth time is $\tau_{\rm ff} \propto \sigma/\Sigma_{\rm g}$, and the instabilities will occur as soon as $Q \propto \sigma\kappa/\Sigma_{\rm g}$ becomes lower than 1. For the same ratio $\sigma/\Sigma_{\rm g}$, a perturbed system with elevated σ and $\Sigma_{\rm g}$ will see the condensations of larger complexes of mass $M \propto \sigma^4/\Sigma_{\rm g}$, in the same time-scale τ. These complexes with larger internal dispersions, and larger gravitational support will be more difficult to disrupt through star-formation, which enhances the star-formation efficiency[54]. The thermal Jeans length is also larger, due to the hotter gas temperature induced by the larger number of stars in the clouds, and high mass stars are favored. This explains the existence of giant and dense star clusters observed with HST in merging galaxies[156,122]. These clusters are estimated to have masses $10^7 M_\odot$, and ages 3×10^8 yr. They could evolve into typical globular clusters in about 15 Gyr.

6.2 *Gas Flow and Starburst/AGN Triggering*

In major mergers, where the two interacting galaxies are of comparable mass, strong non-axisymmetric forces are exerted on the gas. But contrary to what could be expected, the main torques responsible for the gas inflow are not directly due to the companion. The tidal perturbations destabilize the primary disk, and the non-axisymmetric structures generated in the primary disk (bars, spirals) are responsible for the torques. The self-gravity of the primary disk, and its consequent gravitational instabilities, play the fundamental role. The internal structure therefore takes over from the tidal perturbations on the outer parts. The gas is provided by the primary disk itself. Again it is not the gas accreted from the companion which is the main trigger for activity, until the final merger of course. This is why the first parameter determining the characteristics of the merger event is the initial mass distribution in the two interacting galaxies[124]. The mass ratio between the bulge and the disk is a more fundamental parameter than the geometry of the encounter (Figs. 20 and 21).

The central bulge stabilizes the disk with respect to external perturbations. If the bulge is sufficiently massive, the apparition of a strong bar is delayed until the final merging stages, and so is the gas inflow, and the consequent star-formation activity. But the starburst can then be stronger. When the primary disk is of very late type, without any bulge, the gravitational instability settles in as soon as the beginning of the interaction. There is then a continuous activity during the interaction, but at the end the starburst is less violent, since most of the gas has been progressively consumed before (Fig. 22).

The effects of the geometry are more visible on the direct manifestations of the tidal interaction, i.e. on the tails and debris. For the galaxy experiencing a retrograde collision, no extended tidal tail is formed. There is no resonance effects in the target galaxy disk, and less violent material perturbations; a transient leading tidal arm is instead developed. There is however a tidally-induced two-arm density wave, and the subsequent torques produce gas inflow towards the center, as for the prograde encounter. During the interaction the retrograde disk has more star formation than the prograde one, due to the larger gas content. For coplanar mergers, the retrograde disk accretes a significant fraction of the gas from the prograde disk[124].

In the simulations, about 75% of the gas is consumed during the merger, whatever the internal structure of galaxies, or the geometry of the encounter[124]. This is the most uncertain parameter, however. What is the fate of the rest of the gas? In general long tidal tails are entrained in the

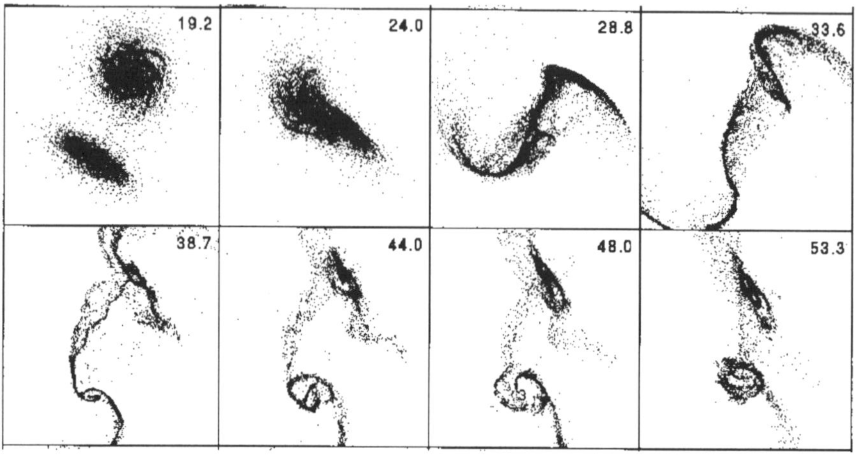

Figure 20. Simulations of the gas and young stellar components in the merger of disk/halo models. Time is indicated in units of 12 Myr. From Mihos & Hernquist[124] ©1996 AAS.

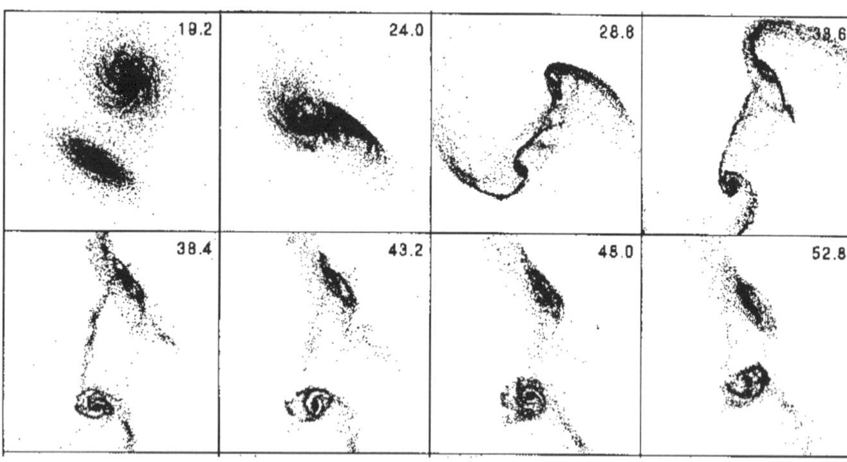

Figure 21. Same as Fig. 20 for the merger of disk/bulge/halo models. From Mihos & Hernquist[124]©1996 AAS.

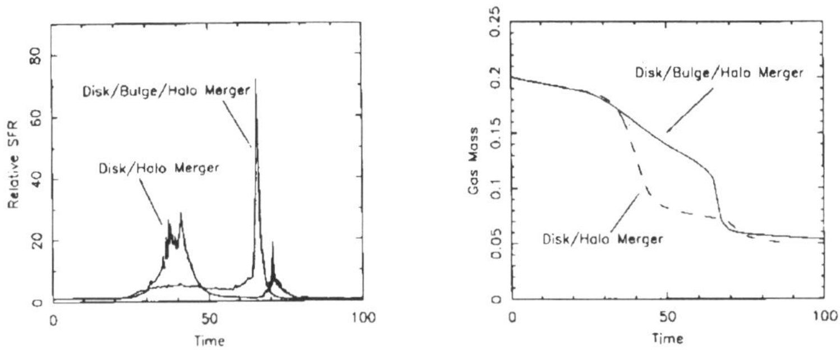

Figure 22. *Left:* evolution of the global star formation rate (relative to isolated disks) for the two models of Figs. 20 and 21. *Right:* evolution of the total gas mass. From Mihos & Hernquist[124] ©1996 AAS.

outer parts, especially in neutral hydrogen, since this is the most abundant component in external parts of galaxies. But most of the material of the tails is still bound to the system, and will rain down progressively onto the merger remnant[77]. Due to phase-wrapping processes, it gives rise to shells and ripples. The HI fraction in the tails appears to increase as the merger enters into more advanced stages. This might be due to the transformation of HI into H_2 in the central part. Also some of the gas is heated to the coronal phase (seen in X-rays).

During minor mergers, where the mass ratio between the two colliding galaxies is at least 3, the same features can be noticed: the first relevant parameter is the mass concentration in the galaxies before the interaction[76]. The torques responsible for the gas mass inflow are exerted by the non-axisymmetric (essentially $m = 2$) potential developed in the disk. The torques are much stronger towards the center; the inflow time is of the order of the rotation time-scale at each radius.

7 Observational Evidence

The role of large-scale dynamics and interactions of galaxies is clear for the starburst activity in galaxies[97,153]. IRAS ultra-luminous galaxies are all mergers[152], and the fraction of interacting galaxies is increasing with L_{IR}/L_B. But the influence of global dynamics is less clear for AGN activity. The fraction of Seyfert and quasars is however increasing with infrared luminosity[98], from 4 to 45%.

A large fraction of galaxies with radio jets (FR-II) are interacting[72,11]. QSOs appear to interact with companions[83,84] more than usual, and low-z quasars are mergers on HST images. Interactions are therefore efficient, but it has been very difficult to find any correlation between bars and nuclear activity. Are bars actually responsible for the fueling?

7.1 Search for Correlations between Bars and AGN

There have been several observational works revealing a correlation between nuclear activity and bars[40,163,128]. But the correlation is weak and depends on the definition of the samples, their completion and other subtle effects. Near-infrared images have often revealed bars in galaxies previously classified unbarred, certainly due to gas and dust effects. However, Seyfert galaxies observed in NIR do not statistically have more bars nor more interactions than a control sample[121,129].

It is however evident from observations that bars are efficient to produce radial gas flows: barred galaxies have more H_2 gas concentrations inside their central 500 pc than un-barred galaxies[151]. Also, the radial flows level out abundance gradients in barred galaxies[118].

Peletier et al.[136] and Knappen et al.[100] have recently re-visited the question, and took a lot of care with their active and control samples. Their Seyfert and control samples are different at 2.5σ, in the sense that Seyferts are more barred. They also measure the bar strength by the observed axial ratio in the images. In Seyferts, the fraction of strong bars is lower than in the control sample[161]. Although a surprising result a priori, this is not unexpected, if bars are believed to be destroyed by central mass concentrations (Sec. 4).

Regan & Mulchaey[146] have studied 12 Seyfert galaxies with HST-NICMOS. Out of the 12, only 3 have nuclear bars, but a majority show nuclear spirals. However their criterium for nuclear bars is that there exist leading dust lanes exist along this nuclear bar. This is not a required characteristic, since these secondary bars in general are not expected to have ILRs themselves.

A morphological study of the 891 galaxies in the Extended 12 μm Galaxy Sample (E12GS) has confirmed that Seyfert galaxies and LINERs have the same percentage of bars as normal spirals; contrary to HII/Starburst galaxies, that have more bars[80]. However, active galaxies show rings significantly more often than normal galaxies or starbursts. LINERs have more inner rings (by a factor 1.5), while Seyferts have more outer rings (by a factor 3-4) than normal galaxies. This might be due to the different time-scales for bar and ring formation: bars form relatively quickly, in a few 10^8 yr; they can drive matter to the central regions and trigger a starburst there, in the same time-scale. Outer rings form then, also under the gravity torques of the bar, but in the dynamical times of the outer regions, i.e. a few 10^9 yr. Since Seyferts are correlated with them, they would be associated to delayed consequences of the starburst, or of the bar, which by this time begins to dissolve.

The percentage of AGN in the E12GS sample is 30%. As for interactions, 25% of the Seyferts are "peculiar" (disturbed). There is also a correlation between AGN and morphological types along the Hubble sequence. Seyferts tend to lie in early-types[174,128]. This has to be related to the existence of inner Lindblad resonances in early-types, favoring the fueling of the nucleus[33] (e.g. Sec.4).

In summary, there might be some evidence of the role of large-scale dynamics on AGN fueling, but it is in general weak, except for the most powerful AGN. A good correlation between bars/interactions and AGN is not expected, from several arguments:
- there must be already a massive black hole in the nucleus, and this might be the case only for massive-bulge galaxies (not all barred galaxies);
- again a large central mass concentration (bulge) is necessary to produce an ILR and drive the gas inwards (early-types);
- other parameters, like geometrical parameters, control the fueling efficiency of interaction;
- time-scales are not fitted: the AGN fueling is postponed after the interaction/bar episode;
- there are other mechanisms to fuel AGN such as a dense nuclear cluster.

7.2 Galaxy Interactions and Nuclear Activity

Huge starbursts require galaxy interactions and mergers[152]. For AGN of low luminosity, external triggering appears less necessary, since only $0.01 M_\odot$/yr is required for a Seyfert like NGC 1068 for example, during 10^8 yr. However, once interactions drive gas to the nucleus, some activity must be revived. Time-scales may be the reason why the actions are not simultaneous. Large-scale gas has to be driven at very small scales in the center, and the whole process requires several intermediate steps.

Even for the good starburst/interactions correlations, there are exceptions for the low- luminosity samples. Optically selected interacting galaxies (not IRAS galaxies) don't often have enhanced star formation[22,104]. Only obviously merging galaxies, like the Toomre sample[178], are truly a starbursting class; it is difficult to reveal a progression along a possible evolutionary sequence[73]. There are too many determining parameters: geometry, distance, gas content, etc...

A complication comes from the time-scales involved: the starburst phase is short, of the order of a few 10^8yr, similar only to the end of the merging phase. Morphological distortions are more correlated with activity than the presence of nearby companions. More than 50% of ULIRGs possess multiple nuclei[25,66].

Are Seyfert galaxies preferentially interacting? According to Dahari[40], 15% have close companions, compared to 3% in the control sample. But the Seyferts with or without companions have the same Hα or radio power[41], although they may be more infrared-bright with companions[42,120]. According to Keel et al.[95], there are 5% of Seyferts in control sample, and 25% in the close pairs of Arp Atlas. But it is possible that the Arp Atlas galaxies suffer from selection effects. Bushouse[22], on the contrary, finds a deficiency of Seyferts in interacting galaxies.

In summary, if the environment influence is evident for powerful QSOs, radio-galaxies and BL Lacs, it is not so significant for Seyferts[45].

Surprising observations also involve low surface brightness (LSB) galaxies that are about 6 times less bright than high surface brightness ones. Their low evolution state is generally attributed to their isolated environment. A large fraction of active nuclei have been reported in their category: 65% in about 50 LSB instead of 1% expected for these low luminosities[101,168]. Are there unknown selection effects?

7.3 Radio-Galaxies

Interactions are evident in the more powerful objects only. The FR-I objects have low-power, radio jets declining with distance, and are in general elliptical galaxies. Fewer than 10% show tails, and tidal interactions[165]. The rarer FR-II objects, with high-power, are classical doubles (Cygnus A, Perseus A). They show a high percentage of interactions[187,82], from 32 to 100%; most of them ($> 50\%$) have at least 2 companions, blue colors, star-forming regions[73,67]. Against the central radio-source, HI is seen in absorption, revealing that gas is most of the time infalling (absorption lines are blue-shifted[184]).

Radio-loud QSOs have 4-5 times more neighbors[188,73], while radio-quiet QSOs have only 2 times more neighbors with respect to a control sample. The morphology of QSO hosts is disturbed in 35-55% of the cases[81,164], while radio-loud QSO are perturbed[164,82] at 70-80%; this is confirmed with HST[47].

The radio power of radio-galaxies has been related to the spin-down of a rotating supermassive black hole[13]. The spin of the black hole may be acquired in galaxy interactions and mergers: either the black hole is spin up through gas accretion from an external disk, or through the merging of two black holes. In this respect, it is significant that radio-galaxies are generally giant ellipticals, since these are expected to be formed essentially by mergers.

7.4 Conclusion

Interactions are an essential mechanism in the fueling of QSOs, high-power radio-galaxies and ULIRGs. But they are not dominant in Seyferts and low-power radio-galaxies.

Observations reveal a close complicity between starbursts and AGN. AGN may be fueled by mass loss during PMS stellar evolution of a coeval stellar cluster of[134] $4 \times 10^9 M_\odot$ within 10pc. Within 10^8yr, mass loss accumulates $1.5 \times 10^9 M_\odot$. Radiation (UV, X) from the inner accretion disk ionizes stellar envelopes, and form the BLR ($V \sim 3000$ km/s in the black hole potential well). The young stellar cluster requires a huge starburst, like those due to mergers. Shocks are then inevitable, model-independent supernovae can help transferring angular momentum. Starbursts and AGN are symbiotic systems[28,137,186,29].

8 Black Hole Evolution

How do the massive black holes in the center of galaxies form and grow? Can we explain the observed peak in quasar activity at $z = 2$ and the decline since then (see Fig. 23)? Why is this curve so parallel to the star formation history? If galaxies are themselves formed through interactions and mergers, what is the fate of the black holes in the merger, and are binary black holes observable?

8.1 Black Hole Growth and Activity Life-Time

It is widely believed that AGN derive their power from supermassive black holes, but their formation history, their demography, their time-scales of activity are still debated. Two extreme hypotheses have been explored: either only a small percentage of galaxies become quasars, and they are continuously fueled, and active over Gyrs; or massive black holes exist in almost all galaxies, but they have active periods of only a few 10^7 yr. In the first hypothesis, black holes with masses 100 times larger than the maximum observed today and accretion rates much lower than the Eddington rate should exist, and this is not supported[26]. Models with a duty cycle of 4×10^7 yr are favored, and

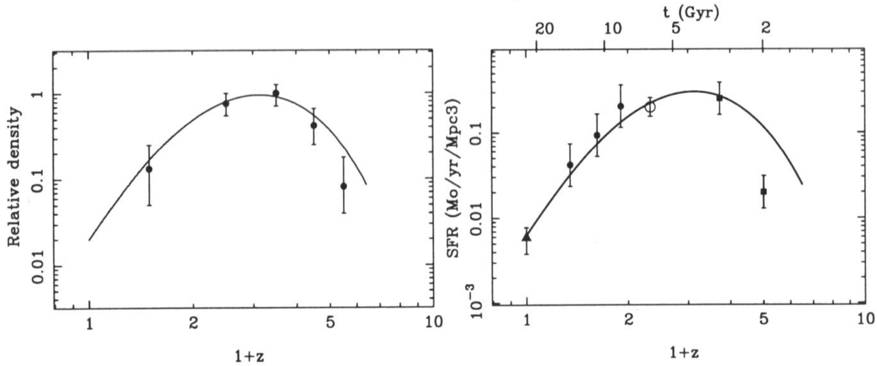

Figure 23. *Left:* space density as a function of redshift, normalized to $z = 2 - 3$, for the Parkes flat-spectrum radio-loud quasars with $P_{11} > 7.2 \times 10^{26}$ W Hz^{-1} sr^{-1}. The optically-selected quasars follow the same curve. *Right:* cosmic history of star formation, for comparison. The various data points, coming from different surveys, give the universal metal ejection rate, or the star formation rate SFR (left-scale), as a function of redshift z. Reprinted by permission from Nature (Shaver et al.[154]) ©1996 Macmillan Magazines Ltd.

many galaxies today should host a starving black hole[68].

The density of black holes derived today from the kinematics of galaxy centers at high spatial resolution[115,57] strongly constrains the duty cycle time-scale. Also the cosmic background radiation detected at many wavelengths constrains the formation history. The volume density of massive black holes today ρ_{BH} can be expressed[69] as a function of the density of bulges Ω_{bul} (normalized to the critical density to close the Universe), and of the observed ratio between black-hole and bulge mass M_{BH}/M_{bul},

$$\rho_{BH} = 1.1 \times 10^6 \frac{M_{BH}/M_{bul}}{0.002} \frac{\Omega_{bul}}{0.002h^{-1}} M_\odot \, \mathrm{Mpc}^{-3} \quad , \tag{20}$$

with h the usual ratio of the Hubble constant to 100 km/s/Mpc. It is interesting to note that this ratio largely exceeds (by a factor 10) the mass density in black holes needed to explain the blue light purely by accretion on AGNs,

$$\rho_B = 1.4 \times 10^5 \frac{f_B \epsilon}{0.01} M_\odot \, \mathrm{Mpc}^{-3} \quad , \tag{21}$$

where f_B is the fraction of light emitted in blue, and ϵ the mass-to-energy conversion efficiency (Sec. 2). This might be interpreted as a strong extinction of the blue light of AGNs, since part of the energy then can be seen in the infrared. From the detected cosmic infrared radiation, and assuming that about 30% is coming from AGN[65], the predicted density is

$$\rho_{\rm IR} = 7.5 \times 10^5 \frac{f_{\rm IR}\epsilon}{0.1} M_\odot \, {\rm Mpc}^{-3} \quad . \tag{22}$$

Part of the radiation is also observed through hard X-rays, and from the background hard-X-ray radiation, the corresponding black hole density would be

$$\rho_{\rm X} = 3.8 \times 10^5 \frac{f_{\rm X}\epsilon}{0.01} M_\odot \, {\rm Mpc}^{-3} \quad . \tag{23}$$

Note however that the sources contributing to the submillimeter and hard-X-ray backgrounds are not the same, as discovered recently with Chandra[158]. This could mean that even the hard-X-rays are extincted and most of the AGN radiation comes essentially in the infrared.

The best interpretation of the observations lead to an AGN activity lifetime of $\sim 4 \times 10^7$ yr. During this time an AGN would accrete and radiate at the Eddington limit. Before, the black hole is not massive enough to accrete efficiently, since the Eddington limit is proportional to its mass. The external fueling rate is larger than the limit during the growth phase, then during its bright phase of $\sim 4 \times 10^7$ yr the AGN radiates the most efficiently at Eddington luminosity, until the fuel is exhausted, and the AGN fades away (see Fig. 24).

In cold dark matter (CDM) hierarchical scenarios, the smallest haloes form first, and they are denser, since the density of the Universe is a strong increasing function of redshift. Then the dynamical time-scale in such structures are much shorter. Due also to a higher gas mass fraction at high redshift (not yet consumed in stars), there are many reasons why the mass of the black holes formed are higher at high redshift (even if they form in smaller-mass systems). This can explain the density distribution and luminosity function observed for quasars[21]: at high z, quasars were both more numerous and their black holes more massive (assuming that we all see them in their bright phase, at Eddington luminosity).

To take these arguments into account, and reproduce the fact that the more massive black holes form in the smaller haloes, Haehnelt & Rees[68] assume the ratio

$$M_{\rm BH}/M_{\rm halo} \propto (1+z)^6 \exp(-v_*/v_{\rm cir})^3 \quad , \tag{24}$$

where the dependence in (1+z) corresponds to the higher efficiency to form a black hole at high z (due to the high densities), the exponential cut-off for low $v_{\rm cir}$ takes into account the difficulty of small mass system to gravitationally retain their gas in their small potential well (v_* is a normalizing value). With

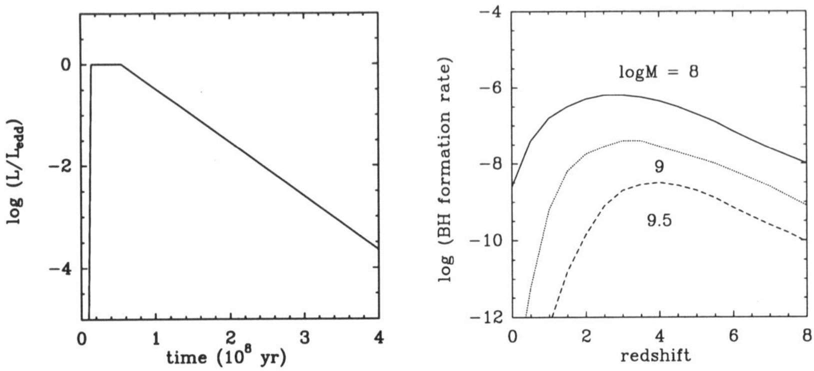

Figure 24. *Left:* a schematic view of the activity phase of a typical quasar. During the growth phase, the AGN does not radiate efficiently, although the fueling rate is larger than Eddington. After its luminous phase of $\sim 4 \times 10^7$ yr, the fuel is exhausted and the AGN fades away. *Right:* the black hole formation rate, or number density of newly formed BH per comoving volume, per time, and $\ln M$, in Mpc^{-3} $(10^7\ \mathrm{yr})^{-1}$ $(\ln M)^{-1}$. From Haehnelt & Rees[68] ©1993 Blackwell Sci.

this model, the result is indeed that at low redshift mostly small black holes are built, while the peak of formation of very massive black holes occurs at $z = 4$ (Fig. 24).

Today, merging of galaxies create a quasar in only small-mass gaseous objects. If the black hole formation process is not efficient at low redshift, the main process creating AGNs now is the tidally triggered nuclear activity, through gas accretion and minor mergers. To explain for instance the 1% of Seyferts in field galaxies[79], it is sufficient to assume a duty cycle of 1% of "normal" galaxies.

8.2 Binary Black Hole Formation

Given the large frequency of galaxy encounters and mergers, if there is a massive black hole in nearly every galaxy, the formation of a binary black hole should be a common phenomenon. The successive physical processes able to brake the two black holes in their relative orbit have been considered by Begelman et al.[12]. Each black hole sinks first toward the merger remnant center through dynamical friction onto stars. A binary is formed; but the life-time of such a binary can be much larger than a Hubble time, if there is not enough stars to replenish the loss cone, where stars are able to interact with the binary. Once a loss cone is created, it is replenished only through 2-body relaxation between stars, and this can be very long (see Sec. 2). Modeling the merger remnant as an elliptical, with a core of radius r_c and mass M_c (and corresponding velocity V_c), the radius where loss cone effects are significant is: $r_{lc}/r_c = (M_{BH}/M_c)^{3/4}$. The various time-scales involved, and corresponding characteristic scales are defined by the following steps:

- the dynamical friction on stars, in less than a galactic dynamical time,

$$t_{df} \sim (V_c/300\,\mathrm{km/s})(r_c/100\,\mathrm{pc})^2(10^8/M_{BH})\,\mathrm{Myr} \quad ; \tag{25}$$

- when the separation of the binary shrinks to a value $r_b = r_c(M_{BH}/M_c)^{1/3}$, the two black holes become bound together;

- the binary hardens, with $r_h \propto (r/r_b)^{3/2}$;

- but when $r < r_{lc} = (M_{BH}/M_c)^{3/4} r_c$, the stars available for the binary to interact with, are depleted through the loss cone effect, and replenished only by 2-body relaxation;

- gas infall can reduce the binary life-time (whether the gas is flung out, or accreted, there is a contraction of the binary) in t_{gas};

- gravitational radiation shrinks the orbit on

$$t_{GR} \sim 0.3\,\mathrm{Myr}\,(10^8/M_{BH})^3(r/0.003\,\mathrm{pc})^4 \quad , \tag{26}$$

  if the two black holes have comparable masses.

All these time-scales are represented on Fig. 25. If the binary life-time is too long, another merger with another galaxy will bring a third black-hole. Since a three-body system is unstable, one of the three black-holes will be ejected by the gravitational slingshot effect.

Since the life-time of the binary is not short, there should be observable manifestations of massive black hole binaries. One of the best tracer is to detect the periodicity of the keplerian motion, with the period $P \sim 1.6$ yr $r_{16}^{3/2} M_8^{-1/2}$. This is the case for the AGN OJ 287 where eclipses have been monitored for a century[173,105,138]. Also, if the black holes are rotating, and their spins have misaligned axes, they precess around the orbital one. Plasma beams (aligned to the hole axis) precess, and curved jets should be observed, with periods between 10^3 to 10^7 yr. This is frequently the case in radio structures observed with VLA and VLBI, modified by Doppler boosting, and light travel time (cf 3C 273, NGC 6251, 1928+738)[91,148]. Finally, pairs

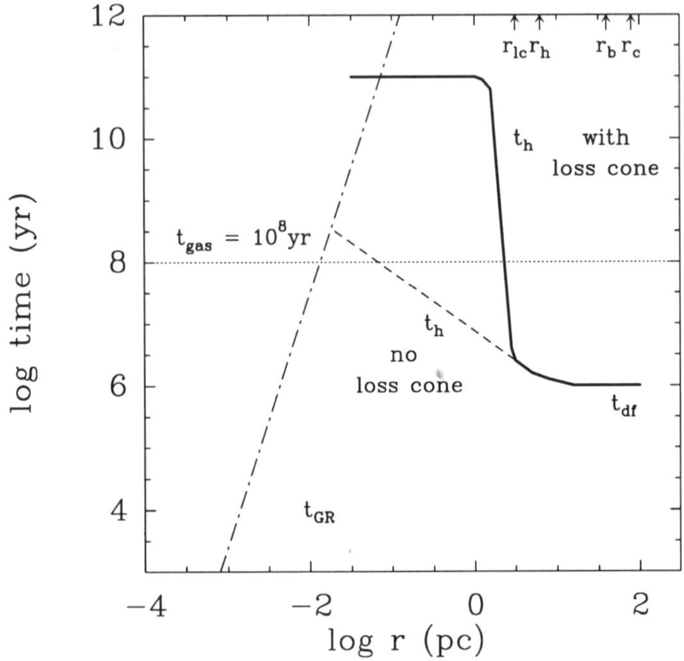

Figure 25. Time-scales relevant to the formation of a black-hole binary, in the merger remnant of two galaxies. Dynamical friction in the core of the elliptical galaxy formed, is acting in $t_{df} \sim 10^6$ yr. Within r_h, the evolution will proceed on $t_h = (r_h/r)t_{df}$ if loss-cone effects are neglected. However, taking into account depletion of stars in the loss cone leads to $t_h = 10^{11}$ yr, a much longer time. If infall of gas is considered, $t_{gas} \sim 10^8$ yr applies. At the end, gravitational radiation will take over (t_{GR}). The core radius r_c, the "bound" radius r_b, the hardening radius r_h and the loss-cone radius r_{lc} are defined in the text. Reprinted by permission from Nature (Begelman et al.[12]) ©1980 Macmillan Magazines Ltd.

of radio galaxies have been observed during their merger with four radio jets (3C 75)[135].

Numerical simulations have brought more precision in the determination of the life-time of the binary, although numerical artifacts have given rise to debates. Ebisuzaki et al.[49] claimed that the life-time of the binary should be much shorter if its orbit is excentric, since then the binary can interact with more stars and release the loss cone problem. The first numerical simulations tended to show that orbit excentricity should grow quickly through dynamical friction[61]. Mikkola & Valtonen[126] and others found that the excentricity in fact grows only very slowly.

Numerical simulations suffer from a restricted number of bodies N, and consequently of a large random velocity of the binary (that should decrease in $N^{-1/2}$). The binary then wanders in or even out of the loss cone, and the effect of the loss cone depletion does not occur[116]. Also the 2-body relaxation time is shorter than in the real system, contributing to replenish the cone. Numerically, the life-time of the binary depends on the total number of particles, i.e. the ratio between the black hole to particle mass,

$$t_{\rm b} \propto (M_{\rm BH}/m*)^{0.3} \quad . \tag{27}$$

To summarize the conclusions of several numerical computations, there is finally little dependence on excentricity e, only in rare cases, when e is large from the beginning[141]. Eventually, the wandering of the binary helps merging the two black holes[142]. The ejection out of the core of stars interacting with the binary weakens the stellar cusp, while the binary hardens. This may help to explain the surprisingly weak stellar cusps in the center of giant ellipticals observed recently with HST. Observations show that bright elliptical galaxies have weak cusps, while faint galaxies have strong cusps, with a power-law slope of density versus radius of up to 2. A way to weaken the cusps is a sinking black hole[131], and this could be the case for giant galaxies that have experienced many mergers in their life.

8.3 Hierarchical Merging Scenario

Semi-analytic models, based on the Press-Schechter formalism, and a CDM hierarchical scenario of galaxy formation[94], can reproduce rather well the essential observations: the proportionality relation between the bulge and black hole mass in every galaxy, the amount of energy radiated over the Hubble time due to accretion onto massive black holes, the past evolution of AGN activity. It is assumed that black holes grow through galaxy merging, both because of merging of pre-existing black holes, and because the infall of gas to the center can fuel the merged black hole. It is also assumed that the cold gas in galaxies decreases with time; this implies that the fueling will also decrease with time, accounting for the observed decline of AGN activity. Finally, the gas accretion time-scale is proportional to the dynamical time-scale, which is shorter at high redshift. Quasars convert mass to energy at a fixed efficiency, and cannot radiate more than the Eddington limit.

The results of such simulations are a strong decrease of the gas fraction in galaxies, from 75% at $z = 3$ to 10% at $z = 0$, corresponding to the gas density decrease observed in damped Lyman alpha systems (cf Fig. 26); this

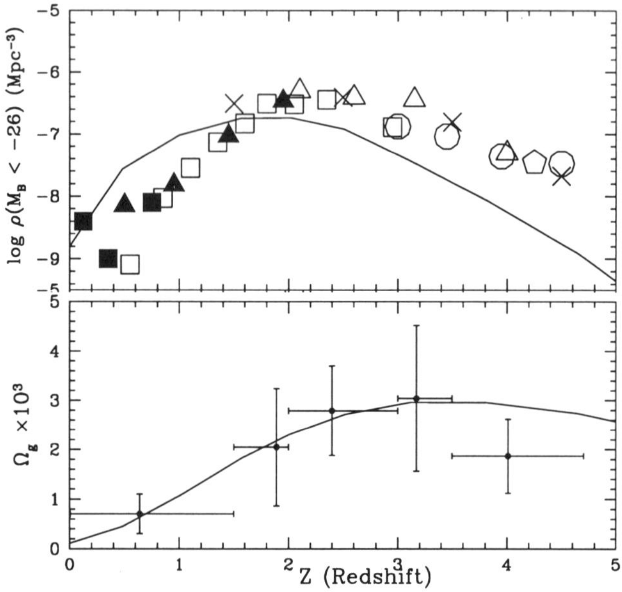

Figure 26. *Top*: the evolution of the space density of quasars with $M_B < -26$ for the ΛCDM model. From Kauffmann & Haehnelt[94]. *Bottom*: the cosmological mass density in cold gas in galaxies as a function of redshift for the ΛCDM model. From Kauffmann & Haehnelt[94] ©2000 Blackwell Sci. (data from Storrie-Lombardie et al.[170]).

implies that the black holes in the smallest ellipticals that formed at high z are relatively more massive, since there was more gas at this epoch. Ellipticals forming today have smaller black holes. Also the brightness of AGN for a given galaxy was relatively higher in the past. The rapid decline of quasars is then due to several causes:

– a decrease in the merging rate (which is also the cause of the decrease of the star forming rate);

– the decrease of the gas amount in galaxies;

– the decrease of the accretion rate (the dynamical processes are slower).

In these kinds of models, it is natural to expect a proportionality between bulge and black hole masses, since they both form with the same mechanisms, the hierarchical merging, and the corresponding dynamical gas concentration. It is interesting to note that the life-time of the quasar phase, a few 10^7 yr is then derived.

9 Conclusions

The source of fueling depends on the strength and luminosity of the AGN:
– for low luminosity AGN and Seyferts, only stars from a dense nuclear cluster are sufficient (through tidal distortions and stellar collisions);
– for high luminosity AGN and quasars, large accretion rates are required, which involve large-scale gravitational instabilities. Those drive gas towards the center that trigger big starbursts, and the coeval compact cluster just formed can provide the fuel through mass loss of young stars and supernovae. This gas must have been driven from galactic disks, through internal gravitational instabilities (bars, spirals), more generally the consequence of interactions and mergers.

Galaxy disks are in general far from stationary equilibrium, most often subject to $m = 2$ and $m = 1$ instabilities. These non-axisymmetric instabilities produce gravity torques, that drive the gas inwards. When the mass concentration towards the center is large enough, a secondary bar can decouple and rotate with a higher pattern speed. Embedded bars or non-axisymmetric structures can take over the gas flow to fuel the nucleus. The fueling is therefore favored in early-type objects.

Interactions and mergers also produce bars and non-axisymmetric structures, that fuel the nucleus through their gravity torques. They first trigger huge starbursts in the merger center, that could afterwards fuel the AGN. Since internal instabilities, external trigger and hierarchical merging both produce the bulges and fuel the nucleus, it is natural to expect a proportionality in their masses.

Acknowledgments

I am very grateful to the organizers for inviting me to give these lectures, in a very friendly and scientifically exciting ambiance.

References

1. Adams, F. C., Ruden, S. P., Shu, F. H., ApJ, 347, 959, 1989.
2. Alexander T., Netzer N., MNRAS, 270, 781, 1994.
3. Alexander T., Netzer N., MNRAS, 284, 967, 1997.
4. Athanassoula, E., MNRAS, 259, 345, 1992.
5. Bacon R., Emsellem E., Monnet, G., Nieto, J. L., A&A, 281, 691, 1994.
6. Bacon R., Emsellem E., Combes F. et al., A&A in press
7. Bahcall, J. N., Wolf, R. A., ApJ, 209, 214, 1976.

8. Barnes J.E., ApJ, 331, 699, 1988
9. Barnes J.E., Hernquist L., ARAA, 30, 705, 1992.
10. Barth A., Ho L. C., Filippenko, A. V., Sargent, W. L., AJ, 110, 1009, 1995.
11. Baum, S. A., Heckman, T. M., van Breugel, W., ApJ, 389, 208, 1992.
12. Begelman M.C., Blandford R.D., Rees M.J., Nat, 287, 307, 1980.
13. Begelman M.C., Nat, 322, 614, 1986.
14. Benedict, F. G., Smith, B. J., Kenney, J. D. P., AJ, 111, 1861, 1996.
15. Bertin G., Romeo A., A&A, 195, 105, 1988.
16. Bertola F., Buson, L. M., Zeilinger, W. W., ApJ, 401, L79, 1992.
17. Bertola, F., Cinzano, P., Corsini, E.M., et al., ApJ, 458, L67, 1996.
18. Block D., Elmegreen, B. G., Wainscoat, R. J., Nat, 381, 674, 1996.
19. Bosma A., AJ, 86, 1825, 1981.
20. Bottema R., A&A, 275, 16, 1993.
21. Boyle B.J., Jones L.R., Shanks T. et al., in "The Space Density of Quasars", ASP Vol 21, p. 191, 1991.
22. Bushouse, H. A., AJ, 91, 255, 1986.
23. Buta R., Combes F., Fundamental Cosmic Physics, 17, p. 95-282, 1996.
24. Buta R., Combes, F., in "Dynamics of Galaxies, From the Early Universe to the Present", ed. F. Combes, G.A. Mamon, & V. Charmandaris, ASP Vol. 197, p. 11, 2000.
25. Carico, D. P., Graham, J. R., Matthews, K. et al., ApJ, 349, L39, 1990.
26. Cavaliere A., Padovani P., ApJ, 340, L5, 1989.
27. Colgate S.A., ApJ, 150, 163, 1967.
28. Collin S., Dyson, J. E., McDowell, J. C., Perry, J., MNRAS, 232, 539, 1988.
29. Collin S., Zahn J-P., A&A, 344, 433, 1999.
30. Combes, F., Gerin, M., A&A, 150, 327, 1985.
31. Combes, F., in "Galactic and Extragalactic Star Formation", ed. R. E. Pudritz and M. Fich, Kluwer, p. 475, 1988.
32. Combes, F., Debbasch, F., Friedli, D., and Pfenniger, D., A&A 233, 82, 1990.
33. Combes, F., Elmegreen, B. G., A&A, 271, 391, 1993.
34. Combes F., in "Mass-transfer induced activity in galaxies" ed. I. Shlosman, Cambridge Univ. Press, p. 170, 1994.
35. Combes, F., Becquaert, J.-F., A&A, 326, 554, 1997.
36. Combes, F., in "Dynamics of Galaxies, From the Early Universe to the Present", ed. F. Combes, G.A. Mamon, & V. Charmandaris, ASP Vol. 197, p. 15, 2000.

37. Comins, N. F., Lovelace, R. V. E., Zeltwanger, T., Shorey, P., ApJ 484, L33, 1997.
38. Contopoulos, G. and Grosbol, P., A&AR, 1, 261, 1989.
39. Courvoisier, T. J.-L., Paltani, S., Walter, R., A&A, 308, L17,1996.
40. Dahari, O., AJ, 89, 966, 1984.
41. Dahari, O., AJ 90, 1772, 1985.
42. Dahari, O., DeRobertis M., ApJ 331, 727, 1988.
43. Dame T.M., Ungerechts H., Cohen R.S. et al., ApJ, 322, 706, 1987.
44. de Grijs, R., Peletier, R. F., van der Kruit, P. C., A&A 327, 966, 1997.
45. deRobertis, M., Hayhoe, K., Yee, H. K. C., AAS, 189, 109.04, 1996
46. Dickey J.M., Hanson M.M., Helou G., ApJ, 352, 522, 1990.
47. Disney M. J., Boyce, P. J., Blades, J. C., et al., Nat 376, 150, 1995.
48. Duncan M.J., Shapiro S.L., ApJ, 268, 565, 1983.
49. Ebisuzaki, T., Makino, J., Okumura, S. K., Nat, 354, 212, 1991.
50. Eckart A., Genzel, R., Hofmann, R., Sams, B. J., Tacconi-Garman, L. E., ApJ 407, L77, 1993.
51. Edwards A.C., MNRAS, 190, 757, 1980.
52. Elmegreen, B. G., Elmegreen, D. M., ApJ, 288, 438, 1985.
53. Elmegreen, B. G., Elmegreen, D. M., ApJ, 342, 677, 1989.
54. Elmegreen B.G., Kaufman M., Thomasson M., ApJ, 412, 90, 1993.
55. Elmegreen, B. G., Elmegreen, D. M., Brinks, E. et al., ApJ, 503, L119, 1998.
56. Englmaier, P., Shlosman, I., ApJ 528, 677, 2000.
57. Ferrarese, L., Merritt, D., ApJ, 539, L9, 2000.
58. Frank J., Rees M.J., MNRAS, 176, 633, 1976.
59. Freeman K. C., ApJ, 160, 811, 1970.
60. Friedli, D., & Martinet, L., A&A, 277, 27, 1993.
61. Fukushige, T., Ebisuzaki, T., Makino, J., ApJ 396, L61, 1992.
62. Galletta, G., ApJ, 318, 531, 1987.
63. Garcia-Burillo S., Guélin M., A&A, 299, 657, 1995.
64. Garcia-Burillo S., Sempere M.J., Combes F., Hunt L.K., Neri R., A&A, in press
65. Genzel, R., Lutz, D., Sturm, E. et al., ApJ, 498, 579, 1998.
66. Graham, J. R., Carico, D. P., Matthews, K., Neugebauer, G., Soifer, B. T., Wilson, T. D., ApJ, 354, L5, 1990.
67. Gubanov A.G., Pis'ma v A.Zh. vol. 17, p. 684, 1991.
68. Haehnelt M.G., Rees M.J., MNRAS, 263, 168, 1993.
69. Haehnelt M.G., Natarajan P., Rees M.J., MNRAS, 300, 817, 1998.
70. Hasan, H., & Norman, C.A., ApJ, 361, 69, 1990.
71. Hasan H., Pfenniger D., Norman C, ApJ 409, 91, 1993.

72. Heckman T.M., Smith E., Baum S. et al., ApJ, 311, 526, 1986.
73. Heckman T.M., in “Paired and Interacting Galaxies” IAU Coll. No. 124 p 359, 1990
74. Heemskerk, M. H. M., Papaloizou, J. C., Savonije, G., A&A, 260, 161, 1992.
75. Heller C.H., Shlosman I., ApJ, 424, 84, 1994.
76. Hernquist, L., Mihos, J.C., ApJ 448, 41, 1995.
77. Hibbard J.E., PhD thesis, Columbia University, 1995.
78. Hills J.G., Nat 254, 295, 1975.
79. Huchra J., Burg R., ApJ 393, 90, 1992.
80. Hunt L.K., Malkan M.A., ApJ, 516, 660, 1999.
81. Hutchings, J. B., Crampton D., Campbell B., ApJ, 280, 41, 1984.
82. Hutchings, J. B., ApJ, 320, 522, 1987.
83. Hutchings, J. B., Neff S. G. AJ, 104, 1, 1992.
84. Hutchings, J. B., Morris, S. C., AJ, 109, 928, 1995.
85. Impey C., Bothun G., ApJ, 341, 89, 1989.
86. Ishizuki S., Kawabe R., Ishiguro M. et al., Nat, 344, 224, 1990.
87. Jog C., ApJ, 390, 378, 1992.
88. Jog C., MNRAS, 278, 209, 1996.
89. Jungwiert B., Combes, F., Axon, D. J., A&AS, 125, 479, 1997.
90. Junqueira S., Combes, F., A&A, 312, 703, 1996.
91. Kaastra J.S., Roos N., A&A, 254, 96, 1992.
92. Kalnajs, A. J., Proc. Astron. Soc. Australia 2, 174, 1973.
93. Kalnajs, A.J., ApJ, 212, 637, 1977.
94. Kauffmann, G., Haehnelt, M., MNRAS, 311, 576, 2000.
95. Keel W.C., Kennicutt R.C., Hummel, E., van der Hulst J-M., AJ, 90, 708, 1985.
96. Kennicutt R.C., ApJ, 344, 685, 1989.
97. Kennicutt R.C., Roettiger, K. A., Keel, W. C., van der Hulst, J. M., Hummel, E., AJ, 93, 1011, 1987.
98. Kim, D.-C., Veilleux, S., Sanders, D. B., ApJ 508, 627, 1998.
99. Kinney A.L., Schmitt H.R., Clarke C.J., Pringle J.E., Ulvestad J.S., Antonucci R.R.J, ApJ in press (astro-ph/0002131)
100. Knapen, J. H., Shlosman, I., Peletier R.F., ApJ, 529, 93, 2000.
101. Knezek, P., Schombert, J., AAS, 183, 4612, 1993.
102. Kormendy J., Bender R., ApJ, 522, 772, 1999.
103. Larson R.B. in "Starbursts and galaxy evolution", ed. T.X. Thuan, T. Montmerle, J. T. T. Van, Ed. Frontières, p. 467, 1987.
104. Lawrence, A., Rowan-Robinson, M., Leech, K., Jones, D. H. P., Wall, J. V., MNRAS, 240, 329, 1989.

105. Lehto H.J., Valtonen M.J., ApJ, 460, 207, 1996.
106. Levine S.E., Sparke L.S., ApJ, 496, L13, 1998.
107. Lewis J.R., Freeman K.C.,AJ, 97, 139, 1989.
108. Lin C.C., Shu F.H., ApJ 140, 646, 1964.
109. Lin D.N.C., Pringle J.E., ApJ, 320, L87, 1987.
110. Ligthman A.P., Shapiro S.L., ApJ, 211, 244, 1977.
111. Lovelace, R. V. E., Jore, K. P., Haynes, M. P., ApJ, 475, 83, 1997.
112. Luminet J-P., in "L'Activité dans les Galaxies", Annales de Physique, vol 12, p. 23 (Les Editions de Physique), 1987.
113. Lynden-Bell, D., Kalnajs, A.J., MNRAS, 157, 1, 1972.
114. Maciejewski, W., & Sparke L.S., ApJ, 484, L117, 1997.
115. Magorrian, J., Tremaine, S., Richstone, D., et al., AJ, 115, 2285, 1998.
116. Makino, J., Fukushige, T., Okumura, S. K., Ebisuzaki, T., PASJ 45, 303, 1993.
117. Mark J.W-K., ApJ, 193, 539, 1974.
118. Martin P., Roy J-R., ApJ, 424, 599, 1994.
119. Masset F., Tagger M., A&A, 322, 442, 1997.
120. McKenty J., ApJ, 343, 125, 1989.
121. Mcleod, K. K., Rieke, G. H., ApJ, 441, 96, 1995.
122. Meurer G.R., in "The interplay between massive star formation, the ISM and galaxy evolution", ed. D. Kunth, B. Guiderdoni, M. Heydari, T. X. Thuan, Editions Frontières, p. 333, 1995.
123. Mihos J.C., Hernquist L., ApJ, 437, 611, 1994.
124. Mihos, J.C., Hernquist, L., ApJ, 464, 641, 1996.
125. Mihos J.C., Richstone D.O., Bothun G.D., ApJ, 400, 153, 1992.
126. Mikkola, S.,Valtonen, M. J., MNRAS 259, 115, 1992.
127. Miller R.H., Smith B.F., ApJ, 393, 508, 1992.
128. Moles, M., Marquez, I., Perez, E., ApJ 438, 604, 1995.
129. Mulchaey J.S., Regan M.W., ApJ, 482, L135, 1997.
130. Murphy, B. W., Cohn, H. N., Durisen, R. H., ApJ, 370, 60, 1991.
131. Nakano T., Makino J., ApJ, 525, 77, 1999.
132. Noguchi M., Ishibashi S., MNRAS, 219, 305, 1986.
133. Norman C., Silk J., ApJ, 266, 502, 1983.
134. Norman C., Scoville N.Z., ApJ, 332, 124, 1988.
135. Owen, F. N., Odea, C. P., Inoue, M., Eilek, J. A., ApJ, 294, L85, 1985.
136. Peletier, R. F., Knapen, J. H., Shlosman, I., et al., ApJS, 125, 363, 1999.
137. Perry J., in "Relationships between active galactic nuclei and starburst galaxies", ed. A. V. Filippenko, ASP Conference Series (ASP, San

Francisco), vol. 31, p. 169, 1992.
138. Pietilä H., ApJ, 508, 669, 1998.
139. Pfenniger, D., & Norman, C., ApJ, 363, 391, 1990.
140. Phinney E.S., in "Mass-transfer induced activity in galaxies" ed. I. Shlosman, Cambridge Univ. Press, p. 1, 1994.
141. Quinlan G.D., New A., 1, 35, 1996.
142. Quinlan G.D., Hernquist L., New A., 2, 533, 1997.
143. Raha, N., Sellwood, J. A., James, R. A., Kahn, F., Nat, 352, 411, 1991.
144. Rand R.J., ApJ, 410, 68, 1993.
145. Rauch K.P., ApJ, 514, 725, 1999.
146. Regan, M. W., Mulchaey, J. S., AJ, 117, 2676, 1999.
147. Richter O-G., Sancisi R., A&A, 290, L9, 1994.
148. Roos N., Kaastra J.S., Hummel C.A., ApJ, 409, 130, 1993.
149. Rubin V.C., Kenney J.D.P., Young J.S., AJ, 113, 1250, 1997.
150. Rubin, V. C., Graham, J. A., Kenney, J. D. P., ApJ, 394, L9, 1992.
151. Sakamoto, K., Okumura, S. K., Ishizuki, S., Scoville, N. Z., ApJ, 525, 691, 1999.
152. Sanders D.B., Mirabel I.F., ARAA, 34, 749, 1996.
153. Sanders, D. B., Soifer, B. T., Elias, J. H., Neugebauer, G., Matthews, K., ApJ, 328, L35, 1988.
154. Shaver, P. A., Wall, J. V., Kellermann, K. I., Jackson, C. A., Hawkins, M. R. S., Nat, 384, 439, 1996.
155. Schwarz, M. P., ApJ, 247, 77, 1981.
156. Schweizer, F., Seitzer, P., AJ, 116, 2206, 1998.
157. Scoville N.Z., Sargent, A. I., Sanders, D. B., Soifer, B. T., ApJ, 366, L5, 1991.
158. Severgnini, P., Maiolino, R., Salvati, M., et al., A&A, 360, 457, 2000.
159. Shields G.A., Wheeler J.C., ApJ, 222, 667, 1978.
160. Shlosman, I., Frank J., Begelman M.C., Nat, 338, 45, 1989.
161. Shlosman, I., Peletier, R. F., Knapen, J. H. , ApJ, 535, L83, 2000.
162. Shu, F. H., Tremaine, S., Adams, F. C., Ruden, S. P., ApJ, 358, 495, 1990.
163. Simkin S.M., Su H.J., Schwarz M.P., ApJ, 237, 404, 1980.
164. Smith E., Heckman T., Bothun G. et al., ApJ, 306, 64, 1986.
165. Smith E., Heckman T., ApJ, 641, 358, 1989.
166. Smith E., Heckman T., ApJ, 348, 38, 1990.
167. Spitzer L., Saslaw W.C., ApJ, 143, 400, 1966.
168. Sprayberry, D., Impey, C. D., Bothun, G. D., Irwin, M. J., AJ, 109, 558, 1995.

169. Statler T.S., ApJ, 524, L87, 1999.
170. Storrie-Lombardi, L.J., MacMahon, R.G., Irwin, M.J., MNRAS, 283, L79, 1996.
171. Taga M., Iye M., MNRAS, 299, 111, 1998.
172. Tagger, M., Sygnet, J. F., Athanassoula, E., Pellat, R., ApJ, 318, L43, 1987.
173. Takalo L.O., Vistas Astron. 38,77, 1994.
174. Terlevich, R., Melnick, J., Moles, M., in "Oservational Evidence of Activity in Galaxies", IAU 121, Kluwer Academic Publishers, Dordrecht, p.499, 1987.
175. Thakar A.R., Ryden B.S., ApJ, 461, 55, 1996.
176. Thakar A.R., Ryden B.S., ApJ, 506, 93, 1998.
177. Toomre A., ApJ, 139, 1217, 1964.
178. Toomre A., ARAA, 15, 437, 1977.
179. Torricelli-Ciamponi, G., Foellmi, C., Courvoisier, T. J.-L., Paltani, S., A&A, 358, 57, 2000.
180. Tremaine S., AJ, 110, 628, 1995.
181. van der Kruit P.C., Searle L., A&A, 95, 105, 1981.
182. van der Kruit P.C., Searle L., A&A, 105, 351, 1982.
183. van der Kruit P.C., A&A, 192, 117, 1988.
184. van Gorkom, J. H., Knapp, G. R., Ekers, R. D., et al., AJ, 97, 708, 1989.
185. Whitmore, B. C., Lucas, R. A., McElroy, D. B., et al., AJ, 100, 1489, 1990.
186. Williams R. J. R., Baker, A. C., Perry, J.J., MNRAS, 310, 913, 1999.
187. Yates, M. G., Miller, L., Peacock, J. A., MNRAS, 240, 129, 1989.
188. Yee H., Green R., ApJ, 280, 79, 1984.

Roberto Terlevich, during one of his lectures.

MASSIVE STARS IN ACTIVE GALACTIC NUCLEI

ROBERTO TERLEVICH
Institute of Astronomy, Madingley Road, Cambridge CB3 0HA, UK
visiting professor at
Instituto Nacional de Astrofísica Óptica y Electrónica, Puebla, Mexico

I will discuss the evidence for the presence of young massive stellar clusters in the central regions of AGN. Some of the possible consequences of massive nuclear Starbursts in AGN will also be discussed

1 Introduction

These lectures are a good opportunity to think once again about the global picture of formation and evolution of galaxies in general and the formation and evolution of the central parts of elliptical galaxies and bulges of spiral galaxies in particular.

I will describe the evidence that proves the existence of large starbursts in the central regions of, at least, some AGN. I will also discuss the possible relationship between star formation and nuclear activity. These lectures do not intend to be a thorough review of the subject, thus the references are not complete.

Rather than trying to address what may be a false controversy, i.e. what powers AGN, starbursts (SBs) or accretion into supermassive black holes (SMBHs), I believe it is better to put AGN phenomenology in the general context of galaxy formation and evolution, and in doing so, to learn what are the roles played by nuclear SBs and accretion into SMBHs in the formation and evolution of galaxies.

Let me briefly summarize some of the major developments that occured in the subject in the last few years:

1 - There is suggestive evidence for the presence of SMBHs in the centres of galaxies[106]. The best examples[46,78] are perhaps the Galactic Center and NGC 4258.

2 - The broad and asymmetric profile of the 6.4 keV Fe K emission line detected in many AGN[35] is an evidence of relativistic orbital motions, and supports the idea that hard X-rays are produced in an accretion disk circling a SMBH.

3 - The gamma-ray experiment GRO failed to detect "normal" AGN[62], which implies that there is a turnover in the X-ray spectrum at about 100 keV. This result indicates that most of the luminosity in non-blazar AGN is pro-

duced at energies between 10 and 500 eV, with hard X-rays and gamma-rays accounting for only a small fraction of the total radiated energy[1]. Thus, the mechanism responsible for most of the AGN energy output has a spectrum that peaks between 10 and 500 eV.

4 - The presence of clusters of massive young stars has been proved in the nuclear/circumnuclear region of Seyfert galaxies and some powerful radio galaxies[100,51,52,48,29,76,58,7]. These massive nuclear SBs emit a substantial fraction of the total energy radiated by the nuclear region. By circumnuclear I denote the central hundred to few hundred pc region only, explicitly excluding the kpc-size star-forming rings sometimes seen in Seyfert galaxies, e.g. NGC 1068. The only exceptions are the two powerful radio galaxies described in section 4.3.

Taken together, these points support a close interaction between star-forming activity and accretion processes in AGN. Given the huge variety of properties spanned by AGN, my feeling is that it may be wrong to try to explain the whole range of observed properties with just one process, either a SB or accretion onto a SMBH.

2 What is a Starburst?

Starbursts are the sites where many and probably most of the high mass stars are formed in the local Universe. They present emission-line ratios that are characteristic of HII regions and their optical/UV continuum is dominated by the spectrum of young massive stars. We usually call SB galaxies those galaxies in which the present rate of star formation can be sustained only for a small fraction ($< 1/10$) of a Hubble time[108]. Because SBs are ubiquitous and appear in disks, bulges, and cores, and have a range of ages of up to 1 Gyr, this convenient definition is difficult to apply. This fact, I believe, has lead to much confusion in the use of the word starburst.

In 1997, I suggested a classification[104] based on a) the relative energy output of the SB (L_{SB}) to that of the rest of the galaxy (L_G), and b) the SB age.

SB galaxies	$L_{SB} >> L_G$
Galaxies with SBs	$L_{SB} \sim L_G$
Normal galaxies	$L_{SB} << L_G$

2.1 *Types of Starburst Galaxies*

The defining characteristic of SB galaxies is that their spectrum is dominated by a young stellar population, either with HII-like emission for very young bursts, or without it for older bursts, but *in all cases with the continuum dominated by young stars.* Taking this into account, I suggest the following three well defined phases:

a) Nebular phase. This is characterized by the presence of strong emission lines from gas photoionized by young massive stars (H-burning stars). The equivalent width of Hα is larger than 100 Å and the SB is less than 10 Myr old. Typical examples are HII Galaxies and IR luminous IRAS SBs.

b) Early continuum phase. The emission lines are relatively weak, only Hα and perhaps [OII]λ3727 are conspicuous. The continuum shows the high Balmer series in absorption. The equivalent width of Hα is smaller than 100Å. The SB age is between 10 and 100 Myr. Typical examples include most Blue Compact Dwarf Galaxies or Zwicky Blue Galaxies, Starburst nuclei, IRAS Galaxies and "weak [OI]λ6300 LINERs"[40].

c) Late continuum phase. The continuum is blue and dominated by Balmer absorptions. Only weak emission lines may be present. Typical ages range between several 100 Myr and ~1 Gyr. Examples include the so-called E+A galaxies[30] and some Zwicky Blue Galaxies.

3 Classification of Emission-Line Objects

When emission-line objects are classified based on their optical spectrum, two stages are used. In a first step, all objects with detected broad permitted lines are classified as type 1 AGN. Objects without broad permitted lines are classified using line ratios, usually applying the method introduced by Baldwin, Phillips & Terlevich[8], i.e. the diagnostic (or BPT) diagrams. These diagrams allow to segregate between AGN-like and HII-like (or SB-like) emission-line ratios. For a general view about AGN classification you should read chapter 2 of Peterson's book[85].

BPT classification is based on combinations of two line ratios, as shown in Fig. 1. In the original paper, we plotted in the same diagram: planetary nebulae, HII regions, Seyfert nuclei and LINERs. We learned that it was possible to segregate between different type of objects. One of the most useful diagrams included [OI]λ6300, [OII]λ3727 and [OIII]λ5007. The [OI]λ6300 line is mainly due to X-rays escaping from the main body of the HII region into the partially ionized region of the nebula, outside the Strömgren radius. For this reason, this diagram is a good estimator of the presence of hard ionizing radiation. This radiation is not normally produced by H-burning massive young stars. Therefore, including [OI]λ6300 in a diagnostic diagram provides a good discrimination between X-loud and X-faint objects or, as it is usually assumed, between AGN and SBs.

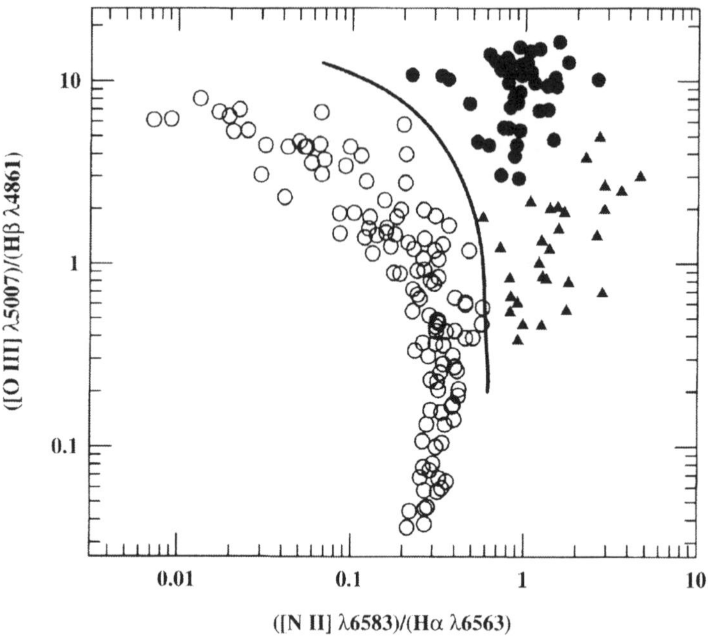

Figure 1. A diagnostic diagram for emission-line objects. Open circles represent HII-type objects ionized by young stars. Filled circles are narrow line AGN (Sy 2 and narrow line radio galaxies) ionized by a harder continuum. The filled triangles represent LINERs. The solid line is an empirical division between the two broad classes of objects, SBs and AGN. From Peterson's book[85] ©1997 Cambridge Univ. Press.

I would like to stress that although diagnostic diagrams provide in principle a quantitative procedure to segregate between different types of mechanisms for photoionizing the interstellar medium (ISM), it is in fact only a method that segregates between ionization by hard and soft radiation.

Few years after the BPT paper, realistic evolutionary models for massive stars including the effect of winds were produced[69], and this represented an important change in the way massive stellar evolution was interpreted. Large mass loss in the form of stellar winds dominates the evolutionary tracks during He burning. The exact dependence of the mass loss with stellar parameters is not properly known, and modellers use scaling-laws to include the winds into the stellar evolutionary models. For massive stars, the scaling-laws should be regarded as semi-empirical. The fundamental fact is that the emitted luminosity of young massive stars corresponds to the Eddington limit, and this is probably the main factor determining the huge mass losses observed in their massive fast winds. Early computations indicated that massive stars migrate in the HR diagram towards higher luminosities and temperatures, spending a substantial part of their He burning life-time in the He zero-age sequence at temperatures well in excess of 10^5K.

The inclusion of this very hot and luminous phase in the early synthesis of the evolution of the spectrum of a young cluster, produces a significant change in the hardness of the emitted radiation[98]. If we use the Geneva tracks, when most massive stars migrate to the He main sequence, the emission-line spectrum of a SB moves in the diagnostic diagram from the region occupied by HII regions to that occupied by AGN. We called these very hot and massive stars *warmers*, but I now believe that we should have called them *sizzlers* instead. Warmers are extreme WC or WO Wolf-Rayet (WR) stars with surface temperatures in excess of 10^5K.

Fig. 2 shows the results[24] of including warmers in the synthesized spectra of star clusters using the Geneva tracks[70] for zero-age main-sequence stars and isochrones for 3 and 4.5 Myr. During the first 3 Myr there is a substantial He-edge at 54 eV, but later, the inclusion of warmers modifies the spectrum considerably. The emitted spectrum in the Lyman continuum mimics a power-law with a slope of about -1.4 to -2, depending on the age of the cluster. Therefore, the associated emitted spectrum changes from that of a classical HII region with relatively weak [OI]λ6300, low excitation and no HeIIλ4686 emission (the HII region abundance is assumed to be solar) to that of a Sy 2 during the early luminous phases, and into a LINER-type spectrum as the ionizing luminosity and the ionization parameter decay.

Subsequent improvements in stellar evolution models suggested a lower luminosity during the warmer phase[89], and the effect of warmers in the global ionizing spectrum was proportionally reduced. It will be interesting to see what the new models with stellar rotation predict for the evolution of the most massive stars. This is only an anecdote that reflects the limitations of the diagnostic diagrams. Objects can migrate and change their type due to the presence of just a few very hot black-bodies contributing to their ionizing spectrum. This reinforces the fact that a *strong* observational definition of SB is really needed, based on the properties of the continuum and not empirically inferred from emission-line ratios. A *strong* definition of a SB should include the presence of absorption features characteristic of young massive stars.

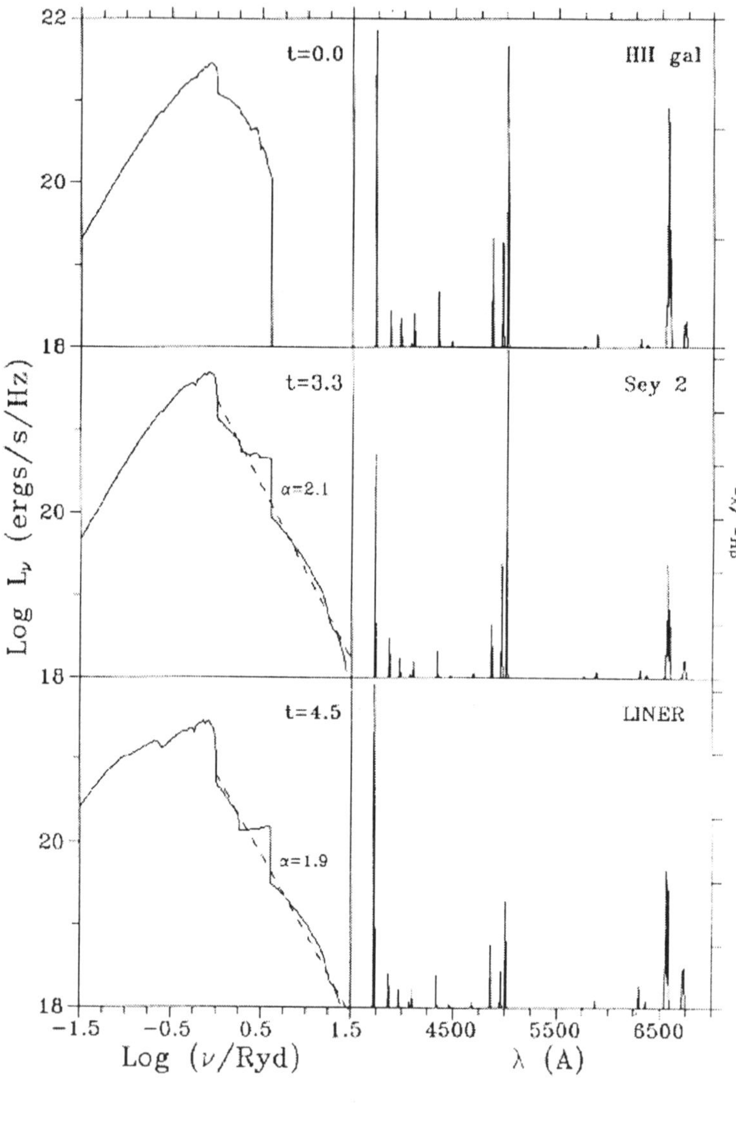

Figure 2. The figure shows the evolution of the theoretical spectrum of a SB computed using the Geneva tracks. The top panel shows the results for the ZAMS, the central one for an age of 3.3 Myr and the bottom one is for an age of 4.5 Myr. The left panels show the evolution of the ionizing spectrum, while the right panels show the computed emission-line spectrum. Note the evolution of the emission lines from those typical of an HII region to those typical of a Sy 2 and finally a LINER. The ionizing spectrum for ages larger than about 3 Myr is well represented by a power law with spectral index ~2. From Cid Fernandes et al.[24] ©1992 Blackwell Sci.

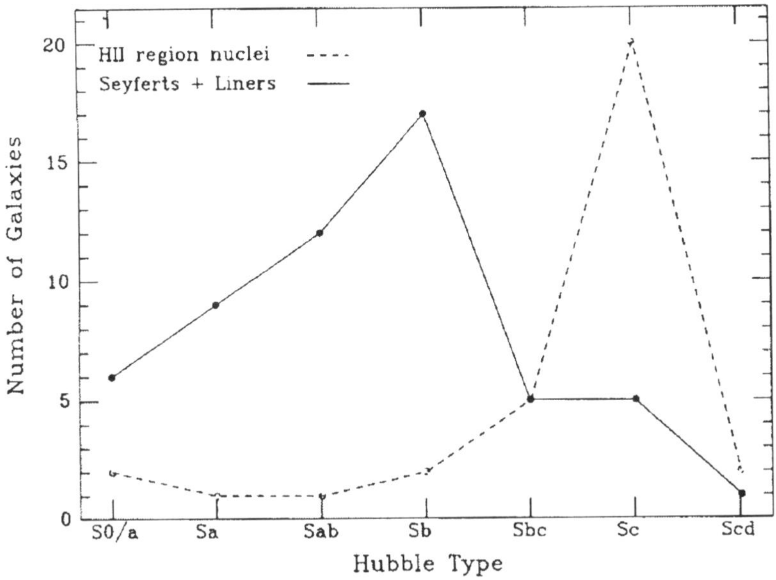

Figure 3. Hubble type distribution of SBs and AGN. Note the strong dichotomy in Hubble type. While most AGN are in galaxies earlier than Sbc, most nuclear SBs are in galaxies later than Sbc. From Terlevich et al.[99] ©1987 Kluwer.

3.1 Selection Effects Introduced by Diagnostic Diagrams

Let me mention now a strong selection effect introduced in samples when diagnostic diagrams are used to segregate between SBs and AGN. Fig. 3 shows the distribution of morphological types of the more than 100 nearby spiral galaxies surveyed by Heckman[50], and classified as SBs or AGN using BPT diagrams[99]. There is a dramatic dichotomy in Hubble type. AGN-like objects reside mostly in early spirals and HII-like objects mostly in late ones, the separation being around Sbc. A consequence of using diagnostic diagrams for classifying nuclear activity, instead of using continuum properties, is to segregate the sample in Hubble type and, therefore, in all the parameters that change from early to late type spirals: disk to bulge ratio, size of the bulge, central velocity dispersion (early types have big bulges with large velocity dispersions), metallicity, gas content, etc.

4 Evidence of Massive Stars in AGN

4.1 The Near-IR CaII Triplet

The early work I did with two very good friends, Elena Terlevich and Ángeles Díaz, that indicated the precence of massive stars in AGN[100], was prompted partially by discussions with Mike Penston during coffee time at the Royal Greenwich Observatory in Herstmonceux more than ten years ago. We followed the hypothesis that some Sy 2s may be powered by SBs. If this was the case, their continuum should show evidence of stellar absorption lines characteristic of young massive stars.

This is a difficult observational experiment because the optical spectrum of young massive stars is characterized by the presence of strong absorptions due to H and He, which unfortunately are exactly the same lines that appear strongly in emission in the associated HII region. It is extremely difficult to detect an absorption line underneath a strong emission line of similar width. In the 80s, with the advent of IR sensitive CCD detectors, it was possible to use the IR CaII triplet (CaT) as a diagnostic for the presence of massive young stars. These lines are very strong in red supergiants (RSGs) and occupy a spectral region that is relatively free of contaminating emission lines. A subreptitious study during several cloudy nights at La Palma included the observation of more than 100 stars with well known parameters: spectral type, temperature, gravity and metallicity. Our study confirmed and extended early results[57] showing that the CaT strongly depends on the gravity of the stars and has little sensitivity to metal content for abundances above 1/3 solar.

Fig. 4 shows the typical optical nuclear spectrum of nearby AGN[56]. The most important feature is the presence of very strong emission lines of H and O on top of a relatively weak continuum. Most absorption features correspond to real stellar absorptions. In particular the Mg complex at about λ5150 should be noticed because we know from the studies of elliptical galaxies, globular clusters and bulges that it is metallicity dependent. It appears weaker in AGN than in normal bulges and ellipticals. The spectrum in the near IR (Fig. 5), on the contrary, is dominated by the continuum and there are almost no emission lines. There is some Paschen and [SIII] emission and in some luminous Sy 1s, [OI]λ8440, but the region occupied by the two strongest lines of the CaT is relatively clean, and accurate measurements can be performed. As the CaT is a good gravity indicator, the masses of the stars can be determined and, therefore, the turn-off point of the cluster can be estimated and, in principle, also the age of the cluster.

After measuring the strength of the CaT index in a sample of normal and

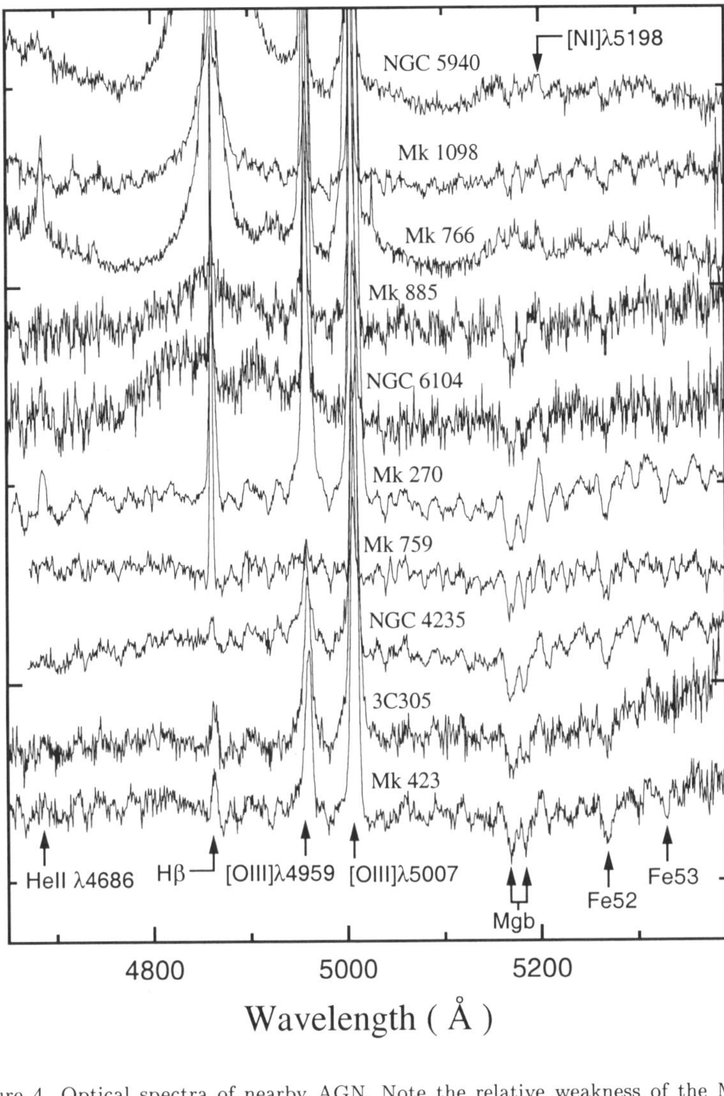

Figure 4. Optical spectra of nearby AGN. Note the relative weakness of the Mg and Fe stellar absorption lines. From Jiménez-Benito et al.[56] ©2000 Blackwell Sci.

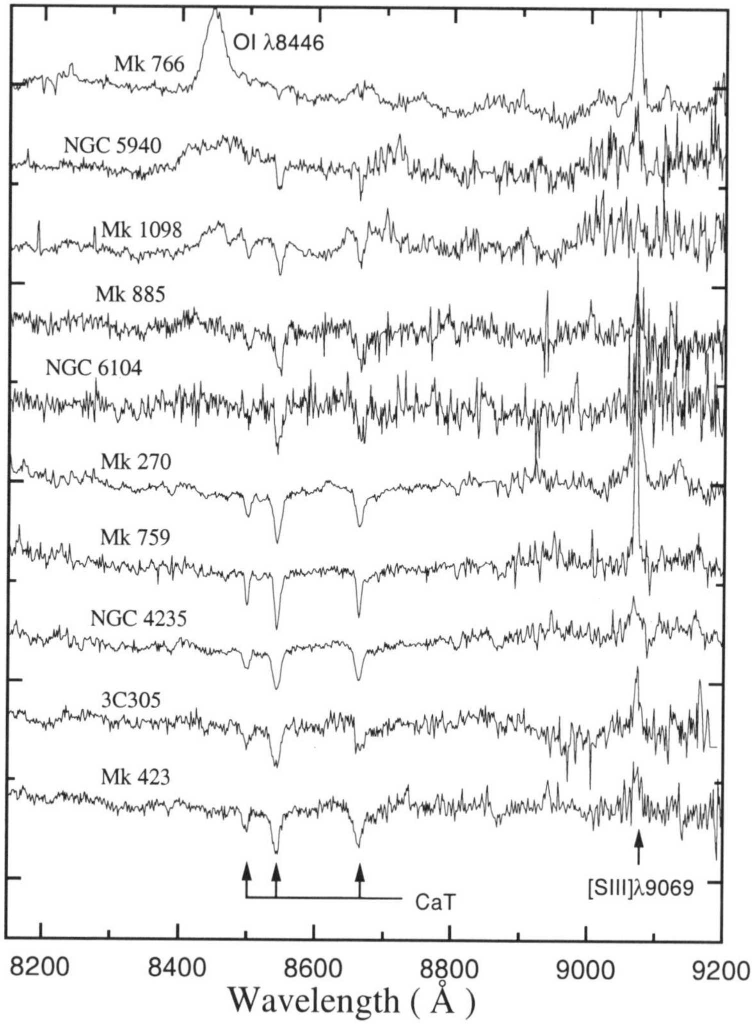

Figure 5. The near-IR spectra of the AGN in figure 4. Note the strength of the Ca triplet absorption and the relative weakness of the emission lines. From Jiménez-Benito et al.[56] ©2000 Blackwell Sci.

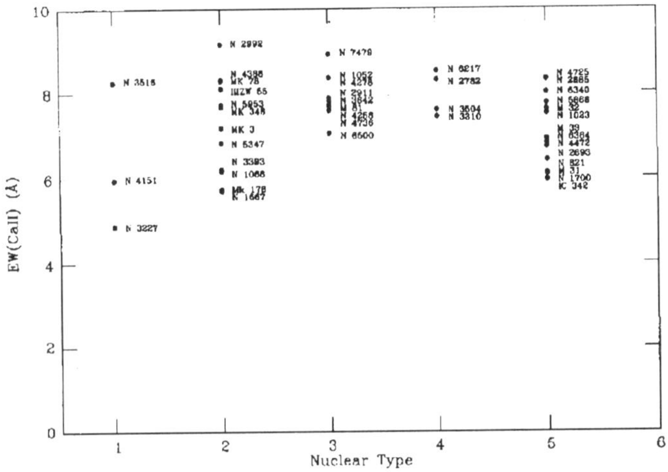

Figure 6. The strength of the CaT is shown for different types of galaxies: Sy 1 nuclei (1), Sy 2 nuclei (2), LINERs (3), SBs (4) and normal ellipticals and spiral bulges (5). Note that the CaT tends to be larger in Sy 2s and LINERs. From Terlevich et al.[100] ©1990 Blackwell Sci.

active galaxies with very blue optical continuum, we reached the conclusion that, unlike the optical metal features, the strength of the CaT in Sy 2s was equal to or larger than that in normal ellipticals. This is illustrated in Fig. 6, where the strength of the CaT is shown for different types of galaxies, from Sy 1s to normal ellipticals and spirals.

The simultaneous presence of a weak Mgb index with a normal intensity CaT is a very strong indication that a dominant part of the continuum originates in a young stellar cluster with age around 10 Myr, and is not due to dilution of an old-bulge light by a featureless power-law continuum as it was believed at the time. This effect is also shown in Fig. 7, where the dilution for Mg is plotted versus the dilution for CaT. It can be seen that the dilution for Mg is much larger than that for CaT and, as the lines indicate, there is no reasonable combination of an old population and power-law that can dilute the Mg without diluting at the same time the CaT. The conclusion is that Sy 2s host an important component of very young stars dominating the optical-near-IR part of the spectrum. Sy 1 nuclei have an additional component: the broad line region (BLR) that dilutes *all* features.

Recently, Nelson & Whittle[80] completed the analysis of a large compila-

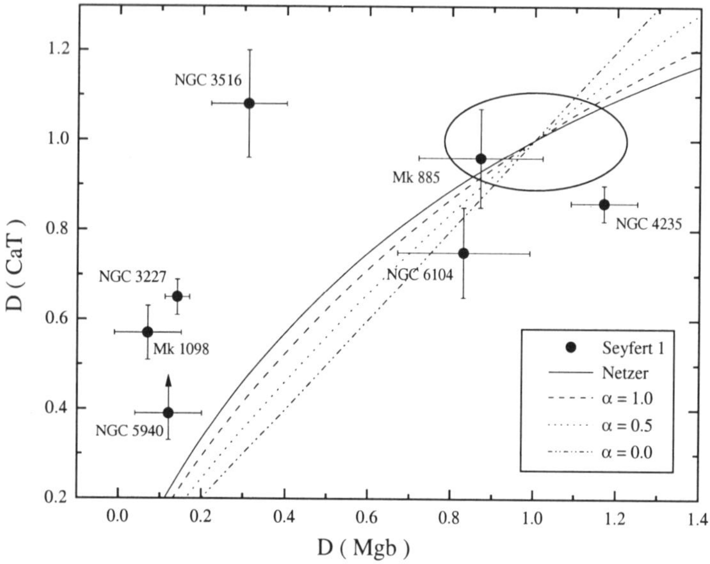

Figure 7. Dilution diagram. The dilution for Mg is plotted versus the dilution for CaT. The dilution for Mg in this sample is larger than that for CaT. There is no reasonable combination of an old population and power-law featureless continuum that can dilute the Mg without diluting at the same time the CaT. From Jiménez-Benito et al.[56] ©2000 Blackwell Sci.

tion of optical and near-IR spectra of Sy galaxies confirming and extending the above results. They have a novel way to show the line dilution in Seyfert nuclei. This is illustrated in Fig. 8, which shows the Mg2 strength versus the velocity dispersion. The strong CaT absorption lines provide an excellent way to measure accurate velocity dispersions (σ) in emission-line galaxies where the optical stellar absorptions are weak or absent. The Mg2–σ relationship is well known in elliptical galaxies and bulges[10,53,55]. Given that both quantities are distance independent, their strong correlation carries information about fundamental properties of the systems. Because the central velocity dispersion is closely linked to the depth of the central potential and ultimately related to the total mass, and because the Mg2 index is mostly a metallicity indicator[79], the common interpretation is that the Mg2–σ relationship is closely linked to the mass-metallicity relationship in normal galaxies. Fig. 8 clearly shows that for a given central velocity dispersion, or total galaxy mass, most Sy galaxies

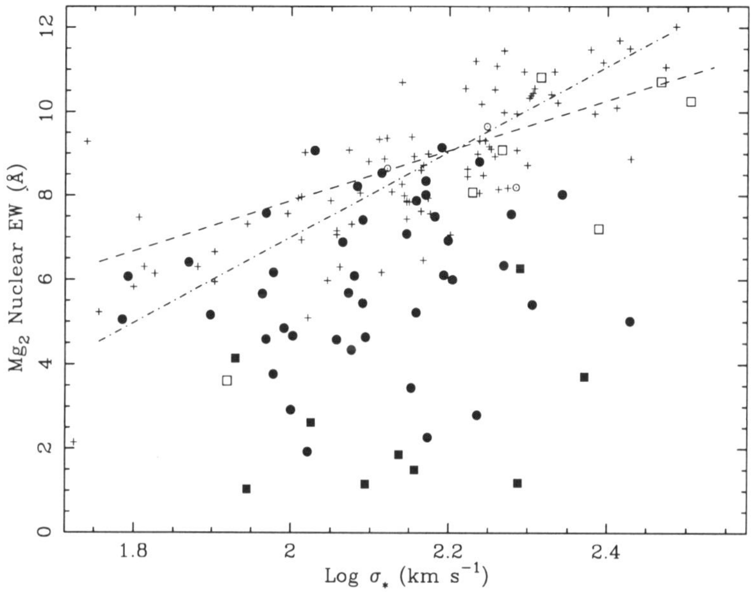

Figure 8. The intensity of the Mg2 index is plotted against the central velocity dispersion for normal spheroidal galaxies (pluses and open circles) Sy 2 (filled circles), LINERs (open squares) and Sy 1 (filled squares). The dashed lines show the best fit to the normal galaxy sample. Most Sy galaxies lie below the lines, showing that for a given velocity dispersion they have weaker optical Mg lines than normal spheroids. Courtesy of C. Nelson.

have weaker Mg2 than ellipticals and spirals. This is clearly a fundamental property.

Fig. 9 shows that unlike the Mg2 versus σ relationship, the CaT strength in normal galaxies does not show any dependence with central velocity dispersion or equivalently, with dynamical mass. This is consistent with the conclusion that the CaT strength does not depend on metallicity in the range of parameters considered. At the same time, the Sy 2s show no systematic shift with respect to normal galaxies, strongly supporting the conclusion that there is no CaT dilution in these systems. Only Sy 1s show dilution, in agreement with the conclusion that they have an additional component: the BLR.

These results can be further visualized by plotting the observed strengths of Mg2 and CaT (see Fig. 10). The solid curve shows the dilution produced by a power-law continuum, where the ticks indicate the relative intensity of

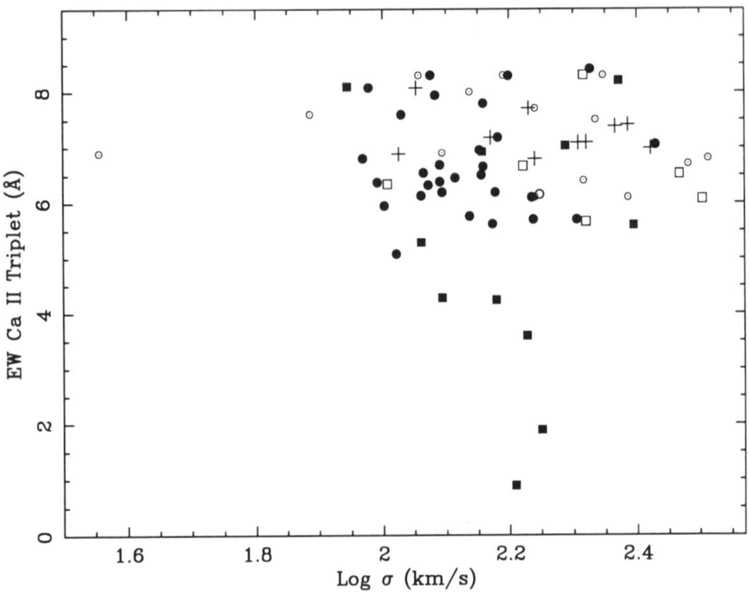

Figure 9. CaT–σ diagram. Symbols are as in Fig. 8. Note than in this diagram there is no relationship between line stregth and velocity dispersion for normal galaxies. This is consistent with the lack of metallicity sensitivity of the CaT. Unlike in Fig. 8, here Sy 2s are located in the same position of the diagram as normal galaxies, indicating the same strength of CaT. Sy 1s are below the region occupied by Sy 2s and normal galaxies, indicating dilution of the CaT. Courtesy of C. Nelson.

the power-law for different relative contributions of power-law and bulge. It is possible to see that the objects with the largest dilution can not be explained by a power-law. Also shown is the evolutionary dilution sequence for a SB modelled with SB99[61]. The lines show different contributions of bulge and SB light, with the ticks indicating the age of the SB. The two tracks illustrate the offset produced by including dust in the models. Clearly, the SB plus old bulge model covers the complete spread of the observations.

Additional information on the presence of luminous stars in AGN comes from the analysis of the M/L relation. Oliva et al.[81] have used the fact that the M/L ratio is lower in stellar clusters in which the light is dominated by RSG stars than in those that are dominated by red giant stars. They were able to show that the M/L ratios observed in SBs and Sy 2s suggest that the

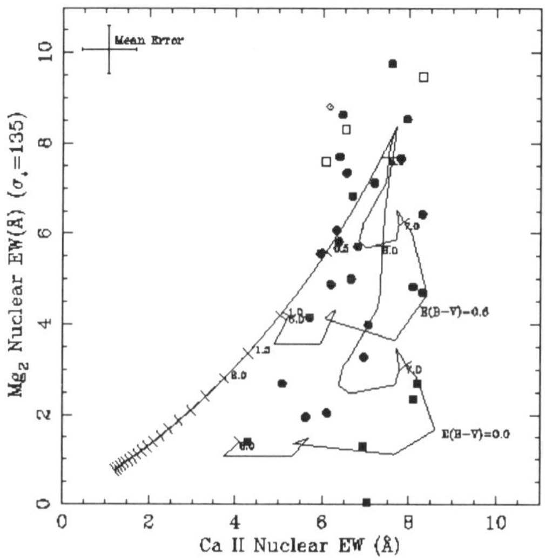

Figure 10. Mg2 index versus CaT index for AGN. Symbols are as in Fig. 8. The solid curve shows the expected strength of the Mg2 and CaT indices for the case of a power-law diluting the spectrum of a normal elliptical. The ticks indicate the relative intensity of the power-law. The other lines correspond to models including a mixture of a young SB and an old elliptical with and without dust extinction. Courtesy of C. Nelson.

IR luminosity is produced by RSG stars.

4.2 A Massive SB in the Luminous Seyfert 2 Mrk 477

Recent work by Heckman et al.[52] has produced a fascinating result. They looked at a critical test for the proposal[100,25] that the continuum in Sy 2s is dominated by massive young stars. To this end, Mrk 477 was selected for a detailed investigation because it is a UV bright, nearby classical Sy 2 (according to the classification using diagnostic diagrams), harbours an obscured Sy 1 nucleus and is also a strong X-ray emitter. Using HST data they detected broad absorption lines associated with the resonant lines of CIVλ1550, SiIVλ1400 and Lyα, typical of massive stars with winds, also found in the UV spectrum of SB. They also found lower excitation lines from CII, SiII, OI that are produced in the ISM and the photosphere of stars and strong IR CaT in absorption. The SiIVλ1400 profile indicates that all of the observed UV continuum can be produced in a nuclear SB. The upper Balmer series was clearly detected in absorption, indicating the presence of massive O-B stars. They also detected broad HeIIλ4686, with an intensity compatible with it being produced by about 30,000 WR stars! Their main conclusion is that from 0.1 to 1 μm, the observed continuum in this Sy 2 galaxy is well represented by a SB of a few Myr of age, mass $M_{\rm SB} \sim 10^8$ $M_\odot$ and luminosity $L_{\rm bol} \sim 10^{11}$ $L_\odot$. Its observed size, $R = 0.25$ arcsec ($\sim$ 90 pc), is very difficult to resolve from ground-based observatories.

All in all, Mrk 477 in not easy to classify, as it is an obscured Sy 1, a canonical Sy 2 according to diagnostic diagrams, a strong X-ray emitter, and a powerful nuclear SB based on its continuum.

One should assess how typical are the stellar properties of Mrk 477 among AGN. The detection of the upper Balmer series in absorption in the nuclear UV spectra of Sy 2s and powerful radio galaxies, that I will present in what follows, are in this respect important results.

4.3 Balmer Absorption Lines in Powerful Radio Galaxies

Aretxaga et al.[7] observed simultaneously the region of the CaT and the upper Balmer series in a sample of powerful nearby radio galaxies (RGs). The CaT was detected in all of them while strong upper Balmer absorptions were detected in two: Hydra A and 3C 285. In both cases the optical continuum is very similar to that of a Sy 2 with weak Mg2 and strong CaT. After subtracting the spectrum of a normal elliptical (scaled using the G band strength), they were able to eliminate most of the metal lines and they ended up with the spectrum of an almost pure SB plus emission lines from the AGN (see Fig. 11). The SB revealed in the nucleus of 3C 285 looks a bit older than the one in Hydra A.

Can we estimate the ages of these nuclear SBs in powerful RGs? One can use the old quantitative classification method for early type stars designed by Barbier and Chalonge that involves the measurement of two independent parameters: the position and amplitude of the Balmer discontinuity[9,17]. The position λ mostly depends on gravity, or luminosity class, while the amplitude D depends on spectral type. The dependence for nearby stars is illustrated by the grid represented in Fig. 12. The scattered spectral types and luminosity classes in the diagram are measurements of nearby stars artificially broadened to reproduce the width of the Balmer lines observed in the RGs. The broadening produces a shift to the left towards lower gravity and higher luminosity types. The values measured for the two RGs are also shown. The spectral type indicated is B or A giant/supergiant stars. These stars have main-sequence masses in the range 15 to 40 $M_\odot$ and main-sequence life-times

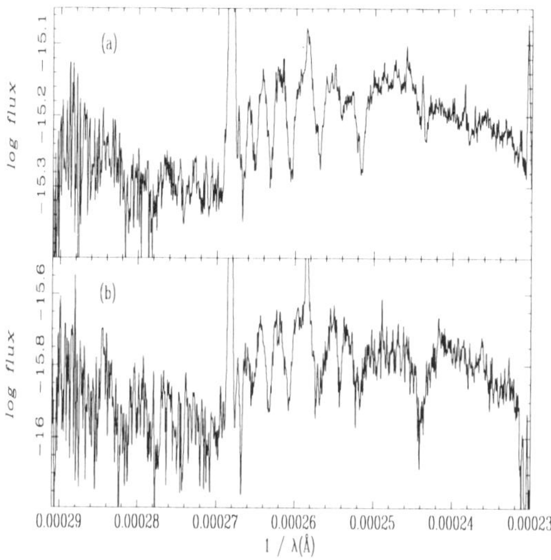

Figure 11. Bulge subtracted optical spectra of Hydra A (a) and 3C 285 (b). The strength and relative narrowness of the upper Balmer series is striking. From Aretxaga et al.[7]

$\sim$ 10 Myr. Thus, this interesting exercise shows that a luminous SB is present in the central regions of two of the most powerful radio sources in the nearby Universe, and that these SBs have ages of about 10 Myr.

The CaT shows similar results to those described above regarding Sy 2 nuclei, i.e. the CaT strength is normal while the optical Mg2 appears diluted with respect to normal galaxies, consistent with a mixed population of an old bulge and a very young SB.

4.4 Balmer Absorption Lines in Seyfert 2 Nuclei

González-Delgado et al.[48] have analyzed data specifically obtained to study the SB-AGN connection and the origin of the UV, optical and near-IR continuum in several nearby Sy 2 nuclei. Their data provide direct evidence of the existence of a central nuclear SB that dominates the UV light, and that seems to be responsible for the origin of the so-called featureless continuum. These SBs are dusty and compact, with sizes up to a few 100 pc, much smaller and closer to the nucleus than those seen in the prototypical Sy 2 galaxy

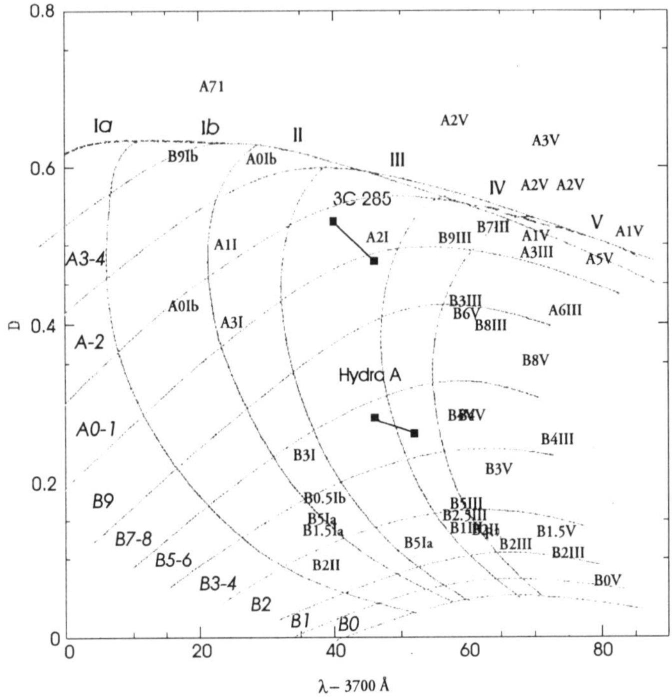

Figure 12. Barbier and Chalonge quantitative classification plane[9,17]. The lines indicate the trend of spectral types and luminosity classes with the position and amplitude of the Balmer jump. The values measured for the radio galaxies 3C 285 and Hydra A are shown by solid squares and connecting lines. SB population is dominated by B or A giants/supergiants. From Aretxaga et al.[7]

NGC 1068.

The UV spectra show photospheric absorption-features from CIII, SiIII, SV and wind features of CIV, SiIV and NV characteristic of massive stars. The comparison of the UV features with evolutionary synthesis models indicates that these are generated in very young SBs with ages between 3 and 5 Myr. Fig. 13 shows the near-UV/optical spectra of these Sy 2 galaxies. The detection of the upper Balmer series in all of them provides direct evidence of the presence of massive young stars in Sy 2 nuclei.

The bolometric luminosity of these SBs is similar to that estimated for their obscured Sy 1 nuclei. The luminosity of these Sy 2s are three to ten

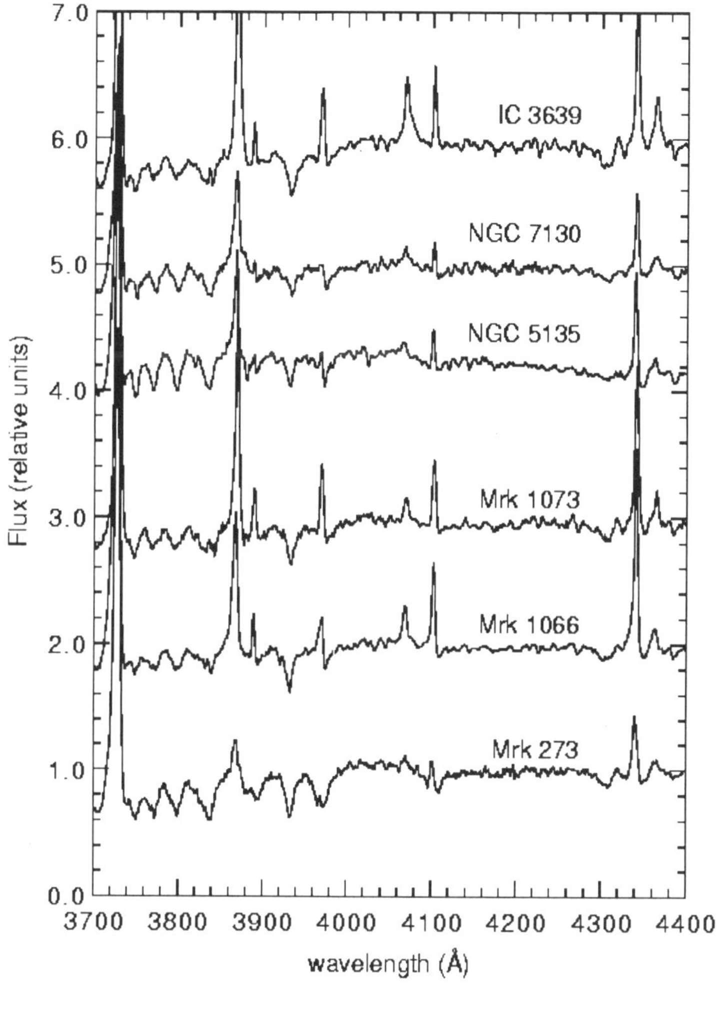

Figure 13. Blue spectra of several Sy 2s showing the clear detection of the upper Balmer series. From González-Delgado et al.[47] ©2001 AAS.

times lower than the bolometric luminosity of the SB and hidden nucleus of Mrk 477, suggesting that more powerful AGN are linked to more luminous SBs.

4.5 Wolf-Rayet Stars in Seyfert 2 nuclei

Kunth & Contini[58] conducted a search for WR features, like the broad HeIIλ4686, in Sy 2 galaxies. They found that about a third of the galaxies of their sample show the signatures of WR stars in their nuclei. The total number of WR stars deduced from the intensity of the bumps is lower than that found for Mrk 477, consistent with the fact that their Seyferts are less luminous.

The ratio of WR to O stars can be estimated from the equivalent width of the WR features, and that in turn can be used to estimate the age of the SB. This method can only be applied during a narrow range of SB ages. One important conclusion from these studies is that, because the WR epoch is short lived and the WR features are weak against the background of O-B stars, the WR features are very difficult, if not impossible, to detect in regions with extended star formation. Thus, the star-formation in those Seyfert nuclei in which WR stars are detected, has to be highly synchronized.

5 Indirect Evidence of the SB-AGN Connection

5.1 The Metal Content of the BLR

Work by Hamann & Ferland[49] has suggested the intriguing possibility that the metal content in the BLR may be higher at higher redshift, perhaps reaching several times the solar value at $z \sim 3$. These results suggest that substantial astration or chemical evolution has already taken place by redshift 2 or 3 in the central regions of luminous AGN. This is an important result that should be explained by models of formation and evolution of galaxies and AGN, although the fact in itself puts no strong constraints on either the adopted cosmology or models of galaxy formation, because only few solar masses of very enriched material are needed to explain the over-solar abundances of the BLR even in high luminosity AGN. It is not difficult to produce metals during the first few hundred million years of evolution of any moderately massive galaxy.

Even before a free-fall time, about a few 10^8 yr, it is possible to generate large amounts of metals, CNO elements and also Fe. The central regions become over-solar in less than a billion years[72,43]. Considering that 10^9 yr represents in present-day favourite cosmologies the age of the Universe at $z \sim 5 - 10$, the over-solar abundances in the BLR are easily explained by these models. Still, the fact that there is not a single AGN known having primordial or near primordial composition (as observed in halo stars and HII galaxies in the near Universe)[59] suggests that star formation and an important level of chemical evolution *precedes* the AGN epoch (but see S. Collin's lecture in this volume).

5.2 Dust in High-z QSOs

A stronger indirect evidence for association between SBs and AGN is based on the fact that mm–submm surveys have detected dust emission from high-redshift QSOs ($z > 3.5$) at the level of few mJy[68]. How large is the mass of dust associated with these high-redshift systems? A helpful observational property of the behaviour of dust emission with redshift is that beyond $z = 1$ the negative K-correction slightly dominates over the distance factor. This implies that the conversion from flux to dust mass is almost independent of distance or redshift, leaving only dust temperature as the main uncertainty.

QSO detections are at the level of a few mJy, which, assuming similar temperatures to that of nearby ultraluminous IR galaxies ($T_d \sim 40$ K), translate into dust masses of 10^{8-9} $M_\odot$ at redshifts well over 3 and up to almost 6. These results imply that a large amount of astration, consistent with the formation of a *complete* massive elliptical has taken place for these objects by $z > 3$. Another important point is that these large masses of dust were produced in less than 1 – 2 Gyr, suggesting that the blue luminosity associated with the massive stars that produced the metals should be detectable.

Lets close this section going back to an important unanswered question: *why has an AGN with primordial or near primordial abundances not been found so far?* Is this high metal content an essential aspect of the AGN phenomenon?

6 Properties of SBs

As discussed in previous sections, the presence of massive SBs in the central regions of some Seyfert galaxies has been proved. In order to explore the possible consequences of having these massive SBs in the central regions of AGN we should first look at the properties of nearby SBs. I would like to present here some relevant properties of optically selected SBs, both nearby and at intermediate redshifts. I will discuss the properties of the stars, estimate the SB sizes, supernova (SN) rates and ISM pressures and how the latter affects the evolution of SN remnants (SNR) at the base of the galactic winds in these nuclear SBs. I will later include these aspects in the discussion of what may shed light into the SB-AGN connection.

6.1 Stellar Make-up

Giant regions of star formation in the Local Group of galaxies have been studied for some time as laboratories for understanding massive star formation, as they, together with SBs, are the places were most of the massive stars in the Universe are formed. One of my favourites is the Tarantula nebula (30 Doradus) in the LMC where the ionized gas extends several hundred pc. Most of the ionizing radiation for the nebula comes from the central half pc of the 30 Dor stellar cluster.

Understanding the intrinsic properties of 30 Dor, its structure and evolution, plays a central role towards the interpretation of violent star formation of more distant and massive systems in general. Despite its importance and a renewed interest in it, much remains to be learned about its remarkable stellar make-up. Early spectroscopic studies[36] reached only the visually brightest stars which, apart from the WR stars and the composite core Radcliffe (R) 136, are mostly late O, B and A supergiants, which do not contribute much to the ionization of the nebula. The ionizing stars are substantially fainter in the optical. Due to their smaller radii, they have high temperatures, emit most of their luminosity in the UV and therefore have large bolometric corrections. The first spectroscopic studies of large number of stars were those by Melnick[73] and Walborn[107]. Recently, fiber-optics and multi-slit techniques have improved the situation considerably[67,13]

I would like to discuss recent work that combines multi-slit spectroscopy of more than two hundred stars with accurate photometry for more than one thousand stars in 30 Dor[13,91,92]. Fig. 14 shows multi-slit stellar spectra obtained with the ESO NTT of some of the most massive and luminous H-burning stars known, O3. This star forming region is hundreds of times more luminous than a typical Galactic star-forming region like that associated with the Orion Trapezium. The slope of the initial mass function (IMF) in 30 Dor is surprisingly close to the Salpeter value: 2.38 ± 0.07 from 120 $M_\odot$ down to 4 $M_\odot$. Stars below 3–4 $M_\odot$ in 30 Dor cannot be accurately classified from the ground due to confusion and contamination by the gas-background emission. This implies that the empirical parameters for the IMF in SBs below 4 $M_\odot$ are not known. It seems that low mass stars are indeed formed, but it is not known in which proportion or which range of stellar masses dominates the cluster mass. This propagates into a large uncertainty when estimating star-formation rates, modelling galaxy formation and evolution, and even chemical evolution. To give some numbers, if the IMF in 30 Dor flattens just below

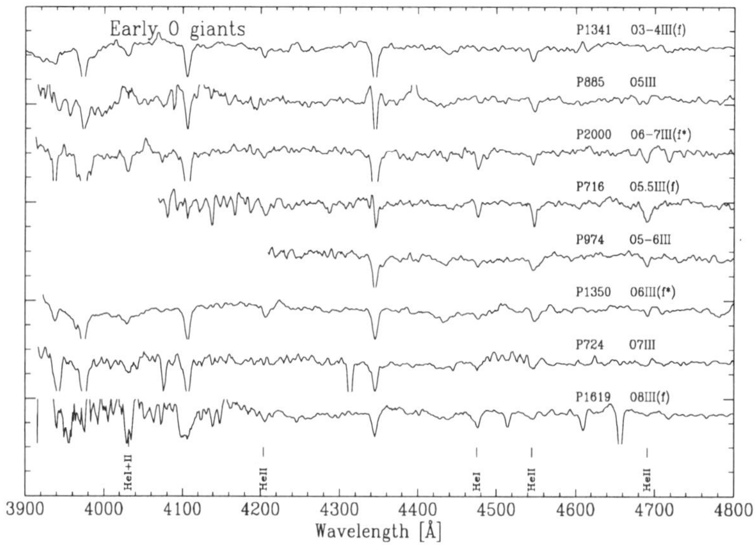

Figure 14. Some of the hottest and most luminous stars in 30 Doradus from multi-slit spectroscopy. Note the narrowness of the Balmer absorption lines. From Bosch et al.[13] ©1999 ESO.

4 $M_\odot$ instead of continuing with the same slope down to 0.1 $M_\odot$, as it is often assumed, the total mass will be an order of magnitude smaller.

Another important finding is that there is substantial mass segregation and probably also kinematical segregation in the stars of 30 Dor. Massive stars tend to occupy preferentially the centre of the cluster while lighter ones occupy a much larger region and have a distribution with an effective radius much larger than that of the most massive stars[13,92]. It is interesting that this segregation represents a radial change in the average mass of the stars and not a radial change of the IMF slope. The kinematics of the cluster suggests also segregation in the sense that the velocity dispersion of the massive stars is smaller than that of the less massive ones[13]. These two facts support quasi-equipartition of energy among the stars in 30 Dor. The fact that this is a very young stellar system suggests that the mass segregation is not the product of dynamical evolution ending in core collapse but rather an initial condition for massive cluster formation.

6.2 The Sizes of Nuclear SBs

What is the size of a nuclear SB? Some of the brightest star-formation regions, giant HII regions and HII galaxies in the local Universe, are very compact indeed[59]. To estimate the optical size of nuclear SBs, I will use results on the structure of 30 Dor and NGC 3603, the two nearest giant bursts of star formation[71,14]. Their core radii are $R_c < 0.07$ pc for 30 Dor and $R_c = 0.024$ pc for NGC 3603. I will adopt a value of 0.1 pc for the core radius of the luminosity profile of 30 Dor. Adopting $(B - V) = -0.3$, the core blue luminosities are $M_B = -11.4$ for 30 Dor and $M_B = -10.2$ for NGC 3603.

Assuming that both clusters are near virial equilibrium (Mass $\propto$ Size) and have similar M/L ratios, the fact that NGC 3603 is about 3 times less luminous than 30 Dor implies that the mass, and therefore the size, of NGC 3603 should be about 3 times smaller than 30 Dor. Allowing for excess of luminosity in the core of 30 Dor with respect to the core of NGC 3603, the expectation is that a *normal* young cluster with core luminosity $M_B \simeq -10$ will have a core radius of $R_c \simeq 0.02$ pc. These estimates can be scaled to the luminosities of nuclear SBs by using the virial theorem and assuming the same M/L ratio for all young clusters. In that case, $L_t \propto M_t \propto R_c \sigma^2$, where σ is the velocity dispersion of the cluster. The velocity dispersion of the ionized gas[75,74] is 16 $\mathrm{km\,s^{-1}}$ in NGC 3603, and 22 $\mathrm{km\,s^{-1}}$ in 30 Dor. If the velocity dispersion of the ionized gas is similar to that of the stars, then the expected size of virialized young clusters would be

$$R_c \simeq 0.02\,\mathrm{pc}\ 10^{-0.4(M_B+10)} \left(\frac{20}{\sigma}\right)^2 . \tag{1}$$

This indicates that a nuclear SB with $M_B = -20$ and velocity dispersion similar to that of normal bulges, i.e. $\sigma = 150\,\mathrm{km\,s^{-1}}$, has a size FWHM $= 2R_c \simeq 7$ pc. If the velocity dispersion of the stellar population in 30 Dor is a factor of two smaller than the velocity dispersion of the ionized gas, as obtained from applying the virial theorem to the total mass obtained from the total luminosity[23], then the predicted size will be proportionally smaller.

There are two important points that should be taken into account when comparing the mass distribution of an old stellar cluster with the light distribution of a very young one. First, mass segregation as detected in 30 Dor implies that the light distribution will be more compact than the mass distribution. In the case of equipartition of kinetic energy, the massive stars will have smaller velocity dispersion and therefore will be even closer to the centre[12]. The second aspect is dynamical evolution. Due to the large mass loss from the central region of the cluster in the form of stellar winds and

SN ejecta, and its short dynamical time scale, the structure will be affected by dynamical evolution in less than a Hubble time. A cluster with Salpeter IMF will lose about half of its initial mass after one Hubble time, $\sim 25\%$ will be lost in the first 100 Myr. This slow mass loss will produce an adiabatic expansion proportional to the mass loss[87],

$$\frac{r}{r_0} = \frac{M_0}{M} \quad . \tag{2}$$

Abrupt mass loss of a small fraction of the cluster mass is twice as effective as the same amount of slow mass loss in puffing up the cluster[11].

The combined effect of mass segregation and dynamical evolution, therefore, can produce a mass distribution that has a core radius that is much larger than the initial light distribution.

6.3 Estimates of the Supernova Rates in Young Clusters

Aretxaga & Terlevich[3] calculated the specific supernova rate, $\nu_{\rm SN}/M_{\rm t}$, and specific blue luminosity, $L^{\star}_{\rm B}/M_{\rm t}$, for a variety of IMFs using the Geneva stellar-evolution models[70]. It was found that both parameters decrease with increasing age of the cluster and are similarly dependent on the assumed IMF. The ratio of the supernova rate per unit mass to the blue luminosity per unit mass, i.e. the supernova rate per unit stellar blue luminosity, $\nu_{\rm SN}/L^{\star}_{\rm B}$, is almost independent of the assumed IMF and basically constant over the whole SN II phase. This is because both quantities, the SN rate and the blue luminosity, depend mainly on the number of stars in a narrow mass range close to the turn-off point, and not on the number of low mass stars that are the bulk of the mass of the cluster. This is a very important property of young clusters. Table 1 lists the SN rate as a function of the absolute blue magnitude of the stars of the cluster, plus some of the relevant parameters for a range of ages and IMFs.

The main result is that a young cluster with age between 10 and 60 Myr and with $M_B \sim -21.5$ produces 1 SN per year, almost independently of the assumed IMF.

6.4 The ISM Pressure in SBs

A central aspect to the evolution of SN remnants (SNRs) is that all parameters depend strongly on the properties of the medium surrounding the explosion. Classical SNRs are the product of the interaction of the SN ejecta with a relatively low density ISM. But before the ejecta reaches the ISM it has to

Table 1. Description of the cluster evolution: (*1*) age of the stellar cluster, in 10^7 yr; (*2*) absolute blue luminosity of a cluster with a rate of 1 SN per year; (*3*) SN rate per unit of stellar blue luminosity, in $\mathrm{yr}^{-1}\ \mathrm{L}_\odot{}^{-1}$; (*4–6*) ratio of stellar blue luminosity to total initial mass of the cluster for several power-law IMFs of slope α=2.35 and lower mass limits (m_l), in solar units; (*7*) initial mass of the stars undergoing explosion, in solar units.

t_7	$M_B^\star$	$\nu_\mathrm{SN}/L_\mathrm{B}^\star$	$L_\mathrm{B}^\star/M_\mathrm{t}$ for $\alpha = 2.35$			M_SN
			$m_\mathrm{l} = 0.3\ \mathrm{M}_\odot$	$m_\mathrm{l} = 1\ \mathrm{M}_\odot$	$m_\mathrm{l} = 3\ \mathrm{M}_\odot$	
1.0	-21.7	1.5×10^{-9}	28.	49.	78.	19.8
2.0	-21.5	1.7×10^{-9}	17.	29.	43.	12.1
3.0	-21.3	2.1×10^{-9}	13.	22.	31.	9.8
4.0	-21.3	2.1×10^{-9}	11.	19.	25.	8.4
5.0	-21.3	2.1×10^{-9}	9.4	16.	20.	7.6
6.0	-21.4	1.8×10^{-9}	8.8	15.	19.	6.9

cross the circumstellar medium (CSM), where it will encounter the material ejected as a stellar wind prior to the explosion. The properties of the wind determine to a large degree the evolution of the SNR[101]. If the density of the circumstellar medium of the progenitor is lower than the critical value $n_\mathrm{cr} \sim 10^5\ \mathrm{cm}^{-3}$, then the interaction produces a classical SNR that goes through the Sedov phase, taking many years before it becomes radiative. If the density is above the critical value, there is no Sedov phase, and the remnant passes directly from the free-expansion phase into the radiative phase. The time-scales involved are short, so these remnants are very luminous. The predicted luminosities are similar to the values observed in the peculiar luminous SN type IIn, also known as compact SNRs or Seyfert 1 impostors[101] The properties of these compact SNRs are discussed in the next section.

Highly active star-forming regions are populated by massive stars radiating at about the Eddington limit, emitting powerful winds and producing high SN rates. If most of the stellar wind and SN energy is thermalized, a high pressure region will develop and a strong wind will be driven out of the star-forming region[97,95]. The evolution of the energy input into the ISM of a solar abundance SB is dominated during the first 10 Myr by the winds of massive stars. Afterwards, type II SNe appear and, if the kinetic energy per SN is 10^{51} erg, they produce a similar level of energy input into the ISM[60].

Classical estimates of the ISM pressure in SBs[20] showed that neglecting gravity and magnetic fields, the pressure of the ISM depends on three parameters: the mass input, $\dot{M}$; the energy input, $\dot{E}$; and the size of the region,

R, in which the mass and energy input are occurring. The resulting ISM pressure inside the active region is proportional to these parameters. During the SN phase, i.e. from $\sim 5 \times 10^6$ to $\sim 5 \times 10^7$ yr, $\dot{M}$ and $\dot{E}$ are both proportional to the SN rate, $\dot{\nu}$. The pressure of the ISM inside the active region, i.e. at the base of the wind, is proportional to the total mass inside the region divided by the square of its size, i.e. the cluster mass surface density: $P \propto \dot{M}^{1/2} \times \dot{E}^{1/2} \times R^{-2} \propto \dot{\nu} \times R^{-2}$. It is interesting that the pressure at the base of the wind is predicted to depend linearly on the surface density of massive stars.

For the central region of M 82, estimates of the ISM pressure at the base of the wind give values[20] of about 10^9 K cm^{-3}. The average surface density of stars in the core regions of elliptical galaxies and spiral bulges is about 10^{5-6} M$_\odot$pc^{-2} . This value is an indication of the surface density of the core regions of massive spheroids at the time when most of the stars we see today were created; since then the effects of dynamical evolution, mass loss and accretion would have modified the initial surface density. However, it is safe to conclude that a *young* cluster with age $t_{cl} < 5 \times 10^7$ yr and average surface density of massive stars about 10^{5-6} M$_\odot$pc^{-2} will have an ISM pressure of $P/k = 10^{10-12}$ K cm^{-3} if all the kinetic energy of the ejecta is thermalized (where k is the Boltzmann constant).

What is the effect that such a high pressure in the ISM has on the slow massive wind associated with the type II SN precursors, i.e. the red supergiants (RSGs)? It is inside this wind that the early expansion of the SN ejecta takes place. The radial density gradient of the wind is central in determining the early evolution of the SN. Therefore, it is important to know how the pressure in the ISM will affect the density of the slow wind. The RSG/SN-progenitor wind is not going to move to infinity, but only up to the stagnation radius were the ISM pressure and the ram pressure of the wind reach equilibrium. At the stagnation radius, all the ejected slow winds from the RSG pile-up forming a dense shell[94,45]. When the SN explodes, its ejecta will first interact with the r^{-2} wind until it reaches the wind shell at the stagnation radius, where it will collide with most of the mass of the slow wind from the progenitor that couldn't go any further because of the pressurized surrounding medium.

These events are happening in a bath of UV flux radiated by massive stars that are still close to the main sequence. Thus, the dense CSM associated with the RSG will be photoionized and will reach electron temperatures of about 10^4K. If we assume pressure equilibrium, this implies that the expected density of the CSM around these stars will be between 10^6 and 10^8 cm^{-3}, depending on the adopted IMF. These are the densities observed in the best studied SNe of type IIn (see Sec. 6.5). In this density range the interaction of the SN ejecta with the shell produces a highly radiative shock.

What is the main message? For students of SBs, it is very important to investigate, not only how the ISM is affected by the input of energy and mass from the massive stars, but also how the stars might be affected by a highly pressurized ISM. We have seen the effect that this may have on the CSM of RSGs and red giants. Other phenomena, like the mass-transfer rate in binaries, may be also affected.

The bottom line is that high pressure $P/k \sim 10^{11}$ Kcm^{-3} is expected in the ISM of massive young clusters for which the column density of massive stars is $\sim 10^{4-5}$ M$_{\odot}$pc^{-2}. The high pressure will stop the expansion of the slow wind from red giants and RSGs and a high density compact circumstellar shell will be formed. When type II SNe explode inside the high-density wind shell, the ejecta will first encounter a wind with a density gradient proportional to r^{-2}, and when it reaches the wind-stagnation radius, it will hit a wall producing a most luminous compact SNR. At the same time, as this is occurring at the base of the wind, a galactic wind will develop. Low pressure and canonical type SNRs occur either in low mass clusters, where the time between SNe is larger than the sound-crossing time inside the system, or in low surface-density clusters. In this case, although normal SNRs develop, the super-wind conditions are still there.

6.5 Compact Supernova Remnants

In this section I will describe aspects of radiative shocks that may be relevant to the physics of the BLR in AGN. I recommend the readers the work by Perry and collaborators[84,110,33,111] that follows a view of the SB/AGN relation that in some sense complements the one presented in theses lectures.

The scenario we will consider is that of a nuclear/circumnuclear instantaneous SB during its first 60 Myr of evolution. The massive SB generates initially a hot, metal-rich bubble. After some time, the bubble expands towards the direction of the highest density gradients, i.e. the poles, and breaks out generating a galactic wind. The high ISM pressure in the core of the SB will generate a large mass flow, and ultimately a galactic wind. The fact that galactic winds are seen in many SB galaxies strongly supports the idea that at the base of the wind, and coincident with the SB, there should be a highly pressurized ISM. The evolution of the slow and massive wind of the red giant and supergiant stars inside this pressurized ISM, as we have seen in the previous section, will differ substantially from that of a star in isolation. If the pre-supernova stellar wind expands in a high-pressure medium, a circumstellar shell of more or less uniform density will form at the stagnation radius[45,94]. At this radius all the mass ejected during the wind phase will accumulate in a relatively narrow shell. Slow high density winds, characteristic of red giant and RSGs will have the smallest stagnation radii and, therefore, the highest density circumstellar shells (CSSs). In the dense CSS scenario, radiative cooling becomes important very early in the evolution of SNRs[101,22,23], when their expansion velocities are still in the range of several thousands of $\mathrm{km\,s^{-1}}$. The radiative phase peaks well before the thermalization of the ejecta is completed, causing the remnants to miss altogether the Sedov evolutionary track [109,93,32,101,103].

Intense cooling leads to luminosities as high as 3×10^9 $L_\odot$ on time scales of only a few weeks or months, with subsequent decrease of the thermal energy of the shocked gas within a few years after the explosion. Such remnants, given their rapid evolution and small final sizes, are called compact SNRs (cSNRs), and are able to account for the most energetic peculiar type II SNe observed.

The post-shock regions of high temperature behind both the reverse and leading shock of a cSNR are sources of ionizing flux with a power-law spectrum[101], and several photoionized regions develop within the remnant. Some of these regions are thought to be responsible for the emission of observed broad lines, as the ionizing radiation is absorbed by rapidly moving shell(s). Ionizing radiation is also absorbed by the decelerated and compressed ejecta exterior to the reverse shock, by the still un-shocked ejecta interior to the reverse shock, and by the unperturbed ambient medium. Both analytical and numerical models of cSNRs are in good agreement with the observations of AGN. In particular, the photoionization of the densest features of cSNRs leads to line ratios matching those observed in BLRs of AGN. Because of these similarities the studies of cSNRs may prove crucial for understanding the BLR of AGN.

6.5.1 Semi-Analytical Solutions

The interaction of gas moving at supersonic velocities with respect to a neutral medium generates a shock with an adiabatic region in which the density behind the shock is exactly 4 times the surrounding density, and the temperature behind the shock is $T_s \propto V_s^2$, where V_s is the velocity of the shock. If the pre-shock densities are larger than 10^5 $\mathrm{cm^{-3}}$, there will be a strongly radiative region, most of the shock energy will be radiated away in some time-scale that is proportional to the cooling time and, after that, the density that is achieved will be proportional to the external density times the square of the Mach number. Hence, for a highly supersonic shock, i.e. large Mach number, the post-shock density will be very high, and the post-shock temperature will be back to the initial condition, because most of the energy would have been radiated away at a typical energy $kT \propto V_s^2$. This means that fast shocks will emit most of their luminosity in the form of X-rays[101,41].

A supernova explosion generates a pair of shock waves: the outward-facing *leading shock*, that sweeps, accelerates and heats up the ambient gas; and the inward-facing *reverse shock*, that decelerates and thermalizes the ejecta. In the uniform density or circumstellar shell case, the leading shock becomes strongly radiative at time t_{sg}, defined by

$$t_{sg} = 0.63 E_{51}^{1/8} n_7^{-3/4} \quad \text{yr}, \tag{3}$$

where E_{51} is the explosion energy in units of 10^{51} erg, and n_7 is the ambient density, n_0, in units of 10^7 cm^{-3}. Throughout the radiative phase, the velocity, radius, and luminosity of the leading shock are given by[109,93,101]

$$V_s = 4600 E_{51}^{1/8} n_7^{-1/4} \left(\frac{t}{t_{sg}}\right)^{-5/7} \quad \text{km s}^{-1} \tag{4}$$

$$R_s = 0.01 E_{51}^{1/4} n_7^{-1/2} \left(\frac{t}{t_{sg}}\right)^{2/7} \quad \text{pc}, \tag{5}$$

$$L_s = 5 \times 10^9 E_{51}^{7/8} n_7^{3/4} \left(\frac{t}{t_{sg}}\right)^{-11/7} \quad \text{L}_\odot, \tag{6}$$

whereas the corresponding equation for the post-shock temperature is

$$T_s = 3 \times 10^8 E_{51}^{1/4} n_7^{1/2} \left(\frac{t}{t_{sg}}\right)^{-10/7} \quad \text{K}. \tag{7}$$

Assuming an even pressure behind the shock, these equations can be combined to obtain indicative values of density (n), ionization parameter (U), mass (M), and column density (σ) of the thin cool shell associated with the leading shock:

$$n_{\text{shell}} = 1.2 \times 10^{12}\, E_{51}^{1/4}\, n_7^{3/2} \left(\frac{t}{t_{sg}}\right)^{-10/7} \quad \text{cm}^{-3}, \tag{8}$$

$$U_{\text{shell}} = 2.0 \times 10^{-4}\, E_{51}^{1/8}\, n_7^{-1/4} \left(\frac{t}{t_{sg}}\right)^{-5/7} \quad , \tag{9}$$

$$M_{\text{shell}} = 1.1\, E_{51}^{3/4}\, n_7^{-1/2} \left(\frac{t}{t_{sg}}\right)^{6/7} \quad \text{M}_\odot, \tag{10}$$

$$\sigma_{\rm shell} = 1.1 \times 10^{23}\, E_{51}^{1/4}\, n_7^{1/2} \left(\frac{t}{t_{\rm sg}}\right)^{2/7} \ {\rm cm}^{-2}. \tag{11}$$

Notice that for $n_7 \sim 1$ and $t > t_{\rm sg}$, the above equations predict total luminosities, sizes and gas velocities close to those typical of low luminosity AGN, while the density, ionization parameter, ionized mass and column density values are similar to those that characterize the BLRs of AGN.

The above formulae assume that the ejecta have been already fully thermalized. As the thermalization time scale is substantially longer than $t_{\rm sg}$, they should be treated only as a guideline. However, these analytical estimates are in good agreement with the results from numerical models[41]. The most important estimate is that of the total luminosity, which in this scenario may peak at, or be even above, 3×10^9 $L_\odot$.

The analysis of the above equations lead to a number of important conclusions. First, $V_{\rm s}$ is still large when cooling becomes important. Thus, the rapidly expanding dense shell which forms due to cooling, and which is photoionized by high-energy photons from the post-shock region, will be a source of broad emission lines. Second, because of the high values of $T_{\rm s}$, for the first several years after the explosion, the remnant will radiate most of its thermal energy in the soft X-ray band. Third, due the decrease of $T_{\rm s}$ with time combined with cooling effects, the post-shock gas will have a broad range of temperatures, resulting in a highly nonthermal spectrum of the ionizing continuum.

To illustrate the latter conclusion, the temperature stratification in the post-shock region can be obtained in an isochoric approximation [41] (only valid in the rapid cooling phase, during which the sound-crossing time for the shell of the post-shock gas is longer than the cooling time). Assuming a constant flow velocity equal to $V_{\rm s}$, the temperature gradient is

$$\frac{dT}{dl} = \frac{16}{9}\frac{n_0 \Lambda_0}{k V_{\rm s}} T_{\rm s} T^{3/2} \quad , \tag{12}$$

where l is the distance from the leading shock, and $\Lambda_0 \simeq 10^{-27}$ erg cm^3 s^{-1} K$^{-1/2}$ in the high temperature regime of $T > 3 \times 10^7$ K, in which free-free transitions dominate the cooling processes. The resulting distribution of T is

$$T(l) = T_{\rm s} \left(1 - l \frac{8 n_0 \Lambda_0}{9 k V_{\rm s} T_{\rm s}^{1/2}}\right)^2 \quad . \tag{13}$$

The radiation emerging from a medium with such a wide range of temperatures will have a power-law-like spectrum at UV and X-ray frequencies[101,103,41].

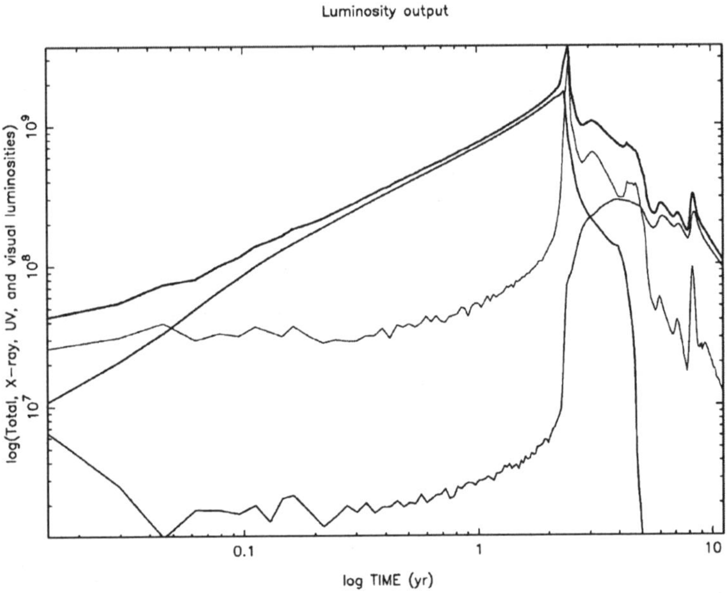

Figure 15. The luminosity output of cSNRs. The bolometric luminosity (upper thick line), X-rays (lower thick line), extreme UV (upper thin line) and optical (lower line) produced by cSNRs throughout their evolution for the case of a CSM of density $n_0 = 10^7\ \mathrm{cm}^{-3}$and $Z_\odot$ abundance.

6.5.2. Numerical Models

Let us now look at some of the results from numerical models of cSNRs and in particular at the details of the light-curve of one of these events. Fig. 15 shows the results of the computation of a model of a 10^{51} erg explosion in a uniform density CSM of $10^7\ \mathrm{cm}^{-3}$. Note that the bolometric light-curve has large amplitude wiggles on a time-scale of months. The luminosity grows as $t^{+0.8}$ reaching a peak of $3.5 \times 10^9\ \mathrm{L}_\odot$. After this maximum, the average or long term decay rate is $t^{-11/7}$, as predicted by the analytical solution above. There is a luminous short event in the bolometric luminosity associated with its maximum. This "flare" is due to the rapid radiative cooling of intermediate temperature gas. Its energy is mostly emitted in the far-UV. The flare, of more than a factor of 2 in luminosity, corresponds to the formation of the thin shell of the forward or leading shock. It occurs behind the leading shock and lasts for about a hundred days.

After the maximum, we see several ups and downs corresponding to thermal instabilities in the shocked gas. This is promoted by the high cooling rate at these high densities. The gas behind the shock, initially at high temperature and pressure, pushes the shock. After a cooling time, less than 1 year for these conditions, the shocked gas cools below 5×10^7K where the cooling rate of gas of solar abundance increases with decreasing temperature, thus generating a catastrophic cooling event. The result is the sudden emission of a UV flare and the drop in the post-shock pressure. This drop in the pressure disconnects the hot cavity from the leading shock, the shock decelerates and the emitted luminosity and temperature of the shock drop. After some time the post-shock temperature recovers and the process repeats in a time-scale of order of a cooling time. This process in both the leading and reverse shocks will generate large amplitude variations in the luminosity of these remnants. Remember that all the parameters scale with the density of the circumstellar medium: the luminosities are higher, and the time-scales are much shorter for high densities.

Let me summarize at this stage the things we believe are relevant for cSNRs:

- cSNRs have predicted line ratios similar to those observed in a typical BLR. Their line-width at maximum luminosity is FWHM$\sim$5000 km s^{-1}, they have small sizes (0.01 pc), the column density in the ionized shells is of the order of 10^{23} cm^{-2}, and the total mass of the emitting shells (collected by the shocks) is between 1 and 10 M$_\odot$. The ionizing spectrum emitted by the shocks from the far-UV to X-rays is a power-law-like continuum. The gas that is emitting the very broad lines is the thin shell gas. This is stable inside the time-scale of evolution of these remnants, but because of the cooling time-scales, rapid UV-variability and also, on longer time-scales, X-ray and broad line variability are predicted.

- cSNRs can reach X-ray luminosities of more than 10^{43} erg s^{-1}, making them the brightest X-ray events known. I believe that the key to the progress in the knowledge of this type of objects lies in high resolution X-ray spectroscopy of young cSNRs.

I would also like to mention what happens with the CSM, after shell formation at the peak of the light-curve. At this point the CSM is pre-ionized to a very high degree and to high temperatures by shock radiation. Therefore, the neutral column density, i.e. the optical depth for X-rays emanating from the shock and traversing the CSM, is going to be very small because the neutral fraction is very low. The CSM is ionized initially to temperatures of about

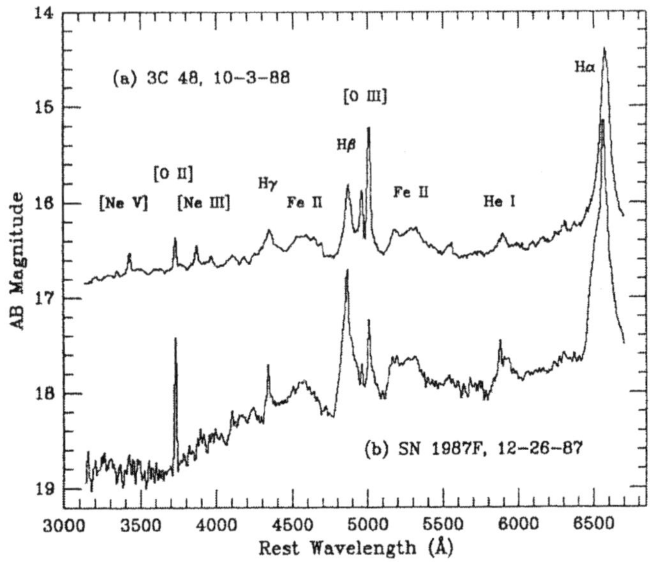

Figure 16. Optical spectra of the quasar 3C 48 (a) and SN 1987 F nine months after discovery. Note the striking similarities between these two spectra. From Filippenko[37] ©1989 AAS.

10^{6-7}K, to an X-ray ionization parameter close to unity. After that, the ionization parameter will drop roughly as t^{-2}. All these plus the additional fact that the CSM covers, as seen by the observer, all the X-ray emitting region, implies that the pre-ionized un-shocked CSM can be identified with an X-ray warm absorber. Following similar lines, the high density cool shell ($n_e \sim 10^{12}$ cm^{-3}, $T < 10^4$ K) behind the leading shock, that reprocesses half of the radiation from the cooling regions of the shock, can be identified with a Compton reflector.

6.5.3 Observations of cSNRs

The optical spectra of SN IIn[90] are characterized by the presence of prominent relatively narrow emission lines seating on top of broad components of FWHM≈ 15000 km s^{-1} at maximum light. The general appearance is extremely similar to those of Sy 1 nuclei and QSOs[39] (see. Figs. 16 and 17). They don't show the characteristic broad P-Cygni signatures of standard SNe, although narrow P-Cygni profiles are detected in some cases at high spectral

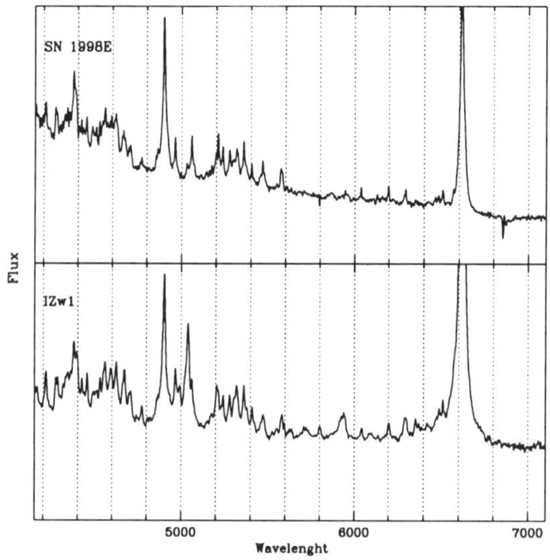

Figure 17. Optical spectra of SN 1998E (top) and the strong FeII emitter Sy 1 galaxy I Zw 1 (bottom). Note the striking similarities between these two spectra.

resolution[88]. SN IIn are normally associated with regions of recent star formation. Despite these general characteristics, SN IIn as a group exhibits some heterogeneity[37,105].

The observed high luminosities, flat light-curves, broad emission lines without P-Cygni profiles, plus in some cases strong radio emission of SNe IIn are interpreted as a result of the interaction between the SN ejecta and a dense CSM with $n_0 \geq 10^5$ cm^{-3} , i.e. as cSNR[18,19,39,96,102,23,42]. In what follows I will assume that SNe IIn, particularly the brightest ones, are cSNRs.

The best studied cSNR is SN 1988Z, which was also the first one detected in hard X-rays[34]. Its optical spectrum shows very broad emission lines of H, He and Fe plus some nebular lines, particularly in the early stages of evolution. Unresolved narrow lines of [OIII] were detected during the early stages[96,5]. These narrow lines are likely to be originated in the CSM. The observed ratio [OIII]λ4363/[OIII]λ5007 ~ 0.5 implies densities in excess of 5×10^6 cm^{-3}. The light-curve of 1988Z followed very nicely the predictions of a radiative shock model[5]. The integration of the radio, optical and X-ray luminosities emitted by SN 1988Z in the 8.5 yr of evolution after discovery gives a total energy

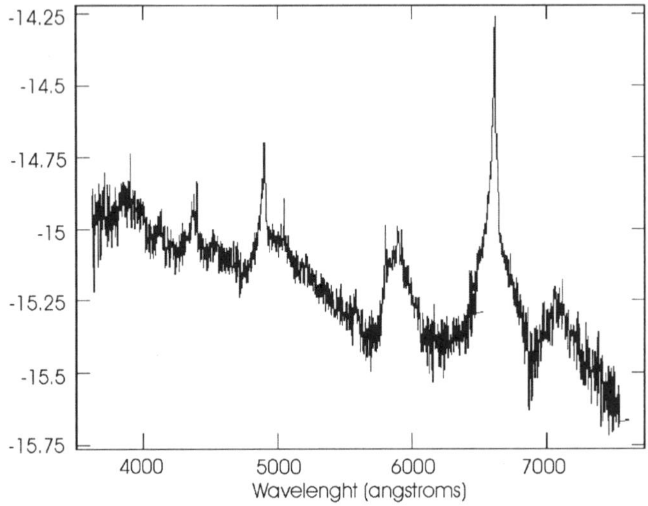

Figure 18. Optical spectrum of SN 1997eg, a broad-line cSNR. Note the extremely broad base of the permitted emission lines (FWZI$\approx$ 30000 km s^{-1}).

close to 10^{52} erg, i.e. larger than the canonical value of the kinetic energy released in a SN explosion, while typical SNe radiate one to two orders of magnitude less energy[82,15,16]. The value of the radiated energy in SN 1988Z is slightly smaller than that estimated for the most energetic kinetic release in a SN explosion observed so far, $(2 - 5) \times 10^{52}$ erg in SN 1998bw[44,54]. SN 1988Z was speculated to be a hypernova that may have been caused by the formation of a black hole in the core of a massive star[5]. Most of the energy in SN 1988Z is radiated in the optical to X-ray interval, with spectra, spectral energy distribution and light-curves departing from those of classical SNe II. The high densities derived from the O forbidden lines and the flat light-curves strongly support the idea of a high-density circumstellar shell reprocessing the kinetic energy through fast radiative shocks.

There is a range of observed spectra of cSNRs that mimics the range of BLRs observed in AGN. This is illustrated in the spectra of four cSNRs (Figs. 16–19), going from SN 1997eg with very broad emission lines to SN 1998es with narrower emission lines and strong permitted FeII emission, reminiscent of the spectrum of the canonical strong FeII emitter I Zw1. This, to me, suggests some link in the physics of cSNRs and the BLRs of AGN.

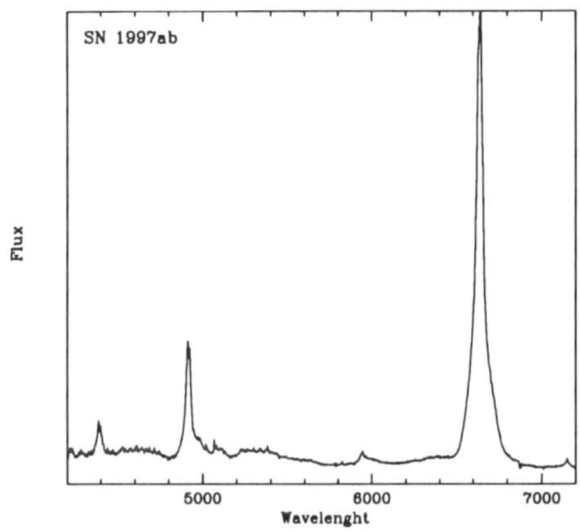

Figure 19. SN 1997ab a "typical" cSNR. The tip of the Balmer emission lines have superimposed narrow (FWHM$\approx$ 90 km s^{-1}) P-Cygni profiles originated in the circumstellar shell[88].

7 Variability in Low Luminosity AGN

Variability is a key aspect that must be explained by any theory of AGN. In the SB scenario, the UV/optical variability of the BLR is due to cSNR activity. The best information regarding variability in AGN is provided by a small group of nearby low luminosity type 1 Seyferts. The optical light-curves of these AGN show two distinct components: an occasional sharp peak variation superimposed onto a long term recurrent modulation[63,64,31,65,66], called rapid and slow components respectively. The peaks of the rapid component last some weeks, while the cycles of the slow one are about two years long. In the SB model, the slow component is associated with the long term behaviour of cSNRs, while the rapid component is mostly due to cooling instabilities in the shells of cSNRs and to SN flashes. For papers dealing with cSNR-generated variability I suggest to read[3,4,5,6,26,27].

A double peak variation is expected in the light curve of a SN evolving in a high density circumstellar medium. The first peak corresponding to the SN flash, and the second one to the time when the cSNR reaches the maximum of its radiative phase[3,4]. As we have seen above, the typical time scale of the radiative phase is t_{sg} (Eq. 3). During the radiative phase, the luminosity of the remnant, can be approximated by

$$L_{bol}(t) \sim 10^{10}\,L_{\odot}\,E_{51}^{7/8} n_{7}^{3/4} \left(\frac{t}{t_{sg}}\right)^{-11/7} , \tag{14}$$

which is emitted mainly in the far-UV and X-rays. cSNR type of events can account[3] for the slow-component variations in the well sampled light-curve of NGC 4151, both in energy and time-scale. The estimated number of cSNRs required to explain the luminosity of the nucleus is also consistent with the number of events expected from the luminosity emitted by the nucleus.

Due to the stochastic nature of the SN activity and the low SN rate in low luminosity AGN, the nucleus can go through transient quiescent phases when any light from cSNRs is faint and, consequently, no broad lines are detected, while the continuum is still blue due to the presence of hot stars. This is likely to be the case of the deep photometric minima of 1981 and 1984 in NGC 4151 during which the nucleus resembled a type 2 Seyfert[2,83,65,3,5]. This characteristic of SN activity gives a simple explanation for the transient stages of type 2 Seyfert observed as LINERs and type 1 Seyfert nuclei.

A strong prediction of the SB model is that the variability amplitude observed in AGN should reach a maximum for those nuclei with luminosity similar to the maximum luminosity of cSNRs, i.e. $M_B \sim -20$. AGN with higher luminosity should show less amplitude of variability due to the superposition of cSNR events, while lower luminosity AGN, i.e. those with $M_B >> -20$ should also show a smaller variability amplitude because they will be representing old and less luminous cSNR which evolve slowly in their spectral properties.

The SN rate is determined by the nuclear continuum blue luminosity. A core SB with a continuum magnitude of $M_B = -16.3$ should have 1 SN event every 100 years. As the cSNRs remain detectable against such a background for about 10 years, one in every 10 low luminosity AGN with $M_B \sim -16.3$ should show broad Hα emission and 1 in every 100 should be found to have a very bright BLR corresponding to a cSNR less than a year old.

Two additional predictions for low luminosity AGN that stem from the properties of old cSNRs are that, the FWHM of the broad lines in these objects should be smaller than in more luminous AGN, and furthermore, the FWHM should decrease slowly with time as FWHM $\propto V_s \propto t^{-\frac{5}{7}}$ while its luminosity should decrease as $L_{bol} \propto t^{-\frac{11}{7}}$. The X-ray spectrum of the BLR should also be much softer than that of typical Seyfert type 1. A relation between the

FWHM of the BLR and the X-rays spectrum is therefore expected in the sense that those AGN with lower luminosity should have also smaller FWHM and softer X-ray spectrum.

8 UV-Variability of the BLR and the Nature of the Lag

Terlevich et al.[102] have analysed the time dependent process that occurs prior to thin shell formation in rapidly radiating cSNRs, as maximum luminosity is achieved. They have found that an inherent delay between the photon emission and the time required to increase the density of the cooling gas leads to a lag between the observed continuum burst and the emission-line response. This delay is intrinsic to the physical processes investigated and not related to the geometry of the system. There are in fact no light-crossing-time arguments involved. The total width of the cold or photoionized region is only about 10^{13} cm (about 300 light-seconds) yet, delays of up to several weeks between continuum and line emission are generated.

Early in the evolution of the cooling wave, large photon fluxes are available, but the UV/optical emission lines are very faint due to the high ionization parameter associated with the still relatively low electron density and the low column density of the cold material. As the cooling wave evolves, however, the column density of cold material soon exceeds $10^{21}\,\mathrm{cm}^{-2}$ and its density starts to increase over the post-shock value. As a consequence, the ionization parameter begins to decrease and lines like NVλ1240 and HeIIλ1640 start to be emitted copiously. By the time the density of the compressed shell reaches a few times 10^9 cm^{-3}, the ionization parameter has dropped to about 0.1 and the electron temperature of the ionized gas is around 10^4K. At this time lines like Lyα and CIV reach their maximum intensity with a delay of several days with respect to the start of the UV burst. CIII shows longer delays in all models. Other lines, like the Balmer series have even longer delays and the longest predicted delays are for MgII and FeII that show lags of more than a month.

Only one lag-producing flare per cSNR is predicted. This flare is the longest and most energetic one. As these luminous flares are produced at the maximum of the light-curve, one important prediction is that flares showing lag between continuum and emission lines are expected to occur close to the maximum of the light-curve. Conversely, no further luminous and lag-producing flare is expected close to the absolute minimum of the light-curve.

Terlevich et al.[102] compared the predicted shape and luminosity of the light curve of the emission lines from the cooling wave models with the observations of NGC 5548. They compared the models with the one flare that

Table 2. Observed and predicted lag in NGC 5548: (*1*) Emission line identification; (*2-3*) Observed and predicted lag in days; (*4-5*) Observed and predicted peak line luminosity in units of 10^{40} erg s^{-1} .

	lag (days)		Peak Luminosity (10^{40} erg s^{-1})	
	Observed	Predicted	Observed	Predicted
Ly$\alpha\lambda$1216	12	10 - 15	100.	90.
NVλ1240	4	2 -10	30.	14.
SiIVλ1440	12 - 34	12 - 17	13.	23.
CIVλ1550	8 - 16	10 - 15	100.	280.
HeIIλ1640	4 - 10	12 - 20	15.	26.
CIII]λ1909	26 - 32	20 - 30	18.	13.
MgIIλ2800	34 - 72	40 - 50	10.	26.
H$\beta\lambda$4861	20	30 - 40	8.	7.5

was strongest and having the best sampling, the second one. Table 2 shows a compilation of the cross-correlation results of the UV and optical campaign of NGC 5548[28,86] together with the results from the models. A quick inspection of Table 2 shows that the observed lags for *all* the observed emission lines are similar to the model predictions. Furthermore the observed and predicted peak luminosities of the emission lines are remarkably similar.

After the first cooling event, dense cold gas will always be present in the outer shell immediately reprocessing the inward moving ionizing radiation produced by further cooling instabilities; little or no lag occurs between continuum and lines as was observed in the 3rd pulse of the NGC 5548 monitoring campaign. These further cooling instabilities are part of an oscillatory instability that produces quasi-periodic modulations in the shock position and velocity.

9 Final remarks

The last few years have produced tantalizing evidence for SMBH in galactic nuclei. Also as we have seen, there is now strong proof that massive nuclear SBs produce a substantial fraction of the emitted energy in many nearby AGN. What happens next?

It is clear how to increase our knowledge of the nuclear regions in type 2 AGN, i.e. where the contamination from the BLR is minimum and the emission lines are narrow. Detailed studies of the stellar populations of type 2 AGN will bring important information about the physics of star-formation in an environment very different from that of a galactic disk.

Detecting and measuring the young stellar components of type 1 AGN will be much more difficult. Clearly, if unification is valid, a substantial fraction of the emitted light in type 1 AGN should also come from a nuclear SB. In order to revoke this last point it is necessary to improve our ability to observationally segregate SBs from BLRs.

There are other critical observations that we can perform. High resolution X-Ray spectroscopy will be crucial in the next few years to assess the possible connection between cSNRs and the BLR. X-Ray line profiles (hot gas kinematics), line ratios (hot gas abundances) and their evolution are very important parameters to help the theoretical understanding of cSNRs, SBs and AGN.

Optical interferometry may solve the problem of the source of the optical/UV variability in the BLR. To assess the possible role played by cSNRs we need to know if the source of variability of the optical and UV continuum is always in the same position of the sky, with a resolution of 0.1 pc or if it jumps as expected from a young cluster with compact supernova remnants, "the supernova dance"[77].

Finally, I would like to thank my colleague speakers, from whom I learned a lot and whose presentations were great fun. I would like also to thank the audience for lively discussions and for their patience, and the organizers for a fantastic experiment in Advanced Schools. Thank you all.

Acknowledgments

This written version of the lectures are a reality thanks to the hard work of Elena Terlevich and the help of Elena and Itziar Aretxaga. To both my warmest thanks. The final version also benefitted from input from Daniel Kunth, David Hughes and William Wall. Thanks to Charlie Nelson and Rosa González-Delgado for providing figures prior to publication and Gabriela López for producing some of the figures. Finally I would like to thank all my colleagues and friends that with enthusiasm have contributed to this work through the years.

References

1. Alloin, D., et al., A&A, 293, 293, 1995.
2. Antonucci, R. R. J. & Cohen, R. D., ApJ, 271, 564, 1983.
3. Aretxaga, I. & Terlevich, R., MNRAS, 269, 462, 1994.
4. Aretxaga, I., Cid Fernandes, R. & Terlevich, R., MNRAS, 286, 271, 1997.

5. Aretxaga, I, Benetti, S., Terlevich, R.J., Fabian, A., Cappellaro, E., Turatto, M., & Della Valle, M., MNRAS, 309, 343, 1999.
6. Aretxaga, I., Joguet, B., Kunth, D. Melnick, J.& Terlevich, R., ApJ, 519, 123, 1999.
7. Aretxaga, I., Terlevich E., Terlevich R.J., Cotter G., Díaz A.I., MNRAS, submitted (available at http://www.inaoep.mx/~itziar)
8. Baldwin J.A., Phillips M.M., Terlevich R.J., PASP, 93, 5, 1981.
9. Barbier, D., in 'Principles fundamentaux de classification stellaire', Coll. Int. Centre Nat. Rech. Sc., vol 55, p. 47, 1955.
10. Bender,R., Burstein, D., Faber, S.M., ApJ, 399, 452, 1992.
11. Biermann, P. & Shapiro, S. L., ApJ., 230, L3, 1979.
12. Binney, J. & Tremaine, S. in Galactic Dynamics, Princeton University Press, 1987.
13. Bosch, G., Terlevich, R., Melnick, J. & Selman, F., A&AS, 137, 21, 1999.
14. Campbell et al., AJ, 104, 1721, 1992.
15. Catchpole et al., MNRAS, 229, 15P, 1987.
16. Catchpole et al., MNRAS, 237, 55P, 1989.
17. Chalonge, D., Ann. d'Ap., 19, 258, 1956.
18. Chevalier R. A., ApJ, 188, 501, 1974.
19. Chevalier R. A. & Imamura J. N., ApJ, 261, 543, 1982.
20. Chevalier R. A. & Clegg A. W.., Nat, 317, 44, 1985.
21. Chu, Y.-H., & Kennicutt, R. C., Jr., ApJ, 425, 720, 1994.
22. Chugai N.N., MNRAS, 250, 513, 1991.
23. Chugai N.N. & Danzinger, MNRAS, 268, 173, 1994.
24. Cid Fernandes, R., Dottori, H. A., Gruenwald, R. B., Viegas, S. M., MNRAS, 255, 165, 1992.
25. Cid Fernandes, R., Terlevich, R., MNRAS, 272, 423, 1995.
26. Cid Fernandes, R., R. Aretxaga, I. & Terlevich, MNRAS, 282, 1191, 1996.
27. Cid Fernandes, R., Terlevich, R. & Aretxaga, I., MNRAS, 289, 318, 1997.
28. Clavel, J. et al., ApJ, 366, 64, 1991.
29. Colina, L., Garcia Vargas, M. L. , González Delgado, R. M., Mas-Hesse, J. M., Perez, E., Alberdi, A., & Krabbe, A. , ApJ, 488, 71, 1997.
30. Couch & Sharples, MNRAS, 229, 423, 1987
31. Dibaĭ, É. A. & Lyutyi, V. M., SovA, 28, 7, 1984.
32. Draine, B.T. & Woods, D..T., ApJ, 383, 621, 1991.
33. Dyson, J. E., Williams, R. J. R., & Perry, J. J., Ap& SS, 237, 187, 1996.
34. Fabian A.C., Terlevich R.J., MNRAS, 280, L5, 1996.
35. Fabian, A.C., et al., PASP, 112, 1145, 2000.
36. Feast, M. W., Thackeray, A. D., Wesselink, A. J., MNRAS, 121, 337, 1960.

37. Filippenko, A.V., ARAA, 35, 309, 1997.
38. Filippenko, A.V. in 'Supernovae', Ed. S.E. Woosley, Springer-Verlag, page 467, 1991.
39. Filippenko, A.V., AJ, 97, 726, 1989.
40. Filippenko, A.V. & Terlevich, R., ApJ, 397, L79, 1992.
41. Franco, J., Miller, W.W., Arthur, S.J., Tenorio-Tagle, G & Terlevich, R., ApJ, 435, 805, 1994.
42. Fransson, C., Lundqvist, P. and Chevalier, R.A., ApJ,461, 993, 1996.
43. Friaça, A. & Terlevich, R., MNRAS, 298, 399, 1998.
44. Galama T.J. et al., Nat, 395, 67, 1998.
45. Garcia-Segura, G., Langer, N. & Mac Low, M. -M., A&A, 316, 133, 1996.
46. Genzel, R., Pichon, C., Eckart, A., Gerhard, O. E. & Ott, T., MNRAS, 317, 348, 2000.
47. González Delgado R. M., Heckman, T., Leitherer, C., ApJ, in press, 2001.
48. González Delgado R. M., Heckman, T., Leitherer, C., Meurer, G., Krolik J., Wilson, A. S., & Kinney A., ApJ, 505, 174, 1998.
49. Hamann F., Ferland G., ApJ, 418, 11, 1993.
50. Heckman, T. M., Crane, P. C. & Balick, B., A&AS, 40, 295, 1980.
51. Heckman, T.M., et al., ApJ, 452, 549, 1995.
52. Heckman, T.M., et al., ApJ, 482, 114, 1997.
53. Idiart, T.P., de Freitas Pacheco, J.A., Costa, R.D.D., AJ, 112, 2541, 1996.
54. Iwamoto K. et al., Nat, 395, 672, 1998.
55. Jablonka, P., Martin, P., Arimoto, N., AJ, 112, 1415, 1996.
56. Jiménez-Benito L., Díaz A.I., Terlevich R.J., Terlevich E., MNRAS, 317, 907, 2000.
57. Jones, J.E., Alloin, D.M., Jones, B.T.J., ApJ, 283, 457, 1984.
58. Kunth D., Contini T. in IAU Symp. 193 'Wolf-Rayet Phenomena in Massive Stars and Starburst Galaxies', eds. K.A. van der Hucht, G. Koenigsberger & P.R.J. Eenens, ASP, p. 725, 1999.
59. Kunth D., stlin, G., A&AR, 10,1, 2000.
60. Leitherer, C., Heckman, T. M., ApJS, 96, 9, 1995.
61. Leitherer, C., Schaerer, D., Goldader, J. D., Gonzlez-Delgado, R.M., Robert, C., Kune, D. F., de Mello, D. F., Devost, D. & Heckman, T. M., ApJS, 123, 3, 1999.
62. Lin, Y.C. et al., , ApJ, 416, L53, 1993.
63. Lyutyi, V. M., SovA, 21, 655, 1977.
64. Lyutyi, V. M., SovA, 23, 518, 1979.
65. Lyutyi, V. M., Oknyanskii, V. L. & Chuvaev, K. K., SovAL, 10, 335, 1984.
66. Lyutyi, V.M. & Oknyanskii, V.L., SovA, 31, 245, 1987.

67. Massey P., Hunter D. A., ApJ, 493, 180, 1998.
68. McMahon, R. G., Priddey, R. S., Omont, A., Snellen, I., Withington, S., MNRAS, 309, 1, 1999.
69. Maeder, A. & Meynet, G., AA, 182, 243, 1987.
70. Maeder, A., AAS, 84, 139, 1990.
71. Malumuth E. M., Heap S. E., AJ 107, 1054, 1994,
72. Matteucci F., Padovani P., ApJ, 419, 485, 1993.
73. Melnick, J., A&A, 153, 235, 1985.
74. Melnick, J., Terlevich, R. & Moles, M., MNRAS, 235, 297, 1988.
75. Melnick, J., Tapia, M. & Terlevich, R., A&A, 213, 89, 1989.
76. Melnick, J., Gopal-Krishna, Terlevich, R., A&A 318, 337, 1997.
77. Melnick, J., Aretxaga, I., & Terlevich, R. in: Science with the VLT Interferometer, Proceedings of the ESO workshop, held at Garching, Germany, 18-21 June 1996, Publisher: Berlin, New York: Springer-Verlag, ESO Astrophysics Symposia, p.112, 1997.
78. Miyoshi, M., Moran, J., Herrnstein, J., Greenhill, L., Nakai, N., Diamond, P. & Inoue, M., Nat, 373, 127, 1995.
79. Mould, J.R., ApJ, 220, 434, 1978.
80. Nelson C.H., Whittle M. ApJ, submitted.
81. Oliva E., Origlia L., Kotilainen J.K. J. & Moorwood A.F.M., A&A, 301, 55, 1995.
82. Panagia N. et al., MNRAS, 192, 861, 1980.
83. Penston, M. V. & Pérez, E., MNRAS, 211, 33P, 1984.
84. Perry, J., Dyson, J.E., MNRAS, 213, 665, 1985.
85. Peterson, B.M., 'An Introduction to Active Galactic Nuclei', Cambrisge University Press., 1997.
86. Peterson, B. et al., ApJ, 368, 119, 1991.
87. Richstone, D. O. & Potter, M. D., ApJ, 254, 451, 1982.
88. Salamanca I., et al., MNRAS, 300, L17, 1998.
89. Schaller, G., Schaerer, D., Meynet, G., Maeder, A., AApS, 96, 269, 1992.
90. Schlegel E.M., MNRAS, 244, 269, 1990.
91. Selman, F., Melnick, J., Bosch, G. & Terlevich, R., A&A, 341, 98, 1999.
92. Selman, F., Melnick, J., Bosch, G. & Terlevich, R., A&A, 347, 532, 1999.
93. Shull J.M., ApJ, 237, 769, 1980.
94. Shull J.M., ApJ, 238, 860, 1980.
95. Silich S. & Tenorio-Tagle G., MNRAS 299, 249, 1998.
96. Stathakis R.A., Sadler E.M., MNRAS, 250, 786, 1991.
97. Suchkov, A. A., Berman, V. G., Heckman, T. M., Balsara, D. S., ApJ, 463, 528, 1996.
98. Terlevich R. & Melnick J., MNRAS, 213, 841, 1985.

99. Terlevich, R., Melnick, J., & Moles, M., in 'Observational Evidence of Activity in Galaxies', Eds. E. Ye Kachickian et al., Kluwer, 1987.
100. Terlevich, E., Díaz A.I. & Terlevich, R., MNRAS, 242, 285, 1990.
101. Terlevich R.J., Tenorio-Tagle G., Franco J., Melnick J., MNRAS, 255, 713, 1992.
102. Terlevich R.J., in Clegg R.E.S., Stevens I.R., Meikle W.P.S., eds, Circumstellar Media in the Late Stages of Stellar Evolution. Cambridge Univ. Press, Cambridge, p. 153, 1994.
103. Terlevich, R., Tenorio-Tagle, G., Różyczka, M., Franco, J. and Melnick, J., MNRAS, 272, 198, 1995.
104. Terlevich, R. , RMAA, 6, 1, 1997.
105. Turatto M., Benetti S., Cappellaro E., Danziger I.J., Mazali P.A., in *'SN 1987A: Ten Years After'*, eds. M.M. Phillips, N.B. Suntzeff, Fifth CTIO/ESO/LCO Workshop, 1999.
106. Van den Marel, R.P., AJ, 117, 744, 1999.
107. Walborn, N. R., in IAU Symp. 116, Luminous Stars and Associations in Galaxies, ed. C. W. H. de Loore, A. J. Willis, & P. Laskarides (Dordrecht: Reidel), 185, 1986.
108. Weedman D., ApJ, 266, 479, 1983.
109. Wheeler J.C., Mazurek T.J. & Sivaramakrishnan, ApJ, 237, 78, 1980.
110. Williams, R. J. R. & Perry, J. J., MNRAS, 269, 538, 1994.
111. Williams, R. J. R., Baker, A. C. & Perry, J. J., MNRAS, 310, 913, 1999.

Brian Boyle at the dinner party,
introducing Mexico to a *Highland Jig*.

THE EVOLUTION OF QSOS

B.J. BOYLE

Anglo-Australian Observatory, PO Box 296, Epping, NSW 1710, Australia

I review the most recent results on the cosmological evolution of the QSO population. I focus on the ensemble *average* properties of the QSO population including their space density/luminosity, clustering, galaxy environment and host galaxies. I show that the cosmological evolution of QSOs closely parallels that of galaxies and that many of the properties of QSOs are consistent with a picture in which QSO activity may be linked to the formation of galaxy spheroidal components in the early Universe.

1 Cosmological Equations

We begin with a brief refresher course on the cosmological equations relevant to the observed properties of QSOs at high redshift. The reader is referred to the excellent book by Peacock[128] for details.

Recall that the Roberston-Walker (RW) metric is given by

$$c^2 d\tau^2 = c^2 dt^2 - R^2(t)[dr^2 + S_k^2(r) d\theta^2] \quad , \tag{1}$$

where r is the distance element and $R(t)$ is the scale factor of the Universe at a given cosmic time t. The relation to the scale factor at the present epoch ($R(t_0) \equiv R_0$) is given by

$$\frac{R(t)}{R_0} = \frac{1}{(1+z)} \quad . \tag{2}$$

This gives rise to the concept of comoving distance ($r_{\rm comov}$) as proper distance ($r_{\rm prop}$) scaled to the present epoch,

$$\frac{r_{\rm prop}}{r_{\rm comov}} = \frac{1}{(1+z)} \quad . \tag{3}$$

The geometry of the Universe is contained in the term $S_k(r)$,

$$S_k(r) = \begin{cases} \sin r & (k=1) \\ r & (k=0) \\ \sinh r & (k=-1) \end{cases} \quad , \tag{4}$$

where k=1,0,−1 for closed, flat and open universes, respectively.

The comoving size $dl_{\rm comov}$ of an object subtending an angle $d\theta$ is then simply given by setting $dt = dr = 0$ in the RW metric,

$$dl_{\rm comov} = d\theta[R_0 S_k(r)] \quad . \tag{5}$$

The comoving volume $dV_{\rm comov}$ is then obtained by considering a volume of space with transverse dimensions $dl_{\rm comov} \times dl_{\rm comov}$ and radial comoving dimension $R_0 dr$, subtending a solid angle $d\omega$,

$$dV_{\rm comov} = [R_0 S_k(r)]^2 R_0 dr d\omega \ . \tag{6}$$

The flux received at a given frequency $S(\nu_0)$ from an object of luminosity $L(\nu_0)$ is given by the inverse square law for an observer on a sphere with radius $R_0 S_k(r)$ from the source,

$$S(\nu_0) = \frac{L(\nu_0)}{4\pi [R_0 S_k(r)]^2 (1+z)^{1+\alpha}} \ . \tag{7}$$

The factor $(1+z)^\alpha$ corrects the observed flux at frequency $\nu_1 = \nu_0/(1+z)$ back to the rest frame frequency ν_0, assuming the object to have a power-law spectral energy distribution between ν_1 and ν_0 of the form $f_\nu \propto \nu^{-\alpha}$. An additional factor $(1+z)$ decrease in energy flux arises from a combination of factors: (a) a factor $(1+z)$ decrease in energy flux due to the redshifting of photon energies between emission and reception (b) a factor $(1+z)$ decrease in energy per unit time due to the redshifting of arrival rates and (c) a factor $(1+z)$ increase energy per unit frequency due to the decrease in bandwidth.

Finally, look-back time is obtained by setting $d\tau = 0$ in the RW metric,

$$dt = \frac{R_0 dr}{c(1+z)} \ . \tag{8}$$

To use these equations, we need expressions for $R_0 S_k(r)$ and $R_0 dr$. These may be obtained from the Friedmann equation (see Peacock[128] for a full derivation), giving

$$R_0 S_k(r) = \frac{c}{H_0} \int_0^z \frac{dz'}{\sqrt{(1-\Omega)(1+z')^2 + \Omega_\Lambda + \Omega_{\rm m}(1+z')^3}} \ , \tag{9}$$

where $\Omega_{\rm m}$ and Ω_Λ are the relative contribution of matter and the cosmological constant to the mean mass density of the Universe (Ω). The integral can only be solved numerically. Trivially, the differential form is given by

$$R_0 dr = \frac{c}{H_0} \frac{dz}{\sqrt{(1-\Omega)(1+z)^2 + \Omega_\Lambda + \Omega_{\rm m}(1+z)^3}} \ . \tag{10}$$

The expressions for length, luminosity, volume element and time can then be obtained for any cosmology by substituting the expressions for $R_0 dr$ and $R_0 S_k(r)$ into the equations for length, volume, luminosity and time above. For a universe with $\Omega_\Lambda + \Omega_{\rm m} = 1$, these simplify to

Length :	$dl_{\rm comov} = \frac{c}{H_0} D_1 d\theta$	(11)
Luminosity :	$L(\nu_0) = 4\pi \left(\frac{c}{H_0}\right)^2 D_1^2 (1+z)^{1+\alpha} S(\nu_0)$	(12)
Volume :	$dV_{\rm comov} = d\omega \left(\frac{c}{H_0}\right)^3 D_1^2 \frac{dz}{\sqrt{(1-\Omega_{\rm m}) + \Omega_{\rm m}(1+z)^3}}$	(13)
Time :	$t(z) = \frac{1}{H_0} \int_0^z \frac{dz'}{(1+z)\sqrt{(1-\Omega_{\rm m}) + \Omega_{\rm m}(1+z)^3}}$	(14)

where the dimensionless distance element D_1 is given by

$$D_1 \equiv \frac{H_0}{c} R_0 r = \left[\int_0^z \frac{dz'}{\sqrt{(1-\Omega_{\rm m}) + \Omega_{\rm m}(1+z')^3}}\right] \quad . \tag{15}$$

For a universe with $\Omega_\Lambda = 0$, we obtain

Length :	$dl_{\rm comov} = \frac{c}{H_0} D d\theta$	(16)
Luminosity :	$L(\nu_0) = 4\pi \left(\frac{c}{H_0}\right)^2 D^2 (1+z)^{1+\alpha} S(\nu_0)$	(17)
Volume :	$dV_{\rm comov} = d\omega \left(\frac{c}{H_0}\right)^3 D^2 \frac{dz}{(1+z)\sqrt{(1+\Omega z)}}$	(18)
Time :	$t(z) = \frac{1}{H_0} \int_0^z \frac{dz'}{(1+z')^2\sqrt{(1+\Omega z')}}$	(19)

In this case the dimensionless distance element D is given by

$$D = \frac{z}{(1+z)} \left[1 - \frac{z(1-\Omega z/2)}{\sqrt{(1+\Omega z)} + 1 + \Omega z/2}\right] \quad . \tag{20}$$

The expressions in physical units for an $\Omega_\Lambda = 0$ universe with $H_0 = 100h$ km s^{-1}Mpc^{-1} are as follows:

$$\textbf{Length}: \left[\frac{dl_{\rm comov}}{\rm kpc}\right] = 14.5 D h^{-1} \left[\frac{d\theta}{\rm arcsec}\right] \tag{21}$$

$$\textbf{Luminosity}: \left[\frac{L(\nu_0)}{\rm WHz^{-1}}\right] = 10^{53.03} D^2 h^{-2} \left[\frac{S(\nu_0)}{\rm Wm^{-2}Hz^{-1}}\right] (1+z)^{1+\alpha} \tag{22}$$

$$\textbf{Volume}: \left[\frac{dV_{\rm comov}}{\rm Mpc^3}\right] = 8.22 \times 10^6 D^2 h^{-3} \left[\frac{d\omega}{\rm deg^2}\right] \frac{dz}{(1+z)\sqrt{(1+\Omega z)}} \tag{23}$$

$$\textbf{Time}: \left[\frac{t(z)}{\rm Gyr}\right] = 9.79 h^{-1} \int_0^z \frac{dz'}{(1+z')^2\sqrt{(1+\Omega z')}} \tag{24}$$

In the optical regime the flux is often quoted in optical magnitudes. Using the relation between AB magnitude ($m_{\rm AB}$) and flux,

$$\left[\frac{S(\nu_0)}{\rm Wm^{-2}Hz^{-1}}\right] = 10^{-0.4(m_{\rm AB}+56.1)} \quad , \tag{25}$$

we can express the luminosity-flux relation in terms of absolute ($M_{\rm AB}$) and apparent magnitude,

$$M_{\rm AB} = m_{\rm AB} - 5\log D - 2.5(1+\alpha)\log(1+z) - 42.39 + 5\log h \quad . \tag{26}$$

In the optical regime, magnitudes are most often quoted in the Johnson B passband. For a QSO with power-law slope $\alpha = 0.5$ the relations between apparent magnitude (B), absolute magnitude (M_B), flux and luminosity become

$$\left[\frac{S(4400\text{Å})}{\rm Wm^{-2}Hz^{-1}}\right] = 3.82 \times 10^{-23} 10^{-0.4B} \tag{27}$$

$$\left[\frac{L(4400\text{Å})}{\rm WHz^{-1}}\right] = 4.59 \times 10^{13} 10^{0.4 M_B} \tag{28}$$

In the optical regime, rather than use a single power-law spectral slope, it is common to use a more complex observationally derived spectral (or K-) correction to convert flux at the observed wavelength to a fixed rest frame wavelength. This is normally expressed as a correction to the observed B (or R) band magnitude as a function of redshift[40,165].

Unless otherwise indicated, throughout this chapter results will be quoted for an Einstein–de Sitter universe ($\Omega_{\rm m} = 1$, $\Omega_\Lambda = 0$) with $h = 0.5$.

2 Evolution of QSO Luminosity

2.1 Introduction

One of the most basic demographic measures of the QSO population is the QSO luminosity function (LF). It can be used to provide information, not only on QSO demographics (e.g. contribution to UV/X-ray backgrounds, QSO redshift cut-offs and comparison with galaxy/star formation evolution), but also, to place constraints on the physical model for QSO evolution (e.g. nature of the central engine, and formation of structure in the early Universe).

The LF is a measure of the comoving space density of QSOs as a function of luminosity and redshift,

$$\Phi(L,z) = \frac{d^2}{dLdV}(L,z) \quad . \tag{29}$$

The evolution of the QSO population is then given by the redshift dependence of the LF. The cartoon in Fig. 1 illustrates the basic features of the LF and its evolution. If a feature is present in the LF, it is possible to determine whether the LF evolves with redshift by: (a) increasing uniformly in space density (PDE: pure density evolution); (b) increasing in luminosity (PLE: pure luminosity evolution); or (c) by a change in shape of the luminosity function (LDDE: luminosity dependent density evolution). To some extent the functional form of the evolution has little bearing on the physics of the QSOs. PLE does not necessarily imply a single generation of long-lived (10^{10}yr) QSOs gradually fading from high to low redshift. It could equally reflect the uniform dimming in the luminosity of successive generations of shorter lived QSOs.

Binned estimates of the LF have been traditionally obtained using the $1/V_a$ statistic[9]. The $1/V_a$ estimator for the LF, $\Phi_{\rm est}$, in the (L,z) interval $([L_{\min}, L_{\max}], [z_{\min}, z_{\max}])$ can be obtained from a combination of $n_{\rm sur}$ surveys comprising $n_{\rm QSO}$ QSOs as

$$\Phi_{\rm est}(L,z) = \frac{1}{\Delta L} \sum_{i=1}^{n_{\rm QSO}\in(L,z)} \frac{1}{V_a(i)} \quad , \tag{30}$$

where,

$$\Delta L = L_{\max} - L_{\min} \quad , \tag{31}$$

and

$$V_a = \sum_{j=1}^{n_{\rm sur}} [V(z_{\max_j}) - V(z_{\min})] \quad , \tag{32}$$

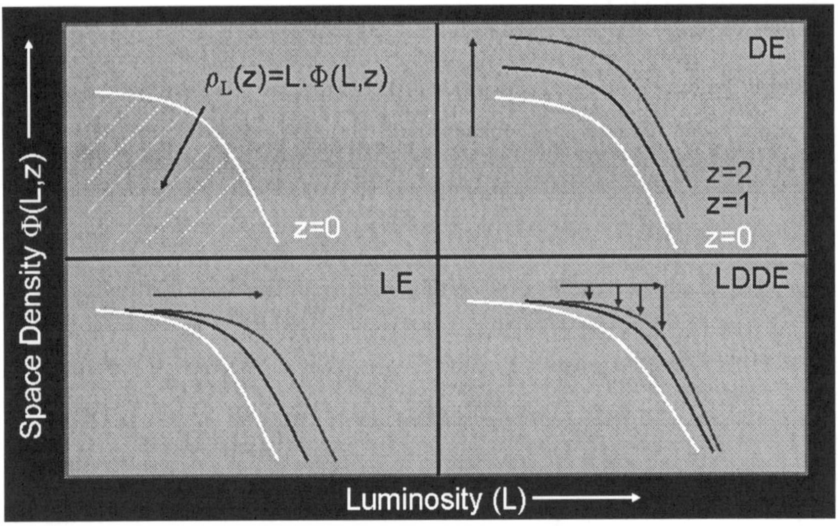

Figure 1. A schematic depiction of the QSO LF and its evolution with redshift. Pure density evolution (PDE: top right-hand panel) corresponds to a uniform shift of the LF along the space density axis. Pure luminosity evolution (PLE: bottom left-hand panel) corresponds to a uniform shift of the LF along the luminosity axis. In both these cases the shape of the LF is preserved. Luminosity dependent density evolution (LDDE: bottom right-hand panel) reflects a more complex form of evolution in which the shape of the luminosity function is not preserved, with different luminosities evolving at different rates. The luminosity density ρ_L associated with the LF at any given redshift z is given by the shaded area under the LF (top left-hand panel) $\rho_L = \int L\Phi(L,z)dL$.

where $V(z_{\max_j}) = \min[z_{\max}, z(L_i, S_{\min_j})]$, i.e. the smaller of the volumes limited by the maximum redshift of the bin or by the maximum redshift an object of given L could have while remaining brighter than the survey flux limit $S_{\min_j}$.

Another useful estimator has recently been developed[127]. In this case the binned estimate of the LF in the (L, z) interval is given by

$$\Phi_{\rm est}(L,z) = \frac{n_{\rm QSO}}{\int_{L_{\min}}^{L_{\max}} \int_{z_{\min}}^{z_{\max}} (L)(dV/dz)dzdL} \ . \tag{33}$$

Although binned estimators provide a good visual impression of the data, they are all biased to some extent. This is largely due to effects of evolution and the presence of flux limits over the small but finite bins used in the analysis. This is removed by using the unbinned likelihood function[111] to test the LF model fits $\Phi(L, z)$ against the observed distribution. The likelihood function is defined as

$$S = -2 \sum_{i=1}^{n_{QSO}} \ln[\Phi(L_i, z_i)] + 2 \int \int \Phi(L, z) \omega(L, z) \frac{dV}{dz} dz dL \ , \qquad (34)$$

where $\omega(L, z)$ is the total solid angle subtended by the surveys under analysis for a given L, z. Goodness of fits can then be derived using the 2D Kolmogorov-Smirnov Test[134].

2.2 Optical Surveys

To date, the largest samples of QSOs suitable for statistical analysis have been obtained in the optical passband. Galactic stars outnumber all QSOs by at least 50:1 (closer to 10^5:1 for $z > 4$ QSOs at $I < 20$). Effective selection techniques therefore need to be applied to lists of stellar images obtained from photographic or CCD images in order to identify a manageable number of QSO candidates for follow-up spectroscopy. A good review of the major selection techniques is provided by Hewett & Foltz[76]. One of the most popular techniques has been to select QSO candidates on the basis of their anomalous broad-band colours. A simple blue $U - B$ colour cut[22] (the UVX technique) provides a relatively complete sample of QSOs at $z < 2.2$. Multi-colour selection based on $U - B/B - R$ colours[23] increases the redshift range to $z < 2.9$, although the completeness in the range $2.2 < z < 2.9$ tends to be lower than at lower redshifts[26]. At the very highest redshifts ($z > 3.5$) a combination of redder colours $B - R/R - I$ or $R - I/I - Z$ can be used to identify suitable candidates[53]. Between $z \sim 2.8$ and $z \sim 3.5$, the colours of QSOs in most optical passbands make them difficult to distinguish from normal galactic stars. Multi-colour selection in this redshift range is therefore subject to significant incompleteness. In the past, QSO candidate selection has been mainly based on photographic datasets, largely due to their uniquely panoramic nature. However, with the increasing availability of large CCD imaging surveys, future optical QSO selection will be based on these new digitally-based catalogues. Such catalogues can also yield much more accurate magnitudes and colours than photographically-based catalogues, and increase the effectiveness and completeness of multi-colour selection at all redshifts.

In most colour-based selection techniques, the requirement QSO candidates also be 'stellar' leads to a bias against low redshift ($z < 0.3$) extended QSOs. This bias is less pronounced in surveys that use slitless spectroscopy (prism/grism) to select QSO candidates[75,89]. Variability[71] and K-band excess[162] have also been utilized as QSO selection techniques. Both

Table 1. Representative Optical Surveys

Survey	Mag Limit	N_{QSO}	z range	Area (deg^2)	Method
Low Redshift $z < 3$					
DMS[125]	$B < 23.8$	55	$0.3 < z < 3.5$	0.83	Multi-colour
BJS[23]	$B < 22.0$	66	$0.3 < z < 2.9$	0.85	UBR
ZM^2B[170]	$B < 22.0$	54	$0.3 < z < 2.9$	0.69	UBR
Durham[22]	$B < 20.9$	420	$0.3 < z < 2.2$	11.0	UVX
HV_1[72]	$B < 21.0$	111	$0.3 < z < 2.2$	4.6	UVX/Vblty
SGP[93]	$B < 20.5$	485	$0.3 < z < 2.9$	24.6	UBR
HV_2[72]	$B < 19.5$	205	$0.3 < z < 3.2$	18.8	UVX/Vblty
HBQS[91]	$B < 18.8$	327	$0.3 < z < 2.2$	555	UVX
LBQS[77]	$B < 18.5$	1055	$0.2 < z < 3.2$	603	Prism
HEQS[166]	$B < 17.5$	415	$0.0 < z < 3.2$	3500	Prism
PG[64]	$B < 16.5$	116	$0.3 < z < 2.2$	10714	UVX
High Redshift $z > 3$					
PTGS[140]	$R < 21.0$	90	$2.7 < z < 5.0$	61.5	Grism
HV_3[73]	$R < 20.3$	16	$3.2 < z < 4.5$	18.8	Vblty/Colour
WHO[161]	$R < 20.0$	86	$2.2 < z < 4.5$	43.0	Multi-colour
DPOSS[88]	$R < 19.6$	10	$4.0 < z < 5.0$	681	Multi-colour
SDSS-ES[53]	$I < 20.2$	22	$3.6 < z < 5.5$	250	Multi-colour
Surveys in progress					
2QZ[26]	$B < 20.8$	25000	$0.3 < z < 2.9$	740	UBR
SDSS[109]	$B < 19.5$	10^5	$0.0 < z < 6.0$	10000	Multi-colour

these latter methods avoid the potential incompleteness of previous multi-colour techniques against any 'red' broad line QSOs which may exist[163]. Some of the major optical QSO surveys are summarized in Table 1.

2.3 The Optical Luminosity Function

Based on the major UVX QSO surveys of the 1980s[111,22], a picture emerged in which the low-intermediate redshift ($z < 2.2$) QSO optical luminosity function (OLF) was modeled by a two-power-law function with a 'break' luminosity at L^*, whose redshift dependence was best fit by pure luminosity evolution (PLE), i.e. a uniform increase in luminosity toward high redshift [110-69].

Later QSO surveys[75,72-166], particularly those that focussed on bright

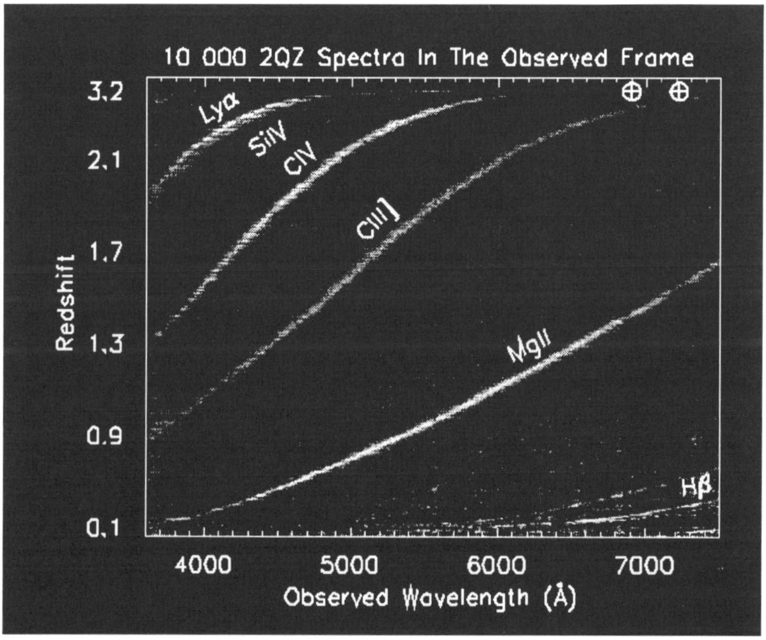

Figure 2. Observer-frame spectra of the first 10000 QSOs in the 2QZ. Spectra are stacked in order of increasing redshift. Note the relative uniformity of the QSO spectra, and the visibility of weak features throughout the spectra (e.g. Fe II emission between CIII] and MgII). Figure courtesy of Robert Smith.

magnitudes ($M_B < -26$) and low redshifts ($z < 0.3$): LBQS[75], HV[72], EQS[63], HBQS[91], HQS[89]; reported evidence for LDDE in which the bright end of the OLF shows a significant flattening with decreasing redshift, from a power-law with slope $\alpha = 3.6$ at $z = 2$ to a much flatter featureless single power-law slope, $\alpha = 2.5$, at low redshift[89,166]. The featureless nature of the low redshift LF is still controversial. A recent independent determination of the $z < 0.15$ LF based on an X-ray selected sample of QSOs[102] has found evidence for a break in the LF at $M_B = -20$.

Surveys used in the analysis of the optical LF have typically comprised from a few hundred up to a thousand QSOs. However, the 2dF QSO survey[26] (2QZ) and the Sloan Digital Sky Survey[109] (SDSS) are set to provide more than an order of magnitude increase in the number of known QSOs. Currently the 2QZ is 50% complete[26] and contains over 10000 QSOs. The spectra of the first 10000 QSOs discovered in this survey are shown in Fig. 2.

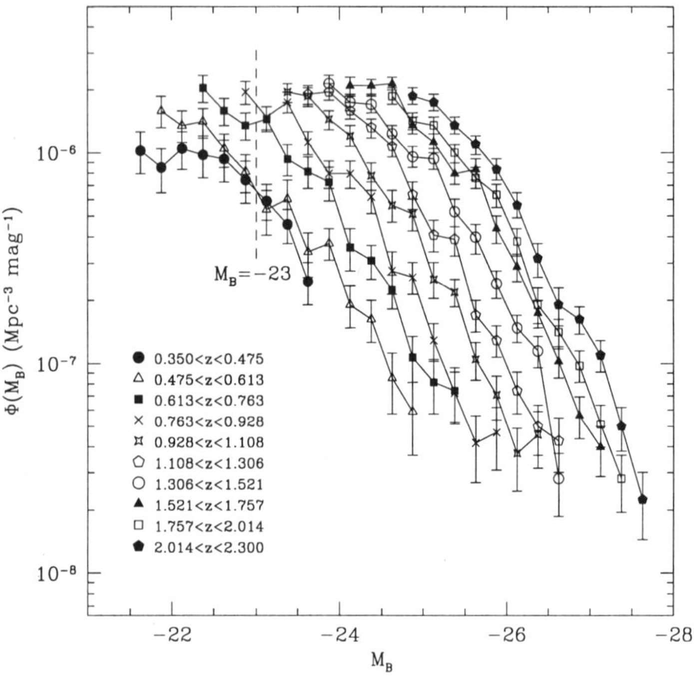

Figure 3. The $0.35 < z < 2.3$ QSO optical luminosity function derived from 6000 QSOs in the 2QZ and LBQS. Bins fainter than $M_B = -23$ may be subject to significant incompleteness. From Boyle et al.[27] ©2000 Blackwell Sci.

Results from the first 5000 QSOs in the 2QZ[27] (Fig. 3) confirm that PLE remains a valid model for the cosmological evolution of QSOs with $-23 > M_B > -26$ over the redshift range $0.35 < z < 2.3$.

The functional form of the optical luminosity function is given by a double-power relation,

$$\Phi(L, z) = \frac{\Phi^*}{\left[\left(\frac{L}{L^*(z)}\right)^{\alpha} + \left(\frac{L}{L^*(z)}\right)^{\beta}\right]} \quad . \tag{35}$$

Based on fitting this form to the 2QZ and LBQS data-sets, Boyle et al.[27] obtain parameter values in the range $3.45 < \alpha < 3.75$, $1.55 < \beta < 1.96$, $-22 < M_B^* < -23$ and $\Phi^* = 10^{-6}$ Mpc^{-3}mag^{-1}. This may be compared

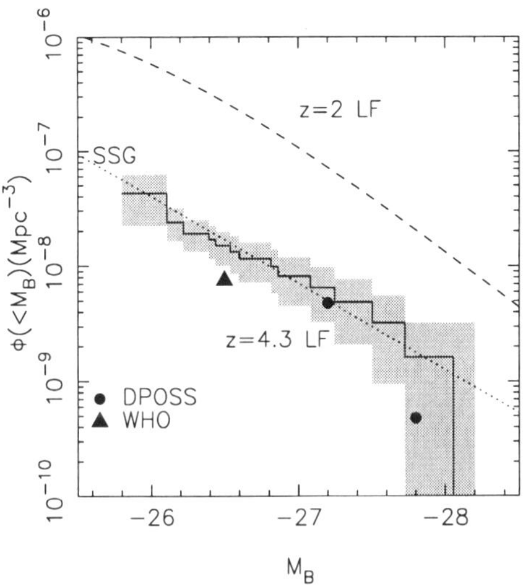

Figure 4. The high redshift ($\bar{z} = 4.3$) luminosity function derived by Fan et al.[54] from the SDSS (solid line). Also shown are points from other high redshift QSO samples (DPOSS, WHO; see Table 1) and the $z = 2$ LF from 2QZ. From Fan et al.[54]

with the LF parameters derived for the 2dF galaxy survey[55]: $\beta = 1.3$, $M_B = -21.2$, $\Phi^* = 0.017$ Mpc^{-3}mag^{-1}. Thus the normalization of the QSO LF is approximately 0.1% that of the galaxy luminosity function at the present day.

Boyle et al.[27] also find that the evolution of the LF at $z < 2.3$ is best fit by a function of the form

$$L^*(z) \propto L^*(0) 10^{1.40z - 0.27z^2} \quad . \tag{36}$$

When extrapolated to higher redshifts, this predicts a strong peak in the evolution of QSOs at $z \sim 2.5$. Such a peak in the QSO evolution has previously been inferred from the analysis of high redshift QSOs. A number of independent studies[140–88] find evidence from multi-colour/grism surveys for a dramatic decline in the space density of luminous ($M_B < -26$) QSOs at $z > 3$. The rate of decline is equivalent to a factor of 3 drop in the space density of QSOs per unit redshift beyond $z = 3$.

While these results have been based on rather small samples of high redshift QSOs (few tens), initial results on the space density of $z > 4$ QSOs from the SDSS[54] confirm the dramatic decline in the numbers of observed QSOs at high redshift (see Fig. 4). These results also suggest that the QSO LF flattens

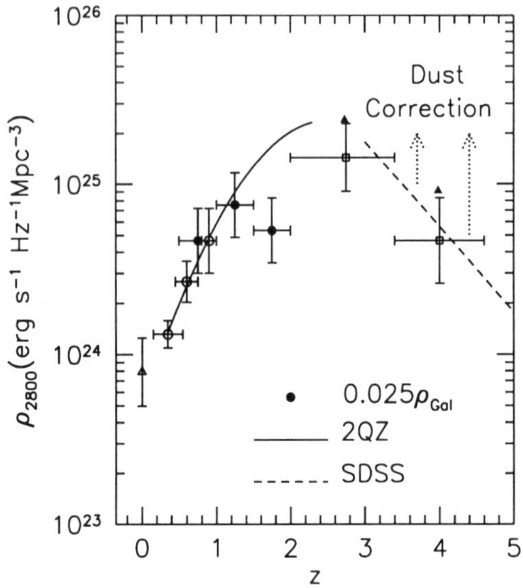

Figure 5. The redshift evolution of the QSO (2QZ: solid line; SDSS: dashed line) and galaxy (data points) UV (2800Å) luminosity density. The galaxy UV luminosity density has been scaled down by a factor 40. Subsequent revision of the galaxy and QSO UV luminosity density at $z > 3$ due to correction for dust obscuration is indicated schematically by the arrows/lower-limits on this diagram. Adapted from Boyle & Terlevich[24] ©1998 Blackwell Sci.

at higher redshifts, from a slope of $\alpha \sim 3.6$ at $z = 2$ to a slope of $\alpha \sim 2.5$ at $z > 4$.

Despite these uncertainties over details, there is general agreement over the overall evolution of the integrated QSO luminosity function or luminosity density, $\rho_L(z) = \int L\Phi(L,z)dL$, as a function of redshift. The integrated optical luminosity of QSOs increases by a factor of approximately 30 between $z = 0$ and $z = 2$, falling precipitously at higher redshifts[54].

Boyle & Terlevich[24] have shown that this mirrors the evolution of the star formation rate derived by Madau et al.[105] (Fig. 5). However, there have been subsequent revisions to the evolution of the star formation rate at $z > 2$, due to the realization that a significant fraction of star formation at high redshift is hidden from optical surveys by dust[82]. The amount of the correction significantly lessens or indeed cancels out any decline in the star formation rate at $z > 2$.

Table 2. Representative Radio Surveys

Survey	$S_{\rm lim}$ (Jy)	$N_{\rm QSO}$	Area(deg^2)
PSR[46]	$S_{2.7\rm GHz} > 0.1 - 2$	171	250–30000
PKS[146]	$S_{2.7\rm GHz} > 0.25$	442	20000
FIRST[164]	$S_{1.4\rm GHz} > 0.001$	636	2682

2.4 The Radio Luminosity Function

The optical luminosity density from QSOs may also be increasingly underestimated at high redshifts[52]. Radio emission from QSOs is unaffected by dust and the number density of radio-loud QSOs at high redshift can provide an important check of the decline in the space density of QSOs at these redshifts.

Unfortunately, the wide optical magnitude range ($16 < B < 24$) for the optical counterparts to even the highest flux radio surveys ($S_{2.7\rm GHz} > 0.2$Jy) have made it difficult to obtain the required complete optical spectroscopic identification for many radio surveys. Most radio surveys therefore impose some pre-selection (magnitude, colour, spectral and/or morphological criteria) on the radio-sources/optical-counterparts to obtain a manageable sample with which to carry out the optical spectroscopy. A representative list of radio-loud QSO surveys with well-defined spectroscopic selection criteria are listed in Table 2.

Based on a sample of 427 radio-loud QSOs in the Parkes Selected Regions (PSR) defined by Dunlop & Peacock[46], the $z < 2$ radio-loud LF is fit by a two-power-law function very similar to that found for the optical LF, with a normalization approximately 1% that of the optical LF. The redshift evolution of the radio-loud QSO LF at $z < 2$ has an almost identical form to that in the optical,

$$L^*(z) \propto L^*(0)10^{1.18z-0.28z^2} \tag{37}$$

At high redshifts, $z > 2.5$, two independent studies[79,145] have found evidence for a decline in the number of QSOs, with a rate equivalent to a factor 3–5 drop in the comoving space density per unit redshift. Thus both radio-quiet and radio-loud QSO LFs exhibit similar evolution over a wide range in redshift in the optical and radio regimes. Shaver et al.[146] have demonstrated that this evolution is also compatible with the evolution in the star formation rate (see Fig. 6).

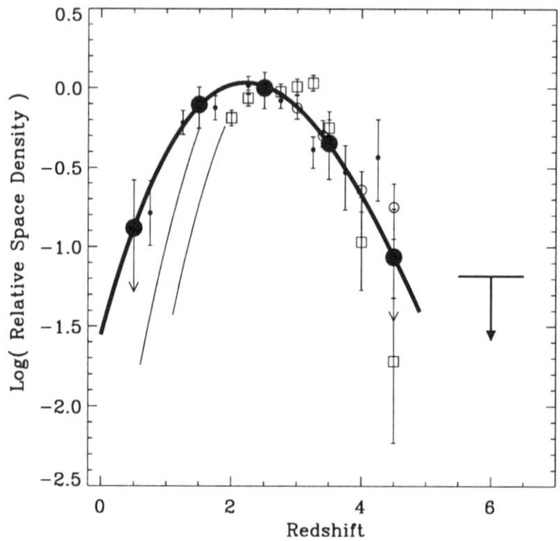

Figure 6. The redshift evolution of the QSO radio-loud luminosity density (bold line), from Shaver et al.[146] ©1999 Astro. Soc. of the Pacific. Open symbols denote scaled results from optical surveys. Thin lines denote the evolution of the $z < 2$ optical luminosity function.

Table 3. Representative X-ray Surveys

Survey	$S_{\rm lim}$ (erg s^{-1}cm^{-2})	$N_{\rm QSO}$	Area (deg^2)
HEAO-I[131]	$S_{2-10\rm keV} > 3.1 \times 10^{-11}$	25	27000
RBS[148]	$S_{0.5-2\rm keV} > 2.5 \times 10^{-12}$	216	20000
EMSS[156]	$S_{0.35-3\rm keV} > 2 \times 10^{-13}$	478	800
SA–N[4]	$S_{0.5-2\rm keV} > 1.3 \times 10^{-13}$	130	685
RIXOS[112]	$S_{0.5-2\rm keV} > 3 \times 10^{-14}$	205	15
DRS[143]	$S_{0.5-2\rm keV} > 5 \times 10^{-15}$	149	1.5
Lockman Hole[139]	$S_{0.5-2\rm keV} > 2 \times 10^{-15}$	68	0.30

2.5 The X-ray Luminosity Function

The soft (< 2 keV) X-ray missions such as Einstein and ROSAT (Table 3) have yielded a dramatic increase in our knowledge of the evolution of the QSO X-ray LF over the past 5–10 years. At low redshifts $z < 2$, many studies have found that pure luminosity evolution is a good fit to the available data[20−86] at $z < 3$ with an evolutionary rate $L_X(z) \propto (1+z)^3$ similar to that observed in the optical and radio regimes. In the X-ray regime, a key issue has been the contribution by QSOs to the X-ray background first detected by Giacconi et al.[61] and now largely ($\sim$ 80%) resolved into discrete sources at soft X-ray energies (2keV) by the deepest ROSAT images[70]. The QSO luminosity function can be extrapolated to provide an estimate of the QSO contribution to the X-ray background, I_{ν_0}[a],

$$I_{\nu_0} = \frac{c}{4\pi H_0} \int_0^{z_{\max}} \int_{L_{\min}}^{L_{\max}} \frac{L_X \Phi(L_X, z)}{\sqrt{1+\Omega z}(1+z)^{2+\alpha_x}} dL_X dz \quad , \tag{38}$$

where α_x is the QSO X-ray spectral index. Recent results have placed the contribution of QSOs to the X-ray background at 2keV in the range 35–50%[20,126].

Recently, luminosity evolution has been challenged by Miyaji et al.[119] Using the deepest ROSAT survey data available they find that more complex LDDE models are required to fit the observed cosmological evolution of QSOs (see Fig. 7). Moreover, they were able to extend the analysis of the X-ray LF to $z > 3$, finding no strong evidence for any decline of the space density of X-ray selected QSOs at these redshifts. This is in contrast to the strong decline seen at optical and radio wavelengths.

At some level, this discrepancy between the Miyaji et al. results and those from previous surveys may rest on the nature of the objects included in the survey. The Miyaji et al. survey includes both broad emission line (Type 1 AGN/QSOs) and narrow emission line (Type 2 AGN) objects in their analysis, whereas previous studies have traditionally included only broad emission-line QSOs[b]. Increasing numbers of high redshift 'Type 2-like' objects are found in soft X-ray surveys at the faintest flux limits[2,114], at harder X-ray energies[123,25] or indeed from surveys in other wavebands (e.g. IRAS09104+4109[57]). These objects have a wide range of neutral hydrogen column densities, from moderate levels ($N_H \sim 10^{21} - 10^{22} \mathrm{cm}^{-2}$) to heavily obscured objects ($N_H = 7 \times 10^{25} \mathrm{cm}^{-2}$) such as IRAS09104+4109[57].

Such objects are predicted to appear at faint X-ray fluxes by models which seek to explain the X-ray background by obscured QSOs or Type 2 AGN. Although broad-line QSOs are capable of producing $\sim$ 40% of the X-ray background at 1 keV, the spectral shape of the X-ray background is much flatter ($f_\nu \propto \nu^{-0.4}$) than that observed for normal QSOs ($f_\nu \propto \nu^{-1}$). Indeed most of the energy in the X-ray background originates at 30 keV[50]

[a]This expression can, of course, be generalized to other passbands for any population of astronomical objects with LF $\Phi(L, z)$

[b]see lecture notes by Bob Goodrich for a complete discussion on AGN classification

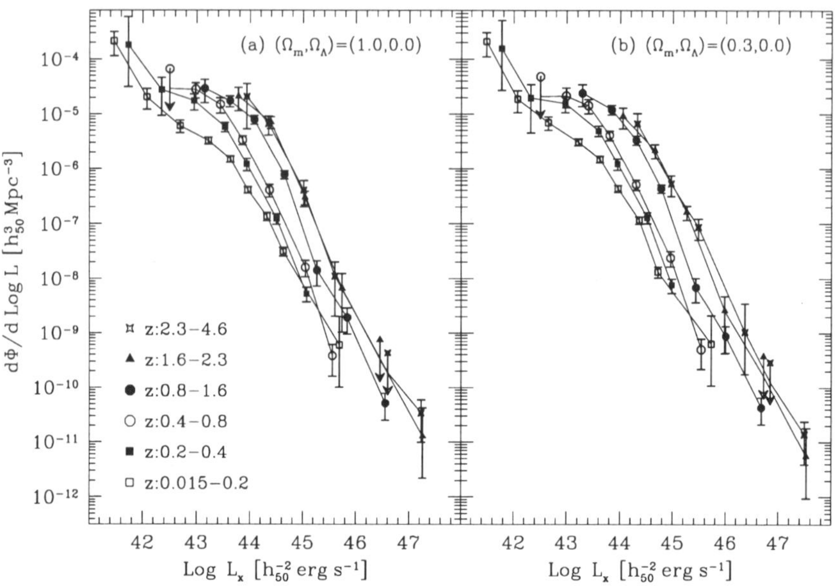

Figure 7. The X-ray (0.5–2keV) QSO LF as a function of redshift for two different universes: (a) $\Omega_m = 1$ (b) $\Omega_m = 0.3$, $\Omega_m = 0.7$. Results based from a compilation of ROSAT surveys by Miyaji et al.[119]

at much harder energies than those studied by most X-ray imaging missions $(0.1 - 10\,\text{keV})$.

By assuming an intrinsic spread in N_H values, Comastri et al.[38] have shown that a mix of unobscured ($N_H < 10^{21}\text{cm}^{-2}$) through to the most heavily obscured ($N_H \sim 10^{25}\text{cm}^{-2}$) QSOs can reproduce the observed X-ray background both in intensity and spectral shape[38].

Such models have recently been vindicated by recent observations with the Chandra X-ray mission[121,28]. The deepest Chandra pointings now resolve over 80% of the 2 – 10keV background into discrete sources. Spectroscopic studies of the optical counterparts to the sources reveal that approximately one-third are Type 1 QSOs, one-third appear 'normal' galaxies (presumably harbouring an obscured AGN), with a further one-third of the sample having no optical counterpart at $I < 25$ (heavily obscured or at high redshift).

The discovery of a large number of obscured AGN could also have a bearing on the derived evolution of the QSO population. For example, it could be used to explain the apparent turn-over in the optical and radio evolution.

While radio emission would be unaffected by dust, obscured AGN lacking broad lines would not be classified as AGN in the optical follow-up, or even rejected as a candidate for follow-up spectroscopy as non-stellar. Thus the evidence for a decline in the numbers of QSOs in the optical and radio regimes may yet be an artifact of dust obscuration associated with the host galaxy at high redshift. Dust obscuration could occur naturally in the formation of the QSOs at high redshift if associated with a significant amount of star formation. A subsequent revision to the evolution of QSO luminosity density with redshift (see Fig. 5) to a near constant value beyond $z = 3$ as implied by the the X-ray results would be consistent with a latest revisions for dust applied to the galaxy luminosity density relation based on the deep sub-mm source counts[82]. In both cases it might appear that dust associated with star formation plays a major role in obscuring our view of QSO and galaxy formation at high redshifts.

2.6 QSO Sub-millimetre Counts

Reprocessing of radiation by dust implies that the heavily-obscured QSOs required to explain the X-ray background or Chandra deep source counts should also show up in the deep sub-millimetre surveys recently carried out with SCUBA[82,12,16].

Almaini et al.[3] have calculated that, for reasonable spectral energy distributions, the Comastri et al.[38] model used to reproduce the X-ray background implies that up to 20% of the sources in the deep SCUBA fields are likely to be QSOs (Fig. 8). This is close to the observed fraction from the most recent SCUBA source identification programs[13]. The far infrared luminosities ($L \sim 10^{12} L_{\odot}$) of these faint SCUBA sources suggest that they are the high redshift equivalent ($1 < z < 3$) of the local ultra-luminous infrared galaxies (ULIGs). To make up the bulk of the SCUBA sources, Barger et al[14]. have shown that the star formation rate in ULIGs must have undergone more than a factor 100 increase in luminosity density between $z = 0$ and $z = 2$. Approximately 20–30% of ULIGs exhibit evidence for an active nucleus[138], and thus the obscured QSO population at $z \sim 2$ is consistent with our current observation of the observed frequency of active nuclei in star forming galaxies at the present day.

The origin of the far-infrared/sub-mm emission (AGN vs. starburst) in these high-redshift QSOs remains an outstanding issue. CO detections in a number of high-redshift QSOs at sub-mm wavelengths[124,31] can be attributed to either a few $\times\ 10^8 M_{\odot}$ dust at 50K or equivalently a star formation rate of a few thousand solar masses per year. The spectral energy distribution of

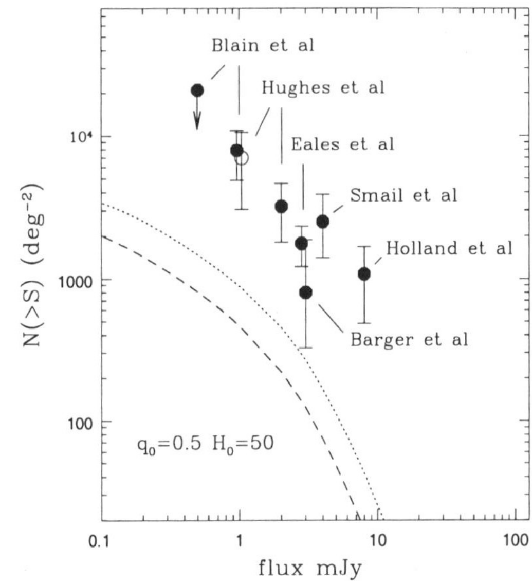

Figure 8. Deep number counts in the 850μm band. The solid line and dashed lines denote the likely 'high' and 'low' predictions of the number counts based on different X-ray LF models. The solid line denotes the predicted QSO contribution assuming that space density of QSOs remains constant at $z > 3$. The symbols denote the observed surface densities in the 850μm band from a variety of different surveys. At the deepest limits QSOs can account for up to $\sim 20\%$ of the sources at 850μm. From Almaini et al.[3] ©1999 Blackwell Sci.

ULIGs is consistent with star formation[137] although 50% also show signs for AGN activity. It should also be noted that the non-detection of the majority of faint SCUBA sources by Chandra[51] further suggest that star formation is likely to be the source of much of the sub-mm background.

2.7 The Physical Interpretation of the QSO Luminosity Function

The luminosity function and its evolution with redshift can be used to provide constraints on a variety of models for the physical evolution of QSOs and their relation to the formation of structure in the Universe.

In the standard supermassive black hole paradigm for QSOs, the luminosity L is given by

$$L = \epsilon \dot{M} c^2 = 5.7 \times 10^{45} \left(\frac{\epsilon}{0.1}\right) \left(\frac{\dot{M}}{1 M_\odot \mathrm{yr}^{-1}}\right) \mathrm{erg\ s}^{-1} \ , \tag{39}$$

where $\dot{M}$ is the mass accretion rate and ϵ is the energy generation efficiency. The Eddington luminosity is the luminosity at which radiation pressure balances gravity,

$$L_{\mathrm{Edd}} = \frac{4\pi G M \mu_e}{\sigma_{\mathrm{T}}} = 1.51 \times 10^{38} \left(\frac{M}{M_\odot}\right) \mathrm{erg\ s}^{-1} \ . \tag{40}$$

The characteristic growth time for a black hole radiating at its Eddington luminosity, τ_{Q}, is therefore given by

$$\tau_{\mathrm{Q}} = \frac{\epsilon \sigma_{\mathrm{T}} c}{4\pi G m_{\mathrm{p}}} = 4 \times 10^7 \left(\frac{\epsilon}{0.1}\right) \mathrm{yr} \ . \tag{41}$$

The fact that the ratio of this time-scale to the age of the Universe is close to the observed ratio between the space density of QSOs and galaxies, i.e.

$$\tau_{\mathrm{Q}} = \tau_0 \Phi_{\mathrm{Q}} / \Phi_{\mathrm{G}} \sim 2 \times 10^7 \mathrm{yr} \ , \tag{42}$$

has been used to argue that every bright galaxy has undergone a QSO phase during its lifetime. The inferred existence of massive black holes in nearby galaxies[107] lends further weight to this picture.

The Press-Schechter (PS) formalism[133] has formed the basis for many studies of the physical interpretation of the QSO LF[48-66]. The semi-analytic approach is used to obtain a halo mass function, $\Phi(M_{\mathrm{halo}})$, at a given cosmic epoch, which then can be related to black hole masses, M_{BH}, via the ratio $M_{\mathrm{BH}}/M_{\mathrm{halo}}$ as a function of redshift.

The time evolution of an individual QSO is assumed to have a form $L(t) = L_{\mathrm{Edd}} \exp(-t/\tau_{\mathrm{Q}})$. There is significant degeneracy between $M_{\mathrm{BH}}/M_{\mathrm{halo}}$ and τ_{Q} in obtaining successful fits to the observed LF. Haehelt et al.[65] successfully reproduce the observed evolution of the QSO LF at $z < 2$ if the $M_{\mathrm{BH}}/M_{\mathrm{halo}}$ falls off dramatically with decreasing redshift, $M_{\mathrm{BH}}/M_{\mathrm{halo}} \propto (1+z)^{5/2}$. Alternatively a constant $M_{\mathrm{BH}}/M_{\mathrm{halo}}$ would require that the mass accretion rate falls by a factor of over 100 between the $z = 2.5$ and $z = 0$. The fall off in black hole formation efficiency or mass accretion rate is not naturally predicted by the PS formalism, but may be driven by local processes near to the black hole[66] or by external influences such as merger activity[32].

However, even models in which QSO activation is achieved via galaxy mergers[87] (see Fig. 9) cannot explain the rapid evolution of the QSO LF at $z < 2$ by the decrease in the merger rate alone. Additional effects such as the

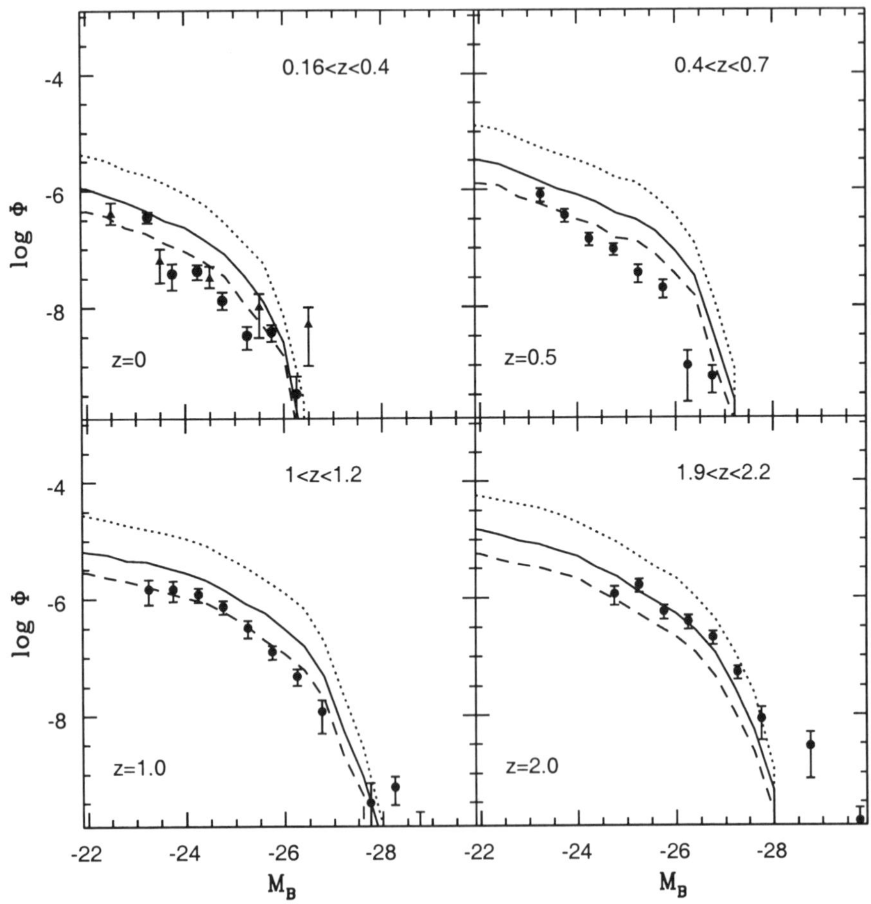

Figure 9. The evolution of the B-band luminosity function of quasars in the galaxy-galaxy merger model of Kauffmann & Haehnelt[87] (see text). The solid lines are for $\tau_Q = 3 \times 10^7$ yr and the dashed and dotted lines are for $\tau_Q = 10^7$ and 10^8 yr respectively. The symbols denote data from a compilation of optical surveys including the Durham/UVX and PG surveys (circles) and the HEQS (triangles). From Kauffmann & Haenelt.[87] ©2000 Blackwell Sci.

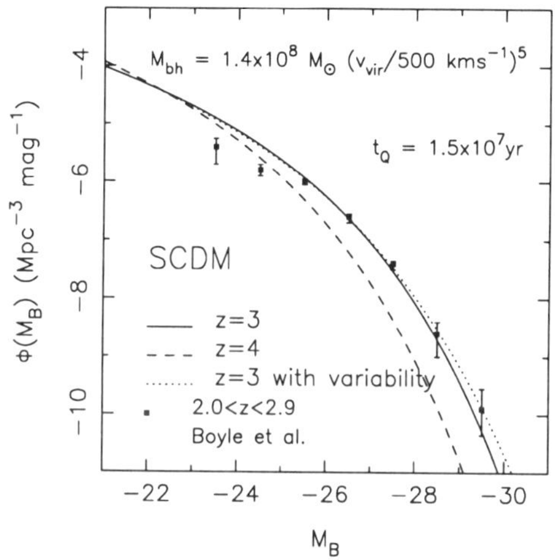

Figure 10. The predicted $z = 3$ (solid/dotted lines) and $z = 4$ (dashed line) optical LF based on the Press-Schechter formalism with $\tau_Q = 1.5 \times 10^7$yr and power-law relation between black hole mass and bulge velocity dispersion. The observed $z \sim 2.5$ LF is indicated by the filled dots. From Haenelt et al[66] ©1999 Blackwell Sci.

decrease in the gas available to fuel the black hole at low redshifts and the assumption that black holes accrete gas more slowly at late cosmic epochs are required.

Haenelt et al.[66] show that good fits to the evolution of the QSO LF at $z < 3$ may be obtained for a range of QSO lifetimes from $\tau_Q = 10^6$ yr for a black hole to halo mass relation of the form $M_{\rm BH} \propto M_{\rm Halo}^{5/3}$ to $\tau_Q = 10^8$ yr for $M_{\rm BH} \propto M_{\rm Halo}$ (see Fig. 10).

The PS formalism has also been used to fit the evolution of QSOs at the very highest redshifts. Haiman & Loeb[67] obtain good fits to the observed QSO LF over the range $2 < z < 4.5$ for a constant $M_{\rm BH}/M_{\rm halo} = 0.0054$ and a $\tau_Q = 6.6 \times 10^5$yr.

The density of accumulated remnant black holes at $z = 0$, $\rho_{\rm acc}$, inferred from the evolution of the QSO LF is given by[66]

$$\rho_{\rm acc} = 1.4 \times 10^5 \left(\frac{f_B \epsilon}{0.01}\right)^{-1} M_\odot {\rm Mpc}^{-3} \quad , \tag{43}$$

where f_B is the fraction of the bolometric luminosity radiated in the B band. This may be compared to the local black hole density inferred from the observation of local bulges[107],

$$\rho_{\rm BH} = 3.3h \times 10^6 \left(\frac{M_{\rm BH}/M_{\rm bulge}}{0.006}\right)\left(\frac{\Omega_{\rm bulge}}{0.002h^{-1}}\right) M_\odot {\rm Mpc}^{-3} \ . \tag{44}$$

Until recently, the typical values assumed for f_B, ϵ and $M_{\rm BH}/M_{\rm bulge} < 0.006$ placed $\rho_{\rm BH}$ approximately an order of magnitude above $\rho_{\rm acc}$. However the recent suggestion that this ratio should be revised downward[136,118] and the significant numbers of obscured AGN required to account for the faint X-ray source population (effectively decreasing f_B) bring these two numbers into closer agreement. In addition, if many low luminosity AGN are powered by Advection Dominated Accretion Flows (ADAFs)[1,122], then ϵ would be reduced further leading to an increase in $\rho_{\rm acc}$.

Much of the recent work on black hole demographics has focussed on the observed correlation between black hole mass and bulge luminosity. However, the latest results suggest that a much tighter relationship exists between black hole mass and velocity dispersion[59], σ,

$$M_{\rm BH} = 1.2 \times 10^8 {\rm M}_\odot \left(\frac{\sigma}{200{\rm km\,s^{-1}}}\right) \ . \tag{45}$$

A relationship of this form provides further evidence for a link between formation and evolution of bulge systems and massive black holes.

Following a different approach, Terlevich & Boyle[159] used scaling laws to demonstrate that the $z = 2$ space density and subsequent evolution of QSOs is consistent with star formation in young ellipticals. A nuclear starburst comprising 5% of the galaxy mass is required to explain the observed space density of $z = 2$ QSOs, where the QSO lifetime, 6×10^7yr, is given by the main-sequence lifetime of an $8{\rm M}_\odot$ star; the lowest mass progenitor for a Type II SN. The evolution rate, $L \propto t^{-2}$, is given by the star formation rate in disipationless models[95]. Such a model also explains the direct correspondence observed between the evolution of the star formation rate and QSO luminosity density[24].

2.8 Luminosity Function: Summary

An overview of the results on the QSO luminosity function and its evolution with redshift are given in Table 4 and Table 5. For simplicity, the evolution is expressed in terms of PLE or PDE models fitted to the data in the various passbands shown. To facilitate comparison between the optical/radio/X-ray regimes, the 'break' luminosity is re-computed in terms of 'break' energy. The luminosity density ($\sim \Phi * L^*$) peaks in the optical/UV.

Table 4. QSO Luminosity Function: parameters ($\Omega_m = 1$)

$$\Phi(L,z) = \frac{d^2N}{d\log L dz} = \frac{\Phi^*}{\left[\left(\frac{L}{L^*(z)}\right)^{\alpha-1} + \left(\frac{L}{L^*(z)}\right)^{\beta-1}\right]}$$

	α	β	νL^* (W)	Φ^* ($d\log L^{-1}\mathrm{Mpc}^{-3}$)
Optical (4400Å)[27]	3.4	1.6	2.0×10^{37}	2.7×10^{-6}
Radio (2.7GHz)[46]	3.0	1.8	4.9×10^{34}	7.1×10^{-9}
X-ray (1keV)[119]	3.3	1.4	2.8×10^{36}	4.0×10^{-6}

Table 5. QSO Luminosity Function: evolution ($\Omega_m = 1$)

	Evolution	
Regime	$z < 2$	$z > 3$
Optical (680THz)[27,53]	$L^*(z) \propto 10^{1.40z-0.27z^2}$	$\Phi(z) \sim \Phi(3)3^{3-z}$
Radio (2.7GHz)[46,146]	$L^*(z) \propto 10^{1.18z-0.28z^2}$	$\Phi(z) \sim \Phi(2.5)5^{2.5-z}$
X-ray (240PHz)[119]	$L^*(z) \propto (1+z)^3$	$\Phi(L,z) \sim$ constant

Over the past decade, a view of QSO evolution has emerged that is remarkably uniform over the different passbands from the radio, through the optical to the X-ray (see Fig. 11). The QSO LF is well represented by a two-power-law form in all regimes. The evolution of the LF is characterized by a strong evolution (a factor 30 increase in luminosity density) between $z = 0$ and $z \sim 2$ at optical, radio, X-ray and infrared wavelengths. The brightest QSOs ($M_B < -26$) appear to exhibit departures from pure luminosity evolution at low redshifts ($z < 0.3$) as do the faintest X-ray AGN. At the highest redshifts, the optical and radio data show evidence for a decline, but this may be an artifact of dust obscuration associated with the galaxies in which these objects are forming.

The evolution in the luminosity density of QSOs as a function of redshift shows striking similarities to the evolution of the star formation rate[24,146,56], providing circumstantial evidence that the epoch of QSO formation is synchronized with that of star formation.

Models based on the PS formalism have proved successful in reproducing

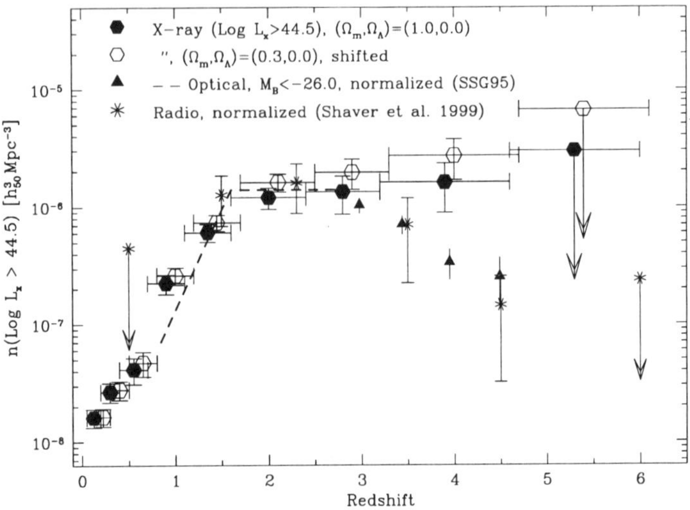

Figure 11. The evolution of QSO space density at a given luminosity (analogous to ρ_L) in the optical, X-ray and radio regimes. Figure reproduced from Miyaji et al.[119] ©2000 ESO.

the observed QSO LF at $2 < z < 3$ from the collapse of $10^{12} - 10^{13}M_\odot$ dark matter halos, although there is still significant degeneracy between the time-scale for QSO activity, τ_Q, and the assumed relationship between the black hole and halo mass. The optical QSO LF may also be reproduced by simple scaling arguments applied to the galaxy luminosity function, where the bulk of the QSO luminosity originates from an initial starburst comprising 5% of the mass of the galaxy.

3 QSO clustering

3.1 Introduction

It is also possible to place limits on QSO formation/evolution from a study of QSO clustering. To date, most studies of QSO clustering have employed the 2-point auto-correlation function statistic, $\xi(r)$, which measures the excess probability, dP of finding 2 QSOs in volumes dV_1 and dV_2 with a mean density ρ_0,

$$dP = \rho_0^2(1 + \xi(r))dV_1dV_2 \quad . \tag{46}$$

The correlation function may also be expressed as the Fourier transform of the power spectrum,

$$\xi(r) = \frac{V}{2\pi^3} \int \delta_k^2 e^{ikr} d^3k \quad . \tag{47}$$

The correlation function is sometimes expressed as a volume average,

$$\bar{\xi}(r) = \frac{3}{r^2} \int_0^r \xi(x) x^2 dx \quad . \tag{48}$$

The comoving separation r_{12} between two objects separated by an angle θ on the sky is given by the spherical cosine rule[128],

$$\begin{aligned} S_k^2(r_{12}) &= S_k^2(r_1)C_k^2(r_2) + S_k^2(r_2)C_k^2(r_1) \\ &+kS_k^2(r_1)S_k^2(r_2)\sin^2\theta - 2S_k(r_1)S_k(r_2)C_k(r_1)C_k(r_2)\cos\theta \quad , \end{aligned}$$

where

$$C_k = \sqrt{1 - S_k^2} \quad . \tag{49}$$

This reduces to the more familiar cosine rule for a flat universe,

$$r_{12} = \sqrt{r_1^2 + r_2^2 - 2r_1r_2cos\theta} \quad . \tag{50}$$

Estimates of the correlation function are obtained by counting QSO pairs $N_{\rm QQ}(r)$ at a given comoving separation $r \to r+\delta r$ and comparing the numbers of random pairs $N_{\rm RR}(r)$ or QSO-random pairs $N_{\rm QR}(r)$ at a similar separation. The most simple correlation function estimator is given by

$$\xi_{\rm est}(r) = \frac{N_{\rm QQ}(r)}{N_{\rm QR}(r)} - 1 \quad . \tag{51}$$

Unfortunately this estimator is sensitive to large-scale gradients in the data. A more frequently used estimator is given by[94]

$$\xi_{\rm est}(r) = \frac{N_{\rm QQ}(r) - 2N_{\rm QR}(r) + N_{\rm RR}(r)}{N_{\rm RR}} - 1 \quad . \tag{52}$$

The data is generally well fit by a power-law correlation function,

$$\xi(r) = \left(\frac{r}{r_0}\right)^{-\gamma} \quad , \tag{53}$$

where r_0 can be interpreted as the characteristic scale-length of QSO clustering.

3.2 Evolution of Clustering

Small density perturbations $\delta\rho/\rho \ll 1$ grow due to gravitational instability as follows

$$\delta\rho/\rho \propto (1+z)^{-1} \quad , \tag{54}$$

while $z > 1/\Omega - 1$. The resulting evolution of the correlation function is given by

$$\xi(r,z) = \frac{\xi(r,0)}{(1+z)^2} \quad . \tag{55}$$

These simple linear evolution models can be extended[41] to incorporate the fact the QSOs are likely to be biased tracers of the matter distribution, where the bias factor $b_{\rm Q}(r,z)$ may be expressed in terms of a formation threshold $\langle\nu\rangle$ compared to the rms mass distribution $\sigma_\rho(t)$

$$b_{\rm Q}(r,z) = \frac{\langle\nu\rangle}{\sigma_\rho(t)} - 1 \quad . \tag{56}$$

At scales where $\delta\rho/\rho \ll 1$ ($r > 10h^{-1}$Mpc) the bias is scale-independent[130]. At these scales the relationship between the QSO and mass correlation function ξ_ρ may then be expressed as

$$\xi_{\rm Q}(r,z) = b^2(z)\xi_\rho(r,z) \quad . \tag{57}$$

Croom & Shanks[41] have developed simple intuitive models for the evolution of QSO bias, assuming that QSO clusters are formed at an arbitrarily high redshift and evolve with the gravitational potential produced by this density fluctuation in the Universe[58]. In this case $b(z)$ may be expressed as

$$b(z) = 1 + [b(0)+1]\, G(\Omega_{\rm m}, \Omega_\lambda, z) \quad , \tag{58}$$

where $G(\Omega_{\rm m}, \Omega_\lambda, z)$ is the linear growth rate of density fluctuations, $G = (1+z)$, for $\Omega = 1, \Omega_\lambda = 0$ (see above). In such models $b(z)$ increases as $z \to 0$, leading to an increase in correlation length with decreasing redshift if $b(z)$ evolves faster than $(1+z)$ (see bias model 1 in Fig. 12).

Alternatively, if the PS formalism is used to identify QSO formation with a specific halo mass[113,120] at any given redshift then the evolution of bias may be expressed as

$$b(z) = 1 - 1/\delta_{\rm c} + [b(0) - (1 - 1/\delta_{\rm c})]\, G(\Omega_{\rm m}, \Omega_\lambda, z)^\beta \quad , \tag{59}$$

where $\delta_{\rm c}$ is the critical over-density for spherical collapse. For $\Omega_{\rm m} = 1, \Omega_\lambda = 0$ we obtain $\delta_{\rm c} = 1.686$ for all redshifts, and this varies by only a few per cent for most reasonable cosmologies[100].

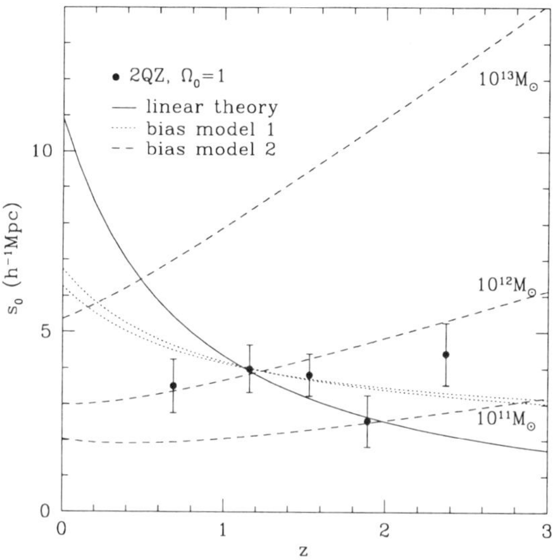

Figure 12. The scale-length of QSO clustering as a function of cosmic epoch. The 'linear' and 'biased' evolutionary models discussed in the text are indicated. Also shown are the observations of the evolution of r_0 from the 2QZ. Figure reproduced from Croom et al.[43]

In this case the QSO correlation length will decrease with time as subsequent generations of QSOs reflect the decreasing level of bias, $b_Q(r, z)$, associated with the collapse of a constant mass as a function of increasing cosmic time (bias model 2 in Fig. 12). Thus the observation of the redshift dependence of QSO scale-length provides an opportunity to break the halo mass/lifetime degeneracy inherent in the interpretation of the LF evolution.

3.3 Observations

Previous studies of QSO clustering[144–92] have been limited to relatively small optically-selected catalogues comprising up to 2000 QSOs with typical L^* luminosities. At a mean redshift $z = 1.4$, a significant (4σ) detection of QSO clustering is found at $r < 10h^{-1}$Mpc, with a scale-length $r_0 \sim 5h^{-1}$Mpc, equivalent to galaxies at the present epoch. Although no significant signal is detected in the QSO correlation function at $r > 20h^{-1}$Mpc, there have been a number of claims[39,35] for superclusters of QSOs with size $r \sim 100h^{-1}$Mpc.

Unfortunately, until recently there was little consensus on the evolution of the QSO scale-length as a function of redshift. Separate studies have found that QSO clustering increases[92], decreases[85] or stays constant[41] with increasing redshift. Support for weak evolution in the clustering of QSOs comes from studies of low redshift ($z < 0.2$) X-ray selected[18] or infrared-selected[60] AGN that also find a scale-length consistent with $r_0 \sim 5h^{-1}$Mpc.

The scale-lengths for QSO clustering are similar to those found for galaxies over a wide range of redshifts: from the present epoch[160–135] ($z \sim 0.05$), $r_0 \sim 5h^{-1}$Mpc; to intermediate[97–34] ($z \sim 0.5$), $r_0 \sim 1.5 - 4h^{-1}$Mpc; and high redshifts[155, 62] ($z \sim 3$), $r_0 \sim 2 - 6h^{-1}$Mpc. The large variation in estimates for r_0 between galaxy surveys at similar redshifts may be attributed to both cosmic variance (at the mean redshift of many surveys, the measured clustering length is equivalent to the angular size of the surveys)[132] and sampling of different types of galaxy with different clustering properties[103].

3.4 The 2QZ Survey

With the increase in QSO numbers provided by the 2QZ and SDSS catalogues, it is now possible to obtain significantly more accurate measurements of the QSO correlation function, and begin to study in detail its evolution with redshift.

Based on a clustering analysis applied to the first 8000 QSOs observed in the 2QZ survey, Croom et al.[43] derive a QSO scale-length $r_0 = 4.5 \pm 1.0h^{-1}$Mpc at a mean redshift $z = 1.4$ ($\Omega = 1$). For an $\Omega_{\rm m} = 1$ universe Croom et al. find that, at small scales, $r < 20h^{-1}$, a correlation function described by a power-law with a slope $\gamma = 1.5$ provides an extremely good fit (see Fig. 13). Moreover there is some evidence for a break in the power-law of the correlation function at $r > 40h^{-1}$Mpc.

For an $\Omega_{\rm m} = 1$, $\Omega_\Lambda = 0$ universe, Croom et al. find weak evidence for a marginal increase in scale-length with increasing redshift, possibly ruling out simple linear (or long-lived) models for QSO evolution. Indeed, the observed evolution of the scale-length for these, typically L^*, QSOs fits the predicted redshift evolution of $10^{12}{\rm M}_\odot$ dark matter halo scale-lengths[113]. For an $\Omega_\Lambda = 0.7$ universe there is a much stronger evolution of scale-length with redshift, more consistent with that predicted for $10^{13}{\rm M}_\odot$ halos[113]. These fits are *not* unique, however, and there are likely to be many other models which the current evolution of the QSO scale-length fits equally well.

3.5 Clustering of Radio-Loud QSOs

All these studies are based on optically-selected QSOs, the vast majority of which ($\sim 95\%$) are radio-quiet QSOs. Given their much lower surface density,

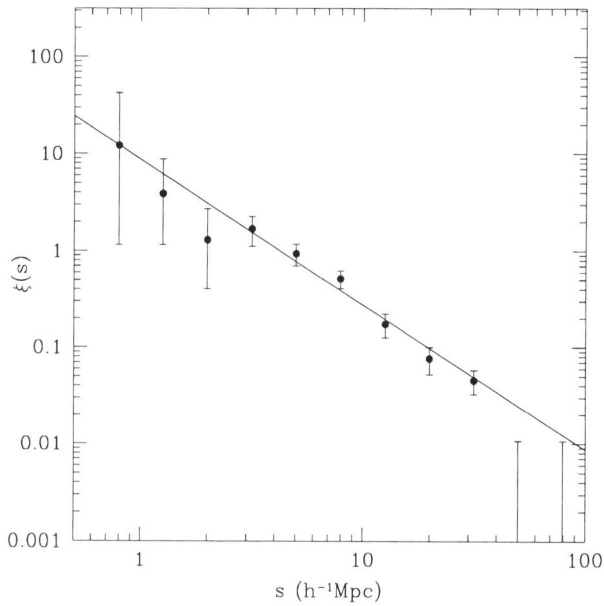

Figure 13. The QSO correlation function determined from the first 8000 QSOs in the 2QZ. A fitted power-law with slope $\gamma = 1.5$ and scale-length $r_0 = 4.5h^{-1}$ is also shown. Figure reproduced from Croom et al.[43]

and the difficultly in procuring a large spectroscopic sample, as yet there are no direct measures for the clustering scale-length of radio-loud QSOs. To date, the existing studies have been limited to the angular clustering of all radio sources (including radio-quasars) on the sky[129,101]. If the clustering of radio sources is assumed to be constant with redshift, then the observed level of clustering on the sky corresponds to a scale-length $r_0 = 12 - 18h^{-1}$Mpc, significantly larger than that for radio-quiet QSOs. On the basis of their host galaxy properties, it has been suggested that the radio-loud/radio-quiet dichotomy is based on spheroid mass[117]. If so, this might naturally explain the difference in clustering properties, with the more massive radio-loud systems originating from larger masses, and hence more strongly biased, dark halos. However, it must be stressed that the scale-length of radio-loud QSOs is still highly uncertain, although this position may change with the new generation of radio-loud QSO surveys (e.g. FIRST[164]).

3.6 QSO Clustering: Summary

The observation that QSO clustering agrees in both form and amplitude with 'normal' galaxies at low redshift ($z < 1$) and galaxies at high redshift ($z \sim 3$) provides further circumstantial evidence that QSOs and galaxies are closely linked in their formation process. The form of the evolution of QSO clustering is consistent with a picture in which QSO activity is associated with the collapse of dark matter halos and the resultant formation of the spheroidal components of galaxies. For the L^* QSOs studied to date, the observed clustering evolution is mostly consistent with model predictions for evolution in the clustering scale-length for halo masses $10^{12} < M < 10^{13} M_\odot$. However, these models for clustering evolution are still rather simplistic and future detailed N-body simulations such as those produced by the Hubble Volume[36] will provide further opportunities to test the observations against theoretical predictions for the growth of structure in the Universe.

4 QSO Environments

4.1 Introduction

Constrained by the prohibitive amounts of telescope time required to obtain redshifts for faint galaxies around large numbers of QSOs, the study of QSO environments to date has largely relied on broad-band imaging studies.

In broad-band imaging surveys the true space density of galaxies around the QSO must be inferred from the observed excess projected on the plane of the sky. The QSO-galaxy spatial cross-correlation function, $\xi_{QG}(r, z)$, may be expressed as a power-law of the form

$$\xi_{QG} = \left(\frac{r}{r_0}\right)^{-\gamma} (1+z)^{-3+\epsilon} \quad , \tag{60}$$

where the term $(1+z)^{-3+\epsilon}$ reflects the redshift evolution of the cross-correlation function. For $\epsilon = \gamma - 3$ clustering stays constant in comoving coordinates, as observed in the QSO and galaxy auto-correlation function (see previous section). The projected angular cross-correlation function, w_{QG}, can than be expressed as[15]

$$w_{QG} = \sqrt{\pi}\frac{\Gamma[(\gamma-1)/2]}{\Gamma(\gamma/2)} B r_0^{\gamma} \theta^{1-\gamma} \quad , \tag{61}$$

where

$$B = \frac{\int_0^\infty g(z) \left(\frac{dN(z)}{dz} \right)^2 dz}{\left[\int_0^\infty \frac{dN(z)}{dz} dz \right]^2} \quad , \tag{62}$$

and

$$g(z) = \frac{dz}{dx}(x)^{1-\gamma} F(x)(1+z)^{-(3+\epsilon-\gamma)} \quad , \tag{63}$$

$$F(x) = 1 + (1 - \Omega)(H_0 x/c)^2 \quad , \tag{64}$$

$$x = R_0 r \quad . \tag{65}$$

Thus, to convert an observed $w_{\rm QG}$ into an inferred QSO-galaxy clustering length, requires knowledge of the galaxy number-redshift relation, $dN(z)/dz$, at the magnitude limit of the galaxy catalogue (typically $R > 21$), and ϵ, the evolution of the QSO-galaxy correlation length. Any study of the QSO environment is therefore limited by our knowledge of the galaxy $dN(z)/dz$. This is a particularly severe problem at faint magnitudes, $R > 22$, and high redshifts, $z > 1$.

An estimate of $w_{\rm QG}$ may be obtained from the observed number of galaxies found at angular distance θ from a sample of QSOs ($N_{\rm QG}$) compared to the number expected from a random distribution of points around the QSO sample ($N_{\rm QR}$),

$$w_{\rm QG}(\theta) = \frac{N_{\rm QG}(\theta)}{N_{\rm QR}(\theta)} - 1 \quad . \tag{66}$$

Since this estimator effectively uses the mean *observed* (rather than the *actual*) galaxy density to normalize $w_{\rm QG}(\theta)$, a constant offset needs to be applied to the measured value of $w(\theta)$,

$$w_{\rm true} - w_{\rm obs} = \int_0^{\theta_{\rm max}} 2\pi w_{\rm QG}(\theta) d\theta \quad , \tag{67}$$

where $\theta_{\rm max}$ is the angular field-of-view. It also requires prior knowledge of the functional form of the correlation function, a significant disadvantage for all correlation function analyses. The only way around this problem is to maximize the field-of-view. A typical clustering scale-length $r_0 = 5h^{-1}$Mpc corresponds to 1 degree at $z = 0.1$ and 10 arcmin at $z = 1$.

4.2 Observations

Much of the pioneering work in this field was carried out by Yee and collaborators[168–49]. Based on imaging studies of both radio-loud and radio-quiet QSOs at $z < 0.6$, they found that the environments of radio-loud QSOs undergo a dramatic evolution at $z \sim 0.4$, from a field galaxy environment (a galaxy clustering scale-length $r_0 \sim 5h^{-1}$Mpc) to an environment consistent with Abell richness class 1 clusters ($r_0 > 10h^{-1}$Mpc). In contrast, the environments of radio-quiet QSOs remained consistent with that of a field galaxy environment throughout the redshift range studied.

Further broad-band imaging studies of low-intermediate redshift QSOs have confirmed this distinction between the galaxy environments of radio-loud and radio-quiet QSOs. At low redshifts ($z < 0.3$), Smith et al.[152] have shown that the correlation length of galaxy clustering around X-ray selected (predominantly radio-quiet) $M_B > -24$ QSOs is identical to that of field galaxies over the same redshift range[106]. More recently Smith et al.[153] have extended this work to higher redshifts ($0.3 < z < 0.7$). They confirm that the level of galaxy clustering around QSOs remains weak $r_0 \sim 2 - 5h^{-1}$Mpc, consistent with the galaxy correlation length at these redshifts[97,34]. Somewhat larger scale-lengths $r_0 \sim 8h^{-1}$Mpc are derived for $0.3 < z < 0.7$ QSOs when colour selection is used to remove the weakly-clustered faint blue galaxy population[30].

At still higher redshifts $z > 1$, most optical[19,42] and near-infrared imaging surveys[158] have found no evidence for any significant excess of galaxies around radio-quiet QSOs. Croom & Shanks[42] demonstrate that their null result is consistent with a $3.5(2)\sigma$ upper limit on the QSO–galaxy clustering scale-length of $r_0 < 6h^{-1}$Mpc for comoving evolution, $\epsilon = -1.2(0.8)$. However, imaging studies carried out by Hutchings[83], cite evidence for significant galaxy excesses around some radio-quiet QSOs at $z \sim 2$. The nature of the discrepancy between the results of Hutchings and other imaging studies remains unclear.

In marked contrast to the environments of radio-quiet QSOs, broad-band imaging optical/near-infrared studies of radio-loud QSOs[78–68], have found an enhanced galaxy surface density of up to 100% around radio-loud QSOs in the redshift range $0.9 < z < 2$. The inferred space densities of galaxies are consistent with Abell richness class 0 clusters or scale-lengths $10 < r_0 < 20h^{-1}$Mpc.

4.3 QSO Environments: Summary

At all redshifts studied to date, there appears to be little evidence that radio-quiet QSOs inhabit anything other than normal galaxy environments. In

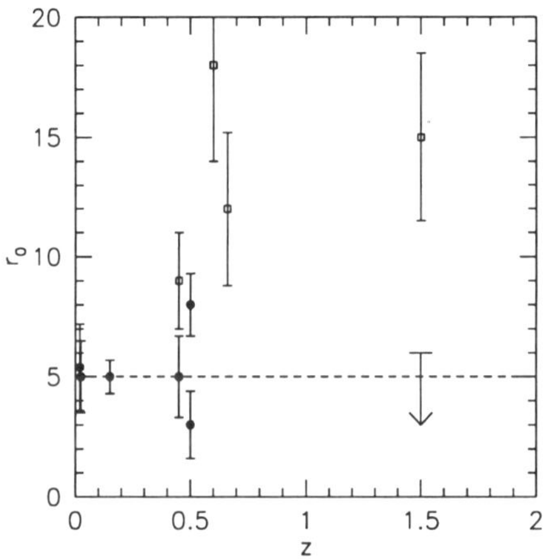

Figure 14. QSO-galaxy scale-lengths derived from optical imaging studies. Filled circles correspond to radio-quiet QSOs, and open squares to radio-loud QSOs. The dotted line denotes the present day galaxy scale-length $r_0 = 5h^{-1}$Mpc.

contrast, the environments of radio-loud QSOs undergo a strong evolution, from a near field galaxy environment locally to a rich cluster environment at $z > 1$. These results are summarized in terms of r_0 in Fig. 14 andTable 6.

The results on QSO environments are therefore very much consistent with those obtained from QSO clustering, with radio-quiet QSOs displaying similar clustering and environmental properties to 'normal' galaxies. This provides strong evidence that radio-quiet QSOs and galaxies exhibit a similar level of bias, consistent with a common formation process. The increased use of photometric redshifts[30] and narrow-band/tunable filter imaging[11] to isolate galaxies at the redshift of the QSO will extend the study of QSO environments to higher redshifts[80,81] and we may look forward to new information on the environmental properties of large samples of QSOs in the key redshift interval $1.5 < z < 3$ in the not-too-distant future.

5 QSO Host Galaxies

5.1 Introduction

At anything other than the lowest redshifts $z < 0.03$, most studies of AGN host galaxies have been limited to broad-band/narrow-band imaging studies.

Table 6. QSO/AGN environment studies

Reference	N_{QSO}	z range	Scale-length
Radio Quiet			
De Robertis et al.[44]	33	$z < 0.04$	$r_0 \sim 5h^{-1}$Mpc
Laurikainen & Salo[96]	–	$z < 0.05$	$r_0 \sim 5h^{-1}$Mpc
Laurikainen & Salo[96]	–	$z < 0.05$	$r_0 > 5h^{-1}$Mpc
Smith et al.[152]	169	$z < 0.3$	$r_0 \sim 5h^{-1}$Mpc
Ellingson et al.[49]	46	$0.3 < z < 0.6$	$r_0 \sim 5h^{-1}$Mpc
Smith et al.[153]	83	$0.3 < z < 0.7$	$r_0 \sim 3h^{-1}$Mpc
Brown[30]	66	$0.3 < z < 0.6$	$r_0 \sim 8h^{-1}$Mpc
Boyle & Couch[19]	27	$0.9 < z < 1.5$	No correlation
Teplitz et al.[158]	30	$0.9 < z < 2.1$	No correlation
Croom & Shanks[42]	150	$0.0 < z < 3.2$	$r_0 < 6h^{-1}$Mpc
Radio Loud			
Yee & Green[169]	10	$0.3 < z < 0.5$	$r_0 \sim 9h^{-1}$Mpc
Yee & Green[169]	9	$0.5 < z < 0.6$	$r_0 \sim 18h^{-1}$Mpc
Wold et al.[167]	9	$0.5 < z < 0.82$	$r_0 \sim 12h^{-1}$Mpc
Hall & Green[68]	9	$1.0 < z < 2.0$	$r_0 \sim 15h^{-1}$Mpc

The basic large-scale properties such as luminosity, size, colour and gross morphology may be determined from profile fitting to the non-nuclear light ('fuzz') surrounding the central unresolved source. Accurate modeling and subtraction of the point source therefore becomes the key to the success of the fitting procedure. Traditionally the galaxy light is fit by either, or both, a bulge component,

$$I_{\rm B}(r_{\rm B}) = I_{\rm B}(0) \exp \left[-7.67 \left(\frac{r_{\rm B}}{r_{\rm e}} \right)^{0.25} \right] , \qquad (68)$$

and a disk component,

$$I_{\rm D}(r_{\rm D}) = I_{\rm D}(0) \exp \left(\frac{r_{\rm D}}{h} \right) , \qquad (69)$$

where $I(0)$ is the central surface brightness, $r_{\rm e}$ is the bulge effective radius and h is the disk scale-length. Integrating the fitted function in both cases gives a total measure of the light in both bulge $I_{\rm B}$ and disk $I_{\rm D}$ components, the ratio of which gives a quantitative measure of morphological classification[149].

Table 7. QSO/AGN Host galaxy studies with HST

Reference	$N_{\rm QSO}$	$N_{\rm RL}$	$N_{\rm RQ}$	z range	M_B range
Bahcall et al.[10]	20	6	14	$z < 0.3$	$M_B < -23$
Boyce et al.[17]	14	6	8	$z < 0.2$	$M_B < -23$
McLure et al.[117]	15	6	9	$z < 0.4$	$M_B < -23$
Schade et al.[141]	76	–	76	$z < 0.15$	$M_B > -24$
Malkan et al.[108]	200	–	200	$z < 0.03$	$M_B > -21$

5.2 Observations at Low Redshift

Ground-based optical imaging studies of the host galaxies of low redshift QSOs over the past 20 years[150–90] have been limited by the spatial resolution attainable from the ground and no strong consensus has been been reached over the general properties of QSO host galaxies from such studies.

Ground-based infrared imaging studies of QSOs have yielded a clearer picture[47–116]. Infrared observations not only benefit from the improved seeing in the H and K bands, but also from the increasing dominance of the red host galaxy against the blue nucleus. Such studies reveal that powerful QSOs inhabit luminous ($L > L^*$) and massive ($r_e > 10h^{-1}$kpc) galaxies. These studies also find that radio-loud QSOs are found exclusively in early-type galaxies. Early-type galaxies have also been found[157] to act as hosts for up to 50% of bright radio-quiet QSOs, challenging the existing orthodoxy that radio-quiet QSOs are predominantly found in spiral galaxies.

With the excellent imaging performance provided by the COSTAR-corrected optics, a number of QSO host galaxy studies have recently been carried out with HST[117,10,17], see Table 7. These observations focus on the brightest ($M_B < -23$) QSOs at low redshift and confirm earlier results that these QSOs are indeed preferentially found in large ($r_{\rm e} > 10h^{-1}$kpc), luminous ($> 2L^*$) galaxies. These studies also find that a significant fraction of QSOs are found in interacting or tidally-disrupted systems[10], providing circumstantial evidence for enhanced merger activity fueling QSOs. However, some authors[117] caution that the incidence of similar levels of merger activity in luminous galaxies without an active nucleus is not itself well known.

McLure et al.[117] also confirm that a significant fraction ($\sim 50\%$) of the radio-quiet QSOs imaged in their study have elliptical hosts, claiming tentative evidence that the fraction of bulge-dominant systems in radio-quiet QSOs increases with luminosity. Indeed, all AGN with $M_B < -24$ in their study were found in bulge-dominant systems. In addition, McLure et al.

demonstrate that the observed correlation between host galaxy and nuclear luminosity is consistent with the observed relation between black hole and spheroid mass[107]. Based on the black-hole/bulge mass ratio of 0.6%, they infer that the most luminous QSOs radiate at up to 20% of their Eddington luminosity, in agreement with results from ground-based infrared studies[116].

In a joint ground-based/HST imaging survey of 76 X-ray selected QSOs, Schade et al.[141] find that 55% of radio-quiet QSOs at low luminosities ($-21 > M_B > -24$) are also found in spheroid-dominant systems. However, Schade et al. find no evidence for any enhanced merger activity in these QSOs. None of the QSOs imaged were in interacting systems (see Fig. 15). Indeed, after accounting for the bias toward bulge-dominant systems, the host galaxies in this survey exhibited identical luminosities and sizes to those of field galaxies (Fig. 16). Schade et al. also find a weak correlation between nuclear and host galaxy luminosity, with inferred nuclear luminosities between 1–5% of the Eddington luminosity. In contrast to the McLure et al.[117] results, Schade et al.[141] found no correlation between the nuclear luminosity and the budge-dominance of the host galaxy.

However, the bulge-dominance amongst host galaxies for radio-quiet QSOs can not extend to the lowest luminosity AGN. An HST study of over 300 $M_B > -21$ active galaxies carried out by Malkan et al.[108] found that the majority of Seyfert 1 nuclei ($\sim 70\%$) inhabit disk galaxies. Malkan et al. also found little evidence for any enhanced merger activity in these AGN, in agreement with the growing body of observational evidence[45,44], that Seyfert 1s do not exhibit any enhanced level of interactions with companion galaxies within $100h^{-1}$kpc. In contrast, there is growing evidence that Seyfert 2 galaxies show evidence for increased interactions and associated circumnuclear star-formation[37]. Such observations are consistent with the 'modified unified model' proposed by Dultzin-Hacyan et al.[45] in which Type 2 AGN arise as a result of an interaction driving dust into the central regions of an AGN and causing obscuration. This would also explain the observation that the heavily-obscured ULIGs are almost exclusively found in interacting systems[138].

Malkan et al.[108] found that 30% of their Seyfert 1 galaxies contained no point-source nuclear component, ascribing this to (patchy?) obscuration of the central source. Schade et al.[141] also a find that 10% of their brighter QSO sample shows no nuclear component at the $M_B > -17$ level. Clearly the nature of these AGN merit further investigation.

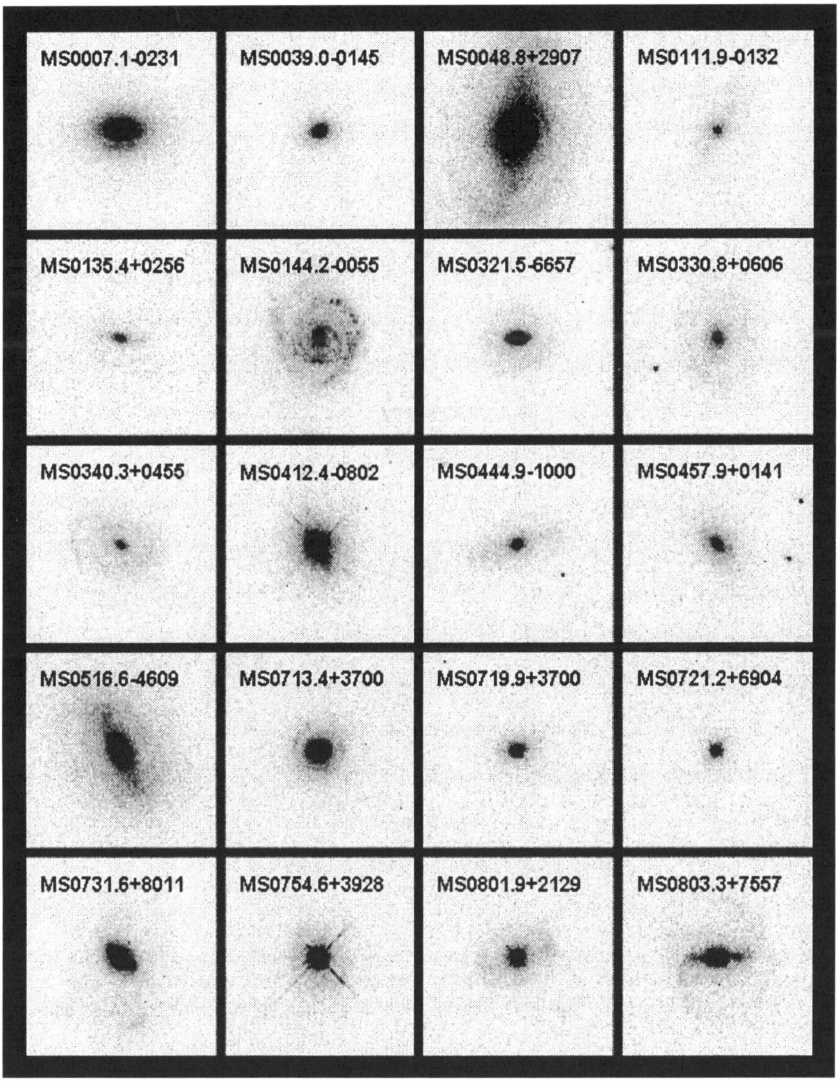

Figure 15: A subset of HST images from the sample of 76 X-ray selected QSOs studied by Schade et al. In contrast to previous studies many QSOs appear to reside in 'normal' galaxies, with few mergers/interactions in evidence. From Schade et al.[141] ©2000 Blackwell Sci.

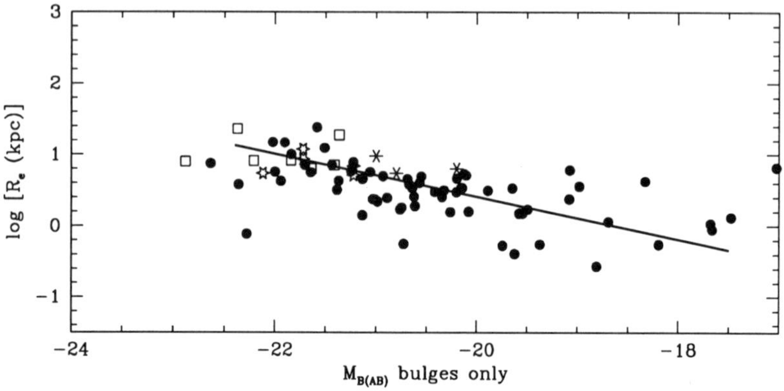

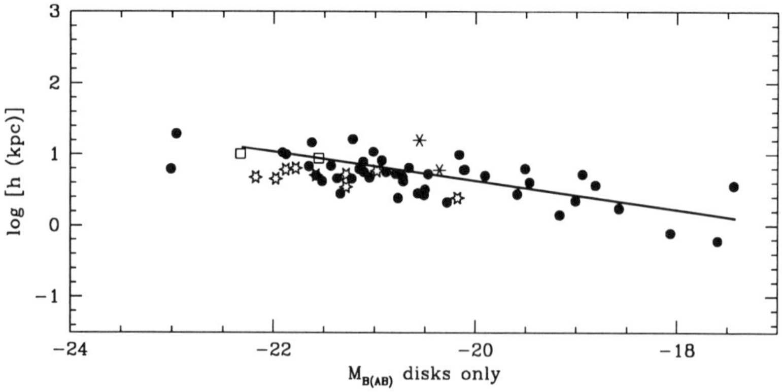

Figure 16: Size-magnitude relation for bulge and disk host galaxies of QSOs in various HST imaging studies. The symbol correspondence is as follows: filled circles to Schade et al.[141], open stars to Bahcall et al.[10], asterisks to Boyce et al.[17], and open squares to McLure et al.[118] Lines denote observed relation for 'normal' galaxies. From Schade et al.[141] ©2000 Blackwell Sci.

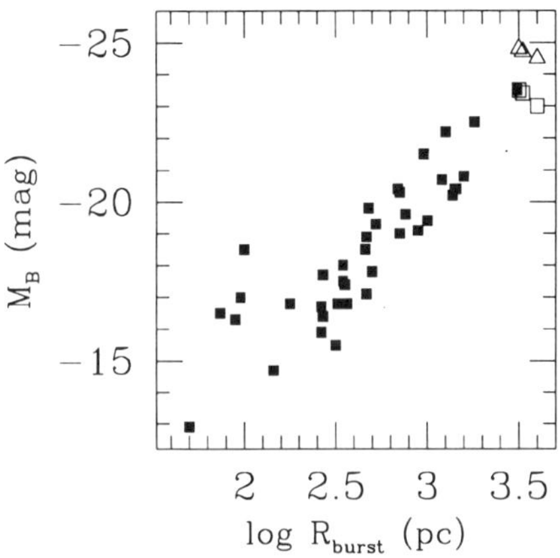

Figure 17. Size-luminosity relation for local HII regions and $z \sim 2$ radio-quiet QSOs. Filled circles correspond to local HII regions, squares to $z \sim 2$ QSOs ($f_\nu \propto \nu^{0.5}$) and triangles to $z \sim 2$ QSOs ($f_\nu \propto \nu^{0.5}$). From Aretxaga et al.[7] ©1998 Blackwell Sci.

5.3 High Redshift

At high redshifts ($z \sim 2$), imaging studies have been largely restricted to the most luminous radio-selected QSOs. Many of these objects show evidence for extended (up to $100h^{-1}$kpc) luminous ($L_{\rm UV} = 10^{12}L_\odot$) galaxies based either on broad-band optical/near-IR[98,99] or narrow-band redshifted Lyα[74] imaging studies of radio-loud objects. Such galaxies comprise between 5 and 40% of the AGN luminosity and exhibit 'knotty' morphologies, indicative of star formation. Typical Lyα luminosities of these galaxies are a few $\times$ 10^{44}erg s^{-1}.

Limited surveys of radio-quiet AGN at high redshift have also been carried out[6−84]. From these studies, comprising no more than 10 QSOs in total, host galaxies have been detected in 2–3 bright ($M_B \sim -29$) QSOs. In common with the host galaxies of radio-loud QSOs at high redshift, the properties of these galaxies are extreme, with up to 35% of the QSO luminosity or $100L^*$. However, the host galaxies of these QSOs appear to have smaller effective radii[7] $r_{\rm e} \sim 10h^{-1}$kpc than their radio-loud counterparts. Moreover, it has been claimed that the properties of host galaxies of radio-quiet QSOs at $z \sim 2$ are consistent with an extrapolation of the luminosity-size relation for H II regions at low redshift[7] (Fig. 17).

At intermediate redshifts, at least one example of a QSO host galaxy has been discovered[29] that exhibits evidence for significant star formation in

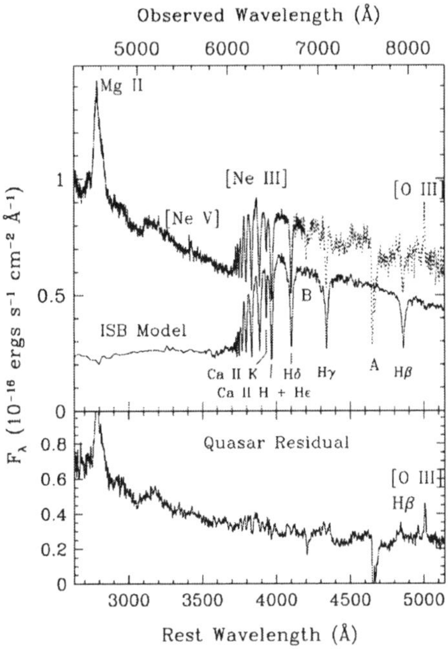

Figure 18. Optical spectrum of UNJ1025–0040 plotted together with the synthetic spectrum of an instantaneous starburst with age 400 Myr. The QSO residual between the observed spectrum and the starburst model is plotted at the bottom of the figure. From Brotherton et al.[29] ©1999 AAS.

the recent past. The optical spectrum of the radio-loud QSO UNJ1025–0040 (Fig. 18) reveals a $10^{11}M_\odot$ 400 Myr post-starburst hosting a $M_B = -23$ QSO at $z = 0.64$. While rare, objects such as this may form a missing link between QSOs and starburst activity, representing a special phase in the lifetime of the QSO. However, other studies of radio-loud QSO host galaxies at intermediate redshifts ($1 < z < 1.5$) reveal a much older (~ 3 Gyr) stellar population[154].

5.4 QSO Host Galaxies: Summary

Overall, imaging studies have shown that at low redshifts QSOs appear to lie in relatively normal galaxies. The correlation between nuclear and host galaxy luminosity results in more luminous QSOs lying in larger and more luminous galaxies. Radio-loud QSOs appear to exclusively inhabit bulge-dominant systems, while radio-quiet QSOs are found in both bulge and disk systems. At least half of the $M_B < -21$ radio-quiet QSOs/AGN are found in bulge-dominant systems. Otherwise the properties (size, luminosity) of QSO host galaxies are consistent with 'normal' galaxies. There is little evidence from recent comprehensive imaging studies for enhanced interaction/merger activity occurring in the vast majority of QSOs/AGN. Merging/interactions may, however, be much more dominant in Type 2 AGN (Seyfert 2s, ULIGs), where the interaction itself is, in part, responsible for the obscured nature of these objects.

These results are entirely consistent with the observed environments of QSOs at low redshifts. Moreover, the suggestion that radio-loud QSOs preferentially inhabit the most massive systems[117] may naturally explain the environmental differences between radio-loud and radio-quiet QSOs.

If there is indeed a luminosity dependence on AGN host galaxy type then (based on the observation that early-type galaxies cluster more strongly than late-type galaxies[103]) we might expect to see a correlation between AGN luminosity and environment. However, environmental studies show little correlation between QSO/AGN luminosity and clustering scale-length. At low redshifts $z < 0.3$, studies of Seyfert 1 galaxies[44] ($M_B > -21$) and more luminous AGN[152] ($M_B < -21$) both yield scale-lengths $r_0 = 5h^{-1}$Mpc.

At high redshifts, studies have been limited to the most luminous QSOs. In all cases, the properties of QSO host galaxies at $z \sim 2$ are consistent with vigorous star-formation. Although the properties of the host galaxies are extreme, sizes and implied star formation rates scale appropriately from H II regions in the local Universe. These observations provide further evidence that QSO activity and galaxy formation are closely linked at the key $z \sim 2$ epoch.

While these results only relate to the most luminous QSOs at these redshifts, diffraction-limited imaging in the near-infrared, offered by the latest generation of 8m-class telescopes, will allow host galaxy studies to extend to more representative samples of QSOs at $z \sim 2$.

6 Summary

I summarize below the key results which have emerged over the past five years in the cosmological evolution of the QSO population:

- The cosmological evolution of the QSO luminosity density at $z < 2$ in the optical/radio/X-ray and infrared bands is remarkably similar in form to that of the star formation rate.

- The luminosity density of QSOs in the optical and radio peak at $z \sim 2.5$. Nevertheless, uncertainties remain over the numbers of potentially obscured QSOs (particularly at high redshift) as revealed by Chandra X-ray observations.

- The evolution of the LF is consistent with QSO formation associated with the collapse of dark matter halos in a Press-Schechter semi-analytic formalism. In these models L^* QSOs correspond to $M_{\rm halo} \sim 10^{12} M_\odot$.

- The clustering length of optically-selected L^* QSOs, $r_0 = 4.5 \pm 1.0 h^{-1}$Mpc remains relatively constant over the redshift range $0 < z < 3$, consistent with the measurement of the galaxy scale over the same redshift interval ($\Omega_{\rm m} = 1$). Such evolution is again consistent with the predictions of the Press-Schechter formalism for $10^{12} - -10^{13} M_\odot$ dark matter halos.

- Radio-quiet QSOs appear to inhabit normal galaxy environments with QSO-galaxy scale-lengths $r_0 \sim 5h^{-1}$Mpc at $z < 1.5$, consistent with the QSO clustering results. Radio-loud QSOs inhabit significant richer $r_0 > 10h^{-1}$Mpc environments at $z > 0.3$.

- QSOs appear to sample normal host galaxies at $z < 0.2$ with a bias towards bulge-dominant systems at high luminosities. QSO luminosity appears to be (weakly) correlated with bulge mass. At $z \sim 2$, the detected hosts are consistent with galaxies undergoing vigorous star formation.

The evolution in the observed properties of QSOs (luminosity, clustering, environments) thus displays many similarities to those of galaxies. This naturally suggests a common origin/formation process for both QSOs and galaxies. A scenario in which QSOs are formed in the initial collapse of a dark matter halo ensures that QSO activity is associated with the accompanying creation of a galactic spheroidal component in a vigorous burst of star-formation. How any supermassive black hole may be formed in this event is unclear.

With the imminent availability of major new QSO surveys such as the 2QZ and SDSS and the advent of 8m-10m class telescopes, the prospects for further disentangling the relationship between QSO activity and star-formation have never been brighter. I hope that many of the participants will play a major role in the exciting times that lie ahead in this key area of astronomical research.

Acknowledgements

I would like to thank the organizers of the Guillermo Haro Advanced School, in particular Itziar Aretxaga and Daniel Kunth, for organizing a very enjoyable and stimulating week. It was a great deal of hard work but immensely rewarding. I would also like to thank the students/postdocs who participated so vigorously with many insightful questions and lively discussion. I am also indebted to Xiaohui Fan, Guinevere Kauffmann, Martin Haehnelt, Takamitsu Miyaji, Robert Smith and Peter Shaver who kindly provided me with postscript versions of their figures reproduced in this chapter. I would finally like to thank Scott Croom of the 2QZ team who provided a number of helpful comments on a first draft of this manuscript.

References

1. Abramowicz M., Chen X., Lasoate J., Reger O., ApJ, 438, L37, 1995
2. Almaini O., Boyle B.J., Griffiths R.E., Shanks T., Stewart G.C., Georgantopoulos I., MNRAS, 277, 31, 1995
3. Almaini O., Lawrence A., Boyle B.J., MNRAS, 305, L59, 1999
4. Appenzeller I., et al. ApJS, 117, 319, 1998
5. Aragon-Salamanca A., Ellis R.S., O'Brien K.S., MNRAS, 281, 945, 1995
6. Aretxaga I., Boyle B.J., Terlevich R.J., MNRAS, 275, L27, 1995
7. Aretxaga I., Terlevich R.J., Boyle B.J., MNRAS, 296, 643, 1998
8. Aretxaga I., Le Mignant D., Melnick J., Terlevich R.J., Boyle B.J., MNRAS, 298, L13, 1998
9. Avni Y., Bahcall, J.N. ApJ, 235, 695, 1980
10. Bahcall J.N., Kirakos S., Saxe D.H., Scheider D.P., ApJ, 479, 642, 1997
11. Baker J.C., Hunstead R.W., Bremer M., Bland-Hawthorn J., Athreya R., MNRAS, in press, 2000
12. Barger A., Cowie L.L., Sanders D.B., ApJ, 518, 1, 1999
13. Barger A., Cowie L.L., Smail I., Ivison R.J., Blain A.W., Kneib J.P. AJ, 117, 2656, 2000
14. Barger A., Cowie L.L., Richards E.A., AJ, 119, 2092, 2000
15. Baugh C.M., Efstathiou G., MNRAS, 265, 143, 1993
16. Blain A.W., Kneib J.-P., Smail I., ApJ, 512, 87, 1999
17. Boyce P.J., et al., ApJ, 473, 760, 1996
18. Boyle B.J., Mo H., MNRAS, 260, 925, 1993
19. Boyle B.J., Couch W.J., MNRAS, 264, 604, 1993
20. Boyle B.J., Shanks, T., Geogantopoulos, I., Stewart G.C., Griffiths R.E., MNRAS, 271, 639, 1994
21. Boyle B.J., Shanks T., Peterson B.A., MNRAS, 235, 935, 1988
22. Boyle B.J., Fong R., Shanks T., Peterson B.A., MNRAS, 243, 1, 1990
23. Boyle B.J., Jones L.R., Shanks T., MNRAS, 251, 482, 1991
24. Boyle B.J., Terlevich R.J., MNRAS, 293, 49, 1998
25. Boyle B.J. Almaini O., Georgantopoulos I., Blair A.J., Stewart G.C.,

Griffiths R.E., Shanks T., Gunn K.F. MNRAS, 297, L53, 1998
26. Boyle B.J., Croom S.M., Smith R.J., Shanks T., Miller L., Loaring N., Phil. Trans. R. Soc. Lond. A., 357, 185, 1999
27. Boyle B.J., Shanks T., Croom, S.M., Smith R.J., Miller L., Loaring N., MNRAS, 317, 1014, 2000
28. Brandt W.N. et al. AJ, 119, 2349, 2000
29. Brotherton M. et al. ApJ, 520, 87, 1999
30. Brown M.J. PhD Thesis, University of Melbourne, 2000
31. Carilli C.L. et al. ApJ, 533, L13, 2000
32. Carlberg R., ApJ, 350, 505, 1990
33. Carlberg R.G., Cowie L.L., Songaila A., Hu E.M., ApJ, 484, 538, 1997
34. Carlberg R.G., et al. in Wide Field Surveys in Cosmology, Editions Frontiers, p143, 1998
35. Clowes R.G., Campusano L., Graham M.J., MNRAS, 309, 48, 1999
36. Colberg J.M. in Wide Field Surveys in Cosmology, Editions Frontiers, p247, 1998
37. Colina L., Vargas M.L.G., Delgado R.M.G. Mas-Hesse J.M., Perez E., Alberdi A. Krabbe A. ApJ, 488, 71, 1997
38. Comastri A., Setti G., Zamorani G., Hasinger G., A&A, 296, 1, 1995
39. Crampton D., Cowley A.P., Hartwick F.D.A., ApJ, 345, 59, 1989
40. Cristiani S., Vio L., A&A, 227, 385, 1990
41. Croom S.M., Shanks T., MNRAS, 281, 893, 1996
42. Croom S.M., Shanks T., MNRAS, 303, 411, 1999
43. Croom S.M., Shanks T., Boyle B.J., Smith R.J., Miller L., Loaring N.S., Hoyle F., MNRAS, submitted
44. De Robertis M.M., Yee H.K.C., Hayhoe, K., ApJ, 496, 93, 1998
45. Dultzin-Hacyan D., Krongold Y., Funtes-Guridi I, Marziani P., ApJ, 513, L111, 1999
46. Dunlop J.S., Peacock J.A., MNRAS, 247, 19, 1990
47. Dunlop J.S., Taylor G.L., Hughes D.H., Robson E.I., MNRAS, 264, 455, 1993
48. Efstathiou G., Rees M.J, MNRAS, 230, 5P, 1988
49. Ellingson E., Yee H.K.C., Green R.F. ApJ, 371, 49, 1991
50. Fabian A.C., Barcons X. ARAA, 30, 429, 1992
51. Fabian A. et al. MNRAS, 315, 8, 2000
52. Fall S.M., Pei Y.C., ApJ, 402, 479, 1993
53. Fan X. et al. AJ, 119, 1, 2000
54. Fan X., AJ, submitted, (astro-ph/0008123.ps), 2000
55. Folkes S. et al., MNRAS, 308, 459, 1999
56. Franceschini A., Hasinger G., Miyaji T., Malquori D., MNRAS, 310, L5,

1999
57. Franceschini A., Bassani L., Cappi M., Granato G.L., Malaguti G., Palazzi E., Persic M., A&A, 353, 910, 2000
58. Fry N.J., ApJ, 461, L45, 1996
59. Gebhart K. et al. ApJ, 539, L13, 2000
60. Geogantopoulos I., Shanks T., MNRAS, 271, 773, 1994
61. Giacconi R., Gursky H., Paolini F., Rossi B.B., Phys. Rev. Lett., 9, 439, 1962
62. Giavalsco M., Steidel C.C., Adelberger K.L., Dickinson M.E., Pettini M., Kellogg M. ApJ, 503, 543, 1998
63. Goldschmidt P., Miller L., MNRAS, 293, 107, 1998
64. Green R.F., Schmidt M., Liebert J., ApJS, 61, 305, 1986
65. Haenelt M., Rees M.J., MNRAS, 300, 817, 1993
66. Haenelt M., Natarajan, P., Rees M.J., MNRAS, 300, 817, 1999
67. Haiman Z., Loeb A., ApJ, 503, 505, 1998
68. Hall P.B., Green R.F., ApJ, 507, 558, 1998
69. Hartwick F.D.A., Schade D.J., ARAA, 28, 437, 1990
70. Hasigner G., Burg R., Giacconi R., Schmidt M., Trumper J., Zamorani G., A&A, 329, 482, 1998
71. Hawkins M.R.S., Veron P., MNRAS, 260, 202, 1993
72. Hawkins M.R.S., Veron P., MNRAS, 275, 1102, 1995
73. Hawkins M.R.S., Veron P., MNRAS, 281, 348, 1996
74. Heckman T.M., Lehnert M.D., van Breugel W., Miley G.M., ApJ, 370, 78, 1991
75. Hewett P.C., Foltz C.B., Chaffee F.H., ApJ, 406, L43, 1993
76. Hewett P.C., Foltz C.B., PASP, 106, 113, 1994
77. Hewett P.C., Foltz C.B., Chaffee F.H., AJ, 109, 1498, 1995
78. Hintzen P., Romanshin W., Valdez F., ApJ, 366, 7 1991
79. Hook I.M., PhD Thesis, University of Cambridge, 1994
80. Hu E., Cowie L.L., McMahon R.G., ApJ, 502, L99, 1998
81. Hu E., McMahon R.G., Cowie L.L., ApJ, 522, L9, 1999
82. Hughes D.H. et al., Nat, 394, 241, 1998
83. Hutchings J.B., AJ, 116, 20, 1998
84. Hutchings J., Crampton D., Morris S.L., Durand D., Streinbring E., AJ, 118, 210, 1999
85. Iovino A., Shaver, P., ApJ, 330, L13, 1988
86. Jones L.R. et al. MNRAS, 285, 547, 1997
87. Kauffmann G., Haenelt M., MNRAS, 311, 576, 2000
88. Kennifick J.D., Djorgovski S.G., de Carvalho R.R. AJ, 110, 2553, 1995
89. Kohler T., Groote D., Reimers D., Wisotzki L., A&A, 325, 502, 1997

90. Kotilainen J., Ward M.J., MNRAS, 266, 953, 1994
91. La Franca F., Cristiani S., AJ, 113, 1517, 1997
92. La Franca F., Andreani P., Cristiani S., ApJ, 497, 529, 1998
93. La Franca F., Lissandrini C., Cristiani S., Miller L., Hawkins M.R.S., McGillivray H.T., A&AS, 140, 351, 1999
94. Landy S.D., Szalay A.S., ApJ, 412, 64, 1986
95. Larson R.B., MNRAS, 166, 585, 1974
96. Laurikainen E., Salo H., A&A, 293, 683, 1995
97. Le Fèvre O., Hudon D., Lilly S.J., Crampton D., Hammer F., Tresse L., ApJ, 461, 534, 1996
98. Lehnert M.D., Heckman T.M., Chambers K.C., Miley G.K., ApJ, 395, 466, 1992
99. Lehnert M.D., van Breugel W.J.M, Heckman T.M., Miley G.K., ApJS, 124, 11, 1999
100. Lilje P.B., ApJ, 386, L33, 1992
101. Loan A.J., Wall J.V., Lahav O., MNRAS, 286, 994, 1997
102. Londish D., Boyle B.J., Schade D.J., MNRAS, 317, in press, 2000
103. Loveday J., Maddox S.J., Efstathiou G., Peterson B.A., ApJ, 442, 457, 1995
104. MacKenty J.W., ApJS, 72, 231, 1990
105. Madau P., Ferguson H.C., Dickinson M.E., Giavalesco M., Steidel C.C., Fruchter A., 1996, MNRAS, 283, 1388, 1996
106. Maddox S., Efstathiou G., Sutherland W., Loveday J., MNRAS, 242, L43, 1990
107. Magorrian J. et al., AJ, 115, 2285, 1998
108. Malkan M.A., Gorijan V., Tam R.T., ApJS, 117, 25, 1998
109. Margon B. Phil. Trans. R. Soc. Lond. A, 357, 93, 1999
110. Marshall H.L., ApJ, 299, 109, 1985
111. Marshall H.L., Tananbaum H., Avni Y., Zamorani G., ApJ, 269, 35, 1983
112. Mason K., et al. MNRAS, 311, 456, 2000
113. Mataresse S., Coles P., Lucchin F., Moscardini L., MNRAS, 286, 115, 1997
114. McHardy I. et al., MNRAS, 295, 641, 1998
115. McLeod K.K., Reike G.H., ApJ, 431, 137, 1994
116. McLeod K.K., Reike G.H., Storrie-Lombardi L.J., ApJ, 511, 67, 1999
117. McLure R.J., Dunlop J.S., Kukula M.J., Baum S.A., O'Dea C.P., Hughes D.H., MNRAS, 308, 377, 1999
118. McLure R., Dunlop J.S., MNRAS, submitted, astro-ph/0009406
119. Miyaji T., Hasinger G., Schmidt M., A&A, 325, 25, 2000

120. Moscardini L., Coles P., Lucchin F., Mataresse S., MNRAS, 299, 95, 1999
121. Mushotsky R.F., Cowie L.L., Barger A.J., Arnaud K.A., Nat, 404, 459, 2000
122. Narayan R., Yi I., ApJ, 452, 710, 1995
123. Ohta K. et al., ApJ, 458, L57, 1996
124. Omont F., McMahon R.G., Cox P., Kreysa E., Bergeron J., Pajot F., Storrie-Lombardi L.J., A&A, 310, 259, 1996
125. Osmer P.S., Kennefick J.D., Hall P.B., Green R.F., ApJS, 119, 189, 1999
126. Page M.J, Mason, K., McHardy I.M., Jones L.R., Carrera F.J., MNRAS, 291, 324, 1997
127. Page M.J., Carrera F., MNRAS, 311, 456, 2000
128. Peacock J.A., Cosmological Physics, Cambridge University Press, 1999
129. Peacock J.A., Nicholson D., MNRAS, 253, 307, 1991
130. Peacock J.A., Dodds S.J., MNRAS, 280, L19, 1996
131. Picinotti G., Mushotzky R.F., Boldt E.A., Holt S.S., Marshall F.E., Serlemitsos P.J., Shafer R.A., ApJ, 253, 485, 1982
132. Postman M., Lauer T.R., Szapudi I., Oegerle, W., ApJ, 506, 33, 1998
133. Press W., Schechter P.L., ApJ, 187, 425 1974
134. Press W.H., Teukolsky S.A., Vetterling W.T., Flannery B.P., Numerical Recipes: 2nd edition, Cambridge University Press, p640, 1992
135. Ratcliffe A., Shanks T., Parker Q.,A., Fong R., MNRAS, 296, 173, 1998
136. Richstone D., Gebhardt K., Pinkney J., BAAS, 195, 9906, 1999
137. Rowan-Robinson M., MNRAS, 316, 885, 2000
138. Sanders D.B., Mirabel I.F., ARAA, 34, 749, 1996
139. Schmidt M., et al. A&A, 329, 495, 1998
140. Schmidt M, Schneider D.P., Gunn J.E., AJ, 110, 68, 1995
141. Schade D.J., Boyle B.J., Letawsky M. MNRAS, 360, 254, 2000
142. Shanks T., Boyle B.J., MNRAS, 271, 753, 1994
143. Shanks T., Griffiths R.E., Georgantopoulos I., Almaini O., Boyle B.J., Gunn K.F., Roche N., Stewart G.C. MNRAS, submitted, 2000
144. Shaver P. A&A, 136, L9, 1984
145. Shaver P.A., Wall J.V., Kellerman K.I., Jackson C.A., Hawkins M.R.S., Nat, 384, 439, 1996
146. Shaver P.A., Hook I.M., Jackson C.A., Wall J.V., Kellerman K.I., in Highly Redshifted Radio Lines: ASP Conf. Ser. 156, eds C.L. Carilli et al., 163, 1999
147. Shepherd C.W., Carlberg R.L., Yee H.K.C., Ellingson E., ApJ, 479, 82, 1998
148. Shwope A.D. et al. Astron. Nach., 321, 1, 2000

149. Simien F., de Vaucouleurs G., ApJ, 302, 564, 1986
150. Simkin S.M., Su H.J., Schwarz M.P., ApJ, 237, 404, 1980
151. Smith E.P., Heckman T.M., Bothun G.D., Romanshin W., Balick B., ApJ, 306, 64, 1986
152. Smith R.J., Boyle B.J., Maddox S.J., MNRAS, 277, 270, 1996
153. Smith R.J., Boyle B.J., Maddox S.J., MNRAS, 313, 252, 2000
154. Spinrad H., Dey A., Stern D., Dunlop J., Peacock J.A., Jimenez R., Windhorst R., ApJ, 484, 581, 1997
155. Steidel C.C., Adelberger K.L., Dickinson M., Giavalesco M., Pettini M., Kellogg M., ApJ, 492, 428, 1998
156. Stocke J., Morris S.L., Gioia I.M., Maccacaro T., Schild R., ApJS, 76, 813, 1991
157. Taylor G.L., Dunlop J.S., Hughes D.H., Robson E.I., MNRAS, 283, 930, 1996
158. Teplitz H.I., McLean I.S., Malkan M.A., ApJ, 520, 469, 1999
159. Terlevich R.J., Boyle B.J. MNRAS, 262, 491, 1993
160. Tucker D.L. et al. MNRAS, 285, L5, 1997
161. Warren S.J., Hewett P.C., Osmer P.S., ApJ, 421, 412, 1994
162. Warren S.J., Hewett P.C., Foltz C.B., MNRAS, 312, 827, 2000
163. Webster R., Francis P.J., Peterson B.A., Drinkwater M.J., Masci F.J., Nat, 375, 469, 1995
164. White R.L., ApJS, 126, 133, 2000
165. Wisotzki L., A&A, 353, 861, 2000
166. Wisotzki L. A&A, 358, 77, 2000
167. Wold M., Lacy M., Lilje P.B., Serjeant S., MNRAS, 316, 267, 2000
168. Yee H.K.C., Green R.F., ApJ, 280, 79, 1984
169. Yee H.K.C., Green R.F., ApJ, 319, 28, 1987
170. Zitelli V., Mignoli M., Marano B., Zamorani G., Boyle B.J., MNRAS, 256, 349, 1991
171. Zitelli V., Granato G.L., Mandolesi N., Wade R., Danese L., ApJS, 84, 185, 1993

Recent Results in Cancer Research 140

Managing Editors
Ch. Herfarth, Heidelberg · H.J. Senn, St. Gallen

Associate Editors
M. Baum, London · V. Diehl, Köln
F. Gutzwiller, Zürich · M.F. Rajewsky, Essen
M. Wannenmacher, Heidelberg

Founding Editor
P. Rentchnick, Geneva

Springer
Berlin
Heidelberg
New York
Barcelona
Budapest
Hong Kong
London
Milan
Paris
Santa Clara
Singapore
Tokyo

H.J. Senn R.D. Gelber
A. Goldhirsch B. Thürlimann (Eds.)

Adjuvant Therapy of Breast Cancer V

With 34 Figures and 71 Tables

Springer

Prof. Dr. med. Hans Jörg Senn
Department of Medicine C, Kantonsspital
CH-9007 St. Gallen, Switzerland

Richard D. Gelber, Ph.D.
Division of Biostatistics
Dana Farber Cancer Institute,
44, Binney Street,
Boston, MA 02115, USA

Prof. Dr. med. Aron Goldhirsch
Ospedale Civico
CH-6900 Lugano, Switzerland

Dr. med. Beat Thürlimann
Department of Medicine C, Kantonsspital
CH-9007 St. Gallen, Switzerland

ISBN-13:978-3-642-79280-9

Library of Congress Cataloging-in-Publication Data. Adjuvant therapy of breast cancer V/H.J. Senn ... [et al.] (eds.). p. cm. – (Recent results in cancer research; 140) Presented at the 5th International Conference on Adjuvant Therapy of Primary Breast Cancer held in St. Gallen, Switzerland, in March 1995. Includes bibliographical references and index. ISBN-13:978-3-642-79280-9 e-ISBN-13:978-3-642-79278-6 DOI: 10.1007/978-3-642-79278-6 1. Breast – Cancer – Congresses. 2. Breast – Cancer – Adjuvant treatment – Congresses. I. Senn, Hansjorg. II. International Conference on Adjuvant Therapy of Primary Breast Cancer (5th: 1995: Saint Gall, Switzerland) III. Series. [DNLM: 1. Breast Neoplasms – therapy – congresses. 2. Combined Modality Therapy – congresses. RE106P v. 140 1996/WP 870 A2354 1996] RC261.R35 vol. 140 [RC280.B8] 616. 99'4 s – dc20 [616.99'44906] DNLM/DLC for Library of Congress 96-25120

This work is subject to copyright. All rights are reserved, whether the whole or part of the material is concerned, specifically the rights of translation, reprinting, reuse of illustrations, recitation, broadcasting, reproduction on microfilm or in any other way, and storage in data banks. Duplication of this publication or parts thereof is permitted only under the provisions of the German Copyright Law of September 9, 1965, in its current version, and permission for use must always be obtained from Springer-Verlag. Violations are liable for prosecution under the German Copyright Law.

© Springer-Verlag Berlin Heidelberg 1996
Softcover reprint of the hardcover 1st edition 1996

The use of general descriptive names, registered names, trademarks, etc. in this publication does not imply, even in the absence of a specific statement, that such names are exempt from the relevant protective laws and regulations and therefore free for general use.

Product liability: The publishers cannot guarantee the accuracy of any information about dosage and application contained in this book. In every individual case the user must check such information by consulting the relevant literature.

Cover design: Springer-Verlag, Design & Production

Typesetting: Best-set Typesetter Ltd., Hong Kong

SPIN: 10474829 19/3134/SPS – 5 4 3 2 1 0 – Printed on acid-free paper

Preface

We often hear physicians, health care professionals, politicians, and patient advocates that "nothing has happened in the treatment of breast cancer," since patients with breast cancer, the most frequent neoplastic condition in women in industrialized countries, are continuing to suffer relapse and succumb to this dreadful disease! This negativistic attitude does not seem to be justified, but, why is the transmission of clinical trial results into general practice, and with it progress, such a slow process?

After many decades of frustrating stagnation of long-term survival expectations, in all stages of early, operable breast cancer treated only by surgery and locoregional radiotherapy, adjuvant systemic therapy (chemo- as well as endocrine treatments) clearly showed to significantly benefit in terms of disease-free and overall survival. This evolution has been extensively expounded on by the Worldwide Oxford Overview and the Expert Consensus Panel at the fourth International Conference on Adjuvant Therapy of Primary Breast Cancer in St. Gallen (Early Breast Cancer Trialists' Collaborative Group 1992; Glick et al. 1992). What has happened since then?

During the past 3–5 years, several new concepts and treatment strategies have emerged and have been studied in various major breast cancer groups and treatment centers worldwide. Some of these can already be considered to assist in the primary treatment of operable breast cancer today, while others are still undergoing clinical trials for better definition of their practical usefulness. Most of these (research) problems were recently discussed during the fifth International Conference on Adjuvant Therapy of Primary

Breast Cancer in St. Gallen, Switzerland, in early March 1995, attended by more than 1000 breast cancer specialists from 52 countries worldwide (Goldhirsch et al. 1995). These new developments include:

- Isolation of "breast cancer genes" (BRCA 1-1-2), which will present new challenges in ongoing and future trials of chemoprevention of breast cancer!
- Biological diagnostics and therapies, involving in particular growth factors and angiogenesis, are still to be studied in solid clinical trials in order to demonstrate their usefulness as adjuvants.
- New factors for prediction of treatment response, such as c-ErbB2 and Ki67-expression for prediction of response to anthracycline-containing adjuvant therapies, and resistance to tamoxifen or CMF: results are interesting, but need to be confirmed in well-designed, randomized clinical trials.
- The unsolved problems of concomitant or sequential adjuvant chemo-endocrine therapies (Osborne 1994) are presently under study in various trials by the International Breast Cancer Study Group and others.
- The concept of the prognostic impact of "full-dose" chemotherapy: based on (scarce) prospective trial evidence, dose reductions of more than 25% are highly detrimental in terms of adjuvant, curative treatment results (Wood et al. 1994). This could greatly influence patient care and breast cancer mortality outside of trials, considering that 80%–90% of patients in most areas are not treated in prospective studies. The use of hematopoietic growth factors has not been systematically analyzed in this regard.
- Very "high-dose" chemotherapy with autologous bone marrow or peripheral blood progenitor cell support is a promising approach for better long-term survival benefit in patients at high risk of relapse, e.g., those with more than four (or at least more than ten) tumor-positive axillary nodes: more prospective trials to define the choice of treatment, clinical effectiveness, and cost-benefit ratios are sorely needed.
- The definition of the usefulness of new, very active single drugs such as taxanes in prospective trials in the adjuvant setting is highly important considering their complete-remission induction potential (especially if combined with anthracyclines) in advanced disease, but also regarding their high toxicity and exorbitant cost.

Is there room for further studies to improve adjuvant therapy of primary breast cancer? We believe the answer is yes!

1. It seems that there are more pertinent questions related to the improvement of adjuvant systemic therapy in breast cancer than interested and logistically able study groups available to resolve these questions in the near future!
2. However, in planning new adjuvant trials, it would be extremely helpful if existing (national and international) breast cancer groups and working parties would not only "work" but rather would work together internationally in designing and conducting common scientifically innovative and clinically relevant studies, which could be completed within acceptable time intervals.
3. The implementation of the concept of "full-dose" chemotherapy outside of clinical trials could mark a major step forward in the curative treatment of early breast cancer, translating clinical research data into clinical and epidemiologic experience.

Although the present curative treatment effects of adjuvant systemic therapies in breast cancer are still modest, they are real and give rise to hope of further improving long-term disease control, especially in younger patients with higher than average risk of recurrence. This bundle of hope and problems is discussed extensively in this volume. The essence of the closing "consensus conference" on the present treatment options for primary breast cancer have also been published in international oncology journals (Goldhirsch et al. 1995).

St. Gallen, Switzerland — Hans Jörg Senn
Lugano, Switzerland — Aron Goldhirsch
Boston, MA, USA — Richard D. Gelber
St. Gallen, Switzerland — Beat Thürlimann
July 1995

References

Early Breast Cancer Trialists' Collaborative Group (1992) Systemic treatment for early breast cancer by hormonal, cytotoxic or immune therapy, 133 randomized trials involving 31 000 recurrences and 24 000 deaths among 75 000 women. Lancet 339:1–5; 71–85

Glick JH, Gelber RD, Goldhirsch A, Senn HJ (1992) Meeting highlights: adjuvant therapy of primary breast cancer. J Natl Cancer Inst 84:1476–1485

Goldhirsch A, Wood CW, Senn HJ, Glick JH, Gelber RD (1995) Meeting highlights: international consensus panel on the primary treatment of operable breast cancer. J Natl Cancer Inst, October 4th, 1995, and Eur J Cancer (in press)

Osborne CK (1994) Interactions of tamoxifen with cytotoxic chemotherapy for breast cancer. In: Jordan VC (ed) Long term tamoxifen for breast cancer. University of Wisconsin Press, Madison, pp 181–198

Wood WC, Budman DR, Korzun AH et al. (1994) Dose and dose intensity of aduvant chemotherapy for stage II, node-positive breast carcinoma. N Engl J Med 330:1253–1259

Contents

*List of Contributors**

---

* The addresses of the authors are given on the first page of each contribution.
[1] Page on which contribution begins.

Introduction

The Control of Breast Cancer: A Look into the Future

U. Veronesi

European Institute of Oncology, Via Ripamonti, 435 Milan, Italy

There is no doubt that the recently published data on the decrease in mortality rates in the United States in the last 5 years, which are in line with similar observations in various European countries [1], are of historical importance. After centuries of increasing rates of mortality from breast cancer, there is now a clear trend in the opposite direction. Even more interesting is the fact that this favorable trend in mortality exists in spite of the continuous increase in incidence rates for breast cancer. This clearly shows that the declining mortality is the result of improved curability of the disease. This trend was first shown by R. Doll in his analysis of mortality rates for various types of tumors in 1990 [2], when he was able to show that the trend in breast cancer mortality in the younger generations appeared to be going in a more favorable direction than in the past.

As soon as the data were made available, a heated discussion immediately started to identify the reasons for the decline, public health officers claiming earlier detection as the main reason for the improved curability and oncologists supporting improved forms of treatment as the real main factor. Looking at the data from both viewpoints, it is extremely likely that new developments both in diagnostic tools and treatment modalities are equally important. In fact, the data showing a reduction in mortality among populations exposed to screening programs are very convincing, as are the results of the meta-analysis on adjuvant chemotherapy.

The encouraging data on mortality reduction will certainly stimulate new plans for a coordinated strategy aimed at better control of breast cancer in the next decade. In the area of primary prevention not much may be expected, as no clear recommendations can be made to the general population as regards lifestyle, mainly because the data on diet are still contradictory. However, in the limited subgroup of women at very high risk for breast cancer due to genetic factors (hereditary cases), some measures can be taken, including a subcutaneous mastectomy in young age, which can now be easily performed as a result of the improvements in plastic and reconstructive surgery.

Chemoprevention represents an interesting area of research, but there is as yet no evidence of efficacy, although the possible preventive role of tamoxifen must be considered with great interest [3]. More promising is the area of breast cancer detection. The extensive use of mammography and its technological improvements, the advances in ultrasound exploration of the mammary gland, and increased awareness among women are changing the characteristics of the population of patients with breast cancer. In advanced centers of Western countries, some 10% of the cases are purely in situ lesions and another 10% are small, nonpalpable carcinomas less than 1 cm in diameter. Moreover, the percentage of cases defined as T1 (less than 2 cm in diameter) is approaching 50% of all cases and will soon become the majority. This trend is focusing attention on problems related to more highly developed treatments of primary carcinoma in terms of good quality of cosmetic results, limitation of radiotherapy, and avoidance of axillary dissection. As the likelihood of distant metastases is low and the prognosis is excellent in patients with early cancer, the need for adjuvant systemic treatments will become less important in these patients with the final results of better accepted and less expensive forms of treatment.

If, therefore, efforts are focused on the highest quality of local treatment, the problem of achieving good cosmetic results and complete local control of the disease will become the main issue. Unfortunately, to avoid local recurrences after conservative surgery, resection of normal mammary tissue around the primary carcinoma needs to be fairly extensive [4], with the result that the breast will ultimately remain somewhat smaller than and unavoidably asymmetric to the other one. The logical solution is a surgical strategy dealing with both breasts during the same session. A few years ago, we introduced the concept of *mirror quadrantectomy*, which is now developing into a larger project, the aim of which is to extend the resection of the affected breast, reducing the other one accordingly. With this new strategy, breast conservation may also be applied to patients with tumors up to 5 cm in diameter, provided that the breast is not small. In the latter case, a mastectomy with immediate reconstruction and contralateral remodeling will become the treatment of choice [5].

The surgical treatment of the axillary nodes is similar. As axillary dissection is considered more a staging procedure than a therapeutic one [6], it appears that if reliable prognostic information can be obtained at the examination of primary carcinoma, axillary dissection may be avoided in patients with clinically negative nodes. In my opinion it is not difficult to predict that, in the future, axillary dissection will be limited to patients with clinically positive nodes. A trial is in progress to test the validity of this hypothesis and, in particular, to evaluate the clinical incidence of axillary metastases in an untreated, clinically negative axilla and the extent of the protection of axillary prophylactic irradiation, while adjuvant treatments will be decided on the basis of prognostic factors derived from primary carcinoma alone (estrogen receptors, grading, proliferative rate).

Although the trend is towards an increasing number of early cases, there is no doubt that many cases of breast cancer of medium or large dimension will be seen for a considerable period of time. These cases will require special efforts to mobilize all our resources in terms of surgery, radiotherapy, and systemic treatments in order to reduce mortality to the lowest possible rates and to reduce, whenever possible, the extent of surgical mutilation. In this context, a new role may be played by preoperative chemotherapy, which represents a stimulating area of research. At present, preoperative (neoadjuvant) chemotherapy is indicated mainly with the aim of reducing the mass of primary carcinoma in order to expand the indications of conservative surgery [7]. Although this is now an important objective, it may become of limited relevance in the future due to the expected increase of detected small carcinomas. A second objective is the reduction of local recurrences in patients treated with breast conservation, an objective which, however, would imply the introduction of preoperative chemotherapy in patients with primary carcinomas of limited extent, which may represent a disputable indication. Finally, a third objective, which might become the most stimulating, is the use of primary chemotherapy as an informative test on chemosensitivity. The response rate of primary carcinoma to a specific type of chemotherapy may represent an important source of information to plan further chemotherapy with the same drugs or with different ones; pilot studies in this field are now in progress. Unexpected data on the importance of the timing of breast surgery has been obtained, whether in the follicular or in the luteal phase, indicating that this may be a factor that influences survival [8]. The available data are somewhat contradictory, and randomized trials are needed to clarify the issue.

The following new areas of research are very promising:

- Genetic studies, which may determine with very high precision the grade of individual susceptibility to breast cancer
- The identification of new drugs which may be useful as chemopreventive agents (toremifene, indole-3-carbinol)
- Research in the field of high-dose chemotherapy and the use of circulating progenitor cells
- Studies in progress to identify tests which may predict the chemosensitivity or the chemoresistance of different types of primary breast carcinoma
- The development of new chemotherapeutic and endocrine agents which will improve the efficacy of systemic treatments

References

1. La Vecchia C, Lucchini F, Negri E, Boyle P, Maisonneuve P, Levi F (1992) Trends of cancer mortality in Europe 1955–1989. III. Breast and genital sites. Eur J Cancer 28A(2/3):927–998

2. Doll R (1990) Are we winning the fight against cancer? An epidemiological assessment. Eur J Cancer 26(4):500–508
3. Jordan VC (1994) Long-term tamoxifen treatment for breast cancer. In: Jordon VC (ed) University of Wisconsin Press, London
4. Veronesi U, Volterrani F, Luini A, Saccozzi R, Del Vecchio M, Zucali R, Galimberti V, Rasponi A, Di Re E, Squicciarini P, Salvadori B (1990) Quadrantectomy versus lumpectomy for small size breast cancer. Eur J Cancer 26(6): 671–673
5. Petit JY, Rietjens M (1991) Deformities after conservative breast cancer treatment. In: Moon (ed) Breast reconstruction, vol 37, pp 455–466
6. Fisher B, Bauer M, Margolese R et al. (1985) Ten-year results of randomized clinical trial comparing radical mastectomy and total mastectomy with or without radiation. N Engl J Med 312:674–681
7. Bonadonna G, Veronesi U, Brambilla C et al. (1990) Primary chemotherapy to avoid mastectomy in tumors with diameters of three centimeters or more. JNCI 82:1539–1545
8. Veronesi U, Luini A, Mariani L, Del Vecchio M, Alvez D, Andreoli C, Giacobone A, Merson M, Pacetti G, Raselli R, Saccozzi R (1994) Effect of menstrual phase on surgical treatment of breast cancer. Lancet 343:1545–1547

I. Epidemiology and Genetics of Breast Cancer

Familial Breast Cancer

B.L. Weber

University of Pennsylvania Medical Center, 422 Curie Blvd., Room 1009,
Stellar-Chance Laboratories, Philadelphia, PA 19104, USA

Introduction

The first documentation of familial clustering of breast cancer in modern times was published in 1866 by a French surgeon who reported ten cases of breast cancer in four generations of his wife's family; four other women in this family died as a result of hepatic tumors (Brocca 1866). Given the strong influence of molecular genetics in medicine in recent years, there is a tendency to assume that familial clustering of disease results from genetically inherited predisposition. However, other explanations for familial clustering of breast cancer are possible and include: (1) geographically limited environmental exposure to carcinogens which might affect an extended family living in close proximity, (2) culturally motivated behavior that alters risk factor profile, such as age at first live birth, and (3) socioeconomic influences that, for example, might result in differing dietary exposures. In addition, multiple complex inherited genetic factors likely influence the extent to which a risk factor for breast cancer plays a role in any one individual; such modifying effects are likely to be shared among genetically similar members of an extended family. Although non-inherited factors certainly play a role in familial clustering of breast cancer, recent advances have provided unequivocal evidence for the presence of breast cancer susceptibility genes responsible for 5%–10% of all breast cancer. Yet much remains to be understood about the number of heritable factors involved, the function of such genes which, when functioning normally appear to prevent the development of breast cancer and the prevalence of such genetic mutations within various populations.

Inherited breast cancer has several distinctive clinical features–age of onset is considerably younger than sporadic cases, the prevalence of bilateral breast cancer is higher, and the presence of associated tumors in affected individuals is noted in some families. Associated tumors may include ovarian, colon, prostate and endometrial cancers and sarcomas (Nelson et al. 1993;

Anderson and Badzioch 1993). However, inherited breast cancer does not appear to be distinguished by histologic type, morphologic grade, metastatic pattern or survival characteristics. While much has been learned about inherited breast cancer in the last 5 years, much more will be elucidated by ongoing and future work based on the foundation of epidemiological and genetic data described below.

Epidemiological Studies

The first attempts to determine the influence of family history on breast cancer risk were published in the first half of this century (Lane-Clayton 1926; Wainwright 1931; Wassink 1935). In general, these studies were performed by comparing breast cancer incidence in relatives of breast cancer cases to healthy controls. Familial clustering was noted in these early studies, however, these studies were often flawed by unverified diagnoses, lack of rigorously defined control groups and the absence of adjustments for family size. These studies were followed by a number of investigations in which control groups were more carefully defined or that used the general population from which families were derived as control subjects. Although many of these studies also have methodological flaws, they consistently demonstrated a two- to three-fold increase in breast cancer risk in mothers and sisters of breast cancer patients, figures compatible with current studies (Sattin et al. 1985; Colditz et al. 1993; Slattery and Kerber 1993).

Using modern epidemiological methodology, several population-based studies have attempted to estimate breast cancer risk associated with a positive family history. The largest of these is a study conducted in Sweden, involving 1330 women with a confirmed diagnosis of breast cancer in a defined geographic region and 1330 age-matched controls without a previous diagnosis of breast cancer (Adami et al. 1981). Within this study cohort, breast cancer in a first degree relative was reported in 11.2% of breast cancer patients as opposed to 6.7% of controls ($p < 0.01$), yielding a standardized relative risk of 1.7. Relative risks of a similar magnitude were found in Canadian population-based study (Lubin et al. 1982) and the U.S. Nurses Health Study (Bain et al. 1980), a large retrospective case-control study. Subsequent studies attempted to stratify relative risk to women with an affected first degree relative based on the age at diagnosis and presence or absence of bilateral disease. By 1980 a significant body of evidence supporting the presence of inherited factors responsible for familial clustering of breast cancer had accumulated and efforts shifted to attempts to determine the inheritance pattern of breast cancer within these families. Several such studies provided evidence for an autosomal dominant breast cancer susceptibility gene with an age-related penetrance (Williams and Anderson 1984; Newman et al. 1988).

Breast Cancer Attributable to Inherited Genetic Factors

Given the overwhelming evidence that at least a portion of breast cancer is inherited, recent studies have focused on determining the frequency of breast cancer-related gene mutations in the general population and thus the percentage of breast cancer in the general population that is directly attributable to inherited genetic factors. Unfortunately, available data addressing this question are limited. Among 225 consecutive breast cancer patients seen at the Creighton University Oncology Clinic, 5% had family histories considered consistent with hereditary forms of breast cancer (Lynch et al. 1976). This percentage increased to 11.4% with a strong family history among women diagnosed before age 50 years. Two groups (Sattin et al. 1985; Claus et al. 1991) have analyzed data from the Cancer and Steroid Hormone (CASH) study, a large case-control study initiated in 1981 by the Centers for Disease Control. This data set contains information on 4730 histologically confirmed breast cancer cases between the ages of 20 and 54 from eight geographic region of the United States and 4688 age-matched controls from the same geographic regions. In this study group, 11% of breast cancer cases reported a first degree relative with breast cancer compared with 5% of controls (Sattin et al. 1985). Fewer than 1% of cases had both a mother and sister with breast cancer. Analysis of this data set provides evidence for the existence of breast cancer-related gene mutations in approximately 0.33% of the population (Claus et al. 1991). The effect of these mutations on breast cancer risk is a function of age; Claus estimated that 36% of breast cancer cases among women ages 20–29 may be attributed to an autosomal dominant susceptibility gene (two of which have now been identified with others sure to follow) with the attributable proportion of breast cancer decreasing to 1% by age 80.

In 1993, two additional studies attempting to calculate risk attributable to autosomal dominant breast cancer susceptibility genes were published. Colditz et al. (1993) analyzed data from the Nurses Health Study and suggested that 6% of all breast cancer occurred as a result of inherited genetic factors while Slattery and Kerber provided an estimate of 17%–19% based on the Utah Population Data Base (Slattery and Kerber 1993). Several explanations for the discrepancies have been noted (Weber and Garber 1993); these include the unique composition and large size of the families in the Utah data base and the small family size and absence of paternal data in the Nurses Health Study cohort. It appears likely that the Nurses Health Study data represent an underestimate while the Utah Population Data Base provides an overestimate of risk of breast cancer attributable to inherited genetic factors and the actual number lies somewhere in between these estimates.

In summary, current estimates place the percentage of breast cancer cases directly attributable to inherited factors between 5% and 10%. It is expected that the identification of the major breast cancer susceptibility genes based on ongoing investigations will eventually allow a direct measurement of the percentage of breast cancers resulting from inherited mutations in breast cancer-related genes.

The work described above provided unequivocal evidence for the importance of genetic factors in the development of breast cancer. Examination of numerous pedigrees has resulted in the recognition of a number of syndromes in which dominantly inherited breast cancer may be featured. The most frequently identified pedigrees contain site-specific breast cancer and are thought to represent the effect of mutations in a single breast cancer susceptibility gene; *BRCA1* and *BRCA2* are examples of such genes and others are likely to be identified in the future.

BRCA1

In 1990 King and colleagues localized *BRCA1*, the first major susceptibility locus for breast cancer, to chromosome 17q21 using genetic linkage analysis in families with a high incidence of early-onset breast cancer (Hall et al. 1990). *BRCA1* is also responsible for a syndrome of inherited susceptibility to both breast and ovarian cancer (Narod et al. 1991; Easton et al. 1993). *BRCA1* has now been isolated (Miki et al. 1994) and, as a result, direct screening for mutations has replaced indirect methods for identifying woment at risk. Inheritance of breast cancer in these families follows the classic Mendelian pattern of autosomal dominant transmission, with each child of mutation carriers having a 50% chance of inheriting a *BRCA1* mutation (Hall et al. 1990). Although estimates of the frequency of *BRCA1* germline mutations in the general population are difficult to determine with accuracy until population based studies can be initiated, it is possible that as many as 1 in 800 women may harbor *BRCA1* germline mutations and the associated increased risk of developing neoplastic disease (D. Easton, personal communication).

Female *BRCA1* mutation carriers are estimated to have an 85% lifetime risk of developing breast cancer and a 40%–60% lifetime risk of ovarian cancer (Ford et al. 1994). *BRCA1* mutation carriers also have an increased incidence of bilateral breast cancer. The Breast Cancer Linkage Consortium data suggest that the cumulative risk of developing a second breast cancer is 65% for mutations carriers by age 70 (Easton et al. 1994). It is estimated that 45% of families with apparent autosomal dominant transmission of breast cancer susceptibility, and approximately 90% of families with dominant inheritance of both breast and ovarian cancer, harbor *BRCA1* germline mutations. The percentage of breast cancer-only families that are attributed to *BRCA1* mutations rises to almost 70% if the median age of onset of breast cancer in the families is less than 45 years (Easton et al. 1993).

A recent study of 33 families with evidence of germline mutations in *BRCA1* was conducted by the Breast Cancer Linkage Consortium (Ford et al. 1994). The incidence of other cancers in individuals thought to harbor germline mutations in *BRCA1* was anlayzed; significant excesses were observed for prostate cancer, with an estimated relative risk of 3.33 for males thought to carry *BRCA1* germline mutations, and colon cancer, with an estimated relative risk to mutation carriers of 4.11. However, the ages at diagnosis of prostate and colon cancer were not particularly early as compared to the general population, as is the case for breast and ovarian cancer in these families. No significant excesses were observed for cancers originating from other anatomic sites. Specifically, male breast cancer does not appear to be a component tumor of this syndrome (Stratton et al. 1994).

After an intensive research effort, *BRCA1* was isolated by Skolnick and colleagues in late 1994 (Miki et al. 1994). *BRCA1* is a large gene, spanning approximately 100 kb of DNA and encoding a very large protein of 1863 amino acids. *BRCA1* is a novel gene with no recognized homology to previously cloned genes except a region of the 5'end which contains a RING finger motif found in several transcription factors. Thus it is expected that *BRCA1* will regulate the expression of other genes, but at present essentially nothing is known about the function of this important gene. The isolation of *BRCA1* makes possible the direct analysis of mutations in the germline of individuals from breast cancer families and other women with breast cancer. To date, published reports include analyses of 1086 unrelated women with a total of 80 mutations described, 38 of which represent invidual mutations so far described in only one kindred (Shattuck-Eidens et al. 1995). Some 80% of the mutations described so far result in the truncation of *BRCA1*, with the remainder being alterations in the DNA sequence that result in an amino acid substitution (missense mutations). Missense mutations in regions of the gene that are not clearly functionally important remain difficult to distinguish from silent polymorphisms. Effort are are underway to develop a predictive test for *BRCA1*; however, the size of the gene and the large number of individual mutations that have been described are significant obstacles in achieving this goal.

BRCA2

As noted above, only 45% of families in whom increased cancer susceptibility is limited to breast cancer appear to harbor mutations in *BRCA1* (Easton et al. 1993). A genome scan by members of the International Breast Cancer Linkage Consortium recently revealed the presence of a second breast cancer susceptibility locus, *BRCA2*, on the long arm of chromosome 13 (Wooster et al. 1994). The original analysis suggests that approximately 70% of breast cancer families that do not carry a *BRCA1* mutation harbor mutations in *BRCA2*, although subsequent work suggests that this may be an overestimate (unpublished data). Women who harbor germline mutations in *BRCA2* have an estimated 87% lifetime risk of developing breast cancer.

In contrast to *BRCA1* families, *BRCA2* families appear to have only a moderately increased incidence of ovarian cancer. Also in contrast to *BRCA1*, male carriers have a 6% lifetime risk of developing breast cancer, a dramatic 100- to 200-fold increased risk as compared to the general population (D. Easton, personal communication). Although direct mutation analysis will not be possible until *BRCA2* is isolated, it is possible to attempt linkage analysis, using genetic markers on chromosome 13, to determine whether *BRCA2* might be responsible for breast cancer in a given family. However, the same difficulties arise as with previous efforts to establish linkage to *BRCA1* in a family (Biesecker et al. 1993). These difficulties include the need for sampling at least four living affected individuals in a family, along with sufficient numbers of unaffected members to make inferences about inheritance patterns within the family. Unfortunately, identification of families with large numbers of living affected members is an infrequent event, the testing is labor-intensive and the analysis requires sophisticated statistical genetic methods.

Other Breast Cancer Predisposition Syndromes

At present, approximately 30% of breast cancer families fit the criteria described above for inherited mutations in an autosomal dominant breast cancer susceptibility gene but have no identifiable *BRCA1* mutation, and evidence for linkage to *BRCA2* is lacking. It is assumed that these families do in fact harbor mutations in similar predisposition genes which have yet to be localized or isolated. Work is currently underway to identify the chromosomal location of additional breast cancer susceptibility genes, which will add to our understanding of the genetics of breast cancer and add to the repertoire of breast cancer susceptibility genes that can be assayed for inherited mutations in women with a family history of breast cancer.

A number of other genetic mutations have been associated with predisposition to breast cancer, but account for a very small proportion of inherited breast cancer. These include: (1) Li-Fraumeni syndrome, resulting from inherited mutations in the *p53* tumor suppressor gene (Li and Fraumeni 1969; Malkin et al. 1990); (2) Cowden disease (multiple hamartoma syndrome); (3) Muir-Torre syndrome, resulting from inherited mutations in the *MSH2* or *MLH1* "mutator" genes (Hall et al. 1994); (4) Peutz-Jeghers Syndrome; and (4) ataxia-telangectasia. These syndromes are often recognizable as a result of other clinical manifestations, and with the possible exception of Li-Fraumeni syndrome account for a very small amount of breast cancer in the general population.

Counseling for Women with Inherited Susceptibility to Breast Cancer

Individuals identified as carriers of germline mutations in breast cancer susceptibility genes have specific needs for education, results disclosure and long-term follow-up that only now are beginning to be elucidated. Individuals in these groups include: (1) men and women with known *BRCA1* mutations, (2) men and women thought to harbor *BRCA2* mutations based on linkage between breast cancer and genetic markers flanking the *BRCA2* region on chromosome 13, (3) individuals in these families with evidence that they did *not* inherit the mutation segregating in their family, and (4) families in which the presence of such a gene appears likely but no molecular evidence exists, precluding identification of families members as mutation carriers and noncarriers. Table 1 provides risk estimates for each of these distinct groups.

Our experience to date with testing for inherited susceptibility to breast cancer suggests that there are several components to this service that are critical to providing support to families undergoing testing. At a minimum, these include: (1) pre-test counseling such that individuals are informed of the nature of the testing being performed, the possibility that no information will be gleaned from the testing, the lack of data to aid in decisions surrounding surveillance vs prophylactic surgery and the risks and benefits of testing before blood is submitted for testing; (2) a multidisciplinary approach to results disclosure that includes the ability to provide information about genetics, the testing process itself, breast and other cancers, options for long-term follow-up and the psychological support that will be needed; and (3) post-test follow-up that includes both medical surveillance and emotional support (Biesecker et al. 1993). Among the most difficult high risk families to counsel are those in whom the existence of a highly penetrant breast

Table 1. Breast cancer risk estimates for members of high risk families

Gene	Sex	Cancer risk[b]
BRCA1	Female	Breast cancer risk 87% Bilateral breast cancer risk 65% Ovarian cancer risk 40%–60% Colon cancer risk 6%
BRCA1	Male	Breast cancer risk negligible Colon and prostate cancer risk 6%–8%
BRCA2	Female	Breast cancer risk 87% Ovarian cancer risk 15%–20%[c]
BRCA2	Male	Breast cancer risk 6% Risk for other cancers has yet to be evaluated
Unknown	Female or Male	Unknown Risk of breast cancer will vary significantly between mutation carriers and noncarriers

[a] Individuals demonstrated to be mutation noncarriers have a cancer risk equivalent to the general population.
[b] Expressed as cumulative lifetime risk to age 80 (Claus et al. 1994).
[c] Personal communication, D. Easton.

cancer-related mutation is suspected but cannot be confirmed. This will include families tested for *BRCA1* mutations and *BRCA2* linkage who enter into testing with the belief that a "yes-or-no" answer will result. These families will require additional support and follow-up, with efforts to keep them informed of developments in the rapidly change field of breast cancer genetics.

Of particular importance is the need to dispel the notion of a 50% risk of developing breast cancer in women from families that have an undefined autosomal dominant pattern of breast cancer inheritance. The 50% risk estimate often offered to these women is actually the risk of inheriting a dominant mutation from a carrier parent if one exists in the family. In fact, none of these women has a 50% lifetime risk of developing breast cancer. With respect to *BRCA1* and *BRCA2*, the risk to mutation carriers is 85%–90% and the risk to noncarriers is that of the general population, between 9% and 11%. This distinction will often prove to be a very important one in counseling a woment from an uncharacterized high risk family who is considering prophylactic surgery. Finally, it must be recognized that, in the rapidly changing field of breast cancer genetics, approaches and recommendations may change as rapidly as the availability of testing and that all of these factors are likely to be heavily influenced by the appearance of commercial testing within the next few years. However, the need for pre-test education, informed consent, post-test counseling and long-term follow-up will remain unchanged regardless of the group providing technical services.

Familial Breast Cancer Not Associated with Major Susceptibility Loci

The large majority of women presenting to a physician for breast cancer risk assessment will not have a family history of breast cancer which is striking enough to suggest the presence of an inherited breast cancer predisposition syndrome. Family history in these situations typically reveals the presence of one or two relatives with breast cancer (often postmenopausal) and an absence of characteristic non-breast tumors. The increased incidence of breast cancer in this group of families is probably heterogeneous at both the molecular and clinical level and may be explained by at least four models. Breast cancer in these families may result from the combined effects of multiple genetic components and/or environmental agents each conferring a small degree of increased risk. This pattern of breast cancer in a family also could be explained by mutations in a single dominant susceptibility gene with low penetrance. This would result in a probability of developing breast cancer which is greater than the general population, but which does not approach the risk seen with highly penetrant genetic mutations such as those in *BRCA1* and *BRCA2*. A third hypothesis explaining the increased incidence of breast cancer in these families is an inherited breast cancer predisposition syndrome, but due to a small kindred or incomplete family history information, the family history does not appear striking. Finally, breast cancer is a common disease and may occur more than once in a large family unrelated to inherited factors.

Despite the lack of a molecular basis for understanding the clustering of breast cancer in families without evidence for muations in dominant breast cancer susceptibility genes, assessment of an individual's risk for developing breast cancer can be undertaken using data from several prediction models based on analysis of large epidemiological studies. Offit and Brown (1994) recently reviewed four major models of risk prediction based on family history and provided a comparison of the different models. Each of the models is derived using a different study design and employs differing risk factors to calculate risk. Thus, risk estimates for a given individual derived from each of the models will differ slightly. Nonetheless, these estimates serve as useful guidelines for counseling patients until more refined predictions based on molecular characterization becomes available.

One of the two most commonly employed models for estimating cancer risk derives from the CASH (Cancer and Steroid Hormone) study data set, as recently analyzed by Claus et al. (1994). The Claus model was derived specifically to predict risk for women with a family history of breast cancer. This analysis allowed construction of detailed tables that predict the cumulative risk of breast cancer at specific ages based on age at onset of various combinations of affected first and second degree relatives (Table 2). An advantage of this approach is that it presents cumulative risks, and it incorporates the age of the affected relative at the time of her diagnosis into the risk calculation. Nonetheless, this model may significantly underestimate the risk of developing breast cancer in a *BRCA1* or *BRCA2* mutation carrier

Table 2. Breast cancer risk estimates based on age and number of affected relatives

Affected relative	Age of affected relative (year)	Cumulative breast cancer risk by age 80 (%)
One first degree	<50	13–21
	>50	9–11
One second degree	<50	10–14
	>50	8–9
Two first degree	Both <50	35–48
	Both >50	11–24
Two second degree[b]	Both <50	21–26
	Both >50	9–16

Risk estimates are derived by including age extremes from the risk tables calculated by Claus et al. (1994). For example, for affected relatives below the age of 50, the lower limit is the calculated risk if the affected relative is in the 40–49 age group and the upper limit is the calculated risk for a relative in the 20–29 age group. Thus, these figures represent the range of risk based on age and are not confidence intervals.

[b] Both paternal or both maternal.

and overestimate breast cancer risk for a woman in such a family who has not inherited a disease-related mutation.

The Gail model (Gail et al. 1989) calculates breast cancer risk using a formula which considers age at menarche, age at first live birth, number of first degree relatives with breast cancer, number of previous breast biopsies and age as pertinent risk factors. This model predicts a cumulative risk from age 20 to age 80 and corrects for other causes of mortality. The Gail model has been used to calculate risk for women considering entry into the NSABP Breast Cancer Chemoprevention Trial. Like the Claus model, the Gail model may underestimate risk for *BRCA1* and *BRCA2* mutation carriers and may overestimate risk for their noncarrier relatives. in addition, the Gail model allows other risk factors to significantly modify breast cancer risk in the presence of a family history, which is currently unsubstantiated. In addition, this model does not take age of diagnosis of breast cancer in a relative into consideration, nor does it consider the presence of breast cancer in second degree relatives, negating any contribution to risk resulting from genetic mutations in paternal relatives. As such, the Gail model may be most accurate when used to predict risk for without a striking family history of breast cancer (Bondy et al. 1994).

Summary

Tremendous progress has been made in recent years in understanding the genetic basis of breast cancer. Epidemiological models have been devised which allow risk estimation for women with one or two relatives with breast cancer. The existence of two autosomal dominant breast cancer susceptibility genes (*BRCA1* and *BRCA2*) have been documented. *BRCA1* has been isolated and we are beginning to catalog *BRCA1* mutations which appear to be etiologically related to the development of breast cancer in certain families. These *BRCA1* mutation studies and the isolation of *BRCA2* will allow development of accurate DNA-based testing for the prediction of individual risk in women with a family history of breast cancer. Work by several groups is progressing rapidly on the isolation *BRCA2*. Ongoing studies investigating the function of *BRCA1*, and eventually *BRCA2*, will lead to a better understanding of the role these important genes normally play in preventing the development of breast cancer. Finally, additional breast cancer susceptibility genes are likely to be identified, further expanding the number of tools we have to study, and eventually prevent, breast cancer.

References

Adami HO, Hansen J, Jung B, Rimsten A (1981) Characteristics of familial breast cancer in Sweden; absence of relation to age and unilateral versus bilateral disease. Cancer 48:1688–1695

Anderson DE, Badzioch MD (1993) Familial breast cancer risks – effects of prostate and other cancers. Cancer 72:114
Bain C, Speizer FE, Rosner B et al. (1980) Family history of breast cancer as a risk indicator for the disease. Am J Epidemiol 111:301–308
Biesecker BB, Boehnke M, Calzone K et al. (1993) Genetic counseling for families with inherited susceptibility to breast and ovarian cancer. JAMA 269:1970–1974
Bondy ML, Lustbader ED, halabi S et al. (1994) Validation of a breast cancer risk assessment model in women with a positive family history. J Natl Cancer Inst 86:620–625
Brocca P (1866) Traite de tumeurs. Asselin, Paris
Claus EB, Risch N, Thompson WD (1991) Genetic analysis of breast cancer in the cancer and steroid hormone study. Am J Hum Genet 48:232–242
Colditz GA, Willett WC, Hunter DJ et al. (1993) Family history, age, and the risk of breast cancer. JAMA 270:338–343
Easton DF, Bishop DT, Ford D et al. (1993) Genetic linkage analysis in familial breast and ovarian cancer-results from 214 families. Am J Hum Genet 52: 678–701
Easton DF, Narod SA, Ford D, Steel M (1994) The genetic epidemiology of BRCA1. Lancet 344:761
Ford D, Easton DF, Bishop DT et al. (1994) Risk of cancer in BRCA1 mutation carriers. Lancet 343:692–695
Gail MH, Brinton LA, Byar DP et al. (1989) Projecting individualized probability of developing breast cancer for white females who are being examined annually. J Natl Cancer Inst 81:1879–1886
Hall JM, Lee MK, Newman B et al. (1990) Linkage of early onset breast cancer to chromosome 17q21. Science 250:1684–1689
Hall NR, Murday VA, Chapman P et al. (1994) Genetic linkage in Muir–Torre syndrome to the same chromosomal region as cancer family syndrome. Eur J Cancer 30A:180–182
Lane-Clayton JE (1926) A further report on cancer of the breast, with special reference to its associated antecedent conditions. In: Rep Min Health No. 32, HM Stationery Office, London
Li FP, Fraumeni JF (1969) Soft-tissue sarcomas, breast cancer, and other neoplasms: a familial syndrome? Ann Intern Med 71:747–752
Lubin JH, Burns PE, Blot WJ et al. (1982) Risk factors for breast cancer in women in Northern Alberta, Canada, as related to age at diagnosis. J Natl Cancer Inst 68:211–217
Lynch HT, Guirgis HA, Brodkey F et al. (1976) Genetic heterogeneity and familial carcinoma of the breast. Surg Gyn Obstet 142:693–699
Malkin D, Li FP, Strong LC et al. (1990) Germline p53 mutations in a familial syndrome of breast cancer, sarcomas, and other neoplasms. Science 250:1233–1238
Miki Y, Swensew J, Shattuck-Eidens D et al. (1994) A strong candidate for the breast and ovarian cancer susceptibility gene BRCA1. Science 266:66–71
Narod SA, Feuteun J, Lynch HT et al. (1991) Familial breast–ovarian cancer locus on chromosome 17q12–23. Lancet 338:82–83
Nelson CL, Sellers TA, Richs SS, Potter JD et al. (1993) Familial clustering of colon, breast, uterine, and ovarian cancers as assessed by family history. Genet Epidem 10:235–244
Newman B, Austin MA, Lee M, King MC (1988) Inheritance of breast cancer: evidence for autosomal dominant transmission in high risk families. Proc Natl Acad Sci USA 85:3044–3048
Offit K, Brown K (1994) Quantitation of familial cancer risk: A resource for clinical oncologists. J Clin Oncol 12:1724–1736
Sattin RW, Rubin GL, Webster LA et al. (1985) Family history and the risk of breast cancer. JAMA 253:1908–1913

Shattuck-Eidens D, McClure M, Simard J et al. (1995) A collaborative survey of 80 mutations in the BRCA1 breast and ovarian cancer susceptibility gene. JAMA 273:535–541

Slattery ML, Kerber RA (1993) A comprehensive evaluation of family history and breast cancer risk: the Utah Population Database. JAMA 270:1563–1568

Stratton MR, Ford D, Neuhausen S (1994) Familial male breast cancer is not linked to the BRCA1 locus on chromosome 17q. Nat Genet 7:103

Wainwright JM (1931) Comparison of conditions asociated with breast cancer in Great Britain and America. Am J Cancer 15:2610–2645

Wassink WF (1935) Cancer et heredite. Genetika 17:103–144

Weber BL, Garber JE (1993) Family history and breast cancer: probabilities and possibilities. JAMA 270:1602–1603

Williams WR, Anderson DE (1984) Segregation analysis of 200 Danish pedigrees. Genet Epidemiol 1:7–20

Wooster R, Neuhausen S, Mangion J et al. (1994) Localization of a breast cancer susceptibility gene, BRCA2, to chromosome 13q-12-13. Science 265:2088–2090

The Food-Derived Heterocyclic Amines and Breast Cancer: A 1995 Perspective

E.G. Synderwine

Laboratory of Experimental Carcinogenesis, National Cancer Institute, Bldg 37, Room 3C 28 Bethesda, MD 20892-4255, USA

Introduction

It is generally recognized that environmental factors influence the incidence of human cancers. Many of these environmental factors are associated with personal lifestyle habits and include smoking, exposure to sunlight, intake of alcohol, and diet (Wynder and Gori 1977). The single most important environmental factor influencing the incidence of human cancer is diet. Upwards of 40% of cancer incidence in males and 60% in females may be associated with diet alone (Wynder and Gori 1977).

Mammary gland cancer in women is an example of one human cancer that appears to be associated with lifestyle factors, especially diet (Wynder and Gori 1977; IARC 1987; Kelsey and Horn-Ross 1993; Wynder et al. 1991; Weisburger 1992). This opinion is supported the finding that the age-adjusted breast cancer incidence in women varies significantly among different countries and by the finding that the offspring of migrants acquire the incidence rates of their new host country. The mammary cancer incidence rates are four- to five-fold higher in the U.S. than in Japan, with the largest differences being observed in the incidence of postmenopausal breast cancer (IARC 1987; Kelsey and Horn-Ross 1993; Wynder et al. 1991). There are striking dietary differences between Americans and Japanese who consume their traditional diet (Weisburger 1992). In the western world, diet is high in fat, rich in calories and high in cooked meats. Dietary fat constitutes about 30%–40% of total calories on average in the U.S. In contrast, with the traditional Japanese diet, carbohydrates are the major source of calories, and dietary fat constitutes about 15% of total calories. In urban areas of Japan, breast cancer rates are now rising, and this rise appears to correspond to a westernization of the Japanese diet (Weisburger 1992).

How diet and nutrition can influence carcinogenesis is a complex question. In the scheme of the multistage model of carcinogenesis, breast cancer may develop in a stepwise fashion from initiation ensuing from mutagenesis in the target cell, promotion to preneoplastic cells, and progression to the malignant phenotype. Various components in the diet might effect the carcinogenic process at different stages (Wattenberg 1990). For example, the diet is a source of carcinogens that form DNA adducts, which if not repaired might lead to mutagenesis and initiation of carcinogenesis. Dietary fat might promote carcinogenesis by several mechanisms including altering hormone levels and responsiveness, altering intercellular communication, and generating lipid peroxy and/or oxygen radicals (Welsch 1985). The diet also contains chemopreventive factors that might block carcinogenesis at several steps by inhibiting uptake (e.g., fiber and chlorophyll) and metabolic activation (e.g., chlorophyll and isoflavonoids) or enhancing detoxification (e.g., indole-3-carbinol) of dietary carcinogens (Wattenberg 1990). Other dietary protective factors such as isoflavonoids (genistein) and B-carotene might inhibit promotion and progression of initiated cells (Wattenberg 1990; Messina et al. 1994). There is growing evidence to support an influence of dietary factors on endogenous hormones that may affect mammary carcinogenesis (Wattenberg 1990; Messina et al. 1994; Michnovicz and Bradlow 1990; Telang et al. 1992; Adlercreutz 1990). For example, dietary fat (Welsh 1985; Adlercreutz 1990), and certain phytochemicals such as indole-3-carbinol (Michnovicz and Bradlow 1990; Telang et al. 1992) and phytoestrogens such as lignans and isoflavonoids (Messina et al. 1994; Adlercreutz 1990) may in part mediate their effects on mammary carcinogenesis by modifying endogenous estrogen levels and estrogen metabolism.

It has been estimated that only about 25% of the patients with breast cancer present with the conventional risk factors such as family and personal history of the disease, and fibrocystic and proliferative breast lesions (Telang et al. 1992). Thus, the etiology of human breast cancer is largely unknown. The large gap in knowledge of breast cancer etiology has prompted the search for causitive agents in this disease, particularly in the realm of diet. With the recognition that etiological factors in human mammary gland cancer may be numerous, this short review will focus on just one class of compounds, the food-derived heterocyclic amines.

Formation, Dietary Levels, and Animal Carcinogenicity of Heterocyclic Amines

The heterocyclic amines comprise a large family of some 20 compounds that have been identified as pyrolysates of amino acids and proteins or purified from cooked meats including beef, chicken, pork, and fish (Wakabayashi et al. 1993; Layton et al. 1995; Felton and Knize 1990; Felton et al. 1986). The five most prevalent of the heterocyclic amines found in the western diet are, in order of approximate descending prevalence: 2-amino-1-methyl-6-phenylimidazo[4,5-*b*] pyridine (PhIP), 2-amino-*9H*-pyrido[2,3-*b*]-indole A(α)C, 2-amino-3,8-dimethylimidazo[4,5-*f*]quinoxaline (8-MeIQx),

2-amino-3,4,8-trimethylimidazo[4,5-*f*]quinoxaline (4,8-DiMeIQx), and 2-amino-3-methylimidazo[4,5-*f*] quinoxaline (IQ) (Layton et al. 1995). These heterocyclic amines are formed in meats cooked at ordinary cooking temperatures by several conventional methods such as broiling, baking, barbecuing, and frying (Wakabayashi et al. 1993; Layton et al. 1995; Felton and Knize 1990; Felton et al. 1986). When meat is cooked it becomes mutagenic in large part owing to the formation of the highly mutagenic heterocyclic amines formed from precursors in the meat. The two most critical precursors of heterocyclic amines in meats are creatine and amino acids. After cooking, approximate concentrations of dietary heterocyclic amines in cooked meats range in the parts per billion (ppb) level, depending on the temperature, method, and duration of cooking. In a well-done fried hamburger, by mass, the most prevalent heterocyclic amine is PhIP (Felton et al. 1986). In one study, about 15 μg PhIP/kg fried hamburger was detected (Felton et al. 1986).

All heterocyclic amines tested to date are carcinogenic to rodents (Ohgaki et al. 1991; Ito et al. 1991; Hasegawa et al. 1993; Kato et al. 1989; Tanaka et al. 1985; Weisburger et al. 1994; Ghoshal et al. 1994; El-Bayoumy et al. 1995). The majority of these carcinogens induce multiple primary tumors when chronically fed to rats or mice at parts per million (ppm) levels. IQ is also a strong hepatocarcinogen in cynomolgus monkeys (Adamson et al. 1990). In one study, IQ induced tumors at several sites including the liver, colon, small intestine, Zymbal's gland, clitoral gland, and skin when given in the diet to Fischer-344 rats (for review see Ohgaki et al. 1991). Other studies have shown that IQ and its structurally related quinoline MeIQ also induce mammary gland cancer in female rats when given chronically in the diet or by repeated gavage (Kato et al. 1989; Tanaka et al. 1985; Weisburger et al. 1994). PhIP, in contrast to IQ, has not been shown to be a liver carcinogen in either mice or rats (Ohgaki et al. 1991; Ito et al. 1991), and PhIP carcinogenicity studies in cynomolgus monkeys are still underway. Notably, PhIP induced only mammary gland and colon tumors in Fischer-344 rats (Hasegawa et al. 1993; Ito et al. 1991). These are the two target sites that might be associated with a carcinogen found in the diet of the western world, where the incidences of breast and colon cancer are highest (IARC 1987; Weisburger 1992; Adlercreutz 1990).

Dietary Fat and Heterocyclic Amine Carcinogenicity

Although the relationship between dietary fat intake and human breast cancer incidence is a subject of some controversy (Hunter and Willett 1994), the results from several epidemiological studies are weighted in favor of the association between dietary fat intake and human breast cancer risk (Carroll et al. 1968; Carroll 1992; Howe et al. 1991). International correlation studies indicate a strong, direct, and linear relationship between the percent of calories consumed as dietary fat and mortality from breast cancer among different countries (Carroll et al. 1968; Carroll 1992). In addition, a large cohort study in Canada reported a positive correlation between breast cancer incidence and fat consumption (Howe et al. 1991).

Numerous experimental studies have demonstrated an association between calorie-rich high fat diets and mammary gland cancer in animals (Ip 1987; Hopkins and Carroll 1979; Chan and Dao 1981). High fat diets fed ad libitum increase the incidence and shorten the latency of mammary gland tumors in female rats exposed to potent mammary gland carcinogens such as 7,12-dimethylbenz[*a*]anthracene (DMBA) and *N*-nitrosomethylurea (NMU) (Hopkins and Carroll 1979; Chan and Dao 1981). In addition, a high fat diet enhanced mammary gland tumorigenesis in transgenic mice carrying mouse mammary tumor virus (MMTV)/v-Ha-ras (DeWille et al. 1993). Generally, diets rich in polyunsaturated fatty acid (e.g., corn oil) have a stronger promotional effect on experimental mammary tumorigenesis than diets rich in saturated fats. However, a diet rich in saturated fatty acids, as found in lard and beef tallow, have an equivalent effect on enhancing tumorigenesis if a minimum of 4.4% of the polyunsaturated fatty acid linoleic acid is also included in the diet (Ip 1987; Hopkin and Carroll 1979).

In a typical western diet, it is likely that individuals consuming a high fat diet are exposed concomitantly to heterocyclic amines, for example, by eating fried hamburgers. In order to better model the human situation, recent studies have examined the mammary tumor incidence in rats treated with IQ or PhIP that were maintained on a high fat diet (Weisburger et al. 1994; Ghoshal et al. 1994; El-Bayoumy et al. 1995). A high fat diet was found to enhance the incidence of mammary tumors induced by IQ given chronically to female rats (Weisburger et al. 1994). We recently observed that a diet rich in linoleic acid (specifically a 23.5% corn oil diet) enhanced the incidence and severity of mammary gland tumors induced by PhIP in female Sprague-Dawley rats (Ghoshal et al. 1994). In rats given ten doses of PhIP (75 mg/kg, po) and subsequently maintained on a defined low fat diet (5% corn oil), mammary tumors were largely benign lesions. In contrast, in rats given PhIP and then maintained on the high fat diet, the majority of mammary gland tumors were carcinomas. It appeared that PhIP in combination with dietary fat participated in the stepwise progression of the normal breast epithelium to the precancerous state, carcinoma in situ, and to infiltrating duct carcinoma. PhIP alone, however, appeared insufficient in causing the malignant phenotypes observed in the mammary gland in our model. These findings are also consistent with the notion that PhIP might serve as an initiating agent in mammary gland cancer and that the initiated cells are subsequently promoted by dietary fat.

Mechanism for the Initiation of Mammary Gland Cancer by PhIP

PhIP is a procarcinogen that requires metabolic activation in order to be genotoxic (Felton and Knize 1990; Snyderwine et al. 1992; Buonarati et al.

1990). The metabolic activation occurs by a two-step process requiring cytochrome P450-mediated N-hydroxylation and subsequent phase II esterification to reactive *N*-acetoxy, *N*-sulfate, and other ester derivatives. These reactive esters form adducts in DNA via the putative arylnitrenium ion. The reactive metabolites of PhIP, like those of several other food-derived heterocyclic amines, form a major adduct at the C8-position of guanine (Snyderwine et al. 1988; Frandsen et al. 1992). Mutation studies using the supF shuttle vector and the endogenous dihydrotolate reductase (DHFR) gene of Chinese hamster ovary cells indicate that PhIP-guanine adducts cause predominantly G to T transversion mutations, and to a lesser extent G to A transition mutations (Carothers et al. 1994; Endo et al. 1994). In vitro studies also indicate that these adducts inhibit gene expression (Fan and Snyderwine 1994), an effect that might have consequences for carcinogenesis.

In female rats administered PhIP, PhIP-DNA adducts are detectable in mammary epithelial cells, the putative target site for mammary carcinogenesis. After ten daily doses of PhIP (75 mg/kg), a dose regimen that induced mammary gland cancer in rats, adducts persisted in the mammary epithelial cells for at least 5 weeks (A. Ghoshal and E.G. Snyderwine, manuscript in preparation). Since PhIP adducts are mutagenic in vitro, it is likely that PhIP adducts in the mammary gland are responsible for the initiation of mammary carcinogenesis in vivo (Felton and Knize 1990; Endo et al. 1994). How PhIP-DNA adducts arise in the mammary epithelium in rats and whether PhIP metabolism in the mammary gland contributes to PhIP-DNA adduct formation are under study in our laboratory. In the rat, the mammary gland has a very low capacity to carry out the first step in the metabolic activation of PhIP, specifically cytochrome P-450-mediated N-hydroxylation. By contrast, the mammary epithelium appears to have a good capacity to further activate N-hydroxy-PhIP, apparently via O-acetyltransferase. In an in vitro assay, the O-acetyltransferase activation in rat mammary gland cytosols was approximately ten-fold higher than that seen in rat liver cytosols. One possible scenario for the formation of adducts in the mammary gland therefore is N-hydroxylation in the liver and transport of the N-hydroxylamine by the bloodstream to the mammary gland where the high O-acetyltransferase activity in the mammary gland would further activate the carcinogen in situ leading to DNA adduct formation.

Recent studies have examined the mutations in various genes including ras and p53 in tumors induced in rodents or monkeys by the heterocyclic amines (Ushijima et al. 1994, 1995; Makino et al. 1992; Fujimoto et al. 1994; Kudo et al. 1991; Takahashi et al. 1993; Kakiuchi et al. 1993, 1995; Herzog et al. 1993; Canzian et al. 1994). In parallel with in vitro studies (Endo et al. 1994), many of the mutations found in heterocyclic amine-induced tumors appear to be associated with the guanine nucleotide, a finding that provides circumstantial evidence for the involvement of PhIP-guanine adducts in tumorigenesis. To date however, the principal genes harboring mutations in mammary gland tumors induced by PhIP in rats remain elusive (Ushijima et al. 1994).

Epilogue

Humans are continuously exposed to low levels of carcinogenic agents from diet and environment, and thus exposure to promotional factors, such as dietary fat, might have a pivotal influence on human carcinogenesis. Several epidemiological studies support a link between cooked meat consumption and breast cancer risk (Iscovich et al. 1989; Matos et al. 1991; Knekt et al. 1994). In view of animal studies showing a strong enhancing effect of dietary fat on PhIP-induced mammary tumors (Ghoshal et al. 1994), further epidemiology studies are needed to assess if the combination of heterocyclic amine exposure and a high fat diet is a risk factor for human breast cancer. Since metabolic activation is critical for the carcinogenicity of PhIP and other heterocyclic amines, metabolic phenotyping might also help to determine whether certain individuals are at higher risk of breast cancer from heterocyclic amine exposure. Analogously, recent studies have supported a higher risk for colon cancer among individuals eating well done cooked meats who show rapid acetylator phenotype and elevated cytochrome P-4501A2 activity (Lang et al. 1994). Various methods for reducing heterocyclic amines from the diet have been described (Adamson 1990) and include, for example, premicrowaving meat and discarding the juices prior to frying or barbecuing. In light of the substantial influence of diet on human cancer incidence (Wynder and Gori 1977), it is nevertheless encouraging that cancer incidence might be significantly reduced soley by modifying diet.

Acknowledgement. The author thanks Drs. Richard H. Adamson and Snorri S. Thorgeirsson for support and helpful discussion.

References

Adamson RH (1990) Mutagens and carcinogens formed during cooking of foods and methods to minimize their formation. In: DiVita VT Jr, Hellman S, Rosenberg SA (eds) Cancer prevention. Lippincott, Philadelphia, pp 1–7

Adamson RH, Thorgeirsson UP, Snyderwine EG, Thorgeirsson SS, Reeves J, Dalgard DW, Takayama S, Sugimura T (1990) Carcinogenicity of 2-amino-3-methylimidazo[4,5-f]quinoline in nonhuman primates: induction of tumors in three macaques. Jpn J Cancer Res 81:10–14

Adlercreutz H (1990) Western diet and western diseases: some hormonal and biochemical mechanisms and associations. Scand J Clin Lab Invest 50 [Suppl 201]:3–23

Buonarati MH, Turteltaub KW, Shen NH, Felton JS (1990) Role of sulfation and acetylation in the activation of 2-hydroxyamino-1-methyl-6-phenylimidazo[4,5-b]pyridine to intermediates which bind to DNA. Mutat Res 245:185–190

Canzian F, Ushijima T, Serikawa T, Wakabayashi K, Sugimura T, Nagao M (1994) Instability of microsatellites in rat colon tumors induced by heterocyclic amines. Cancer Res 54:6315–6317

Carothers AM, Yuan W, Hingerty BE, Broyde S, Grunberger D, Snyderwine EG (1994) Mutation and repair induced by the carcinogen 2-hydroxyamino-1-methyl-6-phenylimidazo[4,5-b]pyridine (N-hydroxy-PhIP) in the Chinese hamster ovary dihydrofolate reductase gene and conformational modeling of the dG-C8-PhIP adduct in DNA. Chem Res Toxicol 7:209–218

Carroll KK (1992) Review: dietary fat and breast cancer. Lipids 27:793–797

Carroll KK, Gammal EB, Plunkett ER (1968) Dietary fat and mammary cancer. Can Med Assoc J 98:590–594

Chan PC, Dao TL (1981) Enhancement of mammary carcinogenesis by a high-fat diet in Fischer, Long-Evans, and Sprague-Dawley rats. Cancer Res 41:164–167

DeWille JW, Waddell K, Steinmeyer C, Farmer SJ (1993) Dietary fat promotes mammary tumorigenesis in MMTV/v-Ha-ras transgenic mice. Cancer Lett 69:59–66

El-Bayoumy K, Chae Y-H, Upadhyaya P, Rivenson A, Kurtzke C, Reddy B, Hecht SS (1995) Comparative tumorigenicity of benzo[a]pyrene, 1-nitropyrene and 2-amino-1-methyl-6-phenylimidazo[4,5-b]pyridine administered by gavage to female CD rats. Carcinogenesis 16:431–434

Endo H, Schut HAJ, Snyderwine EG (1994) Mutagenic specificity of 2-amino-3-methylimidazo[4,5-f]quinoline and 2-amino-1-methyl-6-phenylimidazo[4,5-b]-pyridine in the supF shuttle vector system. Cancer Res 54:3745–3751

Fan L, Snyderwine EG (1994) Inhibition of plasmid reporter gene expression in CHO cells by DNA adducts of 2-amino-3-methylimidazo[4,5-f]quinoline and 2-amino-1-methyl-6-phenylimidazo[4,5-b]pyridine. Mol Carcinog 10:30–37

Felton JS, Knize MG (1990) Heterocyclic-amine mutagens/carcinogens in foods. In: Cooper CS, Grover PL (eds) Chemical carcinogenesis and mutagenesis I. Springer, Berlin Heidelberg New York, pp 471–502 (Handbook of experimental pharmacology, vol 94/I)

Felton JS, Knize MG, Shen NH, Andresen BD, Bjeldanes LF, Hatch FT (1986) Identification of the mutagens in cooked beef. Environ Health Perspect 67:17–24

Frandsen H, Grivas S, Andersson R, Dragsted L, Larsen LJ (1992) Reaction of the N^2-acetoxy derivative of 2-amino-1-methyl-6-phenylimidazo[4,5-b]pyridine (PhIP) with 2′deoxyguanosine and DNA. Synthesis and identification of N^2-(2′ deoxyguanosin-8-yl)-PhIP. Carcinogenesis 13:629–635

Fujimoto Y, Hampton LL, Snyderwine EG, Nagao M, Sugimura T, Adamson RH, Thorgeirsson SS (1994) P53 gene mutations in hepatocellular carcinoma induced by 2-amino-3-methylimidazo[4,5-f]quinoline in nonhuman primates. Jpn J Cancer Res 85:506–509

Ghoshal A, Preisegger K-H, Takayama S, Thorgeirsson SS, Snyderwine EG (1994) Induction of mammary tumors in female Sprague-Dawley rats by the food-derived carcinogen 2-amino-1-methyl-6-phenylimidazo[4,5-b]pyridine and effect of dietary fat. Carcinogenesis 15:2429–2433

Hasegawa R, Sano M, Tamano S, Imaida K, Shirai T, Nagao M, Sugimura T, Ito N (1993) Dose-dependence of 2-amino-1-methyl-6-phenylimidazo[4,5-b]pyridine (PhIP) carcinogenicity in rats. Carcinogenesis 14:2553–2557

Herzog CR, Schut HAJ, Maronpot RR, You M (1993) Ras mutations in 2-amino-3-methylimidazo[4,5-f]quinoline-induced tumors in the CDF_1 mouse. Mol Carcinog 8:202–207

Hopkins GJ, Carroll KK (1979) Relationship between amount and type of dietary fat in promotion of mammary carcinogenesis induced by 7,12-dimethylbenz[a]-anthracene. J Natl Cancer Inst 62:1009–1012

Howe GR, Friedenreich CM, Jain M, Miller AB (1991) A cohort study of fat intake and risk of breast cancer. J Natl Cancer Inst 83:336–340

Hunter DJ, Willett WC (1994) Diet, body build, and breast cancer. Annu Rev Nutr 14:393–418

International Agency for Research on Cancer (1987) Cancer incidence on five continents. IARC Sci Publ 5:882–883

Ip C (1987) Fat and essential fatty acid in mammary carcinogenesis. Am J Clin Nutr 45:218–224

Iscovich JM, Ischovich RB, Howe G, Shiboski S, Kaldor JM (1989) A case-control study of diet and breast cancer in Argentina. Int J Cancer 44:770–776

Ito N, Hasegawa R, Sano M, Tamano S, Esumi H, Takayama S, Sugimura T (1991) A new colon and mammary carcinogen in cooked food, 2-amino-1-methyl-6-phenylimidazo[4,5-b]pyridine (PhIP). Carcinogenesis 12:1503–1506

Kakiuchi H, Ushijima T, Ochiai M, Imai K, Ito N, Yachi A, Sugimura T, Nagao M (1993) Rare frequency of activation of the Ki-ras gene in rat colon tumors induced by heterocyclic amines: possible alternative mechanisms of human colon carcinogenesis. Mol Carcinog 8:44–48

Kakiuchi H, Watanabe M, Ushijima T, Toyota M, Imai K, Weisburger JH, Sugimura T, Nagao M (1995) Specific 5′-GGGA-3′ to 5′-GGA-3′ mutation of the APC gene in rat colon tumors induced by 2-amino-l-methyl-6-phenylimidazo[4,5-b]pyridine. Proc Natl Acad Sci USA 92:910–914

Kato T, Migita H, Ohgaki H, Sato S, Takayama S, Sugimura T (1989) Induction of tumors in the Zymbal gland, oral cavity, colon, skin, and mammary gland of F344 rats by a mutagenic compound, 2-amino-3,4-dimethylimidazo[4,5-f]quinoline. Carcinogenesis 10:601–603

Kelsey JP, Horn-Ross PL (1993) Breast cancer: magnitude of the problem and descriptive epidemiology. Epidemiol Rev 15:7–16

Knekt P, Steineck G, Jarvinen R, Hakulinen T, Aromaa A (1994) Intake of fried meat and risk of cancer: a follow-up study in Finland. Int J Cancer 59:756–760

Kudo M, Ogura T, Esumi H, Sugimura T (1994) Mutational activation of c-Ha-ras gene in squamous cell carcinomas of rat Zymbal gland induced by carcinogenic heterocyclic amines. Mol Carcinog 4:36–42

Lang NP, Butler MA, Massengill J, Lawson M, Stotts RC, Hauer-Jensen M, Kadlubar FF (1994) Rapid metabolic phenotypes for acetyltransferase and cytochrome P-4501A2 and putative exposure to food-borne heterocyclic amines increase the risk for colorectal cancer or polyps. Cancer Epidemiol Biomark Prev 3:675–682

Layton DW, Bogen KT, Knize MG, Hatch FT, Johnson VM, Felton JS (1995) Cancer risk of heterocyclic amines in cooked foods: an analysis and implications for research. Carcinogenesis 16:39–52

Makino H, Ishizaka Y, Tsujimoto A, Nakamura T, Onda M, Sugimura T, Nagao M (1992) Rat p53 gene mutations in primary Zymbal gland tumors induced by 2-amino-3-methylimidazo[4,5-f]quinoline, a food mutagen. Proc Natl Acad Sci USA 89:4850–4854

Matos EL, Thomas DB, Sobel N, Vuoto D (1991) Breast cancer in Argentina: case-control study with special reference to meat eating habits. Neoplasma 38:357–366

Messina MJ, Persky V, Setchell KDR, Barnes S (1994) Soy intake and cancer risk; a review of the in vitro and in vivo data. Nutr Cancer 21:113–131

Michnovicz JJ, Bradlow HL (1990) Induction of estradiol metabolism by dietary indole 3-carbinol in humans. J Natl Cancer Inst 82:947–949

Ohgaki H, Takayama S, Sugimura T (1991) Carcinogenicities of heterocyclic amines in cooked food. Mutat Res 259:399–410

Snyderwine EG, Roller PP, Adamson RH, Sato S, Thorgeirsson SS (1988) Reaction of N-hydroxylamine and N-acetoxy derivatives of 2-amino-3-methylimidazolo[4,5-f]quinoline with DNA. Synthesis and identification of N-(deoxyguanosin-8-yl)-IQ. Carcinogenesis 9:1061–1065

Snyderwine EG, Schut HAJ, Adamson RH, Thorgeirsson UP, Thorgeirsson SS (1992) Metabolic activation and genotoxicity of heterocyclic amines. Cancer Res [Suppl] 52:2099s–2102s

Takahashi M, Minamoto T, Sugimura T, Esumi H (1993) High frequency and low specificity of ras gene mutations in rat Zymbal's gland tumors induced by 2-amino-3-methylimidazo[4,5-f]quinoline. Carcinogenesis 14:1355–1357

Tanaka T, Barnes WS, William GM, Weisburger JH (1985) Multipotential carcinogenicity of the fried food mutagen 2-amino-3-methylimidazo[4,5-f]quinoline in rats. Jpn J Cancer Res 76:570–576

Telang NT, Bradlow HL, Osborne MP (1992) Molecular and endocrine biomarkers in non-involved breast: relevance to cancer chemoprevention. J Cell Biochem [Suppl] 16G:161–169

Ushijima T, Kakiuchi H, Makino H, Hasegawa R, Ishizaka Y, Hirai H, Yazaki Y, Ito N, Sugimura T, Nagao M (1994) Infrequent mutation of Ha-ras and p53 in rat mammary carcinomas induced by 2-amino-1-methyl-6-phenylimidazo[4,5-b]pyridine. Mol Carcinog 10:38–44

Ushijima T, Makino H, Okonogi H, Hosoya Y, Sugimura T, Nagao M (1995) Mutation, loss of heterozygosity, and recombination of the p53 gene in mouse forestomach tumors induced by 2-amino-3,4-dimethylimidazo[4,5-f]quinoline. Mol Carcinog 12:23–30

Wakabayashi K, Ushiyama H, Takahashi M, Nukaya H, Kim S-B, Hirose M, Ochiai M, Sugimura T, Nagao M (1993) Exposure to heterocyclic amines. Environ Health Perspect 99:129–133

Wattenberg LW (1990) Inhibition of carcinogenesis by naturally occurring and synthetic compounds. In: Kuroda Y, Shankel DM, Waters MD (eds) Antimutagenesis and anticarcinogenesis, mechanisms II. Plenum, New York, pp 155–160

Weisburger JH (1992) Mechanisms of macronutrient carcinogenesis. In: Micozzi MS, Moon TE (eds) Macronutrients: investigating their role in cancer. Dekker, New York, pp 3–31

Weisburger JH, Rivenson A, Hard GC, Zang E, Nagao M, Sugimura T (1994) Role of fat and calcium in cancer causation by food mutagens, heterocyclic amines. Proc Soc Exp Biol Med 205:347–352

Welsch CW (1985) Host factors affecting the growth of carcinogen-induced rat mammary carcinomas: a review and tribute to Charles Brenton Huggins. Cancer Res 45:3415–3443

Wynder EL, Gori GB (1977) Contribution of the environment to cancer incidence: an epidemiological exercise. J Natl Cancer Inst 58:825–832

Wynder EL, Fujita W, Harris RE, Hirayama T, Hiyama T (1991) Comparative epidemiology of cancer between the United States and Japan: a second look. Cancer 67:746–763

Menstrual Cycle Timing of Breast Cancer Resection

W.J.M. Hrushesky

Stratton Department of Veterans Affairs Medical Center, 113 Holland Avenue, Albany, NY 12208, USA

Intorduction

In 1836, Sir Astley Paston Cooper observed in his "Practice and Principles of Surgery" that advanced breast cancer regularly waxes and wanes during the patient's menstrual cycle. In 1896, G.T. Beaston, in The Lancet, demonstrated that advanced breast cancer can resolve following oophorectomy of the premenopausal patient. In 1952 and 1953, Huggins and Bergenstal and Luft, respectively, demonstrated tumor resolution following adrenal and hypophyseal resection in women with advanced breast cancer. In the 1960s, Jensen et al. (1967) initiated our growing understanding of the molecular basis of these intricate connections between host and cancer.

Because the fertility cycle constitutes a regularly recurrent, rhythmic, high amplitude fluctuation of hypothalamic, hypophyseal, adrenal and ovarian hormones that are relevant to the balance between a woman and her breast cancer, we decided to extend the observations of Cooper (1836). Experiments with Dr. Fisher's C_3HeB/FeJ breast cancer model demonstrated that the timing within the fertility cycle of potentially curative breast cancer resection determines the metastatic potential of that cancer and the ultimate curability of that mouse (Ratajczak et al. 1988). Follow-up observations associated the time of highest curability, "early luteal" phase, with the time of the cycle when splenocyte natural killer (NK) cell function and interleukin -2 (IL-2) production are greatest (Hrushesky et al. 1988). In October 1989, a similar window for safe breast cancer resection was retrospectively demonstrated in young cycling women with early stage breast cancer (Hrushesky et al. 1989). In our series, women resected in putative early luteal phase of the menstrual cycle have a fourfold better chance of surviving 15 years than those resected during or nearer to menses. At least six subsequent studies, on two continents, have confirmed our observations (Senie et al. 1991; Meyer 1991; Badwe et al. 1991; Ville et al. 1990, 1991; Saad 1992; Spratt et al. 1993) while aditional studies have reported no difference in outcome as a function of resection timing (Powles et al. 1989, 1991; Gelber and Goldhirsch 1989).

Here, the strengths and weaknesses of nine of the retrospective studies repated to date will be reviewed in order to help practitioners decide what to recommend to regularly cycling premenopausal women undergoing breast cancer surgery. A meta-analysis has also been employed to synthesize the data from individual published studies in which data are presented adequately to allow their inclusion.

Preclinical Results

The relationship between the estrous cycle in the mouse and the metastatic potential of a transplantable estrogen receptor-bearing mammary adenocarcinoma was studied first. The C_3HeB/FeJ mouse is an inbred animal that permits growth of a transplanted mammary tumor which originated spontaneously in an old, inbred, retired C_3H breeder mouse. The mammary adenocarcinoma used in this study has been previously shown to be relatively rich in estrogen receptor (Fisher et al. 1982). As is also the case in the human disease, when this mammary adenocarcinoma is surgically removed after several weeks of growth, not every mouse is cured and some die from subsequent metastases (Ratajczak et al. 1986). In an attempt to better understand the basis for this biological heterogeneity, we hypothesized that the stage of the estrous cycle of an individual animal, at the time of primary tumor resection, might influence mestastatic potential. Our results demonstrate that the timing within the estrous cycle of surgical removal of this tumor influences the expression of subsequent tumor cell metastasis (see Fig. 1).

Since the timing within the estrous cycle of surgical removal of a transplanted murine mammary tumor profoundly affects the frequency of pulmonary metastases, we investigated the potential role of the cellular immune response in this phenomenon by measuring splenic NK cell activity and IL-2 production in syngeneic tumor-free mice in each of four estrous stages. Estrous stage was determined by assessment of vaginal smear cellularity immediately prior to killing and spleen harvest. NK splenocyte cytotoxicity against a standard tumor cell target was assessed using a radiolabeled chromium release assay while IL-2 activity was determined in a bioassay utilizing the IL-2-dependent CTLL-2 cell line. A highly statistically significant difference in NK activity was found among the four estrous stages of sacrifice. Splenocyte IL-2 production during the cycle covaried with NK activity ($p < 0.01$). The stages within the featility cycle associated with lowest metastatic potential, roughly corresponding to the early luteal phase, correspond precisely with those of highest splenocyte NK activity (see also Fig. 1). These results indicate that important components of the cellular immune response vary rhythmically during the fertility cycle of the host. In

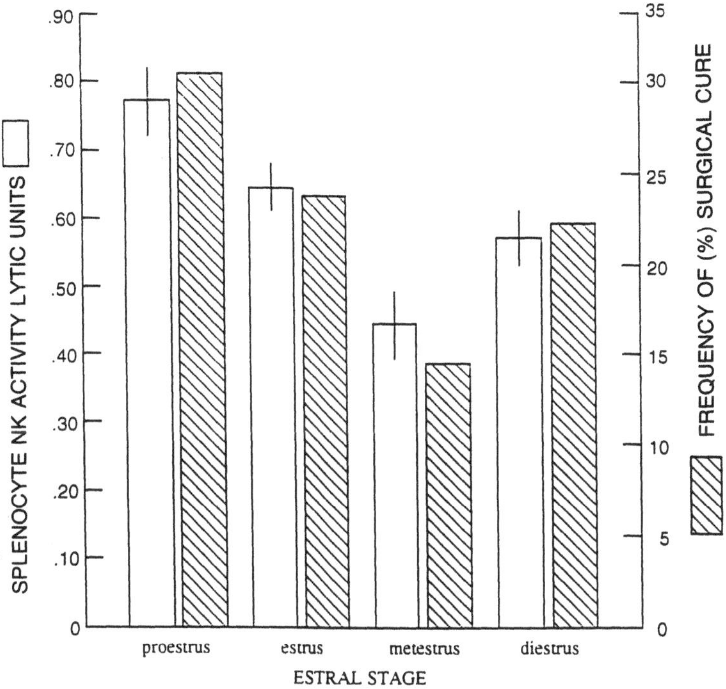

'ig. 1. Three large studies involving several hundred mice have revealed that surgical :urability varies with the estrus cycle phase of breast cancer resection. This figure :ontrasts the likelihood of surgical cure as a function of the estral stage of resection *cross-hatched*) with the average value (± standard error) for natural killer (*NK*) cell :ytotoxicity, estimated by chromium release assay from appropriate target cells after :xposure to splenocyte populations obtained after killing cycling mice at one of four :strous cycle stages. Estrus cycle stage was determined by vaginal cytology. The time)f greatest likelihood of curability coincides with the time within the cycle of highest plenocyte NK activity. This time of the cycle corresponds to the one quarter to one hird of the cycle associated with maximal murine fertility, following ovulation. This s roughly comparable to the early luteal phase of the menstrual cycle which is the ime of the menstrual cycle associated with the best surgical outcome (see text and Table 2)

his model, the rhythmic changes in NK activity may be, in part, responsible 'or the similarly rhythmic frequency of surgical cure. This possibility awaits 'urther study.

Clinical Results

Table 1 summarizes potentially important aspects of each study reported to late. The variables include relevant patient and treatment characteristics, he quality of the fertility cycle data in each study and the types of data ınalysis performed.

Table 1. Clinical characteristics of patients studied retrospectively to date

	N	Median follow-up (years)	Overall survival (10 years)	Minimal surgery	Oral contraceptive exclusion	Perimenopausal exclusion	Menstrual regularity	Menstrual phase balance	Menstrual phase subgroup comparability	Treatment within study	Data analysis
Hrushesky et al. (1989)	41	>10	82%	70%	Yes	Yes	Yes	Balanced	Yes	Consistent	Univariate and multivariate
Senie et al. (1991)	283	~10	74%	>50%	Yes	Yes	Yes	Balanced	Yes	Consistent	Univariate and multivariate
Meyer (1991)	53	>10	>70%	Unknown	Yes	Yes	Yes	Balanced	Yes	Consistent	Univariate
Badwe et al. (1991)	249	9	>70%	30%	Yes	Yes	Yes	Balanced	Yes	Consistent	Univariate and multivariate
Ville et al. (1990, 1991)	165[a]	~5	>80%	Unknown	Unknown	Yes	Yes	Balanced	Unknown	Unknown	Unknown
Saad (1992)	96	>10	60%	50%	Yes	Yes	Yes	Balanced	Unknown	Unknown	Univariate and multivariate
Spratt et al. (1993)	40	10	39%	Unknown	Yes	Yes	Unknown	Balanced	Unknown	Unknown	Univariate and multivariate
Powles et al. (1989, 1991)	81	11	55%	Unknown	No	No	Unknown	Balanced	Not tested	Unknown	Univariate
Gelber and Goldhirsch (1989)	245	9	65%	0%[b]	No	No	No[c]	Unbalanced	Not tested	Not consistent	Univariate

A total of 1253 premenopausal women have been included in these nine studies. Most of the patients were quite mature. Follow-up was a bit brief in Ville's series. The amount of surgery, e.g., lumpectomy vs mastectomy, varied widely from study to study and could contribute to variation in the findings. Oral contraceptive exclusion was a potentially important characteristic and was documented in seven of the nine studies. Exclusion of highly irregular and perimenopausal women was required if menstrual cycle stage was to be accurately assigned according to menstrual history; seven of the nine series did so. The balance of menstrual cycle phase stages indicates that irregularly cycling or noncycling women were probably been excluded. The Goldhirsch study is the most poorly balanced compared to the other seven studies. Menstrual phase subgroup comparability was thoroughly tested in each of the first four studies. These data are not evaluable in the last five studies. Consistency of treatment within study was reasonable in all but the final study. Multivariate data analysis was done in five of the nine studies and each of these analyses identified surgical timing as an independent predictor of outcome.

[a] Serum samples used to assign menstrual cycle phase.

[b] 100% modified radical mastectomy; 20% also had oophorectomy.

[c] Last menstrual period (LMP) range up to 326 days.

In a retrospective study of 41 premenopausal women who underwent resection of a primary breast cancer and were followed for 10–17 years, disease recurrence and metastasis were more frequent and appeared more rapidly in women who had been operated upon during the perimenstrual period (days 0–6 and 21–36) of the menstrual cycle. All irregularly cycling women and those taking oral contraceptives were excluded. Multivariate analysis demonstrated the time of resection in relation to the menstrual cycle proved to be an independent predictor of the likelihood of future metastatic disease. Patients who underwent resection during the perimentstrual span of their cycle had a quadrupled risk of recurrence and death compared with women operated upon during days 7–20 of the menstrual cycle. No deaths or recurrences occurred in women who had their resections during the first half of the luteal phase (13–20) of the cycle (Hrushesky et al. 1989).

Senie et al. have studied disease-free survival and time to death due to breast carcinoma in 283 consecutive premenopausal patients, with last menstrual period data available, treated by mastectomy and axillary dissection. After 10 years of follow-up, the rates of recurrence ($p < 0.03$) and death due to breast cancer ($p < 0.01$) were significantly lower for women treated during the luteal phase than during the follicular phase of their cycles. Survival differences were predominant for patients with axillary lymph node metastases ($p < 0.01$). Breast cancer deaths peaked among patients treated during the late follicular phase, between days 10 and 14 following onset of menses. After controlling for tumor size and number of positive axillary lymph nodes in multivariate analysis, the relative risk of cancer recurrence for patients resected in the luteal phase was two thirds that for women resected in the follicular phase, while the risk of breast cancer death was half as great (Senie et al. 1991).

Meyer (1991) has reported results in 53 cycling women resected for cure in whom menstrual period data were available. Median follow-up in these patients exceeds 10 years. An affect of operative timing within menstrual cycle was found upon patient survival. Patients were divided according to whether they were resected near to or far from the menstrual period. Some 85% of the women resected far from the menstrual period (midcycle) survived 5 years disease-free and 78% survived 10 years without evidence of recurrence. Only 58% and 52% of women resected nearer to the menstrual period survived 5 and 10 years, respectively ($p < 0.01$). Again, the optimal surgical resection timing included the early luteal phase in this series, which precisely reproduced the results of the first study.

Badwe et al. (1991) have examined the records of 249 cycling women not taking oral contraceptives, with last menstrual period data, and treated between 1975 and 1985 and compared overall and recurrence-free survival in those whose operation was 3–12 days after their last menstrual period (n = 75) with those in whom it was 0–2 or 13–32 days after the period (n = 174). Overall and recurrence-free survival was enhanced for those in the latter group who were resected during the luteal phase of their menstrual cycles (p

< 0.001). Actuarial survival at 10 years was 54% vs 84%. This effect was: (1) independent from other factors; (2) of equal importance to nodal status in multivariate analysis; (3) was greatest for node-positive cases and in women with small primary tumors; and (4) was seen in patients with estrogen receptor-positive and -negative tumors. The optimal surgical resection timing clearly included the luteal phase in this as all previous studies.

Ville et al. (1990) report the only series of patients, 165 of them, in whom an objective measure of menstrual cycle phase was obtained. Blood samples were obtained from each woman one day prior to operation, the serum was frozen at −70°C and progesterone, estradiol, and leuteotropin levels were measured. Women were subsequently assigned to one of four menstrual cycle stages: perimentstrual (n = 35); follicular (n = 35); ovulatory (n = 41); or luteal (n = 54). The median follow-up of these patients is currently brief, and clear survival differences by stage are not present. The distribution of the frequency of metastasis is interesting and is consistent with all prior studies. The development of metastasis is two- to threefold less common if surgery was performed in the luteal phase than each of the other cycle phases. A *chi*-squared analysis comparing the frequency with which metastases develop in those resected in the luteal phase (6/54) with all others (29/111) indicates that this difference is significant at the 94% level (x^2 = 3.34; p = 0.06) (Ville et al. 1991).

Saad (1992) recently reported breast cancer outcomes in 96 young women operated upon for stage I or stage II at the London Regional Cancer Centre. Each patient has been followed until death or for at least 10 years from the date of operation. Some 40% of women resected during the follicular phase while 79% of those operated during the luteal phase survived ($p < 0.01$). Overall 10 year survival for women resected during the early luteal phase was 89%.

Spratt et al. (1993) have summarized the outcomes of Louisville premenopausal breast cancer entered into the Breast Cancer Detection Demonstration Project (BCDDP). This national study required investigators to ask and record whether women discovered to have cancer were premenopausal. Upon admission for resection, they also asked the first day of the last menstrual period. The entire national study entered 1087 premenopausal patients. Spratt has reported outcomes of each of the 40 premenopausal women entered in Louisville. Women resected in midcycle days 7–21, which includes the only luteal phase, had a survival probability of 71%. Women resected near the menstrual period (0–6, 21–36) had a 20% 10 year survival probability.

Table 2 graphically portrays the relative surgical curability of the breast cancers in the original murine study and each of the seven confirmatory clinical studies. The table arranges the likelihood of breast cancer cure according to which quadrant of the menstrual cycle the operation was performed. The murine study and the clinical study of Ville objectively documented fertility cycle stage, while all other clinical studies relied totally

Table 2. Surgical curability

Study	Cycle quarter[a] than putative ovulation			
	First	Second	Third	Fourth
Mouse (Hrushesky et al. 1989) $n > 200$	++++	+++	+	++
Women (Hrushesky et al. 1989) $n = 41$	++++	+	+	+++
Women (Senie et al. 1991) $n = 283$	++++	+++	++	+
Women (Meyer 1991) $n = 53$	++++	+	+	++++
Women (Badwe et al. 1991) $n = 249$	++++	+++	++	+
Women (Ville et al. 1990, 1991) $n = 165$	++++	++	+	+
Women (Saad 1992) $n = 96$	++++	+++	++	+
Women (Spratt et al. 1993) $n = 40$	++++	+	+	++++

The fertility cycles of the mouse and woman are divided into relatively comparable quarters across the seven studies providing adequate data to do so. Fertility cycle phase references for mouse were obtained by evaluation of vaginal smear cellularity at the time of attempted surgical cure of a transplanted breast cancer. In the woman, the menstrual cycle phase was estimated by reference to the first day of the last menstrual period prior to primary breast cancer resection in the studies of Hrushesky, Senie, Meyer, Badwe and Saad. In the study by Ville, fertility cycle stage was assigned by reference to a serum sample obtained the day prior to operation in which progesterone, estradiol and leuteotropin were measured. Putative ovulation beginning the putative or actual luteal phase was uniformly the optimum segment of the cycle for maximization of positive outcome regardless of endpoint evaluated. These endpoints included: in the mouse the presence of metastasis; in the Hrushesky, Senie, Meyer, Badwe and Saad studies, disease-free and overall survival; and the appearance of metastasis in the Ville study. Here, the best possible outcome is represented by ++++ and the worst possible is represented by +. While perfect agreement exists for the optimal phase for surgical resection, agreement among studies about the worst quadrant or half of the cycle is somewhat less uniform.

[a] Menses occurs between the second and third quarters; putative ovulation between the first and fourth quarters.

upon the menstrual history. The quarter (7–10 days for humans and one day for mice) at and following putative ovulation, early luteal phase, is clearly the most opportune time for resection. There is, however, greater study-to-study variability in assigning the most inopportune time, although the least favorable time usually appears to be the quarter of the cycle following the first day of the menstrual flow.

At least two nonconfirmatory retrospective studies have been incompletely reported (Powles et al. 1989, 1991; Gelber and Goldhirsch 1989). These two studies were performed in patients with advanced breast cancer and most, if not all, women in them underwent extensive cancer surgery. In the largest of these studies, treatment was not consistent within the study and many of the patients also underwent additional abdominal surgery for oophorectomy. The largest of these studies included many irregularly cycling and/or peri-menopausal patients and neither of them clearly excluded women consuming oral contraceptives. In neither of these studies was subgroup comparability demonstrated nor was multivariate analysis performed. It is interesting that, in the study of Powles, although statistical significance was not realized,

Table 3. Clinical results summary

Author	Number of patients	Optimal span	Endpoint	Ten year comparative advantage
Hrushesky et al.	41	Mid-cycle/early luteal	OS	95% vs 78%
Senie et al. (1991)	283	Luteal	DFS	69% vs 58%
Meyer (1991)	53	Mid-cycle/early luteal	OS	78% vs 52%
Badwe et al. (1991)	249	Luteal	OS	84% vs 54%
Ville et al.	165	Early luteal	DFS	89% vs 62%
Saad (1992)	96	Early luteal	OS	79% vs 40%
Spratt et al. (1993)	40	Mid-cycle/early luteal	OS	71% vs 20%
Powles et al. (1989)	81	No difference	DFS	NS
Gelber and Goldhirsch (1989)	245	No difference	–	NS

OS, overall survival; DFS, disease-free survival; NS, not significant.

those operated at the putative early luteal phase had a 10 year death rate of less than 20% while those resected at other times within the cycle had up to 50% death rates at 10 years. Gelber's data are not reported in a way that will permit evaluation of the behavior of breast cancer in women resected in the luteal phase of their cycles. Table 3 briefly summarizes these retrospective clinical results and lists both the timing of optimal surgery and the 10 year survival advantage associated with optimal resection timing for each study.

Meta-analysis

Meta-analysis is problematic when so few published series exist. Also publication bias causes positive studies to be more likely to be more completely reported. In this particular case the value of meta-analysis may reside more in resolving the two general ways the data have been presented in the literature, rather than determining an accurate level of statistical significance. Two meta-analyses were performed on the published human studies of Hrushesky et al. (1989), Senie et al. (1991), Badwe et al. (1991) and Ville et al. (1990, 1991), (data from Meyer 1991, Saad 1992 and Spratt et al. 1993 are unfortunately inadequate for inclusion in the meta-analysis). The confidence profile method (Eddy et al. 1992) was selected for this analysis for its purely analytical approach to the synthesis of data from multiple studies and for the availability of diagnostics (the ability to quantity the effect of each individual study on the overall parameter estimates). Bayesian methods were employed to estimate the parameter of interest, i.e., the relative change in the rate of disease-free survival of women who underwent resection during different phases of the menstrual cycle.

Luteal vs Follicular

Overall, it was found that women undergoing resection during the luteal phase were more likely to experience disease-free survival than women undergoing resection during the follicular phase ($p = 0.0066$). The effective increase in ≥10 year disease-free survival for women treated during the luteal phase over women treated in the follicular phase was estimated to be 16.82%, with a 95% confidence interval of this increased survival ranging from 3.24% to 30.4%. These data come from analyses of studies of Hrushesky et al. (1989), Senie et al. (1991) and Ville (1990, 1991) (Meyer 1991 and Badwe et al. 1991 did not report survival for luteal and follicular treatments, while Saad 1992 and Spratt et al. 1993 did not provide adequate detail to accurately determine the number of patients treated at each cycle phase).

Midcycle vs Perimenstrual

A second meta-analysis of the difference in the rates of disease recurrence and/or death for women treated perimenstrually vs women treated mid-cycle, including the studies of Hrushesky et al. (1989), Ville et al. (1990), Senie et al. (1991) and Badwe et al. (1991), proved also to be statistically significant ($p = 0.029$), with women treated mid-cycle experiencing a lower rate of disease recurrence and/or death. Both arbitrary divisions of the cycle yielded a statistically significant result because the preponderantly best time within the cycle to resect, early luteal phase, is included in both the luteal and the mid-cycle divisions (see Table 2). Diagnostics indicate that no singular study biased these estimates, and due to the small number of and great similarity among the studies analyzed, no weighting schemes were employed, (the benefits of quality weighting would be hard to substantiate in a case such as this). Variance weighting, a more well known technique of meta-Analysis, was also performed and provided results completely consistent with the above.

Overall, the murine data and seven completely reported retrospective clinical studies with reliable menstrual cycle data or hormonal verification of menstrual cycle phase indicate that limited surgery for early stage breast cancer performed in the early luteal phase results in fewer recurrences and a substantial reduction in the risk of dying from breast cancer. Two incompletely reported data bases fail to confirm these observations. There are some potentially significant differences among the patient populations studied and in the precise way they have been both studied and treated making the value of a meta-analysis suspect for this data base. From a statistical viewpoint, in my opinion, a prospective study is needed to properly investigate this question.

Discussion

There are many questions left open by the retrospective nature of all of the clinical series performed to date. These include, among others: what is precisely the best time within the cycle for operation? How long is this window of opportunity open? What is the relative impact of the stage of disease at resection and the extent of the operation performed? When more than one surgical procedure is employed, what is the relative importance of the timing of an initial incisional or excisional biopsy or lumpectomy compared with follow-up "completion" surgery and/or later node dissection? Each of these and other important questions can be adequately addressed only by a prospective clinical evaluation of breast cancer resection menstrual timing in cycling premenopausal women. When it is performed, such a prospective clinical trial must be limited to those who are not taking oral contraceptives. A careful, general and immedicate past menstrual cycle history must be obtained immediately prior to operation. Either a single-stage primary breast cancer operation must be done or the two procedures must be tightly coordinated. Finally, additional measurements to verify menstrual regularity and ovulation must be employed in such a study.

These findings may also have significant implications for the proper interpretation of diagnostic studies and prognostic indicators. Major changes in breast volume, fat/water ratio and temperature occur within the menstrual cycle. There is every reason to expect that all forms of physical, radiographic and thermographic breast examination may have predictably different signal-to-noise ratios at specific menstrual cycle phases. Prognostic biochemical values, such as estrogen and progesterone receptor concentration in normal and malignant breast epithelium and mitotic index of normal breast epithelium, are known to change rhythmically during the female fertility cycle (Weimer and Donegan 1987; Smyth et al. 1988; Anderson 1984). Wicha's (1991) work demonstrating marked menstrual cycling of serum concentrations of the negative growth modulator of benign and malignant breast epithelium, mammastatin, is a case in point for suspecting that many, if not all, of the biochemical correlates of breast cancer biology may behave similarly. It remains to be determined whether and how much other "hormonally responsive" biochemical correlates of breast cancer aggressiveness, such as cathepsin D concentration, growth factor, oncogene or suppresser gene expression, cycle menstrually. If they do so, each of these and other prognostic indicators may well need to have menstrual cycle phase-specific negative/positive cut-off values and/or normal/abnormal ranges.

In addition to being affected by all the classical prognostic indicators, the ultimate outcome of a premenopausal woman with breast cancer also depends upon the dynamics of the balance between host and cancer at the time of operation. Surgery creates neuroendocrine, humoral and/or cellular signals that affect this balance differently depending upon when in the fertility cycle it is performed. The biology of importance in determining whether a micrometastatic deposit will ultimately succeed in invading and growing to an extent that it is ever clinically detectable and eventually fatal can be grouped within one of three general categories: local support factors, cellular immune defense networks and tumor cell facors (Fig. 2).

Local support structures must provide a framework for tumor growth. This requires adhesion of tumor cells to endothelium, proteolysis of basement membrane, active invasion of tumor cells and the development of new blood vessels. Each of these important steps is potentially sensitive to the different signals resulting from the load of operation at different points in the cyclic relationship of estrogen, progesterone, cortisol, prolactin, FSH and LH represented by the fertility cycle. Angiogenesis is an example of a very tightly and hormonally regulated process, absolutely essential for the development of metastatic cancer (Folkman and Klagsbrun 1987; Folkman 1990; Weidner et al. 1991). Angiogenesis does not occur in the absence of injury or neoplasm in the male. In the female, in preparation for implantation of the fertilized ovum, angiogenesis is turned on within both the ovaries and the uterine lining during each fertility cycle. It is also immediately turned off again if implantation fails to occur resulting in menses.

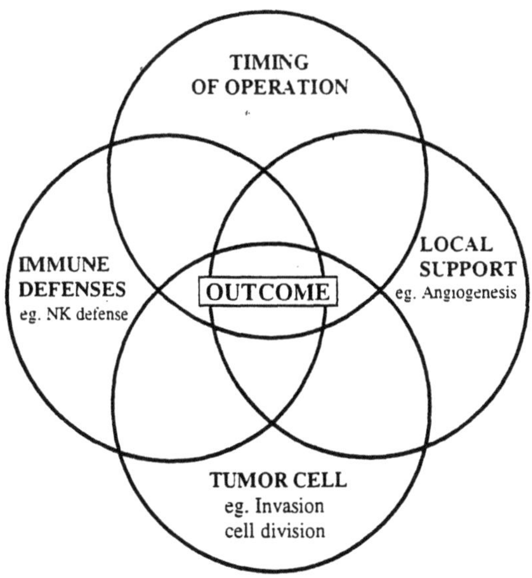

Fig. 2. Whether a woman with breast cancer is cured by a resection or develops metastatic breast cancer and dies from it is determined by a balance among: local support factors, immune defenses, tumor cell biology. This balance is affected differently depending upon when in the womans fertility cycle the hopefully curative operation is performed. This complex interactive relationship is depicted by a four-circle Venn diagram indicating that surgical timing has the capacity for affecting each of the other three sources of outcome variability as well as the bottom line, i.e., outcome

Cellular immune function is known to be important in determining outcome in both murine and human cancer. Lymphocyte helper/suppresser ratios were shown to vary predictably during the fertility cycle in a group of 63 healthy women (Krzanowski 1985). NK cell activity has been shown repeatedly to be important to cancer outcome (Bariozzari et al. 1985; Wiltrout et al. 1985). NK activity is suppressed and tumor metastases are enhanced by estrogenic hormone treatment (Hanna and Schneider 1982). We have demonstrated that NK activity varies rhythmically during the fertility cycle and that it is at its highest level at the time of the cycle associated with highest surgical breast cancer curability (Hrushesky et al. 1988). The specific negative effects of surgery upon the cellular immune networks have been aggressively explored in both murine models and cancer patients (Green and Faist 1988; Pollock et al. 1984, 1987). It is surely reasonable to expect that the fertility cycle stage at surgery might well modulate its subsequent negative impact upon cellular immunity.

The receptivity of the hormonally responsive breast cancer cell to the neuro-humoro-cellular signals created by surgery may well be modulated by the dynamic balance among estrogen, progesterone, cortisol, prolactin, FSH and LH, represented by the fertility cycle (Weimer and Donegan 1987; Smyth et al. 1988). Estrogen and progesterone receptor concentration in nonmalignant breast and uterine epithelium and malignant breast cancer cells very with this cycle. The receptor expression of a patients' breast cancer is well known to translate into significant outcome expectations. The ability of cancer cells to stick to endothelium and move through the basement membranes at either the primary or metastatic site is clearly related to their potential to spread and kill their hosts. Cathepsin D is a protease, the concentration of which in a breast cancer specimen is known to be both hormonally modulatable and relevant to the outcome of the woman bearing that cancer (Rochefort et al. 1988; Spyratos et al. 1989; Tandon et al. 1990). Mammostatin, a negative regulator of breast cell growth and division, is known to be absent or diminished in breast cancer and to cycle with very high amplitude during the menstrual cycle (Wicha 1991). Genetic expression of many important oncogenes and growth factors are responsive to some of the same hormones cycling during each fertility cycle as well as those released in response to stress.

The rhythmic biodynamics of the balance between a woman, her response to surgery and her breast cancer is not a phenomenon which can easily be ignored. If true, these observations can annually save many thousands of young women's lives. Perhaps more importantly, however, these findings demand a fundamental realignment of our thinking about the balance between host and tumor. Fully understanding how these interactions impact cure frequency will generate potential therapeutic strategies to more frequently cure all surgically treated cancers.

Acknowledgement. This work was supported, in part, by NIH RO1 CA31635, "Clinical Applications of Chronobiology to Cancer Medicine," NIH RO1

CA50749, "Chronobiological Investigation of TNF and IL-2," VA RAGS grant, "Chronobiological Investigation of Cytokines and Growth Factors," and VA Merit Review, "Circadian FUDR Pharmacodynamics."

References

Anderson T (1984) Mitotic activity in breast. J Obstet Gynecol 4(12):S114–S118
Badwe R, Gregory W, Chaudary M et al. (1991) Timing of surgery during menstrual cycle and survival of premenopausal women with operable breast cancer. Lancet 1:1261–1264
Bariozzari T, Leonhardt J, Wiltrout RH, Herberman RB, Reynolds CW (1985) Direct evidence for the tole of LGL in the inhibition of experimental tumor metastasis. J Immunol 134:2783–2789
Beatson G (1896) On the treatment of inoperable cases of carcinoma of the mamma: suggestions for a new method of treatment will illustrative cases. Lancet 2:104–107
Cooper A (1836) Practice and principles of surgery. London
Eddy DM, Hasselblad V, Schachter R (1992) Meta-analysis by the confidence profile method. Academic, San Diego
Folkman J (1990) What is the evidence that tumors are angiogenesis dependent? J Natl Cancer Inst 82:4–6
Folkman J, Klagsbrun M (1987) Angiogenic factors. Science 235:442–447
Fisher B, Gunduz N, Zheng S, Saffer E (1982) Fluoresceinated estrone binding by human and mouse breast cancer cells. Cancer Res 42:540
Gelber R, Goldhirsch A (1989) Menstrual effect on surgical cure of breast cancer. Lancet 2:1344
Green DR, Faist E (1988) Trauma and immune response. Immunol Today 9:253–255
Hanna N, Schneider M (1982) Enhancement of tumor metastasis and suppression of natural killer cell activity by β-estradiol treatment. J Immunol 130:974–980
Hrushesky WJM, Gruber SA, Sothern RB et al. (1988) Natural killer cell activity is age, estrous and circadian stage dependent and correlates inversely with metastatic potential. J Natl Cancer Inst 80:1232–1237
Hrushesky WJM, Bluming AZ, Gruber SA, Sothern RA (1989) Menstrual influence on surgical cure of breast cancer. Lancet 2:949–952
Huggins C, Bergenstal E (1952) Cancer Res 12:134
Jensen E, Desombre E, Jungblut P (1967) Estrogen receptors in hormone-responsive tissues and tumors. In: Wissler R, Dao TS, Wood J (eds) Endogenous factors influencing host-tumor balance. University of Chicago Press, Chicago, pp 15–30
Krzanowski M (1985) Low human T4:T8 lymphocyte ratios around ovulation time. Chronobiologia 12:254
Luft R (1953) J Neurosurg 10:301
Meyer K (1991) Season and cycle: variation in presentation and response to therapy in premenopausal breast cancer (Abstr). 77th Annual Clinical Congress, October 21–24
Pollock RE, Babcock GF, Romsdahl M, Nishioka K (1984) Surgical stress-mediated suppression of murine natural killer cell cytotoxicity. Cancer Res 44:3888–3891
Pollock RE, Lotzova E, Stanford SD, Romsdahl MM (1987) Effect of surgical stress on murine natural killer cell cytotoxicity. J Immunol 138:171–178
Powles T, Jones A, Ashley S, Tidy A (1989) Menstrual effect on surgical cure of breast cancer. Lancet 2:1343–1344
Powles T, Ashley S, Nash A, Tidy A, Gazet J-C, Ford H (1991) Timing of surgery in breast cancer. Lancet 337:164
Ratajczak H, Lewis M, Duggal K (1986) Increased chemotaxis of leukocytes from mice bearing tumors. Cancer Res 46:2225

Ratajczak HV, Sothern RB, Hrushesky WJM (1988) Estrous influence on surgical cure of a mouse breast cancer. J Exp Med 168:88–96
Rochefort H, Augereau P, Briozzo P et al. (1988) Structure, function, regulation and clinical significance of the 52K pro-cathepsin D secreted by breast cancer cells. Biochimie 70:943–949
Saad Z (1992) Timing of surgery in relation to menstrual cycle in premenopausal women and its effect on survival. Canad J Surg 35:438
Senie R, Rosen P, Rhodes P, Lesser M (1991) Timing of breast cancer excision during the menstrual cycle influences duration of disease-free survival. Ann Intern Med 115:337–342
Smyth C, Benn D, Reeve T (1988) Influence of the menstrual cycle on the concentrations of estrogen and progesterone receptors in primary breast cancer biopsies. Breast Cancer Res Treat 11:45–50
Spratt J, Zirnheld J, Yancey J (1993) Breast cancer detection demonstration project data can determine if the prognosis of breast cancer is effected by the time of surgery during the menstrual cycle. J Surg Oncol 53:4
Spyratos F, Brouillet J-P, Defrenne A, Hacene K, Rouësse J, Maudelonde T, Brunet M, Andrieu C, Desplaces A, Rochefort H (1989) Cathepsin D: an independent prognostic factor for metastasis of breast cancer. Lancet 2:1115–1118
Tandon A, Clark G, Chamness G, Chirgwin J, McGuire W (1990) Cathepsin D and prognosis in breast cancer. N Engl J Med 322:297–302
Ville Y, Lasry S, Spyratos F (1990) Menstrual status and breast cancer surgery. Breast Cancer Res Treat 16:119–121
Ville Y, Briere M, Lasry S, Spyratos F, Oglobine J, Brunet M (1991) Timing of surgery in breast cancer. Lancet 337:164
Weidner N, Semple JP, Welch WR, Folkman J (1991) Tumor angiogenesis and metastasis – correlation in invasive breast cancer. N Engl J Med 324:1–8
Weimer D, Donegan W (1987) Changes in estrogen and progesterone receptor content of primary breast carcinoma during the menstrual cycle. Breast Cancer Res Treat 10:273–278
Wicha M (1991) Mammastatin: a mammary growth regulator. 82nd Annual Meeting of the American Association for Cancer Research, May 15–18, Houston
Wiltrout RH, Herberman RB, Zhang S-R et al. (1985) Role of organ-associated NK cells in decreased formation of experimental metastases in lung and liver. J Immunol 134:4267–4275

Summary

M.H.N. Tattersall

University of Sydney, Dept. of Cancer Medicine, Blackburn Bldg., Sydney NSW 2006, Australia

There has been considerable progress in identifying the genetic changes which contribute to familial breast cancer risk, but additional information is needed before individual risk can be reliably ascertained. Women with BRCA 1 germ line mutations probably comprise about half of all patients with inherited breast cancer, while mutations to BRCA 2 probably contribute 70% of the remaining patients. Linkage studies in large familial breast cancer pedigrees have indicated that women with BRCA 1 mutation have an 85% life time risk of developing breast cancer, and that approximately half of these women will be diagnosed before age 50. BRCA 1 is a large gene, and the protein consists of more than 1800 amino acids. The initial reports describe seven different mutations, one of which results in complete loss of BRCA 1 m RNA. Many additional mutations are likely to be described, and correlation of genetic studies with outcomes in different familial breast cancer pedigrees are needed before our understanding of the issue of penetrance and the importance of modifying factors will be sufficient to assist counselling of family members.

The complexity of the BRCA 1 gene makes it unlikely that a reliable test for mutations confirming increased breast cancer risk will be developed in the near future. Nevertheless, in particular families in which there is a positive test in a breast cancer patient, genetic screening may allow identification of family members with the relevant germ line mutation.

Interventions which may influence breast cancer risk in BRCA 1 mutation carriers have not been adequately investigated. It is not clear if the reproductive factors important in sporadic breast cancer risk influence breast cancer risk in BRCA 1 mutation carriers. Counselling services and information packages will need to be developed to help women seeking genetic testing to determine breast cancer risk. Women with germ line mutations confirming an increased breast cancer risk are an important resource for the future testing of breast cancer prevention strategies.

Preliminary results do not suggest that BRCA 1 mutations are responsible for sporadic breast cancer occurring without a family history, but additional research is required in this important field. Similarly BRCA 2 germ line mutations, which also contribute to male breast cancer risk, will be studied in the near future.

The contribution of environmental and reproductive factors to breast cancer incidence has been investigated over many years, and large trials investigating the impact on health and breast cancer incidence of dietary modification to reduce fat intake are being undertaken. The importance of vegetables as well as different fats in breast cancer development and promotion is being investigated, but no clear recommendations about dietary modifications can be made. The possible importance of heterocyclic amines which are generated during cooking of animal and fish foods is one area of laboratory research which may lead to interventions to reduce risk. Epidemiological studies have identified dietary components which may contribute to the lower incidence of breast cancer in Asia. In this respect, soybean is of particular interest since it contains the phytooestrogen genistein, which inhibits mammary cancer development in animal studies.

In this part, animal studies investigating physiological changes in hormone levels on breast cancer cell growth were reported. An effect of endocrine status on tumour growth and cell death rates was demonstrated. Tumour inoculation in the early luteal phase of the oestrous cycle in rodents was relatively less conducive to tumour growth, and in contrast tumour growth was maximal during the follicular phase. The concept of biologic time in the menstrual cycle and its effect on tumour behaviour was described. These experiments are relevant to clinical reports from many authors that breast cancer prognosis in premenopausal women is influenced by the time of surgery in relation to the menstrual cycle. Retrospective published studies suggests that risk is maximal if breast surgery is in the follicular phase of the menstrual cycle and minimal in the early luteal or mid-cycle. Prospective studies are required to confirm these results, but possible mechanisms for these interactions have been proposed including hormonal effects on tumour cell growth, and tumour growth factor physiology.

The relevance of these physiological hormone fluctuations to the increasing prevalence of hormone replacement therapy in postmenopausal women diagnosed with breast cancer is unknown.

II. Biology of Breast Cancer

Peptide Growth Factors and Breast Cancer Treatment: Recent Translational Research

M. Pollak

Departments of Medicine and Oncology, Jewish General Hospital and McGill University, Montreal, Quebec, Canada H3T 1E2

Introduction

The term "translational research" refers to investigations at the interface between the laboratory and the clinic. Laboratory research regarding the pathophysiology of peptide growth factors in breast cancer has been ongoing for more than two decades and has led to important insights regarding the pathophysiology of this malignancy. The development of novel treatment strategies based on hypotheses arising from laboratory research concerning peptide growth factors is a relatively recent and encouraging development. In this presentation, only three representative examples of translational research related to peptide growth factors in breast cancer will be reviewed. While many other important examples exist, these three cases should serve to illustrate how laboratory research is yielding concepts suitable for evaluation in clinical trials.

I have chosen three examples that are at different stages of development. The first example, the recent discovery of angiostatin, represents a major step forward in understanding the long-known and somewhat counterintuitive phenomenon whereby resection of a primary tumor can stimulate the proliferation of metastases. This area of research has great potential clinical importance, particularly for adjuvant treatments, but specific clinical trials related to this work have yet to begin. At the other extreme is the second example, the identification of the HER2/neu receptor as a novel therapeutic target. This research has already led to positive phase II trials of a humanized anti-HER2 antibody, and an expanded phase III program will be starting later this year. The third example concerns the hypothesis that certain antineoplastic actions of tamoxifen are augmented by somatostatin analogues; while no definitive clinical trial results are yet available with regard to this hypothesis, preclinical models have been encouraging and clinical trial designs are now being developed.

Clinical Implications of Evidence That Surgical Removal of Primary Tumors Stimulates Growth of Metastases by Facilitating Neovascularization

It has been known for many years that, in a variety of animal models, the presence of a relatively large primary tumor can suppress the growth of a second tumor inoculum injected at a distant site (reviewed in O'Reilly et al. 1994). In humans, there have been reports of relatively rapid growth of distant metastases following the removal of certain tumors, although it has been impractical to study this phenomenon in detail as primary tumors are usually removed prior to the development of measurable metastases. Superficially, the phenomenon has similarities to the well known phenomenon of compensatory hypertrophy following unilateral nephrectomy: The neoplasm appears to compensate for loss of volume caused by surgical resection by accelerating growth of any metastases that are present.

Although mechanisms involving immunological or nutritional factors had been proposed, none of these were supported by direct experimental data. Recently, however, O'Reilly et al. (1994) studied the phenomenon in a rigorous quantitative fashion and hypothesized that the phenomenon might be related to factors controlling angiogenesis in the metastatic lesions. Specifically, it was proposed that primary tumors secrete peptides that control angiogenesis and that the kinetics of release and half-lives of these factors lead to a situation in which angiogenesis is stimulated in the immediate microenvironment of the primary tumor; however, tumor-derived angiogenesis inhibitors, by virtue of their relatively long half-life in the circulation, act to inhibit angiogenesis at metastatic sites as well as in normal organs of a tumor-bearing host.

Subsequent work in Dr. Judah Folkman's laboratory has provided data that strongly support this hypothesis. In a well-defined tumor model, the stimulatory effect of the surgical removal of a primary tumor on the growth of metastatic lesions (as compared to the effect of sham surgery) was quantified. The mean number of pulmonary metastases was more than ten fold higher in animals treated with surgical excision of the primary tumor. From the serum and urine of tumor bearing mice, a protein was isolated that mediated the effect. This protein, called angiostatin, was demonstrated to be a 38 kDa fragment of plasminogen that is a novel and potent inhibitor of angiogenesis. Significant quantities of angiostatin were prepared for further by limited proteolytic digestion of human plasminogen with elastase. Administration of angiostatin at a dose of 25 μg intraperitoneally on the day of tumor excision followed by a dail dose of 12.5 μg could inhibit the growth of metastases as effectively as the presence of a primary tumor.

The implications of these results for the adjuvant treatment of breast cancer may be considerable. However, a priority will be to determine directly the relevance of the experimental results to human breast cancer by assaying angiogenic activity and angiostatin levels prior to and following surgical resection of primary breast cancers in a series of patients. Positive results of such translational research would represent a fundamental advance in our understanding of the pathophysiology of malignancy. Clinical implications would include: (1) a novel rationale for preoperative, as compared to postoperative, chemotherapy, as it would clearly be advantageous to eliminate micrometasteses *prior to* their angiogenic stimulation by primary tumor removal; (2) as tamoxifen has antiangiogenic actions (Furman-Haran et al. 1994) (which may be related at least in part to its inhibitory activity on insulin-like growth factor (IGF)-I gene expression (Huynh et al. 1993), examining the possibility that starting tamoxifen therapy several weeks preoperatively would be more efficacious than current practice would be justified; (3) the development of research programs to evaluate the feasibility and efficacy of using angiostatin itself as an adjuvant treatment would have a high priority; (4) further justification of the concept of targeting the angiogenic process (rather than neoplastic cells themselves) in cancer treatment would be provided.

The HER2/neu Receptor as a Therapeutic Target

The second example provides an instance in which laboratory findings concerning an aspect of peptide growth factors in breast cancer have led to clinical trials which are already ongoing. Briefly, the HER2/neu receptor is a cell surface receptor for certain peptides that are mitogens for breast cancer cells. About 30% of human breast cancers overexpress this receptor, as a consequence of a somatic cell genetic derangement. If a single cell of a tumor anomalously overexpresses a mitogen receptor, it is reasonable to propose that such a cell would become hypersensitive to stimulation by physiological concentrations of mitogen. It would therefore acquire a growth advantage, and its daughter cells would progressively dominate the tumor. In a Darwinian sense, the derangement is "advantageous" to the malignant cell, but from the point of view of the organism, the derangement represents a further escape from physiological regulation, leading to a more aggressive malignant phenotype. We have demonstrated this process experimentally in the case of overexpression of the EGF receptor (Filmus et al. 1987) and also used computer modeling techniques to simulate it theoretically (Pollak et al. 1991).

Following initial observations concerning the HER2/neu receptor (reviewed in Slamon et al. 1987), an approach to interfere with the action of this mitogenic system in laboratory experimental systems by the use of blocking antibodies was developed (Drebin et al. 1988). As this work yielded positive results, efforts were directed to the development of a "humanized" antibody that would not cause a serum sickness syndrome and thus be suitable for clinical trials (Carter et al. 1992). Early trials revealed antitumor activity and acceptable toxicity. An expanded clinical research program is now underway to more accurately define the efficacy of the novel agent and to further document what was initially an unexpected phenomenon, namely, the apparent synergy between the anti-HER2/neu antibody and cytotoxic agents.

Can Somatostatin Analogues Enhance the Activity of Antiestrogens in the Adjuvant Treatment of Breast Cancer?

While the use of adjuvant tamoxifen therapy has clearly been shown to reduce the risk of relapse following surgery, many patients develop overt metastases while receiving adjuvant tamoxifen. The development of pharmacological strategies to improve the efficacy of tamoxifen, without adding toxicity, is therefore an important example of current research directed towards the improvement of adjuvant treatment of breast cancer.

Our recently published results (Weckbecker et al. 1994) clearly show that in the DMBA rat mammary tumor model, the inhibitory effect of tamoxifen (0.5 mg/kg s.c. twice weekly) on the appearance of neoplasms is enhanced by the coadministration of octreotide (10 μg/kg per hour s.c.). At 9 weeks following the initiation of therapy in these experiments, the percentage of control animals with tumors was 80%, while the percentages of tamoxifen-treated and combination (tamoxifen + octreotide)-treated animals with tumors were 20% and 0%, respectively. By 14 weeks, most control animals had to be killed because of extensive tumor growth and 40% of the tamoxifen group had tumors, but only 10% of the combination treatment group had tumors. The mean number of tumors and the mean tumor volume were also significantly reduced ($p < 0.05$ by Student's t test and by Mann-Whitney U test) in the combination group (mean number = 0.4, mean total volume = 200 mm^3) relative to the tamoxifen group (mean number = 2, mean total volume = 3000 mm^3) both during and for 4 weeks after treatment. We also observed that octreotide enhanced the efficacy of ovariectomy in the DMBA system. Further consistent results come from Andrew Schally and colleagues (Szepeshazi et al. 1992), who reported an enhancement of the action of LHRH analogues by the somatostatin analogue RC160 in the MXT mammary carcinoma model.

The mechanism of action underlying these observations has not been established with certainty, but one or both of two processes may be involved. A direct mechanism involves binding of the somatostatin analogue to somatostatin receptors present on tumor cells. Recent imaging studies have confirmed that well over 50% of human breast cancers have somatostatin receptors (Van Eijck et al. 1994). Somatostatin may act as a direct growth inhibitor in somatostatin receptor-positive cells by stimulating intracellular phosphotyrosine phosphatase activity (Lee et al. 1991). Of interest in the present context is the fact that it appears that the ability of somatostatin analogues to act as growth inhibitors is diminished in the presence of estrogens (Setyono-Han et al. 1987). This provides a clear rationale for mproved inhibition when somatostatin is administered with antiestrogens or with ovariectomy.

A further mechanism of action may be indirect in the sense that there is no direct interaction between the somatostatin analogue and the breast cancer cell. Somatostatin is known to reduce the levels of IGF-I, since somatostatin physiologically inhibits growth hormone output by the pituitary and growth hormone up-regulates IGF-I expression. IGF-I is a potent mitogen for breast cancer cells (reviewed in Pollak 1993 and Arteaga et al. 1989). Our group (Pollak et al. 1990) and others (reviewed in Pollak 1993) have shown that tamoxifen also reduces serum IGF-I levels. It also reduces IGF-I gene expression in target organs for metastases (Huynh et al. 1993). We recently demonstrated additive reduction of IGF-I levels and gene expression when the two agents are coadministered (Huynh and Pollak 1994). Thus, reduction of of IGF-I level may also contribute to the observed enhancement of efficacy. It is noteworthy that experience with adult patients with IGF-I and growth hormone deficiencies following hypophysectomy for pituitary tumors shows that these derangements are associated with toxicities considered minor relative to those seen with other cancer treatments. Nevertheless, this issue would have to be addressed carefully in clinical trials.

As there is considerable experience with long-term use of octreotide in cohorts of patients with acromegaly and carcinoid syndrome, data are available which suggest an absence of serious short- or long-term toxicity of this agent. This coincidence may make it more practical to carry out adjuvant clinical trials of the tamoxifen-octreotide combination in cohorts of breast cancer patients with relatively good prognosis (in whom safety is of paramount importance) than it would be to study a compound for which no long-term toxicity data are available. Before such trials can be considered realistic, however, the manufacture of a monthly injectable depot formulation of the somatostatin analogue will have to be achieved, as the current three-times-per-day subcutaneous injection formulation leads to patient compliance problems.

Conclusions

These three examples illustrate only a small subset of ongoing laboratory research that may ultimately lead to improvements in the therapies we have available for our patients.

Many reasonable and interesting hypotheses, when evaluated in appropriate preclinical model systems, are found to be undeserving of further development. We known from the history of the development of new therapies that even of those approaches that appear promising in the laboratory, only a small subset will turn out to be clinically valuable. Potentially promising laboratory results must be evaluated in clinical trials, despite the time and costs involved. There are no short cuts to this process, but it is clear that, in an era of increasing specialization, individuals who have a background in both laboratory research and clinical oncology have facilitating roles to play in academia, the pharmaceutical industry, and in clinical trials organizations.

There sometimes are difficulties in recruiting patients to clinical trials of new therapies based on novel approaches. For progress to occur, we must work to overcome these difficulties. Concerns regarding potential unexpected toxicities are legitimate and must be addressed explicitly, but also must be viewed in the context that our standard therapies are often associated with considerable toxicity, which on occasion tends to be relatively underemphasized simply because we are so familiar with syndromes of hair loss, leukopenia, nausea, mucocytis, etc.

References

Arteaga C, Kitten L, Coronado E et al. (1989) Blockade of the type I somatomedin receptor inhibits growth of human breast cancer cells in nude mice. J Clin Invest 84:1418–1423

Carter P, Presta L, Gorman C et al. (1992) Humanization of an anti-p185-HER2 antibody for human cancer treatment. Proc Natl Acad Sci USA 89:4285–4289

Drebin J, Link V, Greene M (1988) Monoclonal antibodies specific for the neu oncogene product directly mediate anti-tumor effects in vivo. Oncogene 2:387–394

Filmus J, Trent J, Pollak M, Buick R (1987) EGF receptor gene-amplified MDA-468 breast cancer cell line and its non-amplified variants. Mol Cell Biol 7:251–257

Furman-Haran E, Maretzek A, Goldberg I, Horowitz A, Degani H (1994) Tamoxifen enhances cell death in implanted MCF7 breast cancer by inhibiting endothelium growth. Cancer Res 54:5511–5514

Huynh H, Pollak M (1994) Enhancement of tamoxifen-induced suppression of IGF-I gene expression and serum levels by a somatostatin analogue. Biochem Biophys Res Commun 203:253–259

Huynh H, Tetenes E, Wallace L, Pollak M (1993) In vivo inhibition of IGF-I expression by tamoxifen Cancer Res 53:1727–1730

Lee M, Liebow C, Kamer A, Schally AV (1991) Effects of epidermal growth factor and analogues of LHRH and somatostatin on phosphorylation and dephosphorylation of tyrosine residues of specific protein substrates invarious tumors. Proc Natl Acad Sci USA 88:1656–1660

O'Reilly M, Holmgren L, Shing Y, Chen C, Rosenthal R, Moses M, Lane W, Cao Y, Sage E, Folkman J (1994) Angiostatin: a novel angiogenesis inhibitor that mediates the suppression of metastases by a Lewis lung carcinoma. Cell 79:315–328

Pollak M (1993) Effects of adjuvant tamoxifen therapy on growth hormone and IGF-I physiology. In: Salmon S (ed) Adjuvant therapy of cancer VII. Lippincott, Philadelphia

Pollak M, Costantino J, Polychronakos C, Blauer S, Guyda H, Redmond C, Fisher B, Margolese R (1990) Effect of tamoxifen on IGF-I levels in stage I breast cancer patients J Natl Cancer Inst 82:1963–1967

Pollak M, Gora P, Boyarsky A (1991) A mathematical model describing consequences of abnormally high levels of epidermal growth factor receptor on the proliferation of neoplastic cells. Cancer Invest 9:513–520

Setyono-Han B, Henkelman M, Foekens J, Klijn G (1987) Direct inhibitory effects of somatostatin analogues on the growth of human breast cancer cells. Cancer Res 47:156–157

Slamon D, Clark G, Wong S, Levin W, Ullrich A, McGuire W (1987) Human breast cancer: correlation of relapse and survival with amplification of the HER2/neu oncogene. Science 235:177–182

Szepeshazi K, Milovanovic S, Lapis K, Groot K, Schally AV (1992) Growth inhibition of estrogen independent MXT mouse mammary carcinomas in mice treated with an agonist or antagonist of LHRH, an analogue of somatostatin, or a combination. Breast Cancer Res Treat 21:181–192

Van Eijck C, Krenning E, Bootsma A, Lindemans J, Jeekel J, Reubi J, Lamberts S (1994) Somatostatin receptor scintigraphy in primary breast cancer. Lancet 343:640–643

Weckbecker G, Tolcsvai L, Stolz B, Pollak M, Bruns C (1994) Somatostatin analogue octreotide enhances the antineoplastic effects of tamoxifen and ovariectomy on DMBA-induced rat mammary carcinomas. Cancer Res 54:6334–6337

[illegible]

Single-Chain Fusion Toxins for the Treatment of Breast Cancer: Antitumor Activity of BR96 sFv-PE40 and Heregulin-PE40

C.B. Siegall

Molecular Immunology Department, Bristol-Myers Squibb, Pharmaceutical Research Institute, 3005 First Avenue, Seattle, WA 98121, USA

Introduction

Many commonly used anti-cancer drugs have been identified through massive screenings of natural products and chemical compounds. Unfortunately, these screens have resulted in the selection of compounds that are not only cytotoxic to cancer cells, but to most actively proliferating cells as well. The small degree of selectivity usually lies in the fact that cancer cells are amongst the most rapidly proliferating of all cell types in the body. A different approach to the development of anti-cancer therapeutics is through rational design of targeted cytotoxic agents. These molecules are delivered to specific cell-surface receptors or antigens, usually by monoclonal antibodies (MAbs) or growth factor/cytokines, and have been appended to a variety of cytotoxic agents. These molecules include drug-immunoconjugates (Trail et al. 1993), enzyme-prodrugs (Senter et al. 1988), radiolabeled-antibodies (Press et al. 1993), and immunotoxins (Chaudhary et al. 1989) and have been used in vitro, in vivo, and clinically to target cancer cells for elimination. This article will focus on two recently developed immunotoxins, BR96 sFv-PE40 and HAR-TX β2, and will describe their component structure, expression, purification, and preclinical characterization.

Immunotoxins

Immunotoxins are hybrid molecules that combine the cell-binding and modulating ability of MAbs with the cell-killing activity of protein toxins (Brinkmann and Pastan 1994). Immunotoxins also include, in a broader definition, growth factors or cytokines linked to toxins and have been termed oncotoxins, mitotoxins, and hormonotoxins. A multitude of immunotoxins that target cell-surface markers found in carcinomas, autoimmune disease, and hematologic malignancies have been prepared. Many immunotoxins have shown promise in early clinical trials, especially those against hematologic malignancy (Frankel 1993; Ghetie and Vitetta 1994).

The first forms of immunotoxins were prepared by chemically conjugating the components using bifunctional cross-linkers. More recently, they have been engineered and produced in bacterial systems as single-chain immunotoxin (SCIT) fusion proteins in which cloned antibody variable region genes (or growth factor or cytokine genes) are directly connected to DNAs encoding protein toxins (Brinkmann and Pastan 1994). The smaller molecular size of SCITs has some distinct advantages over most immunotoxin conjugate forms including increased ability to penetrate tumor tissue, decreased blood residence time, and faster clearance through liver and kidney. Since SCITs are single polypeptides, they are prepared homogeneously, in contrast to immunotoxin conjugates. Additionally, SCIT can be produced by bacterial fermentation, can be rapidly reconfigured genetically, and have been found to be more potent in vivo than their chemical conjugate counterparts (Siegall et al. 1995a).

Monoclonal Antibody BR96

The Lewis-Y (Le^y) carbohydrate antigen is highly expressed on the surface of many carcinomas types including breast, lung, colorectal, and prostatic. MAb BR96 was obtained using human breast carcinoma cells as the immunogen and binds to an Le^y-related tumor-associated antigen (Hellström et al. 1990). BR96 was found to rapidly internalize into lysosomes and endosomes following its binding to the cell surface (Garrigues et al. 1993). BR96 is reactive with a high percentage of human carcinomas and to a minimal amount of normal tissue, specifically gastrointestinal epithelial cells and pancreatic acinar cells (Hellström et al. 1990). It has also been found to cross-react with rat, dog, and monkey tissue, but not mouse tissue.

Single-Chain Immunotoxin BR96 sFv-PE40

BR96 sFv-PE40 is a single-chain immunotoxin consisting of the heavy and light chain variable regions of BR96 fused to PE40, a binding-defective form of Pseudomonas exotoxin A (PE). PE is a single-chain polypeptide composed of three structural domains, each of which contains specific functions (Allured et al. 1986; Hwang et al. 1987). Domain I encodes the cell-binding activity; domain II, the cytosol entry or translocation mechanism; and domain III, the ADP-ribosylation activity that inhibits protein synthesis. The cell-killing activity of fusion proteins incorporating elements of PE depends upon an intramolecular protease-sensitive region that releases the enzymatic portion of the molecule to enter the cytosol following intracellular cleavage of the toxin. PE40 is a truncated form of PE which is devoid of its binding domain but retains translocation and enzymatic activity (Kondo et al. 1988). The Le^y targeted immunotoxin, BR96 sFv-PE40, is thus cytotoxic by binding and internalizing through the BR96 antigen, translocating through domain II of PE, and ADP-ribosylating elongation factor 2 in the target cell, and thus halting protein synthesis through the enzymatic action of PE domain III.

Production and In Vitro Characterization of BR96 sFv-PE40

BR96 sFv-PE40, encoded in expression plasmid pBW 7.0, is expressed in E. coli using an inducible T7 promoter (Fig. 1A). The fusion protein is expressed as insoluble material in cytoplasmic inclusion bodies, thus necessitating denaturation and renaturation to generate active fusion proteins (Friedman et al. 1993a; Siegall et al. 1994a). BR96 sFv-PE40 has been purified to near homogeneity (Fig. 1B), with yields of up to 100 mg of biologically active fusion protein per liter of fermentation culture.

Using H3396 human breast carcinoma cells in ELISA format, BR96 sFv-PE40 was found to bind as well as BR96 Fab (Siegall et al. 1994a) and approximately 75 times less well than BR96 IgG (unpublished observation) to the BR96 defined antigen. BR96 sFv-PE40 inhibits protein synthesis in vitro (Friedman et al. 1993a) and is cytotoxic towards a variety of human carcinoma cell lines, especially MCF-7 and H3396 breast carcinomas, with EC_{50} values in the picomolar range (Siegall et al. 1994a). The intoxication of cells by BR96 sFv-PE40 is directly related to the amount of BR96 antigen expressed on the cell surface, although other factors, including rate of internalization, intracellular trafficking, and translocation efficiency, can affect the sensitivity of each cell type.

Antitumor Activity of BR96 sFv-PE40

Athymic mice xenografted with L2987 human lung carcinoma cells or MCF-7 breast carcinoma cells responded to BR96 sFv-PE40 with complete regression and in some cases cures of their tumors (Friedman et al. 1993b). The antitumor activity of BR96 sFv-PE40 was found to be both schedule- and dose-dependent. Therapeutic efficacy was determined for BR96 sFv-PE40 against L2987 xenografts with administration schedules of 1, 2, 4, and 7 days between intravenous injections. Higher percentages of long-lasting complete regressions were obtained in groups administered every 4 days than in groups with the other three schedules (Friedman et al. 1993b). Efficacy of BR96 sFv-PE40 using mice xenografted with H3396 breast carcinoma cells was also found to be dose-dependent (Fig. 2). As little as a twofold difference in dose separated partial regression from complete regression of the tumor xenograft.

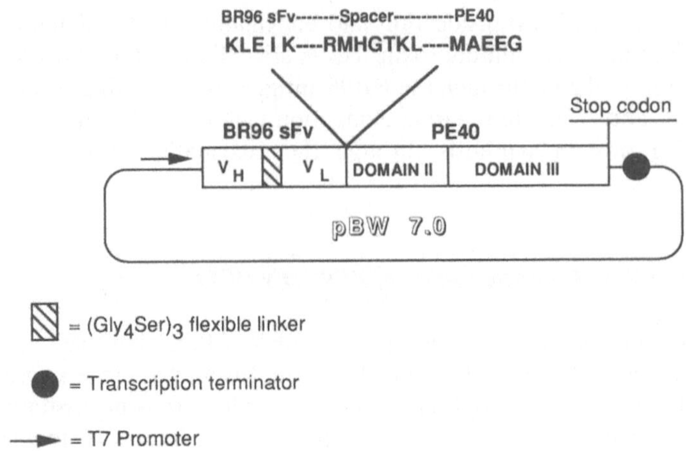

A

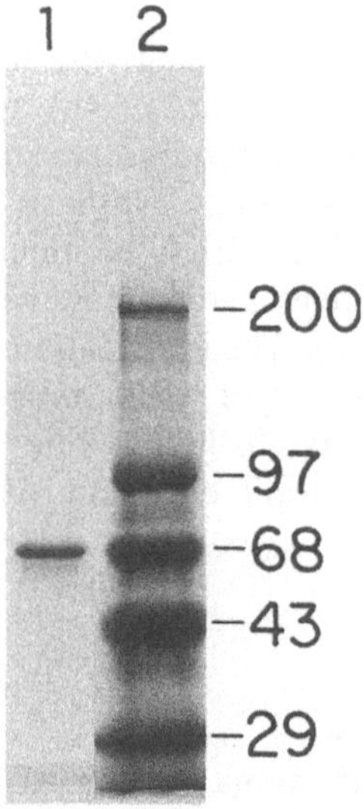

B

Fig. 1A,B. A The expression plasmid encoding BR96 sFv-PE40. **B** Coomassie-blue stained SDS-PAGE analysis of BR96 sFv-PE40 (*lane 1*), molecular mass standards in kDa (*lane 2*)

Unlike mice, which do not express the Lewis-Y antigen, rats express the antigen on normal tissue and have a similar distribution pattern to that found in humans. Thus, rats represent a model in which normal tissue toxicities of BR96 sFv-PE40 would be readily apparent. Complete regression and cures of H3396 xenografts were found in athymic rats upon administration of 0.25–0.5 mg/kg BR96 sFv-PE40 doses (Siegall et al. 1994a). In athymic rats, complete regression of NBT II rat bladder tumor allografts, expressing more than tenfold reduced BR96 antigen than H3396 xenografts, were also obtained (Siegall et al. 1995a).

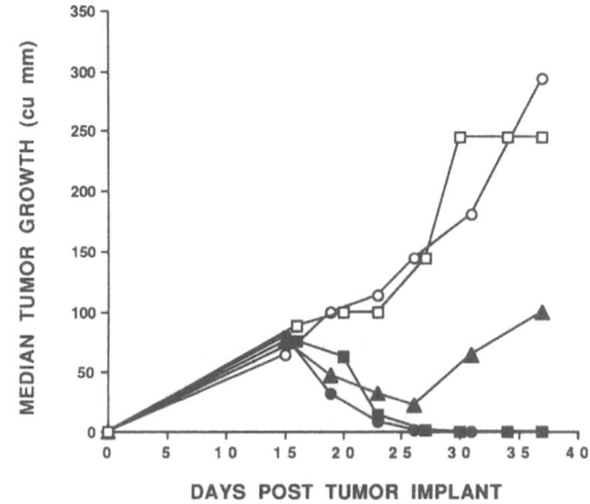

Fig. 2. Antitumor activity of BR96 sFv-PE40. Mice carrying H3396 human breast carcinoma xenografts were treated with BR96 sFv-PE40 using a schedule of q4dx3. Each group consisted of 5 mice. *Filled squares*, 0.44 mg/kg; *filled circles*, 0.33 mg/kg; *filled triangles*, 0.165 mg/kg; *open squares*, 0.088 mg/kg; *open circles*, nontreated control

Dose-Limiting Toxicity of BR96-sFv-PE40

In many immunotoxin trials, efficacy was only observed at doses near or above the maximum tolerated dose, necessitating cessation or reduction of administered dose. Toxicities included hepatotoxicity, vascular (or capillary) leak syndrome (VLS), and myelosuppression. These and other dose-limiting toxicities have hindered immunotoxins from becoming part of standard combination chemotherapy for cancer. As such, the importance of research in this area cannot be overstated. To this end, it is necessary to perform extensive toxicological analysis with each immunotoxin to identify appropriate animal models. While hepatotoxicity may be observed in mice treated with PE-based immunotoxins, no VLS was observed, even up to lethal doses (Siegall et al. 1994b). However, rats respond to 2 mg/kg BR96 sFv-PE40 with VLS in much the same way as is found in humans: that is, with fluid buildup in the thoracic cavity (hydrothorax) and a concomitant decrease in albumin and increase in hematocrit. In contrast, the dose-limiting toxicity in mice, dogs, and monkeys was mainly liver toxicity. Many compounds were evaluated as VLS inhibitors in rats administered with 2 mg/kg BR96 sFv-PE40. Of the agents tested, dexamethasone given prior to administration of the immunotoxin was found to prevent BR96 sFv-PE40 induced VLS in rats without blocking the antitumor activity of the immunotoxin (Siegall et al. 1994b). When dexamethasone was administered concurrently or following the BR96 immunotoxin, hydrothorax was not prevented. Thus, the use of prophylactic dexamethasone should be strongly considered in clinical trials with BR96 sFv-PE40 and other similar immunotoxins.

Heregulin Fusion Toxins

Heregulin is a ligand with structural homology to epidermal growth factor (EGF) and that binds to at least two human EGF receptor (HER) family members, HER4 and HER3 (Plowman et al. 1994; Carraway et al. 1994). Heregulin has also been shown to bind with higher affinity to HER4 than to HER3 (Tzahar et al. 1994). A large percentage of primary breast and prostatic carcinomas express HER3 and HER4 (Siegall and Bacus, in preparation). Due to the potential oncogenic nature of HER3 and HER4 and their expression profile on the surface of certain carcinomas, heregulin was evaluated as a delivery agent for PE40.

Construction, Expression, and Purification of HAR-TX β2

HAR-TX β2, consisting of a chimeric heregulin β2 ligand fused to PE40, was constructed, expressed, and purified in monomeric form (Siegall et al. 1995b). The chimeric heregulin β2 ligand contains the amphiregulin leader sequence genetically fused to heregulin β2. Due to its highly charged nature, the leader sequence is used to enhance the purification of HAR-TX β2. The expression plasmid pSE 8.4, containing the gene fusion that encodes the HAR-TX β2 fusion protein, is diagrammatically shown in Fig. 3A. Similar to BR96 sFv-PE40, HAR-TX β2 localizes to intracellular inclusion bodies following expression in E. coli, necessitating denaturation and refolding to generate active fusion protein. The refolded protein is then selectively purified by cation exchange chromatography, in large part due to the positive charged residues contributed by the amphiregulin leader sequence. Depending on the level of purity following cation exchange chromatography, the HAR-TX β2 fusion protein can be further purified using size exclusion chromatography (Fig. 3B).

In Vitro Characterization of HAR-TX β2 Against Human Breast Carcinoma Cell Lines

HAR-TX β2 specifically binds to breast carcinoma cells which express HER4 and/or HER3 on their surface. HAR-TX β2 binds to low and high affinity sites expressed on MCF-7 and MDA-MB-453 cells (both expressing HER2, HER3, HER4), and H3396 cells (expressing HER2 and HER3, but not HER4) with K_d values ranging from 88 pM to 3.3 nM (Siegall et al. 1995b). Thus, even in the absence of HER4, HAR-TX β2 was able to specifically bind to H3396 cells.

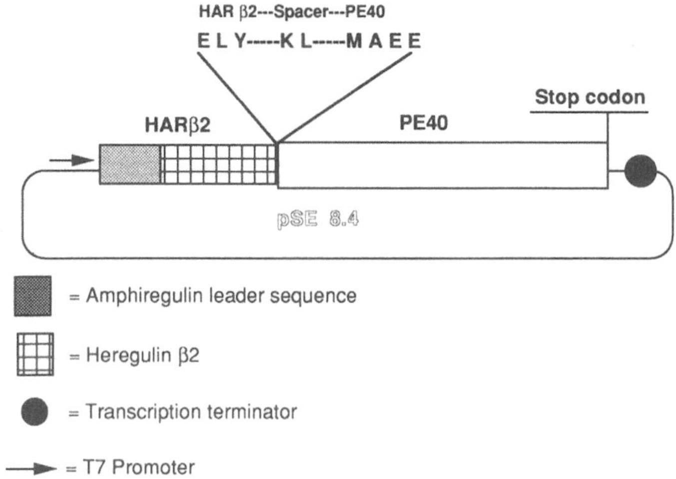

A

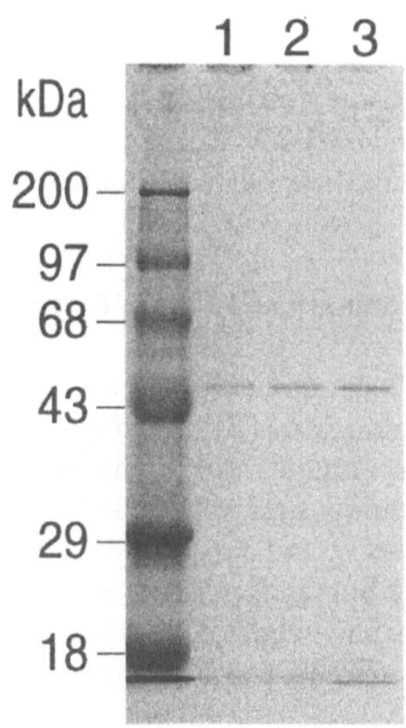

B

Fig. 3A,B. A The expression plasmid enco HAR-TX β2. **B** Coomassie-blue stained SDS-P/ analysis of size exclusion chromatography (S purified HAR-TX β2. *Lanes 1–3*, successive tions of HAR-TX β2 eluted from SEC. Mole(mass standards are in kDa

HAR-TX β2 was cytotoxic towards a variety of human breast cancer cell lines including AU565, SKBR3, MDA-MB-453, T47D, BT474 and MCF-7, all of which express HER2, HER3, and HER4 (Siegall et al. 1995b) A comparison of the cytotoxic activity of HAR-TX β2 with BR96 sFv-PE40 is listed for several breast carcinoma lines in Table 1. However, H3396 cells were not sensitive to HAR-TX β2, even though specific bınding to cell surface HER3 receptors (perhaps HER2/HER3 heterodimers) was observed. Similar observations were also made using a lung carcinoma cell line, L2987,

Table 1. Cytotoxicity analysis of HAR-TX β2 and BR96 sFv-PE40 against breast carcinoma cell lines

Breast carcinoma cell type	EC_{50} (ng/ml)	
	HAR-TX β2	BR96 sFv-PE40
MDA-MB-453	2	40
SKBR3	20	100
T47D	10	5
MCF-7	125	4
H3396	>1000	4

EC_{50} is defined as the amount of fusion toxin required to kill 50% of the target cell line as determined by the cleavage of Calcein-AM, through intracellular esterases found in viable cells, to the fluorescent product Calcein following a 48 h incubation period.

that expresses both HER2 and HER3 but is not sensitive to HAR-TX β2. Therefore, the expression of HER4 correlates with the sensitivity of carcinoma cell lines towards HAR-TX β2. Due to the potent cytotoxic activity of HAR-TX β2 in vitro, experiments have been initiated to evaluate its antitumor efficacy in mice carrying breast tumor xenografts.

Summary and Future Perspective

Most immunotoxins used in the clinic have been formulated as chemical conjugates (Brinkmann and Pastan 1994; Ghetie and Vitetta 1994). Newly developed single-chain forms, including those containing cloned antibody binding sites such as BR96 sFv-PE40, are being prepared for clinical evaluation in the near future and offer exciting opportunities for the treatment of human breast cancer. However, insightful understanding and clinical management of dose-limiting toxicities, including VLS, are needed before immunotoxins become an accepted part of cancer chemotherapeutic regimens.

Acknowledgements. The author thanks D. Chace, S. Gawlak, and B. Mixan for expert technical assistance, Drs. A. Goetze and L. Paul for fusion protein production and purification, Dr. J. Francisco for suggestions to the manuscript, and Drs. P. Fell, I. Hellström, and K.E. Hellström for insightful scientific discussions.

References

Allured VS, Collier RJ, Carroll SF, McKay DB (1986) Structure of exotoxin A of Pseudomonas aeruginosa at 3.0-A resolution. Proc Natl Acad Sci USA 83:320–324

Brinkmann U, Pastan I (1994) Immunotoxins against cancer. Biochim Biophys Acta 1198:27–45

Carraway KL III, Sliwkowski MX, Akita RW, Platko JV, Guy PM, Nuijens A, Diamonti AJ, Vandlen RL, Cantley LC, Cerione RA (1994) The erbB3 gene product is a receptor for heregulin. J Biol Chem 269:14303–14306

Chaudhary VK, Queen C, Junghans RP, Waldmann TA, FitzGerald DJ, Pastan I (1989) A recombinant immunotoxin consisting of two antibody variable domains fused to Pseudomonas exotoxin. Nature 339:394–397

Frankel AE (1993) Immunotoxin therapy of cancer. Oncology 7:69–86

Friedman PN, McAndrew SJ, Gawlak SL, Chace D, Trail PA, Brown JP, Siegall CB (1993a) BR96 sFv-PE40, a potent single-chain immunotoxin that selectively kills carcinoma cells. Cancer Res 53:334–339

Friedman PN, Chace DF, Trail PA, Siegall CB (1993b) Anti-tumor activity of the single-chain immunotoxin BR96 sFv-PE40 against established breast and lung tumor xenografts. J Immunol 150:3054–3061

Garrigues J, Garrigues U, Hellström I, Hellström KE (1993) Le^y specific antibody with potent anti-tumor activity is internalized and degrade in lysosomes. Am J Pathol 2:607–622

Ghetie V, Vitetta E (1994) Immunotoxins in the therapy of cancer: from bench to clinic. Pharmacol Ther 63:209–234

Hellström I, Garrigues HJ, Garrigues U, Hellström KE (1990) Highly tumor-reactive, internalizing, mouse monoclonal antibodies to LeY-related cell-surface antigens. Cancer Res 50:2183–2190

Hwang J, FitzGerald DJP, Adhya S, Pastan I (1987) Functional domains of Pseudomonas exotoxin identified by deletion analysis of the gene expressed in E. coli. Cell 48:129–136

Kondo T, FitzGerald D, Chaudhary VK, Adhya S, Pastan I (1988) Activity of immunotoxins constructed with modified Pseudomonas exotoxin A lacking the cell recognition domain. J Biol Chem 263:9470–9475

Plowman GD, Green JM, Culouscou JM, Carlton GW, Rothwell VM, Buckley S (1993) Heregulin induces tyrosine phosphorylation of $HER4/p180^{erbB4}$. Nature 366: 473–475

Press OW, Eary JF, Appelbaum FR, Martin PJ, Badger CC, Nelp WB, Glenn S, Butchko G, Fisher D, Porter B, Matthews DC, Fisher LD, Bernstein ID (1993) Radiolabeled-antibody therapy of B-cell lymphoma with autologous bone marrow support. N Engl J Med 329:1219–1224

Senter PD, Saulnier MG, Schreiber GJ, Hirschberg DL, Brown JP, Hellström I, Hellström KE (1988) Antitumor effects of antibody-alkaline phosphatase conjugates in combination with etoposide phosphate. Proc Natl Acad Sci USA 85:4842–4846

Siegall CB, Chace D, Mixan B, Garrigues U, Wan H, Paul L, Wolff E, Hellström I, Hellström KE (1994a) In vitro and in vivo characterization of BR96 sFv-PE40. J Immunol 152:2377–2384

Siegall CB, Liggitt D, Chace D, Tepper MA, Fell HP (1994b) Prevention of immunotoxin-mediated vascular leak syndrome in rats with retention of antitumor activity. Proc Natl Acad Sci USA 91:9514–9518

Siegall CB, Wolff EA, Gawlak SL, Paul L, Chace D, Mixan B (1995a) Immunotoxins as cancer chemotherapeutic agents. Drug Dev Res 34:210–219

Siegall CB, Mixan B, Chace D, Bacus SS, Cohen BD, Greene J, Goetze A, Plowman GD, Hellström I, Hellström KE, Fell HP (1995b) HER4 expression correlates with cytotoxicity directed by a heregulin-toxin fusion protein. J Biol Chem 270:7625–7630

Trail PA, Willner D, Lasch SJ, Henderson AJ, Hofstead SJ, Casazza AM, Firestone RA, Hellström I, Hellström KE (1993) Cure of xenografted human carcinomas by BR96-Doxorubicin immunoconjugates. Science 261:212–215

Tzahar E, Levkowitz G, Karunagaran D, Yi L, Peles E, Lavi S, Chang D, Liu N, Yayon A, Wen D, Yarden Y (1994) ErbB-3 and ErbB-4 function as the respective low and high affinity receptors of all neu differentiation factor/heregulin isoforms. J Biol Chem 269:25226–25233

Tamoxifen and Endometrial Cancer: From Experiment to Patient

V.J. Assikis and V.C. Jordan

Robert H. Lurie Cancer Center, Northwestern University Medical School, 303 E Chicago Ave., Chicago, IL 60611, USA

Introduction

Tamoxifen, a nonsteroidal antiestrogen, is the treatment of choice for adjuvant therapy in breast cancer patients. It is currently used both as a palliative treatment for advanced disease and as an adjunct following surgery for node positive or negative disease (Jordan 1994). In addition, tamoxifen's estrogen-like properties have been noted to reduce the incidence of fatal myocardial infarction (McDonald and Stewart 1991; Stewart and Everington 1992) and maintain bone density in postmenopausal patients (Love et al. 1992; Ward et al. 1993; Kristensen et al. 1994). These observations led to the concept of testing tamoxifen as a preventive agent in women with a high risk for breast cancer. Clinical trials that will each recruit about 20 000 women volunteers are now proceeding in the United States, the United Kingdom and Italy (Powles et al. 1994; Vanchieri 1992; Fisher and Redmond 1991). However, in contrast to the beneficial effect on bones and lipids, the estrogenic actions of tamoxifen in the uterus have raised serious doubts about the safety of long-term therapy. Satyaswaroop et al. (1984) first described the partial stimulation by tamoxifen of a human endometrial carcinoma transplanted in athymic mice. Gottardis et al. (1988) cotransplanted a human breast carcinoma and a human endometrial carcinoma into athymic mice to determine whether tamoxifen would control the estrogen stimulated growth of both tumors. Tamoxifen inhibited the growth of the breast tumor but encouraged the growth of the endometrial carcinoma. The demonstration of the target-specific antitumor effects of tamoxifen raised the possibility that patients could develop endometrial carcinoma during long-term adjuvant therapy for breast cancer.

The Stockholm clinical trials group (Fornander et al. 1989) was the first to demonstrate the laboratory principle in a clinical setting. Patients who were randomized to receive tamoxifen had a significant decrease in second primary breast cancers but an increase in the detection of endometrial carcinoma. In fact, patients who were randomized to receive 5 years of tamoxifen had a higher incidence of endometrial cancer. Additionally, a recent report from Yale/New Haven (Magriples et al. 1993) raised considerable concern about the prognosis of endometrial cancers in tamoxifen treated patients. Since thousands of high risk healthy women and hundreds of thousands of breast cancer patients are currently receiving tamoxifen on a daily basis, it is important to address the question: Does long-term tamoxifen treatment cause endometrial carcinoma with a poor prognosis?

Tamoxifen and Endometrial Carcinoma

When evaluating reports of tamoxifen-induced endometrial carcinomas, one should bear in mind that an increased incidence could be due to detection bias. Indeed, autopsy specimens have shown that occult endometrial neoplasia has five times the incidence that is detected in the general population (Horowitz and Feinstein 1986). It is as yet unclear whether tamoxifen's estrogenic properties are promoters of pre-existing cancer growth or cause uterine hyperemia and bleeding and thus lead to an early diagnosis of an existing disease.

Another possible explanation for the increased incidence of endometrial cancer in tamoxifen treated patients is that breast cancer and endometrial cancer share common risk factors: obesity, early menarche, late menopause and high endogenous estrogen levels (Kelsey and Whittemore 1994). Moreover, a Swedish national study documented that breast cancer patients have a relative risk (RR) of 1.72 to develop a uterine tumor. This risk was shown to be age-dependent and rose to 2.4 for patients above 70 years of age (Adami et al. 1987).

The first large scale cohort study to offer evidence of a possible link between tamoxifen and endometrial carcinoma was the Stockholm trial (Fornander et al. 1989). A group of 1846 postmenopausal breast cancer patients was assigned to receive either tamoxifen (40 mg/daily) or placebo. After 5 years of follow-up, 13 endometrial carcinomas were initially reported in the treatment arm and two in the control arm. In a more recent update (Fornander et al. 1993), another four tumors were added to the treatment group and one to the control. The relative risk was reported to be 6.4 ($p < 0.01$). Furthermore, the authors concluded that long-term therapy (>2 years) significantly raised the cumulative frequency of endometrial carcinoma. While demonstrating the benefit of tamoxifen on breast cancer, the Stockholm study focused the clinical community on uterine malignancies.

Prompted by this report, Stewart and Knight (1989) investigated the incidence of endometrial cancers in the Scottish adjuvant tamoxifen trial. There were 1070 postmenopausal patients who were assigned to receive 20 mg/daily or placebo. A 5 year follow-up detected no endometrial carcinomas (but three sarcomas) in the group allocated to tamoxifen while three endometrial carcinomas were reported in the placebo group. An update of this trial (Stewart 1992) added one case of endometrial cancer to each arm.

The discrepancy between the results from the Scottish trial and those found in the Stockholm study was initially attributed to the dose of tamoxifen. This led to the idea that the lower dose of 20 mg daily is safer than higher doses. Indeed, this is the dose currently prescribed by the vast majority of clinicians. In a recent analysis we investigated the role of the dose of tamoxifen in causing endometrial cancer and found that almost half of the patients that developed endometrial carcinoma received 20 mg/daily (Assikis and Jordan 1995). The data we have collected do not support the hypothesis that the higher dose used in the Stockholm study can account for the higher incidence of endometrial cancer.

The second major study that caused concern was conducted by the National Surgical Adjuvant Breast and Bowel Project (NSABP) (Fisher et al. 1994). In adjuvant trial B-14, 20 mg/daily was administered to 1419 out of 2834 randomly assigned breast cancer patients who were followed for an average time of 8 years. In addition, 1220 patients were registered to receive the same dose of tamoxifen and were follwed for more than 5 years. Overall, 15 cases of endometrial carcinoma were reported in the group randomized to tamoxifen. Eight cases occurred in the registered group, one of which was a sarcoma. The RR was determined to be 7.5 (95% CI = 1.7 – 32.7) and the annual hazard rate for endometrial cancer was 1.2/1000 woman years.

The Stockholm and the NSABP trials raised a great deal of concern among patients and clinicians regarding the frequency of endometrial malignancies in tamoxifen treated patients. There was, however, no evidence at that point to demonstrate a worse prognosis of these tumors. Since the carcinogenicity of tamoxifen was attributed to its estrogenic properties, these tumors were expected to have the favorable outcome experienced in patients with a history of estrogen replacement therapy (Grady et al. 1995).

However, a study from Yale published in the Journal of Clinical Oncology in 1993 (Magriples et al. 1993) challenged this belief. The authors conducted a retrospective search of the Yale/New Haven tumor registry for the decade 1980–1990. After screening 3457 patient records, they identified 53 women with a history of breast cancer who later developed a malignancy of the uterine corpus. Fifteen of these patients had a history of tamoxifen use. The dose was 40 mg/daily and the average duration of therapy was 4.2 years. Ten of the 15 tumors (67%) were either grade 3 or MMT (mixed mullerian tumors) or carcinomas with an unfavorable outcome (clear-cell carcinoma, papillary serous carcinoma). Five out of the 15 patients (33%) died of endometrial cancer. The authors concluded that "... women receiving tamoxifen are at risk for high-grade endometrial cancers that have a poor prognosis ..."

Does Long-Term Tamoxifen Induce Endometrial Cancer?

Before addressing the question we posed in the Introduction, we should clarify whether the concept that long-term therapy is worse for the endometrium holds true or not. To investigate this issue we will refer to the data provided by the three aforementioned clinical trials that have a 5-year follow-up.

Although the Stockholm study (Fornander et al. 1993) claimed that long-term tamoxifen therapy enhanced the frequency of endometrial cancer, plotting the actual data leads us to draw a different conclusion. In Fig. 1 we have plotted the duration of tamoxifen therapy in months vs the detection of the 16 endometrial carcinoma cases (one MMT has been excluded). It is apparent that 12 out of 16 patients received 2 years of tamoxifen therapy or less. Figure 1 shows that three times as many endometrial cancers were reported for women that stopped tamoxifen at 2 years as compared to those who continued the drug. Our interpretation of the Stockholm data does not support the position that long-term tamoxifen increases the incidence of endometrial cancer.

Likewise, no evidence of more endometrial carcinoma cases with long-term therapy was shown in the Scottish trial (Stewart and Knight 1989). The update (Stewart 1992) of this trial reports one case of endometrial

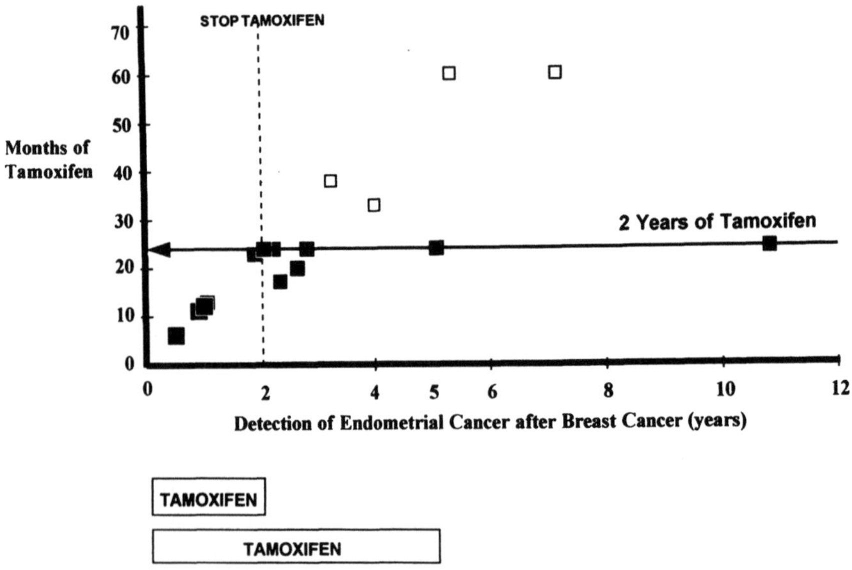

Fig. 1. Endometrial carcinoma cases in the Stockholm study (Fornander et al. 1993): 931 patients were treated with up to 2 years of tamoxifen and then randomized to either stop or continue for 3 more years

carcinoma in the tamoxifen treated group as opposed to three cases in the placebo group.

The NSABP study B-14 (Fisher et al. 1994) reported a total of 22 cases of endometrial carcinoma (we have excluded one sarcoma). Fifteen of these cases were noted in patients (n = 1419) that were randomized to receive 20 mg/day) and were then followed for at least 5 years. One of these patients never received the drug. The placebo group (n = 1424) had two cases. Recurrence of breast cancer in one of these patients and colon cancer in the other prompted their attending physicians to administer tamoxifen to them. Both patients were on tamoxifen therapy when endometrial cancer was diagnosed. Seven cases occurred in patients (n = 1220) registered to receive tamoxifen for at least 5 years. Based on the above analysis of data, we have concluded that the actual total number of patients who received tamoxifen and subsequently developed endometrial carcinoma in the NSABP study was 23. In Fig. 2 we have plotted the endometrial carcinoma cases that occurred in the randomized and the registered group of the NSABP study vs the duration of tamoxifen in years. Eight patients had less than 2 years of tamoxifen therapy, nine had 2–5 years and four had more than 5 years when

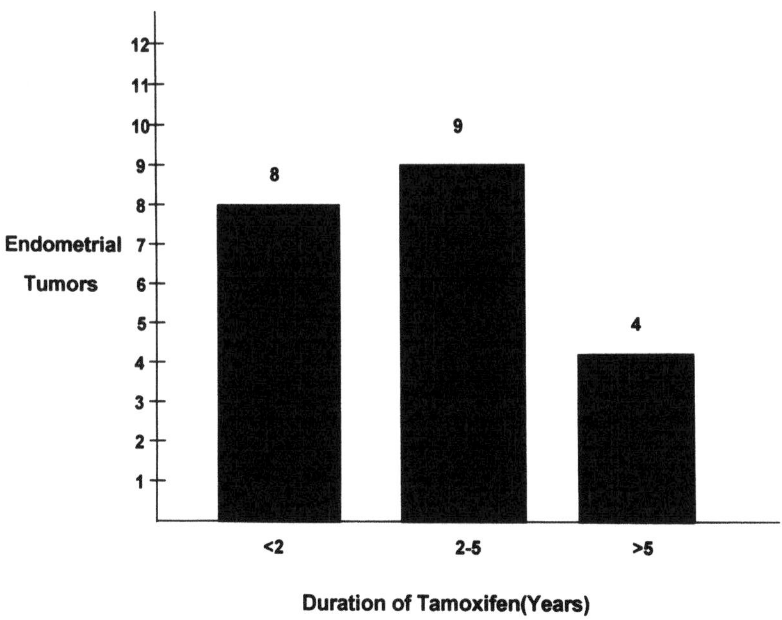

Fig. 2. The occurrence of endometrial carcinoma in the NSABP trial B-14 (Fisher et al. 1994) vs the duration of tamoxifen treatment: 1419 patients were randomized to receive 20 mg/d at of tamoxifen and 1220 were registered to be treated with tamoxifen. Fifteen endometrial carcinomas were reported in the randomized group and eight in the registered group. In our analysis we excluded one of the 15 cases (the patient never took tamoxifen) and one of the eight cases (it was a sarcoma)

endometrial cancer was diagnosed. If we compare the 2 years group to the long-term therapy (2–5 years) group, the RR for long-term therapy is 1.12. Even if the incidence in the NSABP long-term group is much higher than the one noted in the Stockholm study, the RR is only slightly elevated when compared to the up to 2 years group.

Overall, it seems that the 5 year trials do not offer any clear evidence that long-term tamoxifen treatment is worse for endometrial cancer. Moreover, there are data that even a short course of tamoxifen may increase the endometrial cancer cases.

A Danish trial (Andersson et al. 1991) used 30 mg of tamoxifen daily for 48 weeks (1 year) and detected two endometrial carcinomas in patients (n = 846) who received radiotherapy (RT) alone, but seven endometrial carcinomas in patients (n = 864) who received both RT and tamoxifen. The incidence ratio for endometrial cancer was 1.9% (95% CI = 0.8–3.9). The authors reported that after 5 years the cumulative incidence was 0.27% for the control group and 1% for the tamoxifen group. If we compare the incidence in these two groups, we will notice that the ratio of endometrial cases in the (RT + tamoxifen) group (seven cases in 864 patients = 8.1/1000) is 3.5-fold the ratio in the RT group (two cases in 846 patients = 2.4/1000). This would indicate that adding tamoxifen to radiotherapy seriously increases the probability of an endometrial cancer. However, if we look at the control group (11 cases in 1828 patients = 6/1000) who received no therapy, we will notice that comparing (RT + tamoxifen) incidence (8.1/1000) to no therapy (6/1000) shows only a 1.35-fold increase.

We surveyed the world literature and found 250 cases of endometrial carcinoma in patients that had tamoxifen treatment (Table 1). The vast majority of these patients were postmenopausal. There are no data available on the duration of tamoxifen in all of them; however, we do know that 80 patients had less than 2 years of tamoxifen therapy and 100 had more than 2

Table 1. The reported cases of corpus uterine cancer in patients receiving tamoxifen treatment for breast cancer (adapted from Assikis and Jordan 1995)

Endometrial cancer cases	
Endometrial carcinoma	250
Mixed mullerian tumors	13
Sarcomas	7
Patients	
Postmenopausal	173
Premenopausal	1
Duration of treatment	
<2 years	80
>2 years	100

Data are derived from a survey of the literature.

years. The number of patients in these two groups is not significantly different. Our analysis is therefore consistent with the position that long-term tamoxifen (greater than 2 years) is not causing endometrial carcinoma.

It is now time to address our opening question: Does long-term tamoxifen treatment produce endometrial carcinomas with poor prognosis?

Does Tamoxifen Produce Late Stage, High Grade Disease?

In Table 2 we have gathered all the available data on stage and grade of endometrial carcinomas discovered in patients that had tamoxifen treatment. Stage was reported for 156 patients and grade was reported for 165. Some 80% of the reported cases are stage I and good grade (1 or 2). In Table 3 we have compared the data for stage and grade for the Yale study with our survey (we have excluded the Yale cases) and the 1983–1987 Surveillance, Epidemiology and End Results (SEER) data (National Cancer Institute 1993), derived from 12717 endometrial cancer patients in the United States. As far as stage is concerned, there is no significant variation among the three groups but there is a striking difference when we compare the grade of these tumors. Only 20% of cases were of high grade in our database, far less than the 62% reported in the Yale study. Our survey's data are almost identical to the SEER data. It would therefore seem that tamoxifen treated patients demonstrate the familiar favorable pattern seen in most of endometrial cancer cases.

Table 2. Grade and stage of endometrial carcinoma in tamoxifen treated patients

Author	Number of events (n = 191)	Stage (n = 156)		Grade (n = 165)	
		I	I–IV	Good	Poor
Malfetano 1990	7	2	5	6	0
Fornander et al. 1989, 1993	16	13	3	15	0
Segna et al. 1992	11	11	0	9	2
Magriples et al. 1993	13	7	2	5	8
Fisher et al. 1994	23	20	3	17	5
Mignotte et al. 1992	23	8	2	16	2
van Leeuwen et al. 1994	23	17	3	12[a]	11[a]
Silva et al. 1994	12	8	3	3	0
Barakat et al. 1994	27	18	9	16	5
Dallenbach-Hellweg and Hahn (1995)	10	10	0	10	0
Others[b]	26	11	1	18	5

[a] Calculated from a statement made by the authors in the Discussion section of the paper.
[b] Killackey et al. 1985; De Muylder et al. 1991; Seoud et al. 1993; Dauplat et al. 1990; Mathew et al. 1990; Atlante et al. 1990; Spinelli et al. 1991; Cohen et al. 1994; Lahti et al. 1994.

Table 3. Comparison of the stage and grade of endometrial carcinoma in the Yale series[a] with our database[b]

	Yale		Review		SEER[c]
Stage					
I	7/9	78%	118/147	80%	74%
II–IV	2/9	22%	29/147	20%	26%
Grade					
Good (Grades 1, 2)	5/13	38%	122/152	80%	79%
Poor (Grade 3)	8/13	62%	30/152	20%	21%

[a] From Magriples et al. 1993.
[b] From Assikis and Jorday, 1995
[c] The SEER data are used as a standard for grade and stage distribution.

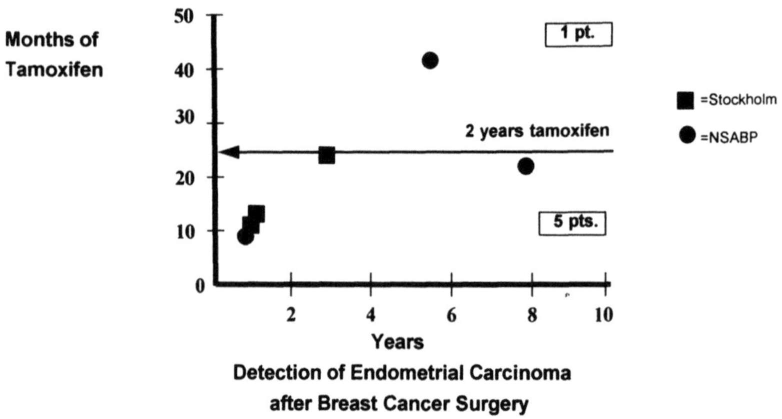

Fig. 3. Deaths due to endometrial carcinoma in the Stockholm (Fornander et al. 1993) and the NSABP B-14 (Fisher et al. 1994) trials. Duration of tamoxifen treatment is plotted against the detection of endometrial carcinoma in patients who eventually died of endometrial cancer

Finally, we have determined the duration of tamoxifen that is associated with poor prognosis endometrial cancer. We have plotted the duration of tamoxifen vs the detection of endometrial carcinoma in patients who are documented to have died of endometrial carcinoma in the Stockholm and NSABP studies (Fig. 3). Five out of six patients developed endometrial cancer after a course of 2 years or less of tamoxifen.

General Conclusions

Tamoxifen is the adjuvant therapy of choice for postmenopausal receptor positive breast cancer patients regardless of nodal status. Its benefit in decreasing breast tumor recurrence and prolonging overall survival has been clearly documented (Early Breast Cancer Trialists Collaborative Group 1992). However, during the past 5 years there has been an ongoing controversy regarding the induction of endometrial carcinoma by tamoxifen. Moreover, during the last 2 years there has been intense debate on whether tamoxifen is associated with poor prognosis endometrial disease.

To address this issue, we searched through the literature to collect all available data on endometrial cancer and tamoxifen. We have found 250 cases of endometrial carcinoma. Based on the fact that there are 6 million woman-years of experience with tamoxifen, the prevalence of endometrial carcinoma seems to be rather low. In the US alone it is estimated that at least 80 000 women will start a long-term of tamoxifen therapy annually. This would translate into a conservative estimate of 1/2 million women currently receiving the drug in the US alone. Since tamoxifen is used worldwide the total number of patients starting tamoxifen annually is enormous. In addition, thousands of healthy, high risk for breast cancer women are already receiving tamoxifen in an effort to investigate the drug's potential to prevent breast cancer. It is therefore crucial to document the link between tamoxifen and endometrial cancer in a definitive way. Out of the 250 cases we collected, we know the stage for 156 patients and the grade for 165 patients. When we compared these data to the data derived from the SEER program, they were almost identical for both stage and grade: 80% of the patients that developed an endometrial carcinoma during or after tamoxifen therapy had early stage and good grade (1 or 2) disease.

It is almost impossible to give a precise rate for the incidence of endometrial cancer in tamoxifen treated patients. Based on the reported incidence in the major clinical trials and on the findings derived from our database, we estimate the risk to be 2–3/1000 women annually.

Clinicians should be aware of the increased incidence of endometrial tumors in tamoxifen treated patients and remain vigilant for early symptoms indicative of endometrial disease. However, the slightly increased incidence of endometrial disease does not outweigh the substantial benefit of long-term tamoxifen in these patients.

References

Adami HO, Kruesmo UB, Bergkvist L, Person I, Pettersson B (1987) On the age-dependent association between cancer of the breast and of the endometrium. A nationwide cohort study. Br J Cancer 55:77–80

Andersson M, Storm HH, Mouridsen HT (1991) Incidence of new primary cancers after adjuvant tamoxifen therapy and radiotherapy for early breast cancer. J Natl Cancer Inst 83:1013–1017

Assikis VJ, Jordan VC (1995) Gynecologic effects of tamoxifen and the association with endometrial carcinoma. Int J Gynecol Obstet 49:241–257

Atlante G, Pozzi M, Vincenzoni C, Vocaturo G (1990) Four case reports presenting new acquisitions on the association between breast and endometrial carcinoma. Gynecol Oncol 37:378–380

Barakat RR, Wong G, Curtin JP, Vlamis V, Hoskins WJ (1994) Tamoxifen use in breast cancer patients who subsequently develop corpus cancer is not associated with a higher incidence of adverse histologic features. Gynecol Oncol 55:164–168

Cohen I, Altaras MM, Lew S, Tepper R, Beyth Y et al. (1994) Ovarian endometrioid carcinoma and endometriosis developing in a postmenopausal breast cancer patient during tamoxifen therapy: a case report and review of the literature. Gynecol Oncol 55:443–447

Dallenbach-Hellweg G, Hahn U (1995) Mucinous and clear cell adenocarcinomas of the endometrium in patients receiving antiestrogens (tamoxifen) and gestagens. Int J Gynecol Pathol 14:7–15

Dauplat J, Le Bouedec G, Achard JL (1990) Adenocarcinome de l' endométre chez 2 malades prenant du tamoxifene. Presse Med 19:380–381

De Muylder X, Neven P, De Somer M, Van Belle Y, Vanderick G et al. (1991) Endometrial lesions in patients undergoing tamoxifen therapy. Int J Gynecol Obstet 36:127–130

Early Breast Cancer Trialists Collaborative Group (1992) Systemic treatment of early breast cancer by hormonal, cytotoxic, or immune therapy: 133 randomized trials involving 31 000 recurrences and 24 000 deaths in 75 000 women. Lancet 339:1–15, 71–85

Fisher B, Redmond C (1991) New perspective on cancer of the contralateral breast: a marker for assessing tamoxifen as a preventive agent. J Natl Cancer Inst 83:1278–1280

Fisher B, Costantino JP, Redmond CK, Fisher ER, Wickerham DL et al. (1994) Endometrial cancer in tamoxifen-treated breast cancer patients: findings from the National Surgical Adjuvant Breast and Bowel Project (NSABP) B-14. J Natl Cancer Inst 86:527–537

Fornander T, Cedermark B, Mattsson A, Skoog L, Theve T et al. (1989) Adjuvant tamoxifen in early breast cancer: occurence of new primary cancers. Lancet i:117–120

Fornander T, Hellstrom AC, Moberger B (1993) Descriptive clinicopathologic study of 17 patients with endometrial cancer during or after adjuvant tamoxifen in early breast cancer. J Natl Cancer Inst 85:1850–1855

Gottardis MM, Robinson SP, Satyaswaroop PG, Jordan VC (1988) Contrasting action of tamoxifen on endometrial and breast tumor growth in the athymic mouse. Cancer Res 48:812–815

Grady D, Gebretsadik T, Kerlikowske K, Ernster V, Petitti D (1995) Hormone replacement therapy and endometrial cancer risk: a meta-analysis. Obstet Gynecol 85:304–313

Horowitz RI, Feinstein AR (1986) Estrogens and endometrial cancer: response to arguments and current status of an epidemiologic controversy. Am J Med 81:503–507

Jordan VC (ed) (1994) Long term tamoxifen treatment for breast cancer. University of Wisconsin Press, Madison

Kelsey JL, Whittemore AS (1994) Epidemiology and primary prevention of cancers of the breast, endometrium and ovary. A brief overview. Ann Epidemiol 4:89–95

Killackey MA, Hakes TB, Pierce VK (1985) Endometrial adenocarcinoma in breast cancer patients receiving antiestrogens. Cancer Treat Rep 69:237–238

Kristensen B, Ejlertsen B, Dalgaard P, Larsen L, Holmegaard SN et al. (1994) Tamoxifen and bone metabolism in postmenopausal low-risk breast cancer patients: a randomized study. J Clin Oncol 12:992–997

Lahti E, Vuopala S, Kauppila A, Blanco G, Ruokonen A et al. (1994) Maturation of vaginal and endometrial epithelium in postmenopausal breast cancer patients receiving long-term tamoxifen. Gynecol Oncol 55:410–414
Love RR, Mazess RB, Barden HS, Epstein S, Newcomb PA et al. (1992) Effects of tamoxifen on bone mineral density in postmenopausal women with breast cancer N Engl J Med 326:885–886
Magriples U, Naftolin F, Schwartz PE, Carcangiu ML (1993) High-grade endometrial carcinoma in tamoxifen-treated breast cancer patients. J Clin Oncol 11:485–490
Malfetano J (1990) Tamoxifen-associated endometrial carcinoma in postmenopausal breast cancer patients. Gynecol Oncol 39:82–84
Mathew A, Chabon AB, Kabakow B, Drucker M, Hirschman RJ (1990) Endometrial carcinoma in five patients with breast cancer on tamoxifen therapy. NY State J Med 90:207–208
McDonald CC, Stewart HJ (1991) Fatal myocardial infarction in the Scottish tamoxifen trial Br Med J 303:435–437
Mignotte H, Sasco AJ, Lasset C, Saez S, Rivoire M et al. (1992) Traitement adjuvant du cancer du sein par tamoxifene et cancer de l' endomètre. Bull Cancer (Paris) 79:969–977
National Cancer Institute (1993) SEER cancer statistics review 1973–1990. NCI, Bethesda (Document no 93-2789)
Powles TJ,. Jones AL, Ashley SE, Obren MER, Tid Y et al. (1994) The Royal Mardsen Hospital pilot tamoxifen chemoprevention trial. Breast Cancer Res Treat 31:73–82
Satyaswaroop PG, Zaino RJ, Mortel R (1984) Estrogen-like effects of tamoxifen on human endometrial carcinoma transplanted into nude mice. Cancer Res 44: 4006–4010
Segna RA, Dottino PR, Deligdisch L, Cohen CJ (1992) Tamoxifen and endometrial cancer. Mt Sinai J Med 59:416–418
Seoud MAF, Johnson J, Weed JC (1993) Gynecologic tumors in tamoxifen-treated women with breast cancer. Obstet Gynecol 82:165–169
Silva EG, Tornos CS, Follen-Mitchell M (1994) Malignant neoplasms of the uterine corpus in patients treated for breast carcinoma: the effects of tamoxifen. Int J Gynecol Pathol 13:248–258
Spinelli G, Bardazzi N, Citernesi A, Fontanarosa M, Curiel P (1991) Endometrial carcinoma in tamoxifen-treated breast cancer patients. J Chemother 3:267–270
Stewart HJ (1992) The Scottish trial of adjuvant tamoxifen in node-negative breast cancer Scottish Cancer Trials Breast Group. Monogr Natl Cancer Inst 11:117–120
Stewart HJ, Everington D (1992) Treatment of elderly patients with breast cancer (Letter). Br Med J 304:1569–1570
Stewart HJ, Knight GM (1989) Tamoxifen and the uterus and endometrium (Letter). Lancet 1:375–376
van Leeuwen FE, Benraadt J, Coebergh JWW, Kiemeney LALM, Gimbrere CHF et al. (1994) Risk of endometrial cancer after tamoxifen treatment of breast cancer. Lancet 343:448–452
Vanchieri C (1992) Breast cancer study initiated in Italy. J Natl Cancer Inst 84: 1555–1556
Ward RL, Morgan G, Dalley D, Kelly PJ (1993) Tamoxifen reduces bone turnover and prevents lumbar spine and proximal femoral bone loss in early postmenopausal women Bone Miner 22:87–94

Summary

A.L. Harris

Imperial Cancer Research Fund, Clinical Oncology Unit, Churchill Hospital, Headington, Oxford OX3 7LJ, UK
Molecular Oncology Laboratories, Imperial Cancer Research Fund, Institute of Molecular Medicine, John Radcliffe Hospital, Oxford OX3 9DU, UK

Introduction

In the last decade there have been major advances in the basic understanding of molecular and cell biology, particularly with relevance to cancer. As these techniques are applied to primary human breast cancer, there has been a better understanding of how conventional therapies work, e.g., tamoxifen and chemotherapy. The biology has also provided many new potential targets for rational development of therapy. Table 1 summarises some of these areas and shows new drugs of more conventional type that are active against breast cancer. It is the cytotoxic drugs that have apparently come through to the clinic much more rapidly than more novel approaches. However, the first taxol studies were carried out 25 years ago and the only novel target shown in Table 1 under cytotoxics is topoisomerase I, which is inhibited by topotecan. Thus, it takes many years to develop even conventional drugs and the new targets raise specific problems in clinical development. One hope for the new approaches is that there will be much greater selectivity and reduced toxicity compared with chemotherapy or radiotherapy.

Specific Problems for New Therapy Approaches

Most patients offered new drugs for breast cancer therapy should already have received optimal conventional therapy which would include chemotherapy and endocrine therapy. Thus, several resistance mechanisms will have developed and clonal evolution for a more aggressive phenotype may have occurred. The response rate may therefore be lower than previously found for conventional agents used to treat therapy – naive populations.

Particularly for drugs used as antiangiogenesis agents and also drugs targeting growth factor receptors (see chapter by Segal, this volume), antitumour effects may be much better on smaller tumours. This was well shown

Table 1. New therapies and targets for treating breast cancer

Cytotoxic drugs
Taxol, vindelbine, topotecan, gemcitabine, infusional 5FU, CI941
Cellular signalling pathways
Growth factor receptor antagonism, kinase inhibitors
Gene therapy
Immunological, prodrug activation
Antiangiogenesis and metastases
Apoptosis
Telomerase

in xenograft experiments in which therapy could be 100% curative or less than 30%, depending on tumour size. Thus, early extrapolation to the adjuvant situation would appear necessary for reasonable therapeutic effect. However, safety is a major consideration and some thought it would be many years before such treatments could be used that way.

However, there are certain situations, e.g., large primary tumours or the population offered high dose chemotherapy with stem cell rescue, in which more innovative approaches could be considered earlier in management.

It will be much more important in the future to assess the relevant target in tumour tissues to select patients for therapy and to develop better ways to monitor anticancer effects on minimal residual disease.

The problem of integrating even existing modalities and their duration of administration is well known; adding many other new treatments will compound these issues.

The size of trials may need to be much larger to detect additional effects of new therapies in combination with existing therapies. International collaborations may therefore be more important in the future.

The issue of patients rights is important, not only to optimise therapy and information, but also to have involvement in clinical trials. In the future, patients may have much more input into trial design to enable improved opportunities for them to have new approaches to therapy.

One area in which early adjuvant administration may be possible is the combination of somatostatin analogues with tamoxifen, since both suppress serum insulin-like growth factor 1 (IGF1) levels (see contribution by Pollak, this volume), there may be a synergistic effect. Octreotide has been in use for many years and its relative safety established. Thus, this would be a new application of an old drug. Phase II trials have shown no additive toxicity but trials involving over 1000 patients are likely to be needed to see an effect in the adjuvant situation.

Understanding How Conventional Therapy Works – Implications for Future Studies

Tamoxifen continues to provide new ideas as we understand more about its mechanisms of action. The endometrial cancer risk is definite but not as high as previously reported by some groups. Although conventional dose progestins could be used to produce regular withdrawal bleeding (see chapter by AssiKis and Jordan, this volume), the effects of this dose of progestogen on breast cancer are unknown and it has been suggested that in advanced disease such a dose may decrease the effectiveness of tamoxifen. A progestogen coil is the most likely way to minimise both risk and side effects of tamoxifen on the uterus and is likely to be well tolerated.

Whether uterine sarcomas may also be increased by tamoxifen was discussed but the numbers of patients involved (seven) are too low to assess at present. Although currently tamoxifen prevention trials are proceeding well, in the future other analogues or synthetic antiestrogens may be used. Some, such as droloxifene, target the osteoblast and may be of use primarily in the management of osteoporosis. Anticancer effects may then be a desirable side effect rather than the main purpose of the drug (see chapter by Assikis and Jordan, this volume). Since cardiovascular mortality and bone fractures contribute to morbidity and mortality far more than breast cancer, new generation antiestrogens with agonist actions, used for the former indications, may become the prevention agents of choice.

Conclusions

Breast cancer biology is a rapidly changing area, and new treatments coming into Phase I trials now were initiated in the laboratory more than 5 years ago. They may take 10 years to be properly evaluated, so optimising existing agents is important as it is something that can be achieved now. This also provides the baseline for integrating new treatment approaches. Cancer centres should include Phase I trials as part of the comprehensive service to their patients, and educating the patient population may help generate the environment for the extra funds required to develop new treatments.

III. Prognosis of Response and Tumor Markers

Review of Known Prognostic Variables

M. Kaufmann

Johann Wolfgang Goethe – University, Department of Gynecology and Oncology, Theodor-Stern-Kai 7, 60590 Frankfurt/Main, Germany

Introduction

Prognostic variables should be used to provide an estimate of recurrence and death in patients with primary breast cancer. Evaluation of prognostic variables have to consider two different aims:

1. Use as a *prognostic factor*, describing groups of low and high risk for recurrence and death. Maximum information is obtained if the favorable group shows an overall survival of approximately 90%–100%, because these patients do not require further adjuvant treatment.
2. Use as a *predictive factor*, describing the responsiveness of endocrine or cytotoxic therapy or both. Treatment could be intensified in patients at high risk and known responsiveness or omitted in patients with known drug resistance.

Useful prognostic and predictive factors have to (a) be reproducible and include quality control; (b) have been proven to be independent variables and validated in clinical trials; (c) be simple and have therapeutic implications. Known established prognostic and predictive variables according to these criteria are: tumor size, axillary lymph node status, histologic type, grade, S-phase, hormone receptor status, and menopausal status and patients' age.

There has been considerable and critically debate (Clark 1994; Dorr 1993; Hawkins 1993; Henderson et al. 1990; Mansour et al. 1994; McGuire and Clark 1992; Osborne 1992) about the utility of these known factors along with many new described variables to answer the questions: *Who* does *not* benefit from therapy? Which *type* of treatment, e.g., surgery, radiotherapy, cytotoxic and/or endocrine (castration, tamoxifen) should be selected? These questions are relevant for the use of local and/or systemic adjuvant treatment.

Tumor Size

The TNM staging system is useful for estimating prognosis; however, the defined categories for describing primary tumor size (T) are not precise enough. It has been well established that there is a strong correlation between tumor size and risk of recurrence and death. The risk increases directly with increasing tumor diameter (Nemoto et al. 1980; Rosen et al. 1989a,b, 1991; Wilson et al. 1984; Carter et al. 1989) (Table 1).

According to SEER data and other reports, tumors ≤1 cm in diameter have a good prognosis, which also means a death rate of less than 10% after 10 years of observation.

It is also well known that there is a strong linear relation between tumor size and involved axillary lymph nodes (Fig. 1) (Nemoto et al. 1980; Carter

Table 1. Relation of tumor size, negative axillary lymph node status and 5-year survival in 24740 patients (Carter et al. 1989)

Tumor size (cm)	Number of patients	Negative nodes	Relative survival (%)
<0.5	339	99.2	96.2
0.5–0.9	996	98.3	94.8
1.0–1.9	6984	90.6	85.8
2.0–2.9	7282	92.3	84.3
3.0–3.9	4329	86.2	77.0
4.0–4.9	2112	84.6	70.3
>5.0	2698	82.2	62.7

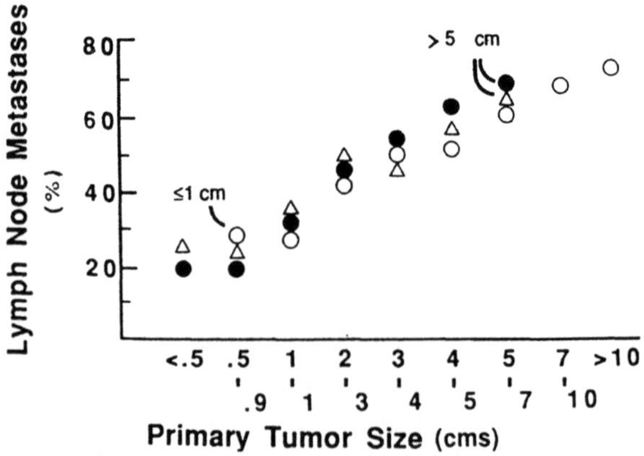

Fig. 1. Primary tumor size versus frequency of axillary lymph node metastases (data from 3 series): *open triangle*, Nemoto et al. (1980); *open circle*, Donegan et al. (1988); *filled circle*, Carter et al. (1989). (From Donegan 1992)

et al. 1989; Donegan 1992). However, the prognostic significance of tumor size was found to act as an independent prognosticator. This suggests that metastases do not occur exclusively via the axillary lymph nodes but also through nonlymphatic routes. Therefore lymph node status mostly serves as an indicator of the tumor's ability to metastasize (Carter et al. 1989; Lin et al. 1993; Seidman et al. 1995).

In general, tumor size in relation to the size of the entire breast is the most important factor in selecting women for breast conservation or primary (neoadjuvant) chemotherapy. Response of chemotherapy, measured by shrinkage of the primary tumor, seems to be the best prognostic factor (Bonadonna 1995).

Large tumors demonstrate lower response rates or even pathological complete response rates compared to small tumors.

In the consensus statement of the NIH Consensus Development Conference (1991) it was concluded: "While all node negative patients have some risk for recurrence, patients with tumors less than or equal to 1 cm have an excellent prognosis and do not require adjuvant systemic therapy outside of clinical trials."

In clinical practice, however, there are also some limitations in measuring tumor size precisely. An accurate measure is often limited by multicentricity and multifocality and by the extent of noninvasive components around the invasive tumor. In small ductal carcinomas the size of only the invasive component, as determined by microscopic measurement, is a better predictor of axillary lymph node status than is the total tumor size (Seidman et al. 1995).

Standardized measurement of tumor response in patients with primary chemotherapy is not yet defined for clinical practice. Pathological invasive tumor size is the most important prognostic and predictive factor for defining risk in patients with primary breast cancer. However, even patients with very small tumors at the time of surgery develop distant metastases.

Axillary Lymph Node Status

Today, the number of involved axillary lymph nodes is still considered to be the main prognostic factor in the systemic treatment of breast cancer. The risk of recurrence and death increases directly with increasing number of involved lymph nodes (risk categories: 0, 1–3, 4–9, 10+ positive nodes) (Table 2) (Nemoto et al. 1980; Wilson et al. 1984; Carter et al. 1989; Donegan 1992). Different risk categories still remain regarding patients treated with systemic adjuvant therapy (Fig. 2) (Kaufmann et al. 1993; von Minckwitz et al. 1995a).

There is general consensus that patients with node-positive tumors will benefit from some form of systemic adjuvant therapy. Absolute benefit for node-positive compared to node-negative patients seems to be twice as high

Table 2. Relation of 5-year recurrence and survival according to number of pathologically positive axillary lymph nodes in 19933 patients (Nemoto et al. 1980)

Number of positive nodes	Number of patients	Recurrence (%)	Survival (%)
0	12299	19.4	71.8
1	2012	32.9	63.1
2	1338	39.9	62.2
3	842	43.0	58.8
4	615	43.9	51.9
5	478	54.2	46.9
6–10	1261	63.4	40.7
11–15	562	71.5	29.4
16–20	301	75.1	28.9
21+	225	82.2	22.2

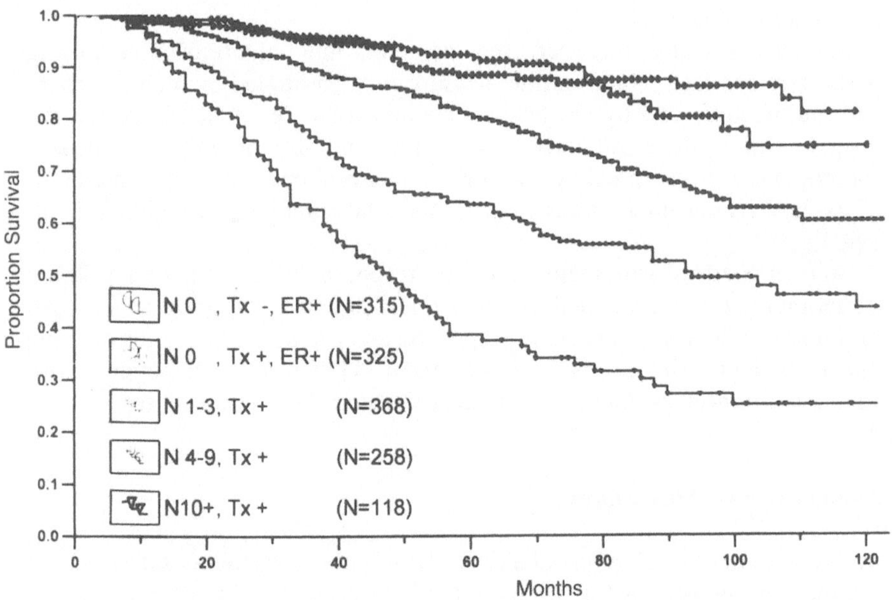

Fig. 2. Axillary lymph node status according different risk categories and survival in pimary breast cancer: GABG trials I and II 1981–1990. Tx^{+} = systemic adjuvant treatment. (From Kaufmann et al. 1993; von Minckwitz et al. 1995)

(Early Breast Cancer Trialists' Collaborative Group 1992). Outside clinical trials, it is still a matter of debate if all patients without involved lymph nodes should have adjuvant systemic treatment (Henderson 1994). Patients with node-negative disease still have a risk of relapse within 10 years of about 20%–30%. This is why, recently, larger proportion of patients with node-negative tumors are being considered candidates for adjuvant therapies, although an exact definition of risk does not exist. The accurate definition of a node-negative axillary lymph node status requires fulfillment of at least two conditions: (1) a sufficient number (at least ten nodes) of nodes have to be examined, (2) nodes must be serially sectioned.

It is well known that serial sectioning of nodes will reveal undetected axillary metastases in about 10%–35% of patients (International (Ludwig) Breast Cancer Study Group 1990).

Information on lymph node involvement may become less important because a down-staging will occur in about 20% of patients treated with primary chemotherapy (Fisher 1995).

Histologic Type

Tumor histopathology also has been shown to be related to different risk categories. Several well-characterized histologic subtypes, e.g., tubular, papillary or mucinous invasive carcinomas demonstrate a relatively good prognosis compared to invasive ductal or lobular carcinomas (Mansour et al. 1994; Rosen et al. 1989a). However, these types (about 15%–20% of cases) are the minority of all invasive breast carcinomas.

Grade S-phase

Nuclear grade is well documented and gives information on tumor differentiation (Rosen et al. 1989a; McGuire 1988). Patients with good grade (Gl) demonstrate low recurrence and death rates.

Besides this prognostic variable, well differentiated tumors may show better response to endocrine treatment than poorly (G3) differentiated tumors. Grading evaluated the phenotypes of tumors, whereas all investigational effects to establish new prognosticators have been directed at the underlying molecular changes. Because intra- and interobserver reliability of grading is low, the value of this highly practical factor is reduced. Therefore pathologists should adopt a uniform and reproducible grading system; however, the detection of S-phase fraction as measured by DNA flow cytometry is thought to represent a more objective and simple method (Mansour et al. 1994; Kaufmann et al. 1988, 1994; Merkel et al. 1993).

S-phase fraction correlates with nuclear grade and hormone receptor status (Clark 1994; Merkel et al. 1993). Good nuclear grade is related to low S-phase fractions and receptor-positive findings in breast cancer. Patients with tumors of high S-phase fractions demonstrate high recurrence and death rates (Clark 1994; Kaufmann et al. 1994). This investigational prognostic factor may have potential clinical value in the near future.

Hormone Receptor Status

Measurement of steroid hormone receptor concentration has been possible for more than 20 years. Estrogen (ER) and/or progesterone (PgR) receptor-positive or-rich tumors demonstrate better prognoses than those with ER- and/or PgR-negative tumors (Mansour et al. 1994; McGuire 1988). There are more data on ER than on PgR. There are also data showing that PgR content is a better determinant of prognosis than is ER status. PgR may be important as a prognostic factor, whereas ER is relevant as predictor of endocrine response.

Intra- and interobserver reliability especially of histologically confirmed receptor results is low. Measurements of receptors can result in misleading data in young women (occupied receptors by estrogens) or in patients with nonfunctional ER (non DNA-binding ER) (Montgomery et al. 1993).

Menopausal Status and Patients' Age

Menopausal status and patients' age are factors which are clinically used for prognostic assessment and treatment decisions. Menopausal status is an excellent predictor for the efficacy of adjuvant systemic treatment and therefor useful to select patients for chemo- or endocrine therapies or both. Chemotherapy appears to be more effective in young (premenopausal) patients, whereas tamoxifen appears to be more effective in older (postmenopausal, elderly) patients. In a randomized trial (GABG-1) CMF chemotherapy was compared to tamoxifen in patients with hode-positive (1–3 nodes involved) and receptor positive tumors. Chemotherapy was found to be superior to tamoxifen in women <50 years old, and tamoxifen appeared to be superior to CMF in patients ≥50 years old (Fig. 3). Castration is effective in premenopausal patients (Early Breast Cancer Trialists' Collaborative Group 1992). So far, it is still unclear if the various effects are related to age alone or to menopausal status, since exact menopausal status data in most clinical trials have been defined by age and have not been confirmed by hormone analyses. Also hormonal replacement therapy interferes with menopausal status, which may be a very important factor in the future.

Young (<35–40 years) patients have a worse prognosis than older patients (De la Rochefordiere et al. 1993), which is only explained in part by a higher frequency of other poor prognostic factors seen in younger women. Moreover outcome in the elderly patient is confounded with concomitant disease.

In addition, age is an important consideration in case of local recurrence after breast conservation. Important tumor factors which predict increased risk of breast recurrence are positive resection margins, lymphatic invasion, and the extension of intraductal components (EIC) (Clarke and Martinez 1992). The most important factor, however, seems to be the age of a

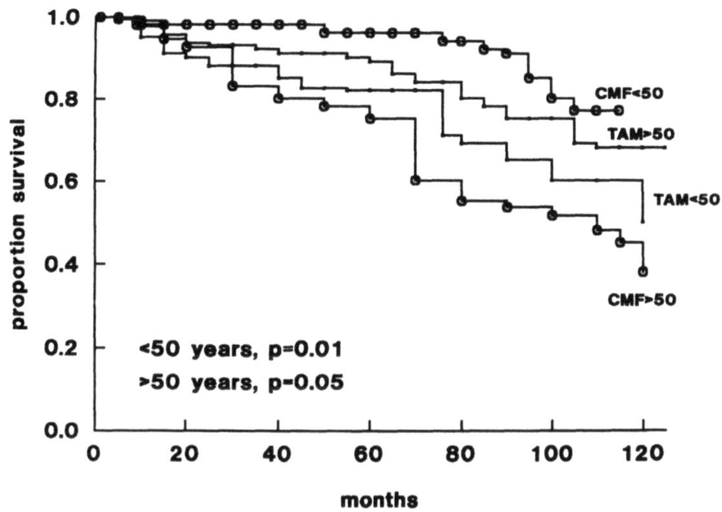

'ig. 3. Adjuvant systemic treatment (10 year results) according to age (menopausal) ubgroups. GABG trial I – low risk patients: 1–3 positive nodes, receptor-positive. From Kaufmann et al. 1993)

›atient. High rates of in-breast recurrences occur in patients <35 years of ıge, despite irradiation, and low rates in women >55 years of age, even vithout radiotherapy (Veronesi et al. 1993). Lack of adjuvant systemic reatment seems also associated with higher recurrence rates.

ssues To Be Resolved Regarding Prognostic/Predictive Factors

n order to gain more information about the above described variables, the 'ollowing issues should be resolved:

.. Measurement of invasive and noninvasive tumor size
!. Standardization of axillary lymph node involvement determination
}. Standardization of nuclear grading system, hormone receptor measurements by immunohistochemistry, and S-phase determination
ł. Definition of menopausal status and influence of hormone replacement therapy
ı. Identification of factors to predict which patients do not require: adjuvant radiation, axillary lymph node dissection, and breast surgery after primary chemotherapy.
ı. Identification of factors to predict drug resistance

nvestigational Prognostic/Predictive Factors

n the future new more relevant prognostic/predictive factors will be needed 'Kaufmann 1993). Recently identified variables can help us to better understand the biology of breast cancer, to evaluate new therapeutic agents, and to achieve more precise interpretation of clinical trials. Our efforts should be concentrated on finding factors which describe the metastatic potential of human breast cancer cells.

Findings such as influence of timing of surgery during menstrual cycle, detection of micrometastases in bone marrow or variants of CD 44 have high prognostic value. Premenopausal patients with positive lymph nodes who have had surgery in the estrogenic (first) phase of the menstrual cycle have a poorer prognosis than those with surgery in the second phase of the cycle (Veronesi et al. 1994); however, our data confirmed this observation only for women who underwent two-step surgery (von Minckwitz et al. 1995b).

Micrometastases detection in bone marrow discriminates between favorable and unfavorable prognostic groups (Diel et al. 1992). For patients with T_1 tumors, the prognostic power of tumor cell detection is superior to that based on axillary lymph node status (Diel et al. 1995) (Table 3).

The presence of variants of CD 44 is related to the length of survival of patients with primary breast cancer (Kaufmann et al. 1995), making discrimination of patients with poor excellent and long-term prognosis possible.

Summary

The prognosis of patients with primary breast cancer depends on prognostic factors per se and on treatment. In certain subgroups of patients, the overall treatment benefits of systemic adjuvant therapy are modest. Therefore, proper selection of patients and treatment modalities is absolutely necessary. Patients with primary breast cancer and a *very good* prognosis can be classified by known prognostic factors such as tumor size $T \leqslant 1\,\text{cm}$, grade 1, hormone receptor positive tumors, and *no* other factors indicating poor prognosis.

Table 3. Relation of time to first metastasis and survival (Cox regression analysis) according to tumor size and tumor cell detection (2 E 11) in bone marrow of primary breast cancer patients (Diel et al. 1995)

Prognostic factor	Recurrence		Survival	
	p value	Risk ratio	*p* value	Risk ratio
Tumor size				
T_1				
Nodal status	0.6	1.2	0.6	1.3
2 E 11	0.0006	7.4	0.08	4.5
T_2				
Nodal status	0.0001	2.2	0.002	2.2
2 E 11	0.01	2.1	0.09	2.2

Table 4. Treatment decisions (type and timing of therapy) according to "classical" known prognostic and predictive factors in primary breast cancer treated outside clinical trials

	Prognostic/predictive factor
Type of therapy	
Surgery	Tumor size
Radiation for breast conservation	Age
Systemic endocrine and/or cytotoxic treatment	Tumor size Menopausal status/age Receptor status Nodal status
Timing of therapy	
Primary and/or adjuvant systemic treatment	Tumor size (experimental)
Surgery	Menstrual cycle (experimental)

Patients with *very poor* prognosis are characterized by young age (<35 years), large tumor size (T > 3 cm), grade 3, hormone receptor-negative tumors, and high S-phase fractions.

Tumor size is the best prognostic factor for the decision to treat (breast conservation surgery, primary preoperative or adjuvant systemic therapy) or not to treat.

Table 4 shows possible treatment decisions (type and timing of therapy) in primary breast cancer based on known prognostic and predictive variables.

The decision to treat or not to treat outside clinical trials should be based on this kind of information. The type and aggressiveness of adjuvant systemic treatment should depend on the receptor status of the tumor and on the age and menopausal status of the patient, as classical predictive factors; however, prognosticators which can describe the metastatic potential of breast cancer cells have yet to be defined.

References

Bonadonna G (1995) Future developments in adjuvant systemic therapy of breast cancer. 5th international conference on adjuvant therapy of primary breast cancer, St Gallen

Carter CL, Allen C, Henson DE (1989) Relation of tumor size, lymph node status, and survival in 24740 breast cancer cases. Cancer 63:181–187

Clark GM (1994) Do we really need prognostic factors for breast cancer? Breast Cancer Res Treat 10:117–126

Clarke DH, Martinez AA (1992) Identification of patients who are at high risk for locoregional breast cancer recurrence after conservative surgery and radiotherapy: a review article for surgeons, pathologists, and radiation and medical oncologists. J Clin Oncol 10:474–483

De la Rochefordiere A, Asselain B, Compana F et al. (1993) Age as prognostic factor in premenopausal breast carcinoma. Lancet 341:1039–1043

Diel IJ, Kaufmann M, Goerner R et al. (1992) Detection of tumor cells in bone marrow of patients with primary breast cancer: a prognostic factor for distant metastasis. J Clin Oncol 10:1534–1539

Diel IJ, Kaufmann M, Solomayer EF et al. (1995) Micrometastatic tumor cells in bone marrow versus nodal status in breast cancer: impact on prognosis in 727 patients with primary breast cancer. 5th international conference adjuvant therapy of primary breast cancer, St Gallen

Donegan WL (1988) Staging and primary therapy. In: Donegan WL, Spralt JS (eds) Cancer of the breast. Saunders, Philadelphia, p 355

Donegan WL (1992) Prognostic factors. Stage and receptor status in breast cancer. Cancer 70:1755–1764

Dorr FA (1993) Prognostic factors observed in current clinical trials. Cancer 71: 2163–2168

Early Breast Cancer Trialists' Collaborative Group (1992) Systemic treatment of early breast cancer by hormonal, cytotoxic, or immune therapy: 133 randomized trials involving 31 000 recurrences and 24 000 deaths among 75 000 women. Lancet 339:1–15, 71–85

Fisher B (1995) The changing role of surgery in a global therapeutic strategy. 5th international conference on adjuvant therapy of primary breast cancer, St Gallen

Hawkins RA (1993) Prognostic factors: seeking a shaft of light or getting lost in the woods? A personal view from Sherwood Forrest. Breast 2:125–129

Henderson IC (1994) Adjuvant systemic therapy for early breast cancer. Cancer 74:401–409

Henderson IC, Hayes DF, Parker LM et al. (1990) Adjuvant systemic therapy for patients with node-negative tumors. Cancer 65:2132–2147

International (Ludwig) Breast Cancer Study Group (1990) Prognostic importance of occult axillary lymph node micrometastases from breast cancer. Lancet 335: 1565–1568

Kaufmann M (1993) Current and new concepts in the systemic adjuvant treatment of breast cancer. Onkologie 16:304–308

Kaufmann M, Feichter GE, Nhila A et al. (1988) Flow-zytophotometrische Parameter, Hormonrezeptoren und axillärer Lymphknotenstatus als Prognosefaktoren beim primären Mammakarzinom. Geburtshilfe Frauenheilkd 48:705–709

Kaufmann M, Jonat W, Abel U et al. (1993) Adjuvant randomized trials of doxorubicin/cyclophosphamide versus doxorubicin/cyclophosphamide/tamoxifen and CMF chemotherapy versus tamoxifen in women with node-positive breast cancer. J Clin Oncol 11:454–460

Kaufmann M, von Minckwitz G, Sinn HP et al. (1994) Combination of grading and new biological factors (S-phase fraction and epidermal growth factor receptor) can predict relapse and survival in patients with node-negative primary breast cancer. Onkologie 17:166–172

Kaufmann M, Heider KH, Sinn HP et al. (1995) CD 44 variant exon epitopes in primary breast cancer and length of survival. Lancet 345:615–619

Lin PP, Allison DC, Wainstock J et al. (1993) Impact of axillary lymph node dissection on the therapy of breast cancer patients. J Clin Oncol 11:1536–1544

Mansour EG, Ravdin PM, Dressler L (1994) Prognostic factors in early breast cancer. Cancer 74:381–400

McGuire WL (1988) Estrogen receptor versus nuclear grade as prognostic factors in axillary node negative breast cancer. J Clin Oncol 6:1071–1072

Mc Guire WL, Clark GM (1992) Prognostic factors and treatment decisions in axillary-node-negative breast cancer. N Engl J Med 326:1756–1761
Merkel DE, Winchester DJ, Goldschmidt RA et al. (1993) DNA flow cytometry and pathologic grading as prognostic guides in axillary lymph node-negative breast cancer. Cancer 72:1926–1932
Montgomery PA, Scott GK, Luce MC et al. (1993) Human breast tumors containing non-DNA-binding immunoreactive (67 kDa) estrogen receptor. Breast Cancer Res Treat 26:181–189
Nemoto T, Vana J, Bedwani RN et al. (1980) Management and survival of female breast cancer: results of a national survey by the American College of Surgeons. Cancer 45:2917–2924
NIH Consensus Conference (1991) The treatment of early-stage breast cancer. J Nat Cancer Inst 80:471–471
Osborne CK (1992) Prognostic factors for breast cancer: have they met their promise? J Clin Oncol 10:679–682
Rosen PP, Groshen S, Saigo PE et al. (1989a) Pathological prognostic factors in stage I (T_1NoMo) and stage II (T_1N_1Mo) breast carcinoma: a study of 644 patients with median follow-up of 18 years. J Clin Oncol 7:1239–1251
Rosen PP, Groshen S, Saigo PE et al. (1989b) A long-term follow-up study of survival in stage I (T_1NoMo) and stage II (T_1N_1MO) breast carcinoma. J Clin Oncol 7:355–366
Rosen PP, Groshen S, Kinne DW et al. (1991) Prognosis in T_2NoMo Stage I breast carcinoma: a 20-year follow-up study. J Clin Oncol 9:1650–1661
Seidman JD, Schnaper LA, Aisner SC (1995) Relationship of the size of the invasive component of the primary breast carcinoma to axillary lymph node metastasis. Cancer 75:65–71
Veronesi U, Luini A, Del Vecchio M et al. (1993) Radiotherapy after breast-preserving surgery in women with localized cancer of the breast. N Engl J Med 328:1587–1591
Veronesi U, Luini A, Mariani L et al. (1994) Effect of menstrual phase on surgical treatment of breast cancer. Lancet 343:1545–1547
von Minckwitz G, Kaufmann M, Schmid H et al. (1995a) Tamoxifen in patients with node-negative and receptor-positive breast cancer: the Heidelberg II and GABG II experience. 5th international conference on adjuvant therapy of primary breast cancer, St Gallen
von Minckwitz G, Kaufmann M, Dobberstein S et al. (1995b) Surgical procedure can explain varying influence of menstrual cycle on prognosis of premenopausal breast cancer patients. Breast 4:29–32
Wilson RE, Donegan WL, Mettlin C et al. (1984) The 1982 national survey of carcinoma of the breast in the United States American College of Surgeons. Surg Gynecol Obstet 159:309–318

Prognostic Variables and Future Predictors of Behaviour and Response

B.A. Gusterson

Institute of Cancer Research, Haddow Laboratories, 15 Cotswold Road, Sutton, Surrey SM2 5NG, UK

Introduction

Recent reviews have clearly defined the present role of prognostic factors in the management of breast cancer (Levine et al. 1991; McGuire 1991; Osborne 1992; Clark 1994; Clark et al. 1994; Gusterson 1995). As adjuvant therapy has become generally used there has been a decreased need for prognostic factors to aid in the optimisation of treatment. It is now accepted that the main application for prognostic factors is to identify those patients who have such a good prognosis that they can be spared any form of adjuvant treatment (Clark 1994). As stated by Page (1991), the "modern" markers of prognosis, based on our understanding of the normal cell cycle and the molecular abnormalities in breast cancer, offer little additional information over and above that of the gold standard, which is a good pathologist with a haematoxylin and eosin section. Thus tumour grade, size of the tumor and lymph node status are the parameters against which everything else should be measured (Fisher et al. 1978; Fisher 1986). S-phase fraction is an objective method of measuring tumour proliferation, but standardisation is still a problem for comparisons between centres. Vascular invasion and angiogenesis (Horak et al. 1992) are recent additions, as predictors of latent metastatic disease. Similarly bone marrow (Cote et al. 1991; Mansi et al. 1991; Neville et al. 1992) and lymph node micrometastases (de Mascarel et al. 1992; International Breast Cancer Study Group 1990) are indicators of tumour burden and thus predictors of relapse. The practical significance of such measurements is, however, unclear as it is not possible to predict relapse and the time of relapse for individual patients based on these measurements.

As interest in prognostic factors has waned focus has turned to analyzing functional attributes of the tumour that may predict response to different treatment modalities and thus to select treatments to match the biology. It has also been of interest to examine some of the known molecular changes in the tumour in relation to therapy to see whether there are correlations that may indicate an interaction between the drug and the molecular abnormality. Thus predictive factors are the vogue. Factors that come under this heading include predictors of drug response and resistance, predictors of malignant potential in benign disease and progression in carcinoma in situ, based on an understanding of their biology, and the identification of molecular abnormalities in tumours that can be used to select them for therapies targeting these abnormalities.

This article discusses the utility of predictive factors and their limitations. In particular the importance of standardisation of methodologies and terminology is emphasized. The need for large multicentre studies and large tissue resources of breast lesions of defined phenotypes is discussed.

Molecular and Biochemical Markers as Surrogate Measures of Cell Proliferation and Differentiation

It is essential with any new biological observation to ensure that its application is addressed to the correct question. All too often a basic research observation is taken up by pathologists for a quick observational investigation of a sample of tumours without apparently first asking: "What is the objective and the hypothesis to be tested?" Any new oncogene product or protein involved in the cell cycle or proliferation pathway is immediately investigated for its prognostic significance. This seems wholly illogical, as we know that prognosis relates to proliferation and different tumours have different abnormalities in their growth control. Within a group of tumours it is likely that any such marker will show a correlation with proliferation and thus prognosis, but as it is only important in a proportion of the tumours it is unlikely to be more important than other nonselective measurements of cell growth.

If a tumour has an increased proliferation rate it accumulates additional molecular abnormalities faster. The tumour progressively dedifferentiates and becomes pleomorphic. Thus any change at the molecular level, whether it be amplification or mutation, is likely to be part of accumulating damage due to genetic instability, perhaps initiated through p53 mutation, and therefore associated with the higher grade tumours. Of course all tumours must have some events which are early, but loss of heterozygosity (LOH) studies have clearly shown that almost any genetic marker will demonstrate LOH in some tumours regardless of their degree of differentiation (Cox et al. 1994; Gusterson et al. 1994). It is thus predictable that any site of LOH will be more frequent in high grade than low grade tumours and will thus be associated with a high proliferation rate and loss of differentiation. Similarly it is likely that there will be an inverse correlation between such genetic abnormalities and estrogen receptors (ERs) and progesterone receptors (PRs) as these are most highly expressed in the well differentiated tumours.

It is possible that biologically there is a correlation between two individual variables such as epidermal growth factor receptor (EGFR) and ER (Sainsbury et al. 1985), but interest is best focused on the biology, not on the general application as a prognostic marker.

In addition to tumour grade, the size of the tumour and lymph node involvement are important prognostic factors. The reason for this is most likely to be due to the time dependency of these two parameters and the fact that there is more time to accumulate genetic damage. If this is true it is to be predicted that those factors that are epigenetic as a consequence of genetic instability will correlate with poor prognosis, but they may not be causally related. The difficulty conceptually is that almost any cellular function could be envisaged as having a positive contribution to the malignant phenotype and thus by loss or overexpression potentially oncogenic. It is therefore going to be difficult to establish which of the many genetic changes seen in any particular tumour are critical and worthy of therapeutic intervention.

It is likely that some of these genetic changes are directly the result of environmental carcinogens and thus specific mutations may be correlated with individual factors in the diet. It has already been shown that DNA adducts accumulate in the breast of normal women (Pfau et al. 1992), suggesting that such damage is potentially occurring before there is any evidence of morphological change.

In view of the increasing complexity of breast cancer at the molecular level (Sato et al. 1990; Callahan et al. 1992; Dickson and Lippman 1992; Gusterson et al. 1994; Cox et al. 1994) it is important that consideration is given as to how we are going to manage this increase in information. It is becoming clear that there may be acquired prenatal factors (Anbazhagan and Gusterson 1994; Hilakivi-Clarke et al. 1994) that influence a woman's susceptibility to develop breast cancer in addition to the inheritance of predisposing genes.

Predictors of Treatment Responsiveness

Evaluation of Treatment Responsiveness?

Many of the features that can be measured in a tumour can be analyzed with respect to different end points. This is most clearly explained by Gelber and Goldhirsch (1992), who state, "Evaluation of treatment responsiveness factors is another way of looking at the same data. However, the distinction in interpretation between risk of bad outcome (prognostic factor) and indication of relative responsiveness to treatment (treatment responsiveness factor) is extremely valuable for getting the most out of a data analysis." This has been recently demonstrated in relation to studies of c-erbB-2, in which different studies have demonstrated that c-erbB-2 overexpressing tumours may have a poor response to treatment (Gusterson et al. 1992; Allred et al. 1992) or the converse (Muss et al. 1994; Tripathy and Henderson 1995) depending upon the therapy given. Thus such observations are hypothesis generating and contribute to our understanding of the biology of the tumour (Goldhirsch and Gelber 1994).

Factors that predict responsiveness or resistance to different treatment modalities are essential for rational patient management. However, reliable indicators of both endocrine and chemotherapy response are lacking (Klijn et al. 1993, 1995). Recently it has been possible to analyze tumours for response to hormonal manipulation based on an improved understanding of the oestrogen response mechanism (Klijn et al. 1995).

Predictors of Hormone Responsiveness and Resistance

This topic has been extensively reviewed recently (Klijn et al. 1995; Tripathy and Henderson 1995; Osborne and Fuqua 1994). The major clinical difficulty is that it is not possible to predict those patients who will be resistant or responsive to tamoxifen. Many possible mechanisms for this have been put forward ranging from individual variations in the metabolism of tamoxifen to mutations in or splice variants of the ER itself. None of these approaches seems to provide the answer and it is most likely that this is a multifactorial problem with different mechanisms in individual tumours. One explanation, yet to be explored in depth, is the possible regulation of the ER by other signalling pathways, such as that of the EGFR or c-erbB-2.

Resistance to Chemotherapy

When considering resistance of a tumour to treatment two patterns occur, namely, acquired resistance and intrinsic resistance. The possible mechanisms involved have been reviewed recently (Klijn et al. 1993; Trimpathy and Henderson 1994). There are multiple mechanisms of drug resistance and for any single drug multiple mechanisms may be operative. There have been excellent recent reviews on this topic including the role of the multiple drug resistance 1 (MDR1) gene in multidrug resistance (Goldstein 1995) and the potential value of topoisomerase inhibitors (Harris and Carmichael 1995). Recently other components of drug resistance have been examined including heat shock proteins (Fuqua et al. 1994).

Predicting Response to Novel Therapies

It is clear that the present modalities of treatment based on current endocrinological and chemotherapeutic strategies do not appear to have very much more to offer in terms of survival benefit. The hope is that by understanding the genetic and biochemical events that are responsible for the malignant phenotype it will be possible to design drugs and molecular strategies that will enable the reversal of some of these processes and the blocking of the pathways essential for growth. The more that is understood about the normal mechanisms of cell cycle control and growth regulation, the better the possibilities for therapeutic intervention; but there are difficulties of producing tumour specificity. There are many examples in which specificity is now a possibility, but often the target for the drug is only expressed in some tumours, so it will be necessary to have methods that will identify those tumours most likely to respond to a particular strategy.

The favoured targets at the present time are growth factor receptors that are overexpressed. These are good targets as they are accessible on the cell surface for antibody targeted therapies. Some antibodies used alone appear to be cytostatic, but it is possible to couple the antibody to a toxin or to use an antibody mediated enzyme/prodrug system (ADEPT). The most obvious targets in breast cancer are the EGFR (Sainsbury et al. 1987; Le Maistre et al. 1994) and c-erbB-2, which is over expressed on approximately 20% of breast cancers (Gusterson et al. 1988). Other targets being considered are some of the receptors for other growth factors. The insulin-like growth factor (IGF) signalling pathway is a good candidate for therapeutic intervention as it does not appear to be essential for normal growth control, but tumours cannot replace its activity by other factors, such as transforming growth factor (TGF)-α (Yee 1994). Other therapeutic strategies being considered are the blocking of the steps in the signalling cascade through the identification of small molecules that are protein tyrosine kinase inhibitors (Lyall et al. 1989; Tapley et al. 1992).

Although most pharmaceutical companies are aiming for drugs that will be specific inhibitors of a defined receptor type (e.g., EGRF but not c-erbB-2) there may be potential in developing inhibitors that have a broader range of activity. It has recently been shown that measurements of cytosolic protein tyrosine kinase (PTK) activity correlate with ER negativity (Romain et al. 1994). This is of interest as 70% of the cytosolic fraction is derived from the c-*src* oncogene product and this is a strong candidate for the development of inhibitors. In the case of IGF there is very little evidence for over-expression in tumours and an alternative strategy may be based on the modulation of IGF binding proteins, which are found in breast cancers (Yee 1994).

For these strategies to be successful it will be essential to first identify those tumours most suitable for a particular treatment on the basis of the cell surface expression of the protein.

Recent Developments in the Prediction of Tumour Behaviour

Are Molecular Markers of Value?

As we learn more about the molecular events that accumulate in a tumour as it evolves (Wolman and Heppner 1992; Gusterson et al. 1994) it becomes necessary to carefully evaluate the cost-effectiveness and added value to be obtained from each additional potential assay that can be developed as a marker of prognosis or as a predictive factor. A strong argument could be made for establishing a few centres that are funded specifically to carry out these important and essential multiparametric studies.

In Situ Disease and Premalignant Lesions

At the present time we do not know the natural history of breast cancer and the relationship of the various types of epithelial hyperplasia to the variant forms of ductal carcinoma in situ and to lobular carcinoma in situ. A major problem with such investigations is that few centres have sufficient cases for detailed studies and in many instances tissue is not available for research purposes owing to local policy in the pathology departments. At the Royal Marsden Hospital in London we have managed to trace 500 cases of in situ disease from files in which tissue is available from patients presenting since 1975. Bearing in mind that this is the largest referral centre for breast cancer in the United Kingdom it is clear that once different variants forms of the disease are taken into account, this is too small a data-base for definitive studies.

Studies of in situ disease and atypical lesions are particularly important as these are most likely to provide information on the key events required for increase in proliferation and invasion (Gusterson 1993). It is to be predicted that in addition to molecular genetic changes that increase proliferation there will be changes in cell motility factors (Litta et al. 1986; Gherardi and Stocker 1991), up-regulation of metalloproteases required for extracellular matrix degradation (Goldberg and Eisen 1990; Waxman and Wasan 1992; Foekens et al. 1994) and alterations in cell/cell (Edelman and Crossin 1991) and cell/matrix (Pignatelli et al. 1991; van Roozendaal et al. 1992; Maemura and Dickson 1994) interactions to enable migration. The question is how this diversity of phenotype is selected for. It seems likely that it takes many years to acquire the necessary combination of events to enable progression to the next stage. By defining the stages in this progression and the sequence in which they occur it should be possible to define the spectrum of molecular changes that identify whether a cell has the potential to invade. Such predictive factors will be very important if we are to produce an objective classification of breast cancer based on biology and identify the molecular profile that will enable us to predict behaviour based on small biopsies and fine needle aspirations.

Pilot studies using microdissection of in situ disease and atypical lesions are producing some interesting results. Using LOH analyses and polymorphic microsatellite markers it has been shown that a similar pattern of loss is seen in the invasive and the in situ component (Stratton et al. 1995). These data confirm the clonal nature of the in situ disease. Using similar methods to investigate in situ lobular carcinoma (LCIS) it has been shown that LCIS and ductal carcinoma in situ (DCIS) share a 16q deletion as a common early event (Lakhani et al. 1995a). This is also an important finding as it confirms that LCIS is a malignant lesion and not just a risk factor for breast cancer. In view of the fact that LCIS can precede both invasive ductal and lobular carvinoma more detailed analyses of the full spectrum of molecular changes will be of interest. It has already been demonstrated that LCIS does not have c-erbB-2 amplification (Gusterson et al. 1988) or p53 mutations (Gusterson et al. 1994). LOH studies in atypical ductal hyperplasia (ADH) have demonstrated that the incidence of LOH is similar to that in DCIS (Lakhani et al. 1995b). This demonstrates that this lesion is monoclonal and thus "neoplastic" and not a hyperplasia. Similar studies carried out in San Antonio have also shown that some of the ductal epithelial hyperplasia are neoplastic and the LOH in the region of BRCA1 is an early event in the evolution of in situ disease (O'Connell et al. 1994).

The application of these methods in conjunction with new techniques such as comparative genomic hybridisation (CGH), (Kalliomieni et al. 1994) will over the next 5 years provide a molecular profile of early breast lesions and the sequence of accumulation of genetic damage that relates to behaviour and prognosis. This should then provide an accurate method of assessing the treatment options for DCIS and LCIS, based on the functional changes that have been acquired.

The Need for Standardisation

In invasive breast cancer and in the investigation of in situ disease there is an urgent need for standardisation of methodology. It is well known that pathologists within a single country will have very differing views on the nomenclature used for the same lesion. While this persists there is very little hope that results will be transferable between centres and it will require much larger numbers of cases to obtain conclusive results. The only way forward is to have internationally agreed upon criteria for each morphological entity. With direct "real time" discussions and teaching via cable and satellite communication (Telepathology), this will be much easier to achieve and should be made a priority. The first step must be to have a nomenclature agreed by all pathologists involved in clinical trials. Only the enthusiasts will be able to establish this, as the majority of non-breast specialists will not rapidly adapt to any proposed changes. It is therefore essential that any resistance to change is directed from above. It is also essential that each new prognostic/predictive factor is assessed as quickly as possible on a large data-base of cases. It is therefore necessary to invest in the these data-bases now. This has been started by the US Army initiative. It will then be possible for the guardians of the data-bases to send out material to investigators working on new markers, so that results can be assessed blindly in an independent group of patients. This will increase the number of parameters in the data-base and thus its power for examining interactions between independent factors. Having established the relative importance of a particular factor it can then be decided whether any added benefit is sufficient to warrant establishing it as a test with utility in clinical practice. The San Antonio group have shown the value of such multiparametric studies (Clark et al. 1994) and have recently focused on the application of neural network analyses to investigate complex interactions between variables (De Laurentiis and Ravdin 1995). The International Breast Cancer Study Group has taken the decision to establish a pathology data-base of material from their large trials and to make it available to researches outside the group for studies approved by their scientific committee. Two of the current active protocols are: (1) to assess the role of angiogenesis with Professor Harris in Oxford (UK) and, (2) to carry out an independent evaluation of the Nottingham Index with this UK group. All such studies are blind to the investigators who then supply these data to the Biostatistics Centre in Boston for analyses against known parameters in the data-base.

The Changing Role of the Pathologist

The role of the pathologist in the management of breast cancer will undergo radical change in the next decade. In addition to the need for standardisation of morphology and the use of Telepathology and the Internet for discussions on particular cases the pathologist will be required to adapt to the needs of the clinicians. This will come at a number of levels, but it is clear that primary chemotherapy will have a major impact if adopted as a standard form of therapy. Even in the absence of primary chemotherapy, the mammographic screening programmes are resulting in detection of smaller tumours and an increase in the detection of in situ lesions. Thus, pathologists are having to deal with a change in the spectrum of disease. An important but unresolved issue is therefore the relative importance of putting through an entire DCIS for histology to exclude microinvasion against retaining some of the tissue for other investigations. The major challenge to the pathologist will be to develop methods to carry out the prognostic and predictive tests referred to above together with the new molecular biological analyses on small cellular samples or punch biopsies. This will firstly require considerable investment on validation of the tests to be used to establish that small samples are representative of the whole tumour. Today we are still in the position that we do not have definitive standardisation of cytology for well established markers such as ER, PR and c-erbB-2. Even with a test as common as S phase fraction, there is considerable variation between laboratories apparently using the same machines and protocols. We have a long way to go.

The pathologist of the future should be both leading the study of breast cancer based on a sound understanding of the morphology and looking for ways to assist the clinician in patient management.

Development of New Assays

It is obvious that it will be necessary in the future to develop new functional assays as predictive tests for drug resistance, to identify those patients that would be suitable for different ADEPT and biomolecular therapeutic strategies. It is not clear at the present which speciality will be responsible for developing these techniques, but it is essential that histopathologists are aware of the importance of these advances and work in collaboration with the biochemists and drug development groups to ensure that maximum utilisation of the patient's material to the benefit of future patients.

Conclusions

The next 10 years will be an exciting time in our understanding of the underlying molecular abnormalities that are responsible for breast cancer. It is to be predicted that we will be frustrated by the complexity of these changes and it is therefore essential that there is maximum cooperation internationally to ensure that the present explosion of small publications on new findings of unproven significance is replaced by large multicentre studies carrying out strict protocols and quality assurance standards. Only in this way will we be able to assess those predictive factors that are of clinical importance.

Acknowledgments. The work in Professor Gusterson's Laboratory is supported by The Cancer Research Campaign, BREAKTHROUGH Breast Cancer, The Leopold Muller Trust and US Army Grant No: DAMD17-94-J-4066.

References

Allred DC, Clark GM, Tandon AK, Molina R, Torney DC, Osborne KC, Gilchrist KW, Mansour EG, Abeloff M, Eudey L, McGuire WL (1992) HER-2/*neu* in node-negative breast cancer: prognostic significance of overexpression influenced by the presence of in situ carcinoma. J Clin Oncol 10:599–605

Anbazhagan R, Gusterson BA (1994) Prenatal factors may influence predisposition to breast cancer. Eur J Cancer 30A(1):1–3
Callahan R, Cropp CS, Merlo GR, Liscia DS, Cappa APM, Lidereau R (1992) Somatic mutation and human breast cancer. Cancer 69:1582–1588
Clark GM (1994) Do we really need prognostic factors for breast cancer? Breast Cancer Res Treat 30:117–126
Clark GM, Hilsenbeck SG, Ravdin PM, De Laurentis M, Osborne CK (1994) Prognostic factors: rationale and methods of analysis and integration. Breast Cancer Res Treat 32:105–112
Cote RJ, Rosen PP, Lesser ML, Old LJ, Osborne MP (1991) Prediction of early relapse in patients with operable breast cancer by detection of occult bone marrow micrometastases. J Clin Oncol 9:1749–1756
Cox LA, Chen G, Le EY-HP (1994) Tumour supressor genes and their roles in breast cancer. Breast Cancer Res Treat 32:19–38
De Laurentiis M, Ravdin PM (1994) Survival analysis of censored data: neural network analysis detection of complex interactions between variables. Breast Cancer Res Treat 32:113–118
De Mascarel I, Bonichon F, Coindre JM, Trojani M (1992) Prognostic significance of breast cancer axillary lymph node micrometastases assessed by two special techniques: reevaluation with longer follow-up. Br J Cancer 66:523–527
Dickson RB, Lippman ME (1992) Molecular determinants of growth, angiogenesis and metastases in breast cancer. Semin Oncol 19:286–298
Edelman GM, Crossin KL (1991) Cell adhesion molecules: implications for a molecular histology. Annu Rev Biochem 60:155–190
Fisher ER (1986) Prognostic and therapeutic significance of pathological features of breast cancer. Monogr Natl Cancer Inst 1:29–34
Fisher ER, Pakelar AS, Rockette H, Redmond C, Fisher B (1978) Pathologic findings from the National Surgical Adjuvant Breast Project (protocol 4). V. Significance of axillary nodal micro and macrometastases. Cancer 42:2032–2038
Foekens JA, Schmitt M, van Puten WLJ, Peters HA, Kramer MD, Janicke F, Klijn JGM (1994) Plasminogen activator inhibitor-1 and prognosis in primary breast cancer. J Clin Oncol 12:1648–1658
Fuqua SAW, Oesterreich S, Hilsenbeck SG, van Hoff DD, Eckarat J, Ostorne CK (1994) Heat shock proteins and drug resistance. Breast Cancer Res Treat 32:67–71
Gelber RD, Goldhirsch A (1992) for the International Breast Cancer Study Group (formerly Ludwig Group) Reporting and interpreting adjuvant therapy clinical trials. Monogr Natl Cancer Inst 11
Gherardi E, Stocker M (1991) Hepatocyte growth factor – scatter factor: Mitogen/Motogen and Met. Cancer Cells 3:227–232
Goldberg FI, Eisen AZ (1990) Extracellular matrix metalloproteinases in tumor invasion and metastasis. In: Lippman ME, Dickson RB (eds) Regulatory mechanisms in breast cancer. Advances in cellular and molecular biology of breast cancer. Kluwer, Boston, pp 421–440
Goldhirsch A, Gelber RD (1994) Understanding adjuvant chemotherapy for breast cancer. N Engl J Med 330:1308–1309
Goldstein LJ (1995) Multidrug resistance in breast cancer. In: Dickson RB, Lippman ME (eds) Drug hormonal resistance in breast cancer: cellular and molecular mechanisms. Horwood, Chichester, pp 219–249
Gusterson BA (1993) The role of the pathologist in dissecting the biology of breast cancer with special reference to in situ disease. Eur J Cancer Prev 2(3):51–58
Gusterson BA (1995) The changing role of the pathologist in the prediction of tumor behaviour and response to treatment. In: Dickson RB, Lipmann ME (eds) Drug and hormonal resistance in breast cancer: cellular and molecular mechanisms. Horwood, Chichester, pp 39–53

Gusterson BA, Machin LG, Gullick WJ, Gibbs NM, Powles TJ, Price P, McKinna A (1988) Immunohistochemical distribution of c-*erb*B-2 in infiltrating and in situ breast cancer. Int J Cancer 42:842–845

Gusterson BA, Gelber RD, Goldhirsch A, Price KN, Säve-Söderborgh, Anbazhagan R, Styles J, Rudenstarn C-M, Golouh R, Reed R, Martinez Tello F, Tilman A, Torhorst P, Gigolato P, Bettelheim R, Neville AM, Bürki K, Castiglione M, Collins J, Lindtner J, Senn H-J (1992) Prognostic importance of c-*erb*B-2 expression in breast cancer. J Clin Oncol 10:1049–1056

Gusterson BA, Mitchell P, Crompton M, Shipley J, Anbazhagan R, Stratton M (1994) Molecular changes in breast cancer. Proceedings of ESMO, Nov 18–22, Lisbon, pp 3–7

Harris AL, Carmichael J (1995) Topoisomerase inhibitors and multiple drug resistance mechanisms in human breast cancer. In: Dickson RB, Lippman ME (eds) Drug hormonal resistance in breast cancer: cellular and molecular mechanisms. Horwood, Chichester, pp 303–323

Hilakivi-Clarke L, Clarke R, Lippman ME (1994) Perinatal factors increase breast cancer risk. Breast Cancer Res Treat 31:273–284

Horak ER, Leek R, Klenk N, Le Jeune S, Smith K, Stuart N, Greenall M, Stepiewska K, Harris AL (1992) Angiogenesis, assessed by platelet/endothelial cell adhesion molecule antibodies, as indicators of node metastases and survival in breast cancer. Lancet 340:1120–1124

International (Ludwig) Breast Cancer Study Group (1990) Prognostic importance of occult axillary lymph node micrometastases from breast cancer. Lancet 335: 1565–1568

Kallioniemi O-P, Kallioniemi A, Piper J, Isola J, Waldman FM, Ray JW, Pinkel D (1994) Optimizing comparative genomic hybridisation for analysis of DNA sequence copy number changes in solid tumors. Genes Chromosomes Cancer 10:231–243

Klijn JGM, Berns EMJJ, Foekens JA (1993) Prognostic factors and response to therapy in breast cancer. Cancer Surv Breast Cancer 18:165–198

Klijn JGM, Berns EMJJ, Dorssers LCJ, Foekens JA (1995) Molecular markers of resistance to endocrine treatment of breast cancer. In: Dickson RB, Lippman ME (eds) Drug hormonal resistance in breast cancer: cellular and molecular mechanisms. Horwood, Chichester, pp 133–168

Lakhani SR, Collins N, Sloane JP, Stratton MR (1995a) Loss of heterozygosity 16q and 17p in carcinoma in situ of the breast: comparison of lobular and ductal variants. J Pathol 175:110A

Lakhani SR, Collins N, Stratton MR, Sloane JP (1995b) Atypical ductal hyperplasia (ADH) of the breast is a clonal proliferation exhibiting loss of heterozygosity at loci on chromosomes 16q and 17p. J Pathol 175:110A

Le Maistre CF, Meneghetti C, Howes L, Osborne CK (1994) Targeting the EGF receptor in breast cancer treatment. Breast Cancer Res Treat 32:97–103

Levine MN, Browman GP, Gent M, Roberts R, Goodyear M (1991) When is a prognostic factor useful? A guide for the perplexed. J Clin Oncol 9:348–356

Liotta LA, Mandler R, Murano G, Katz DA, Gordon RK, Chiang PK, Schiffman E (1986) Tumour cell autocrine motility factor. Proc Natl Acad Sci USA 83:3302–3306

Lyall RM, Zilberstein A, Gazit A, Glion C, Levitzki A, Schlessinger J (1989) Tyrophostins inhibit epidermal growth fact (EGF)-receptor tyrosine kinase activity in living cells and EGF-stimulated cell proliferation. J Biol Chem 264: 14503–14509

Maemura M, Dickson RB (1994) Are cellular adhesion molecules involved in the metastasis of breast cancer? Breat Cancer Res Treat 32:239–260

Mansi JL, Easton D, Berger U, Gazet J-C, Forcl HT, Dearnaley D, Coombes RC (1991) Bone marrow micrometastases in primary breast cancer: prognositic significance after six years follow-up. Eur J Cancer 27:1552–1555

McGuire WL (1991) Breast cancer prognostic factors: evaluation guidelines. J Natl Cancer Inst 83:154–155

Muss HB, Thor AD, Berry DA, Kute T, Liu ET, Koerner F, Cirrincione CT, Budman DR, Wood WC, Barlos M, Henderson IC (1994) c-*erb*B-2 expression and response to adjuvant therapy in women with node-positive early breast cancer. N Engl J Med 330:1260–1266

Neville AM, Bettelheim R, Gelber RD, Säve-Söderborgh J, Davis BW, Reed R, Torhorst J, Garlough R, Peterson HF, Price KN, Isley M, Rudestam C-M, Collins J, Castiglione M, Senn H-J, Goldhirsch A (1992) Factors predicting treatment responsiveness and prognosis in node-negative breast cancer. J Clin Oncol 10:696–705

O'Connell P, Pekkel V, Fuqua S, Osborne CK, Allred DC (1994) Molecular genetic studies of early breast cancer evolution. Breast Cancer Res Treat 32:5–12

Osborne CK (1992) Prognostic factors for breast cancer: have they met their promise? J Clin Oncol 10:679–682

Osborne CK, Fuqua SAW (1994) Mechanisms of tamoxifen resistance. Breast Cancer Res Treat 32:49–55

Page DL (1991) Prognosis and breast cancer recognition of lethal and favorable prognostic types. Am J Surg Pathol 15:334–349

Pfau W, O'Hare MJ, Grover P, Phillips D (1992) Metabolic activation of food mutagens 2-amino-3methylimidazo[4,5-f]quinoline (IQ) and 2-amino-3,4-dimethylimidazo[4,5-f]quinoline (MeIQ) to DNA binding species in human mammary epithelial cells. Carcinogenesis 13:907–909

Pignatelli M, Hanby AM, Stamp GWH (1991) Low expression of β_1, α_2 and α_3 subunits of VLA integrins in malignant mammary tumours. J Pathol 165:25–32

Romain S, Chinot O, Klijn JGM, van Putten WLJ, Guiorou O, Look M, Martin PM, Foekens JA (1994) Prognostic value of cytosolic tyrosine kinase activity in 249 node-positive breast cancer patients. Br J Cancer 70:304–308

Sainsbury JRC, Farndon JR, Sherbert G, Harris AL (1985) Epidermal growth factor receptors and oestrogen receptors in human breast cancers. Lancet 1:364–366

Sainsbury JRC, Farrndon JR, Needham GK, Malcolm AJ, Harris AL (1987) Epidermal growth factor status as predictor of early recurrence and death from breast cancer. Lancet 1:1398–1402

Sato R, Tanigami A, Yamakawa K, Akiyama F, Kasumi F, Sakamoto G, Nakamura Y (1990) Allelotype of breast cancer: cumulative allele losses promote tumor progression in primary breast cancer. Cancer Res 50:7184–7189

Stratton MR, Collins N, Lakhani S, Sloane JP (1995) Loss of heterozygosity in ductal carcinoma in situ of the breast. J Pathol 175:195–201

Tapley P, Lamballe F, Barbacid M (1992) K252a is a selective inhibitor of the tyrosine protein kinase activity of the *trk* family of oncogenes and neurotrophin receptors. Oncogene 7:371–381

Tripathy D, Henderson C (1995) Clinical resistance in breast cancer. In: Dickson RB, Lippman ME (eds) Drug hormonal resistance in breast cancer: cellular and molecular mechanisms. Horwood, Chichester, pp 3–21

Van Roozendaal CEP, van Ooijen B, Klijn JGM, Classen C, Eqqermont AMM, Henzen-Longman SC, Foekens JA (1992) Stromal influences on breast cancer cell growth. Br J Cancer 65:77–81

Waxman J, Wasan H (1992) The architecture of cancer. The tissue metalloproteinases and their inhibitors. Br Med J 305:1306–1307

Wolman SR, Heppner GH (1992) Genetic heterogeneity in breast cancer. J Natl Cancer Inst 84:469–470

Yee D (1994) The insulin-like growth factor system as a target in breast cancer. Breast Cancer Res Treat 32:85–95

Serum (Circulating) Tumor Markers for Breast Cancer

D.F. Hayes

Breast Evaluation Center, Dana-Farber Cancer Institute, Harvard Medical School, 44 Binney St., Boston, MA 02115, USA

Introduction

Tumor markers are defined as substances or processes that can be assayed in order to predict risk, presence, and/or current or future behavior of malignancy (Hayes 1994). These substances or processes differ from the non-malignant state as a result of the malignant process. A variety of specimen types might be assayed for tumor markers for breast cancer (Table 1). In this regard, germ-line genetic markers in all cells of the body might predict a relative susceptibility to the development of breast cancer. Markers of early malignant transformation might be assayed within breast tissue itself, either through biopsy or needle aspirate of epithelial tissue or by aspiration of nipple ductule contents. In patients who have established malignancy, markers can be monitored in breast cancer tissue in order to develop categories of presumed biological behavior or in distant, non-breast tissue to detect the presence of metastatic disease. However, soluble, circulating tumor-associated markers that can be detected in plasma and/or serum might reflect a more immediate status of both tumor burden and biology.

Tumor markers assayed in these specimens might be useful in several clinical situations. These include determination of risk to develop breast cancer, screening for early detection, determination of differential diagnosis, prediction of prognosis (either based on natural propensity of the cancer to metastasize and grow or on the response of the tumor to specific therapies), and/or monitoring patients with established breast cancer for detection of early relapse or estimation of current status during treatment of metastatic disease (Table 2).

Several circulating tumor markers have been commonly associated with breast cancer. The most widely used are carcinoembryonic antigen (CEA), tissue polypeptide antigen (TPA), gross cystic disease protein (GCDP), and a family of mucin-like high molecular weight glycoproteins that are products or are related to products of the MUC-1 gene: CA15-3, CA549, breast

Recent Results in Cancer Research, Vol. 140
© Springer-Verlag Berlin · Heidelberg 1996

Table 1. Specimen sources in which tumor markers might be detected

Tissue/cellular based markers
Cellular material from tissue biopsy
Cellular material separate from tissue biopsy
Fine needle aspiration
Scraping of skin or mucosa for normal cells
Collection of normal circulating cells in blood
Exfoliated cells in secreted contents (nipple aspirate or milk) or in body fluids (effusions, CSF)
Soluble markers
Circulating: serum, plasma
Secretions or body fluids: Nipple aspirate, milk urine, effusions

Table 2. Potential uses for tumor markers

Determine risk of malignancy
Screening for early malignancy
Determination of differential diagnosis
Determination of prognosis
Metastatic potential
Potential benefit from specific therapy (response)
Monitor course
Detect relapse
Follow detectable disease

cancer mucin (BCM), mammary serum antigen (MSA), and mucinous carcinoma antigen (MCA) (Tondini et al. 1989; Hayes et al. 1992). More recently, circulating antibody responses to putative tumor-associated antigens, such as P53 and HER-2/c-neu, have been reported (Davidoff et al. 1992; Disis et al. 1994). However, data regarding these more novel tumor markers are quite preliminary.

Clinical Utility

Detection of abnormal tumor marker levels in patients with breast cancer is not sufficient justification for routine clinical monitoring of that marker. Rather, in order for a tumor marker to be clinically useful, data must exist that knowledge of the marker level leads to a decision in clinical practice that is demonstrated to result in a more favorable outcome than if the marker results were not known. Thus, although circulating levels of a given marker may be more commonly elevated in patients with cancer than in normal subjects, the marker is only useful if therapeutic decisions based on these results lead to improvement in one of four clinical outcomes: overall survival, disease-free survival, quality of life, or cost of care. Indeed, although many claims have been made, few markers have been rigorously shown to have utility in most clinical situations.

For most markers, sensitivities and specificities have been determined in populations of patients with different stages of breast cancer in relation to normal subjects (Tondini et al. 1989; Hayes et al. 1992; Hayes 1993). Table 3 provides rough estimates of these performance characteristics for the commonly used circulating breast cancer markers. These are important considerations when evaluating the potential utility of a marker in each separate clinical use.

Determination of Risk

Prevention, screening, and early therapy might all be applied more efficiently in populations of patients known to be at risk for breast cancer. For example, although males are occasionally afflicted with breast cancer, the disease is predominantly one of females. Thus, current prevention and screening programs are directed exclusively at women. In this case, the presence of a Y chromosome could be considered a marker for *low* risk for breast cancer.

Approximately one of every eight to ten women in Western culture will be diagnosed with breast cancer during her lifetime. Furthermore, although the disease is more common in older women, approximately 25% of all cases and deaths occur in women between the ages of 40 and 50 (Kelsey and Berkotz 1988). Thus, current prevention and screening strategies might be much more effective if markers of extraordinarily high relative risk existed that identified those women who are most likely to develop the disease (as is seen when comparing the risk of breast cancer in males and females).

Recently, investigators have cloned the first gene known to be associated with familial breast cancer (BrCa1) (Shattuck-Eidens et al. 1995; Miki et al. 1994; Futreal et al. 1994). Unfortunately, abnormalities in BrCa1 only account for a small percentage of those patients likely to develop breast cancer. However, it appears that those women who do carry the gene have a very high relative risk of developing the disease, with a lifetime risk approaching 80% in affected families. Therefore, cloning of this gene is a significant step forward in the development of tests for genetic screening of women at risk for breast cancer. Initial assays will be based on detection of mutations/deletions in germ-line DNA. Serologic assays that detect either the abnormal BrCa1 protein product or endogenous antibody response against abnormal BrCa1 protein are likely to be developed as well. The social, ethical, economic, and clinical implications of BrCa1 screening are complex. Whether such assays will have any clinical utility remains to be seen and will require intense and carefully performed investigations (National Advisory Council for Human Genome Research 1994; Shattuck-Eidens et al. 1995; Hoskins et al. 1995).

Table 3. Sensitivity and specificity for circulating tumor markers commonly used in evaluating breast cancer[a]

Tumor marker	Breast cancer stage				Benign conditions		Other malignancies		
	I	II	III	IV/Met	Breast	Liver	Respiratory	GI	Ovarian
CEA	10–15	15–25	35–50	60–70	20–30	60–70	55–60	85–90	50–60
TPA	25–30	35–50	55–65	70–80	15–30				85
GCDP	20–30	30–40	10–30	50–60	30–40		15–20	10–15	0
MUC-1 products									
CA15-3									
CA549	5–30	15–50	60–70	65–90	15–20	50–60	20–70	15–60	40–60
MCA									
others									
MSA	65–70	75–80	75–80	80–90	15–20				

[a] Percent elevated with the relevant condition. Range compiled from available literature (Tondini et al. 1989; Hayes et al. 1992).

Screening

Screening for signs of malignancy that are not detectable by routine clinical means might permit earlier diagnosis and therapy. In order for screening to be effective, the disease of interest must be relatively common, and outcomes must be better if the patient is treated early rather than later in her course. Furthermore, the screening test must be both sensitive and specific. Breast cancer is one of the few diseases for which screening has been demonstrated to reduce mortality (Eddy 1989; Nyström et al. 1993; Kaluzny et al. 1994). However, screening mammography only reduces mortality by 20%–30% at 10 years. An easily performed test with superior sensitivity and specificity might enhance or improve upon the results seem with screening mammography.

Unfortunately, no circulating antigen associated with breast cancer exhibits sufficient sensitivity and specificity to be clinically useful. For example, the sensitivity of MSA has been reported to be as high as 68% in early stage breast cancer (Sacks et al. 1987; Smart et al. 1989). However, as with other circulating antigen assays, the specificity of MSA is too poor to avoid an unacceptably high rate of false-positive tests (Smart et al. 1989). It is possible that detection of circulating antibody responses to tumor-associated antigens might provide an indication of a developing malignancy, but currently available data are too preliminary to suggest routine use of these markers. Furthermore, none of these assays is tissue specific (Table 3). Thus, a rising CEA or CA15-3 might indicate malignant changes in several non-breast epithelial organs, including gastrointestinal tract, lung, and gynecologic organs. Therefore, one is still faced with the daunting task of examining all suspect organs with currently available (and inadequate) diagnostic techniques.

Diagnosis

In Europe and the United States, a "suspicious" abnormality found on screening mammography is generally associated with a positive predictive value of approximately 25% (Harris et al. 1992), and the sensitivity of modern-day film screen mammography is approximately 75%–90% (Donegan 1992). Therefore, in a manner similar to the use of breast sonography, tumor markers could be of great value in helping to differentiate benign from malignant breast masses or mammographic changes. However, as with screening, no circulating tumor marker is sufficiently specific in regards to normal breast epithelial changes to serve as a diagnostic marker.

Markers might also help to determine the tissue of origin in patients who are found to have a malignancy of uncertain site, especially if biopsy confirms that the lesion is an adenocarcinoma. Unfortunately, few if any markers separate breast carcinoma from other types of epithelial malignancies (Hayes et al. 1986; Sekine et al. 1985; Perey et al. 1990; Tondini et al. 1989). Of those available, GCDP antigen appears to be almost exclusively expressed by breast cancers, although it is detected in tumors of the salivary glands (Mazoujian et al. 1983; Cardoso de Almeida et al. 1992). However, only approximately 40% of all breast cancers express GCDP. Therefore, the lack of expression of this antigen can not be taken as evidence that the tissue is not breast in origin. Unfortunately, monoclonal antibodies against specific epitopes present on either CEA or Epithelial Membrane Antigen (EMA) molecules do not confer relative tissue specificity to the extent to be clinically reliable in determining differential diagnosis (Hayes et al. 1986).

Prognosis

Predicting Prognosis of Patients with Newly Diagnosed Breast Cancer

As discussed elsewhere in this symposium, adjuvant systemic therapy confers a proportional reduction in risk of breast cancer recurrence of approximately 25%, regardless of the absolute risk of recurrence (Early Breast Cancer Trialists' Collaborative Group 1992). These data suggest that almost all patients who are destined to develop clinically or radiographically detectable metastasis may benefit from adjuvant systemic therapy.

Clinical staging and the presence or absence of axillary lymph node metastasis are the most important prognostic factors for patients with newly diagnosed breast cancer (McGuire et al. 1992). The relative risk of recurrence for node-positive patients is about twice that for node-negative patients (Elledge et al. 1992). However, approximately 30% of node-negative patients will relapse, and 30% of node-positive patients will not.

Many other putative prognostic factors have been proposed that might be more powerful than nodal status or might further divide node-positive and node-negative patients into those likely or not likely to recur. These factors include demographic features, histologic features, or nonhistologic markers of biologic behavior.

Several studies have suggested that patients with elevated pre- or postoperative CEA and/or MUC-1 product (such as CA15-3) levels have a worse prognosis. However, since these markers are elevated as a result of both tumor burden and biology, the prognostic utility of elevated CEA or CA15-3 has not been independent of clinical, radiographic, and histopathologic staging. For example, Colomer et al. have reported that patients with postoperative CA15-3 levels greater than 40 U/ml have a disease-free survival that is almost half that of patients with normal levels and significantly shorter overall survival (18.3 months vs 25 months, $p < 0.0001$). Similarly, high postoperative CEA levels were also associated with shorter disease-free and overall survival differences, although these differences were not as significant as those seen with CA15-3 (Colomer et al. 1989a,b). None of those studies has determined whether elevated CA15-3 levels are independent of stage as prognostic factors. Since CA15-3 levels are associated with more advanced stages of both primary and metastatic disease, elevated CA15-3 levels may simply reflect the presence of micrometastatic disease and may not be a predictor of biologic behavior. A large, multi-institutional prospective trial assessing this question is currently underway.

Predicting Prognosis of Patients with Newly Diagnosed Metastatic Disease

Currently available therapies do not result in substantial long-erm survival prolongation for patients with metastatic breast cancer (Hayes et al. 1995). However, several studies have suggested that initial prognostic characteristics of a primary tumor remain important for patients who develop subsequent metastasis, especially in regards to predicting response to specific therapy. For example, estrogen receptor (ER) and/or progesterone receptor (PR) content of the primary tumor is predictive of response to hormone therapy for patients who develop metastatic disease (Osborne 1991). However, it has been demonstrated that the discordance between primary and metastatic ER content is approximately 20% (Hull et al. 1983; Raemaekers et al. 1984). Therefore, it would be of value to determine tumor marker content in the metastatic tumor in patients who have relapsed. Unfortunately, collection of tissue from such patients may require difficult invasive procedures.

Preliminary results from recent studies have suggested that certain tumor-associated markers that are reflective of biologic behavior may be elevated in the circulation of patients with metastatic breast cancer. For example, the external domain of the HER-2/neu oncogene product, designated neu-related protein (NRP), is elevated in approximately 40% of patients with metastatic breast cancer when compared to normal subjects (Leitzel et al. 1992; Carney et al. 1991). Elevated circulating NRP levels have been reported to be associsated with worse outcome in patients treated with endocrine therapy (Hayes et al. 1993; Leitzel et al. 1995). These data are consistent with other results suggesting that overexpression of tissue HER-2/neu may be related to resistance to hormone therapy (Wright et al. 1992).

Certain angiogenic factors, such as basic fibroblast growth factor (bFGF), are also elevated in blood and urine of patients with breast cancer (Nguyen et al. 1992; Watanabe et al. 1992). These preliminary results suggest a promising method of predicting response to standard therapies. Since the angiogenic status of a tumor is liekly to change with progressive tumor development, a time-variable circulatory marker might be very valuable. Perhaps more importantly, these findings may be of particular interest as novel therapies directed at process-specific mechanisms are developed, such as anti-HER-2 therapies or angiogenesis inhibitors (Hayes et al. 1995).

Monitoring

Monitoring Patients To Detect Occult Recurrent Disease

Patients with a prior history of invasive breast cancer are at risk for relapse for as long as 20–30 years after initial therapy (Rosen et al. 1989). Therefore, it has been suggested that patients with a prior history of infiltrating breast cancer should be screened for impending relapse throughout their course.

Such screening would require time-dependent marker changes, as opposed to static marker changes such as those that are measured in tissue. Since serum (and/or plasma) and urine are easily accessible, soluble/circulating tumor markers would be ideal for such monitoring. Several studies have shown that CEA and CA15-3 levels may become elevated prior to the development of symptoms or to the detection of metastatic carcinoma by physical or radiographic findings (Table 4) (Ruibal et al. 1987; Thirion et al. 1988; Maigre et al. 1988; Kallioniemi et al. 1988; Colomer et al. 1989a; Safi et al. 1989). In general, approximately 40%–50% of patients who will develop metastases will have a preceding rise in either CEA or CA15-3, with lead times ranging from 3–18 months.

Although impending relapse can be predicted with some degree of certainty using rising tumor markers, the clinical utility of this observation is not clear. Treatment of asymptomatic metastases with potentially toxic systemic therapy has not been demonstrated to result in either increased cure or survival rates, and treatment of patients with asymptomatic metastasis is unlikely to provide improved palliation (Hayes and Kaplan 1991; Stierer and Rosen 1989). The results of two recently completed prospective, randomized studies performed in Italy have suggested that patients who are intensively followed after primary and adjuvant therapy fare no better in regards to overall survival than those who are followed with history and physical examination alone (GIVIO Investigators 1994; Rosselli Del Turco et al. 1994). Neither of these studies included tumor markers as part of the "intensive follow-up" arms. However, since the lead time provided by tumor markers over standard clinical and radiographic evaluations is short (6–9 months), it seems unlikely that these studies would have detected

Table 4. Tumor marker prediction of impending relapse in breast cancer patients who are otherwise free of detectable disease after primary and/or adjuvant therapy

Marker	Patients who relapse with prior elevated marker (%)	Patients with elevated marker who do not relapse (%)	Median lead time (range in months)
CEA	15–80	10–30	6 (1–20)
CA15-3	45–77	9–58	6 (1–18)

Percent range compiled from available literature (Tondini et al. 1989; Hayes et al. 1992).

substantially different results had tumor markers been included. Thus, routine screening of patients following primary therapy should still be considered investigational at this time.

Monitoring Patients with Metastatic Disease

Palliation of patients with metastatic breast cancer may be accomplished with a series of treatment modalities, including surgery, radiation, hormone therapy, and chemotherapy (Hayes et al. 1995). Although each may provide benefit, the risks and toxicities of the separate therapies differ widely. Therefore, accurate knowledge of the status of a patient's disease is essential in deciding whether to continue the current therapy or proceed to an alternate regimen. In this regard, circulating levels of CA15-3 are elevated in approximately 75%–80% of patients with metastatic breast cancer (Table 3) (Hayes et al. 1986, 1992). CA15-3 is a sensitive and reliable marker for clinical course and can be used to supplement other clinical parameters during follow-up of patients with metastatic disease (Hayes et al. 1986). The superior sensitivity and specificity of CA15-3 compared to CEA make the former a preferable marker to follow disease course (Hayes et al. 1992; Tondini et al. 1988). Large-scale studies in the United States are currently under way to precisely determine the independent clinical role of monitoring CA15-3 and other antigens in such patients.

The use of CA15-3 and CEA must be discussed in the context of several important pitfalls. Certain nonmalignant conditions, such as benign breast alterations and benign liver and gynecologic diagnoses, are associated with elevated circulating CA15-3 levels. However, these are usually not important considerations in patients with metastatic breast cancer and therefore do not invalidate the use of the test. Occasional elevation of serial CA15-3 levels early during the course of effective hormone or chemotherapy followed by a fall to or below baseline levels, creating a so-called "spike" effect, is observed in approximately 15%–25% of patients (Hayes et al. 1988). this spike phenomenon must be recognized in order to avoid misinterpretation of a rising tumor marker early in the course of therapy as progression when, in fact, the patient may be responding.

Summary

Many circulating markers have been proposed for breast cancer, with potential utility for identification, screening, prognosis, detection, or monitoring. Of the available markers, those with the greatest promise include circulating tumor markers that provide an indication of clinical course, such as CA15-3 and CEA, and novel markers such as BrCa1, antibodies to p53, antibodies to HER-2/neu, indicators of angiogenesis, and the extracellular domain of

HER-2/neu. However, the precise clinical utilities of all of these markers have yet to be determined. It is especially important that the relative independence of the markers in relation to other available markers be determined so as to avoid the unnecessary cost and expense of redundancy. Moreover, it is important that the clinician be aware of the limitations in both sensitivity and specificity of each marker so as not to over- or under-interpret the predictive value of any test.

References

Cardoso de Almeida PC, Pestana CB (1992) Immunohistochemical markers in the identification of metastatic breast cancer. Breast Cancer Res Treat 21:201–210

Carney W, Hamer P, Petit D, Retos C, Greene R, Zabrecky J, McKenzie S, Hayes DF, Kufe DW, DeLellis R, Naber S, Wolfe H (1991) Detection and quantitation of the neu oncoprotein. J Tumor Marker Oncol 6:53–72

Colomer R, Ruibal A, Genolla J, Rubio D, Del Campo JM, Bodi r, Salvador L (1989a) Circulating CA 15-3 levels in the postsurgical follow-up of breast cancer patients and in non-malignant diseases. Breast Cancer Res Treat 13:123–133

Colomer R, Ruibal A, Salvador L (1989b) Circulating tumor marker levels in advanced breast carcinoma correlate with the extent of metastatic disease. Cancer 64:1674–81

Davidoff A, Iglehart JD, Marks J (1992) Immune response to p53 is dependent upon p53/HSP70 complexes in breast cancers. Proc Natl Acad Sci USA 89:3439–3442

Disis ML, Calenoff E, McLaughlin G, Murphy AE, Chen W, Groner B, Jeschke M, Lydon N, McGlynn E, Livingston RB, Moe R, Cheever MA (1994) Existent T-cell and antibody immunity to HER-2/neu protein in patients with breast cancer. Cancer Res 54:16–20

Donegan WL (1992) Evaluation of a palpable breast mass. N Engl J Med 327:937–942

Early Breast Cancer Trialists' Collaborative Group T (1992) Systemic treatment of early breast cancer by hormonal, cytotoxic, or immune therapy: 133 randomised trials involving 31 000 recurrences and 24 000 deaths among 75 000 women. Lancet 339:1–15, 71–85

Eddy DM (1989) Screening for breast cancer. Ann Intern Med 111:389–399

Elledge RM, McGuire WL, Osborne CK (1992) Prognostic factors in breast cancer. Semin Oncol 19:244–253

Futreal PA, Liu Q, Shattuck-Eidens D, Cochran C, Harshaman K, Tavtigian S, Bennett LM, Haugen-Strano A, Swensen J, Miki Y, Eddington K, McClure M, Frye C, Weaver-Feldhaus J, Ding W, Gholami Z, Söderqvist P, Terry L, Jhanwar S, Berchuck A, Iglehart JD, Marks J, Ballinger DG, Barrett JC, Skolnick MH, Kamb A, Wiseman R (1994) BRCA1 mutations in primary breast and ovarian carcinomas. Science 266:120–122

GIVIO Investigators (1994) Impact of follow-up and testing on survival and health-related quality of life in breast cancer patients: a multicenter randomized controlled trial. J Am Med Assoc 271:1587–1592

Harris JR, Lippman ME, Veronesi U, Willett W (1992) Breast cancer (first of three parts). N Engl J Med 327:319–328

Hayes DF (1993) Tumor markers for breast cancer. Ann Oncol 4:807–819

Hayes DF (1994) Hematology/Oncology Clinics of North America: tumor markers in adult solid malignancies. Saunders, Philadelphia

Hayes DF, Kaplan W (1991) Evaluation of patients following primary therapy. In: Harris J, Hellman S, Henderson I, Kinne D (eds) Breast diseases. Lippincott, Philadelphia, pp 505–525

Hayes DF, Zurawski VR Jr, Kufe DW (1986) Comparison of circulating CA15-3 and carcinoembryonic antigen levels in patients with breast cancer. J Clin Oncol 4:1542–1550

Hayes DF, Kiang DT, Korzun A, Tondini C, Wood W, Kufe D (1988) CA15-3 and CEA spikes during chemotherapy for metastatic breast cancer. Proc Am Soc Clin Oncol 7:38a

Hayes DF, Tondini C, Kufe DW (1992) Clinical applications of CA15-3. In: Sell S (ed) Serological cancer markers. Humana, Totowa, pp 281–307

Hayes DF, Cirrincione C, Carney W, Rodrigue S, Berry D, Younger J, Panasci L, Millard F, Duggan D, Henderson IC, Kufe DW (1993) Elevated circulating HER-2/neu related protein (NRP) is associated with poor survival in patients with metastatic breast cancer. Proc Am Soc Clin Oncol 12:58

Hayes DF, Henderson IC, Shapiro CL (1995) Treatment of metastatic breast cancer: present and future prospects. Semin Oncol 22(suppl 5):5–21

Hoskins KF, Stopfer JE, Calzone KA, Merajver SD, Rebbeck TR, Garber JE, Weber BL (1995) Assessment and counseling for woment with a family history of breast cancer: a guide for clinicians. J Am Med Assoc 273:577–585

Hull DF III, Clark GM, Osborne CK, Chamness GC, Knight WA III, McGuire WL (1983) Multiple estrogen receptor assays in human breast cancer. Cancer Res 43:413–416

Kallioniemi O, Oksa H, Aaran R, Hietanen T, Lehtinen M, Koivula T (1988) Serum CA15-3 assay in the diagnosis and follow-up of breast cancer. Br J Cancer 58:213–215

Kaluzny AD, Rimer B, Harris R (1994) The National Cancer Institute and guideline development: lessons from the breast cancer screening controversy. J Natl Cancer Inst 86:901–903

Kelsey J, Berkowitz G (1988) Breast cancer epidemiology. Cancer Res 48:5615–5623

Leitzel K, Teramoto Y, Sampson E, Mauceri J, Langton BC, Demers L, Podczaski E, Harvey H, Shambaugh S, Volas G, Weaver S, Lipton A (1992) Elevated soluble cerbB-2 antigen levels in the serum and effusions of a proportion of breast cancer patients. J Clin Oncol 10:1436–1443

Leitzel K, Teramoto Y, Konrad K, Chinchilli VM, Volas G, Grossberg H, Harvey H, Demers L, Lipton A (1995) Elevated serum c-erbB-2 antigen levels and decreased response to hormone therapy of breast cancer. J Clin Oncol 13: 1129–1135

Maigre M, Fumoleau P, Ricolleau G, Guillard Y, Cussac A, Vignoud J, Ahkong V (1988) CA15-3 in breast cancer. Comparison with CEA (French). Semin Hop Paris 64:9–13

Mazoujian G, Pinkus GS, Davis S, Haagensen DE Jr (1983) Immunohistochemistry of a gross cystic disease fluid protein (GCDFP-15) of the breast: a marker of apocrine epithelium and breast carcinomas with apocrine features. Am J Pathol 110:105–112

McGuire WL, Tandon AK, Allred DC, Chamness GC, Ravdin PM, Clark GM (1992) Treatment decisions in axillary node-negative breast cancer patients. J Natl Cancer Inst Monogr 11:173–180

Miki Y, Swensen J, Shattuck-Eidens D, Futreal PA, Harshman K, Tavtigian S, Liu Q, Cochran C, Bennett LM, Ding W (1994) A strong candidate for the breast and ovarian cancer suceptibility gene BRCA1. Science 266:66–71

National Advisory Council for Human Genome Research (1994) Statement on use of DNA testing for presymptomatic identification of cancer risk. J Am Med Assoc 271:785

Nguyen M, Watanabe H, Budson A, Richie JP, Hayes DF, Folkman J (1992) Basic fibroblast growth factor (bFGF) is elevated in the urine of patients with a wide variety of neoplasms. Mol Biol Cell 3:234A

Nyström L, Rutqvist LE, Wall S, Lindgren A, Lindqvist M, Rydén S, Andersson I, Bjurstam N, Fagerberg G, Frisell J, Tabár L, Larsson L-G (1993) Breast cancer screening with mammography: overview of Swedish randomised trials. Lancet 341:973–978

Osborne CK (1991) Receptors. In: Harris J, Hellman S, Henderson I, Kinne D (eds) Breast diseases. Lippincott, Philadelphia, pp 301–325

Perey L, Hayes D, Tondini C, van Melle G, Bauer J, Lemarchand T, Reymond M, Mach J, Leyvraz S (1990) Elevated CA125 levels in patients with metastatic breast carcinoma. Br J Cancer 62:668–670

Raemaekers JM, Beex LV, Koenders AJ, Pieters GF, Smals AG, Benraad TJ, Kloppenborg PW (1984) Concordance and discordance of estrogen and progesterone receptor content in sequential biopsies of patients with advanced breast cancer: relation to survival. Eur J Cancer Clin Oncol 20:1011–1018

Rosen P, Groshen W, Saigo P, Kinne D, Hellman S (1989) A long-term follow-up study of survival in Stage I (T1NOMO) and Stage II (T1N1MO) breast carcinoma. J Clin Oncol 7:355–366

Rosselli Del Turco M, Palli D, Cariddi A, Ciatto S, Pacini P, Distante V (1994) Intensive diagnostic follow-up after treatment of primary breast cancer: a randomized trial. J Am Med Assoc 271:1593–1597

Ruibal A, Colomer R, Genolla J (1987) Prognostic value of CA15-3 serum levels in patients having breast cancer. Horm Metab 1:11–15

Sacks NP, Stacker SA, Thompson CH, Collins JP, Russell IS, Sullivan JA, McKenzie IF (1987) Comparison of mammary serum antigen (MSA) and CA15-3 levels in the serum of patients with breast cancer. Br J Cancer 56:820–824

Safi F, Kohler I, Rottinger E, Suhr P, Beger HG (1989) Comparison of CA15-3 and CEA in diagnosis and monitoring of breast cancer. Int J Biol Markers 4:207–214

Sekine H, Hayes DF, Ohno T, Keefe KA, Schaetzl E, Bast RC, Knapp R, Kufe DW (1985) Circulating DF3 and CA125 antigen levels in serum from patients with epithelial ovarian carcinoma. J Clin Oncol 3:1355–1363

Shattuck-Eidens D, McClure M, Simard J, Labire F, Narod S, Couch F, Hoskins K, Weber B, Castilla L, Erdos M, Brody L, Friedman L, Ostermeyer E, Szabo C, King M-C, Jhanwar S, Offit K, Norton L, Gilewski T, Lubin M, Osborne M, Black D, Boyd M, Steel M, Ingles S, Haile R, Lindblom A, Olsson H, Borg A, Bishop DT, Solomon E, Radice P, Spatti G, Gayther S, Ponder B, Warren W, Stratton M, Liu Q, Fujimura F, Lewis CMHS, Goldgar DE (1995) A collaborative survey of 80 mutations in the BRCA1 breast and ovarian cancer susceptibility gene: implications for presymptomatic testing and screening. J Am Med Assoc 273:535–541

Smart YC, Rogers KM, Brien JH, Stewart JF, Burton RC (1989) Factors affecting MSA levels in normal women and women with breast cancer. Aust NZ J Surg 59:805–8–9

Stierer M, Rosen HR (1989) Influence of early diagnosis on prognosis of recurrent breast cancer. Cancer 64:1128–1131

Thirion B, Ricolleau G, Fumoleau P (1988) Result of a prospective study of the CA15-3 reliability in early detection of breast cancer recurrence. 3rd international workshop on monoclonal antibodies and breast cancer, San Francisco, 1988

Tondini C, Hayes DF, Gelman R, Henderson IC, Kufe DW (1988) Comparison of CA15-3 and carcinoembryonic antigen in monitoring the clinical course of patients with metastatic breast cancer. Cancer Res 48:4107–4112

Tondini C, Hayes DF, Kufe D (1989) Circulating tumor markers in breast cancer. In: Henderson IC (ed) Diagnosis and therapy of breast cancer. Saunders, Philadelphia, pp 653–674

Watanabe H, Nguyen M, Schizer M, Li V, Hayes DF, Sallan S, Folkman J (1992) Basic fibroblast growth factor in human serum: a prognostic test for breast cancer. Mol Biol Cell 3:324A

Wright C, Nicholson S, Angus B, Sainsbury JRC, Farndon J, Cairns J, Harris AL, Horne CHW (1992) Relationship between c-erbB-2 protein product expression and response to endocrine therapy in advanced breast cancer. Br J Cancer 65:118–121

Summary

M. Baum

The Royal Marsden Hospital, Fulham Road, London SW3 6JJ, UK

Introduction

At the outset, it is important to make a clear distinction between prognostic factors and predictive indices. Prognostic factors provide us with an estimate of the rate of relapse and death for defined subpopulations of women with breast cancer. Predictive indices are supposed to allow us to determine which subgroup of patients respond preferentially to particular types of treatment. Rational selection of adjuvant systemic therapy is therefore based on two quite separate calculations.

The estimate of prognosis allows one to calculate the absolute benefit to treatments for groups of women assuming a consistent odds ratio for the benefit of adjuvant systemic therapy in unselected groups of patients. In theory, this rational selection could be enhanced if we had biological markers that predicted preferential response to endocrine manoeuvres or preferential response to cytotoxic chemotherapy.

The worse the prognosis or the greater the relative risk of relapse and death over a given period of time, the greater the absolute benefit of the systemic therapy. This allows us to make rational harm-benefit analyses for the individual woman. At one extreme, a patient with only a 10% risk of relapse and death over 10 years may only stand to gain about 2% absolute improvement whereas at the other extreme a woman with say 50% chance of relapse and death might achieve a 12.5% absolute improvement assuming a constant odds ratio of 0.75. Faced with these benefits it is then possible, under the guidance of a wise physician, for a patient to make a rational choice bearing in mind the short-term and long-term hazards of the treatment. Putting it another way, it might be more important to identify a subgroup of patients who have an expectation of survival similar to that of the age-matched population, for whom there is no room for improvement and so adjuvant systemic therapy can safely be avoided.

The number of prognostic variables available and their degree of importance is a subject of great controversy and specialists can be forgiven for their belief that we have more than enough of such indices. The real problem is that we don't yet know how to apply them in a sensible fashion. The other caution is the interpretation of the expression "independent prognostic variables". Multivariant mathematical models can produce conflicting evidence as to what is or isn't an independent prognostic variable. In theory, one could consider time-dependent variables such as tumour size or node status and biological variables that determine the rate of growth and the metastasising capacity of the primary, yet this may be a gross oversimplification in that there could be a phenotypic drift of biological variables with time. In addition, the size of the primary and the lymph node status could also be looked upon as surrogate measures of the rate of proliferation and the metastasising capacity of the primary.

Perhaps more realistically, we could consider independent prognostic variables at the molecular level, for example a p53 mutation on chromosome 17. This could be entirely separate to another mutation on another chromosome that might be related to invasiveness rather than rate of proliferation. If that is the case, then what we might be measuring with variables such as oestrogen receptor status, thymidine labelling index and S-phase could be a series of epiphenomena summating the net effect of a variety of molecular changes at the genomic level. If that is the case, then it would not be surprising that histological grade in the hands of an expert pathologist might provide as much, if not more, information as the sophisticated and apparently more objective measurements at the molecular level.

Passing on now to predictive indices, these might be considered under two headings: those that indicate a qualitative difference in outcome (in other words identifying subgroups of patients who have some benefit vs other groups of patients who have no benefit or in fact a detriment with the treatment) and those quantitative indices which suggest the patients have greater or lesser degrees of benefit from the same treatment. At this point I must introduce a cautionary note on statistical interpretation. We have enough difficulty recruiting sufficient patients into clinical trials to demonstrate with confidence that any one treatment has an overall benefit at a clinical level of significance with statistical power of 90% or more. This might require several thousand patients. However, if we want to study a predictive factor for either the qualitative or quantitative interaction between treatment and a biological index, we would need about twice the number of patients to show this with confidence. Many studies fall into this trap and assume that because they have a conventionally significant result in one subgroup of patients and a nonsignificant difference in another subgroup of patients, this in itself can demonstrate a qualitative interaction. Yet almost invariably, statistical tests for interaction will fail to demonstrate significant differences between these two levels of response.

It is true that there may be quite plausible hypotheses being investigated. For example, one would anticipate that the oestrogen receptor status or the menopausal status of a women would determine responsiveness to endocrine manipulation by tamoxifen or ovarian suppression. I don't wish to drag up the rather tiresome debate about whether or not the oestrogen receptor status can predict which patients will respond or fail on tamoxifen treatment, but there certainly is at least one promising result suggesting that oestrogen receptor status might predict qualitative interactions between chemotherapy for the oestrogen receptor-negative group and ovarian suppression for the oestrogen receptor-positive group (Scottich Cancer Trials Breast Group 1993).

The other plausible hypothesis is that the greater the rate of proliferation of a breast cancer, the more likely it is to respond to adjuvant chemotherapy. One marker that has been proposed in this respect is the c-*erb*B-2 oncogene. However, the literature is somewhat in disarray on this issue, with the IBCSG trial suggesting that the overexpression of c-*erb*B-2 predicts chemoresistance and the CALGB trial suggesting it is predictive of response to high-dose chemotherapy. At the moment it cannot be stated with confidence that the worse the prognosis, the more likely patients are to respond to chemotherapy. After all, those patients with the very worst prognosis who in fact present with metastatic disease show no survival advantage if treated initially with chemotherapy when compared with initial treatment with endocrine therapy proceeding to chemotherapy on relapse or progression.

The authors of the three contributions in this part elaborated on these issues. Professor Manfred Kaufmann, from Heidelberg, Germany, reviewed our knowledge of recognised prognostic variables. Professor Barry Gusterson, from the Royal Marsden Hospital, London, amplified on these from the biological point of view and went on to discuss the potential of future predictors of response. Finally, Dr. Daniel Hayes, from the Dana Farber Cancer Institute, Boston, MA, reviewed the literature on the use of serum tumour markers in the prognosis and monitoring of patients with breast cancer.

Manfred Kaufmann

Age, Menopausal Status

The patient's age and menopausal status are factors which may be clinically useful in the prognostic assessment and treatment decisions. The youngest group of patients, say under the age of 35, have the worst prognosis compared with older premenopausal and/or postmenopausal patients, but this may be explained in part by the increased frequency of poor prognostic factors which are often observed in these younger patients. At the other extreme, the outcome in the oldest cohort of women may be confounded by concomitant disease and competing risks for death. Nevertheless, it is almost self-evident that the menopausal status is the best predictor we have for the efficacy of ovarian suppression and to a lesser extent for adjuvant tamoxifen. There is also some controversy whether a young age itself is a predictor of local recurrence after breast conserving surgery.

Tumour Size

The TNM staging system is still a useful, if somewhat crude, way for estimating prognosis. There is a strong correlation between tumour size and risk of recurrence and death and this can easily be measured by clinical and pathological procedures. Some centres find ultrasound measurements of greater accuracy. In practical terms however, tumour size is the most important factor to select those patients who should be offered an opportunity of breast conserving surgery. In general, there is a strong linear correlation between tumour size and the state of the axillary lymph nodes.

Axillary Lymph Node Status

About 25% of patients who are node-negative at the time of surgery still go on to develop distant metastases, but that aside, the number of involved axillary lymph nodes is still considered to be the most powerful prognostic factor. The risk of recurrence and death increases in a linear fashion with increasing number of involved lymph nodes and there is a general consensus that patients with node-positive tumours will have the greater absolute benefit from adjuvant systemic therapy for reasons described in the introduction. However, the assessment of lymph nodes is limited by the level of axillary dissection decided on by the surgeon, the number of lymph nodes identified by the pathologist and the number of sections taken from each node for study.

Grade/S Phase

Histological or nuclear grade gives evidence of differentiation of the tumour which, as already described, is a summation of molecular phenomena interpreted at the morphological level. In addition, there is quite a good correlation between the degree of differentiation and the presence or absence of oestrogen receptors. Unfortunately, there is extensive intra- and inter-observer variability in the interpretation of histological grade and therefore there have been many efforts made to come up with some reproducible and objective summary factor of epithelial proliferation. The S-phase fraction, measured by DNA flow cytometry, looks promising and correlates well with nuclear grade and hormone receptor status. However, this is dependent on having expensive technology, experts to handle the equipment and some form of quality control. The machinery alone will not guarantee objectivity.

Hormone Receptor Status

Measurement of steroid hormone receptor concentration has been possible for nearly 20 years. Oestrogen receptor status, whether described in qualitative terms, i.e., positive/negative, or quantitative terms based on ligand-binding assays, shows a consistent if rather weak correlation with prognosis. Currently, immunohistochemical tests on frozen or even paraffin embedded sections can provide a relatively quick and cheap estimate. There is considerable difference of opinion as to whether the measurement of the progesterone receptor adds to the prognostic information, although in indicating an intact hormone receptor pathway it could ultimately become a predictive test of endocrine responsiveness. Ligand-binding assay measurements for oestrogen receptors can be misleading in younger women because the receptor sites are occupied by oestrogens. In contrast, the immunohistochemical tests recognise an epitope which is separate from the receptor binding zone.

Barry Gusterson

Professor Gusterson described the study of molecular changes in the development of breast cancer that can explain both the clinical and morphological heterogeneity of the disease. These studies challenge our previous emphasis on the importance of prognostic factors and will in time replace them with predictive factors based on in situ markers of biological behaviour and treatment response. He reemphasised many of the points made in the introduction. There are three areas where prognostic or predictive markers may be of value:

1. Identification of those patients with such a good prognosis that no treatment other than that required for local control is necesary
2. Identification of those patients who have such a poor prognosis that aggressive therapy up to an including bone marrow transplant or stem cell rescue might be justified in the context of a randomised controlled trial
3. Identification of patients whose tumours would be either responsive or resistant to particular therapies

The most significant factors that predict survival are related to the rate of proliferation and metastasising capacity. Any molecular change that contributes to one or other of these aspects of the tumour type can therefore be assumed to have some prognostic significance. As tumours have many diverse ways of producing these phenotypes it is predicted that the reductionist approach will produce numerous subgroups with marginally different prognostic differences. Therefore the most important measurement when applied to a large population will be the summation of these factors which could be assessed, as already stated, by routine examination of an H & E section by a good histopathologist. Thus the molecular dissection of the mechanisms of breast carcinogenesis may have more to offer in the identification of therapeutic opportunities and the prediction of therapeutic response. Many of the molecular abnormalities now being found in tumours are potential targets for new therapeutic strategies, for example to inhibit angiogenesis or specific growth factor pathways or to neutralise the activity of enzymes involved in microinvasion. As our level of knowledge about these molecular events becomes increasingly sophisticated, there is some urgency to develop a new classification of breast cancer that will carry us into the next millennium. It is therefore a priority that the old TNM system be replaced by a new system making use of all the available prognostic and predictive variables, alongside quality control for the measurements of these markers, so that the results can be pooled for use in large multicentre or multinational studies.

Daniel Hayes

The detection and/or monitoring of circulating tumour markers might be useful in several clinical situations including: determination of the risk of developing breast cancer, screening for early detection, determination of differential prognosis, prediction of response to specific therapies or monitoring patients with established breast cancer for detection of early relapse or estimation of current status during active treatment of metastatic disease.

Several circulating tumour markers have been commonly associated with breast cancer. These include carcinoembryonic antigen (CEA), tissue polypeptide antigen (TPA), gross cystic disease protein (GCDP) and the family of mucin-like, high molecular weight glycoproteins (CA15.3, CA549). In general CA15.3 levels are more commonly elevated in patients with metastatic breast cancer than are CEA levels, but both may be elevated by a variety of benign conditions.

Screening

The first gene known to be associated with familial breast cancer has recently been cloned (BRCA1). Abnormalities in BRCA1 only account for a small percentage of those patients likely to get breast cancer, but this is at least a start for the development of tests for genetic screening. Serological assays that detect either the abnormal BRCA1 protein product or endogeneous antibody response against abnormal BRCA1 protein will soon be developed as well. Whether such assays will have any clinical utility remains to be investigated. Unfortunately, currently available circulating tumour markers such as CA15.3 or CEA are insufficiently sensitive and specific to be used to screen the general population.

Diagnosis

Few, if any, markers separate breast cancer from other types of epithelial malignancies. CA15.3 and CEA are frequently elevated in patients with nonbreast epithelial, and even occasional nonepithelial, malignancies. The GCDP antigen is almost exclusively overexpressed by breast cancer. However, only about 40% of such patients express this marker. In contrast, markers associated with ovarian cancer (CA125) or prostate cancer (PSA) are also elevated in patients with breast disease.

Prognosis

Some, but not all, studies suggest that pre- and postoperative elevated levels of CEA and CA15.3 are associated with a poor prognosis. None of these studies has yet determined whether elevated marker levels are independent of the prognostic factors already described. Since they are associated with more advanced stages of the disease, these elevated levels may simply reflect the presence of an extensive micrometastatic tumour burden and not be a predictor of biological behaviour. In contrast, preliminary results from recent studies have suggested that certain tumour associated markers that are reflective of biological behaviour may be elevated in the circulation of patients with metastatic breast cancer. For example, the external domain of the Her-2/neu (c-*erb*B-2) oncogene product is elevated in approximately 40% of patients with metastatic breast cancer when compared to normal subjects. Likewise, certain angiogenic factors such as basic fibroblast growth factor (bFGF) are also elevated in the blood and urine of patients with breast cancer. These preliminary results suggest a promising method of predicting response to anti-angiogenic agents.

Monitoring

Approximately 40% of those patients who will relapse exhibit a rising CEA and/or CA15.3 prior to the emergence of clinically or radiographically detectable disease. The clinical utility of this lead time is not clear. Treatment of asymptomatic metastases with systemic therapy has not been demonstrated to result in either increased cure or survival rates and treatment of such patients is unlikely to provide improved palliation. There is therefore a need for a large, randomised, controlled trial comparing reintroduction of systemic therapy based on a rising tumour marker level with the conventional watch policy in the face of this tumour marker rise. In the meantime, routine screening of patients following primary therapy cannot be recommended.

CEA and CA15.3 are fairly reliable indicators of the clinical course. Circulating levels of CA15.3 are elevated in approximately 75%–80% of patients with metastatic breast cancer. This marker is more sensitive and specific than CEA and could be used to supplement other clinical parameters during follow-up of patients with metastatic disease. However, there are still no data to suggest that this degree of monitoring enhances length or quality of survival. The precise clinical utilities of all these markers have yet to be determined. It is particularly important that we determine whether expression of these circulating markers is in fact independent of expression of the common prognostic or predictive factors which can be measured by readily available methods. Otherwise we will be guilty of generating unnecessary costs. It is also important that the clinician be made aware of the limitations of both the sensitivity and the specificity of these markers.

Reference

Scottish Cancer Trials Breast Group and ICRF Breast Unit, Cuy's Hospital, London (1993) Adjuvant ovarian ablation versus CMF chemotherapy in premenopausal women with pathological stage II breast carcinoma: the Scottish trial. Lancet 341:8856

IV. Screening for Breast Cancer

A Critical Review of Screening for Breast Cancer

M. Rosselli Del Turco

Centro per lo Studio e la Prevenzione Oncologica, V.le A. Volta, 171-50131
Florence, Italy

Introduction

Over the last 30 years numerous studies have been done to evaluate the efficacy of early detection in the reduction of mortality due to cancer.

As is now generally acknowledged, the only reliable results are those of studies with randomized control groups. In fact, these studies have produced positive and consistent results only regarding mammographic screening for breast cancer.

In six randomized studies, mortality due to breast cancer was 24% lower in women 50–69 years of age invited who underwent periodic mammographic examination. The risk ratio was (RR) 0.76, with a 95% confidence interval (CI 95%) of 0.67–0.87 (Wald et al. 1993) compared to the control group. These results cleary demonstrate that mammographic screening is an effective public health instrument for the reduction of mortality due to mammary carcinoma, so long as the quality of the screening program is comparable to that in the controlled studies. Moreover, we know that the efficacy of mammographic screening depends on various factors: the epidemiology of mammary carcinoma, the rate of participation, the sensitivity of the screening test, and, finally, the quality of treatment. These factors are discussed below.

Factors Affecting the Efficacy of Mammographic Screening

The Epidemiology of Mammary Carcinoma

It is apparent that the efficacy of a screening program depends on the level of breast cancer mortality, which, in turn, is related to the expected incidence in the absence of screening and to the survival rate.

It is therefore apparent that the reduction of mortality afforded by the screening program may be just as important in countries with high incidence rates as in countries with lower rates, but with lower average survival of breast cancer patients bacause of a worse stage distribution. For example, in a case control study, Palli et al. (1986) showed that the results of a mammographic screening program conducted in a rural area of the district of Florence, with a low rate of mortality due to mammary carcinoma, were comparable to those obtained in northern European countries (Tabar et al. 1985; Chamberlain et al. 1988; Veerbek et al. 1984).

Participation

Participation in a screening program is a key factor in determining the efficacy of the program. Figure 1 shows the estimated reduction of mortality in an Italian national mammographic screening program on the basis of two different rates of participation. It is estimated that the reduction in mortality after about 30 years (1991–2020) from the start of the program will be 14% if the attendance rate is 70% and 10% if the attendance rate is 55%. In actual numbers this represents a difference of about 400 lives saved in Italy.

Sensitivity of the Test

The sensitivity of the test depends on various technical factors, the professional skills of the personnel, and some choices made regarding the screening

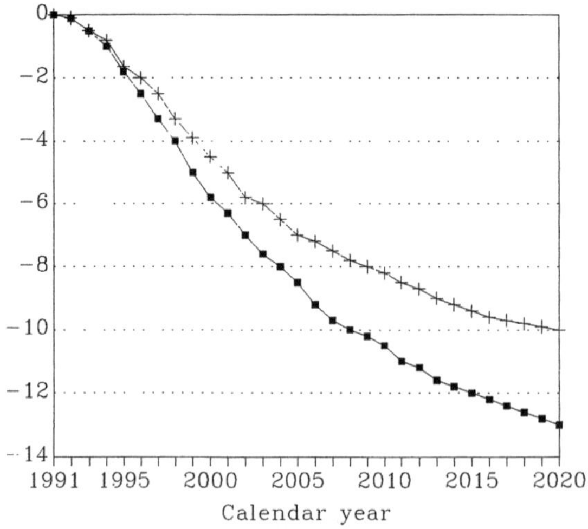

Fig. 1. Mortality reduction with different participation rates by calender year (estimates for Italy 1991–2020). Attendance (%): ■ 70%; + 55%

procedures. It is generally agreed that both the technical and medical aspects of a screening program should be submitted to rigorous quality control (Kirkpatrick et al. 1992). Unfortunately, the experience of the last 10 years suggests that some programs do not meet the recommended European standards because of inadequate personnel training and quality control.

As far as screening procedures are concerned, it is now clear that the use of double views in the first round gives higher sensitivity and specificity. Similarly, independent double readings by two radiologists affords an increase of from 5% to 15% in the detection rate (Bassett et al. 1987; Ikeda and Sickles 1988; Thurfjell et al. 1994; Anttinen et al. 1993; van Dijck et al. 1992). In our experience (Table 1), double readings allow a 4% increase in sensitivity, but it has been confirmed that carcinoma diagnosed by only one of the two radiologists are generally at an earlier stage, and their detection therefore might have a greater impact on the survival rate.

Quality of Treatment

There is no doubt that the efficacy of screening depends upon the ability to provide adequate treatment in the early phase of the tumor's development. Unfortunately the estimates of the efficacy of screening programs are always related to how early the cancer is detected and little information is available on the quality of treatment.

We maintain that more research is necessary with a view to establishing which is the most effective treatment in the preclinical phase and to preparing adequate quality control guidelines for the treatment of screen-detected lesions.

Adverse Effects

It has been established that mammographic screening, like any other means of early detection, has various negative effects such as anxiety, pain and

Table 1. Number of cancer cases detected and of subjects recalled for assessment according to reading radiologist

	Cancers detected (n)	Subjects recalled (n)
First screening	–	–
Both readers	100	524
First reader only	2	76
Second reader only	8	109
Repeat screening both readers	14	28
First reader only	–	2
Second reader only	1	9

discomfort, radiation hazard, false positives, overdiagnosis. In our own program, the high quality of the program and adequate personnel training limits the instances of false positives considerably. In fact, it has been established that the referral rate can be maintained below 5% (Table 2) and the ratio of benign to malignant biopsies below 2:5.

In order to obtain a high predictive value for carcinoma in screen-detected lesions that require open biopsy, two things are necessary: (1) the assessment of screen detected using all the different diagnostic methods available (physical examination, cytology, ultrasound) and in the same session; and (2) the extensive use of citology, either ultrasound or stereotactically guided in the case of nonpalpable lesions.

In our program and in the Swedish programs, the radiologist conducts all the diagnostic examinations at the same place and time, resulting in a predictive value of over 70%.

Cost-Effectiveness

Many studies have been carried out to evaluate the cost-effectiveness of a mammographic screening program (Table 3), but results vary widely. This is partly because the studies employed different methodologies and partly because the programs studied differed in the screening procedures they adopted, the efficacy of their methods, and the local health service costs. We maintain that the estimates of the Dutch study (De Koning et al. 1991) are the most reliable because of the methology adopted, the length of the study period, and the efficiency of the screening procedures.

Using the same methodological criteria as De Koning et al., we have recently evaluated the real costs of the Florence program. As shown in Table 4, the cost per woman examined is about U.S. $40, and the gratest

Table 2. Preliminary results of the Florence program compared to national breast cancer screening programm: first round

	National programs					
	Florence	UK	Netherlands	Sweden	Finland	Iceland
Participation (%)	54.0	71.3	75.0	75.0	89.0	67.0
Referral (%)	4.2	6.2	1.4	5.0	2.9	4.2
Detection rate (%)	8.0	6.2	6.0	6–7	3.8	5.4
Maligant/benign ratio	5:1	2.3:1	1.2:1	4:1	1.6:1	2:1

From Wald et al. (1993); Chamberlain et al. (1993); Ciatto et al. (1994).

Table 3. Cost-effectiveness of mammographic screening (modified from Vanara 1992)

Study	Cost per examined woman (US$ 1990)	Cost per life-year gained (US$ 1990)	QALY (US$ 1990)
UK (Forrest et al. 1986)	39.1	8643	9394
USA (Ota 1987)	104.8	27690	–
Sweden (Jonsson et al. 1988)	38.5	9044	–
Australia (Hall et al. 1990)	155.8	12906	20690
UK (Clarke and Fraser 1991)	32.7	87920	91556
Japan (Okubo et al. 1991)	32.9	14300	–
The Netherlands (van der Mass and de Koning et al. 1989–1991)	110.5	3821	4050

QALY, quality-adjusted life-year gained.

Table 4. Program costs and cost per examined woman in the Florence project: first round (1991–1992)

Program costs	Total costs (US$ 1994)	Percent
Health campaign	3161	0.53
Invitation	39190	6.57
Equipment	32102	5.38
Running costs	97513	16.34
Personnel	358216	60.04
Other	66417	11.14
Total	596599	100
Cost per examined women		
Invitation	3.7	9.2
Screening	30.7	76.0
Assessment	135.9	14.8
Total	40.4	100

single expense is the cost of personnel (60%). If cost-effectiveness is evaluated in a different setting (Germany), with a lower test sensitivity and average rate of participation, the results can be two or three times less favorable than those obtained in Great Britain or in Holland (Beemsterboer et al. 1994).

It is therefore apparent that the cost per life saved due to mammographic screening is acceptable, and lower than that attained by other public health services, so long as high levels of quality and efficiency can be guaranteed.

Research Needs

Effect of Screening Premenopausal Women

Among women under 50, the reduction of mortality attained in the randomized trials was 15% (RR 0.85, 95 CI 0.68–1.08) (Wald et al. 1993). These figures exclude the Canadian trial that recorded an unexplained excess mortality of women under 50 years of age (Miller et al. 1992) and which has been strongly criticized in terms of both technical screening and study quality (Moskowitz 1992; Kopans 1990).

It has been suggested that better results can be obtained with a shorter screening interval (Tabar et al. 1993) and by using high-quality double-view mammography with a double reading. This represents one of the research priorities in the field of screening for breast tumors. In the United Kingdom, a new trial has been undertaken involving about 200 000 women, age 40–41, of whom 65 000 have been assigned at random to the screening group (mammography at annual intervals of 7 years).

In another four European countries (Italy, Holland, Finland, and Sweden) the feasibility of a new trial is being evaluated under the auspices of the International Union Against Cancer. This new trial would follow procedures similar to those in Britain and would enlist 250 000 women age 40–42 years (Eckhardt er al. 1994). Until these results are obtained, we believe that asymptomatic women under the age of 50 who request a mammography "should be fully informed of the lack of proven benefit and the potential adverse effects of mammographic screening," as recommended by the European Group for Mammographic Screening (Rosselli Del Turco 1994).

Screening Interval

The results of randomized studies and evaluation of the incidence of interval cancers in the years following the screening test indicate that, in women over 50, a 2 year interval is optimal and that future studies of the effect of mammography on women under 50 should adopt a 1 year interval.

Natual History of Preclinical Disease

We feel that it is necessary to dedicate a great deal of resources to research in this sector. It has been shown in the Swedish two counties study (Tabar et al. 1992) that screen-detected tumors are associated with more favorable prognostic indicators, as, for some tumors, biological aggressiveness appears only in a later phase and can therefore be avoided by early diagnosis.

A better study of the biological characteristics of the tumor in the preclinical phase and of its appropriate treatment remains one of the most important fields of research. This requires greater cooperation between screening specialists and researchers involved in the study and treatment of mammary carcinoma.

References

Anttinen I, Pamilo M, Soiva M et al. (1993) Double reading of mammography screening films. One radiologist or two? Clin Radiol 48(6):414–21

Bassett LW, Bunnell DH, Jahanshahi R et al. (1987) Breast cancer detection: one versus two views. Radiology 165:95–97

Beemsterboer Petra MM, De Koning HJ, Warmerdam Peter G et al. (1994) Prediction of the effects and costs of breast cancer screening in Germany. Int J Cancer 58:623–628

Chamberlain J, Coleman D, Ellman R, Moss SM (1988) First results of mortality reduction in the UK trial of early detection of breast cancer. Lancet 20:411–416

Chamberlain J, Moss SM, Kirkpatrick AE et al. (1993) National Health Serivce breast screening programme results for 1991–92. Br Med J 307:353–356

Ciatto S, Rosselli Del Turco M, Giorgi D et al. (1994) Assessment of lesions detected at mammographic screening: performance at first or repeat screening in the Florence programme. J Med Screening 1:188–192

De Koning HJ, van Ineveld BM, Van Oortmarssen GJ et al. (1991) Breast cancer screening and cost-effectiveness; policy alternative, quality of life considerations and the possible impact of uncertain factors. Int J Cancer 48:531–537

Eckhardt S, Badellino F, Murphy GP (1994) UICC meeting on breast-cancer screening in pre-menopausl women in developed countries. Int J Cancer 56:1–5

Ikeda DM, Sickles EA (1988) Second-screening with mammography. One versus two views per breast. Radiology 168:651–656

Kirkpatrick A, Törnberg S, Thijssen M (1992) The European guidelines for quality assurance in mammography screening. Commission of the European Communities, Luxembourg, pp 1–54

Kopans DB (1990) The Canadian screening program: a different perspective. AJR Am J Roeutgenol 155:748–749

Miller AB, Baines CJ, To T et al. (1992) Breast cancer detection and death rates among women aged 40 to 49 years. Can Med Assoc J 147:1459–1476

Moskowitz M (1992) Response. J Natl Cancer Inst 84:1368–1371

Palli D, Rosselli Del Turco M, Buiatti E et al. (1986) A case-control study of the efficacy of a non-randomized breast cancer screening program in Florence (Italy). Int J Cancer 38:501–504

Rosselli Del Turco M (1994) Breast cancer screening. News in brief. Lancet 344:813

Tabar L, Fagerberg CJG, Gad A et al. (1985) Reduction in mortality from breast cancer after mass screening with mammography. Lancet 1:829–832

Tabar L, Fagerberg G, Duffy SW et al. (1992) Update of the Swedish two-county program of mammographic screening for breast cancer. Radiol Clin North Am 30(1):187–210

Tabar L, Duffy SW, Burhenne LW (1993) New Swedish breast cancer detection results for women aged 40–49. Cancer Suppl 72(4):1437–1448

Thurfjell EL, Lernevall KA, Taube AA (1994) Benefit of independent double reading in a population-based mammography screening program. Radiology 191(1):241–244

Vanara F (1992) La valutazione economica della diagnosi precoce del carcinoma mammario: una rassegna della letteratura. Epidemiologia a Prevenzione 53:22–31

van der Mass PJ, de Koning HJ, van Ineveld BM et al. (1989) The cost-effectiveness of breast cancer screening. Int J Cancer 43:1055–1060
Van Dijck JAAM, Verbeek ALM, Hendriks, JHCL et al. (1992) One-view versus two-view mammography in baseline screening for breast cancer: a review. Br J Radio 65:971–976
Verbeek ALM, Hendriks JHCL, Holland R et al. (1984) Reduction of breast cancer mortality through mass screening with modern mammography: first results of the Nijmegen project. 1975–1981. Lancet 1:1222–1224
Wald NJ, Chamberlain J, Hackshaw A (1993) Report of the European Society for Mastology Breast Cancer Screening Evaluation Committee. Breast 2:209–216

Ductal Carcinoma In Situ

R.G. Margolese

Introduction

In the past 10 years clinical trials have confirmed the value of breast conserving therapy for invasive breast cancer and many authorities now consider this the preferred and appropriate treatment (NCI Consensus Statement 1992a). However controversy still surrounds the treatment of ductal carcinoma in situ (DCIS). This uncertainty relates in part to a deficiency of prospective studies; such studies have begun only recently and many are not yet concluded. The controversy also relates to the perception that DCIS may be cured by breast ablation, whereas recurrence in a more aggressive form may occur with breast conserving surgery and the patient may develop invasive breast cancer with metastases resulting in death. The real question therefore is how often does this last scenario take place? Also, the expectation mastectomy is 100% curative needs to be critically examined.

With all the publications and discussion of the past decade, the only thing we can say with certainty is that we do not yet have all the answers and we do not yet sufficiently understand the biology of DCIS. There is good evidence that most if not all breast cancers are preceded by a phase of carcinoma in situ but the reverse is not necessarily true. All DCIS may not proceed to invasive breast cancer at least not in a time frame which is realistic in terms of human life span.

We do not yet know whether all of the disease identified histologically is biologically important. With the widespread use of mammography we may in fact be discovering histological abnormalities which would not otherwise have been discovered or become meaningful in that woman's lifetime. A careful rational approach to therapy is necessary to avoid compounding the problem with potential overtreatment.

Historical Review

In the many retrospective series documenting different approaches to treatment, it is clear that total mastectomy does not achieve 100% cure (Lagios et al. 1982; Fisher et al. 1988). There is sufficient evidence of death from breast cancer in such patients that we should acknowledge a 1%–2% likelihood of treatment failure even with modified radical mastectomy.

At the other end of the scale Lagios (1990) has followed a carefully selected cohort of patients who were treated by wide local excision only. These patients were chosen based on small tumor size and favorable histopathology. Considering that they were selected for the most favorable prognostic characteristics, it is somewhat disappointing that there is an 12% local recurrence rate within 10 years, indicating that surgery alone is not likely to be a successful treatment in general and may be inadequate even for a selected group of patients. A similar study by Schwartz et al. (1992) found a 15.3% locla recurrence rate at 4 years follow-up. Subsequently, the National Surgical Adjuvant Breast Project (NSABP) protocol B-17 confirmed this range in a randomized prospective study (Table 1).

Many reports of treating DCIS with lumpectomy and radiation therapy document a relatively low rate of local recurrence (Kuske et al. 1993; Bornstein et al. 1991; Solin et al. 1993; Stotter et al. 1989; Temple et al. 1989; Carpenter et al. 1989; Fisher et al. 1992; Silverstein et al. 1992), but the value of those retrospective series has been clouded by small numbers, uncertain selection features, and difficulty in transferring these outcomes to general recommendations (Table 2).

Prospective Studies

The NSABP has reported on two prospective studies. Protocol B-06 was a trial of breast conserving therapy for invasive breast cancer. Patients were randomized to receive total mastectomy, lumpectomy alone, or lumpectomy and radiation to the breast. Among the 1835 patients randomized in this

Table 1. Ductal carcinoma in situ treated by lumpectomy alone: rates of local recurrence

Author	Year	*N*	Ipsilateral tumor recurrence
Schwartz et al.	1992	70	15
Lagios	1990	79	13
Arnesson	1989	38	13
Fisher[a]	1992	391	16

[a] Randomized trial: B-17.

Table 2. Cases of ductal carcinoma in situ treated by lumpectomy and radiotherapy: rates of local recurrence

Author	Year	*N*	Ipsilateral tumor recurrence	(%)	Invasion
Bornstein et al.	1991	38	8	(21)	1
Kurtz	1989	43	3	(7)	3
Fisher et al.[a]	1986	29	2	(7)	1
Stotter	1990	42	4	(9)	9
Recht	1985	40	4	(10)	2
Zafrani	1986	54	3	(5)	1
Silverstein	1987	51	1	(1)	2
Solin et al.	1993	172	16	(9)	7
Fisher et al.[a]	1993	399	28	(7)	8
Silverstein et al.	1992	103	10	(10)	5
Kuske et al.	1993	70	6	(7)	4

[a] Randomized trial.

protocol, there were 72 patients who on review were determined to have DCIS and not invasive breast cancer. They had been randomized equally into the three treatment groups and analysis indicates that radiation therapy produced a significant improvement in local control (Posner and Wolmark 1992). However, one of the patients treated by total mastectomy eventually succumbed to breast cancer. Although this study is limited by its small size and is essentially an unintentional by-product of B-06, this is the first prospective randomized evaluation of these treatments of DCIS, and the number of patients exceeds that of most of the historical retrospective reports in the literature.

Subsequently, a prospective NSABP trial of lumpectomy or lumpectomy and radiation for DCIS was carried out in protocol B-17. All women had pathologically free margins, defined as no tumor present at the physical margin. The 5 year event-free survival was better in women who received breast irradiation (84%) than in women treated by lumpectomy alone (73%). Overall, the incidence of recurrence or second cancers was reduced by the addition of radiation therpay from 10.4% to 7.5% for noninvasive cancers and from 10.5% to 2.9% for invasive cancers (Bornstein et al. 1991).

These results are early and many authors have cautioned that recurrences may continue over many years. Nevertheless, this is the first large scale prospective study of DCIS and it confirms the preliminary results from the B-06 DCIS patients, that is, that radiation therapy can significantly decrease the likelihood of local recurrence in the breast and especially diminish the prospect of invasive breast cancer developing. It will be important in the future to document the incidence of long-term survival, especially in the small subset of patients who develop invasive breast cancer;

however, the evidence from this protocol suggests that the vast majority of patients with DCIS can be successfully managed with breat conserving surgery.

As to the issues of margins and multicentricity, much confusion remains. Holland et al. (1985) have shown that breast cancers are frequently surrounded by zones of DCIS which may extend for several centimeters away from the primary tumor. Focusing on this tends to bias our thinking anatomically, when consideration of biologic processes might be more important. If 50% of patients have multifocal disease and only 12% develop recurrence, how do we interpret the importance of multifocalness?

Even in protocol B-06 we can see that the issue of margins is confusing. All of the patients in the lumpectomy only group had pathologically clear margins, according to the techniques prescribed in the protocol, yet the 45% local recurrence rate indicates that a substantial number of these patients in fact must have had tumors with positive but undiscovered margins. The majority of these tumors were controlled by postoperative radiation, which explains the difference in local recurrence between the two lumpectomy groups.

The incidence of local recurrence was suppressed even more when adjuvant therapies were used in subsequent protocols (Margolese 1992b). This is an indication that the regrowth of residual cancer at the site of surgery can be influenced by chemotherapy or tamoxifen treatment.

The increased use of adjuvant systemic therapies has changed the expectation for distant disease-free survival and the incidence of local recurrence in invasive breast cancer (Margolese 1992b), and it will be important to consider the effect of adjuvant therapies in DCIS. In this case, the concept of adjuvant therapy is approached from a slightly different direction. Cancer is a problem in control of growth and therefore agents such as tamoxifen, which affect the control of growth, should be useful. Most clinical research on DCIS in the past decade or so has focused on anatomic features: the issue of margins, residual disease, multicentric disease, and concepts for control by surgery and/or radiation therapy. It is now time to combine the understanding we have developed from these studies with knowledge derived from adjuvant therapy protocols designed to suppress the growth of cancer cells or their precursors in order to seek more effective control of this process. The studies of Holland and the Harvard group on extensive intraductal component (EIC) (Holland et al. 1990) of invasive breast cancer suggest that it is often the surrounding DCIS component that contributes to the local recurrence when it does happen. In a situation in which the entire tumor is DCIS, the next logical step in clinical research is to attempt to inhibit the local recurrence with interventions similar to those which were successful in treating invasive breast cancer. Combining tamoxifen with breast conserving surgery and lumpectomy is the object of NSABP protocol B-24 which has reached its target number of patients and closed study entry. This protocol, which awaits analysis, will provide important additional insight because pathologically clear margins were not required.

All patients, based on the findings from B-17, receive radiotherapy and are then randomized to tamoxifen or placebo (Fig. 1). We should therefore have information on the additional effect of tamoxifen on recurrence, especially when pathologically clear margins are not pursued clinically.

Based on the B-06 experience we do not know whether there is a difference between a margin which is apparently negative but undiscovered to be positive and those which are known to be microscopically positive. A recent report from Schniff suggests that there is little or no difference since patients with focally positive margins and extensive intraductal component (EIC) negative tumors have very good rates of local control (Recht 1990).

Some of the above biologic considerations make theoretical sense but remain to be confirmed by clinical trial evidence. The ever increasing use of mammography provides a steadily increasing number of women with DCIS who need management today. Common sense says control of the initial process should be our goal. We have enough evidence to say that breast conserving surgery with appropriate radiation provides effective control. Given the low rate of invasive breast cancer and the probability that most of these will be successfully controlled by a second intervention, it seems that breast conserving therapy will provide the same long-term cure rates that could be obtained with total mastectomy. While results of ongoing clinical trials such as B-24 and the EORTC trial will be awaited with great interest, it is clear that we can reasonably treat the cribriform and micropapillary variants with breast conserving surgery and routine radiation therapy.

The incidence of invasive cancer following lumpectomy and radiotherapy in B-17 is 2.9%. The majority of these tumors are mammorgraphically detected and of small size, suggesting a high cure rate. If we assume a conservative cure rate of 80%, then $0.20 \times 0.029 = 0.58\%$ of lumpectomy/radiotherapy patients will ultimately develop metastatic carcinoma, similar to the outcomes of patients treated by total mastectomy.

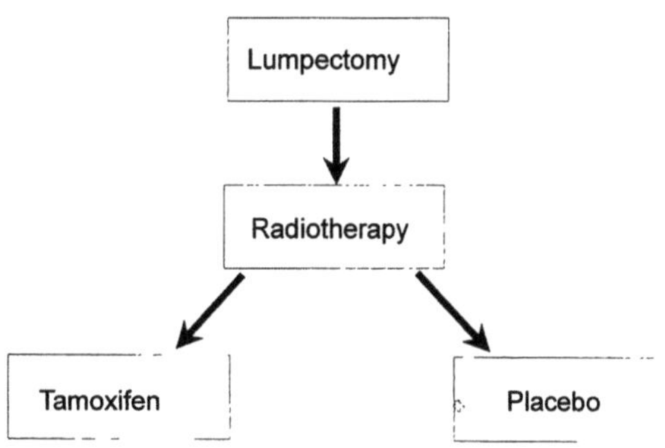

Fig. 1. National Surgical Adjuvant Breast Project (NSABP) protocol B-24: patients with ductal carcinoma in situ treated by lumpectomy and radiotherapy

The comedo variant is reported to have a higher incidence of local recurrence (Schwart et al. 1992; Kuske et al. 1993). There is considerable dispute about this entity, and the experience in the literature does not confirm the relationship (Table 3). It is likely that a small focus of necrosis, not sufficient to quality for the comedo designation, is frequently seen. It can affect statistics depending on how it is interpreted. Perhaps the real problem is poor nuclear grade, since they often coexist. A full-blown comedo subtype with poor nuclear grade may require careful, wide local excision with clear cut negative margins if a reasonably low rate of local recurrence is to be obtained, but this is still uncertain. Even in the presence of these worrisome features it should be remembered that the majority of cases will still be well controlled with conservative surgery.

Conclusions

Prospective trial evidence has confirmed the merit of breast conserving surgery and radiotherapy for DCIS. The long-term survival outcomes reported for treatment with total mastectomy do not provide 100% cure rates. The evidence suggests that similar long-term control can be obtained with either treatment. Extrapolating on the NIH consensus conference for invasive breast cancer, it seems logical to accept breast conserving surgery as standard therapy for DCIS. Whether certain predictive features should indicate need for mastectomy is as yet unproyen; careful follow-up of reported and continuing trials is required for this information.

Table 3. Relationship of comedo histology to local recurrence of ductal carcinoma in situ following lumpectomy and radiotherapy

	Year	Number of cases	Relation to local recurrence	*P* value
Kuske et al.	1993	77	98% vs 88%	0.3
Silverstein et al.	1992	52	13% vs 6%	ns
Solin et al.				
Comedo	1993	172	8% vs 2%	ns
Comedo + N.G.			20% vs 5%	0.005
Bornstein et al.	1991	38	24% vs 22%	ns
Schwartz et al.	1992	70	10 out of 11 comedo	
Lagios[a]	1990	79	19% vs 10%	0.02
Stotter et al.	1990	44	No difference	ns
Temple et al.	1989	109	No identifying features, 12% LR	
Carpenter et al.	1989	38	No features	ns
Fisher et al.	1993	379	No features	ns

NG, nuclear grade; LR, local recurrence.
[a] Lumpectomy only.

The possible beneficial effect of tamoxifen is a future consideration awaiting completion of a clinical trial.

From invasive breast cancer studies it is clear that further research is still necessary to identify predictive factors for local treatments. The same can be said for DCIS but we must keep in mind the difference between clinical and pathologic factors on the one hand and biologic outcomes on the other. Even predictive factors for local recurrence may not be important when the ultimate goal of treatment is considered: long-term survival.

References

Arnesson LG, Smeds S, Fagerberg G, Grontoft O (1989) Follow-up of two treatment modlalities for ductal cancer in situ of the breast. Br J Surg 76:672–675

Bornstein BA, Recht A, Connolly JL, Schnitt SJ et al. (1991) Results of treating ductal carcinoma in situ of the breast with conservative surgery and radiation therapy. Cancer 67:7–13

Bradley SJ, Weaver DW, Bouwman DL (1990) Alternatives in the surgical management of in situ breast cancer. A meta-analysis of outcome. Am Surg 56:428–432

Carpenter R, Boulter PS, Cooke T, Gibbs NM (1989) Management of screen detected ductal carcinoma in situ of the female breast. Br J Surg 76:564–567

Fisher ER, Sass R, Fisher B et al. (1986) Pathologic findings from the National Surgical Adjuvant Breast Project (Protocol B-06). I. Intraductal carcinoma (DCIS). Cancer 57:197–208

Fisher B, Constantino J, Redmond C, Fisher E, Margolese R, Dimitrov Nikolay et al. (1993) Lumpectomy compared with lumpectomy and radiationn therapy for the treatment of intraductal breast cancer. N Engl J Med 328:22:1581–1586

Holland R, Veling SH, Mravunac M et al. (1985) Histologic multifocality of Tis, TI-2 breast carcinomas: implications for clinical trials of breast-conserving surgery. Cancer 56:979–990

Holland R, Connolly JL, Gelman R, Mravunac M, Hendriks JHCI et al. (1990) The presence of an extensive intraductal component following a limited excision correlates with prominent residual disease in the remainder of the breast. J Clin Oncol 8(1):113–118

Kurtz JM, Jacquemier J, Torhorst J, Spitalier J-M, Amalric R, Hunig R et al. (1989) Conservation therapy for breast cancers other than infiltrating ductal carcinoma. Cancer 63:1630–1635

Kuske RR, Bean JM, Garcia DM, Perez CA, Andriole D, Philpott G, Fineberg B (1993) Breast conservation therapy for intraductal carcinoma of the breast.
J Radiat Oncol Biol Phys 26:391–396

Lagios MD (1990) Duct carcinoma in situ. Pathology and treatment. Surg Clin North Am 70(4):853–871

Lagios MD, Westdahl PH, Margolin FR, Rose MR (1982) Duct carcinoma in situ: relationship of non invasive disease to the frequency of occult invasion, multicentricity, lymph node metastases and short-term treatment failures. Cancer 50: 1309–1314

NCI Consensus Statement (1992a) Treatment of early-stage breast cancer. Monogr Natl Cancer Inst 11:1–5

Margolese R (1992b) Surgical considerations in selecting local therapy. Monogr Natl Cancer Inst 11:41–48

Posner MC, Wolmark N (1992) Non-invasive breast carcinoma. Breast Cancer Res Treat 21:155–164

Recht A, Schnitt SJ, Connolly JL, Rose MA, Silver B, Come S, Henderson IC, Slavin S, Harris JR (1985) Prognosis following local or regional recurrence after conservative surgery and radiotherapy for early stage breast carcinoma. Int J Radiat Oncol Biol Phys 16:3–9

Schnitt SJ, Abner A, Gelman R, Connolly JL, Recht A, Duda RB, Eberlein TJ, Mayzel K, Silver B, Harris J (1994) The relationship between microscopic margins of resection and the risk of local recurrence in patients with breast cancer treated with breast-conserving surgery and radiation therapy. Cancer 72:1746–1751

Schwartz GF, Finkel GC, Garcia JC, Patchefsky AS (1992) Subclinical ductal carcinoma in situ of the breast. Cancer 70:2468–2474

Silverstein MJ, Rosser RJ, Gierson ED, Waisman JR, Garnagami P, Hoffman RS, Fingerhut AG, Lewinsky BS, Colburn W, Handel N (1987) Axillary lymph node dissection for intraductal breast cancer–Is it indicated? Cancer 59:1819–1824

Silverstein MJ, Cohlan BF, Gierson ED, Furmanski M, Gamagami P et al. (1992) Duct carcinoma in situ: 227 cases without microinvasion. Eur J Cancer 28(2/3): 630–634

Solin LJ, Yeh IT, Kurtz J, Fourguet A, Recht A, Kuske R et al. (1993) Ductal carcinoma in situ (intraductal carcinoma) of the breast treated with breast-conserving surgery and definitive irradiation. Cancer 71:8:2532–2542

Stotter AT, McNeese M, Oswald MJ, Ames FC, Romsdahl MM (1990) The role of limited surgery with irradiationn in primary treatment of ductal in situ breast cancer. Radiat Oncal Biol Phys 18:283–287

Temple WJ, Jenkins M, Alexander F, Hwang WS, Marx LH et al. (1989) Natural history of in situ breast cancer in a defined population. Ann Surg 210(5):653–657

Zafrani B, Fourquet A, Vilco JR, Legal M, Calle R (1986) Conservative management of intraducatal breast carcinoma with tumorectomy and radiation therapy. Cancer 57:1299–1301

Intraductal Breast Carcinoma: Experiences from The Breast Center in Van Nuys, California

M.J. Silverstein,[1] D.N. Poller,[2*] A. Barth,[3**] J.R. Waisman,[3]
J.A. Jensen,[4] R. Masetti,[1***] E.D. Gierson,[1] W.J. Colburn,[2]
B.S. Lewinsky,[5] S.L. Auerbach,[6] and P. Gamagami[7]

[1]The Breast Center, Division of Surgical Oncology, 14624 Sherman Way, Van Nuys, CA 91405, USA
[2]The Breast Center, Division of Pathology, 14624 Sherman Way, Van Nuys, CA 91405, USA
[3]The Breast Center, Division of Medical Oncology, 14624 Sherman Way, Van Nuys, CA 91405, USA
[4]The Breast Center, Division of Plastic Surgery, 14624 Sherman Way, Van Nuys, CA 91405, USA
[5]The Breast Center, Division of Radiation Oncology, 14624 Sherman Way, Van Nuys, CA 91405, USA
[6]The Breast Center, Fellow in Breast Disease, 14624 Sherman way, Van Nuys, CA 91405, USA
[7]The Breast Center, Division of Breast Imaging (Radiology), 14624 Sherman Way, Van Nuys, CA 91405, USA

Introduction

The Breast Center in Van Nuys, California was the first free-standing, truly multidisciplinary, breast treatment center in the United States (Silverstein et al. 1986a). Housed in a 14000 square foot area, it includes a home-like waiting room, an operating and recovery suite, diagnostic and education center, clinical center, physicians offices, and psychiatric center. Members of the diagnostic and therapeutic teams are always available. The diagnostic center contains four modern mammography machines, a digital stereotactic unit, and ultrasonography equipment. The staff includes surgical, medical and radiation oncologists, a diagnostic radiologist, two reconstructive surgeons, two pathologists and a psychologic team headed by a psychiatrist (psycho-oncologist).

From its inception in 1979, our group has believed that early breast cancer detection by mammography was critical, if mortality was to be reduced. This belief, combined with large numbers of breast screening patients and an aggressive biposy policy, has allowed us to accrue one of the largest

* A visiting fellow from City Hospital, Nottingham, England. Current address: Department of Pathology, United Bristol Healthcare Trust, Bristol, England
** Current address: Department of Medicine (Oncology), The University of Bern, Bern, Switzerland
*** Current address: Department of Surgery, Catholic University, Rome, Italy

single facility series of patients with intraductal breast carcinoma (DCIS) in the world. As of December 31, 1994, the series included 425 patients.

In this article, we would like to update earlier published reports (Silverstein et al. 1990, 1992) of our series and then, discuss four distinct issues. (1) prognostic factors that predict local recurrence, (2) a new prognostic classification for DCIS, (3) the question of whether total excision of DCIS with a procedure less than a true quadrantectomy is possible, and (4) our mastectomy technique when mastectomy is indicated. We have chosen to discuss the various issues raised by each of these questions in relation to our own data.

Patients and Methods

A total of 425 consecutive patients were prospectively entered into the center's computerized database between 1979 and 1994; all were patients with histologically confirmed pure DCIS, without evidence of microinvasion or invasive carcinoma. All pathological specimens were serially sectioned at 2–3 mm intervals at the time of specimen processing. Margins were considered clear if no intraductal carcinoma was found within 1 mm of the margin.

Patients were treated with mastectomy, excision and radiation therapy, or excision only; treatment was not randomized. Patients with large lesions (4 cm and more), multicentricity, or involved margins not amenable to re-excision were treated with mastectomy (usually with immediate breast reconstruction). Patients with smaller lesions (4 cm or less) and microscopically clear surgical margins were treated with breast conservation.

Until 1988, all patients with DCIS who elected breast conservation were advised to add breast irradiation to their treatment. Most patients accepted this recommendation; occasional patients refused and were treated with careful clinical follow-up without irradiation. Beginning in 1989, the physicians within the group were no longer convinced of the value of radiation therapy for DCIS and all breast conservation patients with clear biopsy margins were offered the option of careful clinical follow-up without radiation therapy. Most patients accepted this option; those who refused were treated with breast irradiation.

Because the decision to abandon radiation therapy was arbitrary and not based on clinical factors, such as tumor size, method of detection, etc., patients treated by excision plus radiation therapy or excision alone were clinically similar, other than in median follow-up time. The mean tumor diameter was 18 mm for patients treated with excision and radiation therapy vs 17 mm for patients treated with excision only (p = NS). Some 88% of excision plus radiation therapy patients had nonpalpable lesions compared with 87% of excision only patients (p = NS). The median follow-up time for patients treated with excision and radiation therapy was 97 months; the median follow-up time for patients treated with excision only was 33 months.

Level 1 and 2 axillary dissections were done routinely until 1988 (Silverstein et al. 1986b, 1991), thereafter, a lower axillary sampling was performed only in selected patents treated with mastectomy. Whole breast external beam irradiation (40–50 Gy) was performed on a 4 or 6 MeV linear accelerator with a boost of 16–20 Gy to the tumor bed by iridium-192 implant or external photons or electrons (Silverstein et al. 1990).

The patients were analyzed by 16 prognostic factors, inlcuding treatment, nuclear grade, tumor size, final margins, the presence of necrosis, histologic architecture, estrogen receptor (ER), progresterone receptor (PR), S-phase, ploidy, p53, HER2-neu, microcalcifications, palpability, year of diagnosis, and age. In addition, they were analyzed using the recently developed Van Nuys Classification (Silverstein et al. 1995).

The Van Nuys Classification, a development of an earlier joint Nottingham (England) proposal (Poller et al. 1994), defines three groups of DCIS patients by using the presence of high nuclear grade (nuclear grade 3) to select the most aggressive group (group 3). The remaining non-high nuclear grade lesions (nuclear grades 1 or 2) are then sorted by the presence (group 2) or absence (group 1) of comedo-type necrosis (Fig. 1). Nuclear grade was scored by previously described methods (Lagios 1990; Poller et al. 1994; Poller and Ellis 1995). Essentially, high grade (grade 3) nuclei were defined as nuclei with a diameter greater than two red blood cells, with vesicular chromatin, and one or more nucleoli. Intermediate (grade 2) nuclei were defined as nuclei 1–2 red blood cells in diameter with coarse chromatin and infrequent nucleoli. Low grade (grade 1) nuclei were defined as nuclei 1–1.5 red blood cells in diameter with diffuse chromatin and inapparent nucleoli.

Cases of DCIS traditionally recognized by pathologists as comedo-type DCIS showing central lumina containing necrotic debris surrounded by large pleomorphic viable cells in solid masses were classified as comedo-type necrosis (Poller and Ellis 1995). Other cases of DCIS showing substantial amounts of neoplastic cells of duct origin within duct lumina were also

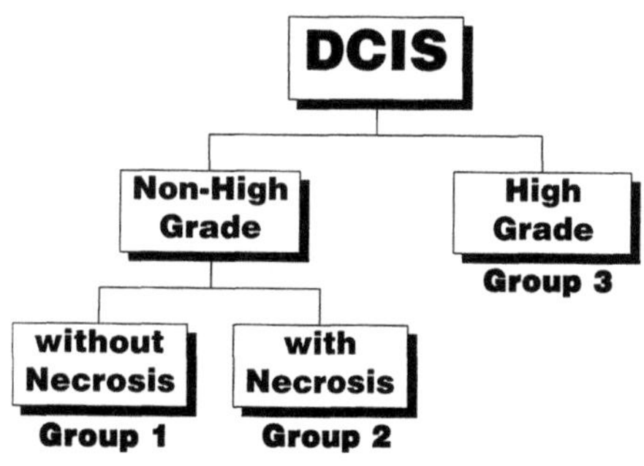

Fig. 1. Van Nuys ductal carcinoma in situ (DCIS) classification. DCIS patients are divided into high nuclear grade and non-high nuclear grade. Non-high nuclear grade cases are then divided by the presence or absence of necrosis

classified as showing comedo-type necrosis. These latter cases were required to contain significant amounts of necrotic cellular debris within duct lumina and included tumors showing cribriform of micropapillary architectural patterns. The assessment of comedo-type necrosis was made without specific reference to the nuclear grade or cell size.

No requirement was made for a specific amount of high grade DCIS, nor was there any requirement for a minimum amount of comedo-type necrosis. Occasional desquamated or individually necrotic cells were ignored and were not scored as comedo-type necrosis.

Group 1 DCIS consisted of non-high nuclear grade DCIS without comedo-type necrosis. Group 2 DCIS consisted of non-high grade DCIS lesions with comedo-type necrosis. Group 3 DCIS consisted of all high nuclear grade DCIS with or without comedo-type necrosis.

Disease-free survival was estimated by the Kaplan-Meier method. The statistical significance between disease-free survival curves was determined by the log-rank test for each variable studied (univariate analysis). Univariate analysis of continuous variables such as age and tumor size was performed with a Cox proportional hazard model using a single variable. Multivariate analysis was performed using the Cox proportional hazard model and included all significant variables found by univariate analysis.

Results and Discussion

The results presented in each section below are based on 425 consecutive patients with histologically confirmed DCIS treated at the Breast Center, Van Nuys, California, USA, between 1979 and 1994. Clinical details reagarding these patients are presented in Table 1. Most lesions were nonpalpable and discovered mammographically. The predominant histologic subtypes were as follows: comedo 45%, cribriform 24%, solid 14%, mic-

Table 1. Characteristics of ductal carcinoma in situ patients by treatment

	Mastectomy	Excision plus irradiation	Excision only
Number of patients (n = 425)	187	139	99
Median follow-up (months)	84	97	33
Size (mm) (n = 425)	39	18	17
Nuclear grade (n = 425)	2.16	2.10	2.15
Nonpalpable (%)	69	88	87
Recurrence (n = 33)	2/187 (1%)	18/139 (13%)	13/99 (13%)
Invasive recurrences (n = 16)	2/2 (100%)	9/18 (50%)	5/13 (38%)
4-year disease-free survival (%)	99	94	81
10-year disease-free survival (%)	98	81	Insufficient follow-up

ropapillary 12% papillary 4%, other 1%. There were 187 patients (44%) who were treated by mastectomy; 238 (56%) were treated with breast preservation, 139 by excison and radiation therapy and 99 by excision alone. Some 1% (2/187) of mastectomy patients had tumor recurrence; 13% (31/238) of breast conservation patients had tumor recurrence. Of these recurrences, 48% (16/33) were invasive. Two patients have died. Both were treated with excision and radiation and both developed inflammatory-type recurrences. A third radiation therapy patient is alive with stable metastatic disease. There have been no deaths in the mastectomy or excision only groups. However, both mastectomy recurrences were invasive and both patients are alive with stable metastatic disease to bone. To date, no patients treated with excision only have developed metastatic breast cancer.

Predicting Recurrence

Of the 425 patients with DCIS, 33 (8%) have had recurrence; 28 with local disease only, four with local plus metastatic disease, and one with metastatic carcinoma only (Table 2). The recurrences have occurred in two of 187 (1%) patients treated with mastectomy, 18 of 139 (!3%) patients treated with excision and radiation, and 13 of 99 (13%) patients treated with excision only. The median follow-up was significantly longer for mastectomy (84 months) and excision plus radiation therapy patients (97 months) than for excision only patients (33 months). Although the absolute recurrence rates are the same for both groups (13%), patients treated by excision only recur more quickly and this is reflected in the actuarial disease-free survival curves (Fig. 2) by a superior disease-free survival for patients treated with excision and radiation than with excision only. The median time to recurrence for excision plus radiation therapy was 4.7 years compared with 1.8 years for excision only patients. The difference between all disease-free survival curves shown in Figure 2 is significant (both p values $\leqslant 0.003$).

Of the 31 (97%) recurrences in breast conservation patients, 30 were at or near (same quardrant) the original DCIS lesion. This suggests that inadequate surgical removal of the primary lesion was the most likely cause

Table 2. Types of recurrences by treatment

	Mastectomy	Excision plus irradiation	Excision only	Total
Local only	0	15	13	28
Local plus distant	1	3	0	4
Distant only	1	0	0	1
Total	2	18	13	33

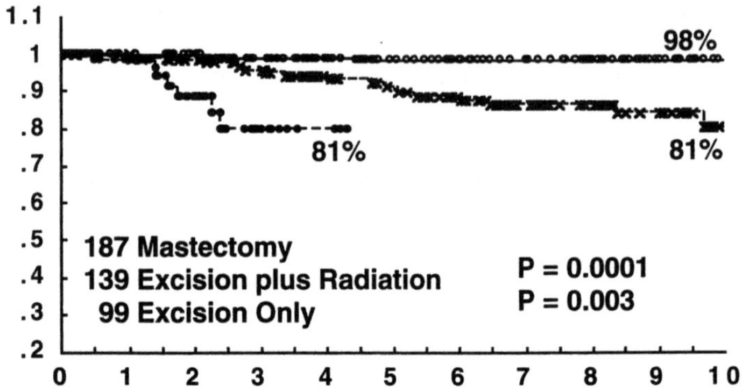

Fig. 2. Eight-year disease-free survival by group: 231 breast conservation patients. The differences between all curves are statistically significant

Table 3. Important predictors of local recurrence: univariate and multivariate analysis

	All 425 patients		238 Breast conservation	
	Univariate	Multivatiate	Univariate	Multivatiate
Treatment	0.00000	0.00001	0.003	0.01
Nuclear grade	0.00003	0.0001	0.00002	0.001
Tumor size	NS	–	0.004	0.03
Final biopsy margins	NS	–	0.0002	0.03
Comedo histology	0.001	NS	0.002	NS
Necrosis	0.001	NS	0.002	NS

of local recurrence. Local recurrence might better be termed "local persistence." We will say more about this below.

In an attempt to predict the likelihood of local recurrence, we evaluated 16 factors by univariate analysis for all 425 patients and for the subgroup of 238 patients who underwent breast conservation. All statistically significant predictors of local recurrence by univariate analysis were then evaluated using a Cox multivariate regression analysis. Ten factors were not significant predictors of local recurrence by univariate analysis for either group: ER, PR, S-phase, ploidy, p53, HER2-neu, microcalcifications, palpability, years of diagnosis, and age. However, ER, PR, phase S-phase, and ploidy were available in less than 80 patients. HER2-neu and p53 were available in only 150 patients.

For the entire group of 425 patients, four factors, including treatment, nuclear grade, the presence or absence of comedo-type necrosis, and histologic architecture (comedo or noncomedo) were significant predictors of local recurrence by univariate analysis (Table 3). However, only the type of treatment and nuclear grade were significant predictors of local recurrence in subsequent multivariate analysis.

Tumor size and final biopsy margin status did not achieve statistical significance as univariate predictors of recurrence for the entire group, because, in general, as tumors became larger or if margins were involved, our patients were more likely to be treated by mastectomy. Since mastectomy patients were unlikely to develop tumor recurrences (only two of 187 patients recurred), tumor size and margin status did not appear as statistically important factors by univariate analysis of tumor recurrence when all patients were considered.

It is instructive to limit the analysis to breast conservation patients who continue to have their treated breast at risk for the development of local recurrence. Here, six factors were important by univariate analysis. They included primary treatment (excision plus irradiation vs excision only), nuclear grade, tumor size, final biopsy margins, comedo-type necrosis, and histologic architecture. Multivariate analysis confirmed that nuclear grade was the most important predictor of local recurrence. However, breast irradiation, tumor size, and final margins also emerged as statistically significant multivariate predictors of recurrence (Table 3). These findings are in accord with the published findings of The National Surgical Adjuvant Breast Project (NSABP) (Fisher et al. 1993), which in a prospective randomized trial showed that breast irradiation decreases the rate of local recurrence in patients with DCIS. In addition, patients with larger tumors or those with margins less than 1 mm are more likely to have residual DCIS and therefore are probably more likely to recur locally than are patients with smaller lesions and widely clear margins.

A New Prognostic Patient Classification

The data generated above can be used to better understand which patients with DCIS are likely to develop recurrence. For more than a year, our group has been looking at combinations of risk factors in an attempt to develop an easy to use, reliable, and reproducible classification for DCIS, a classification that could be used by pathologists and clinicians with little additional training.

If tumor size is an important predictor of recurrence, it should be included in any new classification scheme. Currently, no DCIS classification uses it. Tumor size is difficult to assess and some pathologists do not even attempt to measure it for nonpalpable DCIS. If tumor size is measured, there may be significant interobserver discordance. Conventional blocking techniques, which may cause a large DCIS to be cut into smaller pieces, also make size estimation more difficult. If large block techniques (Jackson et al. 1994) were used, which is admittedly more time-consuming, much more difficult, more costly, and slower than conventional blocking techniques, pathologists would likely be able to make a more accurate size assessment.

We measure DCIS tumor size by direct measurement on the microscopic slide. Since there is approximately 10% shrinkage with formalin fixation and subsequent tissue processing, irrespective of the method of tissue blocking employed, the microscopic tumor size as measured by a pathologist will inevitably be slightly less than that of the size of the fresh tumor prior to tissue fixation. Since every lesion is serially sectioned, we also estimate the maximum diameter by counting the number of sections in which the tumor appears and multiply that number by the thickness of the sections. In addition, the mammographic size is noted. We then record the largest diameter as the size of the DCIS. In spite of this meticulous method, DCIS is extremely difficult to measure and, although our pathologists would like to deny it, there is an obvious rounding bias. Figure 3 shows the histogram for DCIS patients by size. Only patients with tumors 100 mm or smaller are shown. The expected normal distribution is not seen. Rather, there are peaks at 10 mm, 15 mm, 20 mm, etc. Although this is not likely to be too important with lesions such as DCIS, we should not deny its existence.

Margin status should be included in any new classification scheme, but that also is difficult to define. When we excise a DCIS lesion and the pathologist tells us that the margins are free of disease, what does that mean? Does it mean that we have excised the entire lesion? How do we define a clear margin? Is it one fat cell (Fisher et al. 1993), 1 mm (Silverstein et al. 1994), 2 mm (Solin et al. 1991), 1 cm (Van Dongen et al. 1992), the presence of normal breast tissue between the DCIS and the inked margin (Holland et al.

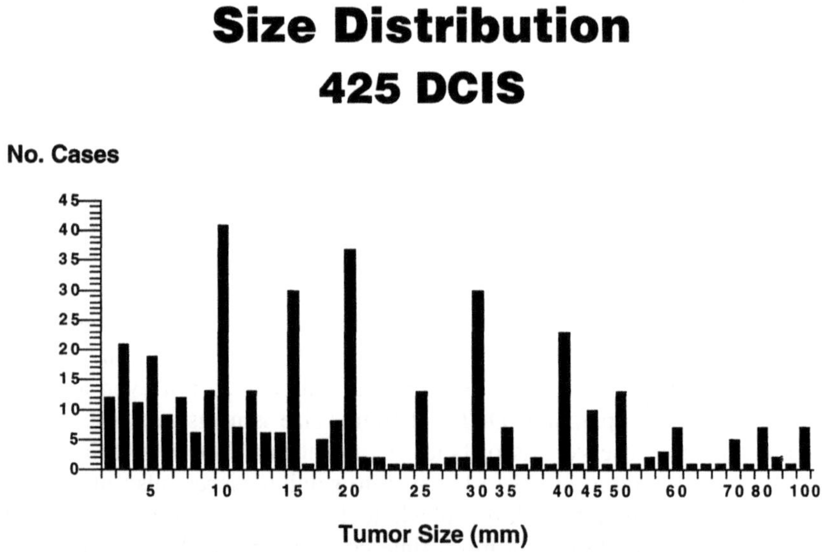

Fig. 3. Size distribution of ductal carcinoma in situ (DCIS) patients showing a rounding bias at 10 mm, 15 mm, 20 mm, etc.

1990)? Or is it something else? Once we define clear margins, will the definition be different for different types of DCIS? For example, will high grade, aggressive looking lesions require a wider margin than low grade lesions?

Nuclear grade appears to be the single most important predictor of local recurrence in patients with DCIS treated with breast conservation. This was agreed upon by participants at a recent EEC funded DCIS classification consensus meeting (Poller and Ellis 1995). There is, however, no international uniform consensus on the definition of nuclear grade. Most pathologists can identify the high and low grade lesions but there is some discordance in recognition of intermediate grade lesions. There is little discordance in the pathologist's ability to recognized comedo-type necrosis; most pathologists are able to recognize it easily (Sloane et al. 1994).

With this in mind, the combination of high nuclear grade and the presence or absence of comedo-type necrosis became the basis of our new DCIS classification system, as described in the Methods section. The Lancet (Silverstein et al. 1995). The Kaplan-Meier calculations of disease-free survival at 8 years revealed 93% for group 1, 84% for group 2, and 61% for group 3 (all $p \leqslant 0.05$).

The Van Nuys Classification defines three distinctly different DCTS groups, each one of which has a different likelihood of local recurrence if treated with breast conservation. We hope that other centers will reclassify cases of DCIS using our classification to test its reproducibility among pathologists and also to confirm whether or not this model can be applied in predicting outcomes in other series of patients with DCIS.

It is important to note that group 1 DCIS patients must be followed-up for prolonged periods of time, since the absence of comedo-type necrosis and, therefore, microcalcifications, will make the diagnosis of recurrent DCIS more difficult for this subgroup. When group 1 patients have recurrence, if they do not develop microcalcifications, their mammographic detection is likely to be delayed.

Can Ductal Carcinoma In Situ Be Completely Excised

The most important cause of local recurrence for conservatively treated patients with DCIS is most likely inadequate resection of the primary lesion (Silverstein et al. 1994) We believe this because 97% (30/31) of local recurrences in breast preservation patients, in this series, recurred at or near the primary DCIS lesion (the same quadrant). In many cases, local recurrence is really probably local persistence. Our statistical analysis of margins does not tell us whether or not tumor was left behind. It merely tells us whether or not margin status, by whatever definintion we are using for clear margins (1 mm in this series), is a statistically important factor.

Serial subgross microscopic evaluation of excised DCIS specimens reveals that for many patients the lesion is larger than expected, often making complete excision impossible with less than a true quadrantectomy (Holland et al. 1990). We currently have data on 277 patients with DCIS in whom the initial excision was performed using a wide local excision, rather than a true quadrant resection, and who subsequently underwent mastectomy (174 patients) or re-excision (103 patients) of the initial biopsy site, allowing us to histologically evaluate whether or not DCIS was left behind (Table 4).

The average size of the initial wide excision was 6.0 cm. As mentioned above, all initial specimens were serially sectioned at 2–3 mm intervals and the margins evaluated microscopically. There were 188 patients with involved margins and 89 with microscopically clear margins after initial excision, all of whom underwent mastectomy or re-excision. At mastectomy or re-excision, the previous biopsy cavity was serially sectioned at 4–5 mm intervals. Overall, combining mastectomy and re-excision, 76% of patients (143/188) with involved margins had residual DCIS; 49% of patients (44/89) with clear margins had residual DCIS.

Today, most DCIS is neither visible nor palpable and is extremely difficult to excise completely using a local excision. When complete excision is desired, we would strongly recommend that multiple bracketing wires (Fig. 4a,b) be used by the radiologist to mark the entire lesion (Silverstein et al. 1989). Unfortunately, this technique, which has been routine at The Breast Center since 1983, does not insure that the entire lesion has been removed. DCIS often extends beyond the area marked by calcifiçations.

If the object is to completely excise DCIS while preserving the breast, a true quadrantectomy may be required in many cases. While we believe that nuclear grade, tumor size, comedo-type necrosis, etc., are important predictiors of local recurrence, inadequate excision of the primary lesion is probably the most important cause of local failure after conservative treatment for intraductal breast carcinoma.

When a multivariate analysis is performed, inadequate excision of the primary lesion cannot really be evaluated. The computer only knows whether or not the pathologist thought the margins were clear by whatever

Table 4. Initial biopsy margins as a predictor of residual ductal carcinoma in situ at mastectomy or re-excision

	Residual DCIS				
	At mastectomy		At re-excision		Total
Initial margins	Involved	Clear	Involved	Clear	
Number of patients	113	61	75	28	277
Residual ductal carcinoma in situ	95 (84%)	31 (51%)	48 (64%)	13 (46%)	187 (68%)

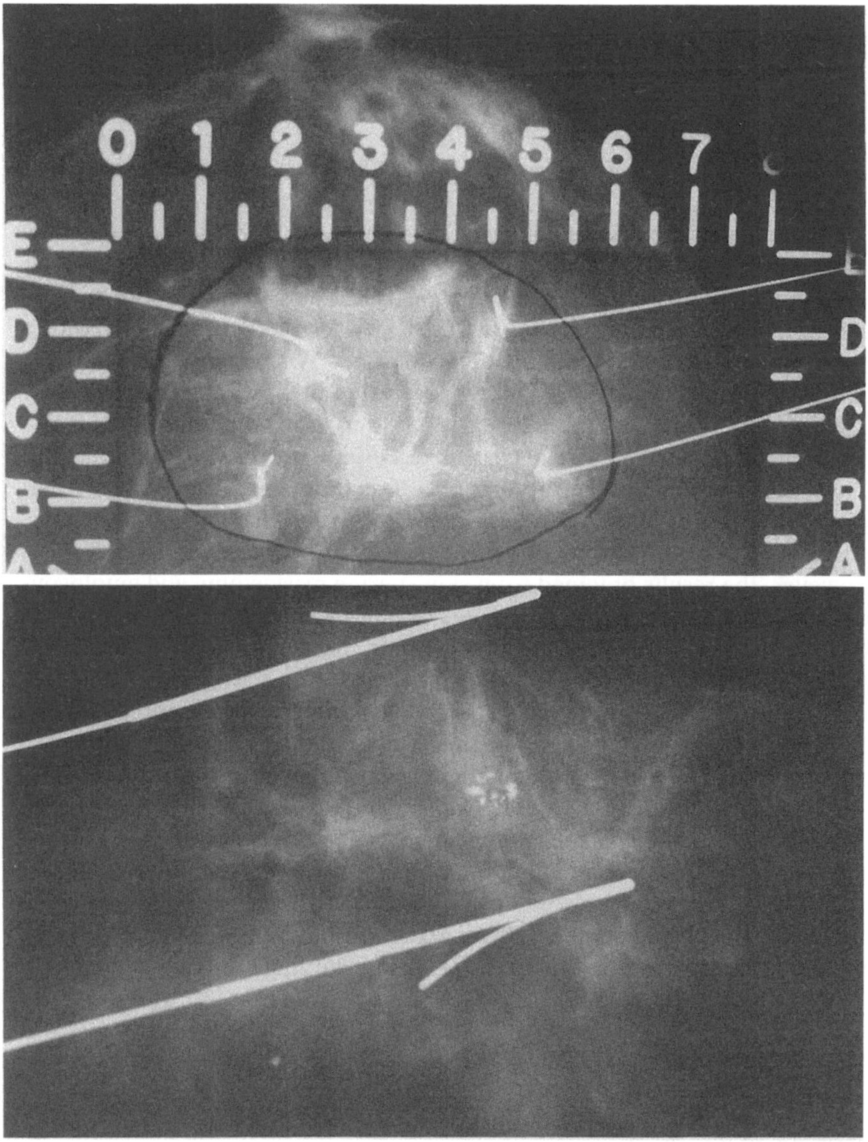

a

b

Fig. 4a,b. **a** Craniocaudal view of a 4-wire directed breast biopsy; **b** specimen radiograph of a 2-wire directed breast biopsy

definition of clear margins was used. Assessment of pathological margins appears, at best, to be an imprecise guide to the persistence of disease. Our pathologists were unable to predict the presence of residual disease, despite meticulous margin assessment, in 49% of patients when 1 mm was used as the criterion for clear margins; 1 mm is clearly an inadequate distance and we would be much happier with margins of 5 or 10 mm. As the distance between the DCIS and the inked margins increases, the likelihood of residual disease will decrease (Holland et al. 1990).

Glandular Replacement Therapy for Intraductal Breast Carcinoma

For those patients whose DCIS would be better treated with mastectomy rather than excision plus radiaiton therapy (XRT), we propose glandular replacement therapy (GRT). We define GRT as a combination of skin-sparing mastectomy and autologous tissue reconstruction, generally with a free transverse rectus abdominis (TRAM) flap. When significant overlying skin is removed, we prefer to call the operation breast replacement therapy (BRT). We have coined these terms because in the United States, medicine is now highly consumer (patient) oriented. Patients need to be fully informed. They often seek multiple consultations and, ultimately, they choose their own therapy. For many patients, breast conservation vs breast removal (mastectomy) seems like a choice between two widely divergent therapies. However, the results obtained from skin-sparing mastectomy and autologous tissue reconstruction are so much better than those obtained from standard mastectomy and reconstruction that we felt the necessity for a new term; hence, GRT for glandular replacement therapy.

Because DCIS does not involve overlying skin, no skin, except the nipple-areolar complex, need be removed. The operation is performed through a circumareolar incision with a linear extension rather than the elliptical skin-removing incision used when performing standard mastectomy (Fig. 5). The skin envelope, including the inframammary sulcus, remains intact allowing autologous tissue reconstruction that yields a new breast that is almost exactly the same size, shape, and texture as the breast that has been removed. The circular defect left by excision of the nipple-areolar complex is filled with abdominal skin trasnsferred on the flap. Initially, this skin is

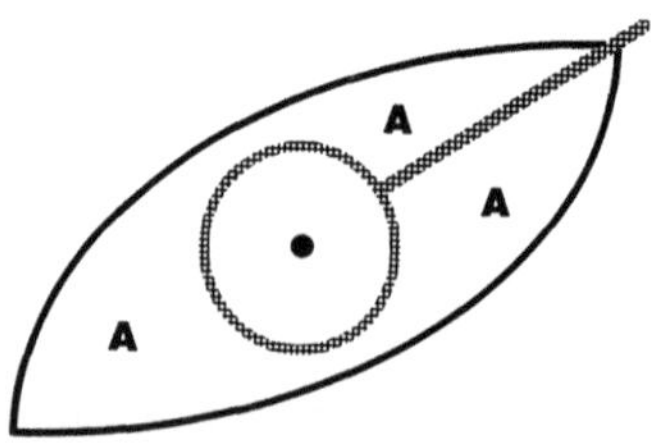

Fig. 5. The elliptical incision for a total mastectomy is shown by the *solid lines*. The lollipop incision for a skin-sparing mastectomy is shown by the *checked line*. The skin that is spared in marked by the letter *A* and amounts to approximately 70 cm^2 in the average patient

used to monitor the flap, and later to reconstruct the nipple-areolar complex (Fig. 6).

From November 1993 – December 1994, we performed 20 of these procedures. There has been no tissue loss in any patient and there have been no recurrences of DCIS. Cosmetic results were excellent in 18 patients and good in two. Initially, GRT is more expensive than XRT, but this is a

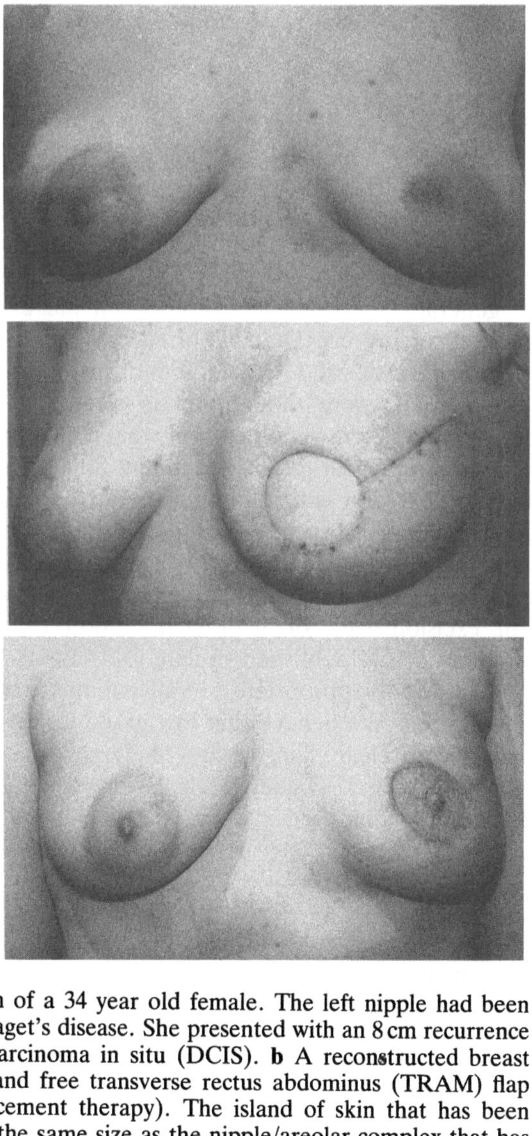

a

b

c

Fig. 6a–c. **a** Pre-op photograph of a 34 year old female. The left nipple had been removed two years earlier for Paget's disease. She presented with an 8 cm recurrence of high grade comedo ductal carcinoma in situ (DCIS). **b** A reconstructed breast after skin-sparing mastectomy and free transverse rectus abdominus (TRAM) flap reconstruction (glandular replacement therapy). The island of skin that has been replaced is circular and exactly the same size as the nipple/areolar complex that has been removed. **c** The nipple/areolar complex has been reconstructed

one-time cost to the patient or health care provider. DCIS patients treated conservatively require life-long mammography. Approximately 40% can expect additional biopsies at some time in the future, and 10%–20% will experience local recurrence (Fisher et al. 1993). Since half of all local recurrences are invasive, there may be a mortality rate of 2%–4% for DCIS patients treated conservatively. GRT allows the patients to accept mastectomy because it minimizes the cosmetic deformity of this treatment. In addition, the chances of local recurrence and mortality are minimized because a complete glandular mastectomy has been performed.

Summary

Intraductal breast carcinoma is a heterogeneous disease, ranging from nonaggressive appearing, low grade lesions, to poorly differentiated high grade lesions. The low grade lesions tend to be diploid and have a low S-phase. They tend not to overexpress p53 or HER2-neu. The high grade lesions are generally the opposite. In between the extremes, there is a complete spectrum of lesions. DCIS is often difficult to excise completely. Not uncommonly, residual DCIS is found surrounding the biopsy cavity, even when margins were thought to be clear. To further complicate matters, there is no uniform agreement on what constitutes a clear margin nor on the definition of nuclear grade. Treatment for DCIS varies widely, from excision alone, to excision with radiation therapy, to mastectomy with and without reconstruction. Most physicians agree that large lesions greater than 5 cm in diameter and those with involved margins not amenable to re-excision are probably best treated with mastectomy. Most smaller lesions can be treated with excision with or without radiation therapy. The only prospective randomized trial published to date (NSABP B-17) has suggested that radiation therapy is appropriate for all patients with DCIS treated with breast conservation. We believe that low grade lesions which are small in size and excised with clear margins are likely to do equally well with or without radiation therapy. Higher grade lesion will likely benefit from radiaiton if the breast is conserved.

References

Fisher B, Constantino J, Redmond C et al. (1993) Lumpectomy compared with lumpectomy and radiation therapy for the treatment of intraductal breast cancer. N Engl J Med 328:1581–1586

Holland R, Hendriks JHCL, Verbeek ALM et al. (1990) Extent, distribution, and mammographic/histologic correlations of breast ductal carcinoma in situ. Lancet 335:519–522

Jackson PA, Merchant W, McCormick CJ, Cook MG (1994) A comparison of large block macrosectioning and conventional techniques in breast pathology. Virchows Arch 425:243–248

Lagios MD (1990) Duct carcinoma in situ: pathology and treatment. Surg Clin North Am 70:853–871
Poller DN, Ellis IO (1995) Ductal carcinoma in situ. In: Kirkham N, Lemoine NR (eds) Progress in pathology, vol 2. Churchill Livingstone, London (in press)
Poller DN, Silverstein MJ, Galea M et al. (1994) Ductal carcinoma in situ of the breast: a proposal for a new simplified histological classification association between cellular proliferation and c-erbB-2 protein expression. Mod Pathol 7:257–262
Silverstein MJ, Handel N, Hoffman RS et al. (1986a) The Breast Center – a multidisciplinary model. In: Paterson AHG, Lees AW (eds) Fundamental problems in breast cancer. Nijhoff, Boston, pp 47–58
Silverstein MJ, Rosser RJ, Gierson ED et al. (1986b) Axillary node dissection for intraductal carcinoma – is it indicated? Proc Am Soc Clin Oncol 5:265
Silverstein MJ, Gamagami P, Colburn WJ et al. (1989) Nonpalpable breast lesions: diagnosis with slightly overpenetrated screen-film mammography and hook wire directed biopsy in 1014 cases. Radiology 171:633–638
Silverstein MJ, Waisman JR, Rosser RJ et al. (1990) Intraductal carcinoma of the breast (DCIS): 208 cases. Cancer 66:102–108
Silverstein MJ, Gierson ED, Colburn WJ et al. (1991) Axillary lymphadenectomy for intraductal breast carcinoma. Surg Gynecol Obstet 172:211–214
Silverstein MJ, Cohlan BF, Gierson ED et al. (1992) Duct carcinoma in situ: 227 cases without microinvasion. Eur J Cancer 28:630–634
Silverstein MJ, Gierson ED, Colburn WJ et al. (1994) Can intraductal breast carcinoma be completely excised by local excision? Cancer 73:2985–2989
Silverstein MJ, Poller DN, Waisman JR et al. (1995) Prognostic classification of breast ductal carcinoma in situ. Lancet 345:1154–1157
Sloane JP, Ellman R, Anderson TJ, Brown CL, Coyne J, Dallimore NS et al. (1994) Consistency of histopathological reporting of breast lesions detected by breast screening: findings of the UK national external quality assessment (EQA) scheme. Eur J Cancer 10:1414–1419
Solin L, Recht A, Fourquet A et al. (1991) Ten-year results of breast conserving surgery and definitive irradiation for intraducatal carcinoma (ductal carcinoma in situ) of the breast. Cancer 68:2337–2344
Van Dongen, Holland R, Peterse JL et al. (1992) Second EORTC DCIS consensus meeting. Eur J Cancer 28:626–629

[illegible]

Summary

John F. Forbes

Department of Surgical Oncology, Newcastle Mater Hospital, Locked Bag No. 7, Newcastle Regional Mail Centre, NSW 2310, Australia

Breast cancer screening and the management of duct carcinoma in situ (DCIS) are closely related. DCIS is now mostly detected by mammongraphy in asymptomatic women and contributes 15%–20% of the cancers found in screening programs. Although the efficacy of mammography screening is well-established from randomized controlled clinical trials (RCT), important questions remain about its optimal use. Dr. Rosselli Del Turco addresses many of these questions, provides answers where possible, and proposes strategies where data are lacking. Two important aspects of DCIS management have been addressed by Dr. Margolese (the surgeon's role) and Dr. Silverstein (predicting recurrence). These two contributions are closely related, as if we knew the true risk of recurrence, we could plan surgery more rationally. Dr. Margolese reports results from an RCT, which should provide unbiased data to answer some questions. Dr. Silverstein has extensive personal experience from which new important questions are identified.

Breast cancer screening is clearly here to stay. The evidence from RCTs to support their efficacy is substantial. A new screening program must reach and maintain standards of quality consistent with early detection of sufficient breast cancers for this to be translated into a reduction in breast cancer mortality in the screened population. In addition, it is essential that women who have breast cancer diagnosed get access to optimal treatment; it is not sufficient to simply diagnose the cancer, and sufficient women must attend for screening to ensure a worthwhile gain for the targeted population. Tabar et al. (1992) have recommended standards to be met to ensure efficacy of screening in terms of (i) detection of small ("curable") tumors (50% <15 mm, >30% Grade 3 <15 mm); (ii) nodal status (>70% node negative); and (iii) detection rates (first screen prevalent detection rate at least three times expected incidence). Dr. Rosseli Del Turco endorses this approach as sensible to evaluate the efficacy of a screening program outside clinical trials.

Should women under 50 years be screened? The data are less persuasive than that for older women. The overview of the Swedish trials (Nystrom et al. 1993) suggests a 13% mortality reduction may be achieved for women aged 40–49 years, and this is the most reliable data set available.

There is little difference in prevalence of breast cancer amongst women aged 40–49 years and 50–59 years. The most appropriate measure of efficacy is years of life gained rather than simply mortality reduction. And it has been claimed that screening women aged 40–49 years has equivalent cost-effectiveness to cervical cancer screening. Faced with the current uncertainties, Dr. Rosseli Del Tureo recommends a very large trial to provide answers for this age group. It may be difficult to accrue the large numbers required, and problems may arise from controls having mammograms. A clear majority of the audience in St. Gallen, however, agreed that the trial should be done. The issues of screening intervals, and screening of special groups such as those at special risk and those aged over 70 years remain important. A 70-year-old women can expect to live about 17 years, and almost 30% of breast cancers occur in women aged 70 and over, so they are clearly an important group. Further trials are now also required to investigate how best to screen women aged 50–70 years, rather than simply whether to screen. Whether to screen still remains, however, the important issue for women aged <50 and over 70 years.

DCIS has emerged as a new management problem through wider use of mammography. DCIS should be completely cured if it is treated by mastectomy and there is no invasive component. Survival is not compromised when invasive disease is treated by breast preservation, but this is because some patients already have disease outside their breast. The issues are different for DCIS. It is not yet known whether patient survival will be worse if DCIS is treated by breast preservation as new invasive disease could develop later and spread outside the breast. Dr. Margolese provides helpful evidence from a RCT. Radiotherapy (RT) added to "lumpectomy" to treat DCIS reduces subsequent recurrence of both invasive and in situ cancer without any survival difference (median follow-up, 43 months). No data are yet available on the important pathology subtypes described by Dr. Silverstein. The RCT data provide no conclusion about mastectomy or the use of tamoxifen for treating DCIS. Ipsilateral recurrence required mastectomy in 64% in the RT arm, but in only 44% of the lumpectomy arm, giving breast preservation rates of 95.7% with RT and 93.1% without RT, and without any survival difference (Fisher et al. 1993). This is a very small gain for RT. The real gain from RT, however, is likely to be the reduction in invasive recurrence (10.5% versus 2.9%), which raises the question of whether RT might actually *prevent* invasive breast cancer from developing. Perhaps the RT lethally damages cells with malignant potential to make them nonviable when they undergo malignant change and divide. This is a plausible explanation of the reduced invasive recurrence rate and has important implications. Current trials (UK, Australia, EORTC) are addressing the issue of tamoxifen and RT both alone and together after a complete local excision (CLE) of the DCIS.

Dr. Silverstein has extensive personal experience with treatment of DCIS and presents data to suggest that grade and morphology can be used to predict outcome and hence also to select optimal local treatments. Patients with high-grade DCIS have very high recurrence rates if the initial DCIS is not treated by mastectomy, and these patients should have either mastectomy or breast preservation with RT. Patients with tumors that are not high grade but which show necrosis apparently have a substantial reduction in local recurrence if RT is added to tumor excision. If DCIS is not high grade and has no necrosis, local recurrence rates seem similar with or without RT (Silverstein et al. 1995). These two presentations on DCIS highlight the role of RCTs in answering questions, are a large personal serices, in indentifying new questions. Both are important, but it is stressed that there is a potential for bias in uncontrolled series.

Finally, it is salutary to note that the only three quantum steps forward in breast cancer management have each come only because rigorous RCTs were done. These have been (i) the introduction of chemotherapy to reduce mortality; (ii) the introduction of breast preservation without compromise of survival; and (iii) the demonstration that mammography screening can lead to lower breast cancer mortality. The next quantum gain would be the demonstration through RCTs that breast cancer can be prevented. Continued rigorous research using RCTs for both screening and DCIS will certainly produce further progress.

References

Fisher B, Costantino J, Redmond C, Fisher E, Mangolese E, Dimitrov H, Wolmark N, Wickerham DL, Deutsch M, Ore L, Mamounas E, Poller W, Karanah M (1993) Lumpectomy compared with lumpeetomy and radiation therapy for the treatment of intraductal breast cancer. N Engl J Med 328:1581–1586

Nystrom L, Rutqvist L, Wall S, Lindgren A, Lindqvist M, Ryden S et al. (1993) Breast cancer screening with mammography: overview of Swedish randomised trials. Lancet 342:973–978

Silverstein MJ, Poller DN, Waisman JR, Colburn WJ, Barth A, Gierson ED, Lavinsky B. Gamagami P, Slamon DJ (1995) Prognostic classification of breast ductal carcinoma in-situ. Lancet 345:1154–1157

Tabar L, Fagerberg S, Duffy S, Day NE, Gad A, Grontoft O (1992) Update of the Swedish two-country program of mammographic screening for breast cancer. Radiol Clin N Am 30(1):187–210

V. Surgery for Breast Cancer: Special Issues

Surgical Treatment of Breast Cancer Relapse

R.G. Margolese

Surgical Oncology, McGill University, 3755 chemin Côte-Ste-Catherine, Montreal H3T 1E2, Canada

Introduction

The widespread acceptance of breast conserving therapy has brought with it the need to consider treatment of ipsilateral breast tumor recurrence (IBTR). In the case of total mastectomy, patients with so-called local recurrences have long-term outcomes similar to those of patients with more distant recurrences, confirming the concept that these are metastases and will need systemic approaches to therapy. For breast conserving therapy however, there is still a controversy about contributing factors to IBTR. Even local retreatment choices may be important. The word "salvage" has been commonly used to indicate the inherent curability of these recurrences.

An important consideration in planning therapy for IBTR following breast conserving therapy is the understanding of the cause of the IBTR. We can think of three main types:

1. Residual disease: inadequate removal of tumor; radioresistant cells
2. Metastatic disease (of which IBTR is one manifestation)
3. New primary cancer

Since virtually all authors report IBTR to be most frequent at or near the site of the original resection site, explanations 1 and 2 seem more likely, since a new primary would realistically more often be represented by tumor elsewhere in the breast.

There are two main patterns of local breast recurrence. Type I is a single focal density more suggestive of regrowth of residual disease and is clearly the more amenable to re-excision either by repeat lumpectomy or by mastectomy. The pattern of type II recurrence is diffused throughout the breast, may involve the skin and is frequently not controllable by surgery alone (Fig. 1). Sometimes even mastectomy will not be sufficient since skin involvement implies presence of tumor in the dermal lymphatics which will often extend beyond the limits of resection.

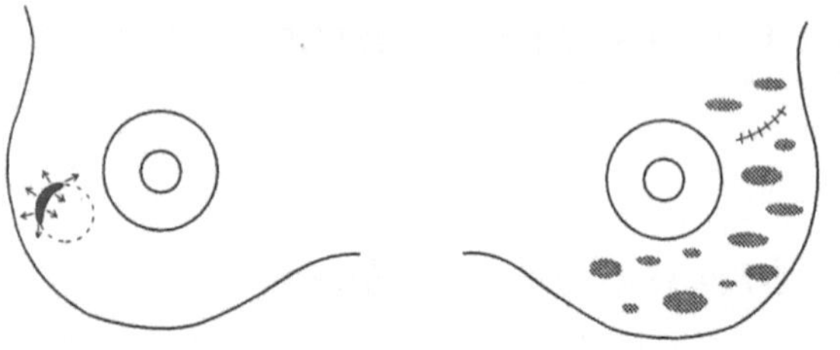

Fig. 1. Patterns of local recurrence. *Panel I*, demonstrates the regrowth of a tumor at the site of microscopically residual disease not removed at the first operation. In *panel II*, a diffuse pattern permeating all or much of the breast indicates a hematogenous origin

Amongst the various clinical features predicting increased risk for local recurrence are tumor size, age of patient, nuclear grade, vascular invasion, and extensive intraductal component (EIC) (Borger et al. 1994; Schnitt et al. 1984; Kurtz 1992; Clarke and Martinez 1992; Harris and Gelman 1994), but some of these features actually predict for increased expectation of metastases (vascular invasion, young age, tumor size), so a consideration of the biology behind IBTR is in order. While there is extensive variability in the conclusions of these analyses some trends are apparent. In cases in which pathological margin assessment is properly performed, tumors with negative margins have a low incidence of local recurrence following lumpectomy and traditional radiation therapy (Margolese 1992). Even the presence of an extensive intraductal component does not change this outcome (Schnitt et al. 1994).

Many of the classic reports on local recurrence date from the time when a standardized approach to pathological margin assessment was not always performed. These have less relevance to the present review and may provide misleading conclusions. Another development which makes the issue of IBTR more difficult is the evolution of systemic treatments during the same time that early studies on breast conserving therapy were being carried out. It is now clear that IBTR is decreased when adjuvant therapies are used.

Effect of Adjuvant Therapy

The National Surgical Adjuvant Breast Project (NSABP) protocol B-06 (Fisher et al. 1985) compared lumpectomy with and without radiation against total mastectomy. After 10 years, the probability of remaining IBTR-free after lumpectomy alone was 57% compared to 88% for patients who received lumpectomy and radiation therapy (Fisher et al. 1991). However,

Table 1. Efect of adjuvant therapies on local recurrence rate in patients undergoing lumpectomy and radiation in NSABP protocols (from Margolese 1992)

Trial number and treatment	Number of patients	Number of events	Average annual failure rate (%)
B-06			
Lumpectomy − radiation	572	193	6.81
Lumpectomy + ratiation	558	47	1.38
B-13			
Lumpectomy + radiation	121	9	2.78
Lumpectomy + radiation + Chemotherapy (methotrexate 5-fluorouracil)	177	2	0.57
B-14			
Lumpectomy + radiation	537	19	1.24
Lumpectomy + radiation + tamoxifen	537	8	0.5

in subsequent years other NSABP protocols employing adjuvant therapy indicated a significantly lower overall rate of local recurrence in systemically treated patients (Margolese 1992). Although true cross-protocol comparisons cannot legitimately be made because of differing selection and eligibility criteria, the annual hazard rate for IBTR can give an indication of the risk for the different treatments.

The average annual hazard rate of IBTR for lumpectomy only patients in B-06 was 6.6%. For lumpectomy and radiation (LpRT) patients it was 1.4%. In protocol B-13, a trial of chemotherapy vs observation in node negative patients with low estrogen receptor (ER) values, the average annual risk of IBTR for LpRT patients decreased to 0.6% compared to 2.7% in the nonadjuvant group (Table 1). In protocol B-14, a trial of tamoxifen vs placebo in node negative ER positive patients, a similar outcome was seen, 0.5% vs 1.2%. Note that in these studies the control groups had essentially the same rate of IBTR as the LpRT patients in B-06.

Rose et al. (1989) studied a group of patients with IBTR and found the 5 year cumulative local failure rate to be 5% in patients receiving adjuvant therapies vs 17% for those who received radiation therapy only. The presence of EIC was the same in both groups. Therefore, IBTR is a less frequently encountered problem in a modern cohort receiving adjuvant systemic therapy than in the classical B-06 and Joint Center studies, where such therapies were not used except in some node positive patients.

Management

Although we cannot individually identify specific causes for IBTR for individual patients we can make some sense of the problem. Since most IBTRs are type I, a localized nodule at the site of original surgery, it is very likely that they represent persistence of unremoved cancer cells. The fact that there is a dramatic reduction in IBTR with more extensive surgery or when local radiotherapy is given supports this idea. However, Cox regression models indicate that IBTR after lumpectomy is a significant independent predictor of distant metastatic disease. The risk of distant disease is 3.4 times greater than in patients who do not develop an IBTR and is higher in patients with earlier recurrences than in those with later ones (Fisher et al. 1991).

Most important, the overall rate of distant treatment failure is no different for patients treated by mastectomy than by lumpectomy with or without radiation. This indicates that IBTR is not the source of the metastases but simply an indicator, or marker, of the increased risk of developing distant metastases. While some IBTR actually represent metastases, the best explanation for most is that factors which contribute to distant metastases also contribute to local recurrence. Thus, patients whose breasts are conserved can display this increased risk in the ipsilateral breast as a marker-like event. Patients treated by mastectomy have the same risk of distant disease but lack the marker since the previously affected breast is absent.

Taken together, it is clear that there is no simple response to the problem of local recurrence. We can speculate that some local recurrences represent residual disease, others signify radioresistant cells, some IBTRs may be new primary tumors and some represent metastatic disease which manifests in the breast.

The concept of salvage surgery implies that we can consider at least some cases of IBTR as a product of incomplete initial excision. These would more aptly be labeled residual tumor. Therefore the initial step in management of IBTR is a careful systemic search to exclude other detectable metastases. We can then adopt a stepwise approach which asks if the recurrence is amenable to surgical treatment. If it is localized, solitary and excisable then local control of this lesion can be accomplished with a repeat lumpectomy. If not, a mastectomy may be necessary (Table 2).

In seeking criteria for the selection of patients for repeat lumpectomy instead of salvage mastectomy, the same selection considerations that exist for primary surgery must be considered: tumor size, absence of grave signs, cosmetic feasibility, and pathological predictors. The main problem concerns this last criterion. The Harvard group has catalogued the role of EIC through many studies (Schnitt et al. 1984, 1994) but it would appear that with modern techniques for margins assessment, appropriate radiation therapy, and use of adjuvant chemotherapy it is difficult to choose patients on the basis of pathological markers for exclusion from lumpectomy at initial surgery. Indeed, in the NSABP protocol B-06 there are no pathological indicators that would exclude patients from initial lumpectomy. Are the considerations for second lumpectomy any different if the lesion is technically resectable?

Table 2. Management plan for ipsilateral breast tumor recurrence

Metastatic search is negative
- Consider lumpectomy
  - Localized?
  - Solitary?
  - Size?
  - Excisable?
- Consider mastectomy
- Consider systemic therapy

Metastases exist
- Consider local control
- Systemic therpay

Salvage Surgery

There are many reports (Recht et al. 1989; Kurtz et al. 1988; Deutch 1993; Veronesi et al. 1990) from retrospective series detailing a generally favorable outcome after salvage surgery following IBTR (Table 3). There are problems in interpreting these results, similar to those encountered in trying to understand the usefulness of predictive factors for IBTR. These are many potentially confounding variables, listed in Table 4. The common point, however, is that, unlike IBTR following traditional mastectomy, a substantial

Table 3. Salvage mastectomy or lumpectomy for local breast cancer recurrence following original breast conserving surgery

	Year	*N*	Metastases	Disease-free survival	Overall survival
Salvage mastectomy (retrospective studies)					
Osborne and Simmons	1994	46	46	55	76
Cajucom et al.	1993	25	25	51	65
Clarke et al.	1985	24	10	93	81
Abner et al.	1993	166	123	59	79
Kurtz et al.	1988	118	118	85	72
Kurtz et al.	1990	71	71	84	–
Kurtz[23]	1989	171	159	88	69
Recht et al.	1989	84	65	63	–
Fourqet et al.	1989	56	45	73	–
Fowble et al.	1990	65	52	59	84
Wide excision as salvage					
Kurtz et al.	1991	50	50	58	–
Kurtz et al.	1988	52	52	79	79
Abner et al.	1993	16	16	59	–

Table 4. Factors confounding retrospective analyses of outcomes for salvage mastectomy

Number inoperable – selection factors?
Number refusing surgery
Number of solitary or multiple breast tumor recurrences
Numter combined with axillary and/or distant metastases
Calculate disease-free survival from relapse or original surgery data?
Calculate for invasive vs noninvasive or combined?
Reported as actuarial or crude survival
Calculate for all recurrences, or only operable?
Surgeon's perference

fraction of patients with IBTR have long-term disease-free survival following retreatment. Almost all such patients were treated by mastectomy, but there are small series documenting similar control following lumpectomy as a second procedure.

Kurtz et al. (1988) reported on 118 patients with isolated recurrences of whom 52 were treated by repeat wide excision instead of mastectomy. Local control was 79% at 5 years. Of 12 patients with second local recurrences 10 were still treatable by further surgery.

The same arguments and considerations that allow breast conserving surgery for primary cancer can be invoked at the time of treatment of IBTR if it is a localized solitary mass. The issue of multicentricity continues to be a distraction. Multicentricity is demonstrable histologically more often than recurrences become evident biologically, and the few reports in the literature indicate local control after conservative salvage operations approximates that observed in the original B-06 trial.

Data from Deutch (1993) indicate that repeat radiation therapy can be accomplished but it is not clear that this is necessary. The use of radiation therapy boost at initial treatment has not been discussed since the low IBTR rates quoted above for the NSABP trials involve no boost. It is uncertain whether a patient with an excised IBTR and who did not have original boost should now receive treatment to the equivalent fields.

Reports on outcomes following salvage mastectomy or repeat lumpectomy both demonstrate the overall feasibility of these approaches. The real question is when and where systemic adjuvant therapies should be employed. If IBTR is considered to represent, for a large number of patients, an increased risk of distant metastatic disease, then this can be considered as an indication that these patients should now receive systemic therapy, especially if they did not receive such therapy at the time of surgery. With the demonstration that adjuvant therapies decrease the incidence of IBTR, patients can be considered good candidates for such therapy based on local considerations alone. No clinical trial has been done to confirm the usefulness of this type of "late" adjuvant therapy aimed at local recurrences, but it makes sense simply because of the increased risk of metastases and the evidence tht it decreases IBTR per primum.

Conclusion

Evidence from the classic NSABP (Fisher et al. 1985) and Milan (Veronesi et al. 1990) studies indicates that breast conserving surgery is equivalent to mastectomy in terms of overall expectation for cure.

Patients undergoing breast conserving therapy are at risk for IBTR. It is clear that initial technical factors (extent of surgery, use of radiotherapy) can decrease the incidence of IBTR. In addition, use of systemic adjuvant therapies also helps control local recurrence rates and in appropriately treated patients the risk is of IBTR is only 0.5% per year.

The search for histopathological or clinical predictors for IBTR is of limited use since the event is uncommon even in the presence of predictive factors, and the vast majority of patients do not experience IBTR. These predictors assume more importance when surgical and pathological techniques are less exact.

Since at least some IBTRs represent increased risk for metastatic disease, it should follow that patients with IBTR would do worse than those who remain disease free. Based on protocol B-06 and others, the consensus is that initial mastectomy would not have changed the likelihood of distant failure or resulted in better survival. Therefore, reports in the literature identifying worse outcomes amongst patients with local recurrence confuse effect with cause and can be misleading. Proper treatment of IBTR requires surgical removal when feasible since lcoal control is always a goal of surgical therapy. Many IBTRs can be managed by repeat surgery, either lumpectomy or mastectomy. Since IBTR may frequently indicate the presence of subclinical metastases elsewhere, additional systemic therapy should be considered at the same time, but clinical trial data do not exist to give us specific guidelines.

References

Abner AL, Recht A, Eberlein T, Come S, Shulman L, Hayes D, Connolly JL, Schnitt SJ, Silver B, Harris JR (1993) Prognosis following salvage mastectomy for recurrence in the breast after conservative surgery and radiation therapy for early-stage breast cancer. J Clin Oncol 11:44–48

Borger J, Kemperman H, Hart A, Peterse H, Van Dongen J, Bartelink H (1994) Risk factors in breast-conserving therapy. J Clin Oncol 12:653–660

Cajucom CC, Tsangaris TN, Nemoto T, Driscoll D, Penetrante RB, Holyoke ED (1993) Results of salvage mastectomy for local recurrence after breast-conserving surgery without radiation therapy. Cancer 71:1774–1779

Clarke DH, Martinez AA (1992) Identification of patients who are at high risk for locoregional breast cancer recurrence after conservative surgery and radiotherapy: a review article fo surgeons, pathologists, and radiation and medical oncologists. J Clin Oncol 10:474–483

Clarke DH, Le MG, Sarrazin D, Lacombe MJ, Fontaine F, Travagli JP, May-Levin F, Contesso G, Arriagada R (1985) Analysis of local-regional relapses in patients with early breast cancers treated by excision and radiotherapy: experience of the Institut Gustave-Roussy. Int J Radiat Oncol Biol Phys 11:137–145

Deutch M (1993) Repeat lumpectomy and radiotherapy for breast recurrence following lumpectomy and radiotherapy. NSABP Annual Meeting

Fisher B, Bauer M, Margolese R, Poisson R, Pilch Y, Redmond C, Fisher E, Wolmark N, Deutsch M, Montague E, Saffer E, Wickerham L, Lerner H, Glass A, Shibata H, Deckers P, Ketcham A, Oishi R, Russell I (1985) Five-year results of a randomized clinical trial comparing total mastectomy and segmental mastectomy with or without radiation in the treatment of breast cancer. N Engl J Med 312:11:665–673

Fisher B, Anderson S, Fisher ER, Redmond C, Wickerham DL, Wolmark N, Mamounas EP, Deutsch M, Margolese R (1991) Significance of ipsilateral breast tumor recurrence after lumpectomy. Lancet 8763(338):327–331

Fourquet A, Campana F, Zafrani B, Mosseri V, Vielh P, Durand JC, Vilcoq JR (1989) Prognostic factors of breast recurrence in the conservative management of early breast cancer: a 25-year follow-up. Int J Radiat Oncol Biol Phys 17:719–725

Fowble B, Solin LJ, Schultz DJ, Rubenstein J, Goodman RL (1990) Breast Recurrence following conservative surgery and radiation: patterns of failure, prognosis, and pathologic findings from mastectomy specimens with implications for treatment. Int J Radiat Oncol Biol Phys 19:833–842

Harris JR, Gelman R (1994) What have we learned about risk factors for local recurrence after breast-conserving surgery an irradiation? J Clin Oncol 12:647–649

Kurtz JM, Amalric R, Brandone H, Ayme Y, Spitalier JM (1988) Results of wide excision for mammary recurrence after breast-conserving therapy. Cancer 61: 1969–1972

Kurtz JM, Amalric R, Brandone H, Ayme Y, Jacquemier J, Pietra JC, Hans D, Pollet JF, Bressac C, Spitalier JM (1989) Local recurrence after breast-conserving surgery and radiotherapy, frequency, time course, and prognosis. Cancer 63;10: 1912–1917

Kurtz JM, Spitalier JM, Amalric R, Brandone H, Ayme Y, Jacquemier J, Hans D, Bressac C (1990) The prgonostic significance of late local recurrence after breast-conserving therapy. Int J Radiat Oncol Biol Phys 18:87–83

Kurtz JM, Jacquemier J, Amalric R, Brandone H, Ayme Y, Hans D, Bressac C, Spitalier JM (1991) Is breast conservation after Local Recurrence Feasible? Eur J Cancer 27:240–44

Kurtz JM (1992) Factors influencing the risk of local recurrence in the breast. Eur J Cancer 28:660–666

Margolese R (1992) Surgical considerations in selecting local therapy. Monogr Natl Cancer Inst 11:41–48

Osborne MP, Simmons RM (1994) Salvage surgery for recurrence after breast conservation. World J Surg 18:93–97

Recht A, Schnitt JS, Conolly JL, Rose MA, Silver B, Come S, Henderson IC, Slavin S, Harris JR (1989) Prognosis following local or regional recurrence after conservative surgery and radiotherapy for early stage breast carcinoma. Int J Radiat Oncol Biol Phys 16:3–9

Rose MA, Henderson IC, Gelman R et al. (1989) Premenopausal breast cancer patients treated with conservative surgery, radiotherapy and adjuvant chemotherapy have a low risk of local failure. Int J Radiat Oncol Biol Phys 17:711–717

Schnitt SJ, Connolly JL, Harris JR, Hellman S, Cohen RB (1984) Pathologic predictors of early local recurrence in stage I and I breast cancer treated by primary radiation therapy. Cancer 53:1049–1057
Schnitt SJ, Abner A, Gelman R, Connolly J, Recht A, Duda RB, Eberlein TJ, Mayzel K, Silver B, Harris RJ (1994) The relationship between microscopic margins of resection and the risk of local recurrence in patients with breast cancer treated with breast-conserving surgery and radiation therapy. Cancer 74:1746–17541
Veronesi U, Banfi A, Salvadori B, Luini A, Saccozzi R, Zucali R, Marubini E, Del Vecchio M, Boracchi P, Marchini S, Merson M, Sacchini V, Riboldi G, Santoro G (1990) Breast conservation is the treatment of choice in small breast cancer: long-term results of a randomized trial. Eur J Cancer 26:668–670

Primary and Secondary Breast Reconstruction with Special Emphasis on the Use of Prostheses

J.Y. Petit,[1] M. Rietjens,[1] C. Garusi,[1] and D. Capko[2]

[1] European Institute of Oncology, Via Ripamonti 435, 20141 Milan, Italy
[2] Hackenn East ACK Medical Center, Institute for Breast Care, 30, Prospect Street, Alacuennfac, NJ, USA

Introduction

Breast reconstruction (BR) is widely accepted as part of the surgical treatment of breast cancer, although mastectomy is less frequently indicated. Delayed breast reconstruction (DBR) is available for patients having undergone a mastectomy in the past, and immediate breast reconstruction (IBR) can be proposed at the time of mastectomy, especially in case of diffuse or extensive in situ carcinomas. Patients with infiltrating carcinomas who require mastectomy are increasingly treated with adjuvant chemotherapy. Therefore, IBR should be considered as an option only if it does not postpone medical treatment, since delay of adjuvant treatment due to IBR can increase the risk of complications, these being directly related to the IBR technique. An evaluation of the risk should help us to decide whether IBR is safe or not when adjuvant treatment is required.

IBR or DBR can be performed with autologous tissues or with an implant. The risk of cancer and autoimmune disease after silicone implantation led, in 1991, to a number of recommendations by the US Food and Drug Administration (FDA).[3] Today, different studies strongly suggest the absence of carcinogenic effects due to silicone prostheses. However, the possible relationship between silicone implants and autoimmune disease remains uncertain, although a few (<100 cases) published observations have shown an association. In any case, women who seek a plasty with an implant should obtain the most recent information from their surgeon in order to be able to make an educated decision.

The overall evolution of indications and techniques of breast reconstruction during the last decade has been similar in most countries: an increasing number of immediate reconstructions and increasing use of autologous tissue reconstruction, especially in the USA where concern regarding silicone prostheses was reenforced by the media (Trabulsy et al. 1994).

Delayed Breast Reconstruction

Breast reconstruction has been performed since the early 1970s. At that time, oncologists were extremely concerned about the oncologic consequences of the procedure following treatment of cancer. Breast reconstruction was most often indicated for patients with in situ carcinomas, and the technique was limited to prosthesis implantation under the remaining post-mastectomy skin. Cosmetic results were poor due to the frequency of local skin necrosis in the postoperative period and the occurrence of capsular contracture, observed after several months. Improvement of the technique (use of myocutaneous flaps in case of poor local skin conditions and improvement of the silicone material) provided a dramatic change in the final results observed after breast reconstruction.

Today, there are actually no limitations to performing DBR, although, obviously, the general condition of the patient must be considered. The rectus abdominal flap, pedicled or freely transplanted, provides enough tissue to reconstruct the breast regardless of the status of the thoracic area (Fig. 1). Psychologic demand is sometimes strong enough to perform a DBR in case of metastatic disease provided that the reconstruction will not alter the general status of the patient and delay further treatment.

a

b

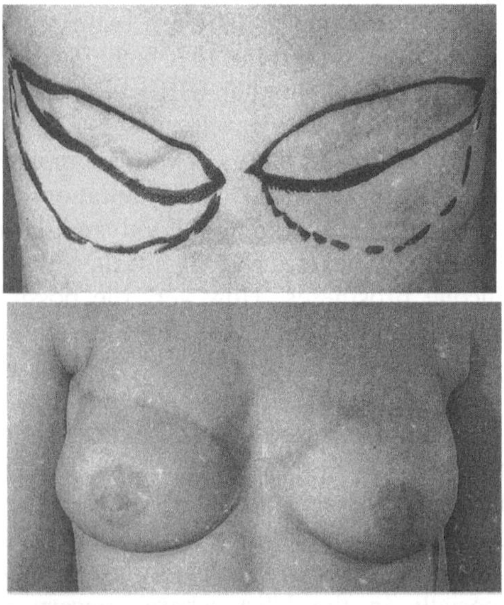

Fig. 1a,b. Delayed bilateral breast reconstruction without implant (**a**) and with a double rectus pedicled flap (**b**)

Immediate Breast Reconstruction

The delay required before undertaking reconstruction has decreased as oncologists have become progressively reassured and convinced of the innocuousness of the procedure. IBR is accepted for patients with extended in situ carcinomas as long as no adjuvant medical treatment is required. The question remains whether IBR is suitable for patients with infiltrating carcinoma. The final answer should be provided by evaluating the rate of immediate complications related to the plastic procedure, mainly infections and the necessity of implant removal. These complications increase the risk of delaying adjuvant therapy and could thus have a deleterious effect on the chances of cure.

Technically, IBR includes many of the same plastic procedures available for secondary reconstruction: implants, expanders with or without flaps, or flaps without implant (Fig. 2). Usually, IBR is performed by a plastic surgery team working together with oncologic surgeons; however, in a few

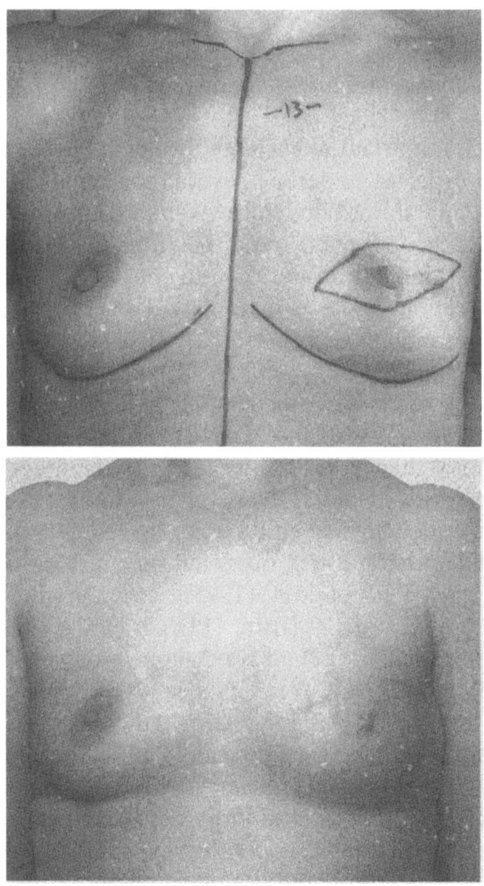

a

b

Fig. 2a,b. Immediate left breast reconstruction with implant

cancer centers in Europe, including Institut Gustave-Roussy and the European Institute of Oncology, the same team is competent in both fields, making coordination of the cancer and plastic surgeries easier. Combination of the two procedures is becoming increasingly frequent, especially in the case of conservative treatment in which the cosmetic result is considered to be an important goal. As shown in our own study (Petit and Rietjens 1991), secondary improvement of poor cosmetic results after conservative surgery is often disappointing. Optimal results are obtained by using plastic procedures at the time of primary treatment. These can best be performed by a team that is competent in both oncologic and breast plastic surgery or by two specialized teams working closely in the same surgical unit.

From the patient's point of view, the advantages of being operated on by a single team include the following: when the patient seeks cosmetic surgery, the surgeon will not forget to perform complete breast cancer screening, and when the patient undergoes cancer treatment the surgeon will perform the less mutilating surgery with partial or total reconstruction.

Complications and Contraindications

It is important to evaluate the morbidity rate related to the plastic procedures in order to underline several contraindications. Immediate morbidity, such as implant or flap failures, could cause delay of adjuvant medical treatment. At the Gustave-Roussy Institute, in a series of 323 IBRs in 306 patients, operated on from 1989 to 1991, the immediate complication rate was: infection (8%), hematomas (3.4%), skin necrosis (7%), seromas (9.5%), and implant removal (4%). In 74% of the patients, no complication was observed.

Most of the complications healed spontaneously in less than 3 weeks. In less than 4% of the cases did an immediate complication lead to a delay of more than 3 weeks in starting chemotherapy. Thus, IBR, in the great majority of cases, does not interfere with cancer treatment. Similar results have been observed by other investigators. A delay before starting adjuvant chemotherapy was observed in one case out of 128 IBRs with autologous tissue by Elliot et al. (1993). Rosen et al. (1990) did not observed any delay in a series of 103 IBRs. O'Brien et al. (1993) did not observe any difference in the rate of immediate complications with or without IBR. Schusterman et al. (1992), however, in a study of 68 IBRs with rectus abdominis flaps observed a delay in 25% of the cases in which adjuvant chemotherapy was required. Such results should be taken into account before using an autologous tissue transfer for an IBR. To minimize the immediate complication rate and thus IBR failure, it is necessary to consider: the experience of the team, whether preoperative radiotherapy has been administered, the prosthesis volume, and the morphology of the patient.

Oncologic or Immunologic Risk Related to Silicone Implants

After breast augmentation, especially when the implant has been inserted under the pectoralis major, the *clinical* detection of a tumor is improved. By contrast, *radiological* detection is more difficult and requires special views, for example, displacement of the gland above the implant (Ecklund technique). Ultrasound provides another opportunity to explore the gland and the prosthesis. Magnetic resonance imaging (MRI) is not reliable due to the lack of guidelines regarding breast cancer screening but looks competent for implant control. Computed tomography is valuable in detecting recurrent breast carcinoma in the chest wall, especially after breast reconstruction. Computerized mammography is one of the best methods to detect local recurrences after prosthetic reconstructions, and it is better than the xeromammogram, which delivers a higher level of radiation (Glicksman et al. 1992) (Fig. 3). The quality of the imaging depends on the breast implant filler material. Several laboratories are working to find a material with minimum radiopacity. Women seeking breast augmentation should undergo a complete clinical and radiological screening. Plastic surgeons should be aware of the different risk factors of breast cancer in order to give their patients the most complete and up to date information.

Several studies are looking at the risk of cancer in women exposed to silicone compared to the risk of cancer in the general population. The largest cohort study, carried out by Deapen and Brody (1992), of healthy women who received a silicone implant for breast augmentation showed that

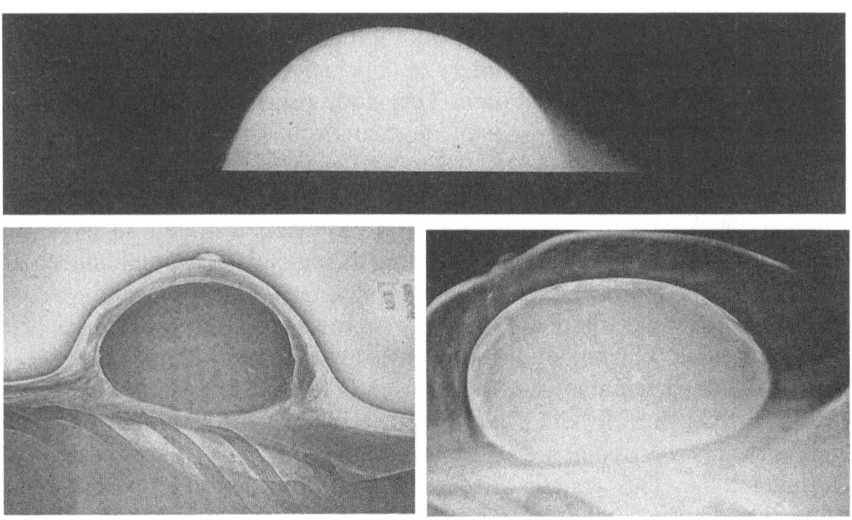

a b c

Fig. 3a–c. Radiological control after breast reconstruction with silicone implant. **a** simple mammography; **b** xeromammography; **c** computerized mammography

the cancer incidence in the silicone group was twofold lower than that in the general population, despite the higher social status.

At the Gustave-Roussy Institute, we performed a study (Petit et al. 1994) comparing the long-term evolution of a group of 146 cancer patients exposed to gel-filled silicone implant after breast reconstruction to a group of 146 matched controls with breast cancer who were treated in the same center by mastectomy without breast reconstruction and were matched for age at diagnosis, year of diagnosis, stage, histologic type of the tumor, histologic grade, and nodal status. The relative risk of death, relapse and second primary cancer were estimated by means of the Cox proportional hazards model stratified with respect to age at diagnosis. The risk of distant metastasis and death due to breast cancer were significantly lower in the breast reconstruction group. The risks of local recurrence, second breast cancer and second primary cancer at a site other than the breast were not significantly different between the two groups of patients. Therefore, our results do not support the allegations of a detrimental effect of gel-filled silicone implants on the course of breast cancer or of an increased risk of death due to another disease.

Another approach to study the risk of cancer or immunological disorders due to silicone implants is presented in an article by Duffy and Woods (1994). They studied the health risks of failed silicone gel breast implants over a 30 year period. Among 2033 patients exposed to silicone, a group of 65 patients with 104 implant failures was analyzed in greater detail, with particular attention given to diagnoses suggesting immune-related disorders or malignancies. The conclusion was that, in the selected subgroup of 65 patients with silicone failures, no excess of expected immune-related disorders or malignancies was identified.

Gabriel et al. (1994) published another population-based retrospective study examining the risk of a variety of connective tissue diseases and other disorders after breast implantation. This study concluded that there was no association between breast implants and connective tissue disease or other disorders.

Rheumatologists are increasingly involved in such studies (Cohen and Rohlich 1994). Wigley and Miller (1992) have demonstrated an increased risk of developing scleroderma following mammaplasty. Goldblum et al. (1992) published the first study describing specific antibodies to silicone elastomer. However, there are presently no specific tests to confirm silicone-related diseases. Further studies will probably help to discriminate between patients with rheumatic manifestations and patients without disease. A recent abstract (Silveira et al. 1992) reported the clinical findings of patients with implants who complained of arthralgias, myalgias, paresthesias and chronic fatigue. This abstract raised new questions requiring further studies to confirm the link between these symptoms and the presence of silicone. However, if the risk exists we must keep in mind that it is very low. Thus, the US and most European countries have declared the use of silicone implants to be legal, provided that the patients are followed regularly.

References

Cohen SB, Rohlich RJ (1994) Evaluation of the patient with silicone gel breast implants and rheumatic complaints. Plast Reconstr Surg 94(1):120

Deapen DM, Brody GS (1992) Augmentation mammaplasty and breast cancer. A 5-year update of the Los Angeles study. Plast Reconstr Surg 89:660

Duffry MJ, Woods JE (1994) Health risks of failed silicone gel breast implants: a 30-year clinical experience. Plast Reconstr Surg 94(2):295

Elliott LF, Eskenazi L, Beegle PH Jr, Podres PE, Drazan L (1993) Immediate TRAM flap breast reconstruction: 128 consecutive cases. Plast Reconstr Surg 92(2):217–227

Gabriel SE, Ofallon WM, Kurland LT, Beard CM, Woods JE, Melton LJ (1994) Risk of connective-tissue diseases and other disorders after breast implantation. N Engl J Med 330(24):1679–1702

Glicksman CA, Glicksman AS, Courtiss EH (1992) Breast imaging for plastic surgeons. Plast Reconstr Surg 90(6):1106–1111

Goldblum RM, Pelley RP, O'Donell AA, Pyron D, Heggers TP (1992) Antibodies to silicone elastomers and reactions to ventriculo peritoneal shunts. Lancet 340:510

O'Brien W, Hasselgren PO, Hummel RP, Coith R, Hyams D, Kurzman L, Neale HN (1993) Comparison of post operative wound complications and early cancer recurrence between patients undergoing mastectomy with or without immediate breast reconstruction. Am J Surg 166(1):1–5

Petit JY, Rietjens M (1991) Deformities after conservative breast cancer treatment. In: Noone B (ed) Plastic and reconstructive surgery of the breast. Mosby, St Louis, pp 455–466

Petit JY, Le MG, Mouriesse H, Rietjens M, Gill P, Contesso G, Lehman A (1994) Can breast reconstruction with gel-filled silicone implants increase the risk of death and second primary cancer in patients treated by mastectomy for breast cancer? Plast Reconstr Surg 94(1):115

Rosen PB, Jabs AD, Kister SJ, Hugo NE (1990) Clinical experience with immediate breast reconstruction using tissue-expansion or transverse rectus abdominis musculocutaneous flaps. Ann Plast Surg 25(4):249–257

Schusterman MA, Kroll SS, Weldon ME (1992) Immediate breast reconstruction: why the free TRAM over the conventional TRAM flap? Plast Reconstr Surg 90(2):255–261

Silveira LH, Cuellar ML, Scopelitis L, Martinez-Osma P, Espinoza LR (1992) Evaluation of rheumatic complaints in women with silicone breast implants (Abstr). Arthritis Rheum 35 Suppl 9:S347

Trabulsy PP, Anthony JP, Mathes SJ (1994) Changing trends in post mastectomy breast reconstruction: a 13-year experience. Plast Reconstr Surg 93:1418

Wigley FM, Miller R (1992) Augmentation mammoplasty in patients with systemic sclerosis: data from the Baltimore Scleroderma Research Center and Pittsburg Scleroderma data bank (Abstr). Arthritis Rheum 35 Suppl 9:S46

The Primary Use of Chemotherapy for Operable Breast Cancer: Does Systemic Therapy Make Surgery Superfluous?

C.-M. Rudenstam

Surgical Department, Sahlgrenska University Hospital, 413 45 Göteborg, Sweden

Introduction

Primary chemotherapy, in combination with radiotherapy and/or surgery, has been used in the treatment of advanced carcinoma of the breast for several decades. The term often used for this type of treatment is "neoadjuvant chemotherapy." More recently, less advanced cases of breast carcinoma have been treated by primary chemotherapy and many studies have been published describing optimistic results; however, very few randomized controlled investigations have appeared, and questions such as: "Should surgery remain the initial treatment of 'operable' breast cancer?" remain (Kurtz 1991).

The theoretical background motivating primary chemotherapy was extensively discussed at an international symposium in Zürich 1983, and the proceedings have been published in Recent Results in Cancer Research, 1985, vol. 98. Especially relevant are chapters by FM Schabel, AN Papaioannu and J Ragaz and the reader is referred to particularly these chapters (Papaioannou 1985; Ragaz et al. 1985; Schabel 1985).

Primary chemotherapy has several theoretical advantages compared to chemotherapy given after surgery or radiotherapy of the primary breast carcinoma:

1. Primary chemotherapy might reduce the volume and the extent of the primary tumor and make an inoperable tumor operable.
2. The size of the primary tumor could be reduced to permit breast saving surgery.
3. Primary chemotherapy is given at a time when the mass of micrometastases in different organs is smaller than when chemotherapy is given after treatment of the primary tumor.
4. The chemosensitivity of the primary tumor can be tested in vivo. If a good tumor response is registered, postoperative or postradiological chemotherapy with the same regimen could be administered.

5. Cells in the primary tumor with the potential to form metastases if released into the circulation by surgery might be killed by primary chemotherapy.
6. In experimental animal models primary chemotherapy followed by surgery of the experimental tumor gave a higher cure rate than surgery followed by chemotherapy. This could, in the clinical setting, improve disease-free and overall survival.

As is evident from this overview of primary chemotherapy, despite an abundance of chinical data, it remains to be seen whether the treatment effects described support the hope that systemic therapy will make surgery superfluous.

Clinical Results

Inflammatory Breast Carcinoma

Clinical experience using primary chemotherapy, mainly combination chemotherapy, was first gained in patients with very advanced breast carcinoma, e.g., inflammatory breast carcinoma. Table 1 presents data from a recent overview by Singletary et al. (1994). Radical mastectomy or simple mastectomy resulted in poor median survival and few patients with 5-year survival, as did the combination of primary radiotherapy with surgery. The introduction of primary chemotherapy followed by radiotherapy or surgery and thereafter, in some studies, by additional chemotherapy and postsurgical radiotherapy was followed by a marked increase of median survival and an improved 5-year survival. In their own series of patients using three cycles of FAC or FACVP followed by mastectomy or irradiation and thereafter adjuvant chemotherapy up to 2 years, Singletary et al. (1994) reported a disease-free survived rate of 58% and overall survival of 72% at 5 years. There was also a better survival rate in patients with complete clinical

Table 1. Management of inflammatory breast cancer

Treatment	Number of studies	Median survival (months)	Five-year survival (%)
Radical or simple mastectomy	8	12–19	0–3.5
Primary radiotherapy + surgery	9	13–29	0–20
Primary chemotherapy + radiotherapy ± adjuvant chemotherapy	6	24–36	40–55
Primary chemotherapy + surgery + adjuvant chemotherapy + radiotherapy	5	49–>56	35–75

response to primary chemotherapy than in patients with partial and minor responses. Surgery before radiotherapy did not improve disease-free or overall survival compared to radiotherapy after primary chemotherapy but gave important histological information of the response to primary chemotherapy, permitting lower doses of irradiation. However, it was pointed out that clinical response to primary chemotherapy is not an accurate prediction of histological response. Of nine patients who had complete clinical response, two had residual tumor in the breast or in the axilla.

Thus primary chemotherapy has improved the outcome of patients with inflammatory carcinoma of the breast if combined with radiotherapy and/or surgery and adjuvant chemotherhapy.

Primary Chemotherapy: Effect on Response of the Primary Tumor and on Method of Treatment

What are the results of primary chemotherapy in patients with less advanced breast carcinoma? Several large nonrandomized studies (>100 patients in each) have been published in which primary chemotherapy was followed by either radiotherapy of the primary tumor or surgical removal. Six of these studies are presented in Table 2. All patients were defined as having "operable" tumors and the results presented in the table refer to the effect on the primary tumor of primary chemotherapy and the method of local treatment. In the investigation by Bélembaogo et al. (1992), 126 patients "who otherwise would have undergone modified radical mastectomy" (MRM) received six cycles of AVCF or AVCFM every 3 weeks before local treatment of the tumor, which was primary radiotherapy in patients with complete remission (CR), tumorectomy in cases with partial response (PR) and MRM in patients with poor response. Mauriac et al. (1991) randomized 272 patients with operable breast carcinomas into two groups: Group A underwent total mastectomy and axillary node dissection (N = 138); group B underwent primary chemotherapy consisting of three cycles of EVM followed by three cycles of MTV every three weeks (N = 134). Group B is presented in Table 2. After primary chemotherapy, patients with CR were treated by radiotherapy alone, patients with PR (tumor <2 cm) with breast conservation surgery and axillary dissection. The remaining patients underwent MRM. Calais et al. (1994) treated 158 patients with operable large breast carcinomas with three cycles of NVCF or EVCF every 4 weeks. Patients with CR were treated by radiotherapy alone while patients with PR and tumors <3 cm were treated by tumorectomy and limited axillary dissection in N_1 patients. The remaining patients received MRM. Local-regional radiotherapy was given to all patients. In a study by Jacquillat et al. (1990), primary chemotherapy consisting of VTMFP ± A was given to 250 patients at intervals of 7–14 days and with three to six cycles according to the stage of the disease. Thereafter, *all* patients received radiotherapy, which always included iridium implants. Complete tumor regression was achieved clinically in all patients.

Table 2. Primary chemotherapy in operable breast cancer

Study	Number of patients	Primary chemotherapy	Primary radiotherapy	Conserving surgery + radiotherapy	Breasts conserved (%)	Clinical response CR (%)	PR (%)	Pathologic CR (%)	TNM classification (UICC 1988)
Bélembaogo et al. (1992)	126	AVCF/3 weeks ±M (six cycles)	41	64	83	36	49	1	T_1–T_3, N_1–N_2
Mauriac et al. (1991)	134	EVM/3 weeks + MTV/3 weeks (3 + 3 cycles)	44	40	62	33	37	?	T_2 (>3 cm) T_2 N_0–N_2 (?)
Calais et al. (1994)	158	NVCF (three cycles)	32	45	49	20	41	?	T_2 (>3 cm) T_3 N_0–N_1
Jacquillat et al. (1990)	250	VTMFP ± A (three to six cycles)	250	0	100	30	41	–	T_1–T_4 N_0–N_3
Schwartz et al. (1994)	189	CMF ± A (mean of seven cycles)	0	55	36	85		10	Stage II, IIA, IIB, IIB inflammaty
Bonadonna et al. (1990)	161	CMF, FAC FEC (three to four cycles)	0	142	88	10	66	4	$T_2T_3T_{4a}$ $N_0N_1N_2$

AVCF, •; VTMFP ± A, •; CR, complete response; PR, partial response.

Two studies in Table 2 evaluated primary chemotherapy followed by surgery. Schwartz et al. (1994) also included 30 patients with inflammatory breast carcinoma among their 189 patients. The basic regime used was cyclophosphamide, methotrexate, 5-fluorouracil (CMF) with the addition of doxyrubicin (A) in patients considered to be nonresponders after three cycles of CMF. If a response then was noted, A was substituted for M in later patients in the study, and responders continued to be treated until there was no additional objective decrease in the size of the primary tumor or axillary nodes. Radical mastectomy was the most common surgery at the start of the study in 1979; but with time and increasing response rates due to the intensified chemotherapy, breast conservation became more frequent and since 1990, approximately 75% of the patients have been treated by conservation surgery. Non-responders were treated by radiation without surgery (15%). As all patients responding to primary chemotherapy were operated on, material was obtained for histological examination in 156 cases. In 16 cases (10%) no residual tumor could be found in the breast and in 11 cases (7%) the diagnosis was ductal carcinoma in situ (DCIS) only. The relation between patients with CR and negative histology was not given.

Bonadonna et al. (1990) were among the first to introduce primary chemotherapy in advanced breast cancer, starting in 1973. A prospective study with primary chemotherapy was started in 1988 with the goal of substituting conservative for mutilating surgery in patients with breast carcinomas having a diameter of ⩾3 cm. Three combination regimes, CMF FAC or FEC, were given to five groups of 33 patients followed by surgery. CMF or FAC were given for three or four cycles, FEC for four cycles. Breast conservation surgery (quadrantectomy) was performed in 142 of 161 patients (88%) and mastectomy in patients with tumors ⩾3 cm in diameter after chemotherapy or in those with multifocal disease. All patients were subjected to axillary clearance. In spite of 10% CR and 66% PR, only seven patients were without evidence of disease based on microscopic examination. This study has been updated by Zurrida et al. (1994) and includes a larger number of patients.

All studies reported in Table 2 confirm that primary chemotherapy induces a high frequency of clinical response. However, in spite of 10%–36% CR, there are relatively few patients with complete histological response. In several of the studies this has not been evaluated mainly due to radiation therapy given to the tumor after primary chemotherapy. It has been possible to use breast conservation surgery with increasing frequency due to down-staging of the primary tumor.

Is it possible using primary chemotherapy to down-stage the disease in locally advanced breast carcinoma to safely permit breast conservation? To answer this question Singletary et al. (1992) carried out a retrospective study of 193 patients administered three cycles of primary chemotherapy (VACP) and treated thereafter by mastectomy and axillary clearance. The response to therapy was evaluated by clinical examination and mammography and correlated to the histological findings in the mastectomy specimen. A total of 33 patients had CR and 110 patients had PR. Only 23% of these (33 patients) were considered in retrospect to have been potential candidates for breast conservation surgery as judged by "clinical, histological and mammographic absence of persistent skin edema, large residual tumour size in relation to the size of the breast, multifocal disease and known extensive intramammary lymphatic invasion". In the noncandidates 50% had tumor in other quadrants.

Some 48% of the candidates had CR but only 6% of the noncandidates; 42% of the candidates had complete histological remission in the breast compared with 0% for noncandidates.

The effect of primary chemotherapy might, however, be increased by more effective chemotherapy combintions and/or increased dose intensity.

Primary Chemotherapy: Effect on the Primary Tumor and on Survival

A few randomized studies comparing primary chemotherapy with other forms of treatment have been performed during the last several years.

Scholl et al. (1994) presented a study in which 414 premenopausal patients with breast carcinomas too large to permit primary wide excision (T_2, T_3, N_0, N_1) were randomized in two groups: (1) Group A received FAC every 4 weeks for four cycles and after evaluation of response treatment by primary radiotherapy; wide local excision or mastectomy was performed in patients with poor response to radiotherapy after 54 Gy. (2) Group B underwent Primary irradiation + boost up to 8000 cGy and thereafter local excision or mastectomy if the tumor persisted. Adjuvant chemotherapy (four cycles of FEC) was then given after completion of the local treatment.

There were 390 patients who were considered as evaluable, 200 in group A and 190 in group B. Breast conservation was possible in 82% in group A and 77% in group B. Of the 154 patients in group A completing four FEC cycles of primary chemotherapy, 30% had CR and 50% PR. After primary irradiation in group B, 41% had CR and 44% PR.

After a median follow up of 54 months, the 5-year survival rates were 86% in group A and 78% in group B ($p = 0.04$), and the 5-year distant disease-free survival rates were 73% and 68%, respectively ($p = 0.09$). There were 73 mastectomies in group A (36 for initial treatment and 37 for local recurrence) and 66 in group B (44 for initial treatment and 22 for local recurrence). The local control rates were the same in both groups. Thus, in this study both methods gave the same rate of breast conservation but survival might be improved by primary chemotherapy. Longer observation time is, however, needed.

One of the studies, by Mauriac et al. (1991), was randomized. Group A ($N = 138$) had initial mastectomy and axillary node dissection (MRM) followed in node-positive patients or estrogen receptor- and progesterone receptor-negative patients by adjuvant chemotherapy (three cycels of EVM followed by three cycles with MTV). Group B (n = 134, presented in Table 2) had the same chemotherapy as during primary treatment. Local therapy was given according to the response to chemotherapy. There were 44 patients with CR who had local-regional radiotherapy only with boost to the breast, 40 patients with PR (tumor <2 cm after chemotherapy) who had breast conservation surgery + axillary node dissection and radiotherapy to the breast alone while the remaining patients (n = 49) had MRM without radiotherapy.

Patients with receptor-negative tumors had a better response to primary chemotherapy than those with receptor-positive tumors.

At a median follow-up of 34 months, there was no statistical difference in local-regional recurrences between the groups but there was a better overall survival in the group given primary chemotherapy (p = 0.04). The significance of this is, however, dubious as 32/136 patients in group A did not get adjuvant chemotherapy.

Response Rate and Dose Intensity

It is generally known that smaller tumors respond better to chemotherapy than larger tumors. Can increasing the dose intensity result in a better response rate of large tumors? In the study by Jacquillat et al. (1990), 250 patients with all stages of breast carcinoma were given primary chemotherapy consisting of VTMFP ± A. The series is described in Table 3. Group 1 was given VTMFP every 14 days for three cycles, group 2 VTMFP every 10 days for four cycles. Group 3 received VTMFPA every 7 days for four cycels and group 4 VTMFPA every 7 days fo six cycles. Due to toxcity delays of later cycles in Groups 2–4 were reqired. In none of the patients was a progression of the primary tumor observed during chemotherapy. As evident from Table 3 there was an impressive level of CR and PR in all groups.

In all groups, chemotherapy was followed by local-regional radiotherapy. In groups 3 and 4, more chemotherapy was given interspaced with radiothe-

Table 3. Neoadjuvant and consolidation VTMFP ± A: stratification in four therapeutic groups according to T and N (increasing dose intensity and number of cycles)

Group		*N*	Complete remission (%)	Partial remission (%)
1	T ≤ 3 cm	49	39	33
2	T > 3 cm or N1b	88	24	46
3	T > 7 cm or N2	41	29	42
4	Inflammation or N3	72	31	43

VTMFP ± A, •

rapy. A boost was given to the breast. Adjuvant chemotherapy was administered over 6.5 months in group 1 and up to 20 months in group 4.

Complete clinical tumor regression was achieved in all patients. At a median follow-up of 5 years, six locoregional metastases in combination with distant metastases and 27 isolated locoregional relapses had been observed in the 250 patients. Breast preservation rate at 5 years was 94%. The study also demonstrates impressive survival results compared to historical controls, results which are difficult to evaluate in the absence of a randomized control group.

Flores et al. (1993) treated 39 patients with T_3–T_4 or N_2 breast carcinomas with methotrexate, vinblastine, doxyrubicin, and cisplatin (MVAC) as primary chemotherapy followed by MRM. MVAC chemotherapy was given as follows: methotrexate 30 mg/m^2, days 1, 15 and 22; vinblastine 3 mg/m^2, days 2, 15 and 22: adriamycin (doxyrubicin) 30 mg/m^2, day 2; and cisplatin 70 mg/m^2, day 2. Folinic acid (10 mg every 6 h × 6) was given 24 h after methotrexate. Of the 36 patients who were followed, 14 had CR and 22 had PR. Two patients refused surgery. Of the remaining 34 patients, eight (24%) had a complete pathological response in breast and lymph nodes.

Brun et al. (1993) treated 19 patients with epirubicin, cyclophosphamide, fluorouracil (ECF), one cycle every 3 weeks for three cycles. The dosaqes were as follows: epirubicin 35 mg/m^2; cyclophosphamide 400 mg/m^2 i.v., days 2, 3 and 4; fluorouracil 750 mg/m^2 as a continuous infusion, days 1, 2, 3 and 4. CR was obtained in five patients and PR in eight. Of these 13 patients, 11 underwent MRM and six had complete pathological responses (31% of the whlol group).

Long-term (6 months) continuous infusion fo 5-fluorouracil (200 mg/m^2 per day) combined with epirubicin (50 mg/m^2 i.v. bolus every 3 weeks) and cisplatin (60 mg/m^2 i.v. bolus every 3 weeks) in 50 patients with large operable breast cancer was studied by Smith et al. (1994). There were 48 patients who had PR (96%), with 34 (68%) achieving CR. Of 39 tumor excised, 28% had comlete pathological remission. Radiotherapy was given to the breast. Surgery is no longer carried out in patients with CR. The regime is now being tested in a randomized trial.

A total of 59 patients who had breast carcinoma, with tumor size >3 cm or situated in the central area of the breast, were treated by primary chemotherapy including navelbine (25 mg/m^2), epirubicin (35 mg/m^2) and methotrexate was given on days 1 and 8 every 28 days for an average number of six cycles (Van Praagh et al. 1995). CR was obtained in 16 patients and PR in 27. Of these 59 patients, 50 had conservative surgery and nine had MRM. Complete pathological remission was achieved in 12 patients (20.3%).

Conventional chemotherapy given as primary treatment induces between 10% and 20% CR and up to 70% PR, but only few pathologically complete responses. More intensified chemotherapy seems to induce higher response rates, in some studies up to a total of 100% for CR and PR. This is followed by a complete pathological response rate of up to 30%.

Long-Term Results with Primary Chemotherapy

In a study with a 10 year follow-up, Valagussa et al. (1990) presented data from 277 patients treated by primary chemotherapy for advanced noninflammatory breast carcinoma considered unsuitable for primary surgical therapy.

In the first 72 patients (group 1), primary chemotherapy consisted of doxyrubiciny (75 mg/m^2 day) and vincristine (1.2 mg/m^2 days 1 and 8 of each cycle of 21 days for four cycles). CR + PR was 68% in this group. Local-regional radiotherapy was then given to all patients, with a boost to the breast.

The remaining 205 patients received doxyrubicin (60 mg/m^2) and vincristine (1.2 mg/m^2) days 1 and 8 every 3 weeks for three cycles. CR + PR was 57% in this group. Some 126 of these patients (group 2) were treated by local-regional radiotherapy with a boost to the breast, and 79 patients (group 3) were treated by radical mastectomy. These 205 patients thereafter received adjuvant chemotherapy consisting of six to seven additional cycles of doxyrubicin (50 mg/m^2) and vincristine (1.2 mg/m^2). Patients with only a minor response to primary chemotherapy were treated with CMF for six cycles.

Local-regional relapse was 52% in group 1, 37% in group 2, and 20% in group 3, indicating the best local control after surgery. Better relapse-free survival and overall survival were obtained by the addition of adjuvant chemotherapy in groups 2 and 3.

In the British Columbia study (Ragaz et al. 1994), the 10 year overall survival was analyzed in 74 patients with stage III breast cancer who were given primary chemotherapy followed by local-regional radiotherapy at 3 months and thereafter MRM. Chemotherapy consisted of ACMLF over 6 months (adriamycin, 40 mg/m^2 i.v. day 1; cyclophosphamide, 200 mg/m^2 po days 2–5; methotrexate, 350 mg/m^2 i.v., and leucovorin day 15; 5-fluorouracil, 600 mg/m^2 i.v. day 15 with cycle 28 days). ER-positive patients had tamoxifen until relapse or indefinitely.

Combined complete and partial responses before MRM was 89% (CR 39 and PR 27 patients). Pathological CR was seen in 16 patients (22%); of these, 25% recurred compared to 60% with less than pathological CR. In the 39 patients with CR, pathological CR was seen in 31%. In this study the pathological CR was more accurate than clinical CR in predicting for relapse. The conclusion was also that mastectomy is important after primary chemotherapy combined with radiotherapy, as 69% of patients with CR will have remaining residual disease in stage III breast cancer.

The use of primary chemotherapy in different stages of breast cancer has evoked a great deal of interest during later years. The majority of these studies are small, with less than 100 patients. Astonishingly, very few studies have been randomized comparing, e.g., primary chemotherapy and adjuvant chemotherapy using the same local-regional treatment. In the randomized study reported by Fischer et al. (1994), primary chemotherapy followed by surgery was compared with primary surgery and adjuvant chemotherapy. Chemotherapy was the same in both arms, i.e., AC every 3 weeks for 4 cycles: adriamycin (60 mg/m^2) and cyclophosphamide (600 mg/m^2).

The study started in October 1988 and by March 1993 included 1313 operable patients. In group A, 549 were evaluated concerning tumor response, with 37% CR and 43% PR. The pathological CR was 15%. The response rate was higher in patients with small tumors (<2 cm; CR 50%) than in those with large ones. Breast conservation surgery was performed in 65% in group A and 57% in group B. A down-staging of axillary nodes was also noted in group A, with 59% negative nodes compared to 42% negative nodes in group B. No data were given for disease-free survival or survival, but such data are awaited with great interest.

Conclusion

Primary chemotherapy reduces the tumor burden in advanced breast carcinoma in the majority of patients. Local therapy, either by radiotherapy or surgery or both, will then be more effective and will reduce the risk for local-regional recurrence in responding patients. Data on inflammatory breast carcinoma are impressive and disease-free survival and survival have been improved without question.

Primary chemotherapy is followed by down-staging of the breast cancer in a large number of patients, making breast conserving surgery possible. Local radiotherapy of the breast tumor might also be possible thereby avoiding surgery, but then important data on the effect of the given chemotherapy on the tumor will not be available (complete pathological response). This might be important for, e.g., the choice of adjuvant chemotherapy after local treatment. Surgery after primary chemotherapy might also reduce the need for high dose radiotherapy with its side effects.

In primary operable breast cancer, the value of primary chemotherapy has not been proven. No reliable data are available concerning effects on disease-free and overall survival. Thus, primary chemotherapy cannot be recommended for general use and should only be given within randomized studies. While the number of breast conservation procedures has increased, in most patients the primary tumor is small at diagnosis often as an effect of screening by mammography. Breast conservation surgery is possible in about 70% of the patients in our clinic, without primary chemotherapy.

The rather low number of pathological CRs (around 15%) after primary chemotherapy (standard combination chemotherapy) might be improved by more effective combinations and treatment schedules and by increasing dose intensity. This low CR rate after primary chemotherapy necessitates local therapy of the breast. In my opinion the form of therapy should still be surgery, if possible conserving the breast, as important data on biological properties of the tumor can be evaluated on the excised tissue. Radiotherapy should be given after surgery in the majority of patients if the breast is preserved.

Before treatment of breast carcinoma with primry chemotherapy, it is essential to have a confirmed diagnosis, preferably a biopsy with enough material to permit analysis of prognostic and predictive factors in the primary tumor.

References

Bélembaogo E, Feillel V, Chollet P et al. (1992) Neoadjuvant chemotherapy in 126 operable breast cancers. Eur J Cancer 28A:896–900

Bonadonna G, Veronesi U, Brambilla C et al. (1990) Primary chemotherapy to avoid mastectomy in tumors with diameters of three centimeters or more. J Natl Cancer Inst 82:1539–1545

Brun B, Poitevin A, Kassem M et al. (1993) High dose ambulatory neoadjuvant chemotherapy in poor prognosis breast cancer patients. Proc ASCO 12:87

Calais G, Berger C, Descamps P et al. (1994) Conservative treatment feasibility with induction chemotherapy, surgery, and radiotherapy for patients with breast carcinoma larger than 3 cm. Cancer 74:1283–1288

Fisher B, Rockette H, Robidoux A et al. (1994) Effect of preoperative therapy of breast cancer (BC) on local-regional disease: first report of NSABP B-18. Proc ASCO 13:64

Flores MR, Morrell L, Richman S et al. (1993) Neoadjuvant methotrexate, vinblastine, doxorubicin and cisplatin (MVAC) for locally advanced breast cancer (LABC). Proc ASCO 12:102

Jacquillat C, Weil M, Baillet F et al. (1990) Results of neoadjuvant chemotherapy and radiation therapy in the breast-conserving treatment of 250 patients with all stages of infiltrative breast cancer. Cancer 66:119–129

Kurtz JM (1991) Should surgery remain the initial treatment of "operable" breast cancer, Eur J Cancer 27:1539–1542

Mauriac L, Durand M, Avril A, Dihuydy J-M (1991) Effects of primary chemotherapy in conservative treatment of breast cancer patients with operable tumors larger than 3 cm. Ann Oncol 2:347–354

Papaioannou AN (1985) Preoperative chemotherapy: advantages and clinical application in stage III breast cancer. In: Metzger U, Largiadér F, Senn HJ (eds) Perioperative chemotherapy. Rationale, risk and results. Springer, Bertin Heidelbary New York, pp 65–90 (Recent results in cancer research, vol 98)

Ragaz J, Baird R, Rebbeck R, Goldie A, Coldman A, Spinelli J (1985) Preoperative adjuvant chemotherapy (neoadjuvant) for carcinoma of the breast: rationale and safety report. In: Metzger U, Largiadér F, Senn HJ (eds) Perioperative chemotherapy. Rationale, risk and results. Springer, Bertin Heidelbery New York, pp 99–105 (Recent results in cancer research, vol 98)

Ragaz J, Manji M, Plenderleith IH et al. (1994) Impact of the residual disease in stage III breast cancer patients treated with preoperative (neoadjuvant) chemotherapy (CT), radiotherapy (XRT) and mastectomy. The 10 year analysis of the British Columbia study. Proc ASCO 13:71

Schabel FM (1985) Rationale for perioperative anticancer treatment. In: Metzger U, Largiadér F, Senn HJ (eds) Perioperative chemotherapy. Rationale, risk and results. Springer, Bertin Heidelbery New York, pp 1–10 (Recent results in cancer research, vol 98)

Scholl SM, Fourquet A, Asselain B et al. (1994) Neoadjuvant versus adjuvant chemotherapy in premenopausal patients with tumours considered too large for breast conserving surgery: preliminary results of a randomised trial: S6. Eur J Cancer 30A:645–652

Schwartz GF, Birchansky CA, Komarnicky LT et al. (1994) Induction chemotherapy followed by breast conservation for locally advanced carcinoma of the breast. Cancer 73:362–369

Singletary SE, McNeese MD, Hortobagyi GN (1992) Feasibility of breast-conservation surgery after induction chemotherapy for locally advanced breast carcinoma. Cancer 69:2849–2852

Singletary SE, Ames FC, Buzdar AU (1994) Management of inflammatory breast cancer. World J Surg 18:87–92

Smith IE, Jones AL, Prendiville J, Walsh G (1994) Primary continuous infusional chemotherapy as an alternative to mastectomy or any surgery for large operable breast cancer. Proc ASCO 13:69

Valagussa P, Zambetti M, Bonadonna G, Zucali R, Mezzanotte G, Veronesi U (1990) Prognostic factors in locally advanced noninflammatory breast cancer. Long-term results following primary chemotherapy. Breast Cancer Res Treatment 15:137–147

Van Praagh I, Feillel V, Cure H et al. (1995) Neoadjuvant NEM chemotherapy regimen for operable breast cancer: results of a co-operative study on 59 patients. 5th international congress on anti-cancer chemotherapy, Paris, O531

Zurrida S, Greco M, Veronesi U (1994) Surgical pitfalls after preoperative chemotherapy in large size breast cancer. Eur J Surg Onc 20:641–643

Summary

W. Wood

Department of Surgery, Emory University School of Medicine, Emory University Hospital, 1364 Clifton Road, N.E., Atlanta, GA 30322, USA

Dr. Fisher's review of the events of the last 2 years from the standpoint of the NSABP illustrates graphically the complexities of performing clinical trials in the public arena. It is important to note that extensive auditing of the NSABP data set for their clinical trial of breast conservation therapy (B-06) was performed by an independent company. The data were validated by this arduous investigation to a remarkable degree. The results of this audit and its review by an independent panel of experts will soon be published by the National Cancer Institute, Bethesda. The NCI held a conference in November, 1994, in which these results and the results from the other randomized trials of breast conservation therapy that formed the basis of the 1990 Consensus Development Conference recommendations were updated. With additional follow-up the same conclusion was sustained, i.e. that there was no difference in survival between women with stage I and stage II breast cancer treated by mastectomy or by breast conservation therapy (involving tumorectomy, axillary staging, and breast irradiation).

With the increasing use of breast conservation therapy more women are seen with breast cancer in a previously treated breast. This represents a recurrence of residual tumor in some women, a new primary breast cancer in others. Dr. Margolese reviewed the importance of restaging such patients. Those who have no evidence of metastatic disease should, again, be treated with curative intent. This goal is apparent. The approach to therapy, on the other hand, is by analogy to primary breast cancer at first treatment. If breast irradiation was never performed, it can now be used together with surgical excision to clear margins. Mastectomy will salvage some women who have had prior breast irradiation. The role of excision alone in the previously irradiated breast is being reported in anecdotal fashion. The case reports presently available do not suggest that this will generally succeed in controlling local disease. The value of a second program of adjuvant systemic therapy is also debated based on first principles rather than any comparative data.

There is a puzzling disparity in reports of survival after local failure in these randomized trials. In the EORTC trial the survival curves after local failure in the breast after BCT or on the chest wall after mastectomy are similar. Memorial Sloan Kettering has reported similar results, but not from randomized trials. The NSABP and the NCI, Bethesda, report more favorable survival when local failure follows BCT rather than following mastectomy. That would suggest that there are new primary cancers and recurrent local disease within the breast that are susceptible to control in addition to those chest wall tumors and the local evidence of systemic disease that present after mastectomy.

There are now sufficient numbers of women treated with immediate reconstruction after mastectomy to conclude that there is no evidence of detrimental effect. Performing the reconstruction at the same time as the mastectomy spares the patient a second operation but also allows skin-sparing mastectomy. This provides a superior cosmetic result. There is no evidence that reconstruction with silastic prostheses is better or worse than reconstruction with native tissue in terms of later health and freedom from recurrent disease.

Current research in breast cancer surgical techniques is focused on two major issues. The first is striking the optimal balance between cosmetic outcome and local control of tumor. Prevention of recurrence is accomplished by wider margins of resection, higher doses of irradiation, and aided by adjuvant systemic therapy. These are the same factors that adversely influence cosmetic outcome. Improved biologic characterization of the primary tumor may allow more accurate discrimination among tumors that require wider margins, those most susceptible to irradiation, and those that require both. The relative cosmetic effects of boost irradiation with a small surgical margin compared with less irradiation and a wider surgical margin relate to breast size and shape but are just beginning to be studied in a meaningful way. Breast reshaping as part of breast conservation is also still in evolution.

The second focus of research is the role of axillary nodal surgery. Sentinel lymph node biopsy, as described by Morton, is under investigation. This involves injecting a tracer (colored dye or radioisotope) into the primary tumor. Only the initial draining lymph node is then excised and biopsied. It is proposed that if this "sentinel" node is not involved, the axilla will be free of nodal metastases. Clinical trials are underway to evaluate this hypothesis. If it is confirmed, it is possible that axillary dissection may be used only in patients found to have involved lymph nodes by this technique. Integrations of the phenotypical characteristics of the primary tumor (prognostic factors) are being studied as a technique to replace the staging value of axillary lymph node dissections. At present the latter have more prognostic power than any combination of descriptors of the primary tumor.

VI. Adjuvant Systemic Therapy

The Role of Anthracyclines in Adjuvant Chemotherapy of Breast Cancer: A Critical Appraisal

H.T. Mouridsen

Dept. of Oncology 5074, Rigshospitalet, 9 Blegdamsvej, 2100 Copenhagen, Denmark

Introduction

Analysis of results of ongoing and future clinical trials will answer many questions regarding the optimal cytotoxic chemotherapy in primary breast cancer. These questions include: optimal time of chemotherapy in relation to primary surgery, the dose-response relationship, whether to use sequential or concomitant cytotoxic and endocrine therapy, and the role of newer drugs such as the taxoids and vinorelbine. This review will focus on the role of the anthracyclines in the adjuvant setting, as assessed from the data of published randomized trials.

Cytotoxic Treatment of Advanced and Primary Breast Cancer

With present available cytotoxic therapies response can be achieved in 50%–60% of patients. Median time to progression is 6–12 months and median survival 12–24 months. Only a small minority of patients will experience survival exceeding 5–10 years (Mouridsen 1992).

In the adjuvant setting, cytotoxic therapy can induce a relative reduction of mortality at 10 years of 25% in premenopausal and 12% in postmenopausal patients (Early Breast Cancer Trialists' Collaborative Group 1992). It still remains to be analyzed whether the observed differences in relation to age may be ascribed to differences in dose intensities administered to the different age groups.

Thus the major objective of treatment of advanced disease is to achieve palliation, applying knowledge gained from the results to develop new regimens to be used in the adjuvant setting. Although the objective of the adjuvant treatment is cure, it is evident from the available data that we are still far from having achieved a dramatic improvement of the prognosis of primary breast cancer.

As a consequence, the most effective among the available chemotherapy regimens should be offered to patients in the adjuvant setting. The curative intent of the adjuvant therapy, however, also requires careful consideration of any clinically relevant long-term adverse events associated with the treatment, problems which has less relevance in the treatment of advanced disease.

Anthracyclines in Advanced Disease

In treating advanced breast cancer the anthracyclines are among the most efficient single agents and three of four major trials have demonstrated that regimens of CAF (cyclophosphamide, doxorubicin, 5-fluorouracil) were significantly superior to similar combinations of CMF (cyclophosphamide, methotrexate, 5-fluorouracil) with regard to time to progression (10–12 months vs 6–8 months), response rate (55%–82% vs 37%–62%) and survival (19–26 months vs 14–20 months) (Muss et al. 1978; Brincker et al. 1984; Tormey et al. 1984; Aisner et al. 1987). In addition two large randomized trials have demonstrated CAF and CEF (cyclofosfamide, epirubicin, 5-fluorouracil) to be similar in terms of time to progression (9–10 months vs 7–9 months), response rate (52%–57% vs 50%–54%) and survival (18–20 months vs 15–20 months) (French Epirubicin Study Group 1988; Italian Multicenter Breast Study with Epirubicin 1988).

Introducing Anthracyclines in Adjuvant Treatment

A logical consequence of the findings regarding treatment of advanced disease is the introduction of anthracyclines in the adjuvant setting. However, many investigators have been reluctant to take this step, mainly because of the potential risk of cardiotoxicity associated with long-term administration of these agents. The demonstration of lack of benefit when adjuvant chemotherapy is prolonged beyond approximately 6 months, and the introduction of epirubicin, which at hematologically equitoxic doses is less cardiotoxic than doxorubicin (Mouridsen et al. 1990), however, has changed this situation.

Analyses of the risk of congestive heart failure following treatment with anthracyclines have demonstrated a slight but steady increase to approximately 5% at cumulative doses of doxorubicin and epirubicin of 550 mg/m^2 and 1000 mg/m^2, respectively (i.e., a ratio of 1:1.8), followed by a rapid increase with higher doses (Praga et al. 1991). As a consequence, 550 mg/m^2 doxorubicin and 1000 mg/m^2 epirubicin have generally been recommended as the maximum cumulative doses to be administered in the treatment of advanced disease.

If the drugs are used as adjuvant therapy in combination with cyclophosphamide and 5-fluorouracil, administered at standard dose intensity every 3 weeks for nine cycles, the cumulative doses would be 360 mg/m^2 (9 × 40 mg/m^2) for doxorubicin and 540 mg/m^2 (9 × 60 mg/m^2) for epirubicin, equivalent to 65% and 54% of the maximum recommended doses, respectively.

Problems in Interpreting the Data of Available Randomized Trials

To interpret the specific role of anthracyclines, whether as a single agent or in combination, compared to other available therapies would optimally require data from randomized trials comparing regimens which differ only with regard to this specific variable. However, as will be discussed later in this review, many of the randomized trials compare regimens that, apart from the anthracycline, differ with regard to other variables such as number of drugs, schedules, and dose intensities.

Furthermore the majority of the trials have included an insufficient number of patients to enable safe conclusions. In the treatment of advanced disease, the response rate achieved with CMF is approximately 50% and with CAF or CEF approximately 60% (Mouridsen 1992). In the adjuvant setting, therefore, an extremely optimistic guess regarding benefit with an anthracycline containing regimen would be at most a 20% relative reduction in mortality. To demonstrate this improvement in a clinical trial using an α value of 0.05, a β-value of 0.10 and thus a power of 90% would require approximately 1000 patients. The published randomized trials only exceptionally include that number of patients, giving rise to a significant risk of false negative and false positive results. This problem may also contribute substantially to publication bias.

Trials Indicating Superior Efficacy with Adjuvant Anthracycline

Four published trials have reported a benefit concerning recurrence-free survival or survival or both with the use of anthracyclines (Table 1). However, the design of some of these trials does not justify safe conclusions as to the role of the anthracyclines.

The NSABP study (Fisher et al. 1989) included patients with node-positive tamoxifen "nonresponsive" tumors, i.e. patients $\leqslant$49 years except those with tumors having both estrogen receptor (ER) and progesterone receptor (PgR) levels $\geqslant$10 fmol and patients 50–59 years with a tumor PgR level of 0–9 fmol. These patients were randomized to PF (L-pam + fluorouracil) for 2 years vs PF with the addition of doxorubicin until a maximum cumulative dose of 300 mg/m^2 was reached. In this trial the benefit with the anthracycline may be associated with the use of three rather than two drugs.

The study from Milan (Buzzoni et al. 1991) recruited patients $\leqslant$70 years who had tumors with more than three positive nodes, irrespective of

Table 1. Trials demonstrating superior efficacy with adjuvant anthracycline

Treatment	Number of patients	Follow-up (years)	Reference
PF/2 years vs PF/2 years + dox (≤300 mg/m²)	697	7	Fisher et al. (1989)
A/4 cycles, CMF/8 cycles vs (CMF/2 cycles, A/1 cycle) × 4	338	2	Buzzoni et al. (1991)
CMF/12 cycles vs AVCF/12 cycles	249	10	Misset et al. (1992)
MMFC/4 cycles vs MMAC/4 cycles	121	8	Beuzeboc et al. (1992)

PF, L-pam, fluorouracil; A, anthracycline (doxorubicin); CMF, cyclophosphamide, methotrexate, 5-fluorouracil; AVCF, anthracycline, vincristine, cyclophosphamide, 5-fluorouracil; MMFC, methotrexate, mitomycin C, 5-fluorouracil, cyclophosphamide; MMAC, methotrexate, mitomycin C, anthracycline (doxorubicin), cyclophosphamide.

receptors. The patients were randomized to four cycles of doxorubicin followed by eight cycles of CMF vs the sequential use of two cycles of CMF followed be one cycle of doxorubicin, the sequence being repeated four times. In this trial the benefit achieved with anthracycline could rather be ascribed to the sequence of the drugs rather than to the role of doxorubicin in itself.

The study conducted by Misset et al. (1992) included pre- and postmenopausal women with node-positive tumors. In this trial 12 cycles of classical CMF were inferior to 12 cycles of AVCF (V, vincristine); however, the latter regimen was administered according to a significantly different schedule (doxorubicin day 1, vincristine day 2, CF days 3–6).

The final study, by Beuzeboc et al. (1992), included a subset of high-risk premenopausal women (<35 years, grade III, no ERs or PgRs and > three positive nodes) randomized to receive a combination of MMFC (methotrexate, mitomycin C, 5-fluorouracil, cyclophosphamide) or with MMAC (A, doxorubicin). However, in this comparison 5-fluorouracil was administered five times and doxorubicin on day 1 only during the 7 week cycles.

Trials Indicating No Superior Efficacy with Adjuvant Anthracyclines

So far six trials have been published, but these also compared various regimens or schedules which prevent exact evaluation of the role of the anthracycline.

The NSABP study (Fisher et al. 1990) recruited patients with node-positive tumors that were tamoxifen "nonresponsive" (≤49 years and those

Table 2. Trials demonstrating no superior efficacy with adjuvant anthracycline

Treatment	Number of patients	Follow-up (years)	Reference
PF + TAM vs PAF + TAM	1093	7	Fisher et al. (1989)
AC/4 cycles vs CMF/6 cycles	1542	3.5	Fisher et al. (1990)
CMF/12 cycles vs CMF/8 cycles, A/4 cycles	486	5	Moliterni et al. (1991)
CMF/9 cycles vs MTV/3cycles, EVM/3 cycles	228	5	Mauriac et al. (1992)
CMF (i.v.)/6 cycles vs CAF (i.v.)/ 6 cycles	528	5	Carpenter et al. (1994)
CMF (classical)/ 6 cycles vs CEF (i.v.)/8 cycles	760	4.5	Marty et al. (1994)

PF, L-pam, fluorouracil; TAM, tamoxifen; PAF, L-pam, anthracycline (doxorubicin), fluorouracil; AC, anthracycline (doxorubicin), cyclophosphamide; CMF, cyclophosphamide, methotrexate, 5-fluorouracil; A, anthracycline (doxorubicin); MTV, mitomycin C, thio-TEPA, vindesine; EVM, epirubicin, vincristine, methotrexate; CAF, cyclophosphamide, anthracycline (doxorubicin), 5-fluorouracil; CEF, cyclophosphamide, epirubicin, 5-fluorouracil.

50–59 years having tumors with a PgR less than 10 fmol/mg cytosol protein, regardless of ER status). Although no significant differences were observed, the authors concluded that AC (doxorubicin, cyclophosphamide) is preferable due to the shorter time spent with chemotherapy and based on the comparative evaluation of toxicities.

The French trial (Mauriac et al. 1992) included premenopausal and postmenopausal patients with node-positive tumors that were smaller than 3 cm and with no ERs or PgRs. The trial compared nine cycles of i.v. CMF with three cycles of MTV (mitomycin C, thio-TEPA, vindesine) followed by three courses of EVM (epirubicin, vincristine and methotrexate). The completely different schedules and the small number of patients enable no safe conclusions.

Another NSABP trial (Fisher et al. 1989) recruited patients with node-positive tumors who were tamoxifen responsive (women 50–59 years with PgRs ≥10 fmol regardless of ERs and women 60–70 years regardless of ERs or PgRs; in addition, those ≤49 years with tumors having ERs and PgRs ≥10 fmol). The patients were randomized to 2 years of PF + tamoxifen 20 mg daily continued until treatment failure vs the same regimen with the addition of doxorubicin to a maximum cumulative dose of 300 mg/m^2. The authors argued that the lack of benefit with doxorubicin might be related to tamoxifen antagonism of the doxorubicin, which needs to be clarified in clinical trials.

The criteria of eligibility in the trial from Milan (Moliterni et al. 1991) included age <70 years and up to three positive nodes. The patients were randomized to 12 cycles of i.v. CMF vs eight cycles of i.v. CMF followed by four cycles of doxorubicin. The authors concluded that, with the expense of minimal toxicity, full dose CMF remains the adjuvant treatment of choice for that specific population of breast cancer patients.

Another two studies compared CMF with CAF or CEF. The Southeastern Cancer Study Group (Carpenter et al. 1994) randomized premenopausal and postmenopausal patients with positive nodes to six i.v. cycles of either CMF (600, 40, 600 mg i.v. day 1 q 3 weeks) or six i.v. cycles of CAF (600, 40, 600 mg i.v. day 1 q 3 weeks). With equal myelotoxicity similar efficacy was achieved with the two regimens. The study conducted by the International Collaborative Cancer Group (Marty et al. 1994) recruited premenopausal patients with node-positive tumors; the patients were randomized to two different schedules of CMF or CEF. Either six classical courses of CMF q 4 weeks vs eight i.v. courses of CEF every 3 weeks, or i.v. CMF days 1+8 q 4 weeks vs CEF day 1 and CF day 8 repeated every 4 weeks. The CMF and CEF regimens were equally effective but, based on preliminary analyses of the schedule–dependent activity, the authors concluded that anthracycline-containing regimens may lead to highly variable results in the adjuvant treatment of breast cancer.

Finally, two major trials are ongoing in high risk patients. One randomizes patients to six classical courses of CMF or CEF (Levine et al. 1993) and one randomizes patients to nine courses of i.v. CMF or CEF (Danish Breast Cancer Cooperative Group 1989). Results from these two studies are not yet available.

Long-Term Toxicity with Adjuvant Anthracycline

Concern has been expressed about potential, clinically significant long-term toxicities associated with adjuvant treatment with anthracyclines, which include cardiotoxicity and acute myeloid leukemia.

Less than 5% of patients treated for advanced disease develop congestive heart failure provided cumulative doses of less than 550 mg/m^2 of doxorubicin and 1000 mg/m^2 of epirubicin are administered (Praga et al. 1991). In patients who develop congestive heart failure following treatment with anthracycline, specific risk factors other than the cumulative dose have often been identified, such as prior mediastinal irradiation, age over 70 years and prior abnormal left ventricular function.

However, endomyocardial biopsy studies have demonstrated that pathological heart changes occur after administration of significantly lower cumulative doses and before any clinical symptoms or signs develop (Billingham et al. 1978). These early changes are likely to occur during adjuvant treatment of breast cancer, but their clinical relevance, in terms of late complications such as congestive heart failure or arrhythmias, remains to be analyzed.

Data reported from Milan (Bonadonna and Valagussa 1992) showed occurrence of cardiac events in 6.4% of patients who had received adjuvant CMF compared to 8.1% in those who had received CMF + doxorubicin. Radiotherapy in patients with carcinoma of the left breast and age >55 years were significant risk factors for development of cardiac events. A recent update from the same group, involving 501 patients who had received a mean dose of doxorubicin of 259 mg/m^2, demonstrated congestive heart failure in only four (0.8%) of the patients (Valagussa et al. 1994). However, recent investigations have raised concerns for all patients with prior exposure to anthracyclines (Robison 1994). Thus, according to experience from long-term survivors of childhood cancer and Hodgkin's disease, about 10% of the patients who survive 10 years after treatment may experience clinically significant congestive heart failure or arrhythmia (Sprinz et al. 1988; Lipschultz et al. 1991; Steinherz et al. 1991; Aviles et al. 1993). This subject has to be carefully analyzed in ongoing and future trials of anthracycline-based adjuvant therapy of breast cancer.

Acute myeloid leukemia (AML) is known to occur in cancer patients treated with DNA-targeted antiproliferative drugs which are known to exert their cytotoxic activity through alkylation (such as cyclophosphamide, melphalan, cis-platinum) or topoisomerase-2 inhibitors (such as etoposide, anthracyclines) (Curtis et al. 1992; Levine and Bloomfield 1992; Ratain and Rowley 1992). However, according to the available data, it is very rare event which has to be balanced against the significant gain in survival achieved with adjuvant chemotherapy of breast cancer.

The clinical data available for breast cancer have recently been reviewed (Borg 1994). In the cases reported so far, anthracyclines were administered either in combination with DNA-targeted drugs or following prior treatment with alkylating agents, which makes it very difficult to evaluate the role of the anthracyclines per se in the genesis of AML. The same relates to recent data indicating that topoisomerase inhibitors, in combination with cis-platinum or other alkylating agents, can also induce acute lymphoid leukemia (Pedersen-Bjergård 1992).

Conclusions

Based on presently available data, no firm conclusions can be drawn about the specific role of anthracyclines in the adjuvant therapy of breast cancer. This can mainly be ascribed to the fact that many of the published trials were not specifically designed to evaluate the role of anthracyclines. Thus, we have to await the data from ongoing randomized trials designed specifically to analyse that question. Furthermore, the demonstration of the superiority of the sequential use of doxorubicin followed by CMF should be confirmed. The long-term evaluations should also include careful analyses of potential late adverse events, especially cardiotoxicity.

References

Aisner J, Weinberg V, Perloff M et al. (1987) Chemotherapy versus chemoimmunotherapy (CAF v CAFV v CMF each +/− MER) for metastatic carcinoma of the breast. A CALGB study. J Clin Oncol 5:1523–1533

Aviles A, Areevila N, Diaz Maqueo JC et al. (1993) Late cardiac toxicity of doxorubicin, epirubicin and mitoxantrone therapy for Hodgkin's disease in adults. Leuk Lymphoma 11:275–279

Beuzeboc P, Mosseri V, Dorval T et al. (1992) Adriamycin based combination chemotherapy significantly improves overall survival in high risk premenopausal breast cancer patients (Abstr). 4th International Conference on Adjuvant Therapy of Primary Breast Cancer, Feb 26–29, St Gallen, p 72

Billingham ME, Mason JW, Bristow MR, Daniels JR (1978) Anthracycline cardiomyopathy monitored by morphologic changes. Cancer Treat Rep 62:865–872

Bonadonna G, Valagussa P (1992) Breast cancer: adjuvant treatment with anthracyclines. Ann Oncol 3:417–418

Borg KO (1994) Doxorubicin-epirubicin and secundary acute myeloid leukemia in breast cancer patients. Company position paper. Pharmacia, Milan

Brincker H, Rose C, von der Maase H, Dombernowsky P (1984) A randomized study of CAF + TAM (tamoxifen) versus CMF + TAM in metastatic breast cancer. Proc Am Soc Clin Oncol 3:113

Buzzoni R, Bonadonna G, Valagussa P, Zambetti M (1991) Adjuvant chemotherapy with doxorubicin plus cyclophosphamide, methotrexate and fluorouracil in the treatment of resectable breast cancer with more than three positive axillary nodes. J Clin Oncol 9:2134–2140

Carpenter JT, Velez-Garcia E, Aron BS et al. (1994) Five-year results of a randomized comparison of cyclophosphamide, doxorubicin (adriamycin) and fluorouracil (CAF) vs. cyclophosphamide, methotrexate and fluorouracil (CMF) for node positive breast cancer: a Southeastern Cancer Study Group study. Proc Am Soc Clin Oncol 13:66

Curtis RE, Boice JD Jr, Stovall M et al. (1992) Risk of leukemia after chemotherapy and radiation treatment for breast cancer. N Engl J Med 326:1745–1751

Early Breast Cancer Trialists' Collaborative Group (1992) Systemic treatment of early breast cancer by hormone, cytotoxic, or immunetherapy. Lancet 339:71–85

Fisher B, Redmond C, Wickerham DL et al. (1989) Doxorubicin-containing regimens for the treatment of stage II breast cancer: the National Surgical Adjuvant Breast and Bowel Project experience. J Clin Oncol 7:572–582

Fisher B, Brown AM, Dimitrov NV (1990) Two months of doxorubicin-cyclophosphamide with and without interval reinduction therapy compared with 6 months of cyclophosphamide, methotrexate and fluorouracil in positive-node brease cancer patients with tamoxifen-nonresponsive tumors: results from the National Surgical Adjuvant Breast and Bowel Project B-15. J Clin Oncol 8:1483–1496

French Epirubicin Study Group (1988) A prospective randomized phase III trial comparing combination chemotherapy with cyclophosphamide, fluorouracil and either doxorubicin or epirubicin. J Clin Oncol 4:679–688

Italian Multicenter Breast Study with Epirubicin (1988) Phase III randomized study of fluorouracil, epirubicin and cyclophosphamide vs. fluorouracil, doxorubicin and

cyclophosphamide in advanced breast cancer. An Italian multicentre trial. J Clin Oncol 6:976–982
Levine EG, Bloomfield CD (1992) Leukemias and myelodysplastic syndromes secondary to drug, radiation and environmental exposure. Semin Oncol 19:47–84
Levine MN, Bramwell V, Pritchard K et al. (1993) A pilot study of intensive cycloprosphamide, epirubicin and fluorouracil in patients with axillary node positive or locally advanced breast cancer. Eur J Cancer 29A:37–43
Lipshultz S, Colan S, Gelber RD et al. (1991) Late cardiac effects of doxorubicin therapy for acute lymphoblastic leukemia in childhood. N Engl J Med 324:808–815
Marty M, Bliss JM, Coombes RC et al. (1994) Cyclophosphamide (C), methotrexate (M), fluorouracil (F)(CMF) versus F-epirubicin (E)-C(FEC) chemotherapy in premenopausal women with node positive breast cancer: results of a randomized trial. Proc Am Soc Clin Oncol 13:62
Mauriac L, Durand M, Chauvergne J et al. (1992) Randomized trial of adjuvant chemotherapy for operable breast cancer comparing i.v. CMF to an epirubicin-containing regimen. Ann Oncol 3:439–443
Misset JL, Gil-Delgado M, Chollet P et al. (1992) Ten years results of the French trial comparing adriamycin, vincristine, 5-fluorouracil and cyclophosphamide to standard CMF as adjuvant therapy for node positive breast cancer. Proc Am Soc Clin Oncol 11:54
Moliterni A, Bonadonna G, Valagussa P et al. (1991) Cyclophosphamide, methotrexate and fluorouracil with and without doxorubicin in the adjuvant treatment of resectable breast cancer with one to three positive axillary nodes. J Clin Oncol 9:1124–1130
Mouridsen HT (1992) Systemic therapy of advanced disease. Drugs 44 Suppl 4:17–28
Mouridsen HT, Alfthan C, Bastholt L et al. (1990) Current status of epirubicin (farmorubicin) in the treatment of solid tumours. Acta Oncol 29:257–285
Muss HB, White DR, Richards F et al. (1978) Adriamycin versus methotrexate in five-drug combination chemotherapy for advanced breast cancer. A randomized trial. Cancer 42:2141–2148
Pedersen-Bjergård J (1992) Acute lymphoid leukemia with t(4;11)(q21;q23) following chemotherapy with cytostatic agents targeting at DNA-topoisomerase II. Leuk Res 16:733–735
Praga C, Trave F, Petroccione A (1991) Anthracycline-induced cardiotoxicity and its relevance in cancer treatment. In: Nimmo WA (ed) Clinical measurement in drug evaluation. Wolfe, London, pp 131–142
Ratain MJ, Rowley JD (1992) Therapy-related acute myeloid leukemia secondary to inhibitors of topoisomerase II. From the bedside to the target genes. Ann Oncol 3:107–111
Robison LL (1994) Late occurring sequelae in survivors of childhood cancer. ASCO Educational Book, pp 115–117
Sprinz PG, Miller JD, Prigent PM, Toledano SR (1988) Do survivors of childhood malignancies treated with anthracyclines have long-term cardiotoxicity? Blood 72:229a
Steinherz SJ, Steinherz PG, Tan CTC et al. (1991) Cardiac toxicity 4 to 20 years after completing anthracycline therapy. JAMA 66:1672–1677
Tormey DC, Weinberg VE, Leone LA et al. (1984) A comparison of intermittent vs continuous and of adriamycin vs methotrexate 5-drug chemotherapy for advanced breast cancer. A Cancer and Leukemia Group B study. Am J Clin Oncol 7:231–239
Valagussa P, Zambetti M, Biasi S et al. (1994) Cardiac effects following adjuvant chemotherapy and breast irradiation in operable breast cancer. Ann Oncol 5:209–216

Adjuvant Systemic Therapy: The Issues of Timing and Sequence

M. Castiglione-Gertsch[1], R.D. Gelber[2], and A. Goldhirsch[3]

International Breast Cancer Study Group [1] Coordinating Center, Konsumstr. 13, 3007 Bern, Switzerland
International Breast Cancer Study Group [2] Statistical Center, Dana Farber Cancer Institute, 44 Binney Street Boston, MA 02115, USA
[3] Servizio Oncologico Cantonale, Ospedale Civico, 6900 Lugano, Switzerland

Introduction

Adjuvant systemic therapy in breast cancer has been widely used during the last three decades and has been shown to reduce relapses and mortality (Early Breast Cancer Trialist's Collaborative Group 1992). The rationale for its introduction was that adjuvant treatment may kill tumor cells released into the circulation at the time of primary surgery. Subsequent hypotheses ascribe the high risk for systemic relapse to micrometastases already present at the time of breast cancer diagnosis. Although the interval from breast cancer diagnosis to postoperative adjuvant therapy is short compared to the presumed preclinical duration of tumor growth, it appears that early chemotherapy onset may be advantageous in terms of tumor cell kill and that a delay in starting adjuvant treatments may result in increased resistance to cytotoxics (Goldie and Coldman 1979).

Preoperative Chemotherapy

Surgical removal of the primary tumor may result in an enhanced growth rate for the residual metastatic disease. Pre/perioperative chemotherapy may abrogate this phenomenon (especially if the increased growth rate is due to recruitment of resting cells into the cell cycle). In animal models, cyclophosphamide, tamoxifen and radiotherapy delivered before surgical removal of the primary tumor prevented the increased proliferation of residual tumor cells, suppressed tumor growth and prolonged survival (Fisher et al. 1983). The early administration of adjuvant therapies may also prevent development of drug resistance, which could be responsible for impaired efficacy of an adjuvant treatment program.

Primary (preoperative) chemotherapy has been tested in large, operable tumors (Table 1) (Bonadonna et al. 1990, 1993; Forrest et al. 1986; Smith et

Table 1. Trials with preoperative chemotherapy in breast cancer

Reference	Patient selection	Number of patients	Treatment	Complete remission (%)	Clinical partial response (%)	Progressive disease	Long-term results
Jones et al. 1993	Large tumors (median 6 cm)	64	CMF or MMM	11 (17)[a]	32 (50)	1 (2)	
Jones et al. 1994	Large tumors (median 6 cm)	50	ECF infusions	33 (66)[a]		16 (32)	
Bonadonna et al. 1990, 1993	Tumors > 3 cm (median 4.5 cm)	227	CMF, CAF, CEF	8 (3.5)[b]	201 (89)[c]	7 (3)	At 3 years, 73% of partial response and 47% of minor remissions relapse-free
Jacquillat et al. 1990	All tumor sizes	250	TVMP ± A	74 (30)[a]	178 (75)	0	
Mauriac et al. 1991	Tumors > 3 cm	272	EVM + MTV vs surgery	44 (33)[a]			DFS at 34 months similar; OS better for primary chemotherapy ($p = 0.04$)

CMF, cyclophosphamide, methotrexate, 5-fluorouracil; MMM, mitomycin-C, methotrexate, mitoxantron; ECF, epirubicin, cisplatin, 5-fluorouracil; CAF, cyclophosphamide, adriamycin, 5-fluorouracil; CEF, cyclophosphamide, epirubicin, 5-fluorouracil; TVMP ± A, thiotepa, vinblastine, methotrexate, 5-fluorouracil, prednisone, ± adriamycin; EVM + MTV, epirubicin, vincristine, methotrexate, followed by mitomycin, thiotepa, vindesine; PD, during preoperative chemotherapy; DFS, disease-free survival; OS, overall survival;

[a] Clinical.

[b] Pathological.

[c] Breast-conserving surgery possible.

al. 1993; Jacquillat et al. 1990; Mauriac et al. 1991; Fisher and Wickerham 1991). This approach enables assessment of the degree of tumor response prior to surgery and may increase the opportunity for breast conservation. However, down-staging of the tumor and its regional spread could lead to the selection of inadequate postoperative adjuvant treatment for some subsets of patients. In the series published by Bonadonna et al. (1990, 1993), only eight out of 227 patients achieved a histopathological complete remission after primary chemotherapy.

Data on the comparison between pre- and postoperative adjuvant treatment are limited, although large trials on this subject are being conducted by the NSABP and the EORTC. The issue of whether preoperative use of chemotherapy will improve survival or disease control when compared to stage-tailored, postoperative administration of cytotoxics is still unresolved. Thus, preoperative adjuvant treatments should be considered experimental for the time being (Fisher and Mamounas 1995). A suggestion by G. Bonadonna, that "a modern sequential multidisciplinary approach could transform the role of surgery such that an operation would aid systemic treatment in achieving curability," might become reality.

Perioperative Adjuvant Systemic Therapy

Adjuvant systemic therapy in breast cancer is generally started a few weeks after removal of the primary tumor to allow healing of the surgical wound. It is likely that micrometastases continue to grow during this time between surgery and the initiation of chemotherapy. Data suggest that the growth fraction in micrometastases is increased after removal of the primary tumor (Fisher et al. 1983).

Tow studies, the Scandinavian and NSABP trials, published almost 20 years ago (Nissen-Meyer et al. 1978; Fisher et al. 1975), tested a perioperative course of cyclophosphamide and thiotepa, respectively. The results showed a survival benefit for the treated patients. (In the NSABP trial the statistically significant benefit was seen only in a subgroup of premenopausal patients with more than four involved axillary lymph nodes.) These trials were the basis for several other clinical trials studying immediate administration of adjuvant cytotoxic therapy after surgery.

The largest trial studying perioperative timing of adjuvant chemotherapy was started in 1981 by the International Breast Cancer Study Group (IBCSG, formerly the Ludwig Group) and included more than 2500 patients. Patients were randomized to start a course of chemotherapy with the CMF regime within 36 h from mastectomy or to no adjuvant perioperative therapy (Table 2). The 1275 patients with node-negative disease received no further treatment. After a median follow-up of 10 years, no significant disease-free survival (DFS) benefit for perioperative chemotherapy-treated patients could be observed (Goldhirsch et al. 1994a). These up-dated results differ

Table 2. IBCSG (Ludwig) study V: study design for all operable breast cancer patients (perioperative randomization): 1981–1985

Randomized surgery	Treatment
Node-negative	● Perioperative therapy is versus
	● No perioperative therapy
Node-positive	▲ Perioperative therapy only is versus
	▲ Perioperative therapy + 6× conventionally timed therapy is versus
	▲ 6× conventionally timed therapy only

Perioperative therapy started within 36 h from mastectomy and consisted of cyclophosphamide (600 mg i.v. days 1, 8), methotrexate (40 mg i.v. days 1, 8), 5-fluorouracil (600 mg i.v. days 1, 8). Conventionally timed therapy started within 25–36 days from mastectomy and consisted of cyclophosphamide (100 mg po days 1–14), methotrexate (40 mg i.v. days 1,8), 5-fluorouracil (600 mg i.v. days 1, 8) repeated every 28 days.

from those reported at a median follow-up time of 5 years, indicating a significant DFS benefit for the single perioperative course ($p = 0.04$) (Ludwig Breast Cancer Study Group 1988). In the up-dated results, the subpopulation of postmenopausal patients with tumors defined as estrogen receptor-negative (ER-negative: <10 fmol/mg protein) continued to show a significant benefit with the use of perioperative chemotherapy (Table 3). This might indicate that patients with a low tumor burden (node-negative) and a high degree of chemotherapy responsiveness (ER-negative) are likely to benefit even from a short course of perioperative treatment. Younger patients with similar tumor characteristics might require a longer duration of cytotoxic treatment. Interesting also is the observation that perioperative chemotherapy was particularly effective in reducing the number of loco-regional and soft tissue relapses (Table 4), a feature which was previously reported by our group to be typical for effective systemic adjuvant treatments (Goldhirsch et al. 1994b).

For patients with node-positive disease, the single course of perioperative CMF was found to be significantly less effective than a conventionally timed prolonged treatment consisting of six cycles of CMF (Ludwig Breast Cancer Study Group 1988). Furthermore, the combination of a perioperative cycle of CMF with six courses of CMF did not increase treatment effectiveness in terms of disease-free or overall survival. Figures 1 and 2 illustrate DFS for premenopausal and postmenopausal patients, respectively.

Four randomized trials evaluating perioperative chemotherapy, which included more than 6000 patients, were reported so far and were evaluated in an overview performed by the EORTC Meta Analysis Unit. The results of the individual trials were not conclusive. In the overview, despite the expectation for a better control of micrometastatic growth, overall survival was not improved by the perioperative application of cytotoxics. Interesting, however, was the observation that patients with ER-negative primaries were

Table 3. IBCSG (Ludwig) study V: Disease-free survival results for node-negative patients at 10 years median follow-up

	Patients (n)	Failure (n)	Disease-free survival (%)		p-value
			5 year	10 year	
Node-negative					
PeCT	848	318	74 ± 2	61 ± 2	0.28
No chemotherapy	427	169	69 ± 2	59 ± 2	
Node-negative, ER+					
PeCT	431	174	73 ± 2	58 ± 3	0.75
No chemotherapy	209	84	70 ± 3	58 ± 4	
Node-negative, ER−					
PeCT	271	93	73 ± 3	63 ± 3	0.008
No chemotherapy	138	65	60 ± 4	53 ± 4	
Premenopausal, ER−					
PeCT	169	66	70 ± 4	59 ± 4	0.46
No chemotherapy	80	35	63 ± 5	56 ± 6	
Postmenopausal, ER−[a]					
PeCT	102	27	79 ± 4	71 ± 5	0.0006
No chemotherapy	58	30	57 ± 7	48 ± 7	

PeCT, perioperative chemotherapy; ER, estrogen receptor.
[a]The overall survival difference was also statistically significant for postmenopausal patients with ER-negative tumors ($p = 0.01$).

Table 4. Sites of failure for node-negative patients in IBCSG (Ludwig) study V at 10 years median follow-up

	Premenopausal		Postmenopausal	
	PeCT	No chemotherapy	PeCT	No chemotherapy
Total number of patients	457	235	391	192
Total failures (%)	38.1	39.6	36.8	39.6
Sites				
Local alone	5.5	6.4	6.7	6.8
Contralateral breast	6.1	4.3	4.6	4.2
Regional ± local	2.6	7.7	2.3	4.2
Soft tissue	0.2	1.3	0.8	2.1
Bone	5.7	6.8	3.8	6.3
Viscera	14.9	11.1	11.5	9.4
Second malignancy	2.4	1.3	3.8	3.7
Death without recurrence	0.4	0.9	3.3	3.1
Unknown	0.2	–	–	–

PeCT, perioperative chemotherapy.

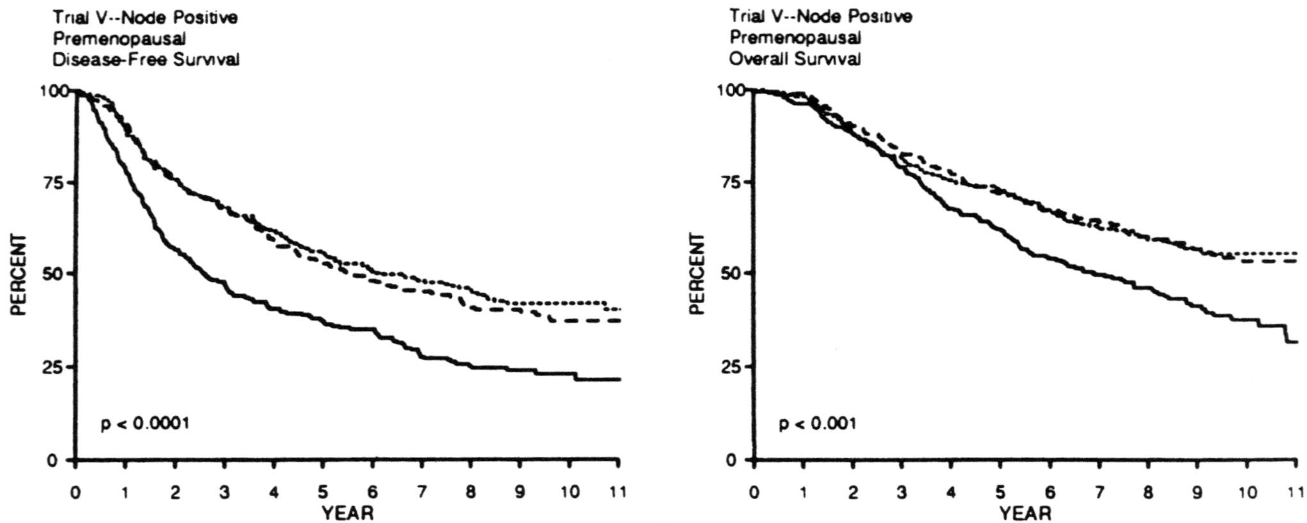

TREATMENT	FAIL	TOTAL	10-YR%	(S.E.)
—— PeCT	180	240	23	(3)
······· PeCT+ConCT	136	239	42	(3)
– – – ConCT	143	236	37	(3)

TREATMENT	DEAD	TOTAL	10-YR%	(S.E.)
—— PeCT	142	240	37	(4)
······· PeCT+ConCT	101	239	55	(3)
– – – ConCT	103	236	53	(4)

Fig. 1. IBCSG trial V: disease-free survival for node-positive premenopausal patients at 10 years median follow-up

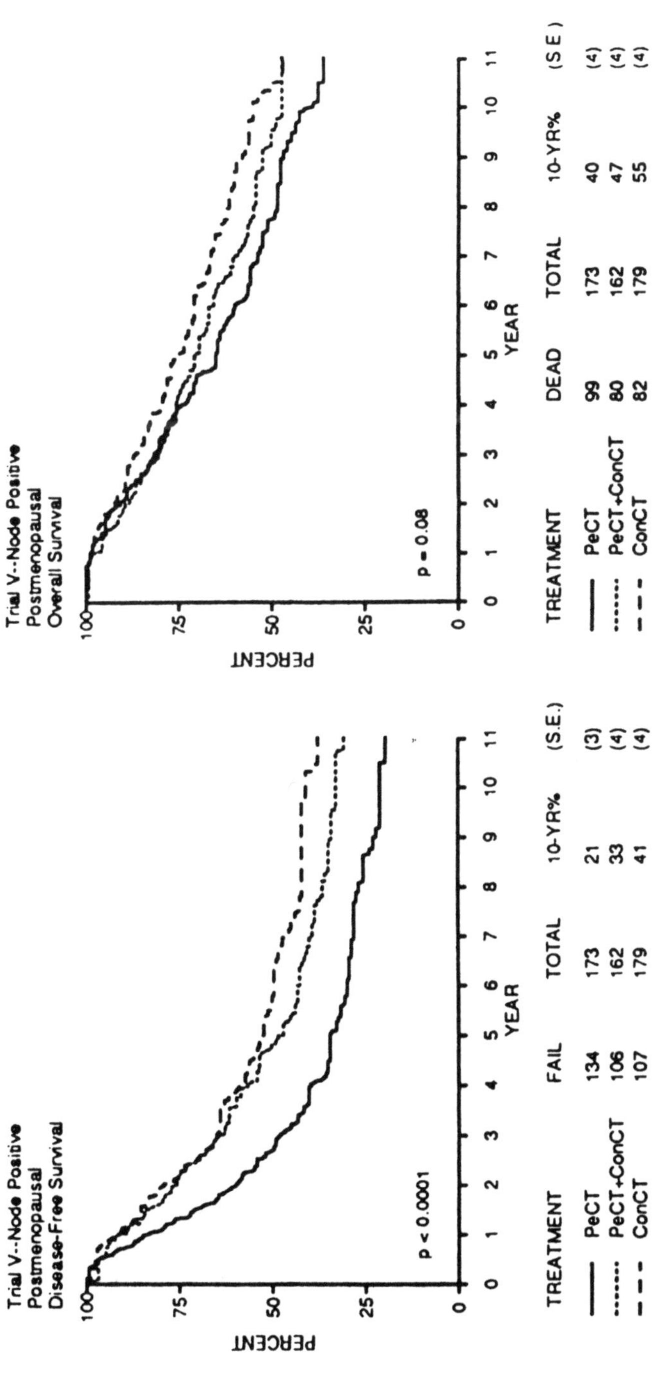

Fig. 2. IBCSG trial V: disease-free survival for node-positive postmenopausal patients at 10 years median follow-up

more likely to benefit from perioperative adjuvant cytotoxics and that locoregional and soft tissue recurrences were influenced the most by this therapeutic approach.

Postoperative Chemotherapy

In 1984, Speer and Retsky (Speer and Retsky 1984) proposed a stochastic numerical model of breast cancer growth-simulating clinical data. Their hypothesis was that growth and regression of breast cancer do not follow an exponential curve, but are more likely to be described by randomly occurring growth phases followed by resting periods. Also, the observation that recurrent breast cancers respond to the same chemotherapy applied during the adjuvant setting (Castiglione, unpublished data), led the IBCSG to investigate the role of reintroduction of chemotherapy after several chemotherapy-free intervals. From 1986 to 1993 two studies (Tables 5, 6) were conducted using this strategy. The data from these two IBCSG trials at a median follow-up time of 52 months are being evaluated and will be published in the near future. A logistical difficulty in studying delayed administration of cytotoxics relates to the ability of patients to complete a protracted course of therapy. Preliminary data on treatment and toxicity in the premenopausal patients show that only 72% completed the treatment regimen, which

Table 5. IBSCG trial VI: study design for node-positive premenopausal patients (1986–1993)

Randomized
CMF × 6 is versus
CMF × 6 + CMF × 1 (months 9, 12, 15) is versus
CMF × 3 is versus
CMF × 3 + CMF × 1 (months 6, 9, 12)

CMF, cyclophosphamide (100 mg/m^2 po, days 1–14), methotrexate (40 mg/m^2 i.v. days 1 and 8 q 4 weeks), 5-fluorouracil (600 mg/m^2 i.v. days 1,8).

Table 6. IBSCG trial VII: study design for node-positive postmenopausal patients (1986–1993)

Randomized
Tamoxifen for 5 years is versus
Tomoxifen for 5 years and CMF × 1 (months 9, 12, 15) is versus
Tamoxifen for 5 years and CMF × 3 is versus
Tamoxifen for 5 years and CMF × 3 followed by CMF × 1 (months 9, 12, 15)

Tamoxifen 20 mg/day for 5 years; CMF, cyclophosphamide (100 mg/m^2 po, days 1–14), methotrexate (40 mg/m^2 i.v. days 1 and 8, 5-fluorouracil (600 mg/m^2 i.v. days 1,8) q 4 weeks).

included an overall duration of 15 months (CMF × 6 followed by three cycles of CMF at months 9, 12, 15). In contrast, 90% of the patients allocated to six continuous courses of CMF completed the treatment program. In the postmenopausal patients, 74% of the patients completed the chemotherapy programs which included reintroduction of CMF. It is noteworthy that, in these trials which entered exclusively patients with node-positive disease, one third of the women who did not complete the delayed treatment program had a breast cancer relapse before month 15 as a cause of therapy withdrawal.

Toxicity of chemotherapy was obviously greater in patients with longer chemotherapy regimes. The most frequently observed side effects were hematological, gastrointestinal, mucositis, eye disorders and depression. Tables 7 and 8 summarize the percentages of patients showing a severe or worse toxicity (WHO grades 3 and 4) in both studies.

Sequencing Adjuvant Cytotoxics and Endocrine Therapies

Laboratory data and clinical evidence (Fisher et al. 1986; Osborne et al. 1989; Goldenberg and Froese, 1985), especially from the NSABP trials, indicate that the concomitant administration of tamoxifen (delaying G_1 transition and increasing the number of cells in G_0) together with cytotoxics (more cytocidal in dividing cells) might result in decreased efficacy, especially for subpopulations of patients with less hormone-responsive malignancies.

Table 7. Severe or worse treatment toxicity (WHO grades 3 or 4) in IBCSG study VI

	Treatment[a]			
	CMF × 6	CMF × 6 + 3	CMF × 3	CMF × 3 + 3
Total number of patients	259	262	243	242
Toxicity (%)[b]	12.0	20.2	6.2	12.0
Percent of patients with:				
Leukopenia	0.4	2.3	0.8	1.2
Neutropenia	3.5	3.1	2.5	2.5
Thrombocytopenia	0	1.5	0	2.1
Nausea, vomiting	5.0	9.2	2.5	4.1
Depression	0.4	2.7	0	1.2
Eye disorders	0.4	1.5	0.4	0.4
Mucositis	1.2	0.6	0	1.2
Weight gain	0.4	2.3	0	0

CMF, cyclophosphamide, methotrexate; 5-fluorouracil.
[a] See Table 5.
[b] Patients with at least one severe or worse toxicity.

Table 8. Severe or worse treatment toxicity (WHO grades 3 or 4) in IBCSG study VII

	Treatment[a]			
	T	T + 3 late	T + 3 early	T + 3 + 3
Total number of patients	194	180	204	199
Toxicity (%)[b]	1.5	6.7	11.3	21.1
Percent of patients with:				
Leukopenia	0	0.6	1.5	1.5
Neutropenia	0	1.7	2.5	5.0
Thrombocytopenia	0	0.6	2.5	3.5
Nausea, vomiting	0	0.6	1.5	5.5
Diarrhea	0.5	0	0	1.0
Depression	0	0	0.5	1.0
Eye disorders	0.5	0	0	0.5
Mucositis	0	0	1.5	3.0
Weight gain	0.5	0.6	0	0.5

[a] See Table 6.
[b] T, tamoxifen. Patients with at least one severe or worse toxicity.

Thus, the subsequent use of both modalities is hypothesised to best overcome a negative interaction between cytotoxics and endocrine agents. Two large clinical trials were specifically designed to assess the concomitant vs sequential administration of chemotherapy and endocrine therapy: the *Intergroup Trial INT 0100 Study* (node-positive postmenopausal, ER-positive or progesterone receptor-positive) started in May 1989 and the *International Breast Cancer Study Group Trial 12-93* (node-positive, postmenopausal, ER-positive) started in May 1993. The study designs of these trials are shown in Fig. 3.

Unpublished data from IBCSG trial VII also suggest that, for patients with ER-negative tumors, the administration of CMF chemotherapy *after* beginning of tamoxifen is detrimental. The issue of concomitant tamoxifen and cytotoxics may be less relevant for anthracycline-based regimens.

Conclusions

Several studies have shown that a large proportion of patients with primary tumors larger than 3 cm can be treated with breast-conserving procedures after use of preoperative chemotherapy. Some studies suggested that in selected cases surgery might be avoided after treatment with chemotherapy and radiation.

Concerns have been expressed that down-staging of tumors due to primary chemotherapy might cause an inadequate choice of postoperative adjuvant

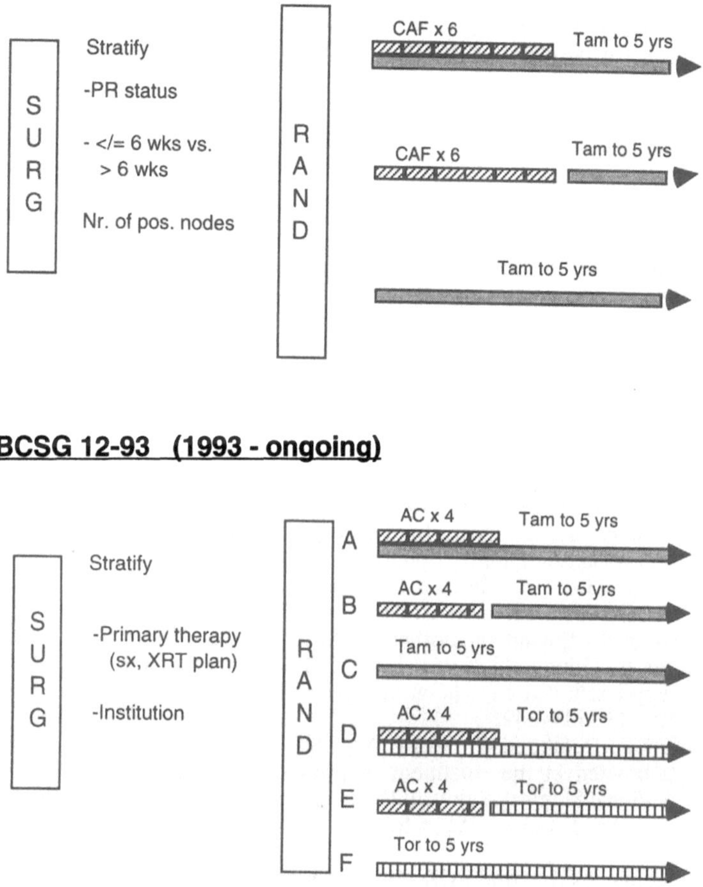

Fig. 3. Intergroup INT 0100 and IBCSG 12-93 trials: study designs for node-positive postmenopausal patients with receptor-positive tumors. *Tam*, tamoxifen; *Tor*, toremifene

systemic therapy for patients found to be node-negative after response to preoperative cytotoxic therapies.

Perioperative chemotherapy was effective in delaying recurrence compared with no adjuvant chemotherapy in node-negative patients, but was insufficient in node-positive patients when compared to a longer duration adjuvant program.

The delayed reintroduction of chemotherapy was postulated to provide some additional reduction of risk of relapse, based upon a mathematical model for tumor cell growth. The role of additional courses of cytotoxics after therapy-free intervals needs to be confirmed in clinical trials.

Clinical and laboratory evidence exist suggesting a negative interaction between tamoxifen and some cytotoxic drugs. Ongoing trials will clarify this relevant issue of a potential antagonism and will shed light on the best timing for combining the two modalities.

References

Bonadonna G, Veronesi U, Brambilla C (1990) Primary chemotherapy to avoid mastectomy in tumours with diameters of three centimeters or more. J Natl Cancer Inst 82:1539–45

Bonadonna G, Valagussa P, Brambilla C, Ferrari L (1993) Preoperative chemotherapy in operable breast cancer. Lancet 3341:1485

Early Breast Cancer Trialists' Collaborative Group (1992) Systemic treatment of early breast cancer by hormonal, cytotoxic or immune therapy: 133 randomised trials involving 31 000 recurrences and 24 000 deaths among 75 000 women. Lancet 339: pp 1–5 and pp 71–85

Fisher B, Mamounas E (1995) Preoperative chemotherapy: a model for studying the biology and therapy of primary breast cancer. J Clin Oncol 13:537–540

Fisher B, Wickerham DL (1991) Preoperative systemic therapy for the treatment of primary breast cancer. In: Powles TJ, Smith IE (eds) Medical management of breast cancer. Dunitz, London, pp 281–286

Fisher B, Slack N, Katrych D, Wolmark N (1975) Ten year follow-up results of patients with carcinoma of the breast in a co–operative clinical trial evaluating surgical adjuvant chemotherapy. Surg Gynecol Obstet 140:528–534

Fisher B, Gunduz N, Saffer EA (1983) Influence of the interval between primary tumor removal and chemotherapy on kinetics and growth of metastases. Cancer Res 43:1488–1492

Fisher B, Redmond C, Brown A (1986) Adjuvant chemotherapy with and without tamoxifen in the treatment of primary breast cancer: 5–year results from the National Surgical Adjuvant Breast and Bowel Project trial. J Clin Oncol 4:459–471

Forrest APM, Levack PA, Chetty U (1986) A human tumour model. Lancet ii:840–842

Goldenberg GJ, Froese EK (1985) Antagonism of the cytocydal activity and uptake of melphalan by tamoxifen in human breast cancer cells in vitro. Biochem Pharmacol 34:763–770

Goldie JH, Coldman AJ (1979) A mathematical model of relating the drug–sensitivity of tumours to their spontaneous mutation rate. Cancer Treat Rep 63:1727–1733

Goldhirsch A, Gelber RD, Castiglione M, Price KN, Rudenstam C-M, Lindtner J, Collins J, Senn H-J, Brunner KW, Galligioni E, Cavalli F, Gudgeon A, Cortes-Funes H, Tattersall M, Marini G, Byrne M, Snyder R, Forbes J, Hürny C, Coates A (1994) For the international Breast Cancer Study Group. Present and future projects of the International Breast Cancer Study Group. Cancer 74:1139–1149

Goldhirsch A, Gelber RD, Price KN, Castiglione-Gertsch M, Coates AS, Rudenstam CM, Collins J, Lindtner J, Hacking A, Marini G, Byrne M, Cortés-Funes H, Schnürch G, Brunner KW, Tattersall MHN, Forbes J, Senn HJ (1994) Effect of systemic adjuvant treatment on first sites of breast cancer relapse. Lancet 343: 377–381

Jacquillat C, Weil M, Baillet F (1990) Results of neoadjuvant chemotherapy and radiation therapy in the breast–conserving treatment of 250 cancer patients with all stages of infiltrative breast cancer. Cancer 66:119–129

Jones AL, Smith IE, O'Brien MER, Talbot D, Walsh G, Ramage F, Robertshaw H, Ashley S (1994) Phase II study of continuous infusion fluorouracil with epirubicin and cisplatin in patients with metastatic and locally advanced breast cancer: an active new regimen. J Clin Oncol 12(6):1259–1265

Ludwig Breast Cancer Study Group (1988) Combination adjuvant chemotherapy for node-positive breast cancer. Inadequacy of a single perioperative cycle. N Engl J Med 319:677–683

Ludwig Breast Cancer Study Group (1989) Prolonged disease–free survival after one course of perioperative adjuvant chemotherapy for node-negative breast cancer. N Engl J Med 320:491–496

Mauriac L, Durant M, Avril A, Dilhuydy JM (1991) Effects of primary chemotherapy in conservative treatment of breast cancer patients with operable tumours larger than 3 cm: results of a randomized trial in a single center. Ann Oncol 2:347–354

Nissen-Meyer R, Kjellgren K, Malmio K, Mansson B, Norin T (1978) Surgical adjuvant chemotherapy: results with one short course with cyclophosphamide after mastectomy from breast cancer. Cancer 41:2088–2098

Osborne CK, Kitten L, Arteaga CL (1989) Antagonism of chemotherapy-induced cytotoxicity for human breast cancer cells by antiestrogens. J Clin Oncol 7:710–717

Smith IE, Jones AL, O'Brien MER, McKinna JA, Sacks N, Baum M (1993) Primary medical (neo–adjuvant) chemotherapy for operable breast canncer. Eur J Cancer 29:592–595

Speer JF, Petrosky VE, Retsky MW, Wardwell RH (1984) A stochastic numerical model of breast cancer growth that simulates clinical data. Cancer Res 44:4124–4130

Current and Future Roles of Adjuvant Endocrine Therapy in the Management of Early Carcinoma of the Breast

M. Baum and J.R. Benson

The Royal Marsden Hospital, Fulham Road, London SW3 6JJ, UK

Introduction

The world overview of adjuvant trials for early breast cancer published in *The Lancet* in January 1992 (Early Breast Cancer Trialists' Collaborative Group 1992) provided us with secure data to allow the rational employment of adjuvant systemic chemotherapy and endocrine therapy, but like the process of science in any other walk of life, in answering one set of questions, a new set of questions has emerged. Effectively therefore, the world overview also set the agenda for the next decade of clinical trials in the management of early breast cancer. A whole raft of new questions has been defined concerning the role of endocrine manipulation in both pre- and postmenopausal women.

Adjuvant Endocrine Therapy in Premenopausal Women

The new questions concerning endocrine manipulation in the management of premenopausal primary breast cancer are very complex. The world overview suggested that the benefits of both adjuvant chemotherapy or ovarian suppression were of the same order of magnitude. This observation has fuelled the continuing debate concerning the mechanism of action of adjuvant chemotherapy in premenopausal women which may in part be due to an indirect chemical ovarian suppression. There was also a suggestion from indirect comparisons that combining these two modalities of treatment might produce a summation of effects. The early trials of ovarian suppression were conducted over 15 years ago and involved either surgical castration or the induction of irradiation menopause. In the last few years a new class of compound has been developed which can produce a reversible chemical ovarian suppression. These are the gonadotrophin-releasing hormone (GnRH) agonists. Pharmacokinetic studies of this class of compound demonstrate that after an initial surge in endogenous oestradiol levels, these then rapidly fall to castrate levels which return to normal on withdrawal or exhaustion of the drug (Nicholson et al. 1985). The theoretical advantage of this approach over a permanent ovarian suppression might be the avoidance of the long-term complications of osteoporosis or ischaemic heart disease that can be expected following castration. In addition, the younger premenopausal woman might still wish to retain the potential of a pregnancy after a period of disease-free survival once her primary and systemic therapy is complete. For these reasons current trials are investigating the benefits of GnRH agonists either by direct head-on comparison with an adjuvant chemotherapy regimen such as CMF or in trials comparing chemotherapy alone vs chemotherapy and ovarian surppression. The recruitment into these trials is close to completion. For example, a large European consortium including the Cancer Research Campaign group in the United Kingdom, a group based in Brussels and a Scandinavian group based in Stockholm, Linkopping and Oslo have between them already recruited over 2000 patients, sufficient to address the main effect question. In addition, the power of this combined study will allow a search for qualitative or quantitative interactions between ovarian suppression and the presence or absence of oestrogen receptors in the primary tumour (vid infra).

Adjuvant Endocrine Therapy in Postmenopausal Women

The dominant question for the postmenopausal group of women concerns the optimum duration of adjuvant tamoxifen. A degree of conceptual rationalisation has persuaded many clinical oncologists that tamoxifen is a cytostatic rather than cytocidal agent and therefore should be prescribed indefinitely until relapse rather than for a set period of time. This might be considered a foolish practice which flies in the face of clinical evidence. For example, in the world overview the average exposure to adjuvant tamoxifen was approximately 2 years (Early Breast Cancer Trialists Collaborative Group 1992). Yet, on withdrawing the tamoxifen there was no evidence of a rebound effect and the survival curves continued to diverge out to 10 years. Furthermore, chronic exposure to tamoxifen might contribute to the development of tamoxifen-dependent mutant variants of breast cancer cells (Jordan 1994). Finally, the recently reported increased incidence of endometrial cancer which might be related to dose intensity and duration of tamoxifen must give us pause for thought (Jordan and Morrow 1994). A number of trials are addressing this question worldwide and the Cancer Research Campaign trial in the United Kingdom is probably the largest of its kind. The design of this trial is simplicity itself. All women with operable breast cancer over the age of 50 are eligible. Primary therapy with surgery, radiotherapy or cytotoxic drugs is not defined in the protocol. The patients are all started on adjuvant tamoxifen (20 mg daily) and then registered with the trial centre. Randomisation occurs amongst event-free survivors at 2 years to either stop the drug or continue for a further 3 years making an aggregate of 5 years. To detect a 6% difference in survival at 5 years (67% vs 73%) with 95% power, 3000 patients are required. Allowing for relapse in the first 2 years and the 20% or 30% of patients who refuse randomisation at 2 years, we have set a target for registration of approximately 5000 patients. We are close to achieving that target and the interim data have so far only been inspected by an independent data monitoring committee (DMC) and are not yet available for publication. At a median follow-up of only 3.8 years, there have been 167 events postrandomisation. It is not surprising therefore that the DMC have recommended continued recruitment. Unfortunately, many of our participants are finding it increasingly difficult to randomize patients after 2 years as the patients themselves or their primary healthcare doctors appear to have prejudged the outcome. Furthermore, at an anecdotal level, we have learned that many patients treated off protocol are left on treatment indefinitely. It is in the interests of all patients that trials of tamoxifen duration are able to recruit sufficient numbers to ensure that cost-benefit analyses can be addressed with statistical confidence and that when armed with these data women can be allowed a rational and informed choice (Tripathy 1994).

When tamoxifen was originally developed it was thought to be an anti-oestrogen. Over the last 10 years or so we have learnt much of interest about this enigmatic compound. It is now perceived as an attenuated oestrogen whose mechanism of action is very complicated and only in part mediated by classical receptor mechanisms. For that reason a lot of interest is currently being expressed in a new class of oral aromatase inhibitors (AIs). In postmenopausal women a background level of endogenous oestradiol at approximately 30 pmol/l is synthesised in the adipose tissue dependent on the enzyme aromatase. The new class of AIs are very specific and nontoxic and do not require the additional support of corticosteroids as with aminoglutethimide. The AIs can now be delivered orally reducing oestradiol levels in postmenopausal women to values below the threshold of detection, thus making them ideal candidates for adjuvant studies. Three pharmaceutical companies have products ready for testing, i.e., ZENEKA, Janssen and Ciba Geigy. There is considerable debate whether the oral AIs will be equivalent in their actions to tamoxifen or possibly additive or synergistic. The Cancer Research Campaign group in the UK is likely to launch a trial later on this year with a three-way randomisation between adjuvant tamoxifen alone, adjuvant AI or a combination of the two.

Finally, a new class of compound has just been reported which may have a role in the adjuvant treatment of early breast cancer amongst pre- and postmenopausal women. This is a specific anti-oestrogen, ICI 182,780 (Wakeling et al. 1991). Amongst the possible reasons for resistance to tamoxifen are that the cancer may be reacting to it as an agonist in postmenopausal women or because the central effect in premenopausal women elevates endogenous oestradiol to out-compete with the tamoxifen for the limited number of unoccupied oestrogen receptors. For this reason Zeneka have developed a new compound which is a pure anti-oestrogen and has been shown to have no agonist activity in animal studies. A recent paper (De Friend et al. 1993) has demonstrated the activity of ICI 182,780 amongst a group of patients with metastatic breast cancer resistant to tamoxifen. There were 37% partial responders and 32% with previously progressive disease that remained static. Future trials might involve ICI 182,780 alone or in combination with tamoxifen in both pre- and postmenopausal women. The major limitations for using this agent as an adjuvant long-term is that it still requires a monthly intramuscular injection. As a pure anti-oestrogen, ICI 182,780 may have greater intrinsic anti-tumour efficacy than tamoxifen when employed as a first-line adjuvant. However, it is uncertain whether the agonist activity of tamoxifen compromises its overall specific tumour suppressive effects, and loss of an agonist component may sacrifice incidental benefits of "impure" anti-oestrogens upon the cardiovascular system and bone. Furthermore, certain additional oestragen receptor independent mechanisms of action of tamoxifen (vide infra) may depend upon the triphenylbutene core which is not possessed by ICI 182,780, an oestradiol derivative. Other triphenylethylenes, such as idoxifene, with selective attenuation of agonist properties may offer more promise, especially in the field of chemoprevention (Chander et al. 1991).

Hormone Replacement Therapy

Hormone replacement therapy for women on treatment for early breast cancer has to be considered a form of adjuvant endocrine therapy if only because its prime objective is to improve quality of life. It has long been the received wisdom that hormone replacement therapy (HRT) is contraindicated in women who have had treatment for early breast cancer (Steinberg et al. 1991). The risk of oestrogens promoting the growth of occult malignant cells has been the primary concern of doctors advising breast cancer patients with menopausal symptoms. However, many such patients will have poor quality of life suffering from intolerable hot flushes, depression, vaginal dryness and other symptoms associated with oestrogen deficiency. These symptoms, common enough in the normal population, are exacerbated amongst women treated for early breast cancer with tamoxifen, chemotherapy and ovarian suppression. There is no doubt that HRT can improve on life quality amongst such women and many patients in our own clinic are provided oestrogen replacement therapy on an ad hoc basis because the real benefits might outweigh the theoretical hazards. As yet, there has never been a properly controlled randomised study looking at the outcome of breast cancer amongst women on or off HRT. It is difficult to speculate what effect this treatment will have on disease-free and overall survival as one can marshall theoretical arguments on both sides of this debate. Furthermore, even if there is a modest excess of cause-specific mortality with oestrogen-replacement therapy, this could be offset with a reduction in mortality from ischaemic heart disease and osteoporosis. For these reasons, a feasibility study has been launched at the Royal Marsden Hospital with full ethical approval and we hope that this will be taken forward into a large pragmatic trial under the auspices of the United Kingdom Coordinating Committee for Cancer Research. Other groups around the world have arrived at similar conclusions. At our own centre we have established a Consumers Advisory Group of women with some experience of breast cancer, either themselves or immediate family, who recognise the importance of this question and are helping in the design of the trial and the production of the patient information leaflets.

Prognostic or Predictive Variables

There is a lot of confusion between prognostic and predictive biological variables. For example, it is naively assumed that prognostic variables which suggest a very aggressive cancer necessarily indicate that such patients will be benefitted preferentially by chemotherapy rather than endocrine therapy. There is little evidence to support such a contention. At the same time it is widely assumed that the presence or absence or oestrogen receptor in the primary tumour will necessarily predict those patients who will benefit from adjuvant endocrine therapy. There are serious methodological problems when one attempts to address these questions. To begin with it is important to make a distinction between qualitative and quantitative interactions. A qualitative interaction suggests that one class of patients has a significant benefit from a treatment whereas the other class of patients has a detriment or zero effect on exposure to the therapy. Such qualitative interactions are exceedingly rare in medicine although it is quite reasonable to expect the menopause itself to select patient groups who may or may not respond to ovarian suppression. In the same way it would be reasonable to assume that amongst premenopausal women the oestrogen receptor content of their primary tumour would select out a group of patients where there might be a quantitative, if not necessarily qualitative, differential response to ovarian suppression. The Scottish Cancer Trials Breast Group (1993) recently published a paper describing a trial comparing adjuvant chemotherapy with ovarian ablation. Overall, there was no difference in survival between these two groups, but when split according to oestrogen receptor status, there was a clear suggestion of interaction with the chemotherapy being more beneficial in the oestrogen receptor-negative patients and ovarian ablation more beneficial in the oestrogen receptor-positive group. The trouble with these kind of studies concerns statistical power. We often have difficulty in recruiting sufficient numbers of patients to demonstrate important and clinically significant differences for the main effect analysis of the whole population. In order to enjoy the same statistical power for subgroups of patients assuming a balanced dichotomy we need twice the original number. These kind of studies have seldom been achieved to date and the majority of subgroup analyses that show significant correlations between subclasses of patients and the treatment modality might confuse qualitative interactions with quantitative interactions. This can be illustrated as follows: Say you have developed a pure anti-oestrogen and you recruit 1000 patients to a trial and the main effect analysis compares 500 receiving the anti-oestrogen with 500 controls. This study demonstrates a 7% difference in overall survival with a p value of <0.01 and a statistical power of 85%. You may then divide the patient groups up according to oestrogen receptor levels of say greater or less than 10 fmols/mg cytosolic protein. The receptor-positive group may show a significant difference of the order of 10% and the receptor-negative group may show a nonsignificant difference of the order of 3%. With those numbers and that much reduced statistical power (there is not even a 50:50 split on oestrogen receptor status at this level), it is impossible to determine whether this is a true qualitative interaction or whether the "oestrogen receptor-negative" group have a modest advantage with the hypothetical anti-oestrogen which is not sufficient to emerge in a test of a population of patients now numbering only approximately 400.

These lessons were learnt the hard way when studying the early results from the trials of adjuvant tamoxifen (Nolvadex Adjuvant Trial Organisation 1987). When the European Group set out to test the hypothesis that adjuvant tamoxifen might prolong disease-free and overall survival, their control groups were untreated and the results from these trials are described as unconfounded. In contrast, in the United States the control treatment was polychemotherapy and the effect of tamoxifen was judged against this background. These trials can be described as confounded and have generated materially different qualitative and quantitative results. One of the flaws in the overview mechanism is to pool these two groups of trials. If separated it is possible to judge that tamoxifen has a similar effect in oestrogen receptor-positive and oestrogen receptor-negative cases producing twice the benefit of adjuvant chemotherapy in the postmenopausal group (Early Breast Cancer Trialists' Collaborative Group 1992). If one looks upon tamoxifen as an anti-oestrogen, then conceptually it is difficult to believe that it can be of any value in receptor-negative cases, yet our modern understanding of the mechanism of action of tamoxifen would suggest such an effect is entirely plausible. Tamoxifen can mediate its tumour suppressive activity upon malignant breast epithelial cells indirectly by modulation of both local and systemic levels cytokines (Colletta et al. 1994). Thus the production of the potent epithelial inhibitory growth factor, transforming growth factor-β (TGFβ) by breast tumour fibroblasts is stimulated by tamoxifen both in vivo (Butta et al. 1992) and in vitro (Benson et al. 1993). This locally induced TGFβ can act in a negative paracrine manner upon breast carcinoma cells irrespective of their oestrogen receptor status. Conversely, adjuvant tamoxifen therapy is associated with a 30% reduction in circulating levels of insulin-like growth factor-I (IGF-I), a stimulatory growth factor which is a potent mitogen for breast cancer cells (Myall et al. 1984; Cullen et al. 1992). This effect is mediated via the pituitary gland and modulation of growth hormone levels. In a rat model, tamoxifen can reduce expression of IGF-I in tissues which are targets for breast cancer metastases (Huynh et al. 1993). Moreover, this effect is partly independent of pituitary function, and the modulation of IGF-I expression occurs in the stromal compartment of tissues. Thus tamoxifen can influence both positive and negative paracrine loops by direct effects upon stromal cells, which do not possess any oestrogen receptor protein (Benson et al. 1993). It is interesting to speculate on how many women have been denied the benefits of adjuvant tamoxifen based on conceptual rationalisation rather than a sophisticated analysis of the clinical trials data. Very recently, the South Sweden Breast Cancer Group published a paper in the European Journal of Cancer suggesting that cathepsin D as well as being a prognostic factor might also be a predictive factor for the effect of adjuvant tamoxifen (Fernö et al. 1994). Whilst superficially convincing, we think once again they have fallen into the trap of confusing qualitative with quantitative interactions. Although there is an undoubted and significant benefit for adjuvant tamoxifen amongst patients with high cathepsin D levels and a nonsignificant difference amongst those with low cathepsin D levels, that final subsubset in the trial only included about 150 patients and therefore lacked sufficient statistical power to demonstrate what could be clinically worthwhile responses.

New Thoughts on the Future Role of Endocrine Manipulation in the Management of Early Breast Cancer

Until quite recently the majority of oncologists in the world had formulated their policies on the unfounded assumption that chemotherapy would have a greater and more lasting benefit than endocrine therapy in the management of early breast cancer. We suspect that this confidence was based on two factors: Firstly that in the management of advanced breast cancer you can expect more than twice the number of objective responses with chemotherapy than with endocrine therapy, and secondly the assumption that chemotherapy would be unselective in its activity whereas endocrine therapy would only benefit the minority, say 30% determined by the level of oestrogen receptor expressed. At long last we are all waking up to the limitations of chemotherapy, particularly as there has never been a trial that convincingly demonstrates that chemotherapy contributes to significantly longer survival compared with simpler endocrine manoeuvres in patients presenting with metastatic disease. Also as we become more sophisticated in studying mechanisms of endocrine response and endocrine resistance, we now recognise that there is perhaps a greater potential for successful endocrine manipulation than we may have believed in the past. We can add to that the intriguing possibility that the endogenous endocrine milieu at the time of surgery could have a profound effect on survival. For example there is an increasing body of data which supports the view that the timing of surgery within the menstrual cycle affects the long-term prognosis of premenopausal women with operable breast cancer and the magnitude of this benefit might exceed that achieved by any current adjuvant regimen (Fentiman and Gregory 1994). It has been suggested by the Guy's group that this effect is mediated by the variation in cohesiveness of the primary tumour cells at the time of surgery so that surgical manipulation itself could release viable clonogenic cells into the circulation if operating at the time of unopposed oestrogen rather than the luteal phase of the menstrual cycle (Badwe et al. 1994). We believe that this is a somewhat naive explanation for what may indeed be a real effect. Unopposed oestrogen per se may be less important than changes in systemic levels of cytokines such as IGF-I, whose levels could be linked to the menstrual cycle via pituitary function and growth hormone (GH) levels. We are rather persuaded by the arguments of Fisher et al. (1989), lately supported by the elegant experimental data from Judah Folkman's group, that micrometastatic foci pre-exist in all cases and many of these will remain dormant as long as the primary tumour is in situ (Folkman 1995). The act of surgery itself is far from biologically neutral and the removal of the tumour bulk, together with the timing of its removal could perturb a variety of homeostatic mechanisms. This could disinhibit the pre-existing metastatic foci, activate microangiogenesis, and kick-start these potential secondaries into a growth cycle whereby proliferation exceeds apoptosis and the increasing tumour volume is sustained without central necrosis by the induction of a malignant microvasculature (Folkman 1994). The limit of tumour growth without new blood vessel formation is approximately 1 million cells, and further growth is dependent upon angiogenesis (Folkman 1990). Cancer cells produce and secrete angiogenic factors which stimulate new vessel growth from normally quiescent capillaries. Fibroblast growth factor (FGF) is a potent angiogenic factor and, when the gene coding for this (FGF-4) is transfected into MCF-7 cells, these become hormone independent and can form tumours in nude mice in the absence of oestrogen (Lippman 1994). There is a correlation between survival and microvessel density of tumours (Weidner et al. 1991; Horak and Harris 1992); so, could the "conversation" between cancer cells and endothelial cells be interrupted? Would it be possible to abort the angiogenic kick-start which heralds the end of dormancy in a nidus of tumour cells? One approach is to employ antibodies against FGF (Hori et al. 1991) or peptide antagonists for FGF and vascular endothelial growth factor (VEGF) receptors such as pentosan polysulphate (Lippman 1994). These could be used as an adjunct in adjuvant schedules in which potential synergistic interactions may occur, e.g., tamoxifen induces TGFβ which is also a potent inhibitor of endothelial cells (Shultz and Grant 1991), and tamoxifen has been reported to inhibit angiogenesis in a chick chorio-allantoic membrane model (Gagliardi and Collins 1993). In theory therefore, the cascade of events that favours the outgrowth of pre-existing micrometastases by the act of surgery could be inhibited by priming the host with anti-hormones, progestins or anti-angiogenic factors before the time of surgery. We find it surprising that so much emphasis has been placed on primary (neoadjuvant) chemotherapy when the biological underpinning for such an approach is so poor. We would have much more confidence in the outcome of primary endocrine therapy for the reasons explained above. Exposure of a tumour to a neoadjuvant endocrine regime may alter the biological potential of cancer cells both within the primary tumour and within any concomitant micro-metastases. This modified biological potential could manifest itself as a changed secretory profile of cytokines and angiogenic peptides and off-set any systemic perturbation induced by the act of surgery. Figures 1–3 are schematic diagrams of the postulated microenvironment of dormant micrometastases, activated metastases if surgery is carried out in the follicular phase and the net favourable environment induced by "neo-adjuvant" endocrine therapy.

At The Royal Marsden Hospital in London we are piloting a study of primary adjuvant tamoxifen, and in the longer term The Cancer Research Campaign Group will probably conduct trials on a large scale comparing primary tamoxifen with or without AIs with a control group receiving conventionally timed medical therapy.

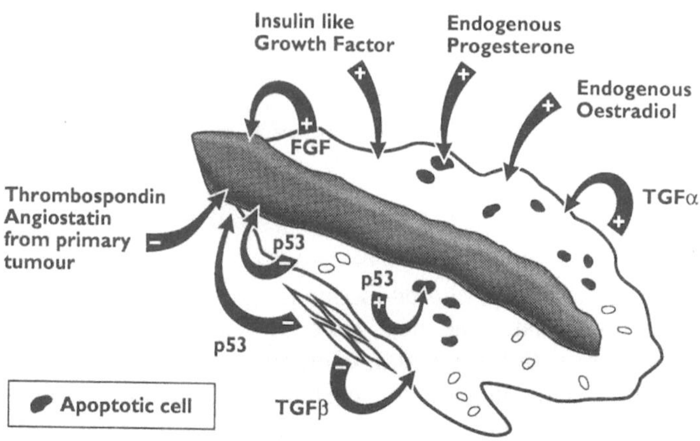

Fig. 1. A dormant micrometastases in dynamic equilibrium – elaboration on the model described by Folkman 1995

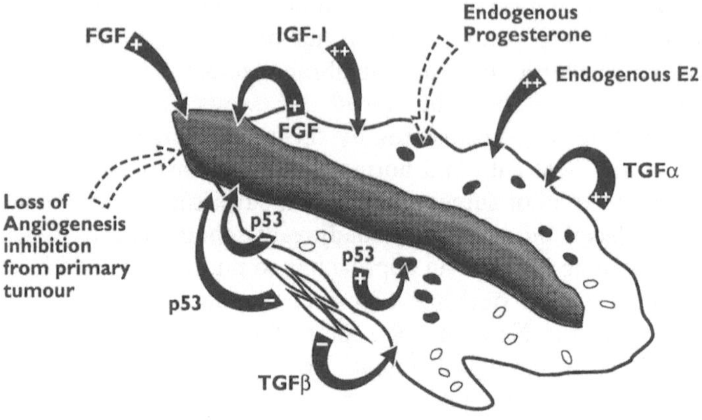

Fig. 2. Hypothetical disturbance of the equilibrium following surgery in the follicular phase of the menstrual cycle

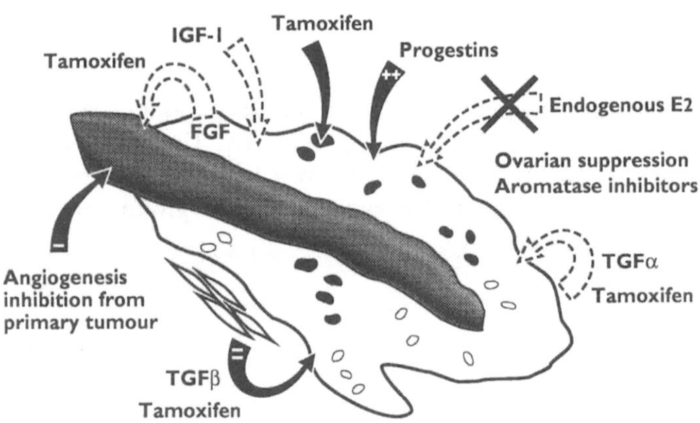

Fig. 3. Theoretical advantages of primary endocrine therapy in maintaining a favourable equilibrium in the perioperative phase

Conclusions

The subject of adjuvant endocrine therapy gets more and more complex. To keep up to date, the next generation of clinical oncologists have to be conversant with biostatistics, endocrinology and cell and molecular biology; they must avoid dogma and constantly keep an open mind. Most of all, they have a responsibility to increase the statistical power of current and future clinical trials. No one has the right to stand on the sidelines or make parasitic use of the data generated by the large collaborative groups. These collaborative groups are now open to all comers and in the long term the best interests of the individual patient and future generations to come will be served by randomisation rather than by the dogmatic insistence on one or another therapeutic intervention.

References

Badwe RA, Gregory WM, Cahudary MA et al. (1994) Timing of surgery during menstrual cycle and survival of premenopausal women with operable breast cancer. Lancet 1:1261–1264

Benson JR, Wakefield LM, Sporn MB et al. (1993) Secretion of TGFβ isoforms by primary cultures of human breast tumour fibroblasts in vitro and their modulation by tamoxifen (Abstr). Breast Cancer Res Treat 27(1/2):163

Butta A, Maclennan K, Flanders KC, Sacks NPM, Smith I, Mackinna A, Dowsett M, Wakefield LM, Sporn MB, Colletta AA (1992) Induction of transforming growth factor beta_1 in human breast cancer in vivo following tamoxifen treatment. Cancer Res 52:4261–4264

Chander SK, McCague R, Luqmani Y, Newton C, Dowsett M, Jarman M, Coombes RC (1991) Pyrrolidino-4-iodotamoxifen and 4-Iodotamoxifen. New analogues of the antiestrogen tamoxifen for the treatment of breast cancer. Cancer Res 51: 5851–5955

Colletta AA, Benson JR, Baum M (1994) Alternate mechanisms of action of antioestrogens. Breast Cancer Res Treat 31:5–9

Cullen KJ, Lippman ME, Chow D, Hill S, Rosen N, Zwiebel JA (1992) Insulin-like growth factor-II over-expression in MCF-7 cells induces phenotypic changes associated with malignant progression. Mol Endocrinol 6:91–100

De Friend D, Blamey RE, Robertson JF, Walton P, Howell A (1993) Response to the pure anti-oestrogen ICI 182,780 after tamoxifen failure in advanced breast cancer (Abstr). Breast Cancer Res Treat 27(1/2):136

Early Breast Cancer Trialists' Collaborative Group (1992) Systemic treatment of early breast cancer by hormonal, cytotoxic, or immune therapy. 133 randomised trials involving 31 000 recurrences and 24 000 deaths among 75 000 women. Lancet 339(8784):1–15, 339(8785):71–85

Fentiman IS, Gregory WM (1994) Peri-operative hormones and prognosis in breast cancer. Eur J Cancer 30A(12):1892–1894

Fernö M, Baldetorp B, Borg Å et al. (1994) Cathepsin D, both a prognostic factor and a predictive factor for the effect of adjuvant tamoxifen in breast cancer. Eur J Cancer 30A(14):2042–2048

Fisher B, Gunduz N, Coyle J (1989) Presence of a growth-stimulating factor in serum following primary tumour removal in mice. Cancer Res 49:1996–2001

Folkman J (1990) What is the evidence that tumours are angiogenesis dependent? J Natl Cancer Inst 82:4–6

Folkman J (1994) Tumor angiogenesis. In: Mendelsohn J, Howley P, Liotta L, Israel M (eds) The molecular basis of cancer. Saunders, Philadelphia (in press)

Folkman J (1995) Angiogenesis in cancer, vascular, rheumatoid and other disease. Nature [Medicine] 1(1):27

Gagliardi A, Collins DC (1993) Inhibition of angiogenesis by anti-oestrogens. Cancer Res 53:533–535

Horak E, Harris AL (1992) Angiogenesis, assessed by platelet/endothelial cell adhesion molecule antibodies, as indicator of node metastases and survival in breast cancer. Lancet 340:1120–1124

Hori A, Sasada R, Matsutani E et al. (1991) Suppression of solid tumour growth by immuno-neutralising monoclonal antibody against fibroblasts growth factor. Cancer Res 51:6180–6184

Huynh HT, Tetenes E, Wallace L, Pollack M (1993) In vivo inhibition of insulin-like growth factor I gene expression by tamoxifen. Cancer Res 53:1727–1730

Jordan VC (ed) (1994) Long-term tamoxifen therapy for breast cancer. University of Wisconsin Press, Madison

Jordan VC, Morrow M (1994) Should clinicians be concerned about the carcinogenic potential of tamoxifen? Eur J Cancer 30A(11):1714–1721

Lippman ME (1994) New approaches in diagnosing and treating breast cancer. Conference organised by Cambridge Healthtech Institute, Philadelphia

Myall Y, Shiu RPC, Bhaumick B (1984) Receptor binding and growth promoting activity of insulin-like growth factors in human breast cancer cells (T47D) in culture. Cancer Res 44:5486–5490

Nicholson RI, Walker KJ, Turkes A et al. (1985) Endocrinological and clinical aspects of LHRH action (ICI 118,630) in hormone dependent breast cancer. J Steroid Biochem 23:843–847

Nolvadex Adjuvant Trial Organisation (1987) Controlled trial of tamoxifen as a single adjuvant agent in the management of early breast cancer. Br J Cancer 54:608–611

Scottish Cancer Trials Breast Group and ICRF Breast Unit, Guy's Hospital, London (1993) Adjuvant ovarian ablation versus CMF chemotherapy in premenopausal women with pathological stage II breast carcinoma: the Scottish trial. Lancet 341(8856):1293–1298

Shultz GS, Grant MB (1991) Neovascular growth factors. Eye 5:178–180

Steinberg KK, Tacker SB et al. (1991) Meta-analysis of the effect of oestrogen replacement therapy on risk of breast cancer. JAMA 265:1985–1990

Tripathy D (1994) How long should adjuvant tamoxifen be continued? Oncology 8(10):25–30

Wakeling AE, Dukes M, Bowler J (1991) A potent specific pure anti-oestrogen with clinical potential. Cancer Res 51:3867–3873

Weidner N, Semple JP, Welch WR, Folkman J (1991) Tumour angiogenesis and metastasis – correlation in invasive breast carcinoma. N Engl J Med 324:1–8

Future Developments in Adjuvant Systemic Therapy for High-Risk Breast Cancer

G. Bonadonna

Division of Medical Oncology, Department of Cancer Medicine, Istituto Nazionale Tumori, Via Venezian, 1, Milan 20133, Italy

Introduction

In an article written by B. Fisher (1984), the following statement can be found: "Competing forces have continuously played a role in determining the management of breast cancer." This statement should remind research physicians that innovative treatment programs will meet with numerous obstacles at all levels before their achievements can be considered as establishing new paradigms. In fact, most clinicians dealing with research are mainly interested in expanding routine science which flourishes as long as the old paradigms remain the standard that governs clinical activity (Fisher 1992).

Current models of breast cancer management are based on seminal biological concepts corroborated by sound clinical results (Fisher 1992; Bonadonna 1992). According to Fisher, in most instances "operable breast cancer is a systemic disease involving a complex spectrum of host-tumor interrelations and . . . variations in local-regional therapy are unlikely to affect survival" (Fisher 1992). Thus, in high risk women, we should consider surgery an "incorrect form of treatment" and only appropriate variations in systemic drug therapy are likely to improve relapse-free and survival.

Building on Prior Experience

The clinical experience achieved over the past quarter of a century, i.e. after the beginning of modern trials with systemic adjuvant therapy, has been considerable and fruitful. The biological principles based on the concept of occult micrometastases and the results of the initial trials were validated through the International Overview by the Early Breast Cancer Trialists' Collaborative Group (1992). This well known meta-analysis, carried out on 75 000 randomized women, confirmed at 10 years a significant reduction in

Table 1. Absolute benefits from the initial trials with systemic adjuvant therapy

Tumor mortality (%)	Pathologic stage	Absolute benefit (%)
10–20	I, good prognosis	4
20–40	I, poor prognosis	8
40–50	II, all patients	12

Data adapted from Henderson (1994).

both relapse and mortality rates in patients treated with surgical adjuvant chemotherapy and endocrine therapy compared to controls given local-regional treatment alone. However, as shown in Table 1, the absolute clinical benefit remained low.

A recent report from the Milan Cancer Institute (Bonadonna et al. 1995a) showed that in node-positive women the administration of a simple, relatively nontoxic combination regimen such as CMF (cyclophosphamide, methotrexate and fluorouracil) could prolong the beneficial effects of surgery by up to 20 years. Even more important for the scope of this article, the authors concluded that the dose of CMF administered was an important factor in achieving the maximum therapeutic benefit regardless of menopausal status, a finding recently confirmed through a prospective American study (Wood et al. 1994). In 1995 another trial from Milan indicated that in women with extensive nodal involvement sequential chemotherapy with doxorubicin (adriamycin) followed by CMF yielded at 10 years superior results compared with the alternating administration of the same regimens or with classical CMF (Bonadonna et al. 1995b). The conclusion from the above mentioned studies is that drug dose intensity and scheduling represent a promising approach for designing future clinical studies.

Recent experience with high dose chemotherapy combined with autologous bone marrow or peripheral blood progenitor cells reinfusion provides further evidence that the dose-response effect is also important in the adjuvant treatment of breast cancer. Table 2 summarizes the most important 5 year results of the pilot study performed at the Milan Cancer Institute with high dose sequential (HDS) chemotherapy in 67 women presenting with at least ten positive axillary nodes. Both relapse free and total survival rates were significantly superior when retrospectively compared with 58 patients having the same tumor characteristics and treated with sequential doxorubicin followed by CMF (Gianni et al. 1995). To properly assess the relative merits of HDS vs doxorubicin/CMF, a prospective randomized trial is being carried out in women with more than three nodes involved, and as of February 1995 a total of 120 patients have been enrolled.

Two important, additional pieces of informations were recently provided by clinical studies. The first deals with current findings of primary (neoad-

Table 2. Essential results from adjuvant high dose sequential therapy in women with ⩾10 positive avillary nodes treated at the Milan Cancer Institute

	Number of patients	At 5 years (%)	
		Relapse-free survival	Total survival
HDS therapy	67	56	78
Doxorubicin followed by CMF	58	41	60

HDS, high dose sequential; CMF, cyclophosphamide, methotrexate 5-fluorouracil.
Data derived from Gianni et al. (1995).

Table 3. Fssential therapeutic results of doxorubicin (adriamycin) plus paclitaxel (taxol) in patients with advanced untreated breast cancer[a]

	N
Evaluable patients	32
Complete response	13 (41%)
Partial response	17 (53%)
Estimated time to disease progression (months)	>12 months

Updated findings from Gianni et al. (1994).
[a] All patients treated at the Milan Cancer Institute.

juvant) chemotherapy in women with large but resectable tumors. The 5-year results from the Milan study clearly indicate that the degree of tumor response (complete vs partial vs minor) is a marker of outcome (Bonadonna et al. 1995c). The second piece of information involves taxol (paclitaxel) given in clinically overt metastatic breast cancer. As shown in Table 3, the combination of a fixed dose of doxorubicin (60 mg/m^2) with various doses of paclitaxel (125–200 mg/m^2) administered through a 3-h infusion proved to be very effective in women previously untreated with chemotherapy (Gianni et al. 1994), including adjuvant chemotherapy. Objective tumor response was documented in all disease sites including bone recalcification of osteolytic lesions, as confirmed by a team of peer reviewers. Transient cardiac effects were recorded in six of 33 patients evaluable for toxicity.

Toward the Year 2000

Collectively, all clinical studies performed during the past two decades have indicated that new drugs, schedules, doses, and strategies improved treatment outcome in most breast cancer patient subsets. As a consequence of better tools and properly conceived trials, medical treatment (in particular chemotherapy) has gained a more central role within the primary treatment of this common malignancy. In the next decade or so we probably have the opportunity to considerably increase the distant control of breast cancer, should more innovative trials be conceived and started. At present, there are many reasons for a persisting palliative treatment mentality in (too) numerous ongoing adjuvant trials for high risk breast cancer. The reasons include the belief that many women are already "cured" by local-regional therapy and treatment by drugs merely delays recurrence without providing cure. Furthermore, many research physicians still feel uncomfortable about programs including the concepts of dose size and dose intensity; they also fear an increased frequency of second malignancies and late cardiac effects, despite recent evidence (Valagussa et al. 1994a,b) that both side effects have a minimal incidence when confronted with tumor mortality due to lack of disease control by ineffective drug therapy. The palliative treatment mentality is also one of the reasons why endocrine therapy alone is still often preferred instead of full dose chemotherapy (with or without hormones), and the effects of adjuvant therapy are often measured in terms of total survival rather than relapse-free survival.

High Risk Patients

In recent years a plethora of new prognostic and predictive indicators (Gasparini et al. 1993) have made patient selection difficult for the clinician, particularly within the node-negative subset. In the vast majority of published studies, the new markers have been evaluated individually rather than by multivariate analysis, retrospectively rather than prospectively; most important, many case series were accrued over a long period of time, and consequently both staging and treatments were heterogeneous.

Table 4. Primary tumor characteristics associated with ≥20% risk of distant micrometastases

Tumor ≥1 cm and negative hormone receptors
Tumor >2 cm and positive hormone receptors
Positive axillary node involvement
Poor nuclear grade
High proliferative activity
Increased angiogenesis
Peritumoral lymphatic vessel invasion
Bone marrow micrometastases detection by monoclonal antibodies
p53 protein overexpression

In spite of this rather confusing picture, clinicians should base their treatment decision on a few tumor characteristics which are associated with a consistent ⩾20% risk of distant microscopic deposits (Table 4). Often two or more such indicators are concomitantly present. The decision about starting any form of adjuvant treatment, particularly in the node-negative subset of patients, should be made after properly informing the patients about the estimated magnitude of risk: more often than not, what is considered by the individual physician to represent a "moderate" risk is actually felt by the individual patient as a "very severe" risk requiring all reasonable efforts to kill micrometastases.

Priority in Strategy

The priority in treatment strategy should involved the search for new effective drug treatments for high risk women affected by invasive breast cancer. In fact, only the incorporation of active new drugs into innovative combination regimens can potentially further ameliorate treatment outcome; modifications of drug sequence are less likely to have a decisive impact in high risk subsets. We now have the possibility to build on the response rate being achieved in treating clinically disseminated disease through the use of doxorubicin plus taxol. Not since the results from using doxorubicin 20 years ago (Bonadonna et al. 1975) has there been a new and promising drug treatment for breast cancer. The combination of the "best" drugs is now providing exciting new results, including complete tumor response, which far exceed the risk of cardiac toxicity. As in the initial experience with doxorubicin, the clinical effects on the myocardium could be easly diminished to <1%, without compromising drug effectiveness, by decreasing the cumulative dose of anthracycline or substituting epirubicin for doxorubicin (Bonadonna et al. 1993). Thus, building on treatment efficacy appears to be a more fruitful strategy than wasting years (as in the past for doxorubicin) in the dissection of cardiac effects.

There are good possibilities that high dose chemotherapy with hematopoietic support may prove to be the most effective adjuvant treatment. Optimization of the dose-response effect, the use of sequential non-cross-resistant drugs, the shortened duration of therapy (about 8 weeks) compared to conventional drug treatment, the reduced incidence of iatrogenic fatalities (about 1%–2%), and the possibility to administer most of the treatment phases in an outpatient clinic all represent important aspects for future innovative trials. Before being adopted on a large scale it is imperative to evaluate the cost-benefit ratio of ongoing randomized trials testing high dose vs conventional dose chemotherapy. Should treatment(s) utilizing high dose sequential chemotherapy emerge as a superior adjuvant modality, the findings will disclose an entirely new avenue of clinical research.

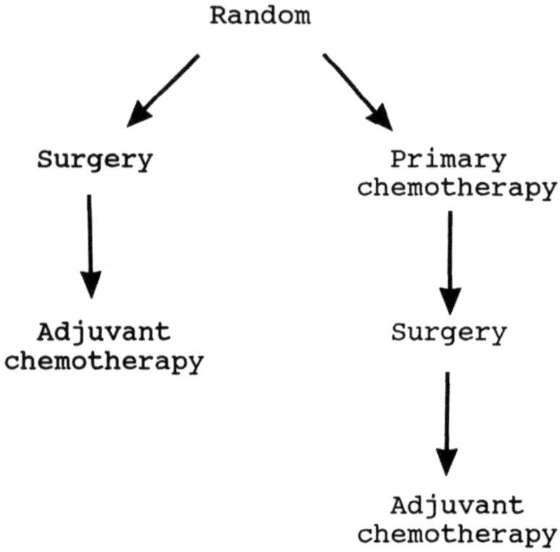

Fig. 1. Proper trial to assess the role of primary chemotherapy in high risk breast cancer

The topic of primary chemotherapy also deserves more attention in the research field. The real question is not whether preoperative drug therapy is, or is not, superior to postoperative administration of the same drugs, but how to properly integrate both modalities in the high risk patient subset (Fig. 1). Since preliminary findings have revealed that the degree of primary tumor response is most probably a simple and useful marker of outcome (Bonadonna et al. 1995c), future studies should focus on how to increase the complete remission rate. This may be achieved by the administration of doxorubicin plus taxol.

Confirmatory Trials

Of great importance are also confirmatory trials aimed at providing answers to specific questions generated by prior experience or not completely clarified through preliminary observations. Examples of such confirmatory trials involve an assessment of treatment efficacy in untreated patients using analogs such as vinblastine and taxotere (docetaxel); whether it is more appropriate to administer chemotherapy and hormones in concomitant vs sequential fashion; whether the combination of goserelin plus tamoxifen is really superior to tamoxifen alone; whether the sequential delivery of doxorubicin (or epirubicin)/cyclophosphamide yields results comparable to those reported after doxorubicin/CMF. Among the potentially innumerable confirmatory trials it remains to be more definitely assessed whether in an adjuvant situation postoperative irradiation should be delivered concomitantly with chemotherapy or after a polydrug regimen to avoid compromising the dose intensity program, as the likelihood of disease recurrence is more frequent in distant than in local sites.

Conclusion

As we are moving toward the year 2000, it is becoming increasingly clear that the scope of treatment of invasive breast cancer is twofold: (1) the eradication of local manifestation(s) of disease while striving to maintain the normal cosmetic appearance of the breast and eradication of systemic disease by effective drug regimens. In a large number of women with tumors ≤2, 0 cm at the largest diameter, breast conserving procedures plus axillary dissection and postoperative irradiation remain an established treatment strategy. Women can be spared routine axillary dissection if the tumor is <0.5 cm while women presenting with involved axillary nodes or other risk factors (see Table 4) should be given adjuvant systemic treatment.

In patients presenting with tumors >2 cm the number of risk factors increases. Thus, more effective combined treatment modalities need to be devised and evaluated by proper trials in order to decrease mortality in spite of current adjuvant treatments. In high risk breast cancer new treatments for distant micrometastases should probably become as intensive as those successfully utilized in the management of bone sarcomas and lymphomas. Such intensive treatments could take into consideration primary chemotherapy with doxorubicin plus taxol for four to five cycles with the specific intent of increasing the frequency of complete remissions; the postoperative adjuvant phase could then include high dose chemotherapy with hematopoietic support, followed by breast irradiation and tamoxifen. The intensive approach must be randomly tested vs a more conventional approach.

In conclusion, current data on new drugs (taxol, taxotere, vinorelbine), regimens (anthracyclines followed by CMF, high dose sequential chemotherapy, chemotherapy followed by hormones) and strategies (e.g., primary chemotherapy) may help reshape the scope and modality of systemic treatment of high risk breast cancer. Innovative controlled studies may conclude that, in high risk women, surgery and radiotherapy are the true adjuvant treatments since the crucial step in disease control is eradication of occult distant micrometastases.

References

Bonadonna G (1992) Evolving concepts in the systemic adjuvant treatment of breast cancer. Cancer Res 52:2127–2137

Bonadonna G, Beretta G, Tancini G et al. (1975) Adriamycin studies at the Istituto Nazionale Tumori, Milan. Cancer Chemother Rep 6:231–245

Bonadonna G, Gianni L, Santoro A et al. (1993) Drugs ten years later: epirubicin. Ann Oncol 4:359–369

Bonadonna G, Valagussa P, Moliterni A, Zambetti M, Brambilla C (1995a) Adjuvant cyclophosphamide, methotrexate, and fluorouracil in node-positive breast cancer: the results of 20 years of follow-up. N Engl J Med 332:901–906

Bonadonna G, Zambetti M, Valagussa P (1995b) Sequential or alternating doxorubicin and CMF regimens in breast cancer with more than three positive nodes. Ten-year results. JAMA 273:542–547

Bonadonna G, Valagussa P, Zucali R, Salvadori B (1995c) Primary chemotherapy in surgically resectable breast cancer. CA 45:227–243

Early Breast Cancer Trialists' Collaborative Group (1992) Systemic treatment of early breast cancer by hormonal cytotoxic, or immune therapy: 133 randomised trials involving 31 000 recurrences and 24 000 deaths among 75 000 women. Lancet 399:1–15, 71–85

Fisher B (1984) The role of science in the evolution of breast cancer management. Heat Memorial Award Lecture. In: Ames FC, Blumenschein GR, Montague ED (eds) Current controversies in breast cancer. University of Texas Press, Austin, pp 1–21

Fisher B (1992) The evolution of paradigms for the management of breast cancer: a personal perspective. Cancer Res 52:2371–2383

Gasparini G, Pozza F, Harris AL (1993) Evaluating the potential usefulness of new prognostic and predictive indicators in node-negative breast cancer patients. J Natl Cancer Inst 85:1206–1219

Gianni AM, Siena S, Bregni M et al. (1995) 5-year results of high-dose sequential (HDS) adjuvant chemotherapy in breast cancer with ≥10 positive nodes. Proc Am Soc Clin Oncol 14:90

Gianni L, Straneo M, Capri G, Villani F, Munzone E, Bonadonna G (1994) Optimal dose and sequence finding study of paclitaxel (P) by 3 h infusion combined with bolus doxorubicin (D) in untreated metastatic breast cancer patients. Proc Am Soc Clin Oncol 13:74

Henderson IC (1994) Adjuvant systemic therapy for early breast cancer. Cancer 74:401–409

Valagussa P, Moliterni A, Terenziani M, Zambetti M, Bonadonna G (1994a) Second malignancies following CMF-based adjuvant chemotherapy in resectable breast cancer. Ann Oncol 5:803–808

Valagussa P, Zambetti M, Biasi S, Moliterni A, Bonadonna G (1994b) Cardiac effects following adjuvant chemotherapy and breast irradiation in operable breast cancer. Ann Oncol 5:209–216

Wood WC, Budman DR, Korzun AH et al. (1994) Dose and dose intensity of adjuvant chemotherapy for stage II, node-positive breast carcinoma. N Engl J Med 330:1253–1259

Summary

H.J. Senn

Department of Medicine C, Kantonsspital, CH-9007 St. Gallen, Switzerland

Adjuvant systemic therapy – the additional treatment of operable breast cancer by some form of cytotoxic and/or endocrine active drugs to decrease relapse rates and increase survival – has always been the "panhandle" of the St. Gallen International Breast Cancer Conferences since their very first meeting in 1979. It is interesting to observe how this "main anthem", while being maintained through the subsequent conferences in 1984, 1988, 1992, and now 1995, has changed its relative position and has increasingly become part of a larger spectrum and more comprehensive vision of how to better cure breast cancer in the future [4]. The initial idea, i.e., to quickly find a simple, effective, and broadly accepted adjuvant chemo- or hormonotherapy for curing breast cancer that is hopefully applicable to most cases, has long since vanished for good.

While making real, but rather modest and very slow progress in ameliorating long-term prognosis of operable breast cancer [3], we are experiencing an ever-increasing segmentalization of breast cancer in prognostic and therapeutic subgroups which in turn requires different forms of adjuvant systemic therapies, from none at all to extremely intensive and expensive ones, depending on ever-more complicated prognostic and therapeutic risk factors! This has led to increasingly more complex, subsequent research questions regarding the optimal timing, dosing, sequencing, and duration of adjuvant systemic therapies, in order to increase effectivity, minimize potential late treatment sequelae, and to avoid compromising the result of subsequent therapies should relapse occur despite prior adjuvant systemic therapies (which is still the fate of too many patients!).

For this reason, *issues of timing and sequence of adjuvant systemic therapies* are analyzed in this session by Monica Castiglione-Gertsch and other exponents of the International Breast Cancer Study Group (IBCSG). Although interesting from the biological and tumor-kinetic point of view, the testing of preoperative (or neoadjuvant) chemotherapy in breast cancer has not yet proven to be superior to conventional timing to date. While in virtually all such studies a relatively large population of women with tumors larger than 3 cm could be subsequently treated with breast conserving measures after primary chemotherapy, true histologic complete remissions by primary chemotherapy are rare [1], and the pharmacological downstaging of tumors might lead to selection of inadequate treatment thereafter. In addition, data from large definitive trials (presently being conducted by NSABP and ECOG in the USA) are still awaited. It is, therefore, correct when the authors stress the point that preoperative adjuvant therapy should be regarded as investigational treatment at the present time.

The same applies to the use of perioperative adjuvant systemic treatments, on which topic the IBCSG has contributed the largest study in the field. After a median follow-up of more than 10 years, the administration of a single cycle of cyclophosphamide, methotrexate, 5-fluoruracil (CMF) is clearly not an effective therapy in node-positive patients, while it partially benefits those with node-negative disease. On the basis of an international overview by the EORTC meta-analysis unit in more than 6000 patients, there is also no clear-cut evidence that perioperative therapy improves the overall survival of patients, with the possible exception of the subgroup with estrogen receptor (ER)-negative N-primaries. In addition, the superiority of delayed re-introduction of adjuvant chemotherapy (based on a mathematical model of tumor growth by Speer and Retsky, first published in 1984 [6]) remains to be shown. One of the problems with prolonged chemotherapy treatment phases clearly is patient (and physician) compliance, and questions of prolonged interference with quality of life. Another yet unresolved area is the clinical relevance of the suggested negative interaction between cytotoxic and endocrine adjuvant therapy, especially tamoxifen. There are presently controversial views on this subject. Hopefully, ongoing large trials will clarify this relevant question of potential antagonism and of the "best combination" in the near future.

The current and future role of adjuvant endocrine therapy in the management of early breast cancer is summarized by Michael Baum and J.R. Benson, mostly based on the Lancet world overview data from 1992 [3], with an interesting theoretical look towards the future. The authors correctly state that adjuvant therapy of breast cancer is becoming ever more complex and "interactive"; to cope with the future evolution, "the next generation of oncologists has to be conversant with biostatics, endocrinology and cell as well as molecular biology". The whole story of ovarian suppression in premenopausal patients has to be revisited as the early trials of surgical and radiation ovarian ablation were conducted over 15 years ago, and since a new class of agents permitting chemical castration have evolved in the past few years, i.e., the gonadotrophin-releasing hormone (GnRH) antagonists. Not only from the oncological, but also from the gynecologic, toxicologic, and osteologic point of view, these agents, which lead to temporary ovarian ablation, might be a desirable alternative for younger women with breast cancer. Recruitment to cooperative studies in which GnRH antagonists are compared with adjuvant therapy or combinations of the two will soon be coming to an end, and the results are awaited impatiently wordwide.

In postmenopausal women, there are few new drugs of interest, but the dominant question at the moment is the duration of the use of adjuvant tamoxifen. The authors criticize the "conceptual rationalism" of most clinicians, who still hang on to the idea that tamoxifen is merely cytostatic and not cytocidal and neglect its direct nonendocrine-mediated effect via insulin-like growth factor (IGF)-1 on breast cancer cells. More sound data about the optimal duration of the use of tamoxifen is urgently needed, due to the potential of selection of tamoxifen-resistant tumor-cell mutants, and especially in the light of recent publications about the reported increased incidence of endometrial cancer, which might be related to dose and duration. The authors stress the necessity of carying out large, meaningful collaborative studies, and urge all surgeons and oncologists treating breast cancer to participate and not leave the responsability for optimizing adjuvant therapy of breast cancer to the "others"! Brief mention is made of a new class of "specific anti-estrogens" [2] which have now entered clinical trials and might become interesting compounds for the as yet open field of chemoprevention. Another unresolved problem in peri- and early postmenopausal patients is hormone-replacement therapy (HRT). Traditionally regarded as contraindicated in women who have undergone breast cancer surgery, the whole issue is now revisited under the presently fashionable viewpoint of quality of life, especially with regard to the fact that there are no sound prospective trial data to prove that HRT is really detrimental to the long-term prognosis of breast cancer patients. The IBCSG and others are finalizing respective study designs, the results of which – although they are much too late – will be greatly appreciated by patients and doctors!

Finally, Baum and Benson confront readers with an interesting hypothesis – schematically outlined in their Figs. 1–3 – showing how the postulated microenvironment of dormant breast cancer cells may be activated by surgery in the follicular phase of the menstrual cycle, and the potential benefit of avoiding this by the use of "neo-adjuvant endocrine therapy". A study into this hypothesis is currently being conducted at the Royal Marsden Hospital in London, UK.

The role of anthracyclines in adjuvant chemotherapy of breast cancer has been critically appraised by Henning Mouridson from the Danish Breast Cancer Study Group in Copenhagen. The whole question has greatly expanded since the NSABP and recently also the IBCSG have included adriamycin (AC) or epirubicin (EC) in their basic chemotherapy regimens (AC × 4 or EC × 4) in virtually all of their trials, and in pre- as well as postmenopausal patients. Their reasons for doing this are not entirely plausible, but are based on identical effects of AC × 4 and CMF × 6, and apparently "better acceptance" of shorter, yet subjectively more toxic (and also more expensive) chemotherapy by the American patients and nurses (NSABP). Furthermore, there is a greater potential for performing studies with dose increases and hematologic growth factors (NSABP, IBCSG, Milan, and others). It is also the stong belief of clinicians as practical oncologists or trialists, that the most effective of the presently available chemotherapy regimens should be available to the patient in the adjuvant setting. However, this unilateral view also requires careful consideration of potential long-term adverse effects by switiching to first-line anthracyclines!

Dr. Mouridsen didactically analyzes all potential trial situations and carefully reviews the role of adjuvant anthracyclines. To summarize, on the basis of presently available data, in his opinion no firm conclusions can be drawn at the moment. This might be due to the fact that most of the previous studies were not primarily designed to elucidate the role of anthracyclines. Furthermore, the demonstration of the superiority of the sequential use of AC followed by CMF (Milan) still awaits confirmation by other trialists in a larger group of patients. While there are a fair number of trials that do not show any superiority of anthracyclines, it seems wise to wait for the results of two hopefully "definitive" and major trials in high-risk patients who were randomized to CMF × 6 or CEF × 6 [5], and the Danish trial DBCG 89D which compared CMF × 9 with CEF × 9, before we jump to any premature conclusions. In addition, it is imperative to carefully reanalyze the long-term cardiac toxicity of anthracyclines in different dose–time regimens – including the potential of self-resolution of these changes in subsequent years. One such study will be started soon by the Swiss Group (SAKK) in the context of current IBCSG studies in pre- and postmenopausal patients treated with and without AC or EC × 4 in addition to CMF and/or tamoxifen or toremifen. Until further evidence is available, every physician treating patients with early breast cancer using adjuvant systemic therapy (with first-line anthracylines!) also has the responsibility of monitoring these patients closely and for long enough in follow-up, regardless of whether they are treated within or outside prospective clinical trials.

The *future developments in adjuvant systemic therapy for high-risk breast cancer patients* is reviewed by Gianni Bonadonna from the Istituto Nazionale Tumori in Milan, Italy. Based on the fact that a relatively simple and conventionally dosed adjuvant combination chemotherapy such as CMF × 6–12 could lead to a prolonged beneficial effect of up to 20 years duration, dose and scheduling of therapies might be a critical factor in achieving maximum therapetic effect, regardless of (menopausal) age, stage of disease, etc. Recent experience from high-dose chemotherapy combined with autologous bone marrow or peripheral blood stemcell retransfusion provide further evidence that the dose–response effect is also important in the adjuvant treatment setting [7]. In addition, there is growing evidence that the degree of tumor response in preoperative, down-staging chemotherapy is a prognostic marker in breast cancer [1]. With regard to the fact that with the advent of taxanes (e.g., paclitaxel, docetaxel) in combination with anthracyclines, we have even more active cytotoxic combination chemotherapy against metastatic breast cancer at our disposal, one may hope for another step forward in adjuvant, curative treatment too.

This leads Dr. Bonadonna to speculate about "breast cancer treatment towards the year 2000". To really achieve progress within a reasonable time, first of all the "palliative treatment mentality" which has greatly marked and also hindered therapeutic evolution in the past 20 years should vanish from the minds of clinicians actively treating early breast cancer! With the logistics and potential of high-dose chemotherapy and growth factors as well as the above-mentioned new drug combinations at hand, the topic of primary chemotherapy urgently deserves more attention in curative breast cancer research. As we approach the year 2000, it becomes increasingly clear that the scope of innovative treatment of invasive breast cancer is twofold in the future: (1) eradication of local manifestation of disease, while maintaining the best possible cosmetic appearance of the breast itself, and (2) eradication of systemic disease by means of more effective drug regimens. More innovative, intensive, and hopefully preoperative chemotherapy approaches must be prospectively compared with more conventional, present-day therapies with regard to effect, acceptance, and costs. In patients with high-risk breast cancer, it becomes ever more apparent that local therapy (breast-conserving surgery and radiotherapy) represents the "adjuvant treatment", while the crucial step in disease control is the more potent pharmacologic eradication of occult distant micrometastases.

More than ever, the timely (and expensive) validation of this vision requires interdisciplinary and international cooperation by open-minded surgeons and physicians in innovative, well-designed trials, to convince the public, the (conservative) medical profession, and also the (sceptical) health-political and health-insurance scene.

References

1. Bonadonna G, Valagussa P, Brambilla C, Ferrari L (1993) Preoperative chemotherapy in operable breast cancer. Lancet 341:1485
2. Chander SK, McCague R, Lugmani Y, Newton C, Dowsett M, Jarman M, Coombes RC (1991) Pyrrolidino-4-iodotamoxifen and 4-iodotamoxifen. New analogues of the antiestrogen tamoxifen for the treatment of breast cancer. Cancer Res 51:5851–5855
3. Early Breast Cancer Trialists' Collaborative Group (1992) Systemic treatment od early breast cancer by hormonal, cytotoxic or immune therapy: 133 randomized trials involving 31000 recurrences and 24000 deaths among 75000 women. Lanct 339:1–15, 71–85
4. Glick JH, Gelber R, Goldhirsch A, Senn HJ (1992) Commentary: adjuvant therapy for primary breast cancer. J Natl Cancer Inst 84:471–475
5. Levine MN, Bramwell V, Pritchard K et al. (1993) A pilot study of intensive cyclophosphamide, epirubicin and fluorouracil in patients with axillary node positive or locally advanced breast cancer. Eur J Cancer 29A:37–43

6. Speer JF, Petrovsky VE, Retsky MW, Wardwell RA (1984) A stochastic numerical model of breast cancer growth that simulates clinical data. Cancer Res 44:4124–4130
7. Wood WC, Budman DR, Korzun AH et al. (1994) Dose and dose intensity of adjuvant chemotherapy for stage II, node-positive breast carcinoma. N Engl J Med 330:1253–1259

VII. Radiation Therapy as Part of Primary Treatment

Radiation Therapy in the Primary Management of Early-Stage Breast Cancer

L.E. Rutqvist

Karolinska Hospital, Radiumhemmet, PO Box 40500, 10401 Stockholm, Sweden

Introduction

The two most important indications for radiation therapy in the primary management of early stage breast cancer are: (1) postoperative treatment of the breast parenchyma following breast-conserving surgery in low-risk patients, and (2) treatment of the breast/chest wall (postmastectomy) and regional nodes in patients with high-risk disease. During the past 2–3 years several controlled clinical trials have contributed important information about the use of radiation therapy in these settings. This article will give a brief overview of some of these studies.

The Role of Radiation Therapy in Breast Preservation

Randomized Trials

A total of four unconfounded randomized clinical trials of radiation therapy following breast-conserving surgery of invasive breast cancer have been published during recent years (Table 1). The total number of randomized patients in these studies is about 3200. "Unconfounded" in this context implies that the treatment arms in the trials have only differed with respect to the radiation therapy. This means that any difference in outcome can be directly attributed to the addition of radiotherapy in one of the treatment groups. Another set of breast preservation trials involving large numbers of patients have compared different types of mastectomy with breast preservation plus radiation therapy. However, these studies studies will not be considered here since they do not directly address the role of the radiation therapy.

The largest and perhaps most well-known study is the NSABP B-06 trial (Fisher et al. 1989). It is a three-arm trial comparing mastectomy, lumpec-

Table 1. Breast recurrence rates in unconfounded randomized trials of radiation therapy after breast-conserving surgery

Trial, stage, follow-up	Breast recurrence rate (%)		P
	Surgery alone	Surgery + RT	
NSABP B-06 (8 years)			
N0	37	12	<0.001
N+	43	6	<0.001
Uppsala-Örebro (T1N0)			
3 years (1990)	8	3	0.06
5 years (1994)	18	2	<0.001
Milan (3 years)	9	0.3	0.001
Ontario (N0, 3.5 years)	26	6	<0.001

RT, radiation therapy.

tomy, and lumpectomy plus radiation therapy to the breast. The two latter treatment arms, involving 1450 patients, permit an unconfounded estimate of the effect of the radiation. At a median follow-up of 8 years there was a significant reduction of breast recurrences among the lumpectomy patients allocated to radiation compared to those treated with lumpectomy alone (Table 1). All patients entered in the trial were required to be clinically node-negative. Those patients who were found to be microscopically node-positive all received adjuvant chemotherapy. Despite this treatment the rate of breast recurrence in the lumpectomy group not receiving radiation therapy was high (43%).

The Swedish Uppsala-Örebro trial was specifically designed to assess whether an adequate local control rate could be achieved without radiation therapy using a strictly standardized surgical technique (sector resection) with meticulous histopathological confirmation of a complete excision (Uppsala-Örebro Breast Cancer Study Group 1990). The total number of patients was 381. The trial only included histopathologically node-negative patients with a tumor diameter less than 20 mm and a 20 mm margin free of in situ or invasive lesions. In addition a perioperative specimen radiography was mandatory. In the first report of the trial, published in 1990 at a median follow-up of 3 years, the breast recurrence rate in both treatment groups was low and not significantly different. However, in the 5-year follow-up, published in 1994, the breast recurrence rate among the control patients had increased to 18% compared to 2% among the unirradiated controls ($p < 0.001$) (Liljegren et al. 1994). It is possible that the short follow-up and the large number of screen-detected patients in the study help to explain the low recurrence rates in the first report of the study. However, the authors suggested that a less meticulous surgery in some collaborating community centers during the later part of the patient accrual period also may have contributed to the difference at 5 years.

The Milan trial was a single institution study including 567 node-negative patients with tumors less than 2 cm (Veronesi et al. 1993). The surgery consisted of a quadrantectomy. At a median follow-up of 3 years there was a significant reduction of breast recurrences among the irradiated patients but the rate among the control patients was also low (9%).

The Ontario trial involved 837 patients with clinically node-negative tumors less than 4 cm (Clark et al. 1992). Ranomization was between lumpectomy and lumpectomy plus radiation therapy. At a median follow-up of 3.5 years there was a highly significant decrease of breast recurrences in the irradiated group (Table 1).

In summary, these trials have all shown that radiation therapy to the breast results in a substantial reduction of breast recurrences. Without radiation, 20%–40% of the patients develop such a recurrence and the risk appears to increase continuously with increasing follow-up. The Milan trial showed an apparently lower breast recurrence rate in the control group than the other trials. However, it should be pointed that that this trial used the most extensive surgery of all studies (quadrantectomy). Moreover, it was the only single institution study.

Can Radiation Therapy Be Avoided in Selected Subgroups?

Despite the substantial reduction of breast recurrences with radiation therapy in the above mentioned trials, it should be acknowledged that the majority of patients would probably have not developed a breast recurrence even without radiation therapy, as evidenced by the results among the unirradiated controls. Therefore, routine radiation therapy to the breast probably constitutes overtreatment in many patients. The Uppsala-Örebro trial was specifically designed to test whether radiation therapy safely could be avoided in a selected subgroup of patients. Obviously, the 5-year results indicate that the entry criteria used in that trial failed to identify such a subgroup.

These observations highlight the importance of prognostic factors for assessing breast recurrence. Many studies have identified nodal status, patient age, and the presence of extensive intraductal cancer as independent prognosticators. However, in the Ontario trial, which only involved node-negative patients, the breast recurrence rate at 3.5 years was 13.5% in the unirradiated subgroup of patients above 50 years without an extensive intraductal component. The authors concluded that they had been unable to identify a subgroup for which radiation therapy safely could be avoided (Clark et al. 1992). These observations underscore the need for an improved understanding of prognostic factors as well as the importance of trials aiming to assess the effect of systemic therapies on the breast recurrence rate. At present, postoperative treatment of the breast parenchyma following breast-conserving surgery continues to be routine at most treatment centers. On the basis of an evaluation of current knowledge about prognostic factors for breast recurrences, Gelber and Goldhirsch suggested in 1994 that the following criteria be used for patient entry into a controlled trial radiation therapy trial aiming to assess whether radiation to the breast safely can be avoided: postmenopausal, node-negative, tumor size less than 2 cm, radical excision, no vessel invasion, no ductal carcinoma in situ (DCIS) component, and no evidence of multicentric disease.

The Role of Systemic Therapy in Breast Preservation

In mastectomized patients adjuvant chemotherapy alone has a marginal effect on local recurrence rates. The same appears to apply to breast recurrences following breast-conserving surgery. The node-positive patients in the NSABP B-06 trial all received adjuvant chemotherapy but the breast recurrence rate after 8 years was 43% among those allocated to no radiation therapy. A similar observation was made in a Scottish trial in which the breast recurrence rate at 4 years exceeded 30% among patients treated with with CMF chemotherapy following breast-conserving surgery without postoperative radiation (Stewart et al. 1989).

In contrast, available information from adjuvant tamoxifen trials indicates that tamoxifen substantially decreases loco-regional recurrence rates in mastectomized patients. Preliminary data from the Stockholm Adjuvant Tamoxifen Trial suggest that tamoxifen therapy results in a substantial reduction of breast recurrences (Rutqvist 1995, unpublished data). In a subgroup of about 400 node-negative women who had been treated with breast-conserving surgery plus radiation therapy, the breast recurrence rate at 10 years was about 10% among the controls compared to 4% among the tamoxifen-allocated patients ($p = 0.02$). No results have as yet been published from the BASO-II study, which is a British, randomized, 2 × 2 factorial breast preservation trial comparing adjuvant tamoxifen, adjuvant tamoxifen plus radiation therapy, radiation therapy, and surgery alone. This study will hopefully permit an assessment of whether the breast recurrence rate with tamoxifen alone is sufficiently low in a defined subgroup of patients so that radiation therapy safely can be avoided.

Optimal Target Volume?

All mentioned breast preservation trials have used radiation therapy encompassing the entire breast parenchyma. However, breast recurrences appear to be most common in the involved quadrant. This observation was he basis for a Christie Hospital trial involving node-negative patients with umors less than 4 cm (Ribeiro et al. 1990; Ribeiro 1994, unpublished data). The randomization was between radiation therapy to the involved quadrant ınd radiation therapy to the entire breast. At 6.5 years the breast recurrence ate in the two treatment groups was 15% and 10%, respectively, for patients with ductal cancers compared to 33% vs 8% for patients with obular cancers. The potential use of a decreased target volume thus appears o be hampered by the current lack of reliable information on prognostic actors for breast recurrence and the lack of trials that have directly compared different surgical procedures involving variations in radicality. It ippears reasonable to assume that different histopathological subtypes of cancer may exhibit differences in the distribution of breast recurrence within he breast parenchyma.

Prognostic Significance of Breast Recurrences and Overall Survival

Many patients with breast recurrences after breast-conserving surgery may be treated by salvage surgery. Survival following such recurrences appears o be substantially better than after locoregional recurrences occurring postmastectomy. In a series from the Stockholm Breast Cancer Group the estimated survival 5 years after a breast recurrence was 60% (Rutqvist 1995, unpublished data). Although a disease recurrence has considerable psychosocial consequences for many patients and often necessitates a mastectomy, the high local control rate following salvage surgery suggests that a "watch policy" may be appropriate provided that the breast recurrence does not compromise the chance of ultimate cure. Two published studies have addressed this issue, that is, the prognostic implications of a breast recurrence.

On the basis of an analysis of the NSABP B-06 trial Fisher et al. (1991) concluded that postlumpectomy breast recurrences are associated with a poor long-term outcome. However, they suggested that the breast recurrence was only a marker of concomitant systemic spread rather than the cause of such spread. In an analysis based on the Ontario trial, Whelan et al. (1994) made similar observations as to the outcome following the breast recurrences but concluded that it was not possible to determine whether the recurrences were the cause of distant dissemination in some patients or if they simply were markers of systemic disease, as suggested by Fisher. The partly conflicting conclusions in these two studies illustrate that the prognostic significance of breast recurrences can only be reliably assessed by analyses of the overall survival in the above mentioned unconfounded radiation therapy trials. If breast recurrences simply are markers and not the cause of distant dissemination, overall survival should be the same irrespective of the widely different breast recurrence rates among the radiotherapy and control groups in the trials.

Table 2. Overal survival in unconfounded randomized trials of radiation therapy after breast-conserving surgery

Trial, stage	Overall survival (%)		p
	Surgery alone	Surgery + RT	
NSABP B-06			
N0	76.6	82.9	n.s.
N+	60.3	68.3	n.s.
Uppsala-Örebro	90.3	91.0	n.s.
Milan	–[a]	–[a]	n.s.
Ontario	91.0	92.1	n.s.

RT, radiation therapy.
[a] No significant difference, exact figures not reported.

Table 2 shows the observed survival by allocated treatment in three of the unconfounded randomized trials. None of the studies has demonstrated a substantial difference between the radiation therapy and control groups and all observed differences are statistically not significant. However, all three studies show slight differences in favor of the irradiated patients. This was particularly the case in the study with the longest follow-up, the NSABP B-06 trial, in which the difference was about 6%–8% at a median follow-up of 8 years. Exact figures on overall survival have not been published from the Milan trial. The authors have simply reported that there were no statistically significant differences. Clearly, the question of whether prevention of breast recurrences through the use of radiation therapy confers a survival advantage remains unresolved. It is probably best addressed in an overview of all available trials and may require longer follow-up than is currently available.

The Role of Radiation Therapy in Ductal Carcinoma In Situ

For many years mastectomy was the only treatment option generally recommended for patients with DCIS. However, with the increasing use of breast preservation for treating invasive breast cancer, many have advocated that similar surgical procedures should be appropriate also in patients with localized DCIS. Unfortunately, data on local recurrence rates and on the radioresponsiveness of DCIS have been scanty. In a histopathological review of patients included in the NSABP B-06 trial some patients were found to have DCIS rather than invasive breast cancer. At a median follow-up of 39 months the breast recurrence rate among patients allocated to lumpectomy alone was five of 22 (23%) compared to two of 29 (7%) among those treated with lumpectomy plus radiation. This observation suggested that DCIS has roughly the same degree of sensitivity to radiation as invasive ductal cancer. This hypothesis was later tested in a prospective randomized trial by the NSABP with included a total of 818 patients (Fisher et al. 1993). At a median follow-up of 3.5 years the breast recurrence rate was 16% among the controls vs 7% among those allocated to breast irradiation ($p < 0.001$). Invasive recurrences were reduced from 8% to 2% ($p < 0.001$) and noninvasive recurrences from 8% to 5% ($p = 0.06$). Onogoing trials in Europe have been designed to assess the effect of tamoxifen with or without radiation therapy and to further study the long-term risk of breast recurrence according to the histologic subtype of DCIS.

Update on Postmastectomy Radiation Therapy

The Cuzick Overviews

In 1987, Cuzick et al. published two overviews including selected, long-term radiation therapy studies. No overall survival benefit with radiation was observed during the first 10 years of follow-up. The radiation therapy used in the trials was even suggested to have had an adverse effect on survival in subgroups of patients after 10–15 years. These results were interpreted to support the assumption that adjuvant radiation therapy is unable to prevent distant dissemination and that a survival benefit only can be achieved with systemic treatments aimed at eradicating distant micrometastases. However, the overviews have been criticized because they included several trials using inadequate treatment techniques, for example, orthovoltage radiation, or with methodological flaws such as biased methods of treatment allocation (Levitt 1988). These circumstance may have obscured a clinically worthwhile benefit with radiation in the more recent trials using modern, megavoltage techniques.

In a recent update there was no longer any significant survival difference between the irradiated and control patients (Cuzick et al. 1994). The reason for this change was that the results for the irradiated patients were, as expected, significantly better in the more recent megavoltage trials, and these studies contributed more long-term data in the updated analysis than they did in the original report. In an analysis of cause-specific mortality among the 10-year survivors there was a significantly reduced mortality due to breast cancer in the more recent trials compared to no such effect in the early Manchester studies which used orthovoltage radiation. However, this benefit was balanced by significantly increased cardiac mortality in most studies, even the more recent megavoltage trials.

The updated Cuzick overview thus confirms previous observations reported in the first Stockholm radiaton therapy trial (Rutqvist et al. 1992). In an analysis of cause-specific mortality according to treatment technique in that study, it was demonstrated that late side effects in the myocardium were related to the distribution of biological dose of radiation in the heart and that an increased cardiac mortality could be avoided through the use of modern megavoltage techniques giving a lower dose to the myocardium.

These observations highlight the importance of appropriate radiation treatment techniques and the significance of local tumor control in the primary management of breast cancer. The updated Cuzick overview confirmed that loco-regional radiotherapy has an impact beyond local control of the disease.

Cardiac Effects from Radiation Therapy to the Breast Parenchyma?

The mentioned cardiac effects of radiation therapy are mostly based on analyses of trials involving treatment of regional nodes particularly the internal mammary nodes. Such treatment generally implies higher radiation doses to the myocardium than is the case when the target volume includes only the breast parenchyma The most common technique for radiation therapy following breast-conserving surgery of low-risk, node-negative patients is tangential photon fields with the patient in a supine position. In node-negative patients, the internal mammary nodes are usually not included in the target volume; this implies that the irradiated heart volume is small. However, depending on the individual anatomy and the extension of the breast parenchyma, a significant proportion of the heart may well be included in the tangential fields in some patients even if no attempt is made to include the internal mammary nodes. Therefore, during treatment planning and patient set-up, care is required in the delineation of the target volume in order to exclude as much as possible of the myocardium (and lung) from the treatment fields. Techniques in which the target volume is defined in relation to fixed anatomical landmarks should probably be avoided. Alternative methods for patient set-up and portal arrangements have also been suggested which would result in lower doses to the myocardium. Treatment planning based on CT or transverse X-ray tomography should be encouraged so that the dose distribution to the heart can be assessed in individual patients.

The 1990 Oxford Overview

The Cuzick overview was based on selected, older studies and did not include more recent trials in which radiation therapy has been tested in the presence of adjuvant chemotherapy or adjuvant tamoxifen. The Early Breast Cancer Trialists' Collaborative Group (EBCTCG) have attempted to conduct a comprehensive overview of all available radiation therapy trials initiated before 1985. However, due to delays because of problems with collection of additional data on patient selection and treatment technique in the available studies, no results have as yet (February 1995) been published from that overview.

Optimal Treatment Schedule and Target Volume?

Despite the fact that randomized radiation therapy trials have been conducted since more than 40 years, there are still several pivotal issues that remain unresolved. One of them concerns treatment schedule. In a survey from the Princess Margaret Hospital in Toronto a total of 48 different radiotherapy schedules had been used following breast-conserving surgery in 551 patients treated outside of controlled trials (Whelan et al. 1993). The total dose varied between 40 and 66 Gy, the number of fractions between 15 and 30, and some patients received a boost dose whereas others did not. In a similar survey of routines at all 15 Swedish radiotherapy centers in 1993, six different schedules were used with a total dose varying from 48 to 56 Gy, the number of fractions ranged from 20 to 27. Routine boost was generally not used (Rutqvist 1995, unpublished data). At present, there are no randomized trials that have assessed the significance of different schedules or of boost treatment. Retrospective studies have, however, conclusively demonstrated that large fraction sizes (>2.5 Gy) and concurrent radiochemotherapy leads to an increased frequency of late side effects and compromises cosmetic results.

In postmastectomy radiation therapy of patients with high-risk tumors the internal mammary nodes are sometimes included in the target volume. These nodes are located close to the heart, and the dose-volume in the myocardium could be diminished if they were excluded. Is it then important to treat these nodes? No randomized radiation trials have specifically studied the relevance of internal mammary node treatment. Nor are there any trials that have specifically addressed the relevance of treatment of other regional nodes, that is, in the axilla and supraclavicular fossa. The only data that are available are based on retrospective analyses.

Metastases in the internal mammary nodes are more common in medially located breast cancers than in cancers located in the lateral part of the breast. The frequency of internal mammary node metastasis in medially located node-positive breast cancers have been reported to be about 50%–55% compared with 20%–25% in lateral, node-positive cancers. Analyses of the survival benefit with radiation therapy in node-positive patients according to the location of the tumor have generally suggested a greater benefit for patients with medial tumors. In a multivariate analysis including more than 1200 patients from the Institut Gustave Roussy, treatment of the internal mammary nodes, either with surgery or radiation, was found to be associated with a significant overall survival benefit (Arriagada et al. 1988). These results suggest that treatment of internal mammary nodes may be important in a subgroup of node-positive patients.

References

Arriagada R, Le MG, Mouriesse H et al. (1988) Long-term effect of internal mammary chain treatment. Results of a multivariate analysis of 1204 patients with operable breast cancer and positive axillary nodes. Radiother Oncol 11:213–222

Clark RM, McCullough PB, Levine NM et al. (1992) Randomized clinical trial to assess the effectiveness of breast irradiation following lumpectomy and axillary dissection for node-negative breast cancer. J Natl Cancer Inst 84:683–689

Cuzick J, Stewart H, Peto R et al. (1987a) Over-view of randomised trials of postoperative adjuvant radiotherapy in breast cancer. Cancer Treat Rep 71:15–29

Cuzick J, Stewart H, Peto R et al. (1987b) Over-view of randomised trials comparing radical mastectomy against simple mastectomy with radiotherapy in breast cancer. Cancer Treat Rep 71:7–14

Cuzick J, Stewart H, Rutqvist LE et al. (1994) Cause-specific mortality in long-term survivors of breast cancer who participated in trials of radiotherapy. J Clin Oncol 12:447–453

Fisher B, Redmond C, Poisson R et al. (1989) Eight-year results of a randomized clinical trials comparing total mastectomy and lumpectomy with or without irradiation in the treatment of breast cancer. N Engl J Med 320:822–828

Fisher B, Anderson S, Fisher ER (1991) Significance of ipsilateral breast tumor recurrence after lumpectomy. Lancet 338:327–331

Fisher B, Costantino J, Redmond C et al. (1993) Lumpectomy compared with lumpectomy and radiation therapy for the treatment of intraductal breast cancer. N Engl J Med 328:1581–1586

Gelber RD, Goldhirsch A (1994) Radiotherapy to the conserved breast: is it avoidable if the cancer is small? J Natl Cancer Inst 86:652–654

Levitt S (1988) Is there a role for post-operative adjuvant radiation in breast cancer? Beautiful hypothesis versus ugly facts: 1987 Gilbert H. Fletcher lecture. Int J Radiat Oncol Biol Phys 14:787–796

Liljegren G, Holmberg L, Adami H-O et al. (1994) Sector resection with or without postoperative radiotherapy for stage I breast cancer: five-year results of a randomized tral. J Natl Cancer Inst 86:717–722

Ribeiro GG, Dunn G, Swindell R et al. (1990) Conservation of the breast using two different radiotherapy techniques: interim report of a clincal trial. Clin Oncol 2:27–34

Rutqvist LE, Fornander T, Lax I, Johansson H (1992) Cardiovascular mortality in a randomized trial of adjuvant radiation therapy versus surgery alone in primary breast cancer. Int J Radiat Oncol Biol Phys 22:887–896

Stewart HJ, Prescott RJ, Forrest PA (1989) Conservation therapy of breast cancer. Lancet 2:168–169

Uppsala-Örebro Breast Cancer Study Group (1990) Sector resection with or without postoperative radiotherapy for stage I breast cancer: a randomized trial. J Natl Cancer Inst 82:277–282

Veronesi U, Luinin A, Del Vecchio M et al. (1993) Radiotherapy after breast-preserving surgery in women with localized cancer of the breast. N Engl J Med 328:1587–1591

Whelan T, Marcellus D, Clark R et al. (1993) Adjuvant radiotherapy for early breast cancer: patterns of practice in Ontario. Can Med Assoc J 149:1273–1277

Whelan T, Clark R, Roberts R et al. (1994) Ipsilateral breast tumor recurrence postlumpectomy is predictive of subsequent mortality: results from a randomized trial. Int J Radiat Oncol Biol Phys 30:11–64

Quality Assurance in Early Breast Cancer Treatment: Clinical Aspects of Postoperative, External, Whole Breast Irradiation

R. Valdagni, C. Italia, P. Montanaro, and M. Ciocca

Division of Radiation Oncology, Clinica S. Pio X, via F. Nava 31, 20159 Milan, Italy

Introduction

"Quality assurance in medicine is the process through which we attempt to monitor the actual quality of care given to an individual patient, to a patient group or to a population. When deficiencies are identified, measures taken for corrective action are formulated" (van der Schueren et al. 1993).

From a theoretical point of view, this definition of quality assurance includes four distinct concepts: (1) the process approach, or the way to assess quality through specific quality control procedures, (2) the standards of reference which define adequate or inadequate care, (3) the identification of deficiencies or quantitative discrepancies between predefined measurements (the standard of reference) and the observed calculated measurements, and (4) the possible corrective action to keep/regain conformance with the standard or to modify/change the standard of reference when identified discrepancies exceed predefined values of acceptance. If the starting point of a specific quality assurance program is the standard of reference, the key event is represented by the quality control process, also known as the audit loop (Fig. 1).

Through the objective and systematic evaluation of the quality and appropriateness of patient care, the audit loop analyzes the way treatment is actually delivered and correctly applied and introduces corrections considered necessary to improve the quality.

In radiation oncology, it is necessary to control and optimize every step of the radiotherapy process in order to achieve an adequate and uniform quality of care given to patients. Three principal areas of investigation have been identified: structure, process and outcome (American College of Radiology 1989; Intersociety Council for Radiation Oncology 1991; van der Schueren et al. 1993).

The structure is assessed by the way a radiation oncology center is equipped and organized: quality control procedures consequently evaluate

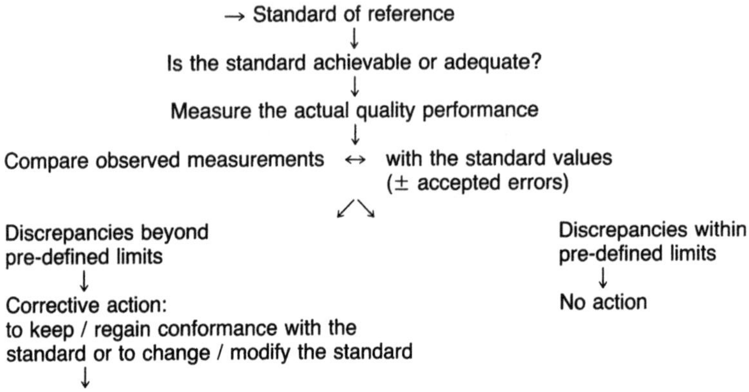

Fig. 1. The quality control process loop or audit loop

equipment, facility and personnel skills. The radiotherapy process is a very complex event: it is assessed by the way treatment is given in each step of the pretreatment evaluation phase, treatment prescription and preparation phases and lastly, in the treatment execution phase. The outcome is assessed by the analysis of treatment results: several end-points can be considered, such as tumor control, survival, palliation, acute and late side effects, cosmetic results, patient satisfaction and quality of life.

In general, it is difficult to document the direct influence of quality of radiotherapy on therapy results because educational, technical, physical and clinical variables are strongly interrelated and conditioned by each qualitative level of radiotherapy structure, process and outcome. Furthermore, in oncology, treatment results are frequently assessed several years later, making identification of possible inadequacies more difficult.

When we deal with quality assurance in early breast cancer (EBC) postoperative, external, whole breast irradiation, several variables, mainly related to the radiotherapy process, have been found to influence treatment outcome. Prescribed total dose, weekly dose, and split course together with the utilization of large fractions per day and the avoidance of boosting the tumor bed when margins are positive have been shown to influence local control rates (Hayward 1974; Osborne et al. 1984; Clarke et al. 1985; Recht et al. 1985; Van Limbergen et al. 1987; Kurtz et al. 1991; Ryoo et al. 1989; Van Limbergen 1993; Fowble 1994). Prescribed total dose, dose per fraction, large volume boost, large breast size and dose inhomogeneity within the target volume have been found to modify cosmetic outcome (Harris et al. 1979; Clarke et al. 1983; Overgaard 1985; Olivotto et al. 1989; Pierce et al. 1992; Wazer et al. 1992; Sarin et al. 1993; Van Limbergen 1993; Borger et al. 1994; Moody et al. 1994; Pezner 1994). For lung and cardiac toxicities, adopted treatment techniques, the use of deep tangents, employed treatment units and dose prescription have been suggested to condition morbidity rates

(Van Dyk et al. 1981; Roberson et al. 1982; Host et al. 1986; Haybittle et al. 1989; Janjan et al. 1989; Lingos et al. 1991; Fuller et al. 1992; Rutqvist et al. 1992; Neugut et al. 1993; Cuzick et al. 1994; Gyenes et al. 1994; Houghton et al. 1994; Inskip et al. 1994; Valagussa et al. 1994).

Guidelines to set standards of adequacy have been recently proposed by the EORTC-EUSOMA Consensus Meeting (Bartelink et al. 1991); this report represents the European standard of reference of quality assurance in EBC postoperative irradiation. Physical and dosimetric aspects of quality assurance are well defined and based on comprehensive programs developed by the medical radiation physics community (ICRU 1978; AAPM 1984; Intersociety Council for Radiation Oncology 1991; Starkschall and Horton 1991). Guidelines are recommended to prevent or correct errors particularly in the radiotherapy process, i.e., in diagnostic patient data acquisition, simulation and treatment unit functioning, treatment aid devices, dose computation and distribution, and treatment verification (Bartelink et al. 1991).

From a clinical point of view, the radiotherapy process includes several complex, interlinked and critical steps: the Consensus report gives guidelines for obtaining good and uniform quality in the four radiotherapy treatment phases. In the pretreatment evaluation phase, clinical data (tumor location, distance from the nipple, volume, etc.), radiological data (tumor location, microcalcifications, dimensions, etc.) and pathological data (tumor size, macroscopic and microscopic involvement of margins) should be checked and confirmed. In the treatment prescription phase (treatment modality and basic treatment technique), specific recommendations were given concerning the application and alignment of tangent beams to reduce the irradiated lung volume with respect to its cranial, ventral and dorsal components (use of collimator rotation, table wedge, dorsal beam alignment, half-beam block, etc.). In addition, guidelines regarding the prescription of total dose and dose per fraction, together with recommendations to accept a dose gradient between *D*max and *D*min in the target volume of less than 15%, were established.

In the treatment preparation phase (patient position and arm position; use of patient position modifiers, immobilization devices, breast shape modifiers; methodology to localize the target volume; etc.) and in the treatment execution phase (patient set-up, use of treatment aids and patient set-up aids, frequency of off-line or on-line verification of geometrical and dosimetric accuracy, evaluation of acute morbidity during treatment, etc.), the recommendations are somewhat less rigid probably because these clinical aspects of the radiotherapy process are frequently technique- and structure-dependent. As underlined in the Consensus report, "The resources and the situations for the treatment techniques in the different institutions vary considerably." Therefore, partially individualized, intrainstitutional, quality assurance programs should be organized in order to assess the quality of radiotherapy given to patients by each center.

Table 1. General and specific routine quality assurance procedures in early breast cancer postoperative external irradiation: physical and clinical aspects

External beam treatment equipment (Linacs)
Daily/weekly/monthly treatment checks:
Mechanical and geometrical checks
Dosimetry checks
Safety system checks
Conventional simulation equipment
Mechanical and geometrical and safety system checks
Image quality consistency checks
CT scanner
Mechanical and geometrical checks
CT number accuracy
Data display, manipulation and transfer checks
Treatment planning system and dose calculations
Hardware and software checks
Patient data transfer
Beam data library and algorithms
Monitor unit calculations
Treatment aids
Half-beam block and/or shielding blocks (geometrical accuracy and beam attenuation)
Immobilization devices
Tissue compensators (bolus, 3-D stepped compensators and related mechanical apparatus)
Patient-related procedures
Pretreatment evaluation
Treatment prescription
Treatment preparation
Treatment execution (daily chart checks, weekly chart rounds, portal imaging, in vivo dosimetry, weekly analysis of acute toxicities)
Outcome
Evaluation at follow up: local control, survival, late effects, cosmesis
Education
New personnel training
In-service training

Within our quality assurance program (Table 1) assessing the quality and appropriateness of whole breast external irradiation in EBC, several areas have been the subject of clinical investigation. Three examples are given below.

Evaluation of Conventional Methods To Localize the Clinical Target Volume (Radiotherapy Process: Treatment Preparation Phase)

Most radiation oncology textbooks or specific reports (Sack 1976; Moss et al. 1979; Fletcher et al. 1980; Lichter and Padikal 1983; Dobbs and Barrett

1985; Chu 1989; Yarnold and Bloom 1989; Bartelink et al. 1991; Fowble et al. 1991; Mendenhall 1991; Ribeiro 1991; McNeese et al. 1992; Perez et al. 1992; Wilson 1994) suggest localizing the target volume (i.e., the residual mammary gland) with conventional methods, i.e., palpation or anatomical reference lines. These two methods consequently represent the standard of reference for clinical target volume (CTV) localization. To evaluate the accuracy of these localization procedures, ultrasound (US) determination of the remaining breast tissue volume in 100 consecutive patients undergoing irradiation has been compared with determinations from conventional methods (Valdagni et al. 1994a). A comparison of these measurements shows: (a) A general overestimation of the CTV using both anatomical reference lines and palpation when mean and median values of each field border are considered. Anatomy overestimates the lateral field border in 74% of patients and palpation in 63% of patients. The medial border is overestimated in 99% of patients with anatomy and in 72% with palpation. (b) An underestimation of at least one field border of the CTV in 55% of patients with anatomy and in 36% of patients with palpation.

The most critical margin to be localized with conventional methods appears to be the superior: an underestimation of the cranial component of the mammary gland has been observed in 51% of patients with anatomy and in 22% of patients with palpation, the problem of "missing the target" more frequently occurring in premenopausal women. The preliminary conclusion from this quality control procedure is that the standards of reference (both conventional methods) are not adequate to localize the target. Due to the discovery of a systematic error, corrective action has become necessary, i.e., inclusion of state of the art imaging techniques to localize the CTV (Fig. 2). The clinical implications of these findings may include a possible reduction of undesired lung/heart irradiation (and related acute and late morbidity) and, through the optimization of CTV localization, a potential decrease in local recurrence rates.

→ Standard of reference:	CTV localised using palpation or anatomical reference lines
↓	
Quality control:	CTV localised using ultrasound
↓	
Comparison of measurements:	The standard is not adequate: geographical miss and overestimation of CTV
↓	
Are corrections needed?	Yes (systematic error)
↓	
Major correction:	CTV localised using US (CT under evaluation)
↓	

Fig. 2. Audit loop on conventional clinical target volume (CTV) localization methods

The Appropriateness of Adding Conventional and Empirical Geometrical Port Margins to the Clinical Target Volume To Determine the Planning Target Volume (Radiotherapy Process: Treatment Preparation and Execution Phase)

The suggestion has often been made (Sack 1976; Moss et al. 1979; Fletcher et al. 1980; Lichter and Padikal 1983; Dobbs and Barrett 1985; Chu 1989; Yarnold and Bloom 1989; Bartelink et al. 1991; Fowble et al. 1991; Mendenhall 1991; Ribeiro 1991; McNeese et al. 1992; Perez et al. 1992; Wilson 1994) to add 1 or 2 cm to the CTV, particularly when the residual mammary gland is localized by palpation. These values should not be considered as an ideal margin (Kijewski 1994). As a matter of fact, they are merely empirical and do not consider the use of immobilization devices, beam characteristics, patient movements, and equipment-related variables (ICRU 1993).

Geometrical port margins are strongly related to the basic treatment technique used and also to the various treatment devices utilized. Consequently, these margins can change when assessed by the different radiation centers. In our department, whole breast irradiation technique is applied using (Valdagni et al. 1992):

1. Two opposed tangential 6 MV photon beams (SSD: 100 cm)
2. A rotatable half-beam block
3. Couch rotation
4. Three-dimensional, stepped, compensating filters
5. An individualized cellulose acetate immobilization cast of bilateral breasts and scapulo-humeral joints, derived from a plaster of paris mold.

Until 1992, the adopted empirical geometrical port margin was 2 cm, which was considered to be the standard of reference. To assess the adequacy of this margin, a comparison of simulation films with biweekly portal films was performed (Valdagni and Italia 1991; Valdagni et al. 1994b). Five variables related to the above mentioned treatment technique were analyzed: cranio-caudal and ventro-dorsal shifts, half-beam block rotation errors and, recently, gantry and couch rotation errors. Comparison of measurements showed: (a) cranio-caudal errors of 2 mm $\pm$ 2 (1 SD) and ventro-dorsal errors of 3 mm $\pm$ 2 (1 SD), and (b) negligible ($<1°$) (Das et al. 1993) half-beam block, gantry and couch rotation errors (Valdagni 1994, unpublished data). As mentioned, these values reflect errors in daily set-up due to equipment- and patient-related variables. Knowing the beam penumbra (approximately 6 mm are related to the presence of the half-beam block and 7 mm to open field borders) and including 1 SD in both cranio-caudal and ventro-dorsal errors, the calculated ideal margins are of about 11 mm for all field borders. Thus, the applied quality control procedure has shown that the empirically adopted standard of reference margin is not adequate and a minor correction should be made to reduce geometrical port margins and consequently the planning target volume (Fig. 3).

→ Standard of reference:	Ideal margin to be added to CTV = 2 cm
↓ Quality control:	Port film analysis
↓ Comparison of measurements:	The standard is not adequate: overestimation of ideal margins
↓ Are corrections needed?	Yes (systematic error)
↓ Minor correction: ↓	Reduce ideal margins to 11 mm

Fig. 3. Audit loop on ideal margins. CTV, clinical target volume

Treatment Morbidity Systematic Evaluation (Radiotherapy Outcome: Acute and Late Side Effects)

Since 1990, acute and late toxicities and cosmetic result have been systematically evaluated using quantitative and qualitative scoring systems (Van Limbergen et al. 1989; Holleb 1992). The preliminary results of quality control procedures that consider treatment outcome aim for a reduction of both acute and late side effects.

Acute Side Effects

Since 1988, three-dimensional, 8 mm thick, stepped brass compensators have been used to improve dose distribution within the breast tissue. Weekly evaluation of acute morbidity in the period 1991–1992 showed that grade 2 toxicity (RTOG/EORTC grade 2: tender erythema or patchy moist desquamation) in the nipple-areolar region and in the inframammary sulcus occurred in 23% of patients. The previously mentioned port film analysis (Valdagni and Italia 1991) showed that the amount of shifts (range of cranio-caudal shift: 0–14 mm; range of ventro-dorsal shift: 0–17 mm) could have caused an undercompensation of these regions. Since 1993 stepped compensators corresponding to free air around the inferior and anterior breast profiles have been added, and grade 2 toxicity has been reduced to 15% ($p = 0.45$).

Late Side Effects

In the period 1989–1991, patients were immobilized using a 2 mm thick, heat and vacuum formed, individualized, cellulose acetate cast. Also, to preserve the build-up effect, windows were opened in the beam area leaving plastic strips to help to maintain the treated breast in a fixed position

(Valdagni et al. 1992). Evaluation of long-term side effects revealed that, after an interval of 10–16 months after completion of treatment, several patients exhibited telangiectases outside the boost region. The morphology of the telangiectasia and a comparison between pictures documenting patient set-up (patients wearing the cast) and those taken to evaluate late effects showed the plastic strips as being the source of the increased skin dose and consequently of the late effects. The casts have been modified and, only in the beam area has a 0.2 mm thick heat- and vacuum-formed, plastic breast mold been substituted for the plastic strips. Outside the field area, the 2 mm thick, cellulose acetate cast is maintained to guarantee patient immobilization. With a follow up of 1–3 years no evidence of telangiectasia has been observed.

As shown by these examples, treatment optimization requires the objective and systematic assessment of radiation therapy quality. It is evident that quality controls in radiotherapy are not theoretical or merely physical-technical procedures. Their clinical impact on standard treatments and on potential technical improvements will surely be relevant and their routine use should therefore be mandatory. "To be effective, quality control must form part of the daily practice of the radiotherapy centre" (Dische 1984).

Acknowledgement. We would like to thank Ms. Tiziana Magnani for her patient and skillful technical assistance and her help in the English version of the manuscript.

References

AAPM (1984) Physical aspects of quality assurance in radiation therapy. American Institute of Physics, New York, pp 5–37

American College of Radiology (1989) Quality assurance program in radiation oncology. American College of Radiology

Bartelink H, Garavaglia G, Johansson KA, Mijnheer BJ, Van den Bogaert W, van Tinehoven G, Yarnold J (1991) Quality assurance in conservative treatment of early breast cancer. Report on a consensus meeting of the EORTC Radiotherapy and breast cancer cooperative groups and the EUSOMA (European Society of Mastology). Radiother Oncol 22:323–326

Borger JH, Kemperman H, Smitt HS, Hart A, van Dongen J, Lebesque J, Bartelink H (1994) Dose and volume effects on fibrosis after breast conservation therapy. Int J Radiat Oncol Biol Phys 30:1073–1081

Chu FCH (1989) Breast cancer. In: Mansfield CM (ed) Therapeutic radiology. Elsevier, Amsterdam, pp 291–325

Clarke DH, Martinez A, Cox RS (1983) Analysis of cosmetic results and complications in patients with stage I and I breast cancer, treated by biopsy and irradiation. Int J Radiat Oncol Biol Phys 9:1807–1813

Clarke DH, Le MG, Sarrazin D, Lacombe MJ, Fontaine T, Travagli JP, May-Levin F, Contesso G, Arriagada R (1985) Analysis of local-regional relapses in patients with early breast cancers treated by excision and radiotherapy. Int J Radiat Oncol Biol Phys 11:137–145

Cuzick J Stewart H, Rutqvist L, Edwards R, Redmond C, Peto R, Baum M, Fisher B, Host H, Lythgoet J, Ribeiro G, Scheurlen H (1994) Cause-specific mortality in long-term survivors of breast cancer who participated in trials of radiotherapy. J Clin Oncol 12:447–453

Das IJ, Cheng CW, Fosmire H, Kase KR, Fitzgerald TJ (1993) Tolerances in setup and dosimetric errors in the radiation treatment of breast cancer. Int J Radiat Oncol Biol Phys 26:883–890

Dische S (1984) Quality assurance in radiation therapy: European experience – present and future clinical efforts. Int J Radiat Oncol Biol Phys 10:55–57

Dobbs J, Barrett A (1985) Practical radiotherapy planning. Arnold, London, pp 119–127

Fisher B, Wickerham DL, Deutsch M, Anderson S, Redmond C, Fisher ER (1992) Breast tumor recurrence following lumpectomy with and without breast irradiation: an overview of recent NSABP findings. Semin Surg Oncol 8:153–160

Fletcher GH, Montague ED, Papley NduV, Barker JL (1980) Radiotherapy in the management of non disseminated breast cancer. In: Fletcher GH (ed) Textbook of radiotherapy, 3rd edn. Lea and Febiger, Philadelphia, pp 527–579

Fowble BL (1994) Conservative surgery and radiation in the treatment of ductal carcinoma in situ and stages I–II breast cancer. 36th annual scientific meeting of American Society for Therapeutic Radiology and Oncology (ASTRO)

Fowble BL, Salin LJ, Schultz DJ (1991) Conservative surgery and radiation for early breast cancer. In: Fowble BL, Goodman RL, Glick JH, Rosato EF (eds) Breast cancer treatment. A comprehensive guide to management. Mosby, St Louis, pp 105–149

Fuller SA, Haybittle JL, Smith REA, Dobbs NJ (1992) Cardiac doses in post-operative breast irradiation. Radiother Oncol 25:19–24

Gyenes G, Fornander T, Carlens P, Rutqvist LE (1994) Morbidity of ischemic heart disease in early breast cancer 15–20 years after adjuvant radiotherapy. Int J Radiat Oncol Biol Phys 28:1235–1241

Harris JR, Levene MB, Svensson G, Hellman S (1979) Analysis of cosmetic results following primary radiation therapy for stages I and II carcinoma of the breast. Int J Radiat Oncol Biol Phys 5:257–261

Haybittle JL, Brinkley D, Houghton J, A'Hern RP, Baum M (1989) Postoperative radiotherapy and late mortality: evidence from the Cancer Research Campaign trial for early breast cancer. Br Med J 298:1611–1614

Hayward J (1974) Conservative surgery in the treatment of early breast cancer. Br J Surg 61:770–771

Holleb AI (1992) Standards for breast-conservation treatment: a landmark report. CA 42(3):133–162

Host H, Brennhovd IO, Loeb M (1986) Postoperative radiotherapy in breast cancer – long-term results from the Oslo study. Int J Radiat Oncol Biol Phys 12:727–732

Houghton J, Baum M, Haybittle JL (1994) Twenty year results from the CRC trial of radiotherapy after mastectomy for breast cancer. Eur J Cancer 30A Suppl 2:S35

ICRU (1978) Dose specification for reporting external beam therapy with photons and electrons. International Commission of Radiation Units and Measurements, Washington (Report no 29)

ICRU (1993) Prescribing, recording and reporting photon beam therapy. ICRU, Bethesda (Report no 50)

Inskip PD, Stovall M, Flannery JT (1994) Lung cancer risk and radiation dose among women treated for breast cancer. J Natl Cancer Inst 86:983–988

Intersociety Council for Radiation Oncology (1991) Radiation oncology in integrated cancer management. Intersociety council for Radiation Oncology, (Report)

Janjan NA, Gillin MT, Prows J, Arnold S, Haasler G, Thorsen MK, Wilson JF (1989) Dose to the cardiac vascular, and conduction systems in primary breast irradiation. Med Dosim 14:81–87
Kijewski P (1994) Three-dimensional treatment planning. In: Mauch PM, Loeffler JS (eds) Radiation oncology: technology and biology. Saunders, Philadelphia, pp 10–33
Kurtz JM, Amalrick R, Brandone H, Ayme Y, Spitalier JM (1991) How important is adequate radiotherapy for the long term results of breast-conserving treatment? Radiother Oncol 20:84–90
Lichter AS, Padikal TN (1993) Treatment planning in primary breast cancer. In: Bleehen NM, Glatstein E, Haybittle JL (eds) Radiation therapy planning. Dekker, New York, pp 639–661
Lingos TI, Recht A, Vicini F, Abner A, Silver B, Harris JR (1991) Radiation pneumonitis in breast cancer patients treated with conservative surgery and radiation therapy. Int J Radiat Oncol Biol Phys 21:355–360
McNeese MD, Fletcher GH, Levitt SH, Kahn FM (1992) Breast cancer. In: Levitt SH, Kahn FM, Potish RA (eds) Technological basis of radiation therapy. Practical clinical application. Lea and Febinger, Philadelphia, pp 232–247
Mendenhall NP (1991) Post-operative irradiation following breast conserving surgical procedures. In: Scandy IK, Copeland EM (eds) The breast. Comprehensive management of benign and malignant disease. Saunders, Philadelphia, pp 781–798
Moody AM, Mayles WPM, Bliss JM, A'Hern RP, Owen JR, Regan J, Broad B, Yarnold JR (1994) The influence of breast size on late radiation effects and association with radiotherapy dose inhomogeneity. Radiother Oncol 33:106–112
Moss WT, Brand WN, Battifora H (1979) The breast. In: Moss WT, Brand WN, Battifora H (eds) Radiation oncology, 5th edn. Mosby, St Louis, pp 289–331
Neugut AI, Robinson E, Lee WC, Murray T, Karwoski K, Kutcher GJ (1993) Lung cancer after radiation therapy for breast cancer. Cancer 71:3054–3057
Olivotto IA, Rose MA, Osteen RT, Love S, Cady B, Silver B, Recht A, Harris JR (1989) Late cosmetic outcome after conservative surgery and radiotherapy: analysis of causes of cosmetic failure. Int J Radiat Oncol Biol Phys 17:747–753
Osborne MP, Ormiston N, Harmer CL, McKinna JA, Baker J, Greening WP (1984) Breast conservation in the treatment of early breast cancer. A 20-year follow up. Cancer 53:349–355
Overgaard M (1985) The clinical implications of non standard fractionation. Int J Radiat Oncol Biol Phys 11:1225–1229
Perez CA, Garcia DM, Kuske RR, Levitt SH (1992) Breast: stage T1 and T2 tumors. In: Perez CA, Brady LW (eds) Principles and practice of radiation oncology. Lippincott, Philadelphia, pp 877–947
Pezner RD (1994) Cosmetic breast fibrosis: it's the local boost! Int J Radiat Oncol Biol Phys 30:1251–1252
Pierce SM, Recht A, Lingos TI, Abner A, Vicini F, Silver B, Herzog A, Harris JR (1992) Long-term radiation complications following conservative surgery (CS) and radiation therapy (RT) in patients with early-stage breast cancer. Int J Radiat Oncol Biol Phys 23:915–923
Recht A, Silver B, Schnitt SJ, Connolly JL, Hellman RS, Harris JR (1985) Breast relapse following primary radiation therapy for early breast cancer. I. Classification, frequency and salvage. Int J Radiat Oncol Biol Phys 11:1271–1276
Ribeiro GG (1991) Breast. In: Pointon RCS (ed) The radiotherapy of malignant disease, 2nd edn. Springer, Berlin Heidelberg New York, pp 255–278
Roberson PL, Lichter AS, Bodner A, Fredrickson HA, Padikal TN, Kelly BA, Van de Geijn J (1982) Dose to lung in primary breast irradiation. Int J Radiat Oncol Biol Phys 9:97–102

Rutqvist LE, Lax I, Fornander T, Johansson H (1992) Cardiovascular mortality in a randomized trial of adjuvant radiation therapy vs surgery alone in primary breast cancer. Int J Radiat Oncol Biol Phys 22:887–896

Ryoo ME, Kagan AR, Wollin M, Tome MA, Tedeschi MA, Rao AR, Hintz BL, Kuruvilla AM, Nussbaum H, Stretter OE, Tabola BR, Miller MJ (1989) Prognostic factors for recurrence and cosmesis in 393 patients after radiation therapy for early mammary carcinoma. Radiology 172:555–559

Sack H (1976) Weibliche und männliche Mamma. In: von Scherer H (ed) Strahlentherapie. Radiologische Onkologie. Springer, Berlin Heidelberg New York, pp 593–615

Sarin R, Dinshaw KA, Shrivastava SK, Sharma V, Deroe SM (1993) Therapeutic factors influencing the cosmetic outcome and late complications in the conservative management in early breast cancer. Int J Radiat Oncol Biol Phys 27:285–292

Starkschall G, Horton J (eds) (1991) Quality assurance in radiation physics. Medical Physics, Madison, pp 311, 364

Valagussa P, Zambetti M, Biasi S, Moliterni A, Zucali R, Bonadonna G (1994) Cardiac effects following adjuvant chemotherapy and breast irradiation in operable breast cancer. Ann Oncol 5:209–216

Valdagni R, Italia C (1991) Early breast cancer irradiation after conservative surgery: quality control by portal localization films. Radiother Oncol 22:311–313

Valdagni R, Ciocca M, Busana L, Modugno A, Italia C (1992) Beam modifying devices in the treatment of early breast cancer: 3-D stepped compensating technique. Radiother Oncol 23:192–195

Valdagni R, Italia C, Montanaro P, Ciocca M, Modugno A, Pellegrino A (1994a) Planning target volume (PTV) in early breast cancer post-operative external irradiation: ultrasound localization of clinical target volume (CTV) and ideal margin identification. Eur J Cancer 30A Suppl 2:S34

Valdagni R, Italia C, Montanaro P, D'Abbiero N, Morandi G, Salvadori B (1994b) Clinical target volume localization in early breast cancer post-operative irradiation: inaccuracy of conventional methods (anatomy and palpation) as demonstrated by ultrasonography. Int J Radiat Oncol Biol Phys 30 Suppl 1:187–188

Van der Schueren E, Horiot JC, Leunens G, Rubens R, Steward W, van Dongen JA, van Oosterom AT, Vantongelen K (1993) Quality assurance in cancer treatment. Eur J Cancer 29A:172–181

Van Dyk J, Keane TJ, Kan S, Rider WD, Fryer CJH (1981) Radiation pneumonitis following large single dose irradiation. A revaluation based on absolute dose to lung. Int J Radiat Oncol Biol Phys 7:461–467

Van Limbergen E (1993) Breast conservative treatment of mammary cancer. A medical decision making model based on qualitative and quantitative clinical experience

Van Limbergen E, Van der Bogaert W, van der Schueren E, Rijnders A (1987) Tumor excision and radiotherapy as primary treatment of breast cancer analysis of patient and treatment parameters and local control. Radiother Oncol 8:1–9

Van Limbergen E, van der Schueren E, Vantongelen K (1989) Cosmetic results: qualitative and quantitative analysis. Radiother Oncol 16:159–167

Wazer DE, Di Petrillo T, Schmidt-Ullrich R et al. (1992) Facts influencing cosmetic outcome and complication risk after conservative surgery and radiotherapy for early-stage breast carcinoma. J Clin Oncol 10:356–363

Wilson JF (1994) The breast. In: Cox JD (ed) Moss' radiation oncology. Rationale, technique, results, 7th edn. Mosby, St Louis, pp 355–401

Yarnold JR, Bloom HJG (1989) Radiation after limited and radical surgery. In: Hoogstraten B, Burn I, Bloom HJG (eds) Breast cancer (UICC). Springer, Berlin Heidelberg New York, pp 181–197

How To Predict the Risk of Local Relapse in the Preserved Breast

J.M. Kurtz

Radiation Oncology Department University Hospital, 24 Rue Micheli-du-Crest, 1211 Geneva 14, Switzerland

Introduction

There are two essential local treatment steps in breast-conserving therapy: local excision of the primary tumor mass along with 1–2 cm of macroscopically normal tissue and adequate megavoltage irradiation of the remaining breast tissue (EORTC Breast Cancer Cooperative Group 1991). Without implementation of other measures, for tumors up to 4–5 cm such an approach is associated with a risk of intramammary failure of about 1.5%–2% per year, at least during the first 10 years. Over the past decade much has been learned about how to significantly improve local control through more effective treatment of the primary tumor area and through the action of systemic therapy, so that local failure rates observed with current multidisciplinary treatment strategies should be well below 1% per year (Fisher and Anderson 1994). Such improvements have in part been due to a better understanding of risk factors and the implementation of this knowledge to optimize treatment of the individual patient. Multidisciplinary collaboration between breast surgeon, pathologist and clinical (radiation and medical) oncologist has resulted in reduction of local failure risk, while at the same time allowing extension of breast conservation techniques to more advanced tumors.

Pathogenesis of Local Failure

A majority of failures occur in the vicinity of the primary tumor and represent remnants of the reference lesion rather than multicentric tumor foci. Tumor growth at a clear distance from the primary tumor is exceptional, at least during the first decade. A small minority of "recurrences" are diffuse, usually involving skin, and may represent hematogenous dissemination rather than true local failure (Kurtz 1993).

Table 1. Parameters which correlate with the risk of intramammary recurrence after conservative sugery (Kurtz 1993)

Breast irradiation
Systemic therapy (chemotherapy, tamoxifen)
Type of excision, resection margins
Patient's age
Tumor size
Nodal status
Histologic or nuclear grade
Estrogen receptor
Extensive intraductal component
Vascular invasion
Necrosis
Mononuclear cell reaction
Multifocality

Most local failures occur as a result of residual cancer remaining in the breast despite the surgeon's impression of having performed a "complete" excision. Risk factors for local failure are thus likely to reflect the mechanisms of local extension, namely interstitial, intraductal, and lymphatic spread. In addition, recurrence risk relates to the ability of residual cancer to persist and progress despite tumoricidal doses of radiotherapy. Thus local failure risk will reflect the number of residual clonogens as well as their radiocurability (intrinsic radiosensitivity, cell kinetics, etc.).

Intramammary tumor recurrence risk has been shown to correlate with numerous clinical and pathologic parameters (Table 1), which have been summarized elsewhere (Kurtz 1993). A detailed discussion of the various factors is beyond the scope of this chapter, which will be devoted to discussing those factors which have emerged as the most useful for assessment of risk. Most factors appear to be reflections of local tumor aggressiveness and presumably correlate with residual tumor burden. In contrast, little is known regarding factors determining radiocurability.

The Influence of Radiotherapy

Most risk factor data have been generated from data sets in which all patients received breast irradiation. Nonetheless, the effect of radiotherapy can be quantitated from four published prospective studies in which conservatively operated patients had been randomized to receive or not to receive breast irradiation (Table 2). Although surgical and radiotherapeutic approaches varied among the studies, it can be seen that the risk reduction associated with radiotherapy is profound. This effect is considerably more important than that associated with any other risk factor. At present little is

Table 2. Risk of local recurrence after conservative surgery and breast irradiation compared with the risk after conservative surgery alone

Study	N	Surgery	Dose (Gy)	Relative risk
NSABP-B6	1140	Lumpectomy	50	0.21
Ontario	837	Lumpectomy	52.5	0.19
Swedish	381	Sectorectomy	54	0.14
Milan-3[a]	567	Quadrantectomy	60	0.03

Radiotherapy reduces the local recurrence risk by a factor of 5 or more.
Data from prospective randomized trials (B. Fisher et al. 1992; Clark et al. 1992; Liljegren et al. 1994; Veronesi et al. 1994).
[a] Calculated from ratio of events.

known regarding which tumors benefit most from irradiation, although there is some indication that radiotherapy may be less effective when cancers are still in an intraductal stage (Fisher an Anderson 1994).

Risk Factors for Residual Disease

Extent of Excision and Resection Margins

It seems axiomatic that extent of excision should influence local risk, since wideness of excision will influence both the proportion of patients having residual cancer and the extent of residual disease. This has been demonstrated by a randomized trial comparing tumorectomy and quadrantectomy (Veronesi et al. 1994), both followed by radiotherapy (albeit with somewhat different techniques). Quadrantectomy appears to reduce local risk by about a factor of 2–3 compared with tumorectomy (19/360 vs 46/345). However, the tumorectomy technique in this study was only considered a "debulking" procedure, without consideration of excision margins. The benefit of quadrantectomy compared with the type of wide local excision commonly employed is likely to be considerably less.

Considering the unfavorable cosmetic effects of very large excisions, surgical strategies are more commonly based on lumpectomy with histopathologic evaluation of inked margins and selective use of re-excision (Solin et al. 1991). In general there is a correlation between recurrence rates and margin status, if radiotherapy is standardized (Table 3). Higher doses of localized radiation may compensate for positive margins (Solin et al. 1991). Resection margin status and the presence or absence of tumor in a re-excision specimen should both be considered useful markers of residual disease. In addition, it may be useful to distinguish between focal and extensive margin involvement (Schnitt et al. 1994).

Table 3. Influence of microscopic resection margin status on local recurrence risk in series employing assessment of inked excision specimens

	Negative	Local recurrence (%)	
		"Close"[a]	Positive
Netherlands Cancer Institute	2	6	16
Harvard Joint Center for Radiation Therapy	0	4	13
University of Pennsylvania	7	11[b]	2[b]

Data from Borger et al. 1994; Schnitt et al. 1994; Solin et al. 1991.
[a] "Close" refers to tumor cells within 1–2 mm of inked margins.
[b] Higher dose given for non-negative margins.

Multifocality

Multiple macroscopic tumor masses and diffuse malignant microcalcifications often indicate nonlocalized involvement of the diseased breast and are generally considered contraindications for breast conservation (Fowble et al. 1993); however, clinical results to support this notion are almost totally lacking, Occasional patients present two or even three apparently separate tumors, either within the same quadrant or at a distance from each other (Kurtz et al. 1990a). Although the inherent local recurrence risk appears to be about twice as great as for unifocal tumors, existing data suggest that satisfactory local control can be obtained if the tumors are excised with clear margins (Table 4).

Tumor Size

Most retrospective series show no effect of tumor size on recurrence risk (Kurtz 1993). This probably reflects the higher risk of lymph node involvement with increasing tumor size and the more frequent use of adjuvant systemic therapies in patients with larger tumors, compensating for what may be a higher inherent risk of local failure. In addition, in patients receiving radiotherapy the number of local events is generally small, and minor differences in recurrence rates would not necessarily be apparent. These notions are confirmed by data from two randomized trials containing unirradiated control arms (Clark et al. 1992; Fisher and Anderson 1994). Chemotherapy was used only in node-positive patients in one of the trials. Tumors larger than 2 cm in diameter appear to have about twice the tendency to local failure of smaller tumors. In practice, this has little practical consequence, as such tumors usually receive more aggressive therapy based on other considerations.

Table 4. Local results of treatment of macroscopically multiple tumors in patients selected for breast-conserving surgery and radiotherapy[a]

	Local recurrence (%)	
	Marseille CI	Rush-Presbyterian Medical Center
Unifocal tumors	11	4
Multiple tumors		
Margins positive or indeterminate	30	–
Margins negative	5	4

Data from Kurtz et al. 1990a; Hartsell et al. 1994.
[a] If microscopic resection margins were negative, local control was comparable to that associated with unifocal tumors.

Vascular (Lymphatic or Blood Vessel) Invasion

As it reflects one of the mechanisms of local extension, vascular invasion should intuitively be an important risk factor. Most studies taking vascular invasion into account suggest that the presence of this factor increases local failure risk by a factor of about 2 (Borger et al. 1994; Fourquet et al. 1989; Locker et al. 1989; Fisher et al. 1986).

Extensive Intraductal Component

This is the risk factor which has been the most extensively analyzed. Extensive intraductal component (EIC) reflects the risk of intraductal extension of the tumor beyond its apparent boundaries. Although the definition initially was based principally on the percentage (>25%) of intraductal carcinoma within the primary tumor, alternative definitions based on examination of peritumoral tissue have been proposed. It is clear that EIC correlates with the quantity of residual carcinoma, particularly intraductal, remaining in the breast after local excision (Holland et al. 1990). Most studies suggest that the presence of EIC augments the local failure risk by a factor of 2–3 (Kurtz 1993).

The Effect of Age

Patient age is a reproducible risk factor for local recurrence, and it is well established that local risk decreases with increasing age (Kurtz 1993). Although risk seems in most series to be a continuous function of age, attention is commonly concentrated on the very young patients who form

Table 5. Relation of very young age to the risk of local recurrence in the breast after conservative surgery and irradiation

	Local recurrence (%)	
	<35 years	>35 years
Marseille Cancer Institute	20/109 (18.4)	165/1574 (10.5)
Harvard Joint Center for Radiation Therapy	15/62 (24)	76/721 (10.5)
University of Pennsylvania	12/64 (18.8)	68/941 (7.2)
NSABP B-06	9/38 (23.7)	33/450 (7.3)
Milan Cancer Institute	9/95 (9.5)	45/1137 (4.0)

The differences shown are all statistically significant.
Data from Kurtz et al. 1990b; Harris and Recht 1991; Fowble et al. 1994; Fisher et al. 1992; Veronesi et al. 1990.

the highest risk group, for example those age 35 or younger (Table 5). Although perhaps an oversimplification, this distinction is convenient, as very young patients also have higher metastatic risk and require more aggressive treatment (de la Rochefordière et al. 1993).

It is unclear whether very young patients have an inherently higher risk of local recurrence, or whether the high risk patients might be able to be identified by a particular morphologic or biologic feature (Kurtz et al. 1990b). It has been documented that breast cancers in younger patients tend to have a prevalence of certain characteristics, such as a higher frequency of estrogen receptor negativity, vascular invasion, EIC, and high nuclear and histologic grade: all factors correlating with a higher risk of local failure (Nixon et al. 1994). No single feature has yet been identified which in itself explains the higher local failure risk in this group. In particular, young age appears to be a risk factor for local failure even in tumors not having EIC (Harris and Recht 1991).

The Effect of Systemic Therapies

Evidence is accumulating from uncontrolled and randomized studies that chemotherapy and tamoxifen both significantly reduce local recurrence risk, probably by a factor of 2 or 3, at least when used in conjunction with breast irradiation (Table 6). Given the increasing prevalence of systemic treatments in current strategies, this could have a major effect on local recurrence rates in the future. Little is known of the possible interactions between systemic agents and radiotherapy, nor of the potential importance of sequencing.

Table 6. Preliminary results of randomized studies suggesting that the addition of systemic therapy reduces local recurrence after conservative surgery and radiotherapy (Rutqvist et al. 1992; Fisher and Anderson 1994)

Study	Treatment	Patients	RR
Stockholm	50 Gy + TAM vs 50 GY	n = 252; N−, postmenopausal	0.31
NSABP B-14	50 Gy + TAM vs 50 Gy	n = 1072; N−, ER+	0.42
NSABP B-13	50 Gy + Chemo vs 50 Gy	n = 239; N−, ER−	0.32

TAM, tamoxifen; chemo, sequential methotrexate and 5-fluorouracil; RR, relative risk (ratio of mean annual hazard functions); N−, node-negative; ER, estrogen receptor.

How Should Margin Status and Other Factors Be Used To Assess Risk?

Although it is unlikely that all of the above parameters are of independent importance, there is presently no coherent approach to the practical application of the notions of risk. Construction of a risk model would ideally require a precisely documented data base of uniformly treated patients and a large number of events. Such a comprehensive model is as yet unavailable, and the existing literature concentrates largely on interactions between only two or three risk parameters. Nonetheless, existing studies provide some notions as to how assessment of certain risk factors might reasonably be implemented.

While to a great extent resection margins are an expression of tumor biology, they are unique among risk factors in that margin status can be modulated by surgical intervention. Nonetheless, in patients who are to receive breast irradiation, there is no rational basis for requiring that margins be histologically negative, as margin status is but one of a number of parameters which appear to influence risk to a similar degree. Without detracting from the value of margin assessment as a measure of residual disease, existing data suggest that studies of the interactions between margin status and other morphologic risk parameters will lead to a more rational basis from which to assess them. In addition, the definition of "positive" margins is controversial, and it is possible that a distinction should be made between focal and diffuse involvement, or between involvement by invasive or by intraductal carcinoma (Schnitt et al. 1994).

The interactions which have received the greatest attention in the literature are those between margin status and EIC. Positive margins are more commonly observed in tumors with EIC (Borger et al. 1994), and the effect of EIC on recurrence risk is most marked in series in which narrow excisions were employed (Table 7). In the NSABP B-06 trial, which required that

Table 7. Correlation of local recurrence in the breast after conservative surgery and irradiation in the presence or absence of extensive intraductal component[a]

	Local recurrence (%)	
	EIC+	EIC−
Narrow excision		
Harvard Joint Center for Radiation Therapy	25	6
Milan-2 (TART)[b]	30	8
Lumpectomy (margins negative)		
NSABP B-06	11	9
Quadrantectomy		
Milan-2 (QUART)[b]	9	4

EIC, extensive intraductal component; TART, tumorectomy, axillary clissection and radiation therapy; QUART, quadrantectomy, axillary dissection and radiation therapy.
From Harris and Recht 1991; Fisher et al. 1992; Veronesi et al. 1994.
[a] The effect of EIC is marked if excision is narrow and becomes insignificant in the face of microscopically negative margins or very wide excision.
[b] 5-year actuarial rates.

inked margins be negative, no influence of EIC on recurrence risk could be demonstrated (Fisher et al. 1992), and in multivariate analyses including inked margin status, EIC tends to lose significance as a risk factor (Borger et al. 1994; Schnitt et al. 1994). In a study from Harvard University EIC only appeared to increase local failure risk if margin involvement exceeded three microscopic low-power fields (Schnitt et al. 1994).

Data from the NSABP B-06 trial suggest that only young age may retain prognostic significance in the face of negative margins (Fisher and Anderson 1994). Nonetheless, it appears that very satisfactory local control can be obtained in this unfavorable group by the use of very wide excision such as quadrantectomy (Table 5). In addition, excellent local control has been reported in very young patients receiving both adjuvant chemotherapy and breast irradiation, suggesting that the increased local risk might now be largely confined to node-negative patients not receiving systemic treatment (Fowble et al. 1994).

A corollary to these studies concerns the significance of margin involvement in the absence of risk factors such as EIC or young age. In patients without EIC, Schnitt et al. (1994) have shown that the 5-year local failure rate in the presence of either close or positive margins was 8%, a value which hardly justifies total mastectomy as as alternative. Similarly, a previous study from the same institution suggested that for tumors without EIC, local recurrence rate depended very little or not at all on the volume of the excision specimen (Vicini et al. 1991). This has been confirmed in a randomized trial by Veronesi et al. (1994), who observed that the local control benefit conveyed by quadrantectomy was more marked for EIC+ than for EIC-tumors.

Summary

In patients treated by macroscopically complete excision and breast irradiation, a small number of parameters appear to be useful in assessing the inherent risk of intramammary tumor recurrence. These factors include microscopic excision margin status, gross multifocality, very young age, EIC, and perhaps lymph vessel invasion. However, if microscopic resection margins are negative, and especially if systemic therapy is administered, local control is adequate, and none of these risk factors represents a contraindication to breast preservation.

Conversely, in the absence of factors indicating an inherently greater risk of local failure, the status of resection margins is likely to be much less important. In such patients existing data suggest that satisfactory local control can be achieved by modern radiotherapy even in the face of microscopically positive margins, especially if such involvement is only focal. Final judgement on this issue awaits availability of additional long-term data from large series in which meticulous evaluation of margin status was prospectively carried out on inked excision specimens.

References

Borger J, Kemperman H, Hart A, Peterse H et al. (1994) Risk factors in breast-conservation therapy. J Clin Oncol 12:653–660

Clark RM, McCollough PB, Levine MN et al. (1992) Randomized clinical trial to assess the effectiveness of breast irradiation following lumpectomy and axillary dissection for node-negative breast cancer. J Natl Cancer Inst 84:683–689

De la Rochefordière A, Asselain B, Campana F et al. (1993) Age as a prognostic factor in premenopausal breast carcinoma. Lancet 341:1039–1043

EORTC Breast Cancer Cooperative Group (1991) Manual for clinical research in breast cancer. EORTC, Leuven, pp 33–51

Fisher B, Wickerham DL, Deutsch M, Anderson S, Redmond C, Fisher ER (1992) Breast tumor recurrence following lumpectomy with and without breast irradiation: an overview of recent NSABP findings. Sem Surg Oncol 8:153–160

Fisher B, Anderson S (1994) Conservative surgery for the management of invasive and noninvasive carcinoma of the breast: NSABP trials. World J Surg 18:63–69

Fisher ER, Sass R, Fisher B et al. (1986) Pathologic findings from the National Surgical Adjuvant Breast Project (protocol 6). II. Relation of local breast recurrence to multicentricity. Cancer 57:1717–1724

Fisher ER, Anderson S, Redmond C, Fisher B (1992) Ipsilateral breast tumor recurrence and survival following lumpectomy and irradiation: pathological findings from NSABP (protocol B-06). Semin Surg Oncol 8:161–166

Fourquet A, Campana F, Zafrani B et al. (1989) Prognostic factors of breast recurrence in the conservative management of early breast cancer: a 25-year follow-up. Int J Radiat Oncol Biol Phys 17:719–725

Fowble B, Yeh I-T, Schultz DJ et al. (1993) The role of mastectomy in patients with stage I-II breast cancer presenting with gross multifocal or multicentric disease or diffuse microcalcifications. Int J Radiat Oncol Biol Phys 27:567–573

Fowble BL, Schultz DJ, Overmoyer B et al. (1994) The influence of young age on outcome in early stage breast cancer. Int J Radiat Oncol Biol Phys 30:23–33

Harris JR, Recht A (1991) Conservative surgery and radiotherapy. In: Harris JR, Hellman S, Henderson IC, Kinne DW (eds) Breast diseases, 2nd edn. Lippincott, Philadelphia, pp 388–419

Hartsell WF, Recin DC, Griem KL et al. (1994) Should multicentric disease be an absolute contraindication to the use of breast-conserving therapy? Int J Radiat Oncol Biol Phys 30:49–53

Holland R, Connolly JC, Gelman R et al. (1990) The presence of an extensive intraductal component (EIC) following a limited excision correlates with prominent residual disease in the remainder of the breast. J Clin Oncol 8:113–118

Kurtz JM (1993) Factors which predict breast relapse. Recent Results Cancer Res 127:137–150

Kurtz JM, Jacquemier J, Amalric R et al. (1990a) Breast-conserving therapy for macroscopically multiple cancers. Ann Surg 212:38–44

Kurtz JM, Jacquemier J, Amalric R et al. (1990b) Why are local recurrences after breast-conserving therapy more frequent in younger patients? J Clin Oncol 8: 591–598

Liljegren G, Holmberg L, Adami H-O et al. (1994) Sector resection with or without postoperative radiotherapy for stage I breast cancer: five-year results of a randomized trial. J Natl Cancer Inst 86:717–722

Locker AP, Ellis IO, Morgan DAL et al. (1989) Factors influencing local recurrence after excision and radiotherapy for primary breast cancer. Br J Surg 76:890–894

Nixon AJ, Neuberg D, Hayes DF et al. (1994) Relationship of patient age to pathologic features of the tumor and prognosis for patients with stage I or II breast cancer. J Clin Oncol 12:888–894

Rutqvist LE, Cedermark B, Glas U et al. (1992) Randomized trial of adjuvant tamoxifen in node negative postmenopausal breast cancer. Acta Oncol 31:265–270

Schnitt SJ, Abner A, Gelman R et al. (1994) The relationship between microscopic margins of resection and the risk of local recurrence in patients with breast cancer treated with breast-conserving surgery and radiation therapy. Cancer 74:1746–1751

Solin LJ, Fowble ML, Schultz DJ, Goodman RL (1991) The significance of the pathology margins of the tumor excision on the outcome of patients treated with definitive irradiation for early stage breast cancer. Int J Radiat Oncol Biol Phys 21:279–287

Veronesi U, Salvadori B, Luini A et al. (1990) Conservative treatment of early breast cancer. Long-term results of 1232 cases treated with quadrantectomy, axillary dissection, and radiotherapy. Ann Surg 211:250–259

Veronesi U, Luini A, Galimberti V, Zurrida S (1994) Conservation approaches for the management of stage I/II carcinoma of the breast: Milan Cancer Institute trials. World J Surg 18:70–75

Vicini FA, Eberlein TJ, Connolly JL et al. (1991) The optimal extent of resection for patients with stages I and II breast cancer treated with conservative surgery and radiotherapy. Ann Surg 214:200–205

Summary

A. Goldhirsch

International Breast Cancer Study Group, Ospedale Civico, 6900 Lugano, Switzerland

Radiation therapy is an important component of the management of breast cancer. New types of radiation equipment, better planning and changes in the local management of breast cancer have led to important developments in radiation therapy. Furthermore, the demonstrated efficacy of systemic adjuvant therapies and the potential benefit, in terms of local disease control, of combining adjuvant and radiation therapies have led to new strategies for improving the care of patients with operable breast cancer.

In the past, chest wall irradiation after mastectomy was routinely carried out in most women with breast cancer. Randomized trials showed that, despite the significant reduction in local recurrence, only a minimal influnce on systemic relapse and overall survival could be obtained (Lichter and Lawrence 1995).

Radiation therapy to the conserved breast after proper tumor surgery was shown to extensively reduce the incidence of relapse in the ipsilateral breast. Disease control in the conserved breast with radiation therapy is dependent upon several features. Adequacy of total tumor excision is related not only to the surgeon's compliance and skills, but also to pathological features of the primary tumor, such as extensive intraductal component and lymphatic vessel invasion (Schnitt 1993). The treatment results of radiation therapy are also often more dependent upon the same pathological features than upon technical details, despite the relevance of the latter and the need for quality assurance programs. These aspects were elegantly described in this chapter.

Several issues associated with local disease control must still be clarified by future research:

1. Screening-detected cancers or very small tumors are still treated in the same way as symptomatic cancers of the breast, despite the fact that data on the adequacy of their treatment with surgery alone are either not available or controversial (Hermann et al. 1993; Gelber and Goldhirsch 1994).

2. The clinical relevance of having clear surgical margins for breast conservation is still an issue for discussion. Obviously, the failure rate in cases of unclear margins, with either in situ or invasive cancer, is increased despite radiation therapy, but the magnitude of this increase is not well defined. The same issue concerning the need to quantify the magnitude of increased risk for local recurrence relates to the presence of extensive intraductal component (Schnitt 1993) in a series of patients with clear margins.
3. The question of adequacy of radiation therapy in the presence of proliferating mastopathy with atypia is also a matter of controversy.
4. The adequacy of radiation therapy to the breasts of young women is an issue for debate. The local failure rate following radiation therapy is higher for younger women, but several features other than age alone were also associated with this increased risk (Kurtz et al. 1990).
5. Elderly patients enjoy nearly the same benefits from proper radiation therapy as younger women. Logistics, however, are often more complex for patients in advanced age resulting in a higher burden of treatment. Cost-benefit considerations for the elderly are likely to yield different results than those expected for a younger population.
6. A small proportion of patients might have an increased susceptibility of normal tissues to radiation therapy. The features required to identify such patients are often ill defined.
7. Some long-term side effects of radiation therapy remain unknown due to the lack of proper longitudinal studies designed to identify late sequelae (Rutqvist and Johansson 1990).
8. The integration of radiation therapy and systemic adjuvant chemotherapy is a matter of controversy. Attempts to first complete the local treatment by delivering radiation therapy immediately after surgery contrast with the logical approach of starting systemic treatment without any delay. Delivering the two modalities simultaneously can result in unsatisfactory effects (Gore et al. 1987; Betzen et al. 1989). However, the International Breast Cancer Study Group showed for the first time, in randomized trials (trials VI and VII), that the incidence of breast recurrence in the conserved ipsilateral breast within 4 years ranged between 8% and 9% for node-positive, premenopausal patients, and between 3% and 6% for node-positive, postmenopausal patients. After 52 months of median follow-up, administering radiation therapy following three or six cycles of chemotherapy for premenopausal women, or following no cycles or three cycles of chemotherapy for postmenopausal women does not influence overall efficacy or local control in this series (Wallgren et al. 1995). At least two other studies will soon be completed to further shed light on this controversy. The effects of a potential interaction between radiation therapy and high-dose chemotherapy are also unknown. Information about this issue will become available from ongoing clinical trials.

Technological advances and quality assurance will most likely further improve radiation treatment in the future. In addition, our knowledge about factors predicting radiation therapy sensitivity or resistance is likely to enhance the accuracy with which the success of breast conserving procedures might be predicted. The three reports in this session are of relevance for the patients of today. They strikingly demonstrate that much more clinical research is needed to answer even very practical questions such as those discussed in this short summary.

References

Betzen SO, Overgaard M, Thames HD et al. (1989) Early and late normal-tissue injury after postmastectomy radiotherapy alone or combined with chemotherapy. Int J Radiat Biol 56:711–715

Gelber RD, Goldhirsch A (1994) Radiotherapy to the conserved breast: is it avoidable if the cancer is small. J Natl Cancer Inst 86:652–654

Gore S, Come S, Griem K et al. (1987) Influence of the sequencing of chemotherapy and radiation therapy in node-positive breast cancer patients treated by conservative surgery and radiation therapy. In: Salmon S (ed) Adjuvant therapy of cancer V. Grune and Stratton, Orlando, pp 365–373

Hermann RE, Esselstyn CB, Grundfest-Broniatowski S et al. (1993) Partial mastectomy without radiation is adequate treatment for patients with stage 0 and I carcinoma of the breast. Surg Gynecol Obstet 177:247–253

Kurtz JM, Jacquemier J, Amalric R et al. (1990) Why are local recurrences after breast conserving therapy more frequent in younger patients? J Clin Oncol 8:591–598

Lichter AS, Lawrence TS (1995) Recent advances in radiation oncology. N Engl J Med 332:371–379

Rutqvist LE, Johansson H (1990) Mortality by laterality of the primary tumour among 55,000 breast cancer patients from the Swedish Cancer Registry. Br J Cancer 61:866–868

Schnitt SJ (1993) Pathologic factors predictive of local recurrence in patients with invasive breast cancer treated by conservative surgery and radiation therapy. In: Fletcher GH, Levitt SH (eds) Non-disseminated breast cancer. Springer, Berlin Heidelberg New York, pp 93–104

Wallgren A, Bernier J, Gelber RD, Goldhirsch A, Roncadin M, Joseph D, Castiglione-Gertsch M (1995) for the International Breast Cancer Study Group (IBCSG). Timing of radiotherapy and chemotherapy following breasts conserving surgery for patients with node-positive breast cancer. Int J Radiat Oncol Bio Phys (in press)

VIII. Women's Health Perception and Breast Cancer: Issues of Fertility, Hormone Substitution, and Cancer Prevention

Reproductive Potential After Adjuvant Chemotherapy for Breast Cancer

P. Valagussa, D. De Candis, G. Antonelli, and G. Bonadonna

Division of Medical Oncology, Istituto Nazionale Tumori, Via Venezian 1, 20133 Milano, Italy

Introduction

The 10-year update of the world-wide overview of breast cancer treatment (Early Breast Cancer Trialists' Collaborative Group 1992) has confirmed that adjuvant systemic therapy can alter the course of breast cancer. In women younger than 50 years of age (i.e. in premenopausal women), highly significant reductions in the annual rates of both recurrence (37%) and death (27%) are produced by adjuvant combination chemotherapy when compared with local-regional treatment alone.

These beneficial effects, however, are in part counterbalanced by the influence that known cytotoxic drugs, and especially alkyating agents, exert on gonadal function. Reports which have appeared in the medical literature dealing with the consequences of treatment of Hodgkin's disease, childhood leukemia and testicular cancer (Myers and Schilsky 1992) have focused mainly on testicular function, since biopsy and semen analysis provide an accurate estimation of male reproductive potential. Data on ovarian function are less frequently reported and usually the recording of menstrual history is taken as a crude measure.

The frequency of amenorrhea following adjuvant combination chemotherapy including cyclophosphamide, methotrexate and fluorouracil (CMF) with or without anthracyclines is well known (Valagussa et al. 1993). Drug-induced amenorrhea is age-related, being an almost universal side effect in women older than 40 years of age, whereas it occurs in approximately 55% of younger patients. In addition, young premenopausal women are not permanently affected by amenorrhea and therefore are potentially able to become pregnant and have children.

Data on pregnancies after adjuvant chemotherapy for breast cancer are scarce (Sutton et al. 1990; Reichman and Green 1994). Apart from the possible consequences on ovarian function, it is usually recommended that breast cancer patients who desire to give birth once treatment is completed defer conception for a few years to avoid the highest risk of recurrence.

In an attempt to quantitate these aspects, we have retrospectively analyzed our series of patients treated in various prospective protocols of adjuvant CMF-based chemotherapy.

Patients and Methods

Between June 1973 and July 1990, a total of 2586 patients were prospectively accrued in various randomized and non-randomized protocols (Bonadonna 1992). Among 1542 women (60% of the case series) who were still menstruating at entry on study, we selected those women who were younger than 38 years of age and therefore potentially able to become pregnant and have children within a few years after the diagnosis of breast cancer.

This subset of young women was followed for a median of 13 years. Fourteen women were treated with local-regional therapy alone and ten of them (71%) presented with new disease manifestations, all within 39 months from surgery.

Present analysis will therefore focus on the 205 patients given adjuvant CMF with or without adriamycin. We report here the tumor characteristics at breast cancer diagnosis, menstrual and pregnancy history before and after adjuvant chemotherapy, and disease status during the first 10 years after surgery. All data were routinely reported at patient entry into the study and during the entire follow-up period.

Results

Within 10 years of surgery, there were 19 pregnancies in 17 patients. Table 1 summarizes the main characteristics of the entire study population and of the 17 women who became pregnant after adjuvant chemotherapy.

As far as reproductive history before breast cancer diagnosis is concerned, 83% of the 205 patients had had at least one pregnancy (median 2, range 1–6). Among the 17 women who became pregnant after chemotherapy, 35% were 30 years of age or younger at the time of surgery and three (14%) had never conceived before.

As far as tumor characteristics are concerned, at pathologic examination 54% of the study population presented with tumors measuring >2.0 cm at their largest diameter; all but three patients presented with axillary nodal involvement, 42% had more than three histologically positive nodes. Estrogen receptor-negative tumors accounted for 36% of the 146 specimens

Table 1. Main characteristics of the patient population

	Total series[a] (n = 205)		Pregnant patients[b] (n = 17)	
	N	(%)	N	(%)
Age at surgery				
≤ 30 years	40	(19)	6	(35)
31–37 years	165	(81)	11	(65)
Regular menses	184	(90)	15	(88)
Use of oral contraceptives[c]	26	(13)	1	(6)
At least one pregnancy[c]	170	(83)	14	(82)
Tumor >2.0 cm	111	(54)	9	(53)
Positive nodes >3	86	(42)	4	(24)
	53	(36)	3	(25)

[a] Median dose of cyclophosphamide: 10 800 mg (2040–27 825).
[b] Median dose cyclophosphamide: 10 800 mg (6960–24 675).
[c] Before breast cancer diagnosis.

assayed. Among the 17 pregnant patients, 24% had more than three positive nodes and 25% estrogen receptor-negative tumors.

Adjuvant CMF was planned for a minimum of six to a maximum of 12 monthly cycles when administered according to the standard regimen and for a minimum of eight to a maximum of 12 courses when given according to the 3-week intravenous administration regimen (Bonadonna 1992). Overall, the median cumulative dose of cyclophosphamide delivered was 10 800 mg both for the total case series of 205 women and for the 17 pregnant women (Table 1).

The 19 conceptions occurred within a median of 26 months from the end of adjuvant chemotherapy (Table 2). There were 13 elective abortions in 12 patients, all performed within the first 5–8 weeks from conception and within a median of 16 months (range 0–29) from the end of adjuvant chemotherapy. Two conceptions occurred during the third and the fifth cycle, respectively, of adjuvant chemotherapy. One of the two women conceived again 3 years later, but this pregnancy was also terminated because of a concomitant breast cancer relapse. One patient had a spontaneous abortion 87 months after the end of CMF, at the age of 43 years.

Five women completed their pregnancies and gave birth to two male and three female babies. Birth occurred within a median of 56 months (range 36–80) from the end of chemotherapy, and only in one instance was there a premature delivery, at the 33rd week, because the mother presented with disease relapse in soft tissues 4 months after conception. Fetal malformations were not recorded at delivery. One woman who had an elective abortion 21 months after the end of chemotherapy became pregnant again and gave birth to a normal baby 20 months later.

Table 2. Pregnancies after CMF-based chemotherapy

	N	From end of chemotherapy (median in months)	Age at conception (median in years)
Total	19	26 (0–87)	36 (23–43)
Elective abortions	13	16 (0–29)	35 (28–38)
Spontaneous abortion	1	87	43
Children	5	56 (36–80)	35 (23–37)

CMF: cyclophosphamide, methotrexate, 5-fluorouracil.

Of the 17 pregnant patients, 11 (65%) presented with new disease manifestations within 10 years from breast cancer diagnosis (median 16 months from abortion or delivery). This percentage, however, is not different from the frequency of disease relapse of the entire series of 205 patients (62%).

In the attempt to assess which fraction of our patient series could potentially become pregnant while in clinical complete remission from breast cancer, we calculated the cumulative risk of disease relapse, the cumulative risk of irreversible amenorrhea or early menopause (before or in the absence of disease relapse), and both risks combined, whichever occurred first.

Table 3 reports the essential results. In this series of young patients the risk of disease relapse was fairly high during the first 5 years after local-regional treatment (53%) and reached 62% at 10 years. A total of 66 women developed early menopause, but half of them also presented with new disease manifestations. Among the 40 patients 30 years of age or younger, three entered menopause (one surgical menopause for gynecological reasons) within 2 years from starting chemotherapy but also developed new breast cancer disease. Two additional women entered menopause 9 years later, at the age of 39 years.

When both disease relapse and early menopause were considered, at 5 years from local-regional treatment only 35% of the 205 women were in continuous clinical complete remission and still menstruating, with no difference between the two major age groups at breast cancer surgery, i.e. 30 years of age or younger vs 31–37 years.

During the subsequent 5 years, we failed to observe any new disease manifestation in the younger age group, but, as previously mentioned, two women entered menopause at the age of 39 years. In the older age group, we documented a continuous risk of disease relapse (from 49% at 5 years to 61% at 10 years). However, among the 58 women who were in clinical complete remission 10 years after breast cancer surgery, 31 had already entered menopause.

Table 3. Cumulative risk of breast cancer relapse and/or early menopause following adjuvant chemotherapy

	Breast cancer relapse (%)	Early menopause (%)[a]	Combined (%)
Within			
3 years[a]	40	18	51
5 years[a]	53	27	65
10 years[a]	62	58	83

[a] Before or in the absence of disease relapse.

Comments and Discussion

In reviewing the medical literature on breast cancer and pregnancy, two important aspects need to be taken into consideration. First, the incidence of breast cancer continues to rise in all age groups and it has been estimated that 10%–20% of the projected new cases in 1993 would occur in women of child-bearing age (Boring et al. 1993). Second, at least in western countries, women have postponed their first childbirth for personal, educational and/or professional reasons (Ventura 1989). It is therefore plausible that, after successful treatment for breast cancer, many young patients will seek medical advice about the safety of pregnancy.

Despite the importance of this topic, data in the medical literature are somewhat meager. In fact, approximately 7% of women not subjected to oophorectomy experienced one or more pregnancies, 70% of which occurred during the first 5 years after breast cancer treatment (Donegan 1977). In reviewing a series of retrospective studies, Danforth (1991) stated that pregnancy subsequent to mastectomy did not alter final outcome of breast cancer and that survival based on disease stage was similar for patients with and without subsequent pregnancy.

Reports on pregnancy after cytotoxic chemotherapy are even fewer. Mulvihill et al. (1987) attempted to evaluate the potential teratogenicity and mutagenicity of different chemotherapeutic regimens and/or irradiation administered in a variety of malignant neoplasms. Data were obtained by means of a postal survey on 133 pregnancies among 66 patients, only four of whom were previously affected by breast cancer. In the 58 pregnancies which occurred in 40 patients after chemotherapy (conceptions after a mean of 27 months from the end of treatment, range 2–104 months), there was an unexpected increase in low birth weight, stillbirths and premature termination of pregnancy, but no excess of genetic disorders. The abnormal outcome was concentrated during the first year after therapy. According to the authors, these data could represent defects in factors (e.g., uterine or hormonal) that normally maintain gestations, and not genetic damage to oocytes.

The only detailed publication available thus far on pregnancies after adjuvant chemotherapy for breast cancer has been provided by investigators from the M.D. Anderson Cancer Center (Sutton et al. 1990). A total of 33 pregnancies occurred in 25 women who were part of a case series of 227 breast cancer patients 35 years old or younger. A total of 19 pregnancies resulted in full-term delivery, with no fetal malformations. The authors suggested that pregnancy did not impact adversely on the clinical course of breast cancer.

Our own experience is in line with the above mentioned data. In a series of 205 women who were 37 years old or younger, there were 19 pregnancies in 17 women. Thirteen pregnancies were terminated, either based on the physician's advice or the patient's wishes, and in these instances conception occurred within a median of 16 months from the end of chemotherapy. Five women gave birth within a median of 56 months after discontinuation of drug treatment, and in no instances were fetal malformations recorded.

In breast cancer patients desiring to give birth after treatment, Danforth (1991) recommended that conception be deferred until 2–3 years after completion of therapy, i.e., until after the period of greatest risk of breast cancer recurrence. We should, however, consider that alkylating agent-containing adjuvant chemotherapy may be responsible for the induction of menopause at an early age (Valagussa et al. 1993).

We attempted to evaluate which fraction of our high risk patient population could potentially become pregnant during the follow-up, while in clinical complete remission from breast cancer. Because of the retrospective nature of this analysis, no complete data were available regarding the use of contraceptives or attempts at conception during the period of observation.

Considering the tumor characteristics of our patient series (54% had breast cancers measuring >2.0 cm and 42% had more than three positive axillary nodes at pathologic examination), the cumulative risk of disease relapse was fairly high at 3 (40%) and 5 years (53%) after local-regional treatment. At the end of the first 5 years, only 35% of our patients were disease-free and still menstruating. During the second 5-year period, an additional percentage of women presented new disease manifestations and/or entered menopause (Table 3); this was especially true in women >30 years old at the diagnosis of breast cancer.

In conclusion, whereas in young premenopausal women pregnancy is still possible after adjuvant chemotherapy, available data are not adequate to provide general guidelines. In counseling women about a subsequent pregnancy, it is important to re-emphasize that a major consideration remains the individual patient's prognosis, which needs to be discussed explicitly with the patients themselves and with their families. In addition, the health, well-being, desire for children, support from partner and family, and other sociodemographic factors need to be considered in this difficult decision-making process (Dow et al. 1994).

References

Bonadonna G (1992) Evolving concepts in the systemic adjuvant treatment of breast cancer. Cancer Res 52:2127–2137
Boring CC, Squires TS, Tong T (1993) Cancer statistics, 1993. CA 43:7–26
Danforth DN (1991) How subsequent pregnancy affects outcome in women with a prior breast cancer. Oncology 5:23–32
Donegan WL (1977) Pregnancy and breast cancer. Obstet Gynecol 50:244–251
Dow KH, Harris JR, Roy C (1994) Pregnancy after breast-conserving surgery and radiation therapy for breast cancer. Monogr Natl Cancer Inst 16:131–137
Early Breast Cancer Trialists' Collaborative Group (1992) Systemic treatment of early breast cancer by hormonal, cytotoxic, or immune therapy. 133 randomised trials involving 31 000 recurrences and 24 000 deaths among 75 000 women. Lancet 339:1–15, 71–85
Mulvihill JJ, McKeen EA, Rosner F, Zarrabi MH (1987) Pregnancy outcome in cancer patients. Experience in a large cooperative group. Cancer 60:1143–1150
Myers SE, Schilsky RL (1992) Prospects for fertility after cancer chemotherapy. Semin Oncol 19:597–604
Reichman BS, Green KB (1994) Breast cancer in young women: effect of chemotherapy on ovarian function, fertility, and birth defects. Monogr Natl Cancer Inst 16:125–129
Sutton R, Buzdar AU, Hortobagyi GN (1990) Pregnancy and offspring after adjuvant chemotherapy in breast cancer patients. Cancer 65:847–850
Valagussa P, Moliterni A, Zambetti M, Bonadonna G (1993) Long-term sequelae from adjuvant chemotherapy. Recent Res Cancer Res 127:247–255
Ventura SJ (1989) First births to older mothers 1970–1986. Am J Public Health 79:1675–1684

Sex Hormones and Breast Cancer: The Issue of Hormone Replacement

K.I. Pritchard, J.-A. Roy, and C.A. Sawka

Medical Oncology/Haematology, Toronto-Sunnybrook Regional Cancer Centre, University of Toronto, 2075 Bayview Avenue, Toronto, Ontario M4N 3M5, Canada

Introduction

The relationship between hormone replacement therapy (HRT) and breast cancer remains controversial. This controversy has been fueled by strong opinions and emotions. There is a large amount of observational data (case control and cohort studies on the relationship between HRT and the development of breast cancer, but virtually no good observational or interventional data (randomized controlled trials) on the effects of HRT in women who have already had a diagnosis of breast cancer.

Hormone Replacement Therapy in Healthy Perimenopausal Women

It has long been appreciated that HRT has some role in alleviating the short-term symptoms associated with the peri- and early postmenopausal period. Over the last 5–10 years, however, increasing data have become available showing that HRT plays a powerful role in reducing long-term disorders associated with the postmenopause, including cardiovascular disease and osteoporosis. It has now become a standard recommendation in many countries that healthy perimenopausal women routinely receive HRT for an indefinite period of time (Society of Obstetricians and Gynecologists of Canada 1994). This recommendation is based on randomized clinical trials showing that HRT can reduce short-term symptoms, on largely observational data suggesting that HRT can reduce cardiovascular disease and/or improve lipid profiles in a way that is believed to be associated with a reduction in cardiovascular disease, and on observational data and now randomized trials, showing a substantial reduction in osteoporotic fractures when HRT is used. Recently, published data from randomized clinical trials are also demonstrating that HRT improves lipid profiles in a way that would be expected to result in a substantial reduction in cardiovascular events.

Relief of Short-Term Symptoms Associated with Hormone Replacement Therapy

For many years the medical management of menopause was focused primarily on the relief of menopausal symptoms. Nonhormonal and hormonal compounds were used in an attempt to achieve these effects. Coope (1976) demonstrated, in a randomized double-blind crossover study, that ERT was superior to placebo for control of hot flashes. Schiff et al. (1980) used a similar design to show the superiority of oral medroxyprogesterone (MPA) over placebo. Estrogen and progesterone have not been directly compared in a randomized design. Veralipride, an antidopaminergic compound, has been widely studied, mainly in Europe, where several randomized trials have shown its superiority over placebo for relief of hot flashes (Wesel et al. 1984; De Cecco et al. 1985). Clonidine has been widely used for the relief of hot flashes when estrogen replacement therapy (ERT) was believed to be contraindicated, but a recent double blind-trial showed no evidence for superiority over placebo (Wren and Brown 1986). Since the decrease in estradiol levels at menopause leads to atrophy of the urogenital tract epithelium, topical estrogen therapy has often been used to relieve urogenital symptoms, but this approach has never been tested in a randomized setting. Certainly it is known that the use of topical estrogens results in considerable systemic absorption (Sitruk-Ware 1990).

Despite the widespread use of ERT, its effects on overall quality of life have been studied only recently. In one trial, ERT with transdermal estradiol for 2 weeks followed by a combination of transdermal norethisterone acetate and estradiol for the next 2 weeks was shown to increase quality of life in postmenopausal women after 3 months of treatment (Wiklund et al. 1992). Recently, Daly et al. (1993) also showed that quality of life may be severely affected by menopausal symptoms and that ERT can substantially improve it. In addition, the randomized trial of Limouzin-Lamothe et al. (1994) comparing HRT to veralipride showed that HRT improved quality of life significantly more than veralipride.

Prevention of Long-Term Problems with Hormone Replacement Therapy

It has been increasingly appreciated that HRT will decrease the rate of bone loss over time and that it likely offers protection from cardiovascular events.

A reduction in bone fractures in postmenopausal women could have major quality of life and economic impact since fracture is a frequent cause of morbidity and even of mortality in postmenopausal women. Observational studies have shown that the use of HRT is associated with the maintenance of bone mass and with a decreased risk of fractures of the hip and lower forearm (Weiss et al. 1980; Cauley et al. 1995). More recently, a randomized trial of ERT has shown a reduction in fractures in postmenopausal women (Lufkin et al. 1992).

In addition, it has long been known that ERT, particularly with unopposed estrogen, is associated with changes in lipid and coagulation profiles that would seem likely to be associated with beneficial cardiac effects. For example, Nabulsi et al. (1993) have shown that current estrogen or estrogen and progesterone users have elevated HDL, reduced LDL and reduced fibrinogen levels in comparison to former users or never users. Eden (1991) however, found that the addition of progesterone to estrogen in postmenopausal women tended to reduce HDL and increase LDL in comparison to estrogen used alone. It has also been postulated by Lobo (1991) that estrogen used in postmenopausal women may, as well as affecting HDL, LDL and fibrinogen, have a direct action on vessel walls, on the myocardium and on platelets.

Stampfer and Colditz (1991), in an internally controlled prospective trial using angiography as an endpoint, showed that estrogen users had as low as a 0.5 relative risk of coronary disease in comparison to nonusers. In addition, Psaty et al. (1994) showed, in a population-based case control study of postmenopausal women with myocardial infarct (MI) as an endpoint, that estrogen or estrogen and progesterone users had relative risks of 0.69 (95% CI: 0.47–1.00) and 0.68 (95% CI: 0.38–1.22), respectively, of developing an MI in comparison to nonusers. Falkeborn et al. (1992) showed a similar relative risk of 0.69 (95% CI: 0.54–0.86) for estrogen users and 0.53 (95% CI: 0.30–0.87) for estrogen and progesterone users compared to nonusers in a prospective cohort study of all women in Upsala County, Sweden.

Now an even more exciting study has been carried out. The PEPI trial, published in JAMA in 1995, represents the first randomized study comparing these interventions. In the PEPI trial, women were randomized to receive premarin 0.625 mg daily, premarin plus 10 mg of medroxyprogesterone acetate 12/28 days, premarin plus 2.5 mg of medroxyprogesterone acetate given continuously, premarin plus 200 mg of micronized progesterone 12/28 days or placebo. After 3 years of follow-up, all groups receiving estrogen, regardless of whether they were receiving progestationals of any type, showed elevation in HDL cholesterol, lowering of LDL cholesterol, lowering of fibrinogen and elevation of triglycerides. Blood pressure rose slightly in all women over the period of the trial, but no more so in those receiving medication than in controls. This study suggests that there may be substantial cardiovascular benefit in women receiving unopposed estrogen or estrogen combined with one of several forms of progesterone. There is little suggestion that the addition of micronized progesterone in particular blunts the beneficial changes in cardiovascular lipid profile.

Risk of Cancer with Hormone Replacement Therapy

Observational studies have shown that endometrial cancer is more common in women who receive unopposed estrogen. Relative risks range from 2.31

for ever users (95% CI: 2.13–2.51) to 4.8 for users of ≥5 years of 0.625 mg of conjugated estrogen, to 8.22 (95% CI: 6.25–10.81) for users of ≥8 years (Grady et al. 1992). Grady's overview has also demonstrated that the risk of endometrial cancer increases with increasing dose as well as increasing duration of estrogen. Fortuitously, the endometrial cancers picked up in women receiving ERT seem generally to be low stage and well differentiated. Thus, it is a real question whether there is an increased risk of death from endometrial cancer in women receiving unopposed estrogen. The risk of developing endometrial cancer is completely abrogated if progesterone is used concurrently for ≥10/28 days of the monthly cycle. The PEPI trial sheds more light on this situation. Some 34% of women who received unopposed estrogen in that trial developed endometrial hyperplasia, while 6% underwent hysterectomy in comparison to women receiving placebo or any combination of estrogen and progesterone, in whom the rate of endometrial hyperplasia and of hysterectomy was only 1% each (PEPI 1995).

The role of HRT in the etiology of breast cancer is much less clear. There is a certain physiologic logic to this connection in that it is known that estrogen and/or progestationals must be present for breast cancer to develop. Physiologically the normal female breast cannot mature without these hormones and men, of course, have a very much lower incidence of breast cancer than women. Furthermore, men who do develop breast cancer often have abnormalities that have resulted in higher estrogen levels (Kleinfelter's syndrome, etc.). Furthermore, the development of breast cancer in women is strongly associated with longer durations of cycling produced by early menarche and late menopause. In addition, there are a number of animal models in which estrogen and/or progesterone are required for the development of maintenance of mammary cancers.

From observational studies including over 30 case control and cohort studies, there is some suggestion that there may be a weak relationship between the use of HRT and the subsequent development of breast cancer. There have been five meta-analyses done of these studies, which suggest a slightly, but nonsignificantly, increased relative risk of developing breast cancer in ever vs never users of HRT (RR 1.0–1.07; NS). A stronger association is suggested however, when one looks at current vs never users. Sillero-Arenas et al. (1992) suggests a relative risk of 1.23 (95% CI: 1.12–1.6) in current users, while Colditz et al. (1993) suggests a relative risk of 1.40 (95% CI: 1.20–1.63). Furthermore, both Steinberg et al. (1991) and Colditz et al. (1993) suggest a relative risk of 1.30 (95% CI: 1.12–1.6, Steinberg et al. 1991; 95% CI: 1.04–1.14 Colditz et al. 1993) in women who have used HRT for ≥15 years. These meta-analyses do not show any relationship between the development of breast cancer and the dose of estrogen; they show no difference between those on unopposed estrogen vs estrogen with progesterone, nor do they show any relationship with family history of breast cancer or with personal histroy of benign breast disease. In conclusion, these five meta-analyses suggest that current estrogen use may be associated with a small increase in risk of breast cancer, that every estrogen use has no significantly increased risk, but that there is a weak association with duration of use. These is no data for those who have received estrogen for ≥20 years.

Hormone Replacement Therapy in Women with Previous Breast Cancer

Since it seems apparent that HRT is, on balance, beneficial for the woman who has not had a previous history of breast cancer, the question arises as to whether HRT can or should be used in women with previous breast cancer. We have traditionally been taught and have told our patients that it is not safe for them to take estrogen therapy, but we have few data to support this point of view. We know that women diagnosed at or around the time of pregnancy tend to have a poorer outcome and we associate this with the hormone increases of pregnancy. It has been shown by Clark and Chua (1989) however, that pregnancy one year or more after a diagnosis of breast cancer does not seem to be associated with a poorer outcome. Vassilopoulou-Sellin and Zolinsk (1992) surveyed 224 women with breast cancer, asking a series of questions concerning menopause, including symptoms related to estrogen deficiency, concerns about osteoporosis or heart disease, and attitudes and perceptions concerning ERT. Some 77% of the women surveyed were postmenopausal at the time of completion of the survey. Of these women, 8% had taken ERT at some point subsequent to their cancer diagnosis. Some 78% were afraid that ERT might precipitate a cancer recurrence but were also concerned about the risk of osteoporosis (70%) and heart disease (72%). Of these women, 44% indicated that they would consider taking ERT under medical supervision. These results suggest that the use of ERT by women with breast cancer may be more common than generally realized and certainly demonstrate that the use of ERT following breast cancer is of interest and concern to women with breast cancer.

Goodwin (1989) performed a decision-analysis of ERT in women made prematurely menopausal by adjuvant chemotherapy. Based on the available data regarding risk of recurrence, risk of death from other causes and menopausal symptoms, it appeared that for women with node-negative breast cancer who had substantial menopausal symptoms, the use of ERT was reasonable.

Three case series of women with breast cancer receiving ERT for menopausal symptoms have been reported. Stoll (1989) published a case series in which 65 postmenopausal women treated previously for breast cancer received conjugated equine estrogen (0.625 mg) and norgesterol given daily for 3–6 months. The women were selected to receive ERT because of either hot flashes which had failed to respond to clonidine or symptoms of severe vaginal or urethral atrophy. All patients treated showed improvement of menopausal symptoms. Patients were followed for at least 2 years with no breast cancer recurrence seen. DiSaia (1993) recently updated a case series of 77 women previously treated for breast cancer who subsequently received ERT for the relief of menopausal symptoms and for the prevention of postmenopausal cardiovascular disease and osteoporosis. The interval between primary therapy and HRT averaged 23.8 months with an average duration of therapy of 27 months. The specific regimen of HRT generally consisted of conjugated estrogen with the majority (64) receiving a concurrent progestagen. Seven patients have had recurrences, five of whom were still taking HRT at the time of recurrence. Of these patients, four stopped HRT at the time of recurrence; one is alive with no evidence of disease, two are alive with disease and one has died of disease. The patient still taking HRT is alive without disease. Powles et al. (1993) recently published a report in which ERT was given together with tamoxifen for the relief of hot flashes in women with breast cancer. There were 35 patients who received 0.625 mg conjugated estrogen together with tamoxifen 20 mg daily. In five patients the estrogen dose was increased to 1.25 mg daily in an attempt to achieve better symptom control. With this approach, 24 patients had either complete or partial relief of menopausal symptoms. The mean duration of HRT was 14.6 months and the mean follow-up was 43 months. Only two patients relapsed during the follow-up.

In addition, two case control studies have been conducted. Wile et al. (1991) matched 24 women with a history of breast cancer who subsequently took HRT to 50 controls using stage, age and duration of observation as matching criteria. The mean interval between breast cancer diagnosis and initiation of HRT in the cases was 25 months. During the average duration of observation on HRT of 2 years, one case and two controls had cancer related deaths. The authors concluded that no adverse effects of HRT on outcome could be detected. Cobleigh et al. (1994) referred in their paper to an unpublished case control study by Eden (1991) in which 90 breast cancer survivors who took HRT were matched to controls on the basis of age and disease-free interval prior to the initiaton of HRT. Significantly fewer tumour recurrences apparently occurred in the HRT group.

Recently, Dhodapker et al. (1995) reported on four women who developed metastatic breast cancer while taking ERT. In all cases, withdrawal of ERT as the sole therapeutic manoeuvre resulted in regression of the metastatic disease. The authors suggested that this manoeuvre would be appropriate as initial treatment for metastatic disease which develops on ERT. It is of interest that three of these four women had been on ERT for 5–15 years prior to developing their metastatic disease. Whether ERT speeded or slowed the development of these metastases is, of course, impossible to determine, but it is clear that these particular women tolerated ERT for many years before developing overt recurrence.

These reports illustrate that data regarding ERT in women with breast cancer are scarce and that any observations must be viewed as preliminary. The total number of women with breast cancer receiving ERT in these published reports is small compared to the number of women with breast cancer who have apparently received ERT, if the survey of Vassilopoulou-Sellin and Zolinski (1992) is at all generalizable. Thus, publication bias may be present. In addition, the mean follow-up the published cases is relatively short, and without randomized data, it is difficult to draw any firm conclusions.

In an attempt to clarify this issue, several randomized prospective studies have recently been initiated or proposed. Vassilopoulou-Sellin and colleagues are conducting a trial for women with disease-free intervals of at least 2 years following resection for estrogen-receptor-negative, and 10 years following estrogen receptor-positive breast cancers. The sample size, however, can detect a difference in recurrence rates of only 10% or greater, and given the high prevalence of this disease, even a small increase in relapse rate due to ERT is of importance. We are also aware of an International Breast Cancer Study Group proposal in which women with node-negative or positive breast cancer will be eligible for randomization to ERT with or without concurrent tamoxifen therapy. Cobleigh et al. (1994) have recently recommended that before any large randomized study is initiated, pilot trials should be conducted in several areas. They feel that appropriate tools must be developed to measure the frequency and severity of menopausal symptoms and their impact on quality of life. Our own group is currently carrying out such a study in which six questionnaires including: (1) a menopause-specific quality of life (Hilditch and Lewis); (2) a survey of attitude and knowledge (Milette); (3) a profile of mood states (POMS); (4) a women's health questionnaire (Hunter); (5) The EORTC quality of life with breast module; (6) a menopausal attitude questionnaire (Lieblum and Schwartzman), and a trade-off task, which are being administered to women with a diagnosis of breast cancer resected with curative intent. These women are selected to be close to menopause by virtue of their age, or of having started or being about to start tamoxifen or adjuvant chemotherapy. The questionnaires and trade-off task will be administered at baseline and the questionnaires at 3 month intervals together with FSH determinations until menopause is entered. The questionnaires will then continue to be administered every 3 months over a 1 year period with the trade-off task being readministered at the beginning of menopausal symptoms and 6 months and 1 year later. We hope thereby to collect data that may be useful in developing adequate endpoints for randomized trials in this area. It will however be most important to ensure that any form of HRT used in this setting does not increase the risk of recurrence of breast cancer. In order to determine this, large studies with good power will be required.

References

Cauley JA, Seeley DG, Ensrud K et al. (1995) Estrogen replacement therapy and fractures in older women. Ann Intern Med 122:9–16

Clark RM, Chua T (1989) Breast cancer and pregnancy: the ultimate challenge. Clin Oncol 1:11–18

Cobleigh MA, Berris RF, Bush T et al. (1994) Estrogen replacement therapy in breast cancer survivors. A time for change. JAMA 272:540–545

Colditz GA, Egan KM, Stampfer MJ (1993) Hormone replacement therapy and risk of breast cancer: results from epidemiologic studies. Am J Obstet Gynecol 168: 1473–1480

Coope J (1976) Double-blind cross-over study of estrogen replacement therapy. In: Campbell S (ed) Management of the menopause and the post-menopausal years. MTP, Lancaster, p 167

Daly E, Gray A, Barlow D, McPherson K et al. (1993) Measuring the impact of menopausal symptoms in quality of life. Br Med J 307:836–840

De Cecco L, Venturini PL, Alibrandi MP et al. (1985) Effet du véralipride sur les taux de LH, FSH et PRL et sur le syndrome climatérique. Rev Fr Gynecol Obstet 80:152–155

Dhodapker MV, Ingle JN, Ahman DL (1995) Estrogen replacement therapy withdrawal and regression of metastatic breast cancer Cancer (in press).

DiSaia PJ (1993) Hormone-replacement therapy in patients with breast cancer. Eur J Cancer Clin Oncol 23:1909–1913

Eden JA (1991) Progestens: an occasional review. Asia Oceania J Obstet Gynaccol 17:289–295

Falkeborn M, Persson I, Adami HO et al. (1992) The risk of acute myocardial infarction after estrogen and estrogen-progestogen replacement. Br J Obstet Gynaecol 99:821–828

Goodwin PJ (1989) Decision analysis (DA) of estrogen replacement therapy (ERT) in women made prematurely menopausal during adjuvant chemotherapy (ADJ/CXT) for breast cancer (BC). Breast Cancer Res Treat 14:147

Grady D, Rubin SM, Petitti DB et al. (1992) Estrogen and progestin for postmenopausal women: a meta-analysis. Ann Intern Med 117:1016–37

Limouzin-Lamother MA, Mairon N, Joyce CRB et al. (1994) Quality of life after the menopause: influence of hormonal replacement therapy. Am J Obstet Gynecol 170:618–624

Lobo RA (1991) Clinical review 27: effect of hormonal replacement on lipids and lipoproteins in postmenopausal women. J Clin Endocrinol Metab 73:925–930

Lufkin EG, Wahner HW, O'Fallon WM et al. (1992) Treatment of postmenopausal osteoporosis with transdermal estrogen. Ann Intern Med 117:1–9

Nabulsi MB, Folsom AR, White A et al. (1993) Association of hormone-replacement therapy and fractures in older women. Ann Intern Med 122:9–16

PEPI (1995) Effects of estrogen or estrogen/progestin regimens on heart disease risk factors in postmenopausal women. The Postmenopausal Estrogen/Progestin Interventions (PEPI) trial. JAMA 273:199–208

Powles TJ, Hickish T, Casey S et al. (1993) Hormone replacement after breast cancer. Lancet 342:60–61

Psaty BM, Heckbert SR, Atkins D et al. (1994) The risk of myocardial infarction associated with the combined use of estrogens and progestins in postmenopausal women. Arch Intern Med 154:1333–1339

Schiff I, Tulchinsky D, Cramer D et al. (1980) Oral medroxyprogesterone in the treatment of postmenopausal symptoms. JAMA 224:1443

Sillero-Arenas M, Delgado-Rodriguez M, Rodrigues-Canteras R et al. (1992) Menopausal hormone replacement therapy and breast cancer: a meta-analysis. Obstet Gynecol 79:286–294

Sitruk-Ware R (1990) Estrogen therapy during menopause. Practical treatment recommendations. Drugs 39:203–217

Society of Obstetricians and Gynecologists of Canada (1994) Canadian Menopause Consensus Conference. J Soc Obstet Gynecol Can 16:4–40

Stampfer MJ, Colditz GA (1991) Estrogen replacement therapy and coronary heart disease: a quantitative assessment of the epidemiologic evidence. Prev Med 20:47–63

Steinberg KK, Thacker SB, Smith SJ et al. (1991) A meta-analysis of the effect of estrogen replacement therapy on the risk of breast cancer. JAMA 265:1985–1990

Stoll BA (1989) Hormone replacement therapy in women treated for breast cancer. Eur J Cancer Clin Oncol 23:1909–1913

Vassilopoulou-Sellin R, Zolinski C (1992) Estrogen replacement therapy in women with breast cancer: a survey of patients attitudes. Am J Med Sci 304:145–149

Weiss NS, Ure CL, Ballard JH et al. (1980) Decreased risk of fractures of the hip and lower forearm with postmenopausal use of estrogen. N Engl J Med 303:1195–1198

Wesel S, Bourguignon RP, Bosumua WB (1984) Veralipride versus conjugated estrogens: a double-blind study in the management of menopausal hot flushes. Curr Med Res Opin 8:696–700

Wiklund I, Berg G, Hammar M et al. (1992) Long-term effect of transdermal hormonal therapy on aspects of quality of life in postmenopausal women. Maturitas 14:225–236

Wile AG, Opfel RW, Margileth DA et al. (1991) Hormone replacement therapy does not affect breast cancer outcome. Proc Am Soc Clin Oncol 10:58

Wren BG, Brown LB (1986) A double-blind trial with clonidine and a placebo to treat the hot flushes. Med J Aust 144:369–370

The Women's Health Initiative: Will It Resolve the Issues?

K.B. Michels[1,2] and W.C. Willett[1,2,3]

[1] Department of Epidemiology, Harvard School of Public Health, 677 Huntington Avenue, Boston, MA 02115, USA
[2] Department of Nutrition, Harvard School of Publich Health, 665 Huntington Avenue, Boston, MA 02115, USA
[3] Channing Laboratory, Department of Medicine, Brigham and Women's Hospital and Harvard Medical School, Boston, MA 02115, USA

Introduction and History of the Women's Health Initiative

Whether fat intake is an important risk factor for the development of breast cancer has been debated over the last decade. Ecological observations, migrant studied and some case-control studies have been interpreted as supporting an association between dietary fat and breast cancer; however, other case-control and most cohort studies have tended not to confirm this suspicion. A compelling mechanism to explain a fat-breast cancer relationship has also not been established.

To further test the fat-breast cancer hypothesis, the U.S. National Cancer Institute, in 1983, began the development of a feasibility study for a 2-year randomized clinical trial among women age 45–69 years, known as the Women's Health Trial (WHT) (Insull et al. 1990). In 1987–1988, two independent reviews of the WHT feasibility study were under taken (Greenwald 1988; Subcommittee for the Women's Health Trial 1988); both groups cited new and conflicting evidence from recent epidemiologic studies recognizing a possible need for modification of the study hypothesis and raised doubts about long-term dietary adherence. In 1991, in a dramatic turnaround, under the influence of Congress, NIH revived the issue of fat reduction trials. The estimated cost of the enterprise was $500 million.

The protocol of the Women's Health Initiative (WHI) (1991) proposed two main goals. The first was to evaluate the effectiveness of specific interventions for the prevention of cancer, cardiovascular disease, and osteoporotic fractures. The interventions to be tested in the randomized trial are hormone replacement therapy, low-fat dietary pattern, and calcium plus vitamin D supplementation. The second goal is to evaluate strategies to achieve healthful behaviors that have established value.

This chapter addresses some of the issues associated with the WHI. Discussion will be restricted to the clinical trial component of the initiative, with special emphasis on the diet-breast cancer component as the central determining factor for size and cost of the study.

Study Protocol

The protocol of the WHI (Women's Health Initiative 1991) proposes a partially blinded, controlled clinical trial of postmenopausal women age 50–79 years. The treatments will be tested in a $3 \times 2 \times 2$ factorial design. The first component will test the efficacy of estrogen replacement or estrogen plus progesterone vs placebo on cardiovascular disease (three arms); the second will test the efficacy of a low-fat vs usual dietary pattern on breast and colorectal cancer, and, as a secondary goal, on cardiovascular disease (two arms); the third will test the efficacy of calcium/vitamin D supplementation vs placebo on fractures (two arms). Safety assessments include breast cancer and endometrial cancer (hormone replacement therapy) and renal calculi (calcium/vitamin D supplementation).

The eligibility and inclusion criteria are as broad as possible. In this way the investigators hope to increase the generalizability of the results to the U.S. population of postmenopausal women. Women will be recruited on the basis of their eligibility and willingness to participate in either the hormone replacement therapy or the dietary modification components, or both. Women with less than 32% intake of fat as a percentage of calories, as determined by a questionnaire, will be excluded from the dietary modification component. It is anticipated that many women who are enrolled in the hormone replacement therapy component also will be enrolled in the dietary modification component and that virtually all women will be randomized into the calcium/vitamin D component. The total clinical trial study population is likely to be in the range of 65000–75000 women. Approximately 40 centers will recruit about 1400 participants each for the trial over a 3 year period. Intervention time will be 9 years and posttrial mortality surveillance for a future 5 years is envisaged so that total follow-up will be an average of 14 years.

The dietary intervention's goal will be to reduce total fat intake to 20% of calories and saturated fat to less than 7% of calories. Complex carbohydrates and fiber-containing foods should be increased to five or more daily servings for fruits and vegetables and to six or more daily servings for grain products.

Chewable calcium carbonate, at a dose of 300 mg two to three times daily with meals, has tentatively been selected. For vitamin D, it is proposed to use 400 IU daily of vitamin D_3.

Effect of Dietary Modification on the Incidence of Breast Cancer and Colon Cancer

Hypotheses

The plan to study whether a reduction in fat intake reduces the risk of breast and colon cancer is based on hypotheses that have been examined in ecological, animal and epidemiologic studies. A secondary goal, the evaluation of the effect of a low-fat diet on coronary heart disease, is based on a hypothesis generated mainly from studies involving lipid-lowering agents. However, there is no evidence that a low fat diet per se reduces the risk of coronary heart disease (Committee on Diet and Health 1989) and the relationship has not been demonstrated in prospective observational studies (Willett 1990).

Dietary fat reduction is recommended by the National Academy of Sciences (Committee on Diet, Nutrition and Cancer 1982), the American Heart Association, and the National Cancer Institute (1984). The Surgeon General's 1988 report on Nutrition and Health states that "substantial epidemiologic and animal evidence supports a relationship between dietary fat and the incidence of both breast cancer and colon cancer" (US Department of Health and Human Services, 1988). Doll and Peto (1981) have suggested that as much as 50% of breast and 90% of colon cancer in the United States might be preventable by changes in diet. Whereas an association between animal fat or red meat intake and colon cancer is relatively well supported (Willett 1989; Willett et al. 1990; Giovannucci et al. 1992; Stemmermann et al. 1984), data on breast cancer are far less consistent.

In some laboratory models, animals fed high amounts of fat experienced an increased incidence of breast cancer (Tannenbaum 1942). This effect occurred with both saturated and unsaturated fats (Carroll 1986). However, no effect was seen in a large trial in which carcinogens were not used as inducing agents (Appleton and Landers 1985). In animal studies, total energy intake has been found to be a more powerful determinant of mammry tumor incidence than dietary fat (Willett 1987; Micozzi 1985). Thus, animal data suggest that total energy balance and growth rates might confound any association of a single nutrient with breast cancer (Welsch 1995; Boissonneault et al. 1986).

International correlation studies have indicated a positive association of fat consumption with breast cancer incidence and mortality rates (Carroll et al. 1968; Armstrong and Doll 1975; Prentice et al. 1988). Industrialized countries have the highest breast cancer rates, so that a multitude of factors, including physical activity, childhood nutrition and growth and reproductive factors, are confounding variables that potentially could distort the international correlations. Breast cancer incidence is correlated with the Gross National Product (Armstrong and Doll 1975) and height (Gray et al. 1979) with comparable magnitude as fat disappearance. Height has been used as a surrogate endpoint for energy restriction in early life; in many case-control and cohort studies a positive relation with risk of breast cancer has been seen (Swanson et al. 1988; Tretli 1989; Vatten and Kvinnsland 1990). Energy restriction might be crucial during childhood and adolescence when mammary tissue might be more susceptible to carcinogenic processes. Also, accelerated prepubertal growth is the main determinant of early age at menarche, a well established risk factor for breast cancer.

Time trends indicating concurrent increases in both breast cancer and fat consumption have been cited to support the fat intake-breast cancer hypothesis (Prentice et al. 1988; Prentice and Sheppard 1990). However, these data on fat consumption were based on the use of food-disappearance figures, that is foods produced and imported, which inaccurately reflects actual intake since food wastage is not taken into account (Willett and Stampfer 1990; Goodwin and Boyd 1987). In contrast, a compilation of 171 studies of actual dietary intake conducted in the U.S. indicates that fat intake has decreased in the last 40 years, rather than increased as suggested by the production and import data, probably due to increased wastage and selective avoidance of fat (Stephan and Wald 1990). As breast cancer incidence rates have increased by about 40% in the last five decades in the United States, the temporal trend contradicts rather than supports a causal association.

Studies of persons migrating from areas with low to those with high rates of fat consumption suggest that environmental factors influence the occurrence of cancer (Buell 1973; Staszewski and Haenszel 1965; McMichael and Giles 1988). However, rates of breast cancer incidence usually take several generations to approach those of the resident population, whereas colon cancer risks increase much faster (Haenszel and Kurihara 1968; Hanai and Fujimoto 1982). Indeed, the extremely slow changes in breast cancer rates among Japanese populations migrating to the U.S., makes highly questionable any assumption that reductions in fat intake by postmenopausal women will, within several years, have the full impact predicted by the international cross-sectional comparisons (McMichael and Giles 1988).

Case-Control and Cohort Studies

Epidemiologic studies on breast cancer in humans have produced inconsistent results. Although some case-control studies provide evidence for a small positive association, such studies are potentially distorted by differential recall of past diets by women with and without cancer, and by uncertainty regarding the appropriate control women. A review of 14 case-control studies did not allow any uniform conclusion (Goodwin and Boyd 1987), whereas a combined analysis of 12 case-control studies showed a weak positive association between breast cancer risk and saturated fat intake in postmenopausal women (Howe et al. 1990).

Little or no relationship between fat intake and breast cancer has been seen in prospective studies (Table 1). In the National Health and Nutrition

Table 1. The prospective studies on dietary fat and breast cancer[a]

Study	Cases/total cohort	Follow-up (years)	Range of total fat intake (as % of calories from fat)	Relative risk (95% confidence interval)
NHANES (1987)	99/5485	10	<30–≥42	0.62 (0.33; 1.19)
NHS 4-years (1987)	601/89 538	4	32–44	0.82 (0.64; 1.05)
Mills et al. (1989)	215/20 341	6	NA	1.21 (0.81; 1.81)
Howe et al. (1991)	519/56 837	5	31–47	1.30 (0.90; 1.88)
Kushi et al. (1991)	408/32 080	4	27–41	1.13 (0.84; 1.51)
NHS 8-years (1992)	1439/89 494	8	<29–>49	0.90 (0.77; 1.07)
Graham et al. (1992)	359/18 586	7	<26–>37	1.00 (0.59; 1.70)
Van den Brandt et al. (1992)	417/62 573	3.3	32–46	1.08 (0.73; 1.59)

[a] The relative risk is given for the highest quartile or quintile of total fat intake compared to the lowest quartile or quintile.

Examination Survey I Epidemiologic Follow-up Study (NHANES) and the Nurses Health Study (NHS), no positive association between fat intake and breast cancer risk was observed (Willett et al. 1987a; Jones et al. 1987). This relationship was confirmed in the 8-year follow-up of the NHS with 1439 incidence cases (Willett et al. 1991). Notably, in the same cohort a clear positive association was observed between animal fat and red meat intake and risk of colon cancer. A cohort study among Seventh-day Adventists concluded that increasing consumption of high fat animal products or percent of calories from animal fat was not associated with increased risk of breast cancer (Mills et al. 1989), and Kushi et al. (1995) found a lack of positive association between dietary fat and breast cancer in the Iowa Women's Health Study. Howe et al. (1991) reported a borderline statistically significant positive association of total fat intake and breast cancer among Canadian women; however, the lowest cancer risk was actually found in the second quartile of total fat intake, which was significantly reduced from the incidence in the lowest quartile. No significant association was found in a New York State cohort of 18 586 postmenopausal women (Graham et al. 1995), or in a Dutch cohort of women (van den Brandt et al. 1993).

Though a potential mechanism for an effect of dietary fat on breast cancer has not been established, an increase in estrogen levels has been suggested (Prentice et al. 1990). However, reports that fat reduction lowers estrogen fractions in women have not included concurrent controls (Prentice et al. 1990), and in carefully controlled animal studies dietary fat does not appear to influence estrogen levels (Welsch and Aylsworth 1983).

Feasibility of the Women's Health Initiative

Is the randomized trial component of the WHI feasible? A key question in a dietary intervention study is whether a significant change in dietary habits can be achieved and maintained during follow-up. In the pilot study for the WHT, a large enough contrast between intervention and control group – although maybe initially achieved – might not have persisted over several years (Insull et al. 1990). During the 2 years of follow-up, the contrast between intervention and control group diminished, leaving the difference between the groups at about 14% in fat from calories. Even this contrast is almost certainly overstated because of the well documented tendency for participants in dietary intervention studies to overstate their compliance (Foster et al. 1990). If the trends observed over the first 2 years were to continue, a progressive increase of about 1% per year in the intervention group and a decrease of the same magnitude in the control group, a meaningful difference between the groups would be lost within a few years.

The stated goal of the WHI is to reduce the fat intake in the intervention group to 20% of calories. Some 7 years ago, when the pilot study was initiated, the contrast to the then average fat intake of almost 40% was plausible. However, the fat intake in the United States is rapidly declining. According to Stephen and Wald, time trends in consumption of dietary fat in the U.S. show a quadratic decline since the late 1950s (Stephan and Wald 1990). Based on these trends, estimates of current fat intake as percentage of total calories would be in the range of 35%, and this has recently been confirmed in preliminary reports of the Third National Health and Nutrition Survey. With the increasing health consciousness of the American population and the recent promotion of low fat products by the food industry, it will become harder to achieve a meaningful contrast between intervention and control group.

Another feature of the WHI, the "MRFIT phenomenon", might add to the problem of compliance. If women are recruited on the basis of their willingness to alter their diet, they might be more likely to change their diet disregarding which group they will be randomized to. In the Multiple Risk Factor Intervention Trial (MRFIT) (1982) study, participants were randomized to risk factor modification, including dietary change to reduce blood cholesterol level and reduction in smoking and hypertension, or to their usual lifestyle. Since potential benefits of the examined risk factors were discussed openly in the media during the study period, many of the participants randomized to the control group changed their lifestyle and sought blood pressure therapy, so that differences in risk factors between intervention and control group were minimal and despite enormous costs, the hypotheses were not adequately tested.

What Could We (Not) Learn from the Women's Health Initiative?

One certainty is that the WHI, as presently designed, will not resolve the hypothesis that reducing dietary fat will decrease breast cancer risk. As noted earlier, the hypothesis is without a well-defined mechanistic basis and is not supported by prospective studies designed specifically to address this relationship. Even if fat intake sometime in life or at some level were related to breast cancer risk, we have little reason to believe that an effect should be observed for a dietary change beginning after ages 50–79 years and with follow-up for only up to 14 years. Indeed, if diet were to have an effect at all, the finding that populations migrating from Japan to the U.S. require several generations to achieve the general U.S. cancer rates argues against finding a major effect of dietary change in older women. For these reasons, a lack of association in the randomized trial would not exclude an effect of diet earlier in life or with longer follow-up. The drift in diets of the intervention and control toward each other within 2 years in the WHT and the failure of other longer term studies to maintain contrasts in dietary intake make it highly uncertain whether even an almost immediate effect of diet on breast cancer incidence could be detected in the proposed study. If an effect of dietary intervention on breast cancer risk were seen in the WHI, despite the enormous obstacles noted above, could we conclude that fat reduction was responsible? Clearly not, because the intervention protocol includes increasing intake of fruit, vegetables and grains. As the epidemiologic evidence for these factors having an influence on breast cancer risk (even though of modest magnitude) is generally as strong or stronger than for fat intake, any reduction in breast cancer could not be ascribed to dietary fat. Even though the WHT feasibility study did not intend to influence alcohol intake, which has been positively related to breast cancer in many studies (Willett et al. 1987b; Schatzkin et al. 1987), women randomized to the low fat diet reported drinking alcoholic beverages less often than the control group (Gorbach et al. 1990). Thus changes in alcohol intake and many other dietary factors will make impossible any inference about fat consumption based on the randomized comparisons.

Even if the design problems noted above, which make any conclusions about fat intake impossible, did not exist, the dietary fat changes proposed in the study do not address questions about fat intake that are of greatest concern. Specifically, the dietary fat intervention proposes to reduce total and saturated fat intake, primarily replacing the fat with carbohydrate. Given the accumulated data of many types and in relation to many diseases, a general consensus has been attained that saturated fat should be reduced in the U.S. diet, so the study does not address an overall public health issue of debate or controversy. An area of genuine concern and uncertainty, however, is whether the saturated fat should be replaced largely with carbohydrate, polyunsaturated fat, or monounsaturated fat. Polyunsaturated fat and monounsaturated fat appear to improve blood lipid patterns relative to carbohydrates, but theoretical concerns exist regarding the relation of long-term, high intake of polyunsaturated fats to cancer. Data to resolve these important issues and to provide dietary guidance are needed, but will not be provided by the WHI.

Finally, the partial factorial design of the trial, although subject- and cost-efficient, might prove additionally problematic for the interpretation of the results. Interactions between the nutritional factors and the hormonal treatment groups are likely if the fat hypothesis were to be correct. Specifically, the leading hypothesized mechanism for a fat-breast cancer association is that reducing fat intake reduces estrogen levels. As the effect, if any, of dietary fat reduction on hormone levels is likely to be modest (Prentice et al. 1990), this change will be swamped by the pharmacologic doses of estrogen to be used in other arms of the trial or taken by women on their own, as estrogen supplementation was not an exclusion criterion for the dietary trial.

Conclusions

Though the cost will be great, randomized trials comprising the WHI would certainly be worthwhile if the questions they purport to address could be answered clearly. The likelihood that some useful information will emerge is substantial because valuable unexpected "spin-off" findings typically emerge from large trials. However, because of the fundamental design, expected difficulties in maintaining long-term compliance and ethical considerations that may require early stopping, the central questions regarding low fat diets and postmenopausal hormone replacement are not likely to be resolved by the WHI. Unfortunately, the large investment required will inevitably divert resources from many promising areas of research.

References

Appleton BS, Landers RE (1985) Oil gavage effects on tumor incidence in the national toxicology program's 2-year carcinogenesis bioassay. Adv Exp Med Biol 206:99–104

Armstrong B, Doll R (1975) Environmental factors and cancer incidence and mortality in different countries, with special reference to dietary practices. Int J Cancer 15:617–631

Boissonneault GA, Elson CE, Pariza MW (1986) Net energy effects of dietary fat on chemically induced mammary carcinogenesis in F-344 rats. J Natl Cancer Inst 76:335–338

Buell P (1973) Changing incidence of breast cancer in Japanese-American women. J Natl Cancer Inst 51:1479–1483
Carroll KK (1986) Experimental studies on dietary fat and cancer in relation to epidemiological data. Prog Clin Biol Res 222:231–248
Carroll KK, Gammal EB, Plunkett ER (1968) Dietary fat and mammary cancer. Can Med Assoc J 98:590–594
Committee on Diet and Health (1989) Diet and health. Implications for reducing chronic disease risk. National Academy, Washington
Committee on Diet, Nutrition and Cancer (1982) Assembly of Life Sciences, National Research Counsil, National Academy of Sciences: Diet, nutrition and cancer. National Academy, Washington
Doll R, Peto R (1981) The causes of cancer. J Natl Cancer Inst 66:1191–1308
Foster JL, Jeffery RW, VanNatta M, Pirie P (1990) Hypertension prevention trial: do 24-h food records capture usual eating behavior in a dietary change study? Am J Clin Nutr 51:253–257
Giovannucci E, Stampfer MJ, Colditz G et al. (1992) Relationship of diet to risk of colorectal adenoma in men. J Natl Cancer Inst 84:91–98
Goodwin PJ, Boyd NF (1987) Critical appraisal of the evidence that dietary fat intake is related to breast cancer risk in humans. J Natl Cancer Inst 79:473–485
Gorbach SL, Morrill-LaBrode A, Woods MN et al. (1990) Changes in food patterns during a low-fat dietary intervention in women. J Am Diet Assoc 90:802–809
Graham S, Zielezny M, Marshall J et al. (1992) Diet in the epidemiology of postmenopausal breast cancer in a New York State cohort. Am J Epidemiol 136:1327
Gray GE, Pike MC, Henderson BE (1979) Breast cancer incidence and mortality rates in different countries in relation to known risk factors and dietary practices. Br J Cancer 39:1–7
Greenwald P (1988) Issues raised by the women's health trial. J Natl Cancer Inst 80:788–790
Haenszel W, Kurihara M (1968) Studies of Japanese migrants. I. Mortality from cancer and other diseases among Japanese in the United States. J Natl Cancer Inst 40:43–68
Hanai A, Fujimoto I (1982) Cancer incidence in Japan in 1975 and change of epidemiological features for cancer in Osaka. Monogr Natl Cancer Inst 62:3–7
Howe GR, Hirohata T, Hislop TG et al. (1990) Dietary factors and risk of breast cancer: combined analysis of 12 case-control studies. J Natl Cancer Inst 82:561–569
Howe GR, Friedenreich CM, Jain M, Miller AB (1991) A cohort study of fat intake and risk of breast cancer. J Natl Cancer Inst 83:336–340
Insull W, Henderson MM, Prentie RL et al. (1990) Results of a randomized feasibility study of a low-fat diet. Arch Intern Med 150:421–427
Jones DY, Schatzkin A, Green SB et al. (1987) Dietary fat and breast cancer in the National Health and Nutrition Examination Survey 1 Epidemiologic follow-up study. JNCI 79:465–71
Kushi LH, Sellers TA, Potter JD et al. (1992) Dietary fat and postmenopausal breast cancer. J Natl Cancer Inst 84:1092–1099
McMichael AJ, Giles GG (1988) Cancer in migrants to Australia: extending the descriptive epidemiology data. Cancer Res 48:751–756
Micozzi MS (1985) Nutrition, body size, and breast cancer. Yearbook Phys Anthropol 28:175–206
Mills PK, Beeson WL, Phillips RL, Fraser GE (1989) Dietary habits and breast cancer incidence among Seventh-day Adventists. Cancer 64:582–590
Multiple Risk Factor Intervention Trial Research Group (1982) Multiple risk factor intervention trial. Risk factor changes and mortality results. JAMA 248:1465–1477

National Cancer Institute (1984) Cancer prevention: good news; better news; best news. Department of Health and Human Services, Washington (DHHS publication no (NIH) 84–2671)
Prentice RL, Sheppard L (1990) Dietary fat and cancer: consistency of the epidemiologic data, and disease prevention that may follow from a practical reduction in fat consumption. Cancer Causes Control 1:81–97
Prentice RL, Kakar F, Hursting S (1988) Aspects of the rationale for the Women's Health Trial. J Natl Cancer Inst 80:802–814
Prentice RL, Thompson D, Clifford C et al. (1990) Dietary fat reduction and plasma estradiol concentration in healthy postmenopausal women. J Natl Cancer Inst 82:129–134
Schatzkin A, Jones Y, Hoover RN et al. (1987) Alcohol consumption and breast cancer in the epidemiologic follow-up of the first National Health and Nutrition Examination Survey. N Engl J Med 316:1169–1173
Staszewski J, Haenszel W (1965) Cancer among the Polish-born in the United States. J Natl Cancer Inst 35:291–297
Stemmermann GN, Nomura AM, Heilbrun LK (1984) Dietary fat and the risk of colorectal cancer. Cancer Res 44:4633–4637
Stephan AM, Wald NJ (1990) Trends in individual consumption of dietary fat in the United States, 1920–1984. Am J Clin Nutr 52:457–469
Subcommittee for the Women's Health Trial (1988) Board of Scientific Counselors, Division of Cancer Prevention and Control, National Cancer Institute, Bethesda
Swanson CA, Torres OY, Schatzkin A et al. (1988) Breast cancer risk assessed by anthropometry in the NHANES. I. Epidemiological follow-up study. Cancer Res 48:5363–5367
Tannenbaum A (1942) Genesis and growth of tumors. III. Effects of a high-fat diet. Cancer Res 2:468–475
Tretli S (1989) Height and weight in relation to breast cancer morbidity and mortality – a prospective study of 570000 women in Norway. Int J Cancer 44:23–30
US Department of Health and Human Services (1988) The surgeon general's report on nutrition and health. Government Printing Office, Washington, p 194 (Publication no 88-50210)
Van den Brandt PA, Wan't Veer P, Goldbohm RA et al. (1993) A prospective cohort study on dietary fat and the risk of postmenopausal breast cancer. Cancer Res 53:75–82
Vatten LJ, Kvinnsland S (1990) Body height and risk of breast cancer: a prospective study of 23, 831 Norwegian women. Br J Cancer 61:881–885
Welsch CW (1995) Relationship between dietary fat and experimental mammary gland tumorigenesis: a review and critique. Cancer Res (in press)
Welsch CW, Aylsworth SF (1983) Enhancement of murine mammary tumorigenesis by feeding of high levels of dietary fat: a hormonal mechanism? J Natl Cancer Inst 70:215–221
Willett WC (1987) Implications of total energy intake for epidemiologic studies of breast and large-bowel cancer. Am J Clin Nutr 45:354–360
Willett WC, Stampfer MJ, Colditz GA, Rosner BA, Hennekens CH, Speizer FE (1987a) Dietary fat and the risk of breast cancer. N Engl J Med 316:22–8
Willett WC, Stampfer MJ, Colditz GA et al. (1987b) Moderate alcohol consumption and the risk of breast cancer. N Engl J Med 316:1174–1180
Willett W (1989) The search for the causes of breast and colon cancer. Nature 338:389–394
Willett W (1990) Diet and coronary heart disease. In: Willett W (ed) Nutritional epidemiology. Oxford University Press, New York, pp 341–379
Willett WC, Stampfer MJ (1990) Dietary fat and cancer: another view. Cancer Causes Control 1:103–109

Willett WC, Stampfer MJ, Colditz GA et al. (1990) Relation of meat, fat, and fiber intake to the risk of colon cancer in a prospective study among women. N Engl J Med 323:1664–1672
Willett WC, Hunter DJ, Stampfer MJ et al. (1991) Dietary fat and breast cancer: an 8-year follow-up. Am J Epidemiol 134:715
Women's Health Initiative (1991) Proposed plan. National Institutes of Health, Washington

The Women's Health Initiative: The Road to Scientific Clarity?

W.C. Willett

Harvard School of Public Health, Epidemiology and Nutrition,
677 Huntington Avenue, Boston, MA 02115, USA

The U.S. National Institutes of Health (NIH) has recently embarked on its most ambitious and expensive trial ever, the Women's Health Initiative (WHI). The primary component of this project involves a randomized factorial trial to evaluate the effectiveness of three interventions for the prevention of cancer, cardiovascular disease, and osteoporotic fractures. The interventions to be tested are hormone replacement therapy, low-fat dietary pattern, and calcium plus vitamin D supplementation. A 10-year period of intervention is planned. A second goal is to evaluate strategies to achieve healthful behaviors that have established value, including smoking prevention and cessation, improved dietary pattern, achievement and maintenance of optimal weight, increased physical activity, and early cancer detection.

The central hypothesis leading to the development of the project has been that dietary fat is the main cause of breast cancer and that reducing fat intake will substantially reduce breast cancer incidence. However, as evidence has accumulated over the last several years, support for this hypothesis, primarily derived from international correlations, has diminished. In each of the six large prospective studies that have examined this association, little or no relation was seen between fat intake and breast cancer incidence. These observational studies have not been able to examine the influence of fat intake below 20% of calories and have not examined the effect of fat before midlife. However, these limitations will also apply to the WHI, particularly since women even over the age of 70 years will be randomized. As in previous large randomized trials, long-term dietary compliance is likely to be problematic; in the pilot study the difference in fat intake between the intervention and control group after 3 years was small: 27% vs 31% of energy. Most fundamentally, the trial cannot answer the question regarding the effect of dietary fat because the dietary intervention has been broadened to include an increase in consumption of fruits, vegetables, and whole

with reduced risk of breast cancer in multiple observational studies, a benefit of the dietary intervention – if of sufficient magnitude and duration – is expected irrespective of any effect of dietary fat.

Whether the trial of hormone replacement therapy (estrogen, estrogen plus progestogen, or placebo) will provide clear guidance for women is uncertain. Evidence that estrogen therapy will reduce risk of coronary disease and fractures is firm, so that the primary question is whether this will be balanced by increases in risks of breast and endometrial cancer. However, the benefit from heart disease and fracture reduction is likely to be rapid whereas the increases in breast cancer are likely to be delayed. Thus the time frame of this trial is not likely to be informative regarding the most important question of whether hormone replacement therapy should be continuous after menopause. The component of the trial relating to calcium and vitamin D is likely to provide clearer information on the benefits of this combination. However, available data more clearly support benefits from vitamin D and this will not be distinguishable from any contribution of calcium in this study.

Although the WHI will provide additional data on health-related issues among women, by design it cannot answer questions about the relation of fat intake to risk of breast and other cancers. Although it will probably contribute to information on the effects of hormone replacement therapy, it will also probably not be able to address the most important practical questions about the optimal duration of treatment. The component of the WHI that seeks to enhance the adoption of lifestyle factors of proven benefit appears well justified.

Design and Current Status of the NSABP Breast Cancer Prevention Trial

C.K. Redmond and J.P. Costantino

Department of Biostatistics, 318 Parran Hall, 130 DeSoto Street, University of Pittsburgh, Pittsburgh, PA 15261, USA

Introduction

When the invitation to contribute to this volume was accepted, in the Fall of 1993, we anticipated that the NSABP Breast Cancer Prevention Trial (BCPT) would have completed or nearly completed accrual and that we, along with the rest of the world's trialists, would be anxiously awaiting for the data to mature. In the intervening time we would continue to conduct an extensive quality assurance program to assure that the trials outcomes were carefully and accurately monitored. In addition, we would maintain an infrastructure to promote the participants and investigators' adherence to the protocol and to maintain enthusiasm for its completion.

Instead, we are reporting today that there remains still about 30% of the original recruitment to be completed. Of concern is the extent to which the unfortunate, tragic disruptions of the NSABP during this past year may have affected the viability of this important trial. Here we will provide a brief review of: (1) the rationale and design of the BCPT; (2) how the events of this past year relate to the design and conduct of the BCPT; and (3) the current status of the trial.

Background and Rationale for BCPT

The systematic investigation of tamoxifen treatment for breast cancer has led to observations important in the rationale and design of the BCPT. The nonsteroidal drug tamoxifen is currently the most widely prescribed treatment for breast cancer. First synthesized in 1966, it has been widely evaluated in both laboratory and clinical studies. Although commonly referred to as an "anti-estrogen," this is somewhat of a misnomer, since the drug exhibits both estrogen antagonist and agonist properties in the body. While the main antitumor effect of tamoxifen is generally attributed to its competing with estradiol for receptor sites in breast tumors, other potentially tumorigenic effects have also been shown, and animal studies provide a basis for the drug's role in preventing either initiation or promotion of mammary cancers. The estrogen agonist properties of tamoxifen are probably responsible for its effect in lowering total and LDL cholesterol and preservation of bone mineral density in postmenopausal women.

The strongest evidence for tamoxifen's effectiveness against breast cancer comes from the numerous clinical trials of early stage breast cancer that have consistently shown a reduction in recurrences and deaths from breast cancer. A recently published overview of randomized clinical trials showed that, on the average, tamoxifen treatment for 1 or more years was associated with a 25% reduction in the odds of recurrence and a 16% reduction in the odds of death (Early Breast Cancer Trialist's Collaborative Group 1992). A concomitant relevant finding for the BCPT was the approximately 40% reduction in the odds of a second primary cancer of the opposite breast, with longer term treatment having a somewhat (not statistically significant) greater effect.

Important for justifying the use of tamoxifen as a breast cancer preventative is the relative rarity of serious adverse effects. As with estrogen therapy there is a risk of thromboembolic disease and endometrial cancer from tamoxifen administration. Although liver cancer has also received much attention as a potential toxicity based on studies of high doses in rats, experience with long-term administration of tamoxifen in clinical trials has not shown this to be a risk for humans at doses used in adjuvant trials. Non-life-threatening side effects, such as hot flashes, appear to be related to the drug's estrogen antagonist properties.

Based on these considerations it was deemed both appropriate and important to evaluate the effectiveness of tamoxifen for primary prevention of breast cancer in a well-designed and conducted clinical trial of women at increased risk for breast cancer.

Design of the BCPT

The BCPT is a multicenter, double masked, placebo controlled trial in women at increased risk for breast cancer. Much attention was devoted to assure that all the clinical and ethical concerns were addressed in the context of a scientifically well-designed protocol that would answer the study's primary hypotheses. The primary aim of the trial is to test the hypothesis that 20 mg daily of oral tamoxifen for at least 5 years reduces incidence and mortality from invasive breast cancer. Figure 1 summarizes the schema for the trial. The original plan was to recruit 16 000 participants during a 2–3 year accrual period with 7 years of additional follow-up required to provide definitive end results. In addition to testing the value of tamoxifen in lowering morbidity and mortality from breast cancer, other specific aims are to evaluate whe-

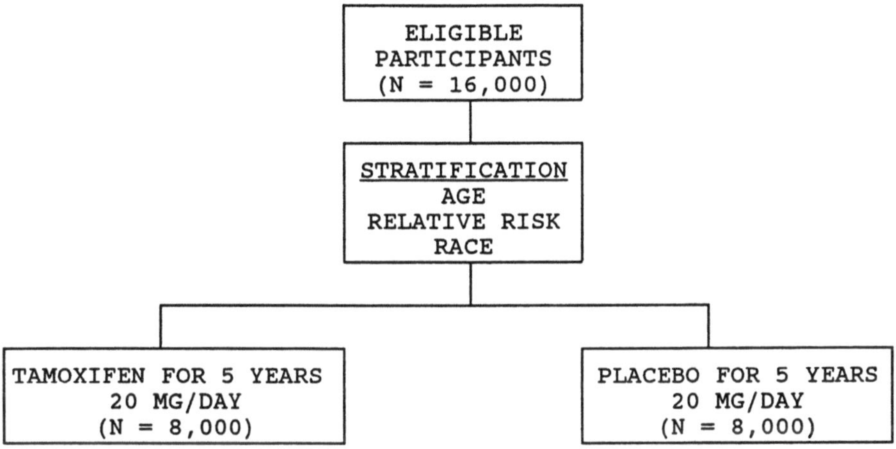

Fig. 1. BCPT schema

ther tamoxifen decreases mortality from coronary heart disease and reduces the incidence of fractures of the hip and wrist in postmenopausal women. Features have been incorporated in the design to reduce the likelihood of serious adverse effects and to monitor carefully all known or suspected side effects. Women in the trial are also monitored for toxicities and side effects by an independent data safety monitoring board, and quality of life is being assessed on a regular basis in all participants.

A major design issue was the weighing of risks and benefits for inclusion of specific groups of women, particularly premenopausal women, in the trial. The criteria for defining the population of women at increased risk for breast cancer who would be eligible for the trial received careful consideration in the design. Although premenopausal women have been included in many cancer treatment trials, their inclusion in a prevention trial was not universally accepted. Based on several justifications it was considered by the BCPT investigators important that the BCPT provide premenopausal, as well as postmenopausal, women with the opportunity to enter the trial. Major justifications for including premenopausal women were:

1. The possibility that the intervention would be more effective if applied earlier in a woman's life. Support for this rationale was provided by NSABP Protocol B-14 (Fisher et al. 1989) for women with stage I breast cancers in which the estimated reduction in incidence of contralateral breast cancers was greater in premenopausal than postmenopausal women.
2. Although increasing age is one of the strongest known risk factors, there are younger women with a family history of breast cancer and/or benign breast diseases who are at increased risk for breast cancer during their premenopausal years and who are subjected to a variety of unevaluated interventions, such as bilateral mastectomy.

3. Known adverse effects appear less frequently in pre- than in postmenopausal women.
4. If the therapy is shown to be effective, it would likely be given to premenopausal women even in the absence of direct information on the risks and benefits of younger women.

Due to ethical concerns about potential reproductive effects and toxicities that might occur with long-term tamoxifen administration, it was felt that well-defined inclusion criteria that related to breast cancer risk must be developed for the trial. The specific statistical strategy utilized was to define women who had a risk for breast cancer that was sufficiently elevated to justify their participation in the trial.

For women interested in participating in the trial a multivariate model incorporating both genetic and nongenetic risk factors is used to estimate for an individual woman the probability of developing breast cancer throughout her remaining expected lifetime. This model, based on work by Gail et al. (1989), provides a prediction of risk that can be compared to the risk of 1.7% over the next 5 years that has been established as the minimum risk qualifying women aged 35–59 for further eligibility screening. (The risk of 1.7% is equivalent to the average 5 year risk of women 60 years of age.) The risk factors used in assessing potential participants include: age, number of first degree relatives with breast cancer, age at first live birth, age at menarche, and number of previous breast biopsies. Table 1 shows the minimum relative risks (RRs) and projected probabilities of developing breast cancer for women of selected ages who meet the eligibility criteria. Because of the higher age-related risk of breast cancer and the increased probability of coronary heart disease, all women 60 years of age or older are considered as potential participants.

In addition to meeting the requirements of a minimum risk of developing breast cancer, from its inception the protocol has included other precautions to reduce the likelihood of fetal harm. Women of childbearing age who wish to participate are counselled and sign a consent form indicating their agreement to use barrier or nonhormonal methods to prevent pregnancy while

Table 1. Minimum relative risk required to enter trial

Age (years)	Relative to women of average risk and the same age	Relative to women of the same age with no risk factors
35	5.07	8.75
40	2.65	4.57
45	1.80	3.10
50	1.52	2.62
55	1.23	2.13

on protocol therapy. Because of a possible increased risk of thromboembolic events with tamoxifen therapy, women with a history of deep vein thrombosis or other thromboembolic events are not eligible. An additional precaution is the requirement that all nonhysterectomized women be queried regarding possible gynecologic symptoms at each 6 month visit and that they have a gynecologic exam on an annual basis.

The NSABP headquarters located at the University of Pittsburgh provides the scientific leadership and overall coordination for conduct of the trial. Recruitment of participants is being conducted through 115 designated BCPT centers which have established over 300 clinical sites located throughout the United States and Canada. The BCPT is funded through the Division of Cancer Prevention and Control of the National Cancer Institute, with supplemental funding from the National Heart Lung and Blood Institute, and the National Institute of Arthritis, Musculoskeletal and Skin Disease.

The study is directed by the Steering Committee, chaired by the Principal Investigator and cochaired by the Director of the Biostatistical Center. The Committee Includes 17 additional members who are principal investigators of the BCPT centers, representatives of the funding governmental agencies and staff of the NSABP headquarters. The Endpoint Review, Safety Monitoring and Advisory Committee (ERSMAC) is the data monitoring committee for the BCPT. ERSMAC is comprised of ten voting members who are completely independent of the BCPT and the University of Pittsburgh. It is chaired by Dr. Theodore Colton. The ERSMAC membership includes clinical trial professionals in the areas of cardiology, gynecology, clinical ethics, biostatistics and epidemiology. The Committee meets in a full day session at least once every 6 months to review unblinded data including: endpoint data, toxicity information, results of blood test and measurements obtained at clinical visits, rates of drug adherence, and compliance with data reporting.

Chronology of the BCPT

Table 2 summarizes important dates for the BCPT. The screening or women for eligibility based upon their risk profile began in April of 1992. Randomization of eligible consenting participants began about 5 weeks later in June. Randomization continued for eligible participants through March of 1994, at which time the National Cancer Institute (NCI) restricted new randomization to only that group of eligible participants who had already signed a consent form. During the next 3 months the BCPT underwent a process of intense scrutiny by several groups including the independent safety monitoring committee for the trial, two NCI advisory boards, and the Food and Drug Administration's Oncologic Drug Advisory Committee. All of these committees recommended that the BCPT should continue essentially as originally designed. Subsequently, the NCI implemented an administrative require-

Table 2. BCPT Chronology

Date	Event
April, 1992	Breast cancer risk assessment initiated
June, 1992	Randomization of eligible consenting participants initiated
March, 1994	NSABP probation began, randomization only of previously consented eligible participants
May, 1994	NCI division of cancer prevention and control board of scientific counselors reviewed and reapproved BCPT
June, 1994	NCI national cancer advisory board reviewed and reapproved BCPT
June, 1994	FDA oncologic drug and advisory board reviewed and reapproved BCPT
June, 1994	NCI subcenter reapproval process initiated
July, 1994	NCI discontinued randomization of new participants pending approval and distribution of protocol amendments
August, 1994	NCI halted risk assessments pending modification
November, 1994	Press release announcing resumption of BCPT randomization
November, 1994	Revised risk assessment profile implemented

ment that the BCPT undergo an extensive reevaluation and modification as part of the process leading to the reinitiation of randomization.

Prior to the NSABP probationary status at the end of March, 1994, revisions to the BCPT consent form were already in process that incorporated updated information on second cancers, especially endometrial cancer. As part of the reinitiation process, additional modifications to the consent form were implemented leading to the necessity of reconsenting all previously randomized participants.

A second modification was the requirement that future participants have a baseline endometrial biopsy as part of the baseline gynecologic examination to assure that women with preexisting uterine cancer or atypical hyperplasia not be entered. Annual follow-up endometrial biopsies were not required but are strongly recommended.

Since over 11 000 participants of the 16 000 projected had been randomized, it was possible to reassess the benefit-risk considerations for the BCPT taking into account more specific information on breast and endometrial cancers. The potential benefit of the trial was reaffirmed. Net benefit was examined both within and across age groups to assure that the risk benefit ratio was positive for all age groups.

Finally, based on an approach that was utilized in the Women's Health Initiative, information was added to the risk profile given to the participant that projected the likely typical age-specific effect of tamoxifen for potential benefits and risks other than breast cancer.

Current Status of BCPT

Table 3 presents our best current projection of the likely benefit-risk pattern among the 8000 participants who will be randomized to tamoxifen, assuming that the remaining participants will have the same age distribution as those already randomized. Table 3 indicates that all age groups are projected to have a positive net benefit if equal weights are assigned to each predicted event. The magnitude of specific beneficial or detrimental effects varies considerably by age. Both heart disease (beneficial effect) and endometrial cancer (detrimental effect) have greater importance in participants 60 years of age or older. Obviously, benefit-risk predictions have a strong subjective component. The individualized risk profiles are an attempt to give potential participants more information to use in deciding whether to enter the trial. It should be noted, however, that, whereas the breast cancer profile takes into account several risk factors other than age, the risks for other events of interest are based only on age-specific rates. The true benefits and risks associated with taking long-term tamoxifen as a breast cancer preventive can only be known by completion of the BCPT.

There has been an increase over the past year in the number of participants who have discontinued taking the protocol medication. Therefore, it became necessary to reestimate the study sample size and power considerations taking into account a higher rate of participant noncompliance than that used in the original estimates. Since over 11 000 participants have already been recruited to the BCPT it was also possible to utilize their predicted risk

Table 3. Updated projected risk-benefit for BCPT tamoxifen-treated patients by age group

Age group	Number of participants	Number of predicted events among BCPT participants[a]					Net benefit of tamoxifen therapy[b]
		Events prevented		Events caused			
		Breast cancer	CHD	Liver cancer	Endometrial cancer	Fatal PE	
35–49	3152	49.8	3.3	0.4	16.8	0.6	35.3
50–59	2456	42.8	8.5	0.6	28.0	1.0	21.7
60+	2392	40.6	32.6	1.3	38.5	2.6	30.8
Total	8000	133.2	44.4	2.3	83.3	4.2	87.8

CHD, coronary heart disease; PE, pulmonary embolism.
[a] Assumes 3 years of accrual and 7 years of follow-up, a 30% reduction in breast cancer, a 20% reduction in CHD, a twofold increase in liver cancer, a twofold increase in fatal pulmonary embolism and a threefold increase in endometrial cancer.
[b] The sum of events prevented minus the sum of events caused.

of invasive breast cancer in the updated power considerations. The group of women recruited have a projected relative risk that is approximately twice the baseline rate from which the sample size of 16 000 was calculated. If accrual is completed at least within a 4 year time frame and if the rate of nonadherence can be brought down to annual rates of 10% of less in succeeding follow-up years, the revised statistical considerations indicate that a sample size of 16 000 with 5–7 years of follow-up beyond accrual should be adequate to achieve sufficient information on invasive breast cancer. Thus, although this past year events have adversely impacted on protocol adherence there is still an opportunity to revitalize the study.

Summary

From its inception the BCPT has been the focus of bitter sustained controversy. In spite of reassessments by numerous advisory groups that have deemed the trial to be scientifically, clinically and ethically sound, its opponents have continued their assaults. That the trial has survived is a tribute to Dr. Fisher's leadership and the dedication of many clinical investigators, who, in spite of the injustice that they have endured, continue to believe in the goals of the NSABP clinical trials. Even greater credit and recognition should be given to the women randomized who have continued their commitment to the study. Their altruism and courage are to be acknowledged. These women undoubtedly understand that the trial affords potential risks, as well as benefits, and that even if the drug is eventually found effective in preventing breast cancer, half of them will have received a placebo in the trial. The ultimate tragedy will be if the study fails to answer definitively the study hypotheses because of what has occurred this past year. We are hopeful that, despite this disruption, the trial will continue to a successful completion.

While larger multicenter randomized clinical trials clearly have vulnerabilities and imperfections, we should not lose sight that they are clearly the best mechanism for rigorously testing new therapies in a manner that will provide reliable, generalizable findings.

Acknowledgement. The authors intend that this manuscript provide updated information regarding the design of the BCPT and the NSABP. The authors are solely responsible for its content. However, we wish to emphasize that the design and conduct of the BCPT has been the collaborative effort of many dedicated individuals within NSABP. Particularly, we wish to recognize the contributions of Dr. Bernard Fisher, Dr. Lawrence Wickerham and Mr. Walter Cronin of the NSABP, and to Dr. Leslie Ford of the NCI to the design, implementation and conduct of the trial. In addition, we wish to recognize the BCPT clinical investigators and the BCPT participants for their personal commitment to maintaining the scientific integrity of the trial. This work was supported by the National Cancer Institute, grant number U10CA37377-70.

References

Early Breast Cancer Trialist's Collaborative Group (1992) Systemic treatment of early breast cancer by hormonal cytotoxic, or immune therapy: 133 randomized trials involving 31 000 recurrences and 24 000 deaths among 75 000 women. Lancet 339:1–15, 71–85

Fisher B, Constantion J, Redmond CK et al. (1989) A randomized clinical trial evaluating tamoxifen in the treatment of patients with node-negative breast cancer who have estrogen-receptor-positive tumors. N Engl J Med 320:479–484

Gail MH, Brinton LA, Byar DP et al. (1989) Projecting individualized probabilities of developing breast cancer for white females who are being examined annually. J Natl Cancer Inst 81:1879–1886

Summary

K.I. Pritchard

Sunny brook Health Science Center, Bayview Regional Cancer Center, 2075 Bayview Avenue, Toronto M4N 3M5, Canada

In a study from the Istituto Nazionale Tumori, Milano, Italy, Dr. Pinucca Valagussa began with a series of 219 patients younger than 38 years of age at diagnosis who were potentially able to become pregnant following their diagnosis. These women came from a series of 2586 women treated between June 1973 and July 1990. Of these, 1542 were premenopausal but only 219 were under the age of 38 at the time of diagnosis. Fourteen were treated with surgery alone, while 205 had adjuvant chemotherapy with CMF in varying protocols. The 205 patients included 43% with more than three positive nodes, 54% with primaries >2 cm and 36% with estrogen receptor (ER)-negative tumours. The women were followed for a minimum of 67 months from surgery.

During follow-up, a total of 62 women (30%) developed permanent amenorrhea or entered early menopause. The risk of entering early menopause over the period of follow-up was 75% in women 31–37, but only 9% in women under 31. During this same follow-up period, 19 pregnancies occurred in 17 women. In 13 cases patients had elective abortion, while one had a spontaneous abortion. Five women gave birth to offspring, none of whom showed fetal malformations. There was a 65% recurrence rate among the patients who became pregnant over the period of follow-up, compared to a 62% recurrence rate in the group as a whole. In this case series, 41% of the patients presented with new disease manifestations within 3 years, 52% within 5 years and 61% within 10 years after surgery. Three major conclusions can be made from this review: (1) There remains a high risk of disease relapse even after the first 3 years following surgery, the maximum time period women are usually advised to wait before becoming pregnant. (2) many patients who may be cured of their disease will become menopausal during a 2–3 year waiting period to become pregnant. (3) No malformations were detected in the five babies born from mothers given adjuvant chemotherapy.

There has been some discussion of these data, including suggestions that women should be fully counselled on their risks of recurrence before considering pregnancy.

Dr. Kathleen Pritchard from Toronto-Sunnybrook Regional Cancer Centre, Toronto, Canada, reviewed the role of hormone replacement therapy (HRT) in healthy women and in women with breast cancer. She discussed the utility of estrogen and of medroxyprogesterone acetate and the superiority of estrogen over veralipride in controlling short-term menopausal symptoms. She also described the literature supporting the role of estrogen replacement therapy (ERT) with or without progestationals in decreasing the rate of bone loss that normally occurs with menopause and in reducing the fracture rate in such patients. In addition, she summarized the studies both observational and randomized, showing that estrogens, with or without progestationals, are associated with increases in high density lipoprotein (HDL), decreases in low density lipoprotein (LDL) and decreases in fibrinogen, all of which are thought to be associated with reduction in cardiovascular events. In addition, she reviewed the observational literature showing that ERT or HRT is associated with a decreased incidence of and mortality from cardiovascular events.

She also reviewed the literature linking ERT to the development of endometrial cancer. This risk is completely obviated by the addition of progestationals. In addition, there is some suggestion that there is a risk of breast cancer developing in healthy women using ERT particularly when it is used for more than 15 years. For breast cancer, there is however no clear dose response, the role of the addition of progestin to estrogens is uncertain, and there appears to be no increased risk associated with ERT in women who have a family history or a history of benign breast disease in comparison to the general population.

Overall it is thought that, in healthy women, the benefits of ERT or HRT in reducing risk of cardiovascular disease and osteoporosis probably outweigh any increased risk of endometrial or breast cancer. This will obviously vary however with the degree of risk of the individual woman for breast cancer or cardiovascular disease in particular.

Women who have had breast cancer previously have traditionally been advised not to receive HRT. It is becoming more common to give such women progestationals, although there is no data showing that this is safe. In addition, ERT is now being given to some women who are at low risk of recurrence. A recent survey suggested that as many as 8% of women with a previous diagnosis of breast cancer may be receiving ERT. Several authors have published case series suggesting that recurrence rates are not dramatically exacerbated by ERT in this setting. There are currently a number of proposals for randomized studies of ERT in women who have had breast cancer; one study is already ongoing and several others are being considered. These trials should be very large in order to rule out even a small increase in risk of breast cancer recurrence, but should also look at other endpoints such as the incidence of and mortality from cardiovascular disease, bone disease, and endometrial cancer as well as quality of life and control of short-term menopausal symptoms.

Dr. Walter Willett, from Harvard School of Public Health in Boston, MA, outlined the Women's Health Initiative (WHI), a 2 × 2 × 3 factorial study in which women are initially randomized to receive estrogen, estrogen plus progesterone or placebo. Each of these three groups then undergoes a second randomization to a low fat diet or a normal diet and each of those six groups then undergoes a third randomization to receive calcium and vitamin D or placebo. These women will be followed for 10 years to look at cardiovascular, osteoporosis and cancer outcomes. Dr. Willet believes that this study may never show clear results for several reasons. First, the accumulating literature suggests that dietary fat may not be a main cause of breast cancer as previously postulated. Six large prospective studies reported in the last few years have seen little relation between fat intake and breast cancer incidence. In fairness they were unable to examine the influence of fat intake below 20% of calories, but this limitation will also apply to WHI. Long-term dietary compliance is likely to be a problem and the diet of the control women may move toward the low fat group as has happened in other lifestyle trials such as MRFIT. In the WHI pilot, the difference in fat intake between the intervention and control groups after 3 years was small (27 vs 31% of energy). In addition, the dietary intervention has been broadened to include an increase in consumption of fruits, vegetables and whole grains which may also, via carotenoids or other mechanisms, be expected to reduce the incidence of breast cancer.

Furthermore, Dr. Willet feels that the hormone randomizations will be probelmatic since estrogen therapy will reduce the risk of coronary disease and fractures in a short time frame (women as old as 70 or more being randomized) whereas the increases in breast cancer are likely to be delayed and may not take place for much less than 10 years. Since the most important question is whether HRT should be continuous after menopause, longer follow-up may be required than has been planned. Thus, although the WHI will give some information, particularly on the role of calcium and vitamin D, it may not clarify the relation of fat intake to risk of breast and other cancers, or address the optimal duration of HRT.

Dr. Carol Redmond from the University of Pittsburgh presented an overview of the design and rationale for the NSABP Breast Cancer Prevention Trial (BCPT) and an assessment of the current status of the trial. The trial opened for randomization of eligible participants on June 1, 1993, and over the ensuing 22 months randomized almost 11000 (79%) of the 16000 participants required for the study. In late March, 1994, the National Cancer Institute (NCI) placed the NSABP on probation. Soon thereafter NCI suspended accrual to all NSABP treatment trials and restricted randomization on the Breast Cancer Prevention Study to that group of eligible participants who had previously signed a consent form. During the following 3 months

(April–June, 1994) the BCPT underwent a process of intense scrutiny by several advisory groups, including the independent data safety monitoring board for the BCPT, two NCI advisory boards, and the Food and Drug Administration (FDA) Oncologic Drug Advisory Committee. All of these reaffirmed that they deemed the study was sound and recommended that the trial continue.

Several protocol modifications were implemented to further assure the safety and informed consent of the participants. At the time the NCI put the NSABP on probationary status, BCPT consent form revisions were already underway that incorporated updated information on endometrial and other second cancers that had emerged subsequent to the initiation of the trial. Additional consent form modifications have resulted in the need to reconsent previously randomized participants. A second protocol modification was a requirement that future participants have a baseline endometrial biopsy to assure that women with preexisting uterine cancer or atypical hyperplasia of the uterus be excluded from randomization. Annual follow-up endometrial biopsies, while not required, were to be strongly encouraged for all women with intact uteri.

The benefit-risk projections for the trial were also reexamined. Since about 11 000 of the 16 000 participants had been randomized, the benefit-risk considerations were revised to take into account more specific information relating to the age and breast cancer risk factor distributions. Of particular relevance is the fact that the predicted relative risks of invasive breast cancer for women entering the BCPT have been on the average about double the baseline risk required to enroll in the trial. (Women are screened for individual breast cancer risk utilizing a multivariate model developed by Gail and colleagues that incorporates age, family history, biopsies, and other hormonally related risk factors) The implication for the trial is that the predicted net benefit overall, taking into account potential benefits and risks, is greater than that originally predicted even if the endometrial cancer is at a threefold risk rather than twofold, as was assumed at the trial's initiation. The net benefit is positive across all eligible age groups (35–49, 50–59, 60+).

Finally, based on the approach that has been adopted in the WHI, estimates have been added to the individual risk profile that projects the likely effect on other potential benefits and risks associated with taking tamoxifen. Whereas the breast cancer profile takes into account several risk factors in addition to age, the projections for other events of interest are age-specific only.

An important concern is the extent to which the criticisms of the NSABP and continued controversy about the BCPT have adversely affected participant adherence and the willingness and capabilities of the trialists to complete accrual to the trial. Although there has been an increase in the number of women who have gone off therapy for nonprotocol indications during the past year, it is remarkable that about 76% of the women have remained adherent. Revised power curves for the trial indicate that there should be adequate sensitivity to detect a clinically important maximal prevention effect of 40%, if the nonadherence rate in successive years is at 10% or less. (A true prevention effect size of 40% will result in a lesser apparent effect of varying magnitude depending on the nonadherence rate)

While the trial has been reopened to randomization since october, 1994, reinitiation of randomization has been slow and not all centers or subcenters are currently actively recruiting. The extent to which the low accrual is attributable to a process of reapproval initiated or a demoralization of staff at participating centers is unclear at present. There is also some concern relating to timeliness of data submission from clinical centers during this past year. It should become apparent over the next 6 months whether the BCPT will be able to recruit the additional 5000 participants in a reasonable time frame (such as 1 or 2 years).

Given the sustained, altruistic commitment of over three quarters of the current participants who remain on therapy and the importance of determining the effectiveness and safety of tamoxifen as a breast cancer preventive drug in women at increased risk, Dr. Redmond hopes that BCPT will move forward without further delays to answer definitively the study hypotheses.

There has been considerable discussion of the relative roles of tamoxifen and tamoxifen plus HRT in both healthy women ans women with a previous diagnosis of breast cancer. The suggestion has been made that tamoxifen might be used in lieu of HRT in healthy women but it certainly does not improve menopausal symptoms, in that it exacerbates hot flashes. In addition, the long-term data, particularly on the effect on bone mass and fractures is not yet available. Tamoxifen does, however, appear to have a cardiovascular benefit very similar to that of HRT. The role of tamoxifen in women with a previous diagnosis of breast cancer is clear in that it reduces recurrence, improves survival, reduces second breast malignancies and may also improve long-term cardiovascular benefit. Once again, however, its role in bone is less clear, although bone density seems improved in comparison to no treatment although perhaps not as much as by estrogen replacement. ERT added to tamoxifen does seem to improve hot flashes and other menopausal symptoms, but it is not clear whether it will further improve cardiovascular or bone outcomes.

Clearly, there is a need for more randomized trials of HRT in well women and women with a previous diagnosis of breast cancer, as well as of tamoxifen as a preventive. It is also to be hoped that the WHI will provide useful information concerning dietary fat reduction and HRT of various types.

IX. International Consensus Conference on Primary Treatment of Breast Cancer

A. Goldhirsch,[1] W.C. Wood,[2] H.J. Senn,[3] J.H. Glick,[4] and R.D. Gelber[5]

[1] International Breast Cancer Study Group, Ospedale Civico, 6900 Lugano, Switzerland
[2] Dept. of Surgery, Emory University School of Medicine, 1364 Clifton Road, N.E., Atlanta, GA 30322, USA
[3] Medizinische Klinik C, Kantonsspital, 9007 St. Gallen, Switzerland
[4] University of Pennsylvania Cancer Center, 6 Penn Towers, 3400 Spruce Street, Philadelphia, PA 19104-4283, USA
[5] Division of Biostatistic, Dana-Farber Cancer Institute, 44 Binney Street, Boston, MA 02115, USA
Other consensus panel participants were K. Antman, M. Baum, B. Fisher, J. Forbes, M. Kaufman, J. Kurtz, H. Mouridsen, K. Pritchard, C.-M. Rudenstam.
* First published in the Journal of the National Cancer Institute, 4 October 1995.

Introduction

Breast cancer is a heterogeneous disease. The choice of primary treatments available today is based on features of the patient, the tumor, and the response to treatment. We have learned that the medical and social environment within which patients are being treated contribute to this heterogeneity with respect to the interpretation of available data on prognosis and cost-benefit of treatments. All of our knowledge about the selection of therapies is derived from results of case series and randomized clinical trials. The latter provide unbiased evidence of relative treatment efficacy on a *selected* patient population. A meta-analysis of randomized clinical trials is extremely effective in reducing statistical uncertainty.

In March 1992 the Consensus Panel of experts at the Fourth International Conference on Adjuvant Therapy of Primary Breast Cancer, in St. Gallen, Switzerland, developed a series of guidelines and recommendations for the selection of treatments in several patient populations (Glick et al. 1992a,b). These guidelines and recommendations were mainly based upon the synthesis of results from individual clinical trials, and the findings from the Early Breast Cancer Trialists' Collaborative Group (1992) worldwide overview that had been made public a few months before the conference. Since then there has been a shift in the proportion of women who are considered candidates for adjuvant therapy as a result of the statistically significant

Table 1. Recent research findings and their implications for patient care

Field or treatment	Status of research/implications for patient care
Genetics	Isolation of *BRCA1*, description of *BRCA2*, evidence for the presence of *BRCA3* and *BRCA4*. Not ready for screening to define individual risk. Will present new challenges to manage severe personal, ethical, and therapeutic dilemmas, when widely available. Ongoing and future chemoprevention trials will have an important impact upon dealing with genetic information.
Timing of surgery for premenopausal women	Some data from retrospective series suggest that surgery on or during the luteal phase of the menstrual cycle is associated with a better prognosis compared with the follicular phase. This issue is controversial due to several studies which do not confirm these findings. Prospective investigations might clarify whether the endocrine environment which characterizes the luteal phase is beneficial for reducing micrometastatic enhancement during and immediately after surgery.
Biological therapies, immunotoxins, antibodies, and gene therapy	In preclinical and phase I/II clinical investigations, mainly in patients with advanced disease; involves particularly *growth factors* and *angiogenesis*; still to enter clinical trials in the adjuvant setting.
New factors for prediction of treatment responsiveness	*c-erbB2* expression for prediction of response to anthracycline-containing chemotherapy and resistance to treatment with tamoxifen or CMF: results require confirmation in prospective randomized clinical trials. *Ki67* expression as marker for proliferation (substitute for tumor grading): results require confirmation and definition of cost and benefit.
Tamoxifen: definition of long-term effects	Increased risk of endometrial cancer by two- to four fold; overwhelmed by proven benefit with respect to breast cancer risk and overall survival advantage.
"Full-dose" chemotherapy	Reduction of the full-dose, except according to objective toxicity guidelines, might be detrimental in terms of treatment results. This should influence patient care outside clinical trials. The use of hematopoietic growth factors is still experimental.
"High-dose" chemotherapy	The administration of a four- to ten fold increase in chemotherapy dose with autologous bone marrow or peripheral blood progenitor cell support (with the aid of hematopoietic growth factors) is a promising approach for patients at high risk of relapse. The use of this modality outside the context of clinical trials is not justified, pending results of randomized, controlled ongoing studies.
Taxanes	Found to be very effective in metastatic and locally advanced disease. The use of taxanes in the adjuvant setting outside the context of clinical trials is not justified, pending results of ongoing studies.

effects of chemotherapy, tamoxifen, and ovarian ablation identified in the overview (Early Breast Cancer Trialists Collaborative Group 1992).

During the past 3 years several new concepts and treatment strategies have been studied. Some of these can be considered to assist in the treatment of patients today, while others are still undergoing clinical investigation for a better definition of their usefulness. Table 1 describes some examples of these findings and their implications or status relative to patient care. These subjects were presented and discussed at the Fifth International Conference on Adjuvant Therapy of Primary Breast Cancer, held in St. Gallen, Switzerland in March, 1995. At the last session a panel of experts discussed prognostic and predictive factors, treatments for DCIS and screening – detected small tumors, and controversial issues related to the adjuvant treatment of node-negative and node-positive breast cancer. In this commentary we describe some areas of ongoing research and update the treatment recommendations presented 3 years ago.

Prognosis and Prediction of Response

Several factors have been identified in attempting to define the prognosis of patients who should not receive any form of adjuvant systemic therapy. The panel agreed that a population of patients having less than 10% mortality at 10 years would not be candidates for receiving routine adjuvant systemic therapy. All other patients could benefit from additional postoperative therapy. Consequently, evaluating the prognosis of patient subgroups today is relevant only within the context of the adjuvant treatments received.

Factors predicting treatment response are those which identify patient subpopulations having a larger or smaller response to a given systemic treatment. Estimating the magnitude of the treatment effect is important for determining the usefulness of a given treatment approach. The p value only indicates the strength of evidence against a null hypothesis and is not a measure of the magnitude of treatment effect. Consequently, a factor which appears to distinguish responders from nonresponders based on the calculated p values may, in fact, predict a quantitative difference in response which is positive for all subpopulations. Evidence of an interaction between treatment group assignment and a potential treatment response prediction factor is required to justify selective withholding of therapy, when an overall result is positive. Randomized clinical trials which include a large number of study patients are required to adequately define these factors for clinical use.

The distribution of initial presentations of breast cancer diagnosed in recent years has changed. The incidence of noninvasive tumors and small, screening-detected, invasive cancers is increasing. Table 2 includes these two types of presentation as distinct categories in addition to the node-negative and node-positive subpopulations. In addition, Table 2 shows the prognostic

Table 2. Presentation of breast cancer; categories with treatment implications

Disease presentation	Prognostic factors used in the definition of risk	Predictive factors used for the selection of adjuvant treatment
Noninvasive tumors: DCIS	Clear margins	None
Small invasive tumors: screening-detected	Nodal status	None
Node-negative (optimally at least ten nodes examined)		
Minimal/low risk	Tumor size	Estrogen receptors
"Good" risk	Grade	Age
"High" risk	Estrogen receptors	Menopausal status
Node-positive		Age
	Number of nodes involved	Estrogen receptors
		Menopausal status

DCIS, ductal carcinoma in situ.

factors used to define candidates for some form of adjuvant therapy and the predictive factors used to "fine-tune" the treatment selection based on considerations of relative effectiveness of available therapeutic regimens.

Ductal Carcinoma In Situ

The primary treatment of noninvasive ductal carcinoma in situ (DCIS) requires clarification. DCIS patients should have breast surgery similar to that of patients with invasive cancer – total removal of the primary tumor with clear resection margins. If it is not possible to achieve this goal, mastectomy is necessary. If there is no invasive component, no additional prognostic information is obtained by removal of the axillary lymph nodes. Therefore, axillary dissection is not indicated. Radiation therapy to the conserved breast after a complete surgical removal of a DCIS lesion showed a significant relative reduction (fourfold) of subsequent invasive tumor growth in the breast, although the number of such events occurring without radiation therapy was small (less than 10% within the first 5 years in the NSABP Trial B-17) (Fisher et al. 1993). Thus, cost-benefit considerations should be used for deciding whether radiation is indicated to reduce the risk of a relatively rare event. Randomized clinical trials are currently underway to better define patient, tumor, and surgical treatment characteristics that could influence the selection of radiation therapy in this setting. There is no indication for systemic adjuvant therapy for patients presenting with DCIS alone at the present time. Ongoing trials will clarify the role of therapy with tamoxifen to reduce the risk of invasion and relapse in this patient population.

Screening-Detected Small Invasive Tumors

The surgical treatment of these tumors (detected by screening mammography only, and usually smaller than 1 cm) should follow the general guidelines for treating breast cancer – total removal of the malignant lesion with clear margins. The issue of whether axillary clearance may be avoided is still controversial since some of the patients will have lymph node involvement. In fact, the prognostic information derived from the histopathological examination of the lymph nodes is important for defining prognosis and treatment selection. Radiation therapy to the conserved breast is indicated, since data are not available to select a subpopulation of patients who will not significantly benefit from the application of this modality. Systemic treatment should be considered for all patients with factors indicating dire prognosis (e.g., node-positive presentation). For all other patients the use of systemic therapy such as tamoxifen is still a matter of clinical research.

Node-Negative Breast Cancer

During the past 3 years, a larger proportion of patients with node-negative breast cancer has been considered candidates for adjuvant systemic therapy. The accurate definition of a node-negative status requires that a sufficient number of axillary nodes be examined. Investigations focused on this staging subject indicated at least ten lymph nodes as appropriate for obtaining the proper prognostic information (Axelsson et al. 1992). Other methods for defining prognosis have not yet been able to provide sufficient information to substitute for axillary lymph node examination. Many of these methods investigate proliferation features of the primary tumor and its metastatic potential (e.g., micrometastases detected in the bone marrow). Neither ploidy nor S-phase was recommended for routine use.

The baseline prognosis of patients with node-negative disease can vary substantially, depending upon several features. Table 3 displays the factors contributing to the allocation of patients to risk categories.

Adjuvant treatment for patients with node-negative disease (Table 4) varies substantially according to the baseline prognosis. For patients considered at "high" risk, the treatment choice follows an algorithm similar to node-positive disease, considering that the prognoses are similar, if the node-negative population is untreated. For these patients the chemotherapy used in clinical trials, especially those included in the overview (Early Breast Cancer Trialists' Collaborative Group 1992), was mainly a

Table 3. Definition of risk categories for patients with node-negative breast cancer

Factors	Minimal/low-risk (all of the listed factors)	"Good" risk	"High" risk (at least one of the listed factors)
Tumor size[a]	<1 cm	1–2 cm	>2 cm
Estrogen receptor status[b]	Positive	Positive	Negative
Grade[b]	Grade 1 (uncertain relevance for tumor <1 cm)	Grade 1–2	Grade 2–3
Age[c]	>35 years		

[a] It was generally agreed that pathological tumor size (of invasive component) was the most important prognostic factor for defining additional risk of relapse.
[b] Estrogen receptor status and grade are expressions of the malignant transformation of the tumor cell. It is difficult to precisely dichotomize these features to indicate a good versus bad prognosis.
[c] Patients who develop breast cancer at young age are considered to be at high risk of relapse, although an exact age threshold for this increased risk has not been defined. While acknowledging this fact, the panel did not accept age as a factor to influence the choice of the type of treatment (chemotherapy or endocrine therapy).

Table 4. Adjuvant treatment for patients with node-negative breast cancer

Patient group	Minimal/low risk	"Good" risk	"High" risk
Premenopausal			
ER-positive	Nil vs Tamoxifen[b]	Tamoxifen[a] Oophorectomy[b] Chemotherapy[b] GnRH analogue[b]	Chemotherapy[a] ± Tamoxifen[b] Oophorectomy[b] GnRH analogue[b]
ER-negative	(Not applicable)	(Not applicable)	Chemotherapy[a]
Postmenopausal			
ER-positive	Nil vs Tamoxifen[b]	Tamoxifen[a]	Tamoxifen[a] ± Chemotherapy[b]
ER-negative	(Not applicable)	(Not applicable)	Chemotherapy[a] ± Tamoxifen[b]
Elderly	Nil vs Tamoxifen[b]	Tamoxifen[a]	Tamoxifen[a]; if ER-negative then Chemothearpy[a] ± Tamoxifen[b]

ER, estrogen receptor; GnRH, gonadotrophin releasing hormone; Nil, no adjuvant treatment.
[a] Treatments accepted for routine use or baseline in clinical trials.
[b] Treatments still being tested in randomized clinical trials.

cyclophosphamide, methotrexate, 5-fluorouracil (CMF)-based regimen. The use of anthracyclines for these patients is currently being investigated. For patients with minimal/low risk disease, the question of whether to treat with tamoxifen or not depends on a cost-benefit analysis which should take into account the low relapse rate within the first 10 years and the potential reduction of the incidence of contralateral breast cancer which has been shown to occur. Patients classified as "good" risk may be assigned to receive an endocrine treatment with tamoxifen. For premenopausal women in this category, other forms of hormonal manipulations, including ovarian ablation and chemotherapy, remain investigational.

Node-Positive Breast Cancer

Treatment selection for patients with node-positive disease was extensively studied in previous years. During the past 3 years, very little has changed with respect to treatment recommendations for this patient population. Table 5 shows the treatments available for patients with node-positive disease.

Specific Aspects of Treatment

Ovarian Ablation. Early Breast Cancer Trialists' Collaborative Group (1992) overview results on ovarian ablation were obtained mainly from trials

Table 5. Adjuvant treatment for patients with node-positive breast cancer

Patient group	Treatments
Premenopausal	
ER-positive	Chemotherapy[a] ± Tamoxifen[b] Ovarian ablation[a] ± Tamoxifen[b] GnRH analogue[b] Chemotherapy[a] ± Ovarian ablation (GnRH analogue)[b] ± Tamoxifen[b]
ER-negative	Chemotherapy[a]
Postmenopausal	
ER-positive	Tamoxifen[a] ± Chemotherapy[b]
ER-negative	Chemotherapy[a] ± Tamoxifen[b]
Elderly	Tamoxifen[a]; if ER-negative then Chemotherapy[a] ± Tamoxifen[b]

ER, estrogen receptor; GnRH, goradotrophin releasing hormone.
[a] Treatments accepted for routine use or baseline in clinical trials.
[b] Treatments still being tested in randomized clinical trials.

comparing this modality (surgical or radiation-induced) to surgery alone. The 15-year follow-up presented in the overview was sufficiently convincing to include ovarian ablation as a clinically relevant treatment option. It had also emerged from data of trials with hormone receptor information that, as expected, ovarian ablation is likely to benefit exclusively patients with estrogen receptor-positive primaries (International Breast Cancer Study Group 1990). The trials of chemotherapy with or without ovarian ablation had a shorter follow-up, and therefore contributed less information to the overview. They seemed to indicate that oophorectomy added to chemotherapy may have an effect greater than chemotherapy alone. The combination, however, remains experimental, since trials comparing it with oophorectomy (or gonadotrophin releasing hormone analogues) alone are ongoing. The relatively small number of patients included into trials investigating oophorectomy (only 1817 patients younger than 50 were included in the overview), the patient's psychological desire to avoid castration, and the concern about its late sequelae are the main reasons that this modality has not been very widely accepted.

Tamoxifen. The benefit from treatment with tamoxifen in terms of reduced incidence of relapse and mortality is clearly overwhelming. The risk related to uterine cancer is increased and varies among countries. It is, however, clear that a threefold increase in the incidence of endometrial cancer, which is below 1%, should be taken into account for postmenopausal women. The duration of the treatment with tamoxifen should be between 2 and 5 years, and current trials compare durations of 5 and 10 years, or lifetime. New data will be available in the near future to establish the best duration of the treatment with this drug.

Chemotherapy Regimen. The treatment of choice for patients with node-negative disease is CMF given on days 1 and 8 every 4 weeks (with cyclophosphamide given either i.v. on the same days as methotrexate and fluorouracil or orally on days 1–14), or a similar combination chemotherapy. The CMF regimen or an anthracycline-based combination, such as doxorubicin and cyclophosphamide (AC) or cyclophosphamide, doxorubicin, and fluorouracil (CAF), is considered to have similar treatment effects for patients with node-positive disease (Shapiro and Henderson 1994). Data indicate that CMF should be given for the duration of six courses, while four courses of an anthracycline combination chemotherapy yield similar results (Fisher et al. 1990a).

Chemotherapy Dose. Reducing the chemotherapy dose is detrimental in terms of treatment results. These are the convincing results of a randomized comparison between the full doses of a CAF regimen chemotherapy and a 50% dose reduction (Wood et al. 1994). However, doubling the dose intensity of the alkylating agents in the AC regimen did not increase therapeutic yield

(Dimitrov et al. 1994). Furthermore, the routine use of marrow growth factors to obtain an enhancement of dose intensity is still experimental.

Chemoendocrine Therapies. The combination of cytotoxic agents, like doxorubicin and cyclophosphamide, and tamoxifen was associated with a better therapeutic outcome when compared with tamoxifen alone, especially in patients with tumors classified as estrogen receptor (ER)-positive (Fisher et al. 1990b). The best way to combine chemotherapy and tamoxifen is unknown, especially if the cytotoxic regimen does not include anthracyclines. In fact, laboratory studies demonstrated that chemotherapy cell kill was inhibited in the presence of tamoxifen (Osborne 1994). Clinical data also suggest a negative interaction between cytotoxics (alkylating agents and 5-fluorouracil) and tamoxifen (Fisher et al. 1986). This issue, however, is still under study.

High-Dose Chemotherapy with Peripheral Blood Progenitor Cell or Bone Marrow Support. There are several randomized trials currently being conducted to investigate this issue. Indirect evidence exists from two single institution phase II studies (Duke University, U.S.A. and Milan National Cancer Institute, Italy) that this approach might be advantageous in terms of treatment results. Although these approaches are promising, they are very costly and should therefore be used only in the context of randomized clinical trials. The validation of high-dose chemotherapy is essential for best defining its indications and yield.

Taxanes. Taxanes have been tested in metastatic and locally advanced disease and their efficacy has been at least partially defined (Rowinsky et al. 1995; this review paper was published after the meeting, during the preparation period of the manuscript, and was added for clarity and completeness). There are, in fact, several open questions about the proper schedule of administration of these agents and about their association with other active cytotoxics, espcially anthracyclines. The testing of taxanes in the adjuvant setting will be carried out in the near future. In the meantime, their use in adjuvant regimens outside of clinical trials is not indicated.

The Elderly. Older patients with a long-term survival expectancy should be considered for adjuvant systemic therapies in the same way as younger postmenopausal patients. Extrapolating from data available from younger cohorts, it is stressed that, if chemotherapy is indicated, a lower dose of cytotoxics should not be given. Lowering the dose intensity may slightly reduce the side effects of the drugs, but unfortunately will compromise the treatment efficacy. Thus, adjuvant cytotoxics should be completely avoided rather than given at a low dose, unless the low dose is given within the framework of a controlled clinical trial.

Hormone Replacement Therapy (HRT). No data are available about the long-term safety of estrogens given for menopausal symptoms to patients with a past diagnosis of breast cancer. Some information on the efficacy of progestins to relieve hot flashes has been generated, but the data are relatively short-term and do not relate to the safety of these compounds (Loprinzi et al. 1994). The use of HRT in this context remains an important matter for future investigation.

Commentary

The panel members attempted to answer many questions related to the best use of treatments investigated in randomized clinical trials. They were convinced, however, that much more progress could be achieved if participation in clinical trials became more acceptable both to the public and to the medical community. Treatment advances, even those of modest magnitude, will have an important influence on breast cancer mortality and/or patients' quality of life because of the high incidence of the disease.

References

Axelsson CK, Mouridsen HT, Zedeler K for the Danish Breast Cancer Cooperative Group (DBCG) (1992) Axillary dissection of level I and II lymph nodes is important in breast cancer classification. Eur J Cancer 28A:1415–1418

Dimitrov N, Anderson S, Fisher B et al. (1994) Dose intensification and increased total dose of adjuvant chemotherapy for breast cancer (BC): findings from NSABP B-22. Proc Am Soc Clin Oncol 13:64 (abstract 58)

Early Breast Cancer Trialists' Collaborative Group (1992) Systemic treatment for early breast cancer by hormonal, cytotoxic, or immune therapy: 133 randomized trials involving 31 000 recurrences and 24 000 deaths among 75 000 women. Lancet 339:1–5, 71–85

Fisher B, Redmond C, Brown A et al. (1986) Adjuvant chemotherapy with and without tamoxifen: five-year results from the National Surgical Adjuvant Breast and Bowel Project Trial. J Clin Oncol 4:459–471

Fisher B, Brown AM, Dimitrov NV et al. (1990a) Two months of doxorubicin-cyclophosphamide with and without interval reinduction therapy compared with six months of cyclophosphamide, methotrexate, and flourouracil in positive-node breast cancer patients with tamoxifen-nonresponsive tumors: results from the National Surgical Adjuvant Breast and Bowel Project B-15. J Clin Oncol 8:1483–1496

Fisher B, Redmond C, Legault-Poisson S et al. (1990b) Postoperative chemotherapy and tamoxifen compared with tamoxifen alone in the treatment of positive-node breast cancer patients aged 50 years and older with tumors responsive to tamoxifen: results from the National Surgical Adjuvant Breast and Bowel Project B-16. J Clin Oncol 8:1005–1018

Fisher B, Constantino J, Redmond C et al. (1993) Lumpectomy compared with lumpectomy and radiation therapy for the treatment of intraductal breast cancer. N Engl J Med 328:1581–1586

Glick JH, Gelber RD, Goldhirsch A, Senn H-J (1992a) Meeting highlights: adjuvant therapy for primary breast cancer. JNCI 84:1476–1485
Glick JH, Gelber RD, Goldhirsch A, Senn H-J (1992b) Adjuvant therapy of primary breast cancer. Ann Oncol 3:801–807
International Breast Cancer Study Group (1990) Late effects of adjuvant oophorectomy and chemotherapy upon pre-menopausal breast cancer patients. Ann Oncol 1:30–35
Loprinzi CL, Michalak JC, Quella SK et al. (1994) Megestrol acetate for the prevention of hot flashes. N Engl J Med 331:347–352
Osborne CK (1994) Interactions of tamoxifen with cytotoxic chemotherapy for breast cancer. In: Jordan VC (ed) Long term tamoxifen for breast cancer. University of Wisconsin Press, Madison, pp 181–198
Rowinsky EK, Donehower RC. (1995) Paclitaxel (taxol). N Engl J Med 332: 1004–1014
Shapiro CL, Henderson IC (1994) Late cardiac effects of adjuvant therapy: too soon to tell? Ann Oncol 5:196–198
Wood WC, Budman DR, Korzun AH et al. (1994) Dose and dose intensity of adjuvant chemotherapy for stage II, node-positive breast carcinoma. N Engl J Med 330:1253–1259

Subject Index

MIX
Papier aus verantwortungsvollen Quellen
Paper from responsible sources
FSC® C105338

If you have any concerns about our products,
you can contact us on
ProductSafety@springernature.com

In case Publisher is established outside the EU,
the EU authorized representative is:
Springer Nature Customer Service Center GmbH
Europaplatz 3, 69115 Heidelberg, Germany

Printed by Libri Plureos GmbH
in Hamburg, Germany

MULTIMEDIA DATABASE MANAGEMENT SYSTEMS

Research Issues and Future Directions

edited by

Bhavani Thuraisingham
The MITRE Corporation

Kingsley C. Nwosu
Lucent Technologies

P. Bruce Berra
Syracuse University

A Special Issue of
MULTIMEDIA TOOLS AND APPLICATIONS
An International Journal
Volume 4, No. 2 (1997)

SPRINGER SCIENCE+BUSINESS MEDIA, LLC